ENCYCLOPAEDIA OF
ANIMAL BEHAVIOUR

ENCYCLOPAEDIA
OF
ANIMAL BEHAVIOUR

(Vol. 1)

Dr. A. Tyagi

ANMOL PUBLICATIONS PVT. LTD.

NEW DELHI - 110 002 (INDIA)

ANMOL PUBLICATIONS PVT. LTD.
H.O.: 4374/4B, Ansari Road, Darya Ganj,
New Delhi-110 002 (India)
Ph.: 23261597, 23278000
B.O.: No. 1015, Ist Main Road, BSK IIIrd Stage
IIIrd Phase, IIIrd Block,
Bangalore - 560 085 (India)
Visit us at: www.anmolpublications.com

Encyclopaedia of Animal Behaviour

First Published, 2006

ISBN 81-261-2675-2 (Set)

PRINTED IN INDIA

Published by J.L. Kumar for Anmol Publications Pvt. Ltd., New Delhi - 110 002 and Printed at ARORA Offset Press, Delhi.

Contents

Preface

Behaviour could be most simply defined as movement or a way one conducts oneself. Animal behaviour is, obviously, the way in which animals-non human-lead their life, such as, feeding, migration to different parts in different seasons, sexual life, gregariousness etc.

Different factors need to be studied to have right knowledge about animal behaviour. One of them is the concept of "Natural Selection" propounded by Darwin. He observes that species compete with each other for survival.

Apart from animals, there also occurs a peaceful contest among birds. For example, the severest rivalry between the males of many species to attract their female counterpart is one of the patterns of animal behaviour. It also includes a genetic component similar to the morphological and psychological traits.

The present multi-volume encyclopaedia is a painstaking research work on animal behaviour. It would certainly prove to be an asset for psychologists and those who seek thorough knowledge on the subject.

Editor

1

Behaviour Evolution

The driving forces of evolution has been natural selection. It will be recalled that natural selection was defined as a process that resulted in the differential reproduction by individuals of different genotypes. There is good evidence in support of the view that behaviour is, at least in part, dependent upon genotype. It follows that as different genotypes reproduce differentially, the behavioural patterns that are dependent upon those differentially-reproducing genes will become differentially represented; i.e., behaviour evolves.

Prerequisite to an understanding of the evolution of behaviour is an understanding of evolutionary processes in general. Some of these processes are being discussed here.

Microevolution

Evolutionary biologists have generally assumed that natural selection produces large scale, qualitative change through the accumulation of small, quantitative increments. These small changes, referred to as micro-evolution, may come about through changes in thresholds in the sensory motivational, or motor system that control behaviour patterns (Manning 1971). Changes in thresholds lead to modifications in frequency or amplitude of displays. For example the long calls of common and herring gulls (genus Larus), which function to advertise territory, are similar, but the two species emphasize different parts. The basic patterns involve jerking the head down, then throwing it back while calling. The herring gull has increased the amplitude of the throwing back of the head (Tinbergen, 1959). Such differences probably function in species recognition. Microevolution also seems important in enabling organisms to adjust to changing environments, but it can lead to the evolution of species as Darwin assumed. Gould and Eldredge (1977) think not, and have argued that new species have arisen in peripherally isolated populations through rapid change, rather than by slow transformation in the central ancestral area.

The Nature of Genetic and Micro-evolutionary Changes : It is worthwhile trying to discover if there is a common thread running through the diversity of genetic and micro-evolutionary changes. We have emphasized the quantitative nature of the changes and many of them can, in behavioural terms, be directly related to changes in thresholds. If, in the same situation, two species or strains differ in the relative frequency with which they perform patterns A and B from a common repertoire, it is reasonable to interpret this as due to threshold differences in the mechanism controlling A and B in each case. There are numerous sites at which the control mechanism could be affected. There might be changes of threshold in the sensory system, in systems affecting ' motivation, in the motor system, or in any combination of these. Threshold changes could be produced by genes acting on the nervous system itself as in the hyperkinetic mutants of Drosophila or more indirectly, by genes affecting metabolic rate or hormone secretion which will in turn affect the nervous system. In no case do we have proof, but sometimes the circumstantial evidence for direct action on the nervous system is strong. Thus, the difference in the performance frequency of various sexual patterns between inbred lines of Guinea pigs remains even after injections of sex hormones or thyroid hormones, which increase the basic rate of metabolism. It seems most probable that the lines differ in the threshold properties of the sexual mechanisms in the brain.

Genetic changes affecting thresholds either within the coordinating centre or one of the subordinate muscle groups will change the form of a fixed action pattern. For instance, muscles might be brought into action earlier or later in a sequence; they might change the intensity and speed of their response as more or less or their constituent fibres are activated. Exhibits a beautiful example of the behavioural results of such changes from Hunsaker's work on lizards. Males of the genus sceloporus show rhythmic head bobbing movements during courtship and whenever they meet other lizards as a kind of species-identity signal. Each species has a characteristic pattern of bobbing which results from a series of contractions in the muscles which extends the front legs and thereby raise and lower the head and shoulders. Micro-evolution has produced several distinct versions by a series of changes of the outlines above. This examples is particularly clear because only two main groups of muscles are involved, but it is not different to envisage the differences in emphasis illustrated.

Evidence for the Evolution of Behaviour : Here we examine the types of evidence used to demonstrate how behaviour changes in a more or less permanent way through generations. The evidence is of three general types:

1. *Phytogeny:* The tracings of the patterns of evolutionary changes through time.
2. *Adaptation:* The way natural selection actually works.
3. *Speciation :* The foundation of new species (Alexander, 1978).

Phylogeny : The most direct evidence about phylogeny is provided by the fossils which, for all their inadequacies represent the only genuinely historical record of the course of evolution. However, since actions normally have no permanent trace to be fossilized, this is not a very fruitful source of information about the evolution of behaviour.

There are some exceptional cases where the consequences of behaviour have survived as fossils. A series of fossil tracks left by feeding organisms from the Cambrian to the lower carboniferous is explained . From their form and position in the fossil series, the following sequence of events in the evolution of foraging behaviour in this group can be proposed; the animals originally fed in loosely meandering paths just below the surface but gradually showed a greater tendency to move sideways and to burrow deeper until the tracks finally took the form of extremely regular spirals. This series of changes probably increased the efficiency of foregoing by reducing the chances of covering the same ground (Seilacher, 1967). The frequency and type of bone fractures observed in excavated human skeletons can be used to provide information about their way of the life (Table) : the incidence of post-cranial bone fractures in American Indians of different periods correlates with their mode of life, deduced from other evidence, post-cranial fractures become less common with the transition from a nomadic, hunting lifestyle to life in more of less modern towns. This coevolution can be used to assess the lifestyle of populations for which only skeletons are available (Steinbock, 1976).

Making Deductions about Behaviour from Fossilized Remains; the Incidence of Post-cranial Fractures in Amerinidian Populations Differing in Age and Way of Life.

Date	*Way of life*	*% of skeletons with post cranial fractures*
4000-1000 BC	Nomadic hunters	10.7
100 BC-1000 AD	Permanent villages	5.4
1000-1600 AD	Agriculture/tours	3.9

Domestication

Artificial selection by humans has changed the behaviour of numerous animal with which humans have associated over thousands of years. Domestication involves more than just taming or socializing animals; humans also control the breeding, care and feeding of domesticated animals. Some ethologists have argued that behavioural plasticity is greater in domestic animals than in their wild counterparts and that neoteny, the persistence of juvenile characteristics in adults is typical in domestic animals. Others have emphasized the degeneracy of domesticants; they cite the laboratory rat. But instance, most carnivores maintain a nest, and urinate and defecate some distance away from it; they are thus suited to living in human habitations. Other factors facilitating the process of domestication include large social groups, the presence of males in the group year round, precodial young, and are omnivorous diet (Hale, 1969).

The evolution of dog behaviour provides the most striking example of selective breeding by Homo sapiens (Scott and Fullar, 1965). The wolf (Canis Lupus), ancestor of the dog, the principle carnivore of the northern hemisphere prior to the arrival of humans has complex, cooperative social system. As the wolf was domesticated and transformed into the dog, it spread all over the world in conjunction with humans and underwent adaptive radiation. Dogs (Canis familiaries) were used for protection for hunting, and for herding other domesticated animals. The relative isolation of human tribes 8,000 to 10,000 years ago, with only occasional genetic interchange, probably encouraged the evolution of distinct dog breeds. As human contact increased with improved transportation, the different breeds of dogs mixed and many breeds, such as those in South Africa and North and South America lost their identity. Only in the last few hundred years have humans practised rigid artificial selection, which has resulted in increased separation of dog breeds. The great diversity of form and behaviour of current breeds illustrates the potential behavioural effects of selection, whether natural or artificial of certain phenotypes.

Scott and Fuller (1965) concluded that the behavioural patterns of the dog and wolf are essentially the same. Selection has modified primarily the agonistic and investigatory systems. For example, the bull dog breed, in the interest of an English sport, was selected for its tendency to attack the nose of a bull and hang on, in contrast with the more typical slashing attack from the rear used by wolves. Terrier breeds have been selected from their tendency to attack prey relentlessly, regardless of any injury suffered. The usual wolf pattern is to snap and then withdraw to avoid injury. Other breeds have been selected for opposing reasons; scent hounds and birds, dogs are examples of peaceable animals that can be kept in groups in a kennel.

Wolves and dogs search for prey rather than lie in wait and therefore have a well-developed investigatory repertoire. Some dogs, such as scent hounds, are bred for the ability to follow a trail. Others, such as bird dogs, use their visual and olfactory senses much more equally; after rapid quartering of ground, they locate prey by scent only when they are a few paces away, and then they freeze. Dogs must be trained for all of these tasks, but in each case artificial selection has produced a phenotype that is predisposed to specific behavioural traits.

Comparative Series : Perhaps the best available evidence for the evolution of behaviour has come from comparative studies of closely-related forms. The presence of a series of intermediate between two extremes suggests what the evolutionary sequence was, but it may not represent the exact route because the living intermediates may have changed from the ancestral type. It is usually assumed that behaviour evolve from simple to complex and such series are often referred to as progressions; but inferring goal-directed change in evolution is unwise.

One comparative series is that of balloon flies, members of the family empididal (Order Diptera). According to Kessel (1955) the story began in 1975 when Baron Osten-Sacken was visiting in the swiss Alps and noticed bright silvery flashes among the shadows of the fir forest. Believing that they were silvery insects his net to capture some, but ended up with dull-coloured flies. Later, he noticed that along with male flies, he had netted packets of filmy material., Osten-sacken assumed that the males carried the material on their books to attract females during the mating swarm. Another observer later discovered that the male flies carried the packet underneath their bodies and suggested that the somewhat flattened balloons serve as aeronautical surfboards on which the flies cavorted among the sunbeams. Another suggestion was that the devices serve as warning signals to birds and predaceous insects; however, no evidence exists that flies seem to use the balloon for attracting females, stimulating mating and reducing the likelihood that the female will try to eat the male once they have coupled.

It is useful, if not critical, to have an independent assessment of phylogenetic relationship with which the behavioural series can be compared. For instance, many wasps (order Hymenoptera) capture prey to provide food on which they lay their eggs on which the larvae subsequently feed. Some species carry the prey anteriorly with the mandibles; other species carry the prey posteriorly, sometimes on a terminal abdominal segment called an 'ant clamp" (Evans, 1962). This series with many intermediate forms, correlates well with other evolutionary evidence and we can be confident of the direction of change. The selection pressure directing the change from anterior to posterior carriage is thought to be ectoparasitism. When carrying prey with the mandibles, the wasp must put the prey down to open or bid a burrow. Other flies or wasps can then lay eggs on the prey as well and thus diminish the food resource.

Among vertebrates, the evolution of pigmentation in the mouth-breeding cichlid fish (Haplochromis spp.) was studied by Wickler (1962). Presumably under the selection pressure of egg predation, some species of cichlid females pick up eggs after spawning (i.e., after the eggs have been laid and externally fertilized) and 'incubate' them in the mouth. The faster the females pick up the eggs after laying, the less the chance of predation. In some species the females take the egg before they are fertilized by the male.

The male presents his anal fin, which bears yellow spot resembling egg; the female snaps at these spots and takes in sperm, which fertilize the eggs in her mouth. Comparative studies have shown that these yellow spots evolved romless conspicuous pearly spots on the vertical fin that have no signalling function. Eible-Eibesfeldt (1975) has pointed out that the following pre-adaptations existed.

1. Female readiness to pick up eggs.
2. Pearly spots on the vertical fins of the male
3. A lateral display by the male during courtship and spawning.

The case in which intermediates are not available is more common and we must draw evidence from related species that have evolved more or less individually from some common ancestor. A weak inference can be made on the basis of the assumption that those behaviour patterns common among the various living species represent ancestral traits.

Tinbergen (1959) compared the behaviours of different species of gulls (family laridae); in particular he examined those patterns that are relatively constant throughout a species and of the fixed pattern type, i.e., they are developmentally stable and resistant to environment change. The basic similarities in displays throughout the family strengthen the conclusion, previously based on morphology that the family is monophyletic, that is, it originated from a single ancestral type. If a species is judged to be closely related to other species because it shares many of their characteristics, any differences can be considered to have been recently acquired.

A variety of animals construct objects for prey capture, shelter, or mate attraction for example, spider webs, caddisfly cases and nets, bee hives, bird nests and bowers. It is relatively easy to examine these semi-permanent structures and to compare closely-related species. Collias and Collias (1963) studied the nests of African weaverbird of the family ploceidae, racing the evolution of the nest from the ancestral cup nest to intricate domed structures with bottom entrances. Primitive members of the family begin the nest as a simple cup or platform; in some species a dome is added later.

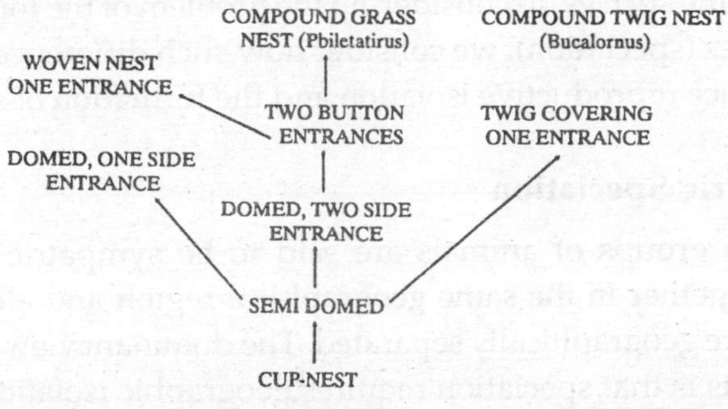

Main trends in evolution of nest building in the water bird (family ploceidae)

In the true weaver (subfamily ploceini) a ring is made from which the roof is built, followed by the floor. In order a species a communal thatched roof is constructed, with brood chambers of many individual pairs underneath. The researchers developed a key to the nests of the true weavers and actually suggested some taxonomic revisions in the family based on nest construction. The techniques used to weave the next from grasses also vary among the weavers, and Collias noted trends toward more regularity in the pattern, tighter, weaving and finer stiches within the subfamily. In general, species making the more 'advanced' nests used the more refined pattern of weaving. In many species the male builds the basic nest and along with neighbouring males, advertises it to the female, who makes the final choice. Males may make many nests before one is chosen for completion by the female, implicating sexual selection in the evolution of nest construction.

Speciation and Reproduction Isolation : An important question in the study of the evolution of behaviour is that of how species are formed. As already defined a species as a group of organisms that are capable of interbreeding freely under natural conditions but which do not interbreed with organisms of other groups. In the study of the evolution of behaviour, it is differences of at least the species level that are generally of interest. There are many quantitative and even qualitative differences among members of different populations of the same species. In considering the problem of the formation of species (speciation), we consider how such differences come to produce reproductive isolation and the formation of species.

Allopatric Speciation

Two groups of animals are said to be sympatric if they occur together in the same geographical region and allopatric if they are geographically separated. The dominant view among biologists is that speciation requires geographic isolation : that is, that speciation must take place in allopatry. However, some biologists maintain that sympatric speciation occurs. Thus while Mayr (1963) maintains that allopatry is necessary for speciation. Scudder (1974) maintains that alternative forms, including sympatric speciation, are possible. We shall limit our discussion to allopatric speciation.

The process of allopatric speciation begins with a single population that clearly represents but one species. Somehow this population must become divided by a geographical and ecological barrier, so that animals from what are now sub-populations are not longer able to interbreed freely. Populations may be separated by the development of a mountain range during a time of glaciation, by being swept up on different islands or by being isolated in different lakes. During the period that gene exchange between the two sub-populations is prevented, there will be changes in the gene pool within each sub-population. Many of these changes will be related to natural selection as the sub-population adapt to slightly different habitats. The extent and rate of these changes will be a junction of phylogenetic inertia and ecological pressure. Still other genetic changes may result from the random processes, such as genetic drift. The latter factors may be particularly important in small sub-populations.

At some point the geographical barrier is crossed or removed and the two sub-populations interact. What happens then depends on the degrees and kinds of changes within the two sub-populations that have occurred during the period of allopatry. If changes in the gene pool have been minimal, the two sub-populations will freely interbreed and eventually merge to become a single population once more. Alternatively, the two sub-populations may have been separated for so long that no breeding can occur, they have already formed species, however divergence may be less extreme than this. The two sub-populations may have diverged to an intermediate degree, so that hybrids could occur but would be at a competitive disadvantage. If hybrid animals are not as healthy or as fertile as those conceived from mating within sub-populations, the fitness of animals breeding with partners from outside their sub-population would be lower than that of animals choosing mates from within the sub-population. In that case, there would be selection against matings between animals from different sub-populations.

It is at this point after two divergent sub-populations have come together, that reproductive isolating mechanism confer selective advantages. Reproductive isolating mechanism are those biological properties of individuals that prevent the interbreeding of sympatric groups. They develop under the present circumstances because of the competitive disadvantage of hybrids between the two sub-populations. Individuals that breed with members of the foreign sub-population would be at a reproductive disadvantage relative to those that breed with members of their own sub-populations. Thus the reproductive isolating mechanism are reinforced and mating between the two sub-populations becomes infrequent or impossible. They now have become two different species.

Typically, sympatry will occur among previously isolated subpopulations only in part of their geographical range. It is in the areas of sympatry than selection will act against inter-population breeding and foster the development of isolating mechanism. The animals from the two sub-populations that occur in the region of sympatry should diverge more from each other than there living in areas of allopatry. This is the phenomenon of character displacement. Character displacement is the phenomenon whereby animals from two newly evolved species interact so that further divergence between the species results in the area of sympatry.

Darwin's Finches

An excellent example of the manner in which a presumed single population can diversify in adapting to new habitats (called adaptive radiation) is provided by a group of finches living in Galapagos Islands off the coast of Ecuador (Lack, 1947, 1953). Because Darwin studied these birds during his voyage on the Beagle, and they are called 'Darwin's finches'. There are currently 14 species of finches on the Galapagos Islands. They have radiated into a number of very diverse forms. This diversification can be seen most easily in an examination of their feeding habits and bills. The bills of some species are adapted for crushing seeds of various sizes; some are adapted for grasping insects and buds; and others are adapted for probing after insects. The appearance of the bill may also function in displays and hence foster reproductive isolation.

Lack proposed that a flock of ancestral finches may have arrived as the first passerine birds on the Galapagos. They found great potential for exploitation of resources because of absence of both competition from other species and of predator. The result was an adaptive radiation in which the original animal becomes isolated on different islands, adapted to local conditions, and formed the total of 14 species from the single-founding population. Many of these species have evolved into distinctly "unfinch like ecological niches" as a result of the unique conditions on the Galapagos. "The geographical isolation from both mainland birds and from each other, together with the opportunity for rapid diversification in adapting to new habits, probably was critical to the adaptive radiation of Darwin's finches.

Experimental Studies of Reproductive Isolation : A number of researches have attempted to recreate artificially the kinds of interactions that must occur when two previously sympatric sub-populations are reunited (Koopman, 1950; Knight, Robertson and Waddington, 1956; Kessler 1966; Crossley 1974). All studies used fruit flies of the genus Drosophila, and 'geneticist's white rat." Flies of the two closely related species or of two similar genotypes differing in some prominent marker gene are intermixed and permitted to mate. Such flies can either mate with partners of their own genotypes or form hybrid matings. The offspring of hybrid matings are removed and not permitted to reproduce. If this procedure is followed over several generations, reproductive isolation between the population can be seen to develop. Whereas a relatively large-proportion of matings in the first few generations and in control lines may be hybrid matings, the percentage of matings in the selected lines that produce hybrids decreases over 20 to 40 generations.

Crossley (1974) analysed the details of the reproductive interactions between her ebony and vestigial flies after selection against hybrids. She found that the courtship interaction of her two population had been changed over the course of her experiment. There occurred (1) an increase in the tendency of females to resist courtship from males of the strange (i.e., heterogametic) strain and (2) a greater tendency of males to respond to the spelling movements from females of the foreign strain by terminating courtship than to respond similarly to the repelling movements of females of their own strain.

These studies differ from the selection studies in that the experimenters did not select for particular behavioural characteristics. Rather, as natural selection appears to do on occasion, the experimenters selected against hybrids. As a result of selecting against hybrids, increased reproductive isolation and evolutionary changes in behaviour developed. Events that occur when two previously allopatric sub-populations first interact may not be too much different from those that occur under experimental conditions with these Drosophila.

Tradition : We have a better understanding of the role of tradition, the passing of specific forms of behaviour from generation to generation by learning. Compared with genetically controlled behaviour, tradition-transmitted behaviour can spread through a population rapidly, sometimes is less than one generation. One common form is dialect in animal communication, bird song has received the most attention. Males learn song during a critical period, when they initiate phrases heard in their neighbourhood. Male immigrants into a local area may adopt the local dialect. The adaptive significance of this phenomenon could enhance communication between male and female in pair bonding. It may also maintain the separateness of locally-adapted populations.

Migratory animals tend to use the same routes and return to the same places as their ancestors. Presumably the young learn the routes from their elders. Water-fowl, for example, use traditional routes and stop at the same places each year during migration, and grazing mammals use the same traits for many generations. Breeding grounds may also be used year after year, in lek species the mating areas may persist for centuries. Tradition is particularly important in species living in closed social groups with prolonged socialisation of young. Most examples of the social transmission of acquired behaviour are from insects, birds and mammals. Among coral reef fish, French grunts (Haemulon flavolineatum) exhibit social traditions of day-time schooling sites and twilight migration routes Helfman and Schultz (1984). Individuals transplanted to new schooling sites and allowed to follow residents used the new migration routes and returned to the new sites even in the absence of resident fish.

Non-human primates can develop a new behaviour patterns through invention and incorporate them rapidly and rather permanently into their social system. The techniques for washing sand from potatoes and wheat grains introduced by a Japanese macaque (Macaca fuscata) and copied by others in the group. Another type of learned behaviour that occasionally spreads through populations is tool use. Tools enable otherwise generalized species to specialize in particular tasks without evolving morphological adaptation; for example, animals may use sticks or similar objects, rather than specialized beaks, to extract insects from crevices in tree bark.

Sound Production in Crickets

Sound production and reception in crickets have been intensively studied and the neurophysiological mechanisms and their genetic bases are well known. Crickets produce sounds by rubbing their elyta together and behavioural observation and experiment show that temporal patterning is a significant aspect of the song. Analysis of wing movements and spike pattern in motor neurons shows that wing beat frequency during flight is the same as that used during stridulation and that the same muscles and motor-neurons are used in the same ways during these two activities. It seems likely that stridulation movements evolved from flight movements, with minor alterations in auxiliary muscles changing the up-and-down movements of flight to the to and fro movements of stridulation.

There is much less information about the evolution of sound reception in females; sense organs which detect air-borne vibration appear to have evolved from proprioceptors monitoring body movement. Further evolution took the form of increasingly complex higher order sensory processing in parallel with the production of more sophisticated songs by males. This comparative study has identified a possible phylogenetic sequence of the song types, demonstrating, the kinds of changes which have taken place in the overt behaviour and suggesting possible physiological bases for these.

Courtship in Peacocks

A constellation of similar evolutionary changes are believed to have occurred in many cases where related species show homologous display movements. These are well-illustrated in the classic study of the evolution of courtship displays in peacocks. The proposed course of evolutionary event is as follows : in the ancestors of the peacock (presumed to resemble the modern domestic; fowl Gallus domesticus), during courtship the male gave a call like that used to attract companions to a new good source, while scratching the ground. Stepping back and picking up pebbles, which served to attract females.

The movements became exaggerated as in the Impayan pheasant (Lophophorus impejanus) in which the male bows rhythmically before the hen with wings and tail expanded and in the peacock pheasant (Polyplectron bicalcaratum) which scratches the gound and bows with spreading wings and tail, offering any available food to the female. Finally, feeding disappears altogether in the peacock itself, which erects it elaborately, coloured tail feathers and shakes these in front of the female, taking several backward steps and bowing.

Dance Displays in Honeybee

Honeybee dancing in such a highly specialized and unusual behaviour that its evolution has understandably been the topic of considerable interest and speculation. Martin Lindaver has used the comparative method creatively to analyse the evolution of this behaviour. Because all populations of the honeybee perform round and waggle dances, the logical group to examine first is the genus Apus to which the honeybee (Apis mallifera) belongs.

The three other Apis species are tropical bees, and all except A. florea exhibit the capacity to perform dance displays on vertical surfaces. This small bee dances on the horizontal surface of the comb built in the open over a limb. To indicate the direction of a food source. A. florea simply orients her waggle run directly at it. Because this is a less sophisticated maneuver than the transposed pointing done on vertical surfaces, it is like a relative ancient form of dance communication. (Moreover, A. florea builds a simpler nest, a single comb, instead of the multiple combs of the other bees).

The transition to vertical surface dancing with coding of sun-compass information in terms of gravity may be favoured initially because there is more vertical surface area on a comb to accommodate more recruiters and their recruits. In the case of A. mellifera, the close packing of vertical combs in a tree trunk or hole in the ground leaves little horizontal surface. Just how the first dancers happened to transpose a sun-oriented dance into an accurate gravity-oriented dance is somewhat puzzling. However, it is known that a number of other insects make crude conversion of this sort. Moreover, moving upwards is moving roughly towards the sun; moving downward leads the bee roughly away from the sun (especially in the tropics). A pioneer bee dancing on a vertical surface might, therefore, be reasonably expected to adopt a code similar to but obviously not nearly as detailed as, the modern display of honeybees.

In any event, because all Apis species have both round and waggle dance displays, one has to look elsewhere for other patterns possibly similar to ancestral behaviours. Not all social bees communicate about foraging sites. However, the stingless bees of the sub-family Meliponini do exhibit an exciting range of communication system.

A worker of the species Trigona droryana that has found high-quality nectar returns to the nest and there performs unstructured excited movements coupled with a high-pitched humming. This arouses its hivemates, who request food samples from the dancer and presumably detect the odour of the source on the dancer's body. With this information they leave the nest and search for similar odours. The actions of the recruitor do not offer any specific cues about the direction and distance to the desirable food, as experiments have provided.

Trigona postica workers, on the other hand, do convey this information, although in a way quite different from the dances of the honeybee. A worker that makes a substantial find marks the area with a pheromone produced by its mandibular glands. As the bee returns to the hive it continues to chew on grasses and rocks very few yards. At the hive entrance there may be a group of bees waiting to be recruited. The forager cooperates by leading these individuals back along the trail it has marked. Martin Lindauer and his associates tested the importance of the scent guided by placing a hive on one side of a pond and then gradually training some foragers to come to a food supply directly on the other side. Now the bees always flew across the water, where they were of course unable to place any odour marks. They did try it, by flying again and again very low over the water surface. They then repeatedly returned to the shore, where they generously bestowed scent marks all over the feeding table, and the notebook and hat of the observer.

During the whole feeding time of two days, not a single newcomer appeared at the feeding places. Then we strung a rope across the water, and hung a few twigs at 2-3 meter intervals. This now served as a bridge for the newcomers. Within 35 minutes, 35 newcomers reached the goal and we could conclusively observe how all the marked collector bees were eagerly placing odour marks on the twig.

Trigona postica lives in tropical areas with dense vegetation, thick tree trunks, and other obstacles between the hive and a source of nectar. Personal leading is a very effective way to overcome these obstacles as well as a useful strategy for attracting newcomers to nectar high in flower-bearing tropical trees. It is instructive that honeybees cannot communicate that a food source is upward rather than outward. Von Frisch trained some would be recruiters to come to a dish of sugar water up in a tower. Although the trained bees returned to the hive to dance, no recruits ever came to the food source, instead, they moved parallel to the ground the 'correct' distance from the hive rather than flying up to the top of the tower. Honeybees have evolved in a more open habitat, where food is generally found close to the ground and rarely in high trees.

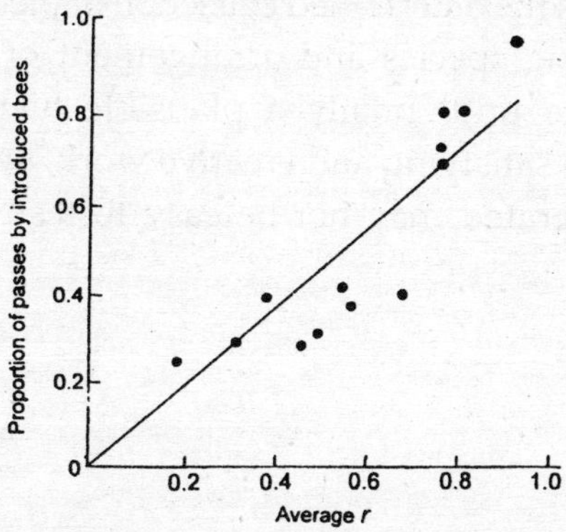

Graphical representation of reproductive success and territory

A number of species belonging to the genus Melipona separate distance and direction information, unlike T. postica. A dancing forager communicates that food is near by producing pulses of sound of short duration. The longer to transmit directional information, the recruiter leaves the nest with a number of followers and performs a short zigzag flight that is oriented towards the source of nectar. The lead bee returns and repeats the flight a number of times before flying straight off to the nectar site with the novice bees in close pursuit. Evidently the zigzag flight alert the recruits to the general direction of the upcoming straight flight.

Thus a comparative approach makes, the dances of honeybees at least somewhat more comprehensible. Like all other behaviour, the dances have an evolutionary past, a gradual history that can be tentatively reconstructed given the necessary comparative material. This does not, however, make the dances of the bees any less wonderful examples of the adaptive effects of the natural selection.

The examples of behavioural evolution presented in the preceding pages are sufficient to illustrate how one can make reasonable guesses about the history of a behaviour with no fossil evidence at hand. In a sense, this kind of enquiry into the origin and subsequent evolution of a behaviour is a detective work. It involves the determined collection of pieces of evidence from many living species and arrangement of the data in a logical sequence until finally a plausible historical pattern emerges. This is satisfying and creative work, even though the hypothesis generated may not be easy to test.

2

Behavioural Studies

Not all behavioural studies take place in well equipped laboratories or in forest, casual observations on pet, domestic or familiar animals also contribute vital information on animal behaviour, therefore, ethology begins from home; it is also a fact that many people watch animals but, few really observe them. Careful or keen observations are different than casual watching. The acclaimed ethologists Karl Von Frisch, Niko Tinbergen and Konrad Lorenz devoted their lives and accumulated important information on behaviour by staying with animals in their natural habitat or by keeping them at their farm houses.

Some other names have become completely associated with the animals they studied e.g. Jane Van Lawick Goodall has spent more than 40 years in the jungles of Tanzania in Gombe National Park to study chimpanzees, Dian Fossy devoted her life studying gorillas in Africa, she used to spend so much time with the gorillas that she was virtually accepted as one of them by the gorilla group. Fossy's extra concern and protectiveness about gorilla population led to her murder by the poachers (the persons involved in illegal hunting). Similarly, George B. Schaller studied tigers and Himalayan fauna, and wrote excellent books, Irwin DeVore carried out exhaustive observations on baboons. Name of John F. Eisenberg is inseparable from the eco-behavioural studies of most of the mammals.

Salim Ali from India has been known all over the world for his most remarkable contribution in Ornithology. Like them not many of us would sit all day in jungles or in front of a cage and spend our entire life studying animal's behaviour. The hazards of encountering carnivores, snakes, insects, and poisonous plants, camping, bad weather, variable food supply in the forests and monotony would be discouraging. Although we all have watched animals around us, inside the Zoological Gardens and Sanctuaries for recreation but scientific observation requires specialized training and procedures. Animal behaviour is a very vast subject, it includes innumerable types of behaviour viz feeding, drinking, fighting fleeing, territoriality, dominance hierarchy, mating, nest building, parental care, social organization, daily or seasonal activity, orientation, migration, mother-infant relationship, inter and intra specific and prey-predator relationships.

The short, specific and scientific methods for studying animal behaviour have been introduced recently. Earlier, people studied animals in a naturilistic way in their farm houses, backyards, and nearby jungles. As mentioned earlier, the great ethologists of earlier times devoted their entire lives studying different animals, though this is the best and most authentic way of studying behaviour but then, one needs a great deal of patience to spend entire life with animals. Not all who have been interested in behaviour get this opportunity. Therefore, quick and reliable methods were formulated by different workers.

Before starting data collection on behaviour the aims,

objectives and subject are ascertained. Study of behaviour can be expanded from body physiology, to individual, to groups, population and species. The first and foremost thing is to decide the behavioural processes of interest. For example, a study may be carried out to know how an animal obtains food and water and which food does it select; another study may focus on how females wean their young ones in monkeys and apes or how adult males take possession of a harem and decide dominance hierarchy in a deer group. One may be interested in knowing how do animals budget their time daily, monthly, seasonally and yearly? When animal behaviour is broken down into such segments of study, observations are easier to make and this makes the study systematic and scientific, easy to quantify and interpret otherwise, ethology is nearly a limitless discipline. This subject can be studied in lab under highly controlled conditions and in field under natural conditions.

Psychology and Ethology both relate to behaviour. The complete understanding of behaviour and its neural mechanism is gained through several methods. The psychologists carry out research mostly in the laboratories on animals like cats, rats, pigeons, and chimpanzees. They keep the animals in confinement, try surgical procedures like removing parts of brain, or pass mild electric current through various regions of the brain or introduce small amounts of chemicals to evoke behaviour. Whereas, ethologists study behaviour of animals in the wild, by allowing them to move freely in their natural habitat, they conduct studies on a wide variety of animals, from weevils to whales.

The study of behaviour can have two different approaches:

(1) With relation to brain and physiology. These studies are carried out in laboratory by utilizing following methods : (i) Neuroanatomical; (ii) Neurophysiological; (iii) Neurochemical

(2) Study of behaviour in wild or in natural habitat in relation to environment and ecology by using following methods: (i) Ad libitum (ii) Focal animal sampling technique (iii) Scan sampling technique (iv) All occurrence (v) One zero

Physiological research is rarely carried out on large sized animals but small sized animals can be used in both instances.

Studies in Laboratory

The first major goal of brain researchers or psychologists was the localization of brain functions, to know which areas or structures of the brain were involved in controlling particular behaviour.

Neuroanatomical or Lesion or Ablation Technique : This is the oldest and also most crude method in neuroethological research for studying the relationship between particular neural structure and behaviour. For this purpose, certain regions of the brain are destroyed and their function is deduced from the abnormal behaviour it causes in the animal. The larger ablation (removal of tissue) or lesion (pathological changes in tissue) may be carried out free hand by means of knife cuts; whereas, the localized and small lesions are produced by passing electric current through platinum, iridium electrode to cauterize the area, small lesions can also be created by neurotoxic kainic acid. Lesion methods are different than stimulation method, in the latter case neurons are activated while in the former they are inhibited or destroyed. This technique helped many workers to form brain atlases or brain maps, which are known as Stereotaxic atlases that are presently available for many mammalian species viz. rat (De Groot, 1959), cat (jasper and Marsan, 1954) and dog (Lin and Coworkers, 1961). The lesions were produced by careful surgery, this particular technique was first practised by medical doctors, who used to examine patients of head injury, these doctors observed that patients showed peculiar behavioural changes due to injury in head, it was rightly correlated by them that behaviour is controlled by brain. While performing the operations, or post-mortem autopsies the destructed areas of brain were identified, and impairment in behaviour was correlated with it, indicating direct influence of brain over behaviour.

Broca (1861) was a doctor who came across a patient with head injury whose speech was defective, this correlation fascinated him and he started observing different patients with similar injuries. After few years, he concluded that there is a definite area in the cortex which is responsible for speech, later using the same technique another scientist-Wernick (1880) established another area in the brain responsible for speech, these two areas are known as Broca's and Wernick's areas respectively. Broca's area controls the infrastructure for speaking a word, this means it has control over neck muscles and larynx, and Wernick's area is word retriever or dictionary. A lesion in Broca's area causes slurring in speech and there is difficulty in understanding the words spoken by such a person though the person would talk sensibly. Opposite to it, if there is a lesion in Wernick's area then the person can talk clearly but the speech will not be articulated or coherent.

Toads have been the major objects for ablation techniques; the experimental finding that there are zones in different areas of the toad brain responsible for the triggering of prey catching (optic tectum) and avoidance behaviour (TP-region) is supported by ablation experiments.

Today many areas in the brain responsible to carry out particular behavioural types have been identified using neuroanatomical techniques. In an experiment, rhesus monkeys were subjected to choose between objects of varying shapes and were rewarded by banana for picking up the triangular one. Monkeys with their visual cortex removed could no longer perform this test, thereby indicating that visual cortex is the part involved in learning to discriminate between various shapes.

The first indications of brain structures that influence aggressive behaviour were obtained by lesion experiments which suggested that amygdala, the medial hypothalamus and periaqueductal grey of the midbrain are responsible for the control of aggression and defence in cats. Classic experiments were done by Lashley (1938) using ablation technique. His experiments proved to be very important in understanding the neural mechanism of learning. Deficits produced by different brain lesions on standard learning tests such as maze learning, object discrimination, avoidance learning and lever pressing have been proved to be very useful one of his famous experiments was performed on rats. The rats were trained to run in a maze, after they learnt it, he removed a small part of cortex and assessed the impairment in learning ability, he kept removing cortex bit by bit and after the completion of experiment he proposed that the impairment in learning was directly proportionate to the amount of cortex removed. 'According to him the majority of learned tasks are equally distributed and stored by cortex. On the basis of this experiment he proposed "the law of equipotentiality".

Discovery of different nuclei of hypothalamus and their role in controlling vital activities like feeding, drinking, mating, aggression and parental behaviour were made using neuroanatomical techniques.

Perhaps the most quoted lesion experiment deals with eating behaviour. Lesions in hypothalamus can lead to adipsia, hyperdipsia, aphagia and hyperphagia, it can make an animal extremely overweight or under weight. In most of the animals lesion in ventromedial hypothalamus increases their hunger to such a great degree that they may eat themselves to death, indicating that the destroyed area had satiety centre, on the other hand, a lesion in the lateral hypothalamus can cause the rat to stop eating for ever so that it eventually starves to death because the destroyed part contained the feeding centre. Experiments conducted using brain lesion technique helped identifying gratification centres of basic needs. This technique has also contributed in locating sensory and motor regions in the cortex. Tedious experimentation has brought to the knowledge that removal of tissue from frontal lobes of brain caused paralysis or inactivity in muscles, and removal of parts from parietal lobe proved to affect sensory activities, these regions of brain were identified as motor and sensory cortex respectively.

In one experiment, bilateral ablation or lesion in amygdala (part of limbic system), enhanced the sexual behaviour of a male cat to such a degree that it tried to mount and copulate with inappropriate mates like dogs and monkeys. This hypersexuality was reduced by forming a lesion in the anterior and ventromedial nuclei of the hypothalamus indicating that there are specific areas in the brain which start and stop the sexual behaviour.

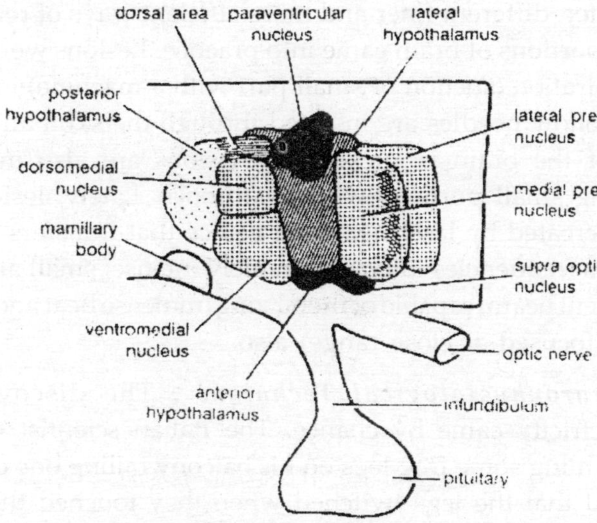

Nuclei of hypothalamus.

In earlier years, neuroanatomical techniques made sufficient contribution to understand the mysteries of the mind but between 1870-1900 ablation technique was dropped by most of the neurologists and was considered crude, because when a particular part of the brain was removed surgically, a portion of surrounding brain tissue was also damaged, it became difficult to rule out whether the impairment in behaviour was due to the specific part removed or due to the damaged surrounding tissue. When critically evaluating lesion experiments, it must also be noted that an animal with a brain lesion is possibly no longer the same animal in many respects apart from the obvious lack of this brain region. The brain represents a complex feed back system, when a component is absent, the whole system is modified. For example, lesion in ventromedial hypothalamus also destroys noradrenergic bundle, lesion in lateral hypothalamus usually damages the nigrostriatal bundle containing dopaminergic fibres, these fibers are involved in mediating sensory and motor functions.

Later, different finer and more efficient ways of removing small portions of brain came into practice. Lesions were made by aspiration (suction of small part with a machine) in which thin hollow needles are inserted through the skull and small part of the brain is sucked out, lesions are also made by freezing small portions with cryoprobes. Lately, lesions are being created by laser beams (a device that transfers light of various frequencies into an extremely intense, small and non-divergent beam; capable of mobilizing immense heat and power when focused at close range) also.

Neurophysiological Technique : The discovery of bioelectricity came by chance. The Italian scientist Galvani (1786) hung some frog legs on his balcony railing one day and noticed that the legs twitched when they touched the metal fence. Galvani at first misinterpreted this phenomenon attributing it to muscles alone. By 1793 Galvani discovered that twitches occurred in the nerve muscle preparations even without any metal; the nerve and muscle were producing electricity by themselves. This finding laid the foundation for electrophysiology.

The discovery that the messages in nervous tissue travel in the form of electric current led to a less destructive and more precise neurophysiological method to study behaviour.

This method involves two ways:

(1) Recording of electrical activities from brain while animal is behaving.

(2) Stimulating the brain region and eliciting behaviour.

During the German-French war of 1870-71, Edvard Hitzig electrically stimulated the exposed brain of a wounded, unconscious soldier, with the result that the soldier moved his eyes. Previously Fritsch and Hitzing (1870) had used same technique on dogs and observed that stimulation of different regions of the cerebral cortex produced contractions in different muscle groups of the body. David Ferrier (1873) performed similar studies on monkeys and Bartholow (1874) on man. Both scientists found that particular motor patterns of movement could be released by stimulation of defined regions of the cerebral cortex and they charted such sites on brain maps. At the end of the 19th century Ferrier was able to roughly localize in the cerebral cortex the sensory centres for vision, auditory, gustatory and olfactory sensations; with the advent of new technologies and scientific knowledge sensory and motor cortex have been mapped with astonishing precision.

It became technically possible to record the electrical potential of active nerves with the aid of implanted electrodes and to amplify this activity by a highly sensitive instrument, where it could be detected and measured. Early in the 19th century scientists knew that nerve tissue responds to electric stimulation; at this time neurophysiological methods came into use for the purpose of evoking behaviour artificially.

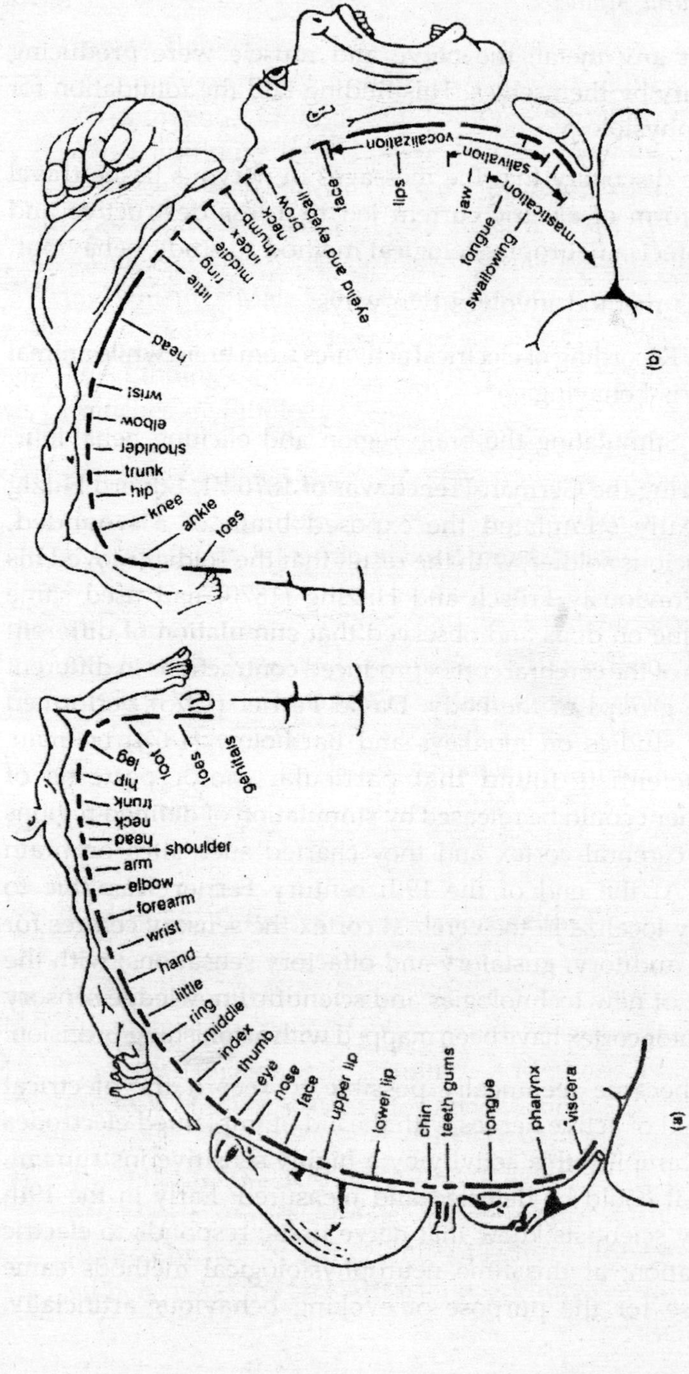

Schematic diagram showing correlation between the body parts and size of cortex controlling its function : sensory (a) and motor (b)

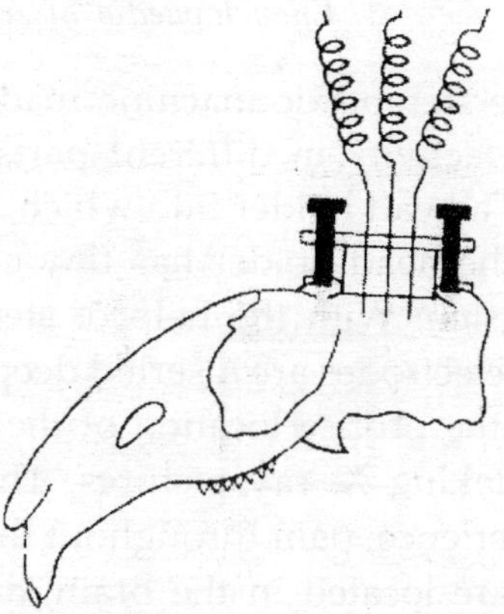

Rabbit Skull with electrodes and holder.

Electrical brain stimulation for deriving the correlation between structures eliciting behaviour was first performed by R. W. Hess (1931) in freely moving cats. This technique has now become a routine tool of neurophysiologists and neuroethologists. The procedure involves cutting the scalp in anesthetised animal, removing portion of the skull by means of special drill, exposing part of the brain surface by cutting dura-mater. Two or more electrodes, each insulated except at the tip are then lowered into the brain region to be examined and fixed to the skull with a quick hardening substance. However, in large animals, a head holder with adjustable electrodes is used. In electrical brain stimulation experiments the threshold current plays an important role. It is presently accepted that different behaviour patterns are often characterized by varying levels of response thresholds in central nervous system e.g. in an experiment conducted on toads the electrical threshold for release of escape behaviour from optic tectum is lower than that for prey catching from caudal thalamic pretactal region. Further more, within a single mode of behaviour, such as the prey catching behaviour, the sequence of the different behavioural components-orienting, snapping, swallowing is assured by different and graded threshold at the same point in the brain.

Invention of the stereotaxic machine made it possible to study the electrical activity in different parts of brain. This machine consists of a head holder into which, the skull of the animal is placed. The head holder has tiny electrodes made up of steel or tungsten. With the help of stereotaxic atlases (brain maps), these electrodes are inserted deep into particular areas of brain and the precise location of their tips can later be ascertained by taking X- ray pictures. The experimental animals do not experience, pain throughout this procedure as no pain receptors are located in the brain, and the surgical wounds in the skull are small and heal quickly. The electrodes are connected with an electric impulse generator. The first successful recording of brain waves from the skull of man was accomplished by Berger in 1929; who named this technique as EEG- Electro Encephalogram. Today, the results of brain stimulation experiments are available from a large number of vertebrates, primarily birds and mammals, and from several invertebrates, especially insects, crabs, and crayfish.

A real break in the field came in 1930 with the work of RW. Hess in Switzerland for his work on cats. His research

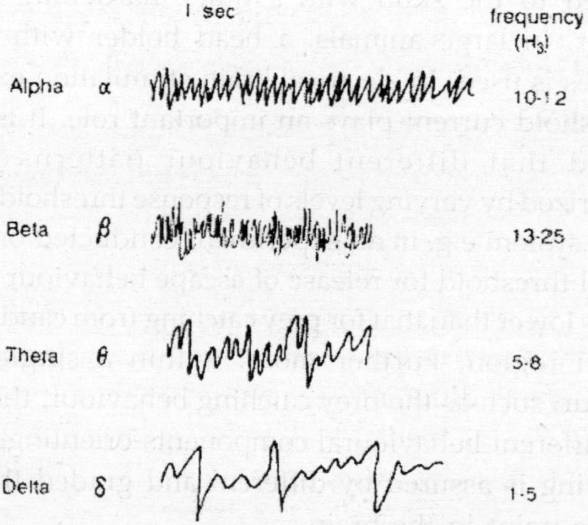

Types of electrical activities in the human brain.

was honoured in 1945 with the Nobel prize. Generally, the neurophysiological experiments can be carried out in two different ways. The first one involves recording of the normal electrical impulses in brain and correlating them with behaviour. Basically four types of electrical brain waves have been recorded in human beings from the brain: Alpha waves from parietal and occipital lobes when the person is at rest with eyes closed; at this time brain is alert but unoccupied. These waves are also produced when we are awake, but in a peaceful, relaxed mind.

Beta waves are produced from frontal lobe when brain is stimulated by sensory input or mental activity, these waves occur at the times of concentration and thought, and in normal daily alertness; Theta, waves are generated from temporal and occipital lobe under emotional stress or when brain receives unpleasant stimuli; theta waves are associated with earliest stages of sleep, hallucination and creativity; and lastly Delta waves are the result of deep sleep.

The second method is the study of behavioural patterns that are initiated as the result of electrical stimulation delivered to a specific brain region by implanted electrodes. For example, by stimulating a particular locus in the diencephalon of a cat, J. Flynn (1929) elicited prey catching behaviour. A. Zukerman (1965) was able to release behaviours like fleeing, courtship and threat by electrically stimulating different regions of a pigeons brain. E. Von Holst and Von Saint Paul (1960) were able to elicit sleeping behaviour by stimulation with an electrode placed into the brain stem of a cock.

Similarly, T. D. Delius (1958) located two areas in the brain of a cat for two different behaviours; one was motor cortex and the other was medial thalamus. Holst and Saint Paul in an experiment observed that when an area in diencephalon is stimulated electrically in chickens by implanting electrodes, the chicken assumed a fighting posture against a toy cat, threatened or attacked it and when the electrical stimulation was stopped, the chicken stopped all these activities on the other hand, if the stimulation continued the chicken flew away.

Another interesting finding made by Von Holst and Von Saint Paul (1960) was that when two separate areas of the brain responsible for different behavioural patterns are stimulated simultaneously, very different responses take place, e.g. suppose two areas in brain X and Y are stimulated, simultaneously the following variation might occur: superposition of both patterns (X + Y), average (X + Y/2), alternate (Y,X,Y or X,X,X), disappearance (X + Y = 0), change to another pattern (X + Y = Z) or suppression of one of the patterns (X or Y). The changed behavioural pattern was shown by T. Brown and his associates (1967) in an experiment with cats, they identified two separate loci in a cat's brain that induced growl and flee, simultaneous stimulation of these two loci produced attack behaviour.

Using the same technique of electrical stimulation by electrodes certain brain areas are identified in human brain, as pleasure centres. During clinical tests C-Sem Jacobson (1967) found that patients sometimes experienced a stimulation through an electrode as pleasant. Some of these patients reported sexual pleasure and some described it as general feeling of Well being. Unfortunately, our hopes of treating depression by administering this kind of brain stimulation remain futile, because certain undesirable side effects like addiction cannot be eliminated. But the diseases can be controlled using this technique.

Since childhood, Ray a boy in USA had suffered from several attacks of epilepsy. During these attacks he used to become violent and unreasonable, even homicidal. Drug treatments were tried and found to be ineffective. Later, a small electrode was surgically implanted in Ray's brain and attached to a stimulation device which he carried with him, the device had a remote control his instructions were to press the button of his brain stimulator whenever he felt an attack coming on. Ray never got a complete attack after the operation. He was able to eliminate his violent obsessions by simply pressing a button and sending electrical impulse to the part of the brain which inhibited the attack of epilepsy.

Though precise to single cell recordings by microelectrodes is possible still the neurophysiological techniques have problems regarding their application in stimulating specific areas, for example,, in rat's hypothalamus electrical stimulation at one point elicit both eating and drinking behaviours though not at the same time. Initially it appeared that the two behaviour patterns were controlled by the same network of neural connections but later it was found that eating and drinking were controlled by two different and independent neural circuits which are very closely set to each other, and therefore the stimulus current can activate both circuits at the same time. Fisher in 1964 found out that chemical stimulation activates only those circuits, for which that particular chemical stimulus is specific. S. Grossman (1966) discovered a method for activating eating and drinking circuits separately. He used a particular canula to inject noradrenaline at one particular point and could evoke only eating behaviour.

The nearby circuit for drinking behaviour did not get excited by the introduction of noradrenaline; this confirmed that there are two different centres for eating and drinking which are very closely set, electrical stimulation can cause excitation of both centres, but if activity of one of the circuits is inhibited by using neuro-inhibitors, it is possible to excite only one behaviour. Electric stimulation frequently activates several neuronal circuit (thin lines,), simultaneously releasing various behavioural responses resulting in overlapping effects. Chemical stimulation, on the other hand, activates mainly the circuit (thick line; for which the chemical stimulus is specific (Fisher, 1964). After such discoveries use of drugs came to practice and opened a new field of research.

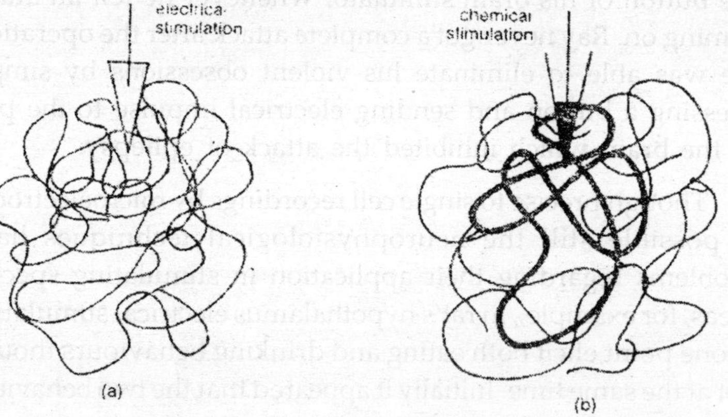

Schematic representation showing (a) several neural circuits excited simultaneously after electrical stimulation and (b) selectivity of chemical stimulation.

Neurochemical Technique : The electrical brain stimulation method by itself has only a limited significance for studying the behaviour, as discussed earlier, electrical stimulation does not unequivocally show whether the stimulated area takes part directly or indirectly in the activation of a behaviour pattern. Therefore, the search for a possible chemical coding of behaviour led to the development of "chemical brain stimulation techniques". In these methods, neurons are affected more selectively in their activity by local application of drugs, neuro-transmitters, blocking agents and hormones.

The use of certain substances such as alcohol, opium, hashish to abolish pain, to induce euphoria (sense of well being), relaxation and sleep has long been practised by men. Addictive use of "bhang" (Cannibis) in India has been age old, festivity, fun, frolic are attached to it. Addictive use of drugs and alcohol has led many to criminal path. For the Yanomamo Indians of northern Brazil and Venezuela, drugs sometimes cause so intense alteration in behaviour that they become violent.

Introduction of certain chemicals and drugs orally or through blood or muscles or directly on brain parts produces behaviour. For example, when people drink alcohol or take opium they develop a feeling of well being and derive much pleasure. These drugs are absorbed and sent to blood, through which the drugs reach brain and when they reach specific target areas in the brain euphoric feeling is developed. To locate the exact centre in the brain the drugs are directly poured or introduced into different places of brain and when the person describes a feeling of well being it is ascertained by the experimenter that this particular part is the centre for pleasure. After the discovery of such centres in the brain and their stimulation studies brought forward many drugs which helped solving problems like insomnia (sleeplessness), depression, etc. The clinical and experimental use of psycho-active drugs viz. tranquilisers, sedatives to fight insomnia and depression is of recent origin. There are other drugs which stimulate mental state, give a feeling of hallucination and are very addicting, lysergic acid (LSD) is one popular example, such drugs are called psychomimetic drugs.

Drugs can be administered via many routes intraperitonially, intramuscularly, subcutaneously or orally. Accurate method of studying behaviour with the help of drugs is the introduction of chemicals directly into specific brain sites. For example Fisher in 1956 showed that sexual and maternal behaviour can be induced by the injection of minute quantities of sex hormones into the regions of hypothalamus. Transmission across synapses is carried out by means of chemicals or neuro-transmitters. There are certain drugs which can block their activates. A portion of brain which is responsible for the release of aggressive behaviour if treated with antineuro-transmitter or neuro-inhibitors will immediately block the aggressive behaviour because of blockage in synaptic conduction.

In order to learn about the glucose sensitivity of the neurons, blood sugar was replaced by goldthoiglucose, (GTG), the organism is unable to distinguish this substance from glucose, as soon as GTG is taken up by cells with glucose receptors it destroys them, it was discovered that cells in ventromedial hypothalamus were glucose receptor and formed a part of satiety centre, after they were destroyed, the animal showed hyperphagia.

In another experiment, minute amount of acetylcholine in rat's hypothalamus produced drinking behaviour. Acetylcholine excited or stimulated the drinking centre and the animal started drinking. R.P. Mitchell (1968) administered oestrogen with a small canula into the brain of female cats when they were not in heat, still the cats responded immediately by inviting tomcats for mating. Oestrogen was introduced in anterior hypothalamus, if it was put in any other area of brain, it did not produce this behaviour. Radioactive oestrogen injected into blood stream was absorbed only by anterior hypothalamus.

Testosterone is a male sex hormone, if in a male or female rat it is injected into the central area of the preoptic region located in front of the hypothalamus, using a fine injecting instrument and a thin canula, the rats of both sexes respond with maternal behaviour, they start building nests. If testosterone is injected slightly lateral to the central area of the same general brain region, male as well as females start exhibiting male sexual behaviour.

The chemicals can affect behaviour in different ways. Depending upon their nature, chemicals and drugs have been classified as:

1. *Physiological Agents:* Adrenalin, histamine, serotonin,

lutamic acid are known as physiological agents. They are present normally in the body but they are considered drugs when administered in concentrated dose.

2. *Foreign Agents:* Reserpine, barbiturates, lysergic acid fall under this category. They are not normally present in the body. According to another classification neuro-active drugs could be identified as tranquilisers, energizers and hypnotisers. Tranquilisers reduce agitation, excitement and emotion; energizers or stimulants include caffeine, amphetamine, they tend to produce hyperexcitability and also facilitate sensory and motor performance, they also act as anti-depressants. Whereas, hypnotisers such as barbiturates are the drugs that produce sensory and motor depression and deep sleep. In large doses, they cause unconsciousness and even death.

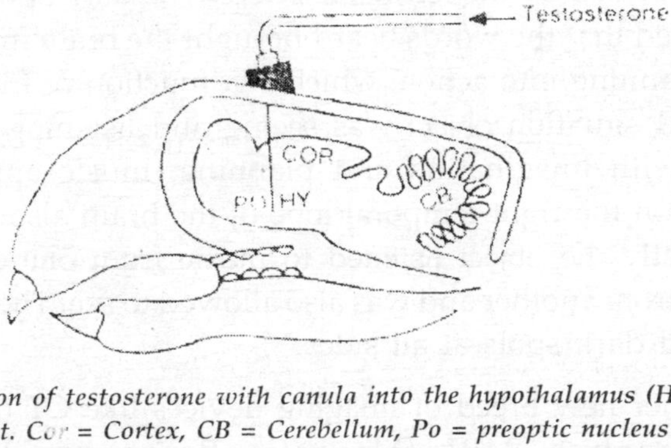

Introduction of testosterone with canula into the hypothalamus (Hy) in rabbit. Cor = Cortex, CB = Cerebellum, Po = preoptic nucleus

Most recently the brain is being explored for its various functions using tomography techniques (any method that produces images of cross section of the body reconstructed by a computer programme, most widely used technique with reference to behaviour is the PET-Positron Emission Tomography). There are millions of unanswered questions regarding brain and behaviour; these techniques are opening new gates to the unreachable and unexplored areas of the brain. It has been investigated by a team of workers that there is more blood supply to the part of the brain which is actively controlling behaviour at that point of time. For this, doctors give the person to be tested a special solution of sugar - this sugar is tagged with a radioisotope brewed in a small, low energy cyclotron, by tracing the radioactive substance, a doctor can pinpoint areas of normal and/or abnormal activities. The PET scanner detects the rate at which different parts of the brain "burn" or use up this special test sugar. A computer attached to the PET scanner shows the activity in different colours on a TV Screen.

In an experiment the object (person) was allowed only to see. Dark spot at the back of the brain indicated that the eyes of the object were open and the visual cortex was active. The object was allowed to see and hear words from friend. A dark spot appeared on the left side of brain. The spoken language was analysed by the left hemisphere. The dark spots in front indicated that the words heard brought the brain imagination and planning into action, which is a function of frontal lobe. In third situation object was seeing and listening to music. Alongwith imagination and planning, music appreciation centre on the right temporal lobe of the brain also got active and lastly the object listened to music from one ear, heard friend from another and was also allowed to see. The PET Scan showed dark spots at all sides.

Other new breed of imaging devices like CT (computed tomography), MRI (Magnetic Resonance Imaging), sonography, DSA (Digital subtraction Angiography) have been developed which enable doctors and researchers to watch vital organs at work along with the brain and behaviours.

Each method described here has its merits and demerits. A comprehensive and balanced understanding of neural function can be gained from the integration of results obtained with different techniques.

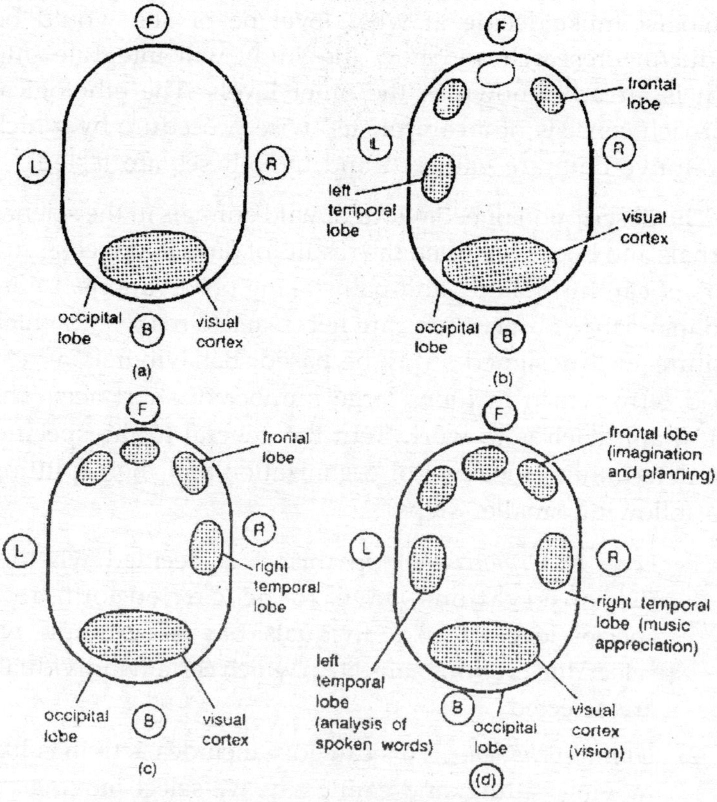

F=Front B=Back L=Left R=Right of brain

Diagramatic representation of PET Scan photographs (a-d).

Studies in Wild

Behaviour can be studied at different levels, it can be studied at the level of whole species (e.g. courtship); population (say migratory pattern in a particular population); family groups (e.g. flying to food source); dyad (agonistic encounter between individuals); individual (swimming pattern in an individual); behavioural type (way of ingestion), behavioural act (catching the prey), body parts (hind legs, their movement and usage), muscles (gastrocnemius), neurons (tibial nerve). These levels make ethology a limitless discipline, therefore an ethologist must decide at what level he or she would be conducting research and how the study will integrate into what is already known at the other levels. The ethological approach consists of a careful step-wise procedure by which descriptive data are gathered and hypotheses are tested.

The description of behaviour of wild animals in the science journals and books are often the results of days and weeks and years of careful stalking and painstaking observations. Initial reconnaissance observations are necessary to make, on which a future well designed study be based. Behaviour is a very broad term which includes large number of short acts. The logical approach is to work from the general to the specific, first determining the level of organization and then splitting into following smaller steps.

1. *Level of Organization:* In this, it is decided whether the observations have to be carried out at a species level or an individuals. Say for example we select ducks as our subject, in which certain individuals are selected.
2. *Unit of Behaviour*: This category includes activities like moving, sitting, and eating say, we select moving i.e. we wish to carry out observations on moving ducks.
3. *Pattern of Behaviour:* Unit of behaviour we have selected can have following patterns — (i) walking, (ii) swimming, (iii) diving, (iv) flying. With this, one can select all of them or anything say, flying is selected.
4. *Act of Behaviour:* With the pattern of flying there can be many acts involved- (i) take off, (ii) flight, (iii) landing, say from this take off is selected.
5. *Component of Behaviour:* This includes - (i) Anatomical (movement of body parts e.g. head, wings, legs), (ii) Neurological. (involvement of nervous system), (iii) Physiological (movement of muscles and bones).

Ethologist mainly concentrate till anatomical components of behaviour and psychologists generally start their work from remaining two components viz. neurology and physiology.

Observations of animal behaviour under natural conditions are extremely interesting. The important steps involved are :

(i) selection of subject,
(ii) what to study,
(iii) what are the questions and objectives,
(iv) stating research hypotheses, design of research,
(vi) chalking out the list of behavioural acts of interest,
(vii) deciding an appropriate methodology,
(viii) reconnaissance observations,
(ix) taking notes without disturbing the animals in their activities,
(x) interpretation of data.

All studies in wild must start by observing animals the whole day from dawn to dusk and later on a working schedule can be set up. Depending upon the animal's activities. The notes could be taken for 10-15 minutes at a time followed by a 3-4 minutes interval, the observer should be close enough to make accurate observations, but not too close to disturb or distract the animals. The clothes one wears also matter, dark flashy colours should be avoided and browns, greens be preferred.

Observing wild animals is not an easy task. One has to be very patient and tough to remain in the field, following and taking notes on animals of interest. Forest, savanna, desert, rain, snow or hot sun, the ethologist has to be in the field. There are many ways to record behavioural activities in the field. First and most interesting is the naturalistic way. That means the person has to spend most of the time with animals or virtually has to live with them and take notes on animal's behavioural repertoir - great studies have been done like this coupled with scientific methods by Konrad Lorenz, Karl Von Frish, NikoTinbergen, Jane Goodall and Dian Fossy.

They lived in large farm houses or protected forest and were naturally surrounded by variety of animals, they learned to identify animals individually, took elaborate notes on them. Konard Lorenz had greylag geese in his farm house, his keen observation and life long experimentation brought forward the concept of imprinting. Niko Tinbergen and Karl Von Frish conducted remarkable studies on a wide variety of animals, their work on digger wasps and bees, respectively is still acclaimed worldwide. Jane Godall has been observing chimpanzees of Gombe National Park, Tanzania since 1957, she has spent her entire life studying them, she recognizes each and every chimpanzee living in the national park by its face and she can trace their family trees; Dian Fossy spent much time with gorillas in Africa, the 400 lbs great ape virtually accepted her as one of their group members. Not many people have this kind of devotion or get the opportunity of living with animals.

Therefore, other methods were adopted to record behavioural acts in a comparatively short time. Notes taken haphazardly can not be quantified, the data must be collected in a systematic way which should be reliable and reproducible. While recording behaviour- states, events and bouts are noted down specifically.

State: The behaviour in which an animal (or group) is engaged in; an on going behaviour (a parrot flying). Basically, a state is a behaviour which can be timed with a stop watch.

Event: A change of state; it approaches an instantaneous occurrence (e.g. a robin taking off or landing) events generally occur so rapidly that they can only be counted.

Bout: The term bout is generally applied to (i) a repetitive occurrence of the same behaviour (e.g. a bout of pecking) or (ii) a relatively stereotyped sequence of behaviours that occur in a burst (e.g. courthip display)

While taking the field notes it is made sure that they are complete, accurate, clear and concise. Ethogram (set of comprehensive descriptions of the characteristic behaviour patterns of a species) of many animals have been prepared using different observational techniques.

Most widely accepted methodology was proposed by Altmann in 1974. He formulated following method:

(1) Ad libitum,

(2) Focal animal,

(3) Scan or Instantaneous sampling,

(4) All occurrences,

(5) Sequence,

(6) One zero.

***Ad Libitum Sampling*:** This is the simplest way of recording behavioural acts. In this, one selects a group of animals and remains with them for some time, which could be hours, days, months or years depending upon the research problem, and whenever the behaviour of interest takes place it is noted. Ad Libitum sampling provides ideas for future research and often reveals rare, but significant, behavioural events. What is recorded are generally the behaviours of those individuals (or groups) that are most easily observed. This technique is used when an ethologist is merely recording as much as he can during (i) an unplanned encounter with a species (ii) during reconnaissance observations for a later study. e.g. If the behavioural observations are being made on mother and infant relationship in deers and occasionally adult males come and interact with them, the notes could be taken on males also. In that case, we would say that interaction of adult males with mother - infant dyad were taken Ad libitum.

Focal Animal Sampling : In this method, one individual is the focus of observation during a particular sample period; that is, a particular individual of interest receives the highest priority for recording its behaviour. Suppose we are interested to note how female langurs wean their infants. Then we should select some adult females from a langur group, identify all those females individually by using body cues; then take notes on only those activities in which we have been interested. To take notes using focal animal technique, the experimenter has to formulate criteria and follow them strictly viz, this should include:

(1) Individual identification of animals of interest.

(2) Prior decision about behavioural acts of interest.

(3) Formation of a check sheet to be carried out in the field. Abbreviation of long descriptive words.

(4) Each animal has to be observed at predetermined time interval (sample period)

(5) Every time the animal observation performs an act of interest its frequency and duration is noted. According to J. Altmann (1974) with proper choice of behaviour units, sample periods and focal individuals this method is the best to use.

Scan Sampling : By this method, several individuals are observed one after another in quick succession in predetermined time interval. Their behavioural acts are noted on a well designed check sheet. One important use of this method is to estimate the percentage of time the individuals of a group spend in various pre-decided activities.

For example, we want to know what the members in a deer group do the whole day or how do they budget their time in a day/month or year? Like other methods following criteria have to be decided :

(i) activities we are interested in

(ii) time duration for scan, depending upon the group size.

(iii) duration of interval.

The scan records activities of all individuals of that group between a predetermined sample time, this task is preferably carried from dawn to dusk. If the notes are taken for a day then at the end you would know how many activities you noted (total number) and out of that percentage of other activities can be calculated. Similarly, if the data is collected for fifteen days, six months or one full year, activity pattern of that particular herd of deers can be known.

All Occurrence Sampling : It may be desirable to focus on one or a limited number of behaviours and record all occurrences of them. This is in contrast with focal animal sampling where the focus is on the individual. For example, we are interested to know if there is any interaction between rhesus and langurs, then select a research site where they co-exist, spend time with them and whenever there is any interaction, note it. In this, the notes could be more specific e.g. one can note what kind of interaction was there, agonistic or amiable? If agonistic then whether in was chase, supplant, tail pulling, threat, or bite? All occurrences of selected behaviours is possible if the following factors exist:

(1) Observational conditions are adequate.

(2) The behaviours have been carefully defined so that they all are easily recognized.

(3) The behaviours do not occur more often (or more rapidly) than the observer can record them.

Sequence Sampling : In sequence sampling the focus is on

a chain of behaviours. These may be performed by a single individual e.g. courtship displays in male ducks, parturition in female, courtship display in sticklebacks. The initiation of a sample period is usually determined by the beginning of a sequence and the sample period terminates when the observed sequence terminates.

One Zero Sampling : It is a sampling method in which the observer scores whether behaviour occurs (one) or not (zero) during a short interval of time (sample period). It is suitable for recording states and or events, this method is also called as Hansen system. This method has the following features

(1) In each sample period the occurrence or non occurrence is scored.

(2) Behaviour of one or two individuals is recorded in each sample period.

(3) Occurrence refers to either an event or a state.

(4) The sample periods are generally short.

(5) Results are presented as frequencies (number of sample periods in which it occurred divided by total number of sample periods).

One has to be very careful in selecting the method because different methods help in generating different data for different interpretation.

Identification and Naming of Individuals : While studying animal's behaviour it is very necessary to learn to identify them age and sex wise, and individually by looking at natural marks, in the absence of such marks the animals are captured and marked either by ink, coloured collar or numbers. Data collection equipment varies from simple note book and pen to the complex on line computer terminal along with clocks, counters, stopwatches, compass, range finder, radio telemetry gadgets event recorders, stenographers, audio tape recorders, still and motion photography are some of the most useful equipment used in behavioural research.

A necessary prerequisite for ethological studies is that the observer be able to recognize individuals, particularly when the studies have to be carried out for a long period (longitudinal studies).

Natural Marks : The best way to identify individuals is by their natural marks e.g. broken horns, cut tails, scars on body, notches or holes in ears, pigmentation, however, these types of markings are generally not sufficiently wide spread in a population to allow large numbers of individuals to be recognized, only some species provide sufficient individual variation for easy recognition e.g. facial patterns in oryx, stripe patterns in zebras, patterns on dorsal fins of bottlenose porpoises, pug marks and body marks of tigers. Rudhai (1973) developed a method of individual identification for African lion (Panthera leo) based on variations in the patterns of vibrissa spots (hair near the nostrils). The other traits for individual identification can be missing limbs, presence or absence and size of canine teeth.

Photography is not the most useful tool, chiefly because of difficulties in obtaining close up photographs of small objects from various angles in addition to poor or varying lighting conditions in the forest, moving animals and expenses reduce photography to a supplemental method of recording. The preferred procedure is to observe the animals with binoculars, prepare drawings, write descriptions on identification cards.

Artificial Marking Technique

1. *Trapping:* Usually it is difficult, expensive and time consuming to capture live animals, there is also a risk to the animal being captured. Many kinds of live traps are commercially available, but they are expensive however, traps can be fabricated less expensively. The

traps can be put at suitable places prohibited with suitable food, as soon as the animal gets trapped it must be removed without traumatizing. The larger animals are anaesthetised by putting a syringe on a long stick, the safest anaesthesia for medium sized animals like monkeys is ketamine (Martin, 1972).

2. *Darting:* Darting techniques are useful for capturing animals under many circumstances when trapping is difficult or impossible, there are species that are trap shy or spend most of their time high in the trees or are too large to trap. A variety of darts and projecting weapons are available. Most commonly used are dartguns, blowguns, and crossbows. Darts can be made from tuberculin syringes with the plungers weighted so that they inject upon impact. The drug or anaesthesia is a crucial part of the darting system, it should have a correct ratio of lethal to effective dose, sernylan and ketamine are the chemicals which are generally used to immobilise the animal. The animals have to be followed after darting because drugs do not work instantaneously, when animals are darted in trees it is desirable to catch them, to prevent injury.

3. *Netting:* This technique has been used successfully in capturing birds, monkeys, apes and deers. Rocket nets and mist nets have been most widely used. The Rocket nets, explosives are attached at one side, ignition makes rockets fly along with the net, and an entire group of deers or animals could come under it.

4. *Ear Notches and Ear Tags:* Ear notches are formed with scissors, by notching the pinnae of the two ears in different ways one can make a variety of distinctly different pattern by which individuals can be recognized. Ear tags are easily applied to and are more easily recognizable marks, the ear tags can bear numbers or colour codes.

5. *Tattoos:* For this, small holes are made in the skin with a fine needle and permanent ink is rubbed into the small holes. It is best to tattoo a relatively hairless place that is lightly pigmented. Clipping and dyeing hairs and putting a collar around the neck are also effective methods for marking animals.

Locating Individuals in Wild

Sometimes it is very essential to follow animals in forest to collect data on various behavioural processes. There are three main ways to locate and follow the animals.

Following on Foot : Locate the animal or group of animals early in the morning before they leave their resting site and remain with them day and night. But at times, it becomes almost impossible to keep animals in sight, the animals could be too shy, they can live in a dense forest where following becomes difficult, the weather can be too harsh for the experimenter to remain in the field. In this case, alternative methods have to be adopted.

Biotelemetry or Radiotelemetry : Radio transmitters have been used for locating a variety of species. This is accomplished by installing a small, radio - transmitter on the body of the animal. This animal is left freely in the jungle. The experimenter keeps a receiver which is tuned with the transmitter. The receiver set has an antenna (like TV antenna), the experimenter rotates the antenna until the maximum signal is received. Simultaneous directional information from two or more receiving stations can then be used to triangulate the location of the animal. The ears of experimenter who has perfected the technique can detect slightest changes in the characteristics of the signal received and is correlated with the behaviour of the animal. Radio-tracking system have been made completely automatic by transferring time and directional information into a computer where it can be stored.

Satellite Telemetry System : It is a recent technique introduced in USA. This system consists of two polar orbiting United States National Oceanic and Atmospheric Administration (NOAH) satellites, along with numerous earth based receiving stations and data processing centres. The purpose of satellite is to collect and process data from radio-collared animals from earth. They have been successfully used to track dugongs, manatees, muskox, polar bears and species of marine mammals. For large species that range widely, satellite telemetry is much less expansive than traditional biometery. The signals which are given out from radio are received by satellite, which directs them to earth to be received by a dish linked to computer for analyses. It forms an important part of Geographic Information System (GIS).

3

Social Evolution

Animals join groups to enhance their foraging, success and gain better protection against predators; but group living also exposes them to competition, diseases and social interference. The costs and benefits of social living vary with the age, sex, experience and physical condition of individuals. Social groups are maintained by chemical, visual, auditary and tactile signals. Signals benefits both receivers and senders, but animals do not always communicate their intentions clearly and accurately. Complex social systems have evolved from offspring remaining with parents to help rear future broods and from cooperation among adults of the same generation.

Every solitary organism must interact to exchange gametes. Among many sessile organisms, the gametes do their association by themselves in the water, but most mobile animals get together to copulate even if they part immediately afterward. From these belief and simple interactions, social behaviour has developed into more and more elaborate systems, culminating in the complex societies of social insects and vertebrates. How these systems have evolved, why it was advantageous for animals to associate with one another and how social groups are maintained are discussed in this chapter.

Costs and Benefits of Social Living

Human tend to think that the highly complex societies of some vertebrates, ourselves included, represent a pinnach of evolution. This self-congratulatory view implies that "highly developed" creatures, like ourselves, represent a form of life superior to the less highly social species. To adopt this attitude is to ignore the broad spectrum of potential disadvantages, as well as possible advantages, that arise from social living (Table). One of the more thorough explorations of the costs and benefits of living in a group is the study of bank swallow behaviour by John Hoogland and Paul Sherman. Bank swallow form nesting colonies of dozens to hundreds of pairs in sand quarries and river banks. Hoogland and Sherman showed that by nesting together the swallows are able to reduce the effectiveness of some of their diurnal predators. When a hunting jay or weasel approaches the colony, the bird mobs the predator, much as gulls do and sometimes succeed in driving their enemy away. In addition, pairs seek to acquire a centrally located nest burrow and so to "use" the other members of the colony as a shield against nest-robbing predators that tend to attack the peripheral members of group first. Moreover, because the birds link their reproduction with the pairs nesting around them, they avoid producing eggs or immature young around them, they avoid producing eggs or immature young when other birds are not reproducing. If the breeding period were extended, a predator could harvest the offspring of the swallows one by one, as they were generated. By synchronizing egg laying a pair of swallows adds its progeny to a large pool of vulnerable offspring that exists for only a short period. Because predators can consume only a few nestings per day, synchronized breeders may have a better chance of getting their clutch through the dangerous period that asychronous reproduces.

Some Major Advantages and Disadvantages of Sociality

Advantages	*Disadvantages*
1. Reduction in predator pressure by improved detection or repulsion of enemies.	1. Increased competition within the group for food, mates, nest sites, nest materials or other limited resources.
2. Improved foraging efficiency for large game or clumped ephemeral food resources.	2. Increased risk of infection by contagious diseases and parasites.
3. Improved defence of limited resources (space, food) against other groups of conspecific intruders.	3. Increased risk of exploitation of parental care by conspecifics.
4. Improved care of offspring through communal feeding and protection.	4. Increased risk that conspecifics will kill one's progeny.

Some workers have suggested that an additional benefit of colonial nesting in this species might be improved foraging success. Birds that had failed to find much food might be able to follow other individuals to ephemeral concentrations of aerial insects these luckier or superior hunters had found. Although Hoogland and Sherman found little evidence to support this possibility in the colonies they observed, it might apply in other years on in other areas.

Animals in groups gain a number of advantages related to defence while a group of animal is bigger than a single one, and so can be spotted from further off. If the predator only eats one prey at a time, it will take longer to find a meal if the preys are clumped than if they are evenly spread out. Furthermore, there are more pairs of eyes in a group, so the predator is less likely to creep up undetected. To some extent, this is offset by the fact that animals in groups are able to be less vigilant than solitary individuals. Lonegene, for instance, raise their heads much more frequently while cropping the grass than they do when feeding in groups. Because there are more pairs of eyes they can afford to keep their heads down and feed for longer periods with the security that others will raise a warming if danger threatens.

A final example of grouping geared to predator defence is a very curious one and this is in the nesting behaviour of ostriches. Several female-ostriches lay eggs in one nest, but only the first to lay incubates. She accumulates a much larger number of eggs in this way than she could lay herself. After the eggs hatch, the female may have around 25 chicks following her as she sets off across savannah, but later she may collect even more as a result of chasing other hens away from their chicks. Brain Bertram, who studied this extra ordinary system, even recorded one group in which two adults were accompanied by 105 chicks. What advantage can there be to a female ostrich in caring for the eggs and chicks of others ?

The answer seems to be that chicks survive better in larger groups so that her own chicks benefit by a 'dilution effect'. Predators such as golden jackals will eat ostrich eggs and chicks, but the more of these that a female has with her the less likely is the one that is taken to be one of her own offspring. This cannot be the whole story, however, otherwise a female would gain whether she cared for her young herself or let another female do so. At the stage of incubation, it seems that there are advantages to the hen that sits, for she can recognize her own eggs and she gives preference to them so that they remain in the centre of the nest and are more likely to hatch. She is not therefore being a generous many helping out others, but is running the creche for her own advantage. It remains to be seen whether females with chicks also gain from giving care themselves rather than leaving them to be looked after by others.

Finding Food

A further range of benefits to group living is related to the exploitation of food supplies. If food is clumped, as is very often the case, animals that feed on it tend to become so too. Seeing another individual feeding may be a good indication of where food is abundant. So groups may form for this reason alone without there being any advantage in being a member of a group. Indeed, the animals may interfere with each other's feeding effort but, if they are conspicuous, there is no way in which they can stop newcomers arriving to share in the spoils.

Not all feeding groups, however, are passive clumps, while many predators, such as frogs, snakes and domestic cats, are solitary hunter, using stealth to catch small prey, some others, like lions, spotted hyaenas and wild dogs, cooperate in the hunt. This may increase the chances of success, because individuals can take turns in chasing and because they can spread out so limit escape routes. It also enables them to tackle far larger prey than could an individual on its own : a pack of hyaenas can pull down a zebra on a wild beest, or some of them can distract a rhino mother while attack her calf.

A final advantage of grouping may also be important in animals that are not carnivorous. This is that groups may be able to make use of resources in a more systematic way than can isolated individuals. The bumble-bee flying from flower to flower may be following close behind another one and so getting less reward than it would otherwise do. If individuals form flocks and move round their range together, as is often true of small birds outside the breeding season, they can be sure that they are all arriving in a fresh area where the food has had a chance to replenish itself since last they came that way.

Many monkey groups occupy quite large territories and more over them in a single party searching for the fruit or buds on which they live. They too may gain from the systematic foraging that group living allows, but their groups are more integrated and the relationship of individuals within them are more varied than in a flock of birds or a herd of deer. The advantages of food finding or predator avoidance may explain why they, like many other animals, come to live in groups, but do little to account for the richness of the interactions between the group members. It is among various facets of the behaviour of individuals within groups towards one another that we shall now turn.

Kinship

Not surprisingly, animals that live in social groups tend to be related to one another. Young ones are born into the group to which their mother and father belong and may often stay in it to become parents themselves. However, if this was always, animals in the group would become more and more inbred. Too much inbreeding leads to young which one less viable, so it is discouraged by natural selection. What usually happens, therefore, is that as maturity approaches, young animals of one sex or the other move away from the group in which they were born and so do not mate with close relatives. In birds, it is usually the females that disperse to different areas, whereas in mammals in the males are more likely to leave, although there are quite a few exceptions to this rule, including species in which both sexes move away.

Which sex disperses may have a profound effect on various aspects of social behaviour. Male white-crowned sparrows sing songs which are almost identical with those of their neighbours and quite different from those of birds a few kilometres away. Song learning usually only occurs in the just three months of life and so is complete long before they set up their territories and start to sing. But they breed only a short distance from where they hatched and thus the song dialects are maintained. Although female birds do not usually sing, their choice of mate is probably influenced by the songs of prospective partners. In white-crowned sparrows, for example, there is evidence that females may prefer males from the same dialect area as themselves, though there is some controversy about this. By contrast with birds, groups of many mammalian species are matriarchal, female staying with their mothers, while males leave to find other groups which they can join. Lions are a case at this point. As already mentioned that males leave their natal group and go in search of another that they can take over. Females, however, stay in the group in which they were born so that the lionesses in a pride are usually closely related to one another. They often give birth at around the same time and it has been found that they will suckle each other's cubs, something one would only expect if they had an interest in them through being related. Furthermore again in line with what kinship would predict, they suckle their sister's offspring more than those of their cousins.

Alarm calling is another behaviour where kinship is important. Making a noise when a predator is around is obviously a risky thing to do, and one would expect to find some advantage to offset this. Many birds only call when they have young in the nest, so saving these from being discovered seems to be the benefit. Using a stuffed badger as a 'standard predator', John Hoogland has shown that black-tailed prairie dogs in America produce alarm calls when they have close relatives in their group than when they do not. Males called a lot in their natal group but ceased to do so after they moved elsewhere, only to start calling again once their own young were born in the new group that they had joined. All these findings fit in well with the idea that animals are more likely to call when relatives may benefit.

Male rats are not very mobile and their burrow systems are isolated from each other. It is likely, therefore, that most of the animals within a colony are relatives and that their close social relationship stems from the common interest that this gives them in the offspring of the breeding female. Unfortunately, we do not know just how closely related they are to each other, and this is very often the case in studies of social behaviour. It is easy to speculate that kinship is involved when we observe animals being altruistic to one another, but only in a few cases, such as that of the social insects, do we know in any detail exactly what the relationship is between the individuals involved.

Leadership

Leadership and social dominance are not one and the same thing. In colonies, of ants no peck order is known, but there are leaders. In flocks of hens the leaders are often found in the middle of the peck order. In some groups of mammals the leader is the alpha number in the peck order. A leader is not necessarily the strongest animal of the group, or the fastest, or the one that is found out in front when the group is moving. Leadership may depend on other attributes.

A leader is one that can furnish guidance to the group. Those animals which are most alert to danger, most concerned with the welfare of the group, and which are the cleverest in devising escapes from danger may become leaders. In some herds of mammals, like those of the red deer of Scotland, old experienced females are the societies of rhesus monkeys, an experienced and aggressive male is the leader. The howler monkeys of the canal zone have a group of male leaders. Leadership is based on more than force. It involves protection and consideration for others. It involves an interdependence between the leader and the followers. There can be no leader if others are not willing to follow. And being willing to follow a leader implies a process of socialisation in the young.

The preparation of the young for their later roles in the group is reflected in their play activities. Perhaps you have seen young lambs playing "follow the leader". Some deer play a game oft "Eag" much like the one you played when you were younger. They chase and touch each other with their front legs.

Such play activities mark the beginning of the establishment of dominance relations and of the roles of the leader and follower. The leader-follower relationship has survival value for the group. This point was illustrated earlier in the description of a troop of baboons. The Russian scientist V.M. Sdobnikov in 1935 described a similar leader-follower relationship in reindeer herds.

He observed that reindeer herds are not simply collections of animals. Instead, each herd is organized into two groups. One group is made up of the "fringe" reindeer at the edge of the herd; the other group are the "middle" reindeer at the centre of the herd. The fringe reindeer are divided into three subgroups depending upon where they travel and graze in relation to the herd. The "vanguard", the "side", and the "rear".

The reindeer in the different parts of the herd behave differently. The vanguard reindeer are the most alert and restless. They include the leaders if the herd has any. The side reindeer are timid and restless and tend to leave the herd. The rear reindeer are usually quiet and seldom wander away from the herd. The middle reindeer are also quiet and peaceful.

The main point we have been leading to is this. The social life of animals is based on cooperative behaviour, for example, the leader-follower relationship.

Roles of the Sexes

Because of physiological difference between them, together with differences in benefits and costs they drive, males and females often play different roles in their social systems. Females invest more energy in gamete production than do males, but they do not always invest more in the care of offspring. Only female mammals have functional mammary glands. Males in most species contribute nothing to feeding of offspring. The exceptions are among carnivores where male capture prey and deliver them to females and their young. Among herbivorous mammals, males may warn of the approach of predators and defend females and young, but they do not provision young. Among birds, except for laying eggs, there is no aspect of reproduction that cannot be carried out equally well by both males and females. Sexual role specialization, when it occurs, is due to differences in opportunities rather than to powerful physiological restrictions. Among fishes, there is little feeding of young by adults. Parental care consists primarily of guarding eggs and young. Guarding differs from feeding because it can more readily be shared among groups of offspring. A male fish can guard a number of clutches of eggs in his nest, but an item of food can be given to only one offspring.

If there is more parental care by an individual of one sex than the other, it is nearly always the male who does more among fishes but the female who does more among mammals and birds. Most fishes guard eggs but do not care for young. When there is guarding of young and leading them to good feeding areas, biparental care is the rule. Greater care of offspring by males than by females unknown among mammals, for the obvious physiological reasons, but does occur occasionally among birds. The most striking examples involve reversal of the normal sexual roles. In most birds, if there is plumage difference, males are brighter than females, and among species with reverse sexual roles (Phalaropes, for example), females are brighter than males. These females complete with one another for males, lay eggs for them, and leave incubation and care of the young to the males. This breeding pattern has evolved a number of times among birds, always among species in which the young are able to around and feed themselves once they hatch. Two parents are less important in this case because food is not being delivered to the young. Therefore, females are able to terminate parental care early and gather more energy to lay additional clutches for the same or other males.

The amount of investment in parental care made by males

also depends on the certainty of paternity. A male of a species with internal fertilization can know with which females he has copulated, but he cannot know if his sperms have been used to fertilize their eggs unless he was with the females throughout their fertile periods. A male that fertilizes eggs externally after a female has laid them, however, is more certain about his paternity. Males with a higher probability of percentage invest more in parental care than males that are less certainly related to the offspring of their mates. Male parental behaviour in fishes and amphibians, groups with species having both internal and external fertilization, illustrate the phenomenon. Male parental care is much more prevalent in species with external fertilization than in species with internal fertilization.

Relationship Between Mode of Fertilization and Male Parental Care in Fishes and Amphibians with some Parental Care

Number of Females containing one or more species with :

	Male Parental Care		*Female Parental Care*	
Taxon	*Internal Fertilisation*	*External Fertilisation*	*Internal Fertilisation*	*External Fertilisation*
Fishes	0	48	15	21
Amphibians	2	13	11	7
Total	2	61	26	28

Information is sent and received by means of one or more sensory modes, such as chemical, visual, auditary and tactile. Each of these modes has specific properties that influence how it can be used for communication.

Chemical Communication

The two main types of chemical communication are contact communication, known as taste, and distance communication known as smell. Smells originate at some distance from the receiving individual and tastes do not, but in practice they may be difficult to distinguish. Food cannot be tasted as strongly if we have a cold and our nostrils are clogged, indicating that much of our taste sensation is really a combination of taste and smell. Chemicals can be produced easily and are energetically in- expensive. By changing the size and structure of a chemical signal, the distance it will carry and the length of time it will persist can be altered. Chemical communication is the most widespread and evolutionary oldest form of communication. It is harder for us to study chemical communication than visual or auditary communication, however, because humans have such a poor sense of smell. Many animals are sensitive to lower concentrations of chemicals that we can detect without best machines. If we had a sense of smell like that of dogs, we would probably know much more about chemical communication than we do.

Chemical signals between conspecific individuals (analogous to the role of hormones within individuals) are called pheromones and they either act by stimulating an olfactory receptor or they are ingested and have their primary influence on the endocrine and reproductive systems. Because of the high sensitivity of animal receptors, a few molecules of a chemical signal can influence a great deal of behaviour. The number of signals that can be produced is large. Even a species with such poor olfactory abilities as ourselves can recognize over 10,000 odours. The long face-out times of pheromones means that chemical signals can be broadcast over great distances and last a long time for every little cost.

Chemical signals are used in trail marking (ants and stingless bees), in territory marking mammals and crocodilians, and in sexual stimulation. Many female insects attract males from long distances with sex signals. Males of some insects, such as bumblebees, deposit scent on vegetation and then patrol the areas to encounter females attracted to the deposits.

Odours are also used to distinguish individuals and classes of individuals. Bees recognize hive members by thin odours and attack bees that do not smell right. Trained dogs can distinguish the odour trails of particular human beings. Most of us can recognize the distinctiveness of the odours emitted by our associates even though we usually conceal those odours with artificial ones judged to be more pleasant.

Visual and Auditary Communication

Visual signals can be sent only to animals with eyes, but vision is widespread and such signals are important in many species. Visual signals are easy to produce, and an endless variety of signals is possible. The position of the signaler is automatically indicated, something that is not usually true for chemical and auditory signals. Visual signals can also be changed very rapidly and are accordingly, excellent for indicating rapid shifts in position and mood. Not all parts of the body are of equal importance of visual signalling. Among fishes, gills and fins are especially important, but among terrestrial vertebrates, the head and face are most important. Relatively minor alternations in these regions influence the responses of associates more than major alternations elsewhere. The identity of a person is concealed by covering only a small part of the face.

Auditory signals possess some of the advantages of visual signals — case of production and variety but they are more difficult to locate. In addition, unlike visual signalling purposes an animal can remain silent, but it cannot avoid sending visual information about its activities if it can be seen by others. This feature of sound has enabled scientists to tease out the factors influence the structure of auditory signals in a more sophisticated way than has been possible with other signals. In addition, the sound spectrograph, a machine originally invented to help deaf people learn to speak by providing them with pictures of sounds, allows us to make precise measurement of the pitch, loudness and duration of the components of sounds. Vocalisations play major roles in identifying species and individuals, in maintaining contact among members of groups, and in communicating the emotional states of the callers.

Mammals, birds and reptiles produce most of their sound by the forced movement of air across internal vocal cords, whose properties are controlled by a sample set of muscles. Fishes produce sounds mainly with their swim bladders or by vibrating one part of their skeleton against another, such as grinding their teeth or vibrating their vertebrate. Cicades have a complex drum that is vibrated by special sets of muscles, other insects produce sounds by rubbing one part of their body against another. However, the majority of animal species are mute and do not communicate vocally.

Tactile and Electrical Communication

Tactile signaling is specially important among animals with poor vision or among those living in places where visual stimulation is difficult or impossible. Many deep sea fishes, living in perpetual darkness, have highly elaborate feelers used for both prey detection and sexual signalling. Tactile mating signals are highly developed among many spiders, particularly web-building species that have poor vision. Among webspinners, the male usually seek out the female on her web, but it is dangerous for him to approach her because food is normally detected by the vibrations it produces. The approaching male may become a meal rather than a suitor. To reduce this possibility, males have evolved elaborate plucking rituals that precede close approach to the female.

The most highly evolved communication system that has an important tactile component is that of the honeybee, so carefully studied for several decades by Karl Von Frisch and his associates in Germany. Dancing bees also make sounds and carry odour on their bodies, but much of the information in their dances is conveyed by movement that is monitored by other bees that follow and touch the dancers in the darkness of the hive. When a foraging bee finds food, she communicates her discovery when she returns to the hive, dancing in the dark on the vertical surface of the honey comb. If the food is less than 80 metres from the hive, she performs a round dance by running rapidly in a circle, reversing her direction after each circumference.

The odour on her body indicates the flower to be looped for, but the dance contains no information about the direction in which to go. If the food source is farther than 80 meters, she performs a waggle dance, which conveys information about both direction and distance of the food source. The direction of the food is indicated by the orientation of the waggle-run of the dance. Straight up means the direction of the sun and the direction left or right of the sun is indicated by the angle to the left or right of vertical on the waggle run. A food source in the opposite direction from the sun is signalled by a downward pointing waggle-run, while a run 9° to the right of the sun is signalled by a waggle run 9° to the right of vertical. The distance to the food source is signalled by the speed of dancing. The farther away it is, the longer the time spent in each waggle run. This communication is unusual because the system is based on a symbolic convention. Straight down could just as well indicate the direction of the sun as straight up. Such symbolic conventions are developed to an extreme degree in human language.

Seven different groups of fishes have evolved the ability to produce electric fields, which they use to detect objects in the surrounding water. An electric field can be used to locate objects with conductivities different from that of water because they deflect the lines of force. Electric fish live in muddy tropical waters where vision is of limited use; and they not only find, and in some cases sting their prey with electrical discharges, but also communicate with one another electrically.

Interspecific Communication

There is clear evidence of only a small number of cases of intraspecific deceit, but deceit is widespread in interspecific signalling. Prey animals mimic leaves, branches, thorns, bark, stones, and other objects. They also mimic individuals of other species that are less desirable as prey. Palatable butterflies mimic distasteful ones. Spiders mimic ants, all sorts of insects mimic bees and wasps. The eggs of brood parasites that are deposited in the nests of other species, where they are incubated and reared by their hosts, mimic those of their hosts. This deceit may work even when the individuals do not resemble the ones they are supposedly mimicking. A young cuckoo does not look anything like the young of its hosts, but it is nonetheless fed by its foster parents. It functions as a supernormal stimulus in much the same way that abnormally large eggs do.

Interspecific deceit is more prevalent than intraspecific deceit because deceiver and deceived are not members of the same gene pool. Within a species, if a mutation raises the relative fitness of its bearers by causing them to deceive others, the mutant allele will spread. However, the more common it becomes, the more likely its bearers are to deceive other individuals with the same allele. When all individuals have the allele, there are no more non-bearers to deceive. Therefore, the value of intraspecific deceit is greater when deceivers are rare. This is not a constraint on intraspecific deceit, because the value of the deceit is not diminished even when all individuals in a population practice it.

Choosing Associates

Social systems are complex, but they are based on a rather small set of choices by their members. Individuals choose their associates, how to interact with them and when to leave them. Variation in those decisions are made which can result in an amazing variety of social systems. The most widespread kind of choice of associates, made by most sexually reproducing species, is choice of mating partner. Selection can be based on the inherent quantities of a potential mate, on the resources it holds, or on a combination of the two. In many species, no individuals control any resources, and traits of the partner are the only relevant criteria for mate selection.

One way of assessing mate quality is to let individuals of the opposite sex flight and then accept the winner. Females have the most to gain from making careful choices, because a poor choice may jeopardize their yearly reproductive output. Therefore, it is usually male that fight while female may make their assessment by going to some place where males are displaying and fighting or they may go to where they will breed and let males come to them. Grouse that display communally are examples of the first system. Males sage grouse gather on traditional dancing grounds where they jostle for position. Females come to the dancing ground and observe the males. It is not known exactly how females make their choices, but they probably compare the vigour of displaying among the males. When females have made their choices, each female mates once then leaves to build her nest, incubate her eggs, and care for her young unaided by the male.

Elephant seals illustrate the second system. Females come ashore on good beaches where they give birth to their calves and nurse them to independence. There are only a few good quality pupping beaches and females congregate there. This gives males an opportunity to fight for possession of the best beaches. The winners have nearly exclusive copulatory rights with the females on the areas they defend, but they could not do so if were not advantageous for females to congregate in those areas. Males cannot control where females come ashore to give birth to their pups.

Many other females pick their mates on the basis of

resources held. A female hanging fly will mate with a mate only if he provides her with a morsel of food. The bigger the food item, the longer she copulates with him and the more of her eggs he fertilises. The female gains from this behaviour she obtains a good supply of energy for egg production. Many birds engage in courtship feeding in which males bring food items to their potential mates. In some of these species there is a considerable amount of divorce at these early stages. A female common tern is more likely to leave a male who is providing infrequent meals for her than a male who is feeding her more regularly. There is good correlation between the ability of a male to bring food during courtship and his ability to provide food for young later in the season.

Males of some species do not bring food to females but control areas that contain important resources. The best known cases involve territorial behaviour in which a male defends an area that contains food, nesting sites, or other resources. The defending individual expels all individuals other than mates or potential mates. When female arrives, they assess both the male and his resources. If a male controls sufficient resources, he may attract more than one female, a phenomenon known as resource defence polygamy. Among red-winged blackbirds, the number of females attracted to a territory does not depend strongly on which male is holding the area. When females first arrive, they spend most of their time low in the vegetation looking for food and nest sites. They pay little attention to the vigorously displaying males. The size of the harem a male gets is strongly correlated with the quality of the resources he holds. Among redwings, these are safe nesting sites and feeding areas.

An important criterion for choosing an associate other than a mate that of genetic relatedness. Relatives share alleles with one another by common descent. The probability than an allele in one individual is an identical copy, by descent, of an allele in another individual is known as their coefficient of relatedness.

An individual can influence its fitness not only by producing its own offspring but also by influencing the survival of relatives that share those alleles. The effect on future allele frequencies of helping the survival of relatives (known as kin selection) will be explored. For the present we need only know that by cooperating with non-relatives an individual cannot increase the proportion in future generations of the alleles it possesses unless the cooperation is of immediate benefit to itself. However, an individual can lower its own survival and reproductive success to help relatives provided the cost imposed on itself and compensated by improvements in survival and reproductive success of the relatives. The relatives need not do anything in return.

Primate Societies

The Social Life of Primates : Different species of primates live in different kinds of societies; indeed, the same species may form different kinds of social groups according to the conditions. The Hanuman langur (Presbytic entellus) forms both harems, with one male and a larger number of adult females. The reason is uncertain, but may be related to population density, for multi male groups are commoner when the total population density of an area is lower. Before considering the altruistic behaviour of primates, let us consider briefly some of the different kinds of social groups.

The white-handed gibbon (Hylobates lar) lives in the trees of South East Asia in monogamous family groups. The male and female are similar in size and appearance, and live as a pair with 0-4 young. The pair defend a territory of about half a square mile. Within the family, there is no dominance or aggression : male and female live together peaceably as equals. Other primates too, such as the spider monkeys (Ateles), also live in family groups. But how different are the societies of hamadryas baboons.

The hamadryas baboon (Papio hamadryas) is a terrestrial species, inhabiting the plains of North East Africa and South West Arabia. The hamadryas baboon society has several levels of organization. The lowest level, the breeding unit, is a harem of one to ten females with a single male. The male treats his females aggressively, often attacking them if they stay away. The male is about twice as large as the female, which influences the dominance relations of the sexes, but the large size of males is probably mainly an adaptation for fighting off enemies, especially rival males. At thê next level, several of these harems may walk around together in larger bands while feeding, and these bands may act as a unit to defend a food source from a rival band. The hamadryas baboon, however, does not defend a territory.

The bands do confine their wandering to a large area of about 12-15 square miles, but different bands may overlap in their use of this area. It is, therefore, called a home range, to distinguish it from a defended area, which would be called a territory. At a higher level, several bands may join into larger groups for sleeping. If suitable shelters for sleeping are difficult to find, as many as 700 hamadryas may sleep together. Whereas in the hamadry baboon the mating unit is a single harem aggressively controlled by a single male, in the howler monkey (Alouatta), several adult males may live peaceably together in a single group. When Ray carpener watched howlers on Barro Colorado Island, in Central America, the group sizes were variable but contained an average of 3 males, 8 females, and 7 young.

The male howler monkey is about 30% larger than the female, and has an enlarged voice box covered by a beard. The males roar daily, making the loudest animal noise in the American forests, a noise which carries for over a mile. This howling serves to space out the different groups. Within each group there is little aggression and non-obvious dominance hierarchy.

These three species live in entirely different kinds of societies. One is monogamous, another polygamous within single male groups, another polygamous with multi-male group. One shows dominance and aggression within groups. The other two do not. Other primate species live in yet other kinds of societies. Ethologists would like to be able to explain why different primate species live in different kinds of societies. Why are some species aggressive, others not ? These question cannot yet be answered satisfactorily. Our understanding of the diversity of primates societies is still limited. One trend which can be explained is as follows. There is a tendency in species in which there are many females per male in the group for the males to be bigger than the females; in polygamous species the male is larger. This had presumably arisen because sexual selection has favoured larger males in polygamous species, because larger males are more successful in fights over females.

Altruism in Primates

Many kinds of altruistic behaviour can be seen in primate groups. There is the feeding, caring and defence of young, not only by their mothers; cooperative searching for hunting and exploitation of food, food-sharing; cooperative group defence against enemies; and most common of all, that favourite pastime of primates grooming. Most, perhaps all, of these habits should probably be explained by kin selection. We, however, have considered the application of that theory in two examples already, and will therefore use a primate example to illustrate reason why altruism may evolve : the theory of reciprocal altruism.

The example concerns 'consorting' which is a habit of males, in many species that live in multi-male groups, whereby a male stays close to a female during the receptive phase of her oestrous cycle, and defends her from the advances of other males. It is an adaptation produced by the male competition component of sexual selection. In the olive baboon (Papio anubis), Craig Packer observed that two males may occasionally cooperate to fight off a single male; clearly two will be stronger than one, which makes the advantage of cooperation clear. The advantage, however, is only for the one male that copulates with the female; he has gained a benefit of as much as one extra offspring; the other male has paid a cost of the risk of injury in a fight, but obtains no benefit. He has behaved altruistically. If what we have seen so far were the end of the matter natural selection should eliminate the altruistic habit. But it is not the end of the matter.

The next stage arrives when another female in the troop comes into oestrous. The roles of the same two males may now be reversed. Packer saw ten cases in which a male who had previously been 'solicited' into cooperating to defend (but did not copulate with her) herself solicited a male into cooperative defence. In nine of the ten cases, the solicited male was the individual whom he had previously helped. It looks, therefore, as if males form cooperating pairs to defend females and take turns in the copulating that is the end of the defence. If so, it would be an example of 'reciprocal altruism'. Altruism can evolve under individual selection, without any need for the animals to be related, if the altruism is more than paid back later. The danger of any such reciprocal arrangement is that they will be cheated on. There is a clear short-term advantage to receiving altruism but then not paying it back; that way the cheat gains the benefit but does not pay the cost. This being so, reciprocal altruism is expected mainly to evolve in species that form stable group, with individual recognition. Then a cheat, after gaining a short-term benefit, can be recognized and excluded from future transactions, the cheating will then not play. Without the opportunity and mechanism of discriminating against cheats, reciprocal altruism is likely to break down. Because reciprocal altruism requires rather special conditions, it may be rarer than selected altruism. It is, however, a theoretical possibility; and in olive baboons it has been realized in fact.

4

Organisation of Society

Studies on social organizations of animals is not a very old field of research, serious observations on animal sociality began with the work of W.C. Allen (1938) and his students and since then studies in this field with respect to ecology and evolution have accelerated rapidly in different animals during last few years and have also given rise to a specialized branch of biology i.e. sociobiology. Workers from various disciplines viz. zoologist, psychologists, anthropologists and sociologists have converged on this subject in the hope that they will ultimately provide information on the origin of human societies, their intricacies, complexity, problems and solutions. Social organization varies considerably from one species to another, it ranges from simple short term interaction between male and female lizards during the breeding season to the complex, stable societies of monkeys and apes.

Social organisation of a group depends on the total number of individuals, their age, sex, sex ratio and number of adult males and these in turn depend on many important factors like abundance, availability, dispersion of food (Jarman, 1974), predation pressure, type of habitat and mating strategies are the other important factors. Not all the groups of animals are truly social. There is a clear distinction between mere aggregation or association and true societies. Alverdes (1927) defined association as - chance gatherings produced solely by external factors e.g. insects gather around a source of light for mating or groups formed by simultaneous attraction to a common source, rather than to each other e.g. earthworms collect under a rock for moisture and protection or human spectators at a sporting event. Association is a temporary union of individuals reacting in the same way to similar environmental stimuli e.g. collection of house flies around food source, collection of animals around a water hole, a migration point where birds, fishes, crabs or waterflies gather. Groups or aggregations are formed largely by non- breeding individuals based on a mutual attraction e.g. school of fishes, large wildebeests and zebra herds are also called as survival groups. Alverdes considered societies to be - genuine communities which exist in virtue of some particular social instincts. An animal society is a relatively permanent union of individuals held together by mutual attraction of its members. Alternatively, the distinction might be defined by stating that an association endures only while a specified set of external stimuli are effective, and that a society persists inspite of great variations in external conditions.

The basis of social life is the interaction of individual members who exchange food, water, body care and sexual favours. According to Tinbergen any interaction between one individual of a species with another member of the same species is known as social behaviour, this includes all those behaviours that influence, or are influenced by other members of the same species. Assemblage of animals can either be chance aggregation of individuals as stated earlier that may never again assemble in the same configuration or an integrated social organization with some degree of permanency. The social organization itself can be temporary or permanent, seasonal or annual.

A true society will involve more than a mated pair i.e. adults, subadults, juveniles, infants of different age-sex classes. It will mean a stable group whose members inter-communicate extensively and bear some relatively permanent social relationship to one another.

Vertebrate social groups are organized differently from insect societies. The most profound difference is in the higher proportion of reproductive individuals and the absence of sterile casts in vertebrate groups.

The social groups have certain properties, Wilson (1975) and Brown (1975) independently proposed following characteristics of social groups and behaviour:

(1) The first and most obvious characteristic of social behaviour involves the number of animals of the same species that actively come together or remain together in a group. The minimum level of sociality or the smallest social group is found between a male and female who interact only during breeding season or between mother and infant.

(2) Social behaviour depends in part on the "length of time or part of the life cycle that the group remain together."

(3) As opposed to the length of time that the members are simply in a group, that is, in physical proximity to each other, this element concerns the "energy actually spent in social behaviour". Many animals, although physically grouped together, nonetheless may spend much of their time or energy in behaviour that would be more properly classed as individual rather than as social behaviour e.g. a group of wildebeests in Africa may be grazing together or move together or remain together but they function at individual level.

(4) "Reciprocal communication" is generally considered necessary as a mechanism for attracting and keeping the members of a group together. The communication may be visual, auditory, olfactory or tactile.

(5) Much social behaviour is marked by a division of labour and "social structure or what is frequently referred to as roles."

(6) A feature of social group in many species is an "overlap of generations;" that is, families or parts of families may stay together. Parents defend or protect their young ones. Among the most highly developed situations, the group may include three or more generations that remain together through out the entire year and for some animals entire lives.

(7) The last point represents the highest level of social behaviour "Altruistic or aid giving behaviour where there is a cost to the altruistic individual" In most extreme forms it includes even the sacrifice of one's life and reproduction. Origin and existence of sterile or non-reproductive castes among ants, bees, wasps and termites poses a difficult problem for natural selection theory, as, such individuals are not maximizing their own reproductive potential, instead, they work for the benefit of the colony. They have probably evolved through indirect or kin selection. In a colony all the related animals live, the sterile casts by helping related animals are helping in perpetuation of some part of their own genes through their kins which is an extreme form of altruism.

Eisenberg (1965) also proposed four properties of organized societies :

1. *Communication:* All organized societies have some form of complex communication system. The members of a social group make gestures, postures, change colour,

raise hair, thev scent mark, communicate through vocalization, they may indicate messages by touching each other, or have some specialized forms like echolocation in bats, tail and waggle dance in bees. Communication in higher animals is generally elaborate and capable of putting across complex messages for example the facial expression alongwith vocalization in chimpanzees communicate various forms of emotions and messages.

2. *Cohesion:* The individuals constituting a society tend to remain in close proximity to one another e.g. all the bees of a group live in one hive, the individuals in a herd of deer, pride of lions and peck of wolves remain in close vicinity in a given home range, the individuals in a troop of baboons while moving tend to remain in close proximity. The adult and dominant males (ADM) remain in the centre close to females, while adult subdominant males (ASM) take dangerous positions at front and rear.

3. *Division of Labour:* In organized societies, animals of different status, sexes or age groups have different functions in maintaining the society. Baboon and macaque young adult males often serve as front or rear guard for group as the group moves, they are the ones to face predators. The old adult, dominant males remain in a more central place and they decide the foraging and resting areas, the function of females is to bear and bring up infants and care for young. In a colony off bees there is one queen, few drones, and workers (fanners, repairers, builders, soldiers and foragers), similarly, termites have a caste system too. In a troop of baboons when confronted with a predator the ADM leave the centre and take up positions in front, followed by ASM, while females and young baboons retreat. The function of females is to

bear and rear infants, juveniles spend time playing and exploring.

4. *Permanence and Impermeability:* The individuals making up a society tend to be same. There is little migration from the group, in most mammals the core of the group is formed by females who are related to each other, the males come and go, otherwise, the membership among females is permanent. Most organized societies resist immigration by outsiders.

Mating Groups or Mating Strategies

Males and females of all organisms form mating relationship in a way that best enhances their reproductive success. However, the mating relationship yielding maximum success is not necessarily the same for both sexes. Mating systems are most commonly classified according to the duration of pair bonding and the number of mates acquired by each sex. Mating groups have been broadly classified into monogamy, polygamy and promiscuity, and there is further variation within each category.

Monogamy

It is the simplest mating system because it involves the smallest number of animals. Prolonged association and essentially exclusive mating relationship between one male and one female at a time is monogamy. It is the condition in which one male and one female join to rear at least a single offspring, it may last for a season or for life time. In other words, two that breed and remain together to rear offspring are called monogamous. 90% birds, less than 4% mammals and only 20% human societies are truly monogamous. Monogamy can include lifelong breeding pairs that spend all time together, if the mate dies then the new mate is accepted e.g. in gibbon. Typically monogamous pairs are formed by dikdik, duikers, steenbok, grysbok (Ungulate, songbirds, eagles, beavers, foxes, coyotes and swans.

Monogamy has been classified into three different ways based on:

(i) The extent of male parental investment, whether it is less, more or equal as compared to the female parent.

(ii) The extend of territoriality (Spatial).

(iii) Temporal relationship.

The category based on male parental investment is further distinguished into three types.

1. *Facultative Monogamy:* In this the male parental investment is low as compared to female parental investment e.g. bewick's swans, wood duck.
2. *Obligate Monogamy:* In this type male parental investment is equal to the female parental investment e.g. pelicans, Canada geese. In herring gulls both male and female incubate eggs, defend the nest and collect food for chicks.
3. *More Male Investment Monogamy:* The males spend more time and energy rearing up the young ones e.g. emperor and golden lion temarins both of which are small, beautiful, highly endangered species found in Brazil. Males invest more in young ones as compared to the females, who generally give birth to twins. The mother takes charge of the infants only to feed them, rest of the time the male takes care of both infants.

The next category is based on territoriality and is further divided into following types:

1. *Territorial Monogamy:* The monogamous pair shares a common territory, e.g. golden lion tamarin, indri, gibbon, both male and female defend territory.

2. *Female Defence Monogamy:* Each male takes possession of a female and defends her instead of defending territory.

Last category is based on the temporal relationship between male and female and is further divided into two categories:

1. *Annual or Serial Monogamy:* Male and female form a bond for one breeding season or for an year e.g. small passerine birds, sparrow, warblers, chickadees.
2. *Permanent or Perennial Monogamy:* Male and female usually remain together for life e.g. kittiwake, a north American bird pair for life and both parents invest equally in offspring; the other examples are swans, geese, eagles, Siberian cranes and gibbons.

Polygamy

In the broad sense it covers any form of multiple mating. Reproductive unit of one individual of one sex and several of another. It has two categories, polygyny and polyandry.

Polygyny: A single male mates with more than one female or prolonged association and essentially exclusive mating relationship between one male and two or more females at a time is known as polygyny. It is the most common sexual strategy among mammals. The males do not stay with females to help rear up young. This has further distinction into:

(a) *Territorial Polygyny:* Some proportion of males form separate pair bonds with each of two or more females and keep them in their territories during the mating season.

(b) *Harem Polygyny:* Males first establish their dominance rank order, social hierarchy or status system. It is the ordered distribution of rights and duties within a group. Strong and experienced animals remain at the top of the dominance hierarchy and have certain advantages (such as priority at the food or drinking site, access to preferred sleeping sites, they have priority for mating too), but high ranking animals may also have to perform certain duties (leadership role, guard duties, fighting, chasing, etc.). The highest ranking group member is designated as the alpha animal, the lowest one the omega animal. This system provides stability of relationship within the group and among themselves. Most dominant males would possess the harem (a group of females). A single male defends access to each social group of females. It can also be called as female resource polygyny, where a male defends a group of females and mates with them. The seals have to leave the water once an year and come to suitable beaches, to mate and give birth to pups, suitable beaches are very few, consequently large number of individuals gather, the males compete for females very aggressively, only the most powerful males are successful and they become harem masters. In langurs, lions, kudus and most of the deer and antelopes, females form small social groups which are visited by males during the mating season. The rest of the males form all male groups.

In some polygynous species males neither defend a food resource nor a group of females, instead, the mating privilege among males is determined by dominance hierarchy that males establish among themselves. This kind of system is called male dominance polygyny, the typical example of which is Lek formation in sage grouse. Leks also occur among insects, fish and mammals, they have been studied most extensively in birds. A Lek is a defined area in which a number of males compete with each other for the possession of very small territories or mating stations, these serve only as mating sites. All the males stay at Lek throughout the breeding season, females visit Lek, mates and leave, lay eggs and rear up offsprings at some other nesting site. Lek formation is also considered as promiscuity by many workers.

(c) *Territorial Harem or Resource based Polygyny:* Males control access to female social groups by defending group home ranges or habitat where female groups congregate or males defend resources and would mate with visiting females, e.g. in impala, which are browsers of African wooded grassland, females tend to form groups, they do not move around much but spend time together in areas where food is abundant, the dominant males takes possession of such harems and defend both females and territory. The males of orange rumped honeyguide, a bird found in Nepal raid the comb or bee hive of giant honeybee, establish their territories around it, defend it and allow only females to enter for mating, in a season, this male may mate with as many as 18 females. In American bird, dickcissel, the males defend territories where vegetation is dense, because that attracts more females, dense vegetation provides more safe nest sites. Even the size of territory can also be an important factor. In three spined sticklebacks, the males with larger territories attract more number of females.

(d) *Successive or Serial Polygyny:* One male is bonded with several females during a breeding season, but acquires only one female at a time e.g. pied flycatcher, or three spined sticklebacks, similarly, the dogs follow the same strategy, they would mate one bitch after another without having them in a group.

(e) *Simultaneous Polygyny:* Polygynous males acquire all their mates at the same time and mate with them after having them in a group e.g. red winged blackbird, fur seals, elk and langurs.

Polyandry: It is the rarest and unusual mating system. It involves a reversal of usual roles of the two sexes. One female bonded with several males or prolonged association and essentially exclusive mating relationship between one female and two or more males at a time, or mating of one female with more than one male is polyandry. It occurs primarily in birds, this strategy also has further variations.

(a) *Territorial Polyandry:* Several males are paired with territorial females, i.e. the females have territories where males visit for mating e.g. jacana.

(b) *Non Territorial Polyandry:* Females desert their first males and pair with new males elsewhere, females do not possess a territory.

(c) *Polygynous :* Polyandrous system. Both males and females form relatively short term pair bonds with several mates e.g. rhea, shorebirds.

(d) *Serial or Successive Polyandry.* Polyandrous females acquire each of their mates in temporal succession (one after another) e.g. species of shorebirds which form temporary pairbonds and lay complete clutches for their first mates and then desert to repeat the cycle with the next male.

(e) *Simultaneous Polyandry:* Polyandrous females acquire all their mates at the same time e.g. jacanas which live on lily covered ponds where their elongated toes enable them to walk on the floating vegetation.

Females defend large territories which are subdivided into smaller territories, each defended by a male and containing a floating nest, females visit male territories, mates and lay eggs, which are incubated totally by respective males. The female is left free from all parental duties; she spends her time in mating, egg laying and defending her large territory against other females, she is larger and dominant over male.

Promiscuity

It refers to species in which no durable mating relationships are formed; members of both sexes mate randomly with several partners. The two sexes do not form long term bonds or associations, the mating may or may not be multiple by promiscuous animals, the examples are grouse, bears, wildebeests, chimpanzees. Promiscuity can also be of different kinds:

1. *Broadcast Promiscuity or Explosive Breeders:* Species which have very short breeding seasons congregate to breed in large numbers, frogs may have a breeding season that lasts only two days, they get together for spawning and mating takes place randomly. Gametes are shed into the water where external fertilization takes place, e.g. fishes and frogs.
2. *Overlap Promiscuity:* Mating occurs between solitary individuals with individuals of overlapping home ranges or during brief visits by one sex to the home range or territory of the other e.g. rodents.
3. *Lek or Area Promiscuity:* Males defend a display area or territory that is used exclusively for attracting females e.g. sage grouse, Uganda kob (antelope) (it has been considered a form of polygyny by some authors).
4. *Hierarchical Promiscuity:* Males establish dominance hierarchies that affect their ability to mate females e.g. cape buffalo, elands, oryx.

Advantages of Being Social

Several advantages have been proposed with reference to ultimate, ecological and social driving forces that lead to grouping and subsequently, to social behaviour.

Antipredation **:** Improved detection of predators - with more eyes, ears, there is an increased chance that one or more individuals will detect a predator before the others and be able to warn rest of the group e.g. in a group of deers or monkeys, one or two individuals will give alarm calls at the sight of approaching predator and the rest of the group members will take advantage of it, in a flock of birds alarm calls are given by few individuals and the rest get the advantage. Similarly, in prairie dogs which live in large colonies, only one would utter a loud alarm call when a coyote approaches, and rest of the prairie dogs would dash into the burrows.

Animals also show "guard behaviour" in which one or a few individuals assume the role of watching for the entire groups; thus freeing the others almost completely from the job of vigilance e.g. adult males of some of the monkey groups e.g. hanuman langur Presbytis entellus, occupy the highest canopy and remain vigilant for intruding members of all-male groups and/or for predators. Male baboons and pata monkeys often act as sentinels while other troop members forage.

Mutual vigilance has also been observed among the members of different species. Social animals living in the same habitat respond to the alarm calls given by the members of other species e.g. baboons, zebras, gazelles often forage together, and each species responds to warning signals of each other.

Erratic flight and explosive scattering, fleeing along an unpredictable path is a common defensive tactic in social or herding animals. Simultaneous erratic scattering of many individuals is likely to be more confusing to a predator than flight by a single individual. Sudden, explosive movement startles and confuses the predator. Most of the deer and antelopes use this tactic. Wildebeests form herds of 500 to 1000 individuals they start running in all directions in a very confusing fashion as the predator approaches, deer too confuse predators by presenting many choices that the predator is unable to decide which one to catch first, and in the mean time, often all of them successfully escape.

Mobbing running away or fleeing is being the most commonly used anti predatory strategy among animals yet there are examples when adult animals, generally males mob a predator or form a defensive screen to scare it away. Soldiers or workers of social insects defend colonies by collectively attacking. Muskox, a large Bovid found in north America when attacked by wolves, adult female muskoxen form a defensive ring around their calves, while adult males form a ring around them and collectively chase wolves in attempt to drive them off, similar defensive formation occurs in elephants also. Groups of baboons have been seen chasing leopard, cheetah and lions. Coatimundis, agoutis, California ground squirrels and various non human primates have been seen mobing snakes. Large Ungulates sometimes charge in a group to repulse lion, hyenas and wild dogs. Many species of colonial or flocking birds also mob predators. Predatory bird, hawk, prey upon wood pigeons, it was observed that attack success of hawk decreased as the flock size of wood pigeons increased. Black headed gulls also protect their nests and chicks from predators by an elaborate system of colonial defence.

Due to geometrical effects it is most likely that predators attack individuals on the periphery of a group, prey animals in a group should therefore, gain protection by remaining in or moving towards the centre of a group. The animal is safer when it can keep another individual between itself and predator, by doing so it follows the "you - first" principle, if there are others around, the chances are less that any one individual will be taken by a predator; that is, chances are that a neighbour will be captured first. In essence, one takes cover in the crowd.

Feeding Efficiency and Information Sharing : It is easier for a group of animals to catch a prey instead of catching it alone or cooperative foraging is beneficial e.g. for a group of monkeys it is easier to locate a bonanza of fruits. Early in the morning, members of a monkey group will spread and roam around in search of food and even if one individual locates the food it communicates vocally to other members of the group who get the benefit of this information. By cooperative hunting even small carnivores like wild dogs, wolves can hunt big animals like elk, moose, mountain sheep and zebras. Lions, hyenas, killer whales and some dolphins also get benefit of cooperative hunting. Nile crocodiles have a unique way of foraging together, they form a semicircle where a temporary water channel enters, they face the in-rushing water, and do not let fish pass by through that temporary channel and feed on them, it wouldn't have been possible for just one crocodile to block the water to get that much food. Lionesses also do most of the hunting together, often in groups, in an organized way, some females will drive prey toward others lionesses lying in wait.

An experimental study involving chaffinches and house sparrows confirmed that flock members observe what others are eating and then begin eating those foods themselves. In England just a few titimice birds started eating cream from milk bottles and gradually it became a major nuisance because hundreds of these birds learned by watching each other to open the milk bottles left on porches by milkmen. Therefore, in a group, watching others eating, can improve foraging in remaining members.

The other advantage is that in a group, mutual vigilance for predators allow each individual to spend more time foraging, without increasing vulnerability to predators. The animals of a group can leave their young ones with other group members and go foraging e.g. the pups of wild dogs, wolves and lions are often attended by few adults while other peck members can go out hunting. In a group of hanuman langurs, the related adult females temporarily take away the infant from mother to enable her to forage, this behaviour is called aunt behaviour.

There are many ways by which animals can maximize foraging success just by living in a group. Cooperation improves the ability of carnivores to protect carcasses from scavengers, a single lion usually cannot prevent a peck of hyenas or wild dogs from stealing a carcass, but two or more lions can. When a baboon comes upon a water hole, it conveys this information to the others by its conspicuous drinking posture, with the hind part and tail sticking up. Similarly, the neighbours are attracted by the excited hand movements with which a baboon digs up a tuber. During a drought, an old male baboon can lead his group to a distant pond which he remembers from a previous visit. One reason why rats have been so difficult to eradicate, is their capacity to transmit information regarding poisonous bait to other members in a group. If the animals were living solitary they would be deprived of these informations.

Facilitation of Reproduction : Group living improves reproductive success. For example, in solitary animals like rhino and orang-utan it is difficult for them to find a mate, they have to cover large areas in forest, spend time and energy to find a suitable mate but it may be easier to find each other in a large social group. In a group, watching others courting and mating initiates sexual behaviour in other members also.

Societies of Lions, Deer, Antelopes, Monkeys and other Mammals

Lions : The lion is a migrant in India. Fossil remains found in England, Germany, France indicate that the Cave lion a primitive form, now extinct, had originated in central Europe in prehistoric times, from where it migrated to Asia, Africa and Egypt. It is believed that lion entered India from Persia about 6000 years ago, and then spread in northern India. It always remained totally absent from south of India. The occurrence of lions in fairly good number was recorded along the banks of the Sabarmati river in Gujarat and around Mount Abu, Rajasthan in 1830's. In Gujarat, lions mainly entered the Gir and its forests, where they are found till date. In India, lion hunting was a great royal sport, which also became the major cause for decreasing their population, and subsequently, habitat destruction, deforestation, competition with tigers and human beings brought them to near extinction. Now there are only 200 lions left in Gir, and they have completely disappeared from rest of the northern India.

No other big cat (tiger, leopard, cheetah) is as social as the lion. Lions live and hunt in families known as "Pride" which comprise of about 20 or more individuals, large and small, of different age sex classes, consisting of upto 3 adult males and 15 to 17 females and their young. They are sociable creatures with strong family ties, they are most noisy and vocal. The lion follow matriarchal system which is a form of society where the core of the group is formed by the related females, the males are accorded with temporary membership, they come and go. These related females stay together with their recent offsprings.

The growing males leave the natal group and may either join to form an all- male group or would wander alone without the pride. The males roam around in large territories. As the breeding time approaches the dominant, adult males start associating themselves with the group of females (harem). Possession of harem is not easy; the males have to first establish their dominance among themselves, ther a dominant male takes over the harem from other weak male. The dominant male along with few subdominant males stay with the females. One or two adult males mate with many consorting females in heat. The lion males extend their association with females over a long time whereas, male tigers generally leave soon after mating. The adult male lions do not actively invest in rearing up the young ones, it is generally the responsibility of mother to feed them and train them for hunting. Lions prey on animals like gazelles, antelope and zebra, but may also cooperate to kill larger mammals such as buffalo and giraffe. Breeding occurs at any time of year. A litter of 1 to 6 young, usually 2 or 3, is born after a gestation of 102 to 113 days. They are suckled for about 6 months, but after the first 3 months increasing proportion of their food comes from the kills of adults, the cubs are left behind with one or two adults while the rest of the pride goes off to hunt, if a kill is made, a lioness returns to take the pups to the hunting site. Once the cubs are over 4 months old, they accompany their mothers everywhere. The cubs get sexually mature by 18-20 months, young males are driven away from the pride at about this age, but the growing females are retained back by the other female family members.

Deer and Antelopes : The deer, antelopes and gazells belong to order Artiodactyla of mammals, this order has 181 species in all. To a commoner, there is hardly any distinction between them, the deer belong to family Cervidae (There are 10 species in India) whereas, antelopes, gazells belong to family Bovidae. The deer such as hangul (Cervus elaphus), thamin (cervus eldi), musk deer (Moschus moschiferus), sambar (Cervus unicolour), swamp deer (Cervus duvauceli) and chital (Axis axis;) have a large opening in the skull below each eye called the lachrymal fissure, canine teeth are present in the upper jaw, absence of gall bladder, solid branching antlers present in males only are some of the main characters. The antlers are temporary, which are shed after the rutting (breeding) period is over, then regrow and remain "in velvet" (covered by soft skin) for sometime.

The antlers branch and give rise to twines or times. As the males get ready for breeding the soft skin or velvet is rubbed against tree trunks to get the hard and sharp antlers exposed. At this time the stags (males) are ready for rutting season and start searching for does or hinds (females). On the other hand, antelopes and gazelles viz. chriv (Pantholops hodgsoni), chinkara (Gazella gazella), black buck (Antilope cervicapra), chowsingha (Tetracerus quadricornis) and, nilgai (Boselaphus aragocamelus) have horns which may or may not be present in females but they are permanent structures in males, with prominent rings. Antelopes have many scent glands situated at different places viz under the eyes, between the digits. The gazelles are characterized by sandy skin colour and a distinctive white streak on each side of the face. Horns are present in both sexes, they have prominent rings, gazelles have tufts of hair growing from their knees.

The social system of most of the deer and antelopes is essentially matriarchal. The social organization among deer and antelopes can vary from monogamy to polygamy, from three individuals to several, they may form large groups. Adult males defend resources which could be water or food or females as the breeding time approaches the fight among adult males establishes the winner male's dominance, who acquires a green grass land, scent marks the territory and defends it. The females in heat visit the male in that territory, mate and stay with that male to bring up the offspring, another female may also enter the same territory. Hence, ultimately, one stag will have a harem of 5-7 such females. The leadership of the herd, care of young is the exclusive role of does or hinds. As the males attain maturity they leave the natal group and associate with other stags to form an-all-male group.

Social organisation, body size, group size among animals is influenced very much by abundance, availability and dispersion of food and the size of home range. Jarman (1974) studied certain African antilopes to establish inter relationship between these variables. He studied many antelopes (including members of Bovidae) found in different vegetations in Africa. He classified vegetations depending on rainfall and its pattern through the year into four types, he identified the antelopes, their body size group size, food habits and derived correlation between variables

1. *Forest:* Heavy rainfall and well distributed through out the year.
2. *Non-deciduous Woodland, Savanna, Grass*: Less heavy rainfall but well distributed throughout the year.
3. *Deciduous Wood Land:* Less heavy rainfall and seasonal.
4. *Scrub:* Light rainfall and erratic

Antelopes are ground dwelling harbivores and eat two kinds of food, grass and young leaves. Grasses provide bulk to the animal but of low nutritive value, whereas, browse provides little in quantity but highly nutritive food. The selection of food in antelopes is such that the plant or plant parts should provide them high protein, digestible carbohydrates, low fibre contents, water and minerals.

The questions which occurred to Jarmen were why some antelopes are small, some are big? Why some live in small groups some in large? Why some have small home range and some large? Why some use concealment as anti-predatory tactic and others can chase away the predator? To answer these questions he studied 74 species of antelopes and classified them on the basis of:

Principal foraging mode i.e. whether grazers or browsers and degree of dietary selectivity.

He observed that some species of antelopes like small sized dikdik select parts of individual plants via. young leaves, flowers or buds and they were browsers. Jarman called these species highly selective, whereas, there were other big sized species buffalo which eat whole strands of plants, they were grazers and were nonselective. He further realized that selectivity was based on nutritional requirement of that species. If a species requires food item of high protein content, it should eat newly emerged leaves or be a browser, if a species did not need high protein content it should eat the entire plant or graze. In other words browsers need more protein, therefore, they are highly selective and grazers needless protein, and hence, they are non-selective. Regarding relationship between body size and feeding styles Jarman (1974) correlated diet with body-size, he said grazers like buffalo, wildebeests, topi, etc. feed on poor nutritive diet (less protein) unselectively, ingest more bulk, so they have big stomach and subsequently big bodies, on the other hand browsers like dikdik and duickers feed on rich nutritive food and are highly selective, ingest little food and hence, have small stomach and small body size. Because grazers eat unselectively their food is rarely scarce-they eat more and become big while because of high selectivity among browsers they eat less-so they remain smaller. Small sized antelopes are physically better suited (small muzzel) to select small parts of a plant as compared to large antelopes, the question arises why do small animals require high protein? Because small sized animals have high metabolic rate, they require more protein than large animals.

Dispersion of food and home range of animals is inter dependent. Duickers and dikdik form small groups of 1-3 and are highly selective in their food habits. Each food item must be precisely located. An intimate knowledge of the resources of an area would promote searching efficiency, decrease time taken, therefore a small territory is preferred, and hence they have small home area. They exploit one vegetation type, territory is fixed may be for years and form monogamous groups,' because food resources are limited.

Reed bucks and bush bucks exploit more than one vegetation type. They are selective feeders have small home ranges, the male stays with one or two females, the group has 3-6 individuals. Greater kudu and water buck exploit a range of vegetation type, have large home range, they are moderately selective, live in large groups of 50-60 individuals. Males establish territories and herd many females, wildebeests and topi also exploit many vegetation types, they roam around in large areas, have poorly defined home range, the group may have 100-150 individuals, buffalos and eland are large sized grazers, wander in large areas, the herd may include 500-2000 individuals.

Relationship between group size and food dispersion or feeding style was also established by Jarmen, he said that feeding style in most species places a limit on the maximum number of individuals that can feed together as a coordinated cohesive group. The effect of one individuals feeding on the food availability to others is important in a herd.

In duickers and dikdik a feeder eats its way through an area and removes all young leaves. These animals take almost all the acceptable item of their feeding area and do not leave much for another member, such a style automatically separates group members. Its better if individuals exploit different areas. Hence the group size remains smaller. Wildebeests and eland eat the grass through an area and reduce the size of item leaving their dispersion the same. Food item will get smaller, but next following animal can still eat it. Hence many animals can feed together and live in big herds. It was also analysed that why do some antelopes conceal from predators while others do not ?

Dikdik who are small sized animals live singly or in pairs, they live in covered areas to hide, conceal, freeze when they see a predator, they scent mark territory, do not give calls because they do not want to attract predators. These small antelopes have more laterally placed eyes to cover more area, whereas, greater kudu, water buck, wildebeest, topi, buffalo and eland have more binocular vision, they live in open grass lands and are big sized animals and they live in big herds and are capable of facing the predator.

All the above mentioned parameters or variables are interlinked and influence each other:

Small sized animals	Big sized animals
↓	↓
have high metabolic rate	have low metabolic rate
↓	↓
they are highly selective in feeding	they are non selective in feeding
↓	↓
therefore they browse on young leaves, buds, seeds, flowers and fruits	therefore they graze on grass
↓	↓
they restrict their movement in small areas for intimate knowledge	they roam around in big areas, with plenty of food
↓	↓
therefore they live singly or in pairs	therefore they live in big groups, they always have something for the other animal
↓	↓
because they live singly they rely on concealment as anti-predatory tactic.	because they live in large number they can face a predator.

Monkeys

1. *Solitary, Except for Mother Infant Pair : Examples* : Microcebus, galago, slow loris, slender loris, aye - aye. These primates are forest dwellers, smaller in size, arboreal. Males are usually found solitary but they can

be seen in following combinations as well :male and female; male and female with infants; female with infants; 1-2 subadults. Since they are mostly alone, their anti-predatory device is concealment. They remain in one home range and may defend it actively. The sexual partners only meet for reproduction. Nothing much is known about these species as being solitary they are shy to intruders and as they inhabit thick forest, remain hidden in foliage.

2. *Monogamous Adult Pairs with Recent Offsprings or Nuclear Families: Examples* : Indri, mentawei island langur, silvery marmoset, gibbon, dusky titi marmosets, tree shrews, red bellied lemur, mongoose lemur, ruffed lemur, night monkey.

Nowhere among the social primates are females accorded more permanently privileged positions than among the monogamous species. There are some 37 of them, most of them are rare and endangered. They are inhabitants of tropical rain forest. Gibbons are found in south America. Their groups include 4-8 individuals consisting of an adult male and adult female and upto 4 young ones. Male-female bonding is permanent and lasts life long. Nevertheless, it is true for gibbons, marmosets and mentawei langurs that partners who disappear are soon replaced. In such groups usually there is not much difference in the body size of male and female. They have equal dominance. Both of them (male and female) involve themselves in all activities with same intensity. Indri, a lemur, found in Madagascar are classically monogamous, 'no other primate comes so close to fulfilling proverb "love your wife as you love yourself, and honour her more". Male indri eat less, they allow their females to eat more. All monogamous groups have definite territory which is defended actively, both partners sing "songs". Dueting is sympatric of monogamy. Titi monkeys and marmosets scent mark their territories, whereas, gibbons do not scent mark but give hoot calls from their territories which can be heard by other groups at a distance. Monogamous primates are usually smaller in size; they feed on high protein diet viz., insects, new leaves and ripe fruits. The parental investment is equal, the female feeds the young and male carries young ones on his back. In titi monkeys father is the primary caretaker of the young. Anti-predatory device is concealment.

3. *Single Male Groups with Bonded Females and Offsprings or Unimale Bisexual Groups*: *Examples* : Erythrocebus patas, hanuman langurs, red howler monkey, redtail monkey, blue monkey.

Typically they live in unimale bisexual groups, their group will have 20 to 100 individuals, there will be just one adult, fully grown, big sized dominant male popularly designated as "overlord" or "resident male", rest of the group is formed by adult females, subadult females, male-female juveniles and infants. Growing males either leave their natal groups wilfully or are chased away to form all male groups. Adult male is the leader and coordinator of group activity, he initiates and determines the direction of group movement and activities like where to go, when to feed, where to sleep. Males coordinate inter-group relations by producing the loud, resonating calls, which is necessary for quick and effective gathering of a group and also to tell the neighbouring group that - "I am here and this area and harem belongs to me". This male is easily distinguished from rest of the individuals because of its large size and agility.

In unimale bisexual groups especially hanuman langur, it is usually the adult male who alone defends the territory, though not very actively because it is a rather big area, he herds females away from intruding males of all male group, generally he alone indulges in fights, very seldom females may also participate. In unimale bisexual groups the males are much larger in size than females. The male is dominant over all the members of his group, there is no rigid dominance hierarchy among females, male parental investment is almost nil. The Anti-predatory strategy is climbing up the tree branches with all agility. Changing of overlord in a unimale bisexual group is of common occurrence. During interaction between adult male of unimale bisexual group and males of all male group, one of the adult males of all male group would chase the resident adult male and take possession of harem, this is known as "takeover".

Erythrocebus patas are found from Senegal to Sudan and Tanzania, they form groups of 20-30 individuals with one adult male and his harem. These monkeys live in dry Savannas and are truly terrestrial. If attacked they run away (fleeing) with great speed. The red howler (Alouatta seniculus) is found in Columbia, Brazil and Bolivia, the male of the group occupies a territory, makes loud calls, and has a group of 10-15 females. Red tail monkey (Cercopithecus ascanius) and blue monkey (Cercopithecus mitis) are found in Africa and they both live in harems which consist of a permanent group of adult females with their young and one temporary adult male, their group size may vary from 30-50 individuals. The special thing about them is that these two species share the same habitat and sometimes interspecific mating takes place and hybrids are produced.

4. *Many Males with Many Females and Offsprings or Multimale Bisexual Groups*: *Examples* : rhesus, gorilla, baboon, spider monkey, bearded saki, uakaris, wolly monkey, squirrel monkey.

Typically there are 3-8 adult males in a group, each of which has 5-7 bonded females who remain with their infants. In a way there are many small units living together thus forming a big group. Multimale bisexual groups are generally larger than other social groups. One such group can have upto 180 or even more individuals.

One of the reasons of origin of multimale bisexual groups is linked to their terrestrial nature. Millions of years ago when competition increased in trees, few of the primate species left the trees and came to the ground, where they became more vulnerable to predators; these primates increased the number of males, they increased the overall number of members in a group and they increased in body size as well, in addition, they developed powerful jaws and became more aggressive, because their Anti-predatory strategy was "fight the predator." There is a variation from the typical type of grouping, the following is the broad category which indicates two major types within multimale bisexual grouping.

Those which do not divide in smaller feeding groups e.g. gorillas and hanuman langurs.

Those which divide daily into smaller feeding groups e.g. macaques and baboons.

In gorillas and hanuman langurs the group is typically a multimale bisexual type with several males and several females, all the members remain together. The males have dominance hierarchy, the most dominant is known as alpha, then beta, gamma and so on. There is a straight line status system among the adult males of a troop, however, there is no clear-cut dominance system among females. In a typical group of gorillas of 26 individuals, the oldest and largest male, develops gray hair on his back and is called "silver back", he is also the most dominant, rest of the males would be lower in status. Dominance consists of possession of right of way on a narrow path, or to a resting place or feeding site. Surprisingly and most unlike other primates, dominant male in gorillas is not very aggressive and all other males also have access to receptive females. There is no conflict for mating. These groups have also been called as "age graded" which means that there is only one oldest male, all other males are younger and are graded age wise.

In the next category of multimale bisexual groups come rhesus, hamadrya baboons, geladas and olive baboons which live in large groups of many males and many females. In these species the group splits daily to form smaller family units of one male and many females. The rhesus is widely distributed in India, lives in large multimale bisexual groups. The males have dominance hierarchy and in them the bonded females acquire dominance from the males they have been affiliated or bonded with. For example, if alpha is the most dominant male his bonded females will enjoy high place in dominance hierarchy among females and rest of the group members, even their infants acquire that dominance. The dominant males can be identified easily by their confident walk and by their long strides, they carry their tails up and a subdominant male walks carefully and will tuck its tail between the hind limbs, if the alpha male goes away from group even for a short while the beta male will raise its tail and as soon as the alpha returns it will again take the tail down. The split in rhesus group is not a common occurrence but temporarily their group may divide into family units for foraging, all family units remain in near vicinity, and can unite at the time of danger, for day resting and every evening for roosting.

Hamadrya baboons (Papio hamadryas) are large sized primates found in Ethiopia and Somalia, the males have heavy mane around neck, have dog like muzzle, they form large troops on cliffs to sleep together. The troops separate into bands before travelling to foraging areas each morning, bands fragment into one male units while foraging and then reunite into bands to travel back to the cliff in evenings and nights to sleep together as a large multimale bisexual group.

Contrary to hamadryas baboons, gelada baboon (Theropithacus gelada) form slightly different kind of group. These animals are found in northern Ethiopia, have heavy mane around the face and on the chest they have prominent three areas of bare rea skin. They are large sized, aggressive monkeys. The adult males are capable of chasing a leopard. These animals sleep separately on cliffs as one male family unit, in the morning, these units get together to form a large troop of many males and females which remains together all through the day while foraging, as the night falls they all get separated once again into smaller one male units to sleep on a common cliff.

The other variation is exhibited by chacma baboons, (Papio ursinus) which are found in Angola and Zambia. They form one male family units of about 20 individuals, many such units start living close by, and may appear as one group of 200 individuals. Their home ranges overlap considerably, therefore, during the day time the entire troop appears as a multimale bisexual group, but as the night falls they may separate into smaller families to occupy their own sleeping areas.

One more type could be exemplified by squirrel monkeys (Saimiri sciureus) found in the northern south America. Their group size vary from 35 to 100. Adult females with their young, form the core of the group, adult males intermingle in the group for few months during mating season, turning the group into many male and many female type then take more peripheral position during the remaining year.

5. *Diffused Social Parties*: *Examples* : pygmy chimpanzee (Pan paniscus) are found in Zaire and Chimpanzee (Pan troglodytes) are found from Guinea to Zaire, and in Uganda and Tanzania, form diffused social parties. Their social structure is more variable than other primates. They follow no strict social organization. They may be seen in bands of males, groups of females, with or without offspring, or form large troops of males and females with young, they may also be seen wandering alone.

Social Organisation among Mammals

Platypus are solitary animals therefore they do not wander long distances and hence are territorial, they communicate with other platypus by pheromones which are produced from cloacal gland. Spiny anteaters are not territorial but live solitary lives. The majority of marsupials are solitary, living in independent home ranges, as is the case of the American opossum and tasmanian devil. The sugar glider are monogamous, live in family groups of male and female occupying the same nest. The most sociable of the marsupials are kangaroos and wallabies which live in multimale/ multi-female groups called herds or mobs, scent-marking is very extensive in them through the deposition of saliva, urine, faeces, and through the secretions of cloacal, chin, cheek, ventral and sternal glands. Insectivora is a large order which has 344 species and the majority of them are believed to be solitary like the short-tailed shrew, the only species known to be social is the streaked tenrec which is colonial, living in complex burrow systems, Eisenberg (1981) has indicated that this species is polygynous, with a colonial harem system; social communication in them is largely by chemicals and vocalization. Territorial marking with faeces is common and communication includes mutual sniffing of scent glands on meeting. Elephant shrews are monogamous and territorial, territorial boundaries are scent-marked with the subcaudal gland and perineal and sternal glands.

The tree shrew are monogamous and territorial too, male territories generally overlap with that of a single female but some males, which have larger territories, are polygynous and have territory including those of up to three females. Male tree shrews use chin, anal and sternal glands to scent-mark their territory on conspicuous objects such as tree trunks, fallen trees, stones. Order Xenarthra is comprised of three distinct groups, the anteaters, sloths and armadillos, the majority are solitary species, which prefer living close by in their small territories in a large home range, armadillos found in South America, often nest in colonial burrows, nothing much is known of their mating strategy; two-toed sloth also found in South America marks its home range with secretions from its cloacal gland and nostrils, is a solitary, sluggish and slow moving creature. The members of Logomorpha viz. hares are solitary animals, pikas are colonial animals in which a male's territory may include one or more females. Rabbits are colonial too but live in multimale/multifemale colonies. Logomorphs have a well-developed chemical communication system, using chin marking, cheek marking, paw scraping, urine and faeces to mark their territories.

American agouti are usually monogamous, male and female territories completely overlap and each member of the pair defends it, depending upon the density, polygynous sexual strategy may also be followed. Monogamy has been identified in a number of rodents such as the beaver and is also believed to exist in the prairie dog and the old field mouse. Prairie dogs and belding's ground squirrels live in multimale/multifemale colonies made up of territorial female-bonded clans. Some rodent may have variable social organization dependent on the nature of their environment and population density. The house mouse (Mus musculus) ranges from being solitary and territorial to forming multimale/multifemale colonies. Rodent Heterocephalus glaber possesses a female reproductive caste and three other castes, non-workers, occasional workers and frequent workers, female members of these castes appear to be sterile and it is probable that the male non-workers mate with the fertile female.

Perissodactyls range in social structure from the solitary browsing tapirs through the slightly more sociable rhinoceroses to the highly social zebras and horses. Breeding male rhinoceroses are generally territorial. Male rhino form dominance hierarchies. Male ranges overlap and adult males are hostile to one another and compete for oestrous females, related females form long-term associations. The horse family (Equidae) contains the most social species of Perissodactyls. Two types of social organization exist; in one, breeding males are territorial and oestrous females mate only with these males. Territorial males allow non-territorial males to enter their territory and permit them to approach and sniff anal and nasal, regions before chasing them off for 50-100m. Non-breeding males and females may be sociable and form herds, but there are no permanent associations between individuals e.g. asses and grevy's zebra. Territorial ownership is advertised by roaring and by scent-marking with dung piles and urine. The second type of social organization in Equidae is, one in which there are permanent groups of females (harems) and young who are closely related, these harems are taken over by a male who defends the group against other males; non-breeding males live in hierachically-organised bachelor herds. Harem owners abandon their females if successfully challenged by a younger male. In addition, use of urine and faeces for territorial marking, personal recognition appears to involve perineal, anal and circum-oral scent glands.

Among elephants the related females along with their young, form the stable core group. These groups are believed to be matriarchal, i.e. females remain in their natal herd to breed while males are driven out at puberty or leave of their own accord. Females form kinship groups which, on meeting, show elaborate greeting ceremonies. When a male African elephant leaves the nursery herd he may join other males in a bull group. Bull groups appear to be temporary aggregations but there is evidence that the bulls in an area know one another and have hierarchical organization. High-ranking males have intermittent periods of heightened sexual activity known as 'musth'. During this period they leave the male group and compete for females which are in oestrus., Only one male from any bull area is in musth at any one time so that direct sexual competition occurs between bulls from different areas. Each bull has his own large home range which is undefended, but may include female groups. If a cow is in oestrus, bulls in the area attach themselves to it. The highest-ranking male mates with the oestrous female. Male elephants have an orbital gland which produces a copious secretion during the period of musth, which is rubbed on trees.

All Pinnipeds come ashore to breed and, during this period, show a rich repertoire of social behaviour. After leaving the breeding grounds they return to the sea. Arctic and Antarctic cold waters provide the richest food sources so that the largest populations of Phocids are found in these areas and there is a tendency for species with the highest density of population to be the most highly polygynous. Pack ice provides plentiful breeding sites with ready access to the sea in channels between floes; solid ice, however, offers less access to water so that the breeding area is more restricted and leads to the animals crowding together with very large harems. The ringed seal breeds on pack ice and is monogamous. The common seal is either monogamous or has a small harem of 1-4 females. The grey seal breeds on rocky coasts and islands and has a harem of 1-10 females. Walruses are all polygynous. The male defends a territory which includes a group of females. Males challenge one another at boundaries of the territory and attempt to prevent oestrous females from leaving. Males have a dominance hierarchy, while females are hostile to one another but seldom attack. Males do not feed for the entire breeding season of 2-3 months, their sole function is herding females and mating.

There are two living genera of aquatic herbivores, the manatees and dugong. Both form herds, but the only social tie is between mother and offspring. In Florida manatee, an oestrous female is competed for by anything up to seventeen males known as 'the oestrous herd', members of which may have an age-graded rank order. The humpbacked whale generally lives alone or in groups of two or three individuals; large breeding assemblies of up to 500 individuals, however, come together in the area of Hawaii and Berumda. The mother-calf bond is strong and often a mother with young is accompanied by an adult escort who may play some role in defending them. Ganges river dolphin are solitary, or found in very small groups. Species found in both riverine and coastal environments, such as the Common porpoise live in slightly larger groups other species of porpoise may form all-male and nursery herds. In the killer whale the stable group appears to be an adult male and a group' of females with young, thus suggesting a mobile harem structure. Common dolphin, spotted dolphin and spinner dolphin do not form a definite social bond but travel in large schools.

The bats are extremely diverse in their social behaviour, ranging from solitary to highly gregarious forms. Solitary bats such as Micropteropus pusillus, Pipistrellus nanus have independent hunting territories and roost in isolation. Four main types of social organization have been recorded for bats, namely, monogamy, harems, bisexual groups and leks. A number of species of bat are believed to be monogamous on the grounds that they usually occur in pairs. In the false vampire, a South American bat, the male protects and feeds the female and young. The Australian grey-headed flying fox is a colonial bat; some males are polgynous, but males which mate with a female who already has an infant are monogamous. Phyllostomus discolor harems consist of 1-15 females per male, though there are also all-male groups. The South American Phyllostomus hastatus, an omnivorous, colonial bat which roosts in caves has harems of 10-100 females per male. The Indian flying fox Pteropus giganteus forms bisexual societies, both sexes being highly integrated. There is a male hierarchy which confers breeding success on high-ranking individuals. At parturition, females become segregated and form all-female nursery groups.

5

Human Behaviour

The human, Homo sapiens is the animal which is best equipped to overcome challenges imposed by the environment. Largely as a result of his superior mental abilities, the human is less restricted than other species by environmental problems. The human is the most highly developed of all creatures in his greater independence of environment. Not only does he cope with existing environmental problems, but he is also capable of changing the environment to suit his needs. Thus he influences the course of his own evolution as well as that of other species. The human uses his superior brain to manipulate the environment which is a direct reflection of human intelligence. Intelligence is reflected in the many facets of human behaviour which has the potential to influence the very existence of all life forms. Since the human nervous system is so complex, it should come as no surprise that analysis of the hereditary and environmental compounds which influence human intelligence and behaviour is the most difficult of all genetic investigations. Sufficient data may accumulate to permit a breakthrough into a fuller appreciation of the human mind and the reasons why the acts as he does. The ability to communicate through speech and the written words associated with oral language makes it possible for each generation to build on the culture of its ancestors to utilize it, and improve it, thus becoming more and more independent of environmental problems.

Communication Symbols

The behaviour of human parents in rearing their young has specific features which are expressed during the long period of development required by the very helpless human infant. While incapable shortly after birth of performing feats such as walking, swimming, or feeding himself, the potential of the human infant exceeds by far that of any other animal. The human's genetic endowment has provided him with a central nervous system of the utmost complexity. The countless nerve connections and relays in the brain itself provide him with the machinery to accomplish something no other animal can do. This is the ability to deal with abstractions. To receive symbols, arrange and interpret them in a meaningful coherent fashion. There is evidence from studies with chimpanzees with symbols, but it does not approach the facility of the human in this respect. Moreover, the human is unique in the ability to communicate symbols freely through a spoken language. Again, chimpanzees have been taught after labour to utter a few simple words. The inability of the ape to speak freely may not be entirely a reflection of limited intelligence. It is to a large degree a consequence of the fact that the ape is not equipped anatomically to handle spoken words. Speech demands a voice box, palate, nose and mouth cavities of a certain construction. No other animal possess the type of anatomy required for the free formation of the words.

Language : Normal hearing apparatus is essentially associated with proper speech. This is the receiving part of the system which delivers the sound stimuli to the brain. It is here that these stimuli are arranged and interpreted into a meaningful whole. Before any human can speak properly, he must be provided with these sound stimuli. These enable him to form his own sound by association of symbols in the complexes of his brain. Ability to speak a language, therefore, is a complex characteristic. It depends in large part on anatomy, the hearing apparatus to receive the sound stimuli, the brain cells to interpret them, and the vocal apparatus to form words. The anatomical features necessary for language communication have a large hereditary component. Certainly, very fertile human must carry in the sperm or egg information required for the embryo to develop these anatomical requirements. A defective gene may cause upsets in the differentiation of one of these structural parts. Without sound reception, a human will never be able to speak clearly and freely, even though the brain and vocal apparatus have developed perfectly.

Intelligence : The ability to deal with symbols is a reflection of the intangible is known as intelligence. The increase in intelligence which the human possesses over other animals is seen in the greater ability of memory retention, to think, and to arrange information stored in the brain in such a manner that foresight is possible. This foresight enables the human to act in ways which permit him to avoid potentially harmful environmental influences with foresight, the human can plan and transmit his knowledge and culture. This thinking, learning and foresight depend, of course, on a brain with billions of neurons (nerve cells). Increase in intelligence thus depends to a large degree on increase in the number of brain cells and thus increase in the size of the brain. It is well known that differences in intelligence among different animal species is related to differences in brain size. However, within a species, there is no evidence to indicate that a large brain reflects greater intelligence. Nevertheless, the species-specific behaviour of the human in regard to the ability to think, plan and speak depends on an important genetic component which provides the information to construct the brain of a certain size as well as the other anatomical features essential for the expression of the brain's full potential. However, the environment in which these genetic factors operate is a paramount importance and may carry genetic factors for normal speech or above average intelligence, the environment can deprive him of the full benefits which they provide. We see that behaviour, like any other phenotypic characteristic, is thus the result of the interaction of many genes and the environment. It should also be evident that species-specific behaviour has been subjected to the operation of the forces of evolution. Natural selection has provided each species with a particular behaviour which enables it to cope best with its environmental challenges. Behaviour, therefore, has an advantage just as any other characteristic of a species of individual. The human's behaviour as noted here, has enabled to make use of the greatest number of environmental niches, and to manipulate the environment for his own end.

Heredity and Environment

While much of human behaviour depends on heredity, it must not be forgotten that a great deal of it depends on learning. Human behaviour is distinguished by its plasticity, its ability to be changed and molded. A normal child is born with genetic factors which make it possible for him to receive environmental stimuli. Among the most important of these are the ones provided by other humans. An infant is associated with a family, and as he continues to develop, he is exposed to more and more members of his particular cultural group or society. Certain stimuli will determine what language he will speak. He carries the genetic information which provides the ability to speak, but whether he will speak and what language he will become expert in depends on other humans about him. The amazing inherent talent which a child possesses to learn a language is seen by the youngster's ability to become expert in several tongues if he receives the appropriate sounds during his formative years. No one must tell him what is English, Chinese or French. He sorts them out automatically without being told a thing about rules of grammar. But alas, this genetic ability is lost about the age of puberty, so that no adult will ever have the facility of the child who can so effortlessly master two or more tongues. Not will most adults ever speak a language perfectly if they have learned it after they have reached sexual maturity. There is nothing that dictates what language a child must speak. The great plasticity of this aspect of human behaviour is quite apparent, as in the interaction of its hereditary and environmental components, few human behavioural traits are based exclusively on genetic factors. The plasticity provided by the environmental interaction with the genetic component is seen in a child's general conditioning as he becomes integrated into his peculiar society. He learns the customs of his society and tends to behave according to the expectations of his specific cultural group. Reared in a completely different society with different values and expectations, another child with the same genetic potential as the first one will come to behave very differently as he acquires different habits, skills, and beliefs prescribed by the second group.

Evolution of Cultures

Human brains appear to have properties that facilitate certain kinds of learning. The classic example is language learning : there is considerable evidence that the brain contains neural mechanisms that facilitate language acquisition and use. Simply because there is cultural variation in the languages spoken around the world is not the proof of independence of this behaviour from genetic influence. Genes are deeply involved in the production of the neural foundation of language, and it seems inescapable that alleles have been selected on the basis of how well their brains have been able to contribute to language use. If selection can produce humans with great skill at language, it should also be able to shape the design of the human brain so that human can readily learn which individuals are relatives, how to use tools, how to behave in certain social settings, how to extract economic gains from the environment, and so on. As in the case of language, the successful genes need not produce neural mechanism that are restrictively designed to permit only the learning of a specific cultural trait (such as Turkestan of Jivaro or English language). Instead genes may endow their brains with more open learning abilities permitting a more flexible (but still not completely plastic) accommodation to whatever cultural variables dominate an individual's social environment. By this view, cultural evolution involves selection for various learning abilities that permit individuals to adopt that cultural practices of their societies. Thus genetic and cultural evolution complement one another rather than compete with each other.

But why have individuals with genes that indirectly helped them acquire a culture enjoyed a selective advantage in the history of our species ? To answer this question it may be helpful to consider a second basic behavioural principle discussed earlier in the book that social organization is a reflection of the ecology of the species. If a species confronts varied ecological conditions, it may involve the ability to alter its social structure to match the conditions of the movement. If this is true for humans, cultures may have evolved because the adoption of cultural practices enabled humans to invade a wide variety of environments each with its own special ecological pressures. Many authors have noted that cultural adaptation permits a more rapid adjustment to variable ecological circumstance than biological adaptation. If this hypothesis is correct, we should predict that cultural variation is not random, instead, the cultural practices of one group should permit individuals to exploit their environment more successfully than other cultural options. We should also expect that cultural changes elevate the inclusive fitness of the individuals that invent or adopt the change. People that practice the wrong option (one that results in lowered relative reproductive success) will be replaced by those who practice the superior option, transmitting this option culturally to their offspring. It seems obvious that the spread of tool use and the perfections of tool manufacture that occurred during human evolution were related to improve resource acquisition and the ability to produce and support more offspring. Moreover, the two major cultural innovations of the past 10,000 years, the invention of agriculture and the development of, the use of fossil fuel as an energy source have:

(1) spread rapidly around the world and

(2) been associated with a phenomenal increase in the human population. Both inventions, in effect have given participating individuals access to more calories and nutrients for conversions into offspring than non-participating individuals.

But what about those many peculiar cultural practices that actually appear to reduce the fitness of individuals rather than raise it ?

In some American culture it is customary to cook cornmeal in a 5% solution of lime, pour off the solution, cool the mash, was it, and finally use the prepared material to make tortillas. This involves a considerable expenditure of time and energy as well as the loss of some of the thiamine, niacin, nitrogen, and fat content of the corn. Although it would be easy to interpret this method of corn preparation as maladaptive, S. H. Katz and his colleagues have shown that alkali cooking actually improves the nutritional quality of cornmeal. It does so by breaking down the indigestible proteins of the corn kernel and thereby freeing for human use substantial quantities of certain essential amino acids, including lysine, which is deficient in those parts of the corn seed that can be digested without this system of preparation. A diet that is dependent on cornmeal as a staple would be dangerous without alkali cooking, we lack the metabolic pathways to manufacture lysine, an amino acid that appears in many vital proteins. Moreover, alkali-treated corn has a lower ratio of the amino acids leucine to isoleucine, which improves its nutational value. It is hard to imagine how was invented this cooking procedures, but it is striking that all south and central American cultures that rely on maize as the basic food stuff have adopted the practice.

Effect of Alkali Cooking on the Availability of Certain Essential Amino Acids

	Milligrams of amino acid per gram of nitrogen in	
Amino Acid	*Uncooked corn*	*Alkali-treated corn*[a]
Histidine	0.012	0.028
Isoleucine	0.074	0.158
Leucine	0.309	0.358
Lysine	0.045	0.126
Threonine	0.145	0.381
Ratio : Leucine/ Isoleucine	4.2	23

[a]Com subjected to enzymatic digestion for 12 hours.

Sacred-cow Tradition

Another cultural trait that has baffled outside observers is the sacred-cow-tradition, which forbids or greatly restricts the sale or slaughter of cows for meat. In India the Hindu reverence for cows is such that they are allowed to roam freely, interfering with traffic, and they are almost never killed, even in times of famine. Why are not emaciated beasts sold or killed and eaten by their owners ? Marvin Harris has argued persuasively that far from being a liability, cow love is almost certainly to the economic benefit of the Indian peasants who maintain the tradition. Emaciated or not, cows provide their owners with a certain amount of milk and with quantities of dung, which is used for fertilizer, cooking and construction; most importantly they occasionally produce offspring, the males of which can be converted into oxen. Oxen are the Indian farmer's tractor, and without them he faces economic doom. This is powerful incentive not to dispose of your cow, even if she is a decrepit beast. The sacred-cow tradition is a proximate mechanism that encourages individuals to do those things which are economically and biologically adaptive. If this argument is correct, we should expect to find that other groups that rely on cattle for their subsistence have a similar attitude toward cows.

The cattle-owning Karamojong of Uganda and Masai of Kenya have exactly the same view. The tribesman try to accumulate large numbers of cattle, which provide their owners with milk and blood, important elements of their diet. Despite pressure in recent times to sell portions of their herds to prevent overgrazing, the Karamojong and Masai are extremely reluctant to do so, in part because cattle exchanges are currency for forming protective alliances with other men and for the purchase of wives. The herdsmen feel a strong affinity for the cattle they own and absolutely refuse to trade or kill female animals which may provide them with additional herd members. Thus, although Western observers of Indian and African cultures may feel that these people's attitudes are irrational, from the perspective of individual cattle owners their practices are entirely sensible.

From the perspective of evolutionary theory, the cultural traditions of these groups help rather than hinder individual genetic success.

Evolution of Human Welfare

Let us examine the complementarity of biology and culture in the context of one of the most dramatic of all human cultural activities, violent group aggression. The frequency and degree of aggressiveness exhibited by different cultures vary from the essentially pacifist societies of some Eskimos and Pygmies to the combativeness of the Yanomavo Venezuela, many New Guinea tribes in the recent past, and the societies of the United States and Russia which are either constantly engaged in war or armed to the teeth in anticipation of the next round. This variation has been interpreted by some as evidence that war is largely a capricious features of certain cultural environments and could easily be eradicated.

Certainly the evidence indicates that the genotypes of humans do not blindly program them to engage in warfare. But this does not mean that our genes have not endowed but nervous systems with special properties that provide humans with the capacity to engage in group aggression under Certain Conditions. From a evolutionary perspective it is possible that some ecological circumstances may so elevate the benefits of warfare to individuals that they exceed the costs to individual inclusive fitness. An argument that the abilities that underlie the capacity for warfare are generally advantageous does not imply either that warfare is inevitable or that is morally desirable. This does not prevent us from accepting as a hypothesis that the traits that may be employed in group aggression can help maximize the inclusive fitness of the humans that possess these traits. Here we shall examine the plausibility of the hypothesis, taking this example to illustrate how (i) evidence about our history as a species, (ii) comparative behavioural data from other living animal species (iii) ecological comparisons of human cultures may be brought to bear upon the question.

Evidence from Human History

Our hominid forebears were group-living primates, closely related to the great ape lineage, whose key ecological invention was to become a hunter of living animals in addition to gathering vegetable foods. The use of tools became a central adaptation that enabled protohominids to capture and process game efficiently. The importance of this culturally-transmitted adaptation may be related to the dramatic increase in the brain size of members of the human line over time and is surely reflected in alterations in the anatomy of the hand. The relation between the thumb and other fingers of Australopithecus africanus (a fossil hominid thought to be in the direct human evolutionary time or close to it) differs from that of modern humans. The changes that had occurred by 500,000 years ago permitted better power and precision grasping, capabilities of the greatest importance in tool making and tool using. The evolution of a longer straighter thumb with a broader tip meant that human could apply more surface area to an object held between the thumb and other fingers, thus producing area a firmer grip. There is a corresponding increase in the abundance diversity, and quality of tools found in hominid campsites over their evolutionary time.

The improvement in tool's quality is correlated with transition of the hominid diet from its early reliance on small game lizards, birds, fishes and antelopes to a focus on really large animals. Campsites of our ancestors living as long as 300,000 years ago have been found that contain great quantities of the bones of the largest mammals then alive mastadons rhinoceros, and cave bears. Bit-game hunting may have been the central adaptation of most human populations from that time until recent past.

Both logic and the examples provided by the modern hunting gathering tribes suggest that big-game hunting favours a division of food-collecting labour among men and women, a division that reflects the constraints placed on women by childbirth and infant care. Selection pressure on human females has produced traits that facilitate bearing and rearing children. But a wide pelvis that makes the birth of a large-brained infant possible also makes locomotion slightly less efficient and therefore interferes with hunting (just as selection exerted on male lions to be effective territory defenders has favoured the evolution of a large conspicuous mane, which makes prey capture more difficult for miles). Dividing the gathering of food resources between the sexes permits specialization and thus presumably greater efficiency in exploiting a diversity of food.

Hunting large and sometimes dangerous animals should favour individuals capable of cooperation in planning and executing a successful hunt. Moreover, the savannah habitat in which large herbivores are found and which our hominid ancestors in search of game also is home to potential predators of man. The danger of isolation and the advantage of group defence may have been major selective agents promoting tight social cohesion among the members of a band. In addition, the cooperative abilities useful in hunting big game and repelling predators could also be employed to kill or drive off other competitive species the large herbivores of the savannah. Finally, the great likelihood that members of a band were probably close relatives would have further elevated the benefits of intraband cooperation. An individual capable of working with others for a common goal would not only be improving his or her chances of reproducing, but might also be helping increase the fitness of other humans with many shared genes.

Evidence from other Animals

The plausibility of this sketch of the evolutionary events and selective forces that set the stage for human warfare can be examined by making some careful comparisons of human behaviour with that of other animals. A common complaint of critics of an evolutionary analysis of human behaviour is that socio-biologists make caplicious use of comparative material, selecting whatever species help support the hypothesis favoured by socio-biologist. However there are species to gain insight into the ecological and historical bases of a behavioural characteristic of interest. First, we can look at closely related species to determine which of their behaviour are similar to our own. This information can help us determine the history of a behaviour pattern. Second, we can examine unrelated species that experience similar ecological pressures to determine if they have evolved convergent behavioural solutions to these shared problems. For humans, therefore, the species whose behaviour may be most useful for an understanding of our own abilities include :

(1) other primates, especially our closest relatives, the chimpanzee, gorilla, orang-utan, and gibbon and

(2) the mammalian social cornivores, which are not closely related but which face the same obstacle associated with capturing big game.

Human Sex Roles in Society

The relative roles of the hereditary and environmental components in the expression of human behaviour has been a matter of heated debate in relation to the topic of human sex roles in society. Many persons expect a female to behave in a certain manner and a male to respond to a particular situation in a characteristic way which is distinct from that of a female. Generally, the female is expected to be more retiring, the male more aggressive. For a male or female to behave in a fashion considered inappropriate to his or her sex is often labelled unnatural. According to the ideas of some students of behaviour, characteristic male and female behaviours are fixed in the species. There is a genetic basis, according to this thesis, which is responsible for the expression of male social behaviour and of female social behaviour.

The argument of the components of this idea is based mainly on studies of the fossil record, observation made on the behaviour of various species of primates (the order to which the human belongs), and observations of behaviour patterns in various human societies today. Proponents of the concept contribute the fixation of sex roles to the advantage. This behaviour imparted to the human species during its evolution in the relatively recent past. Social behaviour in which sex roles are distinct would have given a decided benefit to the human when he becomes a hunter. Once the human species assumed hunting activities, so the argument goes, an assignment of roles to each sex would have given insurance to the success of these activities and hence would have increased the survival of the group. Any genes or genetic combinations which would magnify the differences in sex role behaviour would be selected under the force of natural selection. As the hunter, the human male would require strength, endurance, speed, alertness, and prowess among other attributes.

Such hunting effort would benefit from improvement of these traits, and so any genetic determinants in this direction would be selected. However, such traits would benefit directly only the active hunter, the male. Interaction of genes for excellence in these traits with the hormonal environment would fix them in the male. On the other hand, any hormonal controls to make the female more different from the male would have value. The changes in the female during this part of human evolution would have occurred only in rotations to the requirements of the male in his role as the hunter. In her role as child-bearer and rearer, intelligence would not be at a premium cunning, strength and swiftness would not be needed for good gathering activities. Genetic controls to make the female subordinate and docile would benefit the needs of the male after the demands of the hunt and would in turn make him a more sufficient hunter. Hormonal regulations in the female to make her sexually receptive at any time would have value.

They would have been selected in the relation to the need of the hunter who would be assured that the female would be available to him whenever, he so desired after return from the hunt. Strong bonds would form between males. Such bonding would be of advantage since it would assure cooperation during hunting activities and would reduce hostility among the males. No such bonding among females would be required, since the female's activities did not depend on major group activities. Therefore, sex roles became more and more distinct as the success of the hunt became more and more a certainty.

Even today, mainly the men perform decision-making activities. Woman accept these decisions. The male is aggressive whereas the female is the homebody concerned primarily with child-rearing and pleasing the male. All of the behaviour patterns which we today consider male or female are largely a direct result of the genetic factors which control these activities and which became established in the past in relation to the hunting activities of Homo sapiens.

Evolution of Human Reproductive Behaviour : Warfare is behaviourally highly complex, and its relation to inclusive fitness is not immediately clear. Reproductive behaviour is much more obviously linked to individual fitness, and therefore, the imprint of selection should be readily apparent on it. Yet it is well known that there is considerable variation within and among cultures with respect to many reproductive practices. Moreover, there are actions, including mate sharing, adoption of another's child, and the practices of birth control, which on the face of it, appear to reduce fitness.

The family unit is universally accepted by anthropologists and others as the basic components of human societies. Human males and females have the capacity, shared with many other animals, to form a family in which male and female live together and cooperate in the rearing of the progeny they produce. The selection of a mate should have a great deal to say about the genetic success of individuals. What evidence is there than human males and females behave in ways that promote their genetic success in this phase of the reproductive process ?

Mate Selection

One significant feature of mate selection is the culturally reinforced prohibition against copulation between some sexually mature individuals. Incest taboos appear to be a universal feature of human social behaviour. Thus parent and their children are forbidden to copulate, and people are usually not permitted to marry their siblings (and often first cousins). Incest taboos are often culturally extended to forbid marriage between individuals belonging to particular categories, such as the same clan. The strength of the incest taboo has been demonstrated by the refusal of children reared together in Israeli Kibbutzim to marry one another when the kibbutz system was devised, the social scientists in charge envisioned that communally-raised children would eventually marry within the Kibbutz. Such marriages were actually encouraged, and yet not a single one has occurred; the children of one Kibbutz always fall in love with members of another Kibbutz. It is as if the brain had a program that reads : "Do not become sexually attracted to humans with whom you have been reared". (When unrelated cactus mice were taken from different litters and experimentally reared together, they too were reluctant to copulate with nestmates. Over evolutionary time juvenile mice, and humans, that grew up together were almost always siblings or very close relatives.)

The inability to become sexually attracted by close relative contrasts sharply with due ability of humans to fall in love with complete strangers. This ability the incest taboo, and the culturally extended proscription against copulating with or marrying certain classes of individuals all contribute to outbreeding. The occurrence of similar mechanisms in other animal species suggests that a major advantage of these traits is due avoidance of inbreeding. This prevents the accumulation of deleterious recessive alleles in double close in one's progeny. Humans from incestrous matings exhibit a substantially higher rate of genetic defects and congenital abnormalities than members of outbred population. Thus human should benefit genetically to the extent that they avoid in breeding. An added benefit to outbreeding among humans is the formation of alliances between different genetic lineages. The progeny produced a marriage between two separate family groups creates a tie between the two bands that may promote cooperation between them. Without a mutual interest in the children of the marriage, the two groups night have no other shared concerns and no motivation to help one another.

Female Choice

The incest taboo applies to both males and females, but there is some indication that male parents break the taboo (by copulating with their daughters) more often than females. This may reflect the lower threshold for sexual excitation among males than females that is characteristics of animal species in which males may pass on their genes at little expense relative to the female. Female humans, like all female mammals, must make a large investment in each progeny while it is in the embryonic stage and during the early years of its life when it is nourished at the breast. Because a male's investment in an offspring may be as little as the sperm used to fertilise the egg, evolutionary theory predicts that females should be more selective than males in the choice of copulatory partners. There is ample anecdotal evidence as well as scientific data collected by Masters and Johnson that females of our culture are less easily sexually aroused than males. Female should be more cautious because they have much more to lose than a male. If a woman becomes pregnant at an inopportunate time (e.g. when she is unmarried and her partner has no intention of helping her rear her child), her large physiological investment may be poorly rewarded. At least in the past, the child of a husbandless woman would have had a reduced chance of surviving to maturity; if it did reach reproductive age, it would probably have a lower fitness than average because of juvenile malnourishment or the possession of few material goods, with which to attract mates. Moreover, a single mother may reduce her long-term chances to acquire a helpful mate, either because potential husbands would consider her a poor risk or because they are not eager to help care for another man's offspring.

Thus in cultures in which males may provide material resources for their children, a female should permit copulation to occur only if her partner has demonstrated both a willingness and an ability to make a large parental investment in their progeny.

Male Parental Investment

It is easy to see why females might prefer mates willing and able to provide them with food and protection. If a division of labour between the sexes has been the rule throughout much human evolution, males have controlled highly valuable resources (such as protein-rich game in many societies), which when used by a female and her progeny would greatly enhance their fitness. But, as discussed earlier, the male animal typically enjoys reduced benefits relative to the female from equal parental investments per offspring. First, the reliability of paternity is often less than the reliability of maternity, and a male therefore runs the risk of helping someone else's genes. Second, the cheapness of sperm means that males can potentially inseminate many females. This elevates the benefits of a reproductive strategy that emphasizes effort to copulate with many females rather than invest heavily in one or a few mates. Male mammals, including most primates, rarely are monogamous and rarely provide much in the way of parental care for the progeny of females with which they have copulated. Why are human males an exception to this rule ?

The following explanation is speculative. We begin with the hypothesis that males have consistently been more successful hunters than females. A female protohominid in estrus might attract a consort male who would guard her during the time she was receptive and most likely to ovulate. During the consort phase a male might share prey he had recently captured with the female because it would be likely that the progeny produced by the female would bear his genes; protein gifts to the female might, therefore, have genetic gain for the male. (As usual, there is no requirement that individuals have to be aware of the genetic consequences of their acts). If gift-giving became an established component of courtship and consort guarding, then a female might gain if she could extend the period of guarding, and thus receive more presents. Instead of blatantly advertising the line of ovulation by becoming receptive just prior to this event, female protohominids might have partly hidden the time of ovulation by lengthening the period of receptivity.

The advertisement of ovulation strategy is practised by female mammals as an incitation display to encourage competition among males for access to them, thus increasing the probability of mating with a dominant male. But if parental investment is offered by a male, the quality of this investment may become more important to a female than signals of male genetic quality. By taking it difficult for a male to determine when his chances of fertilizing an egg were highest, females that retained their receptivity could make it advantageous for at least some males to practice prolonged guarding. The more hidden the ovulation time and the more prolonged the period of receptivity, the longer a male must remain with a female (copulating with her at regular intervals) if he is to increase his chances of fertilizing her eggs when they do become available for fertilization. A widespread use of a strategy of permanent guarding (marriage) reduces the availability of fertilizable females and reduces the time a married male has available to find these females (because of the guarding time required to protect a mate). This being so, it may become advantageous to be a devoted husband and father and to expend resources on one's presumptive offspring, which are highly likely to bear one's genes. This may increase their chances of surviving and reproducing, providing a genetic gain that may often (but not always) exceed a casanova strategy of attempts at multiple sexual conquests.

Because human males can offer useful parental services to their offspring, one can expect females to choose husbands partly on their potential ability to provide superior investments, whereas males should make the parental investment in ways that tend to promote their own genetic success. A female should choose a husband who appears to be genuinely devoted to her and whose gifts and reputation are evidence that he is willing to help her. Males can and do deceive some females, seducing them with the intent of deserting them at some point. Female coyness and caution about copulation have the function the probability of making mistakes about males.

Women should also be concerned about the economic status of males, present and potential. The fact that females in a great many human societies willingly enter into polygynous marriages is at first glance puzzling. One might predict that given the potential role of males in child rearing, female should prefer to mate with a single male. Carnivorous mammals generally show a greater tendency toward monogamy than herbivorous ones. In fact, the vast majority of human marriages have surely been monogamous. Nevertheless, why should there be any polygynous marriages fully entered into, given the male's potential contribution to the welfare of his wife and children ? Perhaps human men are rather like redwinged blackbird males. Most human societies offer opportunities for truly exceptional performance in various fields, with commensurate material rewards. Outstanding leaders, hunters, weapon makers, and healers have probably always secured large amounts of food directly or indirectly. Because the difference between successful, and unsuccessful men is fairly obvious, marriageable women have probably had no difficulty in discriminating between the two. Given a choice between an outstanding provider and an inferior one, it may be to the women's advantage to choose the former, even though this means entering into a polygamous relationship. In general the economic status of a potential husband should be at least as important as his physical attractiveness (and those male traits that are generally considered attractive—a well proportioned body, broad shoulders, tallness, strength and intelligence are the kinds of attributes useful in hunting and in competing with other males both within and outside the group).

Characteristics of Males that are of Potential Interest to Females in Selecting Mates

Degree of relatedness

Fathers, brothers, prescribed relatives not acceptable mates.

Willingness to provide parental investment

Quality of attention given female by suitor

Gifts given to female by suitor

Reputation of suitor

Ability to provide parental investment

Quality of gifts given to females by suitor

Resources known to be controlled by male

Dominance status of male within the society

Intelligence

Male Choice

The reproductive interest of males and females are not necessarily convergent and may often be at odds. This possibility is believed to be related to the low threshold for suspicion and fear of exploitation that characterizes many interactions between the sexes (and which is well developed in the feminist movement and in the hostile response of many males to militant feminism). Males, like females appear to behave in ways that promote their own special reproductive goals because of the selective about helping children. This is because the reliability of paternity can never equal the reliability of maternity in our species, husbands should be somewhat more cautious than their wives about providing materials resources for their children. We predict the willingness of a male to maintain with his wife and help her children should be proportional to his certainty of paternity.

Suspicion of adultery and fear of cuckoldry are extremely widespread phenomena in cultures in which males take care of their children. These emotions often play a major role in the social relations of husbands and wives. The ultimate cause for male concern about these matters is the severity of the genetic consequences of female adultery for providing a husband. If the child he helps, lacks his genes, he is not only failing to promote his reproductive success but actually elevating the fitness of a competitor. From a female's genetic view point, a child, whether the product of an adulterous liaison or not, will have 50% of her genes. In many societies, male and female adulteress receive harsh treatment at the hands of aggrieved husbands. Among the relatively peaceful bushmen of the kalahari desert, most murders are committed by men is disputes over women, especially if adultery is involved. Even if the reduction of certainty of paternity is not the female's fault, as in the case of rape, a husband may still divorce his wife, a response accepted or even encouraged by a variety of religious groups and cultures. Raped women often do not report the incident for fear of the reaction.

The protectiveness of husbands about their wives and their reluctance to continue to care for adulterous (or raped) wives and their progeny seem to confirm evolutionary predictions about how males should behave to maximize their inclusive fitness. But what about wife-sharing, a practice not unknown in our society and fully sanctioned in certain other cultures ? A speculative interpretation of wife-sharing is that it may be an extreme example of reciprocal altruism. If the sharing is truly equal, each husband stands the same chance as the other of eventually providing support for offspring fathered by the other individual : Therefore, neither male necessarily reduces his fitness if he sees to it that his generosity is fully reciprocated. Moreover, if the exchange of wives helps cement an alliance between the males that enables them to cooperate more effectively in the acquisition of resources, the fitness of both males might be elevated in the long run. Males that exchange the right to copulate with their mates do not lack the emotion of jealousy on the sense of possessiveness about their wives, as is shown by the violent treatment or death that adulterous (non-reciprocating) males receive in some Eskimo groups and other tribes in which wife-sharing occurs.

Any number of cultures are far more sexually permissive than our own, and in them adulterous sexual relationships may be common place. The sexually open south Sea Island societies of the Pacific are a classic case. In human groups of this sort, the reliability of paternity may be very low. The standard practice within such cultures is for males to with hole parental care from the children of their wives and instead to adopt the father role for the children of their sisters. In our society we consider it natural for mothers and fathers of their sisters. In our society we consider it natural for mothers and father to love and care for their own offspring. There are many proximate mechanisms that encourage individuals to do this.

For example, it is now recognized that a mother, usually forms a strong, loving attachment to her baby in the first few hours after it is born. Incidentally, the recent hospital practice of separating the neo-natal infant from the mother immediately after it is born and keeping contact to a minimum for the first few days of its life almost certainly interferes with the development of the bond. If a strong bond is not established at this time, this may have permanent negative effects on the relationship of the mother to her child. More generally, parents, male and female, are disposed to find their children truly delightful and charming creatures, even when more dispassionate observers are hard-pressed to see why. This is especially true of helpless newborns and younger children, who possess certain facial creatures (for example chubby cheeks and large eyes) that are exceptionally attractive to adults.

From an evolutionary perspective, parents may offer their offspring love and care for many years because these individuals usually have a night proportion of their genes. But a male whose wife's children were fathered by other males does not share genes in common with these offspring. If most or all of his wife's children are genetic strangers, it may become adaptive to share resources with his relatives instead of his wife and her progeny. Richard Alexander has shown that if, on the average a male fathers only one of four children produced by his wife, he would gain genetically by diverting his parental care to the children of his sister. Under these conditions, a male shares 5/32 of his genes on the average with his nephews and nieces, but only 4/32 with his wife's children. Importantly in this society three times out of four a male helping his wife's offspring will be assisting an individual with no shared genes at all. In contrast, by becoming an uncle father, he will be helping a sister's progeny. All of whom have at least some of his genes by common descent. Because humans everywhere display an enormous interest in and aptitude for learning "who is related to whom and by how much", they are in a position to regulate their distribution of their resources and services in ways that could help them maximize their inclusive fitness. The ability to assume the uncle-provider role in certain circumstances is merely one example of the genetic utility of knowing who is related to you and being, predisposed to provide parent like-assistance in accordance with the degree of relatedness.

Adoption

Two aspects of human reproductive behaviour provide what may be the most difficult challenges to an evolutionary approach to over behaviour. They are the practices of adoption and birth control. We all know people in our society who have adopted into their families the child of a complete stranger, which they treat with the same affection and generosity that is customarily offered to one's biological children. These adoptive parents are, in effect, subsidizing competitor genes. How is it possible to account for such a maladaptive response in evolutionary terms ?

Advocates of an evolutionary approach tackle problems of this sort in a special way. If the action is highly unlikely to increase inclusive fitness, as is the case for the adoption of a complete stranger, there are two alternative evolutionary hypothes is that may help explain the behaviour:

1. The action is one of the costly effects of an ability whose benefits, on average, exceed the costs.
2. The action is currently disadvantageous solely because of the radically altered setting in which it is now performed; in the past, the disadvantageous response would not have occurred.

These explanation for traits that seem to harm an individual's chances for genetic success have been used throughout this text. For example, the copulatory response of a male has been to an orchid flower or the flight of a moth to a bola spider were held to be disadvantageous aspects of the generally adaptive ability of these individuals to find receptive females of their species. A hypothesis of the second type was employed to explain why robins sometimes hurl their banded offspring from the nest. The silvery band stimulates the nest cleaning response, which is used in a way that would never have occurred in the past because human bird banders are a recent, novel addition to the bird's environment.

Similar hypothesis may help explain the modern practice of the adoption of strangers. First adapting a non-relative may be a genetic cost of the usually adaptive desire of adult humans to have children and raise a family. Although it is true that adults who adopt strangers are reducing their cause them to behave this way are no average beneficial. Because these traits usually elevate the fitness of the humans that exhibit them, they are therefore maintained within human population. The point is that all adaptation have costs and benefits, and one cost is that they may be misused, from a genetic perspective, especially in rase or unusual circumstances. In Joel Welty"s the Life of Birds, there is a photograph of a cardinal feeding a beakfull of insects to a goldfish that has longed out of the water, mouth gaping to receive the food. This act certainly could not benefit the cardinat's genes. The bird, however, had recently lost its brood and was employing its food gathering and brood feeding in a rare and inappropriate manner. The ability to collect food for young and the strong desire to feed them surely elevates the fitness of most parent cardinals.

The misplaced parental care hypothesis generates a testable prediction, which is that husband and wives who fail to produce children themselves should be especially prone to adopt strangers. Their frustrated desire to be parents could find inappropriate expression in the adoption of a biological child substitute, such as an animal pet or an unrelated infant. However, the risk that parental drive will be misused is relatively slight, and the overall genetic benefit of the eagerness to have children is obvious.

Another prediction is that if the adapting parents proceed to have children of their own (not an uncommon occurrence) then they will be more likely to withhold care and resources from an adapted child of strangers that from their own biological children. One way to test this prediction would examine the victims of child abuse in a family with both natural and adopted children.

The second basic alternative hypothesis is that the adoption of strangers is a novel, recent phenomenon that occurs primarily because of changes in the typical human environment over the past several hundred years. In large industrial societies children are made available to people who do not know the parents of the adoptee. Throughout the vast majority of human evolution in which people lived in small bands or villages, this event would be exceptionally rare. Adoption in this environment may have been practised regularly to the genetic benefit of the adopter. There are at least three possible kinds of adoptions that could raise the inclusive fitness of the adopting individual.

1. Almost certainly acceptance of the children of relatives was the most common kind of adoption in the past. A direct way to elevate the fitness of a relative and thereby perhaps to raise one's own inclusive fitness would be to assist the children of that relative. In effect, males that practice the uncle, father strategy are adopting their sister's children as their own. Adoptive assistance given related persons might also occur frequently when the biological parents died or experienced some disastrous turn of fortune that greatly reduced their ability to care for their children. By stepping in to help, an individual could salvage nieces and nephews whose reproductive potential would otherwise be nil or very low.

2. In case in which an individual adopted a stranger, the adoptee might be a captured child from an enemy group. The adopting family would probably not treat the child in an altruistic fashion, but instead exploit him or her as a servant or slave.

3. There are some well-documented cases of self-sacrificing adoption in pre-industrial societies of individuals known not to be relatives of the adopters. Marshall Sahlins, who believes this phenomenon cannot

be explained by recourse to evolutionary theory, also paints out that "adoption is, like marriage, a made of alliance between groups", thereby providing a possible reciprocal altruism hypothesis predicts that for adoption of nonkin, the adopting individuals, although temporarily lowering their fitness by adoption of a non-relative, in the long run elevate their fitness by establishing a mode of alliance with the parents (and kin) of the adopted child. In the Pacific island cultures where this practice was widespread, warfare was also highly developed. Adoption among non-relatives may have increased the possibility of alliances between villages for mutual benefit in defensive or offensive war manoeuvres. Such alliances are often critical in determining the survival of a group in cultures where warfare is a major activity. Adoptive practices may also have decreased the probability that allies would turn on each other during or after the war because they would have a direct genetic stake in the survival of the other group. Finally, if one could adopt the child of an enemy tribe, one might reduce the likelihood of continuing aggression between the opponents and so end or moderate hostilities. In this context it is significant that the war-like tahitians were willing to adopt the child of a person they had killed in a battle or a child of a close relative of the decreased person.

6

The Ethology

Ethology is the study of animal behaviour and it is one of the recent branches of Zoology. Initially it was studied by non professionals, generally by nature lovers, only for past few decades it has drawn attention of scientists.

Ethology is the scientific, biological and specific study of behaviour of animals. The word behaviour is used in a very broad sense in ethology. It refers to all externally recognizable changes that bring about communication and which can release behaviour patterns in other animals, viz movement patterns of whole body or portions, facial expressions, body postures, vocalization, as well as changes in colouration, release of odour, etc. are all included in behaviour.

The behaviour is not only restricted to movements, even the freeze reaction in a pigeon after seeing an approaching cat; or standing motionless on specific area by male antelope indicating his ownership of that area as its territory; and remaining still by prey towards a passing predator are also part of behaviour. Chemicals play an interesting role in behaviour e.g. a female butterfly releases an odour (pheromone) which attracts a male is also an integral and important part of their reproductive behaviour. The term Ethology (from the Greek word Ethos habit, convention), as is used today for the biological study of behaviour, is quite old. It first appeared in the middle of the 18th century in publications of the French Academy of Sciences, later by 19th century, the term was used to express the life styles of animals. The use of the term ethology or the study of behaviour in its correct perspective was introduced in 1950 by Niko Tinbergen.

In early years, instead of ethology, the synonym-Animal Psychology (in German: Tierpsychologie) was also used. In USA and UK ethology is usually referred to by the term Animal Behaviour. However, presently the most widely accepted term is Ethology.

Animal behaviour is a continuing source of awe and mystery that sparks the imagination of scientists and public alike. This subject, though recently introduced as a scientific branch, has been known to a commoner for many years and to some extent all of us are experts of behaviour. The way a person walks, talks or the minutest movement of a muscle in the face like frown, smile, blush can tell so much about the mood he or she is in. If we own pets we learn to interpret their actions; we can distinguish a friendly dog from an angry one, a hungry cat from the one that wants to be out in search of mate. The behaviour of pets and domestic animals are observed carefully by their owners.

The farmers know much about farm animals and pests. Most people's knowledge ends at about this point because learning about the behaviour of animals in greater details is an extremely time consuming and scientific process. When we enjoy the fascinating wildlife films on television, it is worth knowing that, it may have taken hundreds of hours of careful watching before these points of interest were captured on film. We have always been interested in the behaviour of the animals that surround us for our benefit and recreation and also because we are inherently very inquisitive. The sense of wonder has always been at the root of scientific enquiry.

A question arises that why is it necessary for us to study animal behaviour? It is because, the vast majority of people, wherever they live and whatever may be their occupation, come in contact with animals. It is obvious that the hunter has to know the ways of his quarry (animal which is hunted); that the farmer must be aware of the habits of his farm yard animals. Cattle breeding, poultry, piggery, fishery, apiculture and sericulture have been developed as industries for the benefit of human beings, only after understanding behaviour of respective animals. Knowledge of behaviour of creatures like insects and rodents that damage our crop has helped mankind in controlling them.

Even the modern city dwellers have close encounters with animals; they may want to ward off the cockroaches and termites from their houses or keep a dog, cat or birds for recreation; for healthy keeping of pets their behaviour must be understood. Man also visits zoo or amusement parks to get pleasure of observing animals. Lastly, we must share our world with fellow creatures and that is possible only by understanding their ecology and behaviour to ensure that animals can be preserved and managed for our future generations at least for them to see, this effort in turn would automatically maintain delicate ecological balance on this earth, which has already entered into an era of environmental crisis.

The other questions that come to our mind are how do animals behave, why do animals behave and why behaviour is important for animals ? We all know that the conditions are by no means favourable everywhere on earth and every living being has somehow adapted for maintaining itself in suitable surroundings, for example, small creatures like bacteria will not enter into a drop of acid placed in their culture medium, they will move away from it; the whole surface of amoeba is sensitive to negative or positive stimuli; and it would respond accordingly. All animals behave for their benefit and survival. Behaviour can be observed right from the simplest, single-celled protozoans to most intelligent and highly developed primates. Behaviour constitutes an extremely important aspect for species survival.

The male stickleback (a small fish) habitually builds a tubular nest of grass and weeds on the bed of the river, after inducing one or more females to spawn in it, the male fertilizes the eggs and guards them by swimming around the nest and intermittently would be seen fanning the eggs with the help of pectoral fins. By doing this, it directs water towards the fertilized eggs to ensure fresh supply of adequate dissolved oxygen. As the eggs grow they require more oxygen, so the male increases its intensity of fanning and forms a number of extra openings in the roof of the nest to make ventilation more efficient. If the male is removed, the eggs will die. This is a simple and direct example to indicate how behaviour is necessary for survival. In fact, animals feed, drink, fight and flee to survive; fleeing from danger, fight for limited resources, caring for young ones, feeding fledglings, helping conspecifics and mating to leave the progeny, are all important behavioural forms for species survival.

There are many questions which remain unanswered. Nature is simply fascinating and some of the things that animals do, are curious, striking and mind boggling. Why do animals migrate ? How do they find their way ? Why do bees visit some flowers more than others? Do animals sacrifice their lives for others? At one point study of behaviour appears very simple but actually it takes years to come to a conclusion regarding a behaviour.

Two causations or explanations can be assigned to a particular behaviour of animal. One is the proximate causation and the other is ultimate, if there is a question like why does a dog eat ? The proximate causation would be "because it hungry, it must eat" which is an immediate short and narrow sighted explanation whereas, the ultimate causation has the evolutionary explanation that the dog eats, because it must not die of hunger, it must survive to reproduce and leave its progeny for the perpetuation of species.

It has not been easy to define behaviour in totality. It has been defined differently by various workers:

(1) Observable and quantifiable events constitute behaviour. It is a directional activity necessary for survival.

(2) The behaviour of an animal is defined as the totality of its movements, sound emissions and body postures; also the externally noticeable changes such as colour change, secretions of odorous substances that serve bilateral communication and can therefore, release other behaviours in the partner.

(3) Behaviour is the response of living matter to some form of stimulus.

(4) Behaviour includes all those processes by which an animal senses the external world and adopts itself to that environment.

(5) Movement of whole animal because of some external stimulus is known as its behaviour or activities of an animal's effector organs (muscles) is called behaviour.

(6) Behaviour is always a combined result of the actions of many body organs.

(7) Behaviour includes all movements, gestures, postures, changes of colour and vocalization displayed by an animal.

A synthesis of above mentioned definitions could be that, behaviour is the scientific study of the totality of visible movements (including freeze or arrest reaction or standing still), body position (posture) and its orientation in space, emotions, expressions (gestures), vocalization, production of odorous substances, colour change and pilo-erection (raising of hair) brought about by coordinated activities of nervous, endocrine and muscular systems of an organism. Behaviour is actually a consequence of stimulus and response, the receptors identify the stimulus, conductors coordinate the response and effectors carry out behaviour.

The science of animal behaviour is still considered in its infancy, it has been studied extensively for last 50 years only. It still has a very long way to go because it is not so easy to observe the behaviour of various animals, they behave in a bewildering variety of ways; in fact, the range of animal behaviour patterns is as great as the variety of animals found on this earth with many different shapes, sizes and colours which took generations of Zoologists to describe and classify them. None of the species are identical nor do they behave alike. Not only this, there can be many different types of behaviours in one individual of a species.

Many trained researchers all over the world are spending their energy and time in carrying out research work, and although they are making good progress, still there is much more to observe. Earlier we had to rely on what we immediately saw, but now we have still and video cameras with time lapse and slow motion. Scientists have recorded amazingly precise films on tigers, cheetah, elephants, chimpanzees, gorillas, bees and ants. Sounds can not only be heard but also recorded on tape to be heard again and again. Scientists have successfully recorded and studied the songs of cricket and birds, songs of humpback whales have attracted much attention.

The complexity of animal behavioural study does not depend on elaborate mathematical treatments, delicate instruments and expensive chemicals or giant computers, the paraphernalia that people usually associate with science. Although these devices have their place, they are after all only a means of writing facts from nature. An experienced student of animal behaviour is equipped with binoculars, stopwatch and camera, hidden in a blind and can gather facts about his subject in few days time, to keep him pondering for an year.

One thing that must be brought to your knowledge here is that while writing or describing about animal's behaviour, we the human beings have the tendency of using the expressions of our own emotions, or try to attribute human characteristics to animals for example, we would use words like, angry, sad or amused to explain the moods of our pet. This is an anthropomorphic way of describing animal's behaviour and is not considered proper in scientific writings.

This is also not a scientifically consistent line of enquiry, we can not say that an animal fights because it is angry or that it escapes because it is afraid, but the same thing can be described as : animal attacks to express its aggression and it retreats accepting its failure to face the aggressor. Earlier workers like Edmund Brehm (1712) in his book titled Brehms Tierleben (Brehms Animal Life) explained animal behaviour using anthropomorphic terms like clever, mean, sly, gentle and proud. Though anthropomorphic terms make the presentation interesting, yet their use is not recommended.

Due to inquisitive human mind, dedication, and elaborate research it has become possible to define several areas of ethology. Today ethology has many specialized branches and there is much overlap with other areas of biology such as ecology, physiology, and genetics.

Branches of Ethology

Ecoethology is a comparatively new branch of ethology in which the relationships between the behaviour of a species and other living and non-living components of environment are investigated. Ecoethology can proceed in one of two ways; it can emphasize or focus either on a group of species or on a particular habitat. In focusing on the habitat, one would be interested in the parallel behavioural adaptations that are found in certain habitats, for example, deserts or tropical rain forests.

This is of interest even in species that are not closely related, with an interest in how the various species differ from one another and how such differences can be considered adaptations to various habitats, i.e., what is the "biological" significance of a behaviour or study of similar adaptations by animals of different species living in the same habitat. Especially interesting results can be expected when within a related group of species there is one that lives in an entirely different habitat and whose behaviour deviates substantially from the behaviour typical for the group or study of difference in behaviour within a single species found in different habitats.

Behavioural Physiology Or Ethophysiology deals with the physiological basis of behaviour. Two of its main branches are concerned with the two large control systems of an organism that are also important in the area of behaviour. Neuroethology deals with the sensory process and the central nervous system that underlie a particular behaviour. Ethoendocrinology deals with the relations between hormones and behaviour.

Behaviour Genetics is called Ethogenetics, and it investigates the genetic basis of behaviour, using simple concepts of single or multiple gene inheritance. Its goal is to prove empirically the relationship between genetic factors and their influence on behaviour.

Two areas of ethology deal with the change of behaviour over time. The Phylogeny of Behaviour traces the evolutionary origin and development of behavioural characteristics, and the Ontogeny of Behaviour studies the development of behaviour in an individual. A sub area of this latter branch is Behavioural Embryology, which is concerned with the prenatal development of behaviour patterns.

The most recent area is Human Ethology, whose goal is to study human behaviour with ethological methods. It emphasizes phylogenetically transmitted and genetically determined regularities and variability of human behaviour.

A revolution happened in biology during the 1960's and 1970's. Behavioural biologists began to realize how remarkably well animal social behaviour is attuned to environmental and social circumstances. This led to an explosive surge of theoretical and empirical investigations devoted to the adaptive significance of complex animal social systems, this formed Sociobiology which is a fascinating area in ethology and has already had a substantial influence on contemporary biological thoughts. With respect to subject matter, sociobiology stands between ethology and population biology, a research area that deals with the temporal and spatial distribution of individuals and their relationships to the living and non-living environment.

Like ethology, sociobiology is a comparative science. It examines the biological and evolutionary bases of various aspects of social behaviour and deals primarily with the selective advantage of specific social structures. To some extent, Sociobiology deals with the mechanisms that determine the social organization of a species or population, and/or how it is maintained.

Sociobiology is an amalgamation of behavioural science, modern population ecology, and evolutionary theory. Its central concern is to understand how and why animal social behaviour has evolved. Although sociobiology focuses on ultimate causation, proximate mechanisms do not lie entirely beyond its purview. Sociobiology does not have discrete boundaries, it is interwoven with many disciplines, including physiology, embryology and ultimately anthropology, psychology and even history. Sociobiology has its root in Darwin's (1859) contention that behaviour, like morphology and physiology, evolved through natural selection. David Lack (1968) also proposed the view that many aspects of behaviour could be analysed and understood from an evolutionary perspective. E.O. Wilson (1975) wrote an excellent book "Sociobiology : The New Synthesis" which laid the cornerstone for an interdisciplinary approach to both animal and human behaviour.

Sociobiology is a new and exciting field of study that promises to go far in answering the question of why animals behave as they do? Combined with cultural anthropology and other social sciences, it could also go far in explaining why people do what they do? Sociobiology has caused ethologists to pay more attention to the intraspecific variation of behaviour and to individual behaviour patterns. Beyond that it has offered biological explanations for those behavioural characteristics that had not previously been explained by the theory of natural selection, such as the existence of altruistic behaviour (sacrificing behaviour).

Two other fields that border ethology are Bioacoustics and the study of Biological Rhythms. The goal of bioacoustics is the investigation of sound production with the help of high quality recording equipment and tape recorders. The ability to produce sound spectrograms resulted in an objective and reproducible presentation of animal sounds which discontinued the use of old and subjective description by words or musical notes. More generally, bioacoustics include the investigation of vocal and instrumental acoustics, auditory sense organs and the sound-producing organs like vocal cords, stridulating organs, and instruments like feathers, wings, legs (modified to produce sounds), etc. In all, it includes the study of production and perception of sounds, their mechanical and physiological processes. It also includes the relationships between the Vocalisations of the animal and its habitat. Bioacoustics have contributed important knowledge to ethology in the area of the ontogeny of behaviour. The study of Biological rhythms is concerned with recurring, rhythmic events of living organisms and their underlying processes. Since such periodicity is so prominently expressed in the behaviour of an animal, biorhythms form an integral part of ethology.

Behaviour is studied by many people throughout the world and its fundamental principles are taught in almost all courses of biology and psychology in western countries, it is a recently introduced subject in our country. Ethologists are formally trained in biology and approach behaviour from the view point of its relation with nature and ecology. They study a wide variety of behavioural repertoire in an equally wide variety of free living animals. By contrast, most psychologists are formally trained in psychology or medical science and are mainly interested in the behaviour of people and for that reason, may carry out research on animals. Their studies are generally carried out on dogs, cats, rats, pigeons and chimpanzees in laboratory conditions, they concentrate most on learning feats of which animals are capable, aggression, reproduction, and maternal behaviours.

The ethologists prefer to remain behind a hide looking out at the freely moving animal, while the psychologists move about freely around his laboratory collecting results from an animal enclosed in a cage. Initially, after World War II, there were two main schools of animal behaviour, the first consisted largely of Europeans and they called themselves ethologists. The other group was mainly of American psychologists. World renowned ethologists are K. Lorenz, Karl V. Frisch and Niko

Tinbergen. Their work is mentioned later in this chapter. Among psychologists, B. Frederic Skinner became the founder of "behaviourism" a branch of psychology founded early in this century in USA, which regards learning as of the supreme importance, Skinner studied learning by "reward" and his work has given rise to the school of "Skinnerian psychology". This method has general laws of learning through the study of animals, usually rats, trained to do tasks such as to press a lever which is arranged to deliver them a pellet of food. In this "Operant conditioning" the amount of a particular behaviour pattern, like pressing of lever known as the operant, in upward or downward direction, is associated with reward or punishment.

Skinner believed that the principles emerging from such experiments could be generally applicable to other animals and even to human beings. Harry F. Harlow, one time Director of University of Wisconsin's Primate Laboratory, observed the intricate relationship between mother and infant and effect of deprivation of maternal care on growing infants in rhesus (Macaca mulatta).. His work has got him world wide recognition. He has proved that if the infants are kept in isolation, they show abnormal behaviour even in their adult life i.e. early childhood experiences have profound effect on remaining life. I.P. Pavlov was a psychologist, born in Ryazan in Russia. He is best known for his studies on conditioned reflexes which has formed the basis of much subsequent work on learning. In his best known experiment, he studied the salivation of dogs when they were presented with food and showed that, if a bell is rung everytime the food appeared, the dogs will ultimately come to salivate to the sound of the bell alone. He was of the view that the building up of such associations is an important part of learning. Besides this, other well known psychologists, like Hess, Hinde, Thorndike, Sherrington also greatly contributed in exploring animal behaviour.

Few basic differences between the ethologists and psychologists could be summerised as follows:

Ethologists	*Psychologists*
1. They work in forests, in national parks and in reserves using camera, binocular and stop watch.	1. They work in laboratory using specially designed boxes, mazes and other instruments.
2. Watch animals in their natural habitat.	2. Watch animals in captivity.
3. Study wide variety of animals from ants to elephants, from wevil to whales and from protozoa to primate.	3. Study few selected animals viz rats, cats, pigeons and chimpanzees.
4. Deal with visible actions which includes innumerable types of behavioural acts.	4. Deal with feelings and emotions, also concentrate on learning processes.
5. Correlate behaviour with environment or vice-versa. Persons specializing in it are called ecoethologists.	5. Correlate behaviour with body physiology specially nervous system; persons specialising in it are called neurobehaviourists.
6. Let their animals roam around freely in natural habitat, they rarely kill animals.	6. Let their animals remain in cages and may kill them for experimental use.

It is interesting to note that at first, ethologists and psychologists took little notice of each other's findings, but when they did, cross fire began and the best thing was that both sides scored good points and their dispute generated more light than heat and they were more enlightened than annoyed.

History : The history of ethology is as old as prehistoric man because at that time human beings were hunters and they had to be careful students of behaviour, a fact revealed in the paintings on the walls of excavated caves of that time. It was actually the great Greek philosopher Aristotle (372 BC) who for the first time wrote excellent descriptions of animal behaviour in his book - Historia Animalium meaning "the history of animals". He collected a great deal of information on animals and wrote that animals also possessed insight, and love for families. William Harvey, a scientist of seventeenth century, studied breeding, nesting and incubating behaviour of many birds. Gallen (1840) carried out extensive observations on life of animals and wrote complete theory in details in a book consisting of two large volumes.

Scientific study of animal behaviour has its origin in the work of eighteenth century naturalists such as Gilbert white (1720-1793), of England who made meticulous observations of many aspects of behaviour of swallows. But it was Charles Darwin (1809-1882) who must be regarded as the first one to make scientific study of animal behaviour due to three reasons (i) his theory of natural selection gave background to understand animal behaviour in evolutionary terms (ii) his views on instincts formed firm basis for other theories to develop (iii) and his observations on behaviour and their compilation in his book "The expression of the emotions in Man and the Animals" (1872) gave a detailed and vivid description of animal behaviour, this book is still considered a classic book because of its contribution in understanding the emotional expression and their development in animals and human beings. He wrote in his book on origin of species by natural selection in 1859 that phylogeny of behavioural development would be subjected to the same laws as that of morphological changes and perhaps all instinctive motor sequences have emerged gradually, developing from many small and useful modification. Darwin could not live to see his theories being appreciated and accepted. Later, Darwin's assumptions got support from Mendel's Laws of Inheritance. In 1882, the year Darwin died, John Romanes, a student and friend of Darwin wrote an excellent book on "Animal Intelligence" which is also considered a classic book on animal behaviour. Charls O. Whitman, (1908) a Zoologist at the university of Chicago, USA studied the behaviour of pigeons and doves.

He wrote some of his observations in his book entitled "Animal Behaviour". After his death, his student O. Riddle published his remaining work. Earlier people started their work by studying pets, domestic and farmhouse animals. Fundamental turning point came in the works of a German Zoologist Oskar Heinroth (1871-1945) who devoted his life studying waterfowls and wrote a book "Ethology of Anatidae", his findings and dedication fascinated many of his colleagues and students. One of his most devoted students was Konrad Lorenz. Ethology reached its peak of popularity in the works of three great workers Konrad Lorenz, an animal psychologist in Vienna, Niko Tinbergen, a Zoologist in Holland and Karl Von Frisch from Germany. These three shared a nobel prize in 1972-73 for their remarkable contribution in the field of Animal Behaviour. They also got the credit of introducing an absolutely new branch of science to Zoology and making it popular among common people.

Konrad Lorenz (1903-1989) is regarded as the founder father of ethology, he was a Lecturer in Animal Psychology at the University of Vienna, Austria. He got initial inspiration from his teacher Oskar Heinroth, his close interaction with this great teacher proved to be extremely useful in building up firm basis for future studies. Lorenz used to live in a farm house, where he had many pets (his wife had to keep their son into the cage). He observed shrews, frogs, monkeys, dogs, jackdaws and mainly greylag geese on which he carried out extensive studies and developed a theory of Imprinting (Pragung in German) or childhood learning during the critical period . He wrote popular books like "King Solomon's Ring" (1952), "Man Meets Dog" (1954), and "On Aggression" (1962). All these books showed his deep understanding of animals, and hence received wide attention. He is rightly known as founder father of ethology because his work created enthusiasm in many others for the study of behaviour. Although between Darwin and his time several others carried out studies on this subject but his contribution was more wide ranging.

Another major force in behavioural studies was Karl Von Frisch (1886-1982;) who was born in Vienna but spent his life in Germany. He also had the opportunity of living in a house surrounded by a big farm, with many trees, flowers and pet animals, his main interest was in bees. With great patience he conducted most tideous experiments on this tireless social insect, by carrying out a series of experiments with coloured cards. He observed that bees perceive colours, that they can locate the sun even during a heavy overcast because they can perceive ultraviolet rays, his most significant discovery was that bees have an extensive communication systems, bees have two types of dances-waggle and circle to communicate to their conspecifics, the exact position of sun from the hive, flowers from the hive, and the type of flowers. He also studied behaviour of bees during total solar eclipse. His marvellous work on bees got him the laurel of Nobel prize. He also carried out work on the hearing and colour vision of fishes.

The third most important name in forming the history of ethology is Niko Tinbergen (1907;) he was born in Holland, but lived in England. He devoted many years observing and teaching behaviour; he was a Zoologist by profession. He studied variety of animals from butterflies and diggerwasos to three spined sticklebacks and gulls in their natural habitat. He is an author of many books; his book on "Animal Behaviour" a LIFETIME publications has helped many young readers to get acquainted with the subject. His other books "The Study of Instincts" (1951), "The Animal in its World" (1972) and "Social Behaviour in Animals" (1965) have been milestones in ethology. Niko Tinbergen's contribution on the details of sign stimuli which are needed to elicit a specific instinctive behaviour is still appreciated. He found out that, in addition to colours and light-dark contract relationship, the size and position of sign stimulus also plays an important role.

Study of Animal behaviour has its roots in European countries, Lorenz (1970), Frisch (1970), Tinbergen (1972), Hess (1957), Thorpe (1958), Ramsay (1954) and many students of Lorenz contributed significantly for ethology, this new branch in Science grew and took an advanced, developed shape in USA. Pioneer American workers in Animal Behaviour and Psychology are Beach (1967), Scott (1970), Griffin (1971), Wilson (1973), Eiesenberg (1976), Harlow (1951), Skinner (1967) and Carpenter (1940). Presently, extensive research on Animal Behaviour has been undergoing in USA, Europe, Africa, South America, Australia and India.

7

Genetics of Behaviour

The study of relationship between genes and behaviour begins with a fundamental question that whether heredity directly affects behaviour or merely defines the stage on which behavioural patterns may be molded by environmental factors. Regarding this issue, both biologists and psychiatrists have taken opposite views but now the two groups recognize that environment and heredity, both are relevant for virtually all behaviour patterns. At every point in development, environmental factors are interwoven with inheritance mechanism, essential is to identify the role of each and to evaluate their relative importance in specific situation. Some of the cases are there where genetic programming simply states that animals have specific learning capabilities at certain developmental periods. For example, in zebra finches, the physical equipment for the learning process have been inherited but they can never learn to differentiate males and females of their own species if they are isolated from species members early in life. Other aspects of behaviour, such as the calls of some birds are less learned and more inherited.

H. F. Harlow and his associates have shown that young female Rhesus monkeys separated from their mother at birth were deficient in basic patterns of maternal behaviour, social play and sexual activity. Again, structural and physiological characteristics that provide the tangible framework for these behavioural patterns must follow the DNA "blue prints" for the organism. But through remaining in contact with the environment, an important component of the behaviour itself must be learned. There are basically ranges of modiflability which are inherited, with the segregation of genes and the forces of natural selection which account for observing individual's differences in behaviour patterns within a given population.

Genetic Mechanism

The genetic mechanism of some behaviour traits have proved surprisingly simple in contrast to the common complex investing of environment and heredity. There are some specific examples which depend on few genes and respond to a limited range of environmental stimuli.

The first concerning example is the honeybee. It lives in the hives in large societies including nearly 15,000 bees. Bees are able to build their hives in any convenient hollow, for example, inside of an old tree. In a hive there are series of vertical wax honeycombs in which the bees store their food and rear their offspring. Now-a-days, most of the bee hives are made by humans and the vertical shelves on which the bees built their combs are so built that they are able to slide in and out of the hive.

Larvae of honeybee are susceptible to disease. There is particularly one disease called American foul brood which is caused by the bacterium bacillus larvae. A larva which catches this foul brood disease does not survive. The dead, rott larvae than become a source of infection to any other larvae in nearby combs. There are few strains of bees which by removing the rotting larvae help in preventing the spread of the disease. These strains are known as "hygienic". This "hygienic bee' performs two functions —firstly it eliminates the cap of the cell containing a rotting larva and secondly it throws the larva away. All the bees strains are not hygienic, therefore W. C. Rothenbuhler experimented on the inheritance of the habit and showed that the difference between hygienic and non-hygienic strains is due to their possession of different genes.

Rothenbuhler, first of all, did an experiment, he infected combs of larvae with the help of foul brood. These were then slid into different hives and it was simply observed that these hives contained hygienic bees. He then crossed a hygienic and non-hygienic strain and their hybrid progeny was tested with combs of infected larvae. All the hybrid were found to be non-hygienic.

Therefore, the genes for hygienic behaviour must be recessive. Rothenbuhler did a backcross of the hybrid with the hygienic parental strain in the next stage. As a result from these backcross, he reared 29 colonies in all. Six of the 20 were hygienic (6/20 is nearly one quarter). About another quarter (9/20) behaved very strangely. Bees of these nine colonies took the caps off the larval cells, but then left the dead larvae inside. The remaining half (14/20) of the colonies were non-hygienic they left cap and larva alone. Rothenbuhler thought that hygienic behaviour might be controlled by two recessive genes. One gene would control uncapping, the other, throwing out of the larvae.

The nine colonies which uncapped but left the dead larvae inside had one of the genes, that for uncapping, but not the other for removing. Therefore, Rothenbuhler said that out of the 14 colonies which were apparently non-hygienic, about seven would be having genes for eliminating dead larvae whereas seven other might be lacking the gene for uncapping.

The genes for removing would not be expressed. He then removed the caps from the combs of foul brood-ridden larvae accordingly, and put these combs into his 14 hives. Thus, according to his predictions, six of the colonies promptly removed the dead larvae. This experiment performed by Rothenbuhler provides a clear example of the genetic control of behaviour. There is only one gene which controls each of the two behavioural stages of hygiene.

Genes and Evolution

Gene Frequency : We have studied a number of ways in which genes affect behaviour, but here is another aspect to the gene-behaviour relationship. In a population, behaviour may significantly affect the frequency and expression of certain genes. If any change occurs in the gene pool than it is one of the outcome of natural selection which in turn affect behaviour.

For example, variations occurring in courtship behaviour—the process of selection of mate and synchronization of activities that results in mating are found to alter gene frequencies. In order to test these possibilities, Merrell (1949, 1953) used wild-type and mutant strains of fruit flies. Four mutant strains—designated yellow, cut, raspberry and forked—had sex-linked recessive alleles, expressed only in the male phenotypes and detected by examination of external characteristics. He started populations of flies in bottles with a mutant gene frequency of 0.5.

Thus for each generation of flies, a departure from random mating would be indicated by an increase or decrease of the mutant gene in males. The number of raspberry, cut and yellow mutants decreased after several generations but the number of forked mutants does not exhibited any significant shift. Other tests demonstrated that although fertility or viability did not differ in any of the mutant types, mating success differed significantly. Thus, this test implicated some aspects of courtship and mating behaviour as the critical factor leading to differences in productivity and, hence, to shifts in gene frequencies.

Adaptations

It has been demonstrated by other investigators that gene-controlled behaviours may vary between groups of animals of the same species. Some of these differences are adaptive to habitat characteristics and others are adaptive to stages of a population cycle.

Habitat Population : Woodland deermice (P. M. gracilis) inhabit only forested areas in South-central Canada and the extreme northern and north-eastern regions of the United states. In contrast Prairie deermice (Permyscus maniculatus bairdi) of the mid western United States inhabit only fields are never caught in forested areas. Woodland deermice prefer a temperature of 29.1 °C; grassland deermice select a temperature of 25.8 °C in agreement with the cooler habitat of the letter subspecies. In addition, the two subspecies have different reactions to sand and in a one-day test, grassland mice removed a median of 5.9 lb of sand in contrast to 0.1 lb for woodland mice. This result again seems to be logical because grassland deermice are terrestrial living in the ground; woodland deermice are semi-arboreal, sometimes found resting in trees.

Population Cycles : The populations of meadows voles (Microtus Pennsylvanicus) rise and fall in cycles that last three or four years. During the low phase of the cycle in a particular area, voles are difficult to find. In contrast, at the time of peak populations, there may be over 200 voles per hectare. Trapping data indicate that many animals emigrate during the increase phase of population cycle. However, at peak density few animals emigrate and losses in number must be attributed to deaths. In voles, where the specific gene loci regulate the production of blood proteins could be measured by electrophoresis (a technique in which substances are separated from one another on the basis of their molecular weights and electric charges, is called electrophoresis).

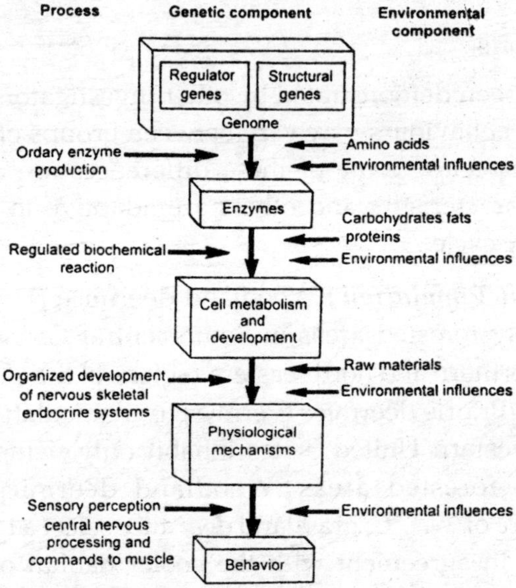

Model illustrating relationship between genes and environment in control of behaviour.

Breeding Behaviour in Insects : J. Hirsch and Th. Dobzhansky and their associates have done their experimental work with Drosophila and have shown that fruit flies exhibit different kinds of inherited behaviour pattern. Several techniques make it possible to select for specific types of behaviour patterns.

Selection Experiments : Usually most individual laboratory-reared species of Drosophila are negatively geotactic. When the flies are placed in a vertical tube, they move upward against gravity. A maze has been devised to sort out from a population of flies those having different degrees of response to gravity. Individuals exhibiting the most positive geotaxis were mated together, as were those showing the most negative geotaxis. The initial population of D. pseudo-obscura used by Dobzhansky in his experiments contained individuals with positive, neutral and negative responses in approximately equal proportions. After selection and inbreeding for 15 generations sub-populations were obtained all of whose members were either definitely negative or definitely positive.

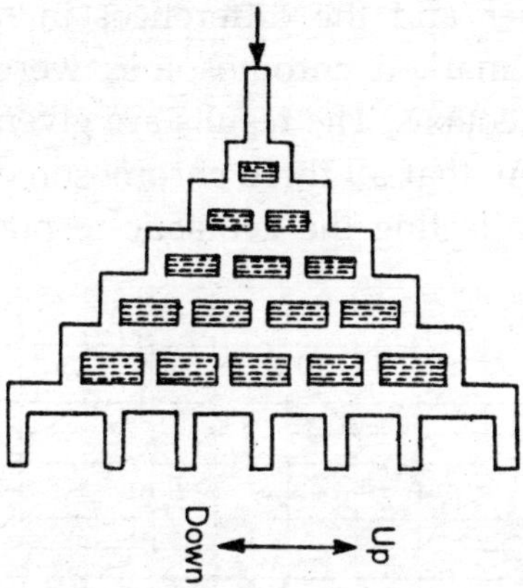

Diagram of the type of maze used to separate flies with negatively geotactic responses from those with positive responses. Flies are introduced where indicated by arrow.

A similar result was obtained by using a maze that segregated flies according to the degree to which they are attracted to light. The data supporting the genetic control of response to light are even more striking than those obtained for response to gravity.

Hirsch and Erlenmeyer Kimling have obtained very similar results with populations of D. melanogaster and have also been able to show that the response to gravity is determined by a number of genes.

The mating pattern illustrated in Fig. was used. It enables one to test the effects of chromosomes X, II, and HI on geotaxis. The three chromosomes carrying the dominant markers B, Cy and Sb also carry inversions that suppress crossing over. The tester females were crossed to males from the following three different lines : (1) inbred for positive geotaxis, (2) inbred for negative geotaxis and (3) an unselected control stock.

Only female flies carrying heterozygous B, Cy, Sb markers were used in the back cross. The eight types of expected female progeny are listed in the fig. Flies heterozygous for each of the three chromosomes were compared to those homozygous for the chromosomes and the differences in response to light measured. The marked chromosomes were assumed to be unselected for geotaxis. The results are given in the table and they clearly show that all three chromosomes must have one or more genes affecting the geotactic response.

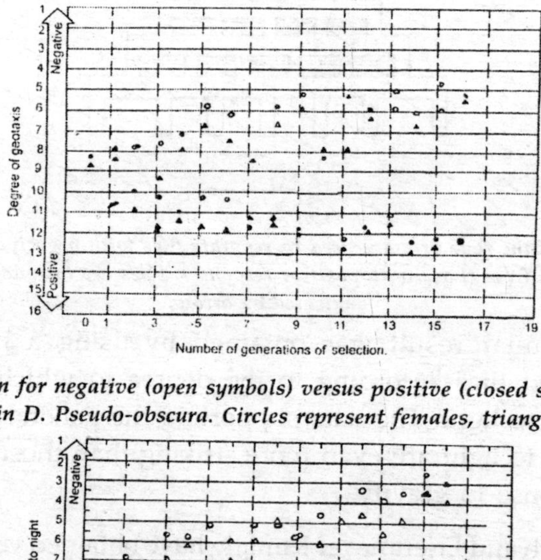

Selection for negative (open symbols) versus positive (closed symbols) geotaxis in D. Pseudo-obscura. Circles represent females, triangles males.

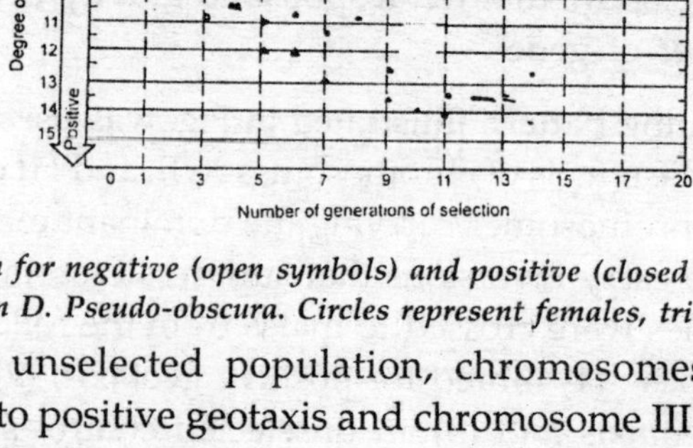

Selection for negative (open symbols) and positive (closed symbols) phototaxis in D. Pseudo-obscura. Circles represent females, triangles males.

In the unselected population, chromosomes X and II contribute to positive geotaxis and chromosome III to negative geotaxis. The flies from the population selected for positive geotaxis show a change in sign from negative to positive for chromosome III while selection for negative geotaxis produces marked changes in all three chromosomes.

This is what is to be expected for phenotypes as complex as responses to such stimuli as light and gravity. The fact that in a given population there are individuals with different degrees of response going from very negative to very positive is also understandable. These differences are another example of natural diversity, which confers on the species a degree of flexibility and plasticity to enable it to survive environmental changes.

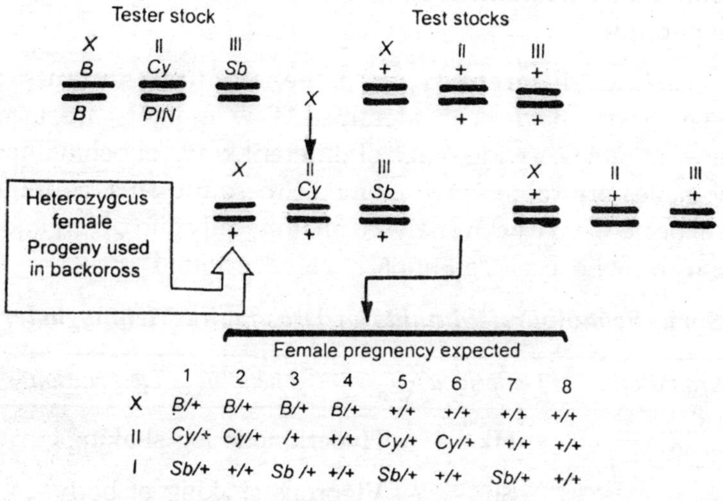

Genetic analysis of geotactic response in Drosophilia melanogaster.

Estimates of cumulative effect on Geotaxis of Genes located on the X, II and III chromosomes of D. melanogaster

	Chromosomes		
Phenotype	X	*II*	*III*
Positive	1.39	1.81	.12
Unselected	1.03	1.74	–29
Negative	47	33	– 1.08

Mutations Affecting Behaviour in Drosophila **:** Experimental results do not pinpoint the role of specific genes. It may be assumed that there are genes that have a primary effect in determining a behaviour pattern by their control of the development and functioning of some specific components of the nervous system.

Several single gene mutants of D. melanogaster demonstrate anomalous behaviour some of them exhibit marked deviations from the normal male courtship pattern; others will not fly, although their wings are well-developed; and some become paralysed when shocked either mechanically or by changes in temperature.

Various different types of behavioural mutants of melanogaster have been identified. Few mutants are listed below in table. A wide range of different kinds of behavioural anomalies are represented. Some of these mutants, as might be expected, have been analysed anatomically and divergences from the wild type condition have been found.

Some Behavioural Mutants of Drosophila Melanogaster

Chromosome	*Gene locus*	*Phenotypic Description*
X	Hk	Hyperkinetic leg shaking
	Sh	Vigorous shaking of body
	Eag	Shaking of body
	Para	Paralysis at 29°C, but not at 22°C
	per	disturbance of circadian rhythm
	drd	Drop dead beginning a day after eclosion
	tko	Become immobile when shocked
II	sps	Unable to walk or fly

Contd...

Chromosome	*Gene locus*	*Phenotypic Description*
	ap	Become immobile day after eclosion from pupa
	es	Hypersensitive to diethyl ether and chloroform
III	fty	Males court each other even in presence of females
	Sc	Male have difficulty separating from females after copulation

Mosaic flies provided with some mutant and some normal tissues have been produced. By using these aberrant individuals, the effect of one tissue type on the other can be observed. Mosaic has been constructed with the technique already described using X chromosomes carrying both morphological markers and behavioural mutations. Because of the presence of morphological marker on the X-chromosome, the mutant tissues are easily recognizable. One significant finding made with behavioural mosaics is that the mutant tissue behaviour is generally unaffected by the presence of wild-type tissue immediately adjacent to it.

This autonomous behaviour is not unexpected because it is also generally true for morphological mutant phenotypes in Drosophila. Autonomous action by genes refers to the fact that the effects of their activity are confined to the cell within which they reside. Thus, a mutant gene in a single cell surrounded by cells with the wild-type allele will not affect the phenotype of the surrounding cells, nor will the mutant cell be affected by its wild-type neighbours. However, sometimes the gene products are diffusible through the cell membranes. In this case, the action of the genes will not be autonomous. However, rather than rule this is an exception.

Mating behaviour in Drosophila also has genetic components. Discrimination of mates leading to non-random mating can greatly affect the array of genotypes in the population. Pheromones, the chemicals that elicit courtship activity and mate selection, have been shown to allow differentiation of mates within a population, and to reduce the incidence of mating among close relatives of D. melanogaster. Furthermore, many instances have been demonstrated whereby individuals with a genotype in low frequency in a population will mate more frequently than effected under random mating. Some of these instances have been related to phenomenon differences.

Reproductive isolation also seems to occur due to behavioural modifications. For example, closely- related species of some Hawaiian Drosophila use different markings on their wings to signals one's species identity between individuals. Sexual dimorphism often reflects sexual selection, whereby courtship displays of the male are enhanced by the development of extreme phenotypes. Sexual dimorphism in Hawaiian Drosophila is found frequently, as one might expect, if males must visually attract the attention of receptive females. The morphological differentiation which enhances the courtship behaviour might play an important role in sexual isolation and hence in evolution.

In crickets each species possess its own characteristic pattern of chirp. D. Bentley and R. Hoy have shown that these distinctive song patterns are not learned, but are inherited. Hybrids made between two different species chirp with a pattern intermediate to the parents. Several different kinds of songs are common to males of each cricket species. These may be identified as:

(1) the "Calling Song" used by the male to call sexually receptive females,

(2) The "Courtship Song" that facilitates copulation,

(3) the triumphal song following copulation, and

(4) a kind of song sung by fighting males; male crickets are very chauvinistic and will usually attack another male who invades their territory.

Genetic and Environmental Interactions in Dog : Among dogs, personality difference occurs even within a breed. Some are timid and other are confident, some are gentle and others are aggressive. Those that are "socialized" (allowed to interact with people) early in life function as friendly and understanding companions of humans. On the other hand, the same breed which are not given similar experiences while young may become fearful or even hostile toward people.

Extensive observations on genetic and environmental factors have been made by J. P. Scott and J. L. Fuller in the building behaviour patterns of dog. In some of their work, daily observations were continued until the dogs were 16 weeks of age. Every effort was made to observe the earliest manifestation of hereditary difference before the effects of experience were noted. Of course, during the first few days after birth, there was very little behaviour to observe in the pups. As soon as recognizable behaviour become apparent, interactions between hereditary and environmental influences were already present. The puppies changed markedly in reactions from day to day in response to learning and increasing physical and mental maturity. The environment was optimal for learning, and innate faculties became active at progressive developmental stages. Genetic control was acting on a very different animal with each new developmental stage, and environmental influence was apparent.

Scott and Fuller obtained quite unexpected and significant results. During the very early stages of development when behaviour is minimal, genetic differences had few opportunities to be expressed. When behaviour pattern did appear all at once early in development, to be modified by later experience. Instead, they developed under the influence of environmental factors. Scott and Fuller concluded that the raising and lowering of response thresholds is one of the most significant behaviour in dogs which is affected by heredity. These studies showed that although heredity is an important fact in dog behaviour but sometimes the actual appearance of specific patterns depend upon varying degree on the individual's experience.

Generally, dogs' breeds have managed to retain a great deal of genetic flexibility, despite man's intensive selection over time. Through selection of genetic qualities and training, remarkable behaviour patterns have been made available in some breeds of dogs. For example the Basenji, commonly called as the African non-barking hound, is a 'scent' hunter by nature. This dog is used in Africa to find and drive wild game. He is basically intelligent and can be taught such feats as the advanced American kennel club obedience program, which includes retrieving a dumbbell over a high jump. The Basenji's inherent intelligence is put to a relatively artificial use in this demonstration.

In England, retrieving dogs have been selected and trained for more than 100 years by enthusiasts of the waterfalling sport. Several breeds exhibited some of the characteristics necessary for an excellent retriever, such as strength, moderate size, endurance, enthusiasm, aquatic ability, keen scenting ability, courage, favourable temperament and trainability. By intercrossing the most favourable breeds—Newfoundland, setter and spaniel — and selecting progeny, a letter of four puppies with a favourable combination of traits was obtained in 1868.

From this beginning, the golden retriever stock was developed. These dogs are light yellow in coat colour. They also have aquatic ability, pleasing temperament, keen nose, tracking ability and tenacity to retrieve under severe conditions. It is natural for a pup from this stock to want to retrieve. Training merely perfects and polishes the performance.

Environment and Development in Animals : According to the central hypothesis of the famous studies of Margaret and Harry Harlow, the social experiences of young primates may be critical for the development of normal social behaviour. According to this view, the abnormal social behaviour in primates may be the product of social deprivation. The Harlow's standard experimental technique involves separation of the infant rhesus from its mother shortly after birth. The baby is placed in a cage with an artificial surrogate mother which may be a wire cylinder or terry-cloth figure with a bottle from which the baby is able to nurse.

The young rhesus gains weight normally and develops physically in the same way that non-isolated rhesus infants do. However, it soon begins to demonstrate signs of behavioural abnormality, which may eventually develop into continuous crouching in a corner, rocking back and forth, and various forms of self-mutilation. If confronted with a strange object or another monkey, the isolated baby is especially likely to react fearfully. It may withdraw as completely as possible from the object or playmate by hudding in a corner or by clinging tightly to a terry cloth surrogate mother.

Monkeys reared in total isolation for the first 6 months of their lives sometimes exhibits permanent disturbed behaviour in late life, when rocking, head banging and self-biting may be almost the only activities of the adult animal. Harlows stress the significance of receiving maternal care for normal behavioural development. But they also point out that young rhesus monkeys reared alone with their mothers do not develop truly normal sexual play, and aggressive behaviours. When the animals encounters their own age later in life, they are likely to behave very inappropriately, reacting with excessive fear or violence. Evidently contact and interaction with peers as well as a mother early in life are essential ingredients for normal behavioural development in the rhesus monkey.

Development Homeostasis

Paradoxically, these experiments, although they do show the importance of early experience for normal behavioural development also demonstrate how difficult it is to alter the course of behavioural development through environmental manipulations. In order to produce a rhesus with behavioural disorders, Harlows find it necessary to isolate an infant in a small wire mesh cage at birth or to keep it in a similar environment with just its mother. Neither event has even occurred under natural condition for rhesus infants. An abandoned (i.e., isolated) baby would be dead in a short time; and because rhesus monkeys always live in groups, a young animal is never entirely alone with his mother for prolonged periods of time, to say nothing of 6 months.

Thus, no animal in the history of the species has had to cope with the bizarre environmental situations devised by the Harlows and other researchers in this field. It is not surprising that the animals's developmental system break down under these conditions. What is surprising and revealing is that otherwise totally isolated infants will develop relatively normal behaviour if they are introduced into a cage with three other infants for just 15 minutes each day. At the outset, the young rhesus monkey simply cling to one another, but later more time is spent in play. In their natural habitat, rhesus babies, beginning at about one month, start to play more and more with others about their age. By six months, they are spending practically every waking moment in the company of their peers. Yet in the laboratory, despite the barren existence of isolated infants, a few moments of playtime a day is sufficient to counteract the effects of extreme deprivation. Rhesus monkeys, therefore, exhibit Development Homeostasis, the ability of developmental system to compensate for environmental and genetic deficits and so produce key adaptive traits in animals exposed to less than optimal conditions.

Another unexpected demonstration of the resilience of the developmental system of the rhesus is provided by tests of isolated infants reared on a swinging, instead of immobile, surrogate mother. These infants develop more typical behaviour in many regards than those reared on non-moving surrogates. Again, stimulation that only approximates when a baby rhesus would receive from its active mother nevertheless provides the basis for the relatively normal development of some behaviour.

If we were to extrapolate from these studies to human beings an evolutionary approach would suggest that young children are probably not high susceptible to various environmental deficits. The human genetic-developmental system should be as protected against environmental deprivation as a rhesus monkey's development program. Yet a large majority of developmental psychologists and the general republic, at least in the United States, take it for granted that experienced young children have in the first few years of life an enormous impact on the development of their intelligence, personality and other aspects of their behavioural repertoire. This viewpoint is reminiscent of the philosophy of the Englishman John Locke as expressed in 1686 in his "Essay of Human Understanding". Locke proposed that the human infant's mind is like a blank sheet of paper on which experience is written and that human are, in effect, entirely products of what happens to them. The advice often given to young parents presumes that Locke's view is correct and, moreover, that the human infant is highly impressionable and sensitive to its initial experiences in life.

There is growing evidence, however, that human genetic programs can compensate for severe environmental deficits even if these occur early in life, as studies of the mental performance of a group of Dutch teenagers have shown. These adolescents were born or conceived at a time when their mothers were being starved during the Nazi transport embargo, which prevented food from reaching the large Dutch cities during the winter of 1944-45. Deaths from starvation were common, and many persons lost one fourth of their body weight. For most of the famine period, the average caloric intake was about 750 calories (roughly one half of the absolute minimum standard) and at one time as low as 450 calories per day. Pregnant women in large cities were subjected to famine. The birth weights of their babies were much less than those of the infants born at the same time to women living in rural areas and smaller towns in Holland; the towns and country side were less dependent on food transported to them.

It is commonly observed that poor nutrition during pregnancy will inevitably result in intellectual damage to the offspring because of the critical nature of this period for the development of the brain. However, the famine babies did not exhibit a higher incidence of mental retardation at age 19 than those born or conceived at the same time in non-famine areas. Nor did these deprived areas score more poorly that relatively well-nourished infants on the Dutch intelligence tests administered to men of draft age.

The children were in some way protected from this environmental insult and despite being born at a lowered weight, suffered no permanent damage. These data indicate that an improper prenatal environment can be compensated for through the resilience of the developmental program. This makes sense, humans, including pregnant women, have in all probability, been regularly exposed to food shortages throughout evolution. Genetic-developmental systems in embryos that were highly sensitive to food-deprivation would be likely to produce aberrant individuals with lowered fitness. Those women whose embryonic offspring were relatively insensitive to nutritional deficits would tend to produce normal functional individuals with normal to superior fitness. It is not said that humans are impervious to all environmental influences or that it is a good idea for pregnant women to starve. Still, just because there is an environmental factor that potentially could effect development is no guarantee that it will, in fact, have an irreparable effect.

Currently, little is known about genetic-developmental systems which are able to keep on even keel in the face of great genetic and environmental variation. But it seems likely that as more is learned about developmental homeostasis. Perhaps there is considerable redundancy in genetic control mechanisms. Just as the designers of manned space vehicles provide these machines with a number of backup systems so that one failure is not fatal, so too the genotype may have the capacity to provide developing cells with alternate metabolic pathways for the production or disposal of key substances. Should a mutant gene's product interfere with the manufacture of degradation of a material, other genes may become activated and provide substitute enzymes for these tasks. The maintenance of metabolic homeostasis within cells should promote the regulation of cell function and development. In this way, genes can maintain control of the course of development of nervous systems and so stear behavioural development along certain lines. The detailed analysis of the role of genetic redundancy and other properties of the genotype that contribute to developmental homeostasis is a formidable but enormously exciting challenge for behavioural workers.

8

The Orientation

The life on earth is existing under the protective covering of Biosphere. Biological diversity goes hand in hand with the physical diversity of the planet. This physical variation has created a well defined area of living for different species, the habitat. Every living being has to find optimal localized living conditions for its survival. It would always like to avoid adverse conditions and find a suitable environment. The sense organs play an important role in guiding the animal through the environment and achieve this goal by random or actively controlled movement in space. The active maintenance or change of position is orientation behaviour (Rudolf Jander). This behaviour is common in all mobile animals and one can attempt an understanding of this from different point of views (i) by studying its mechanisms and (ii) in the ecological context.

To understand the position and movement of a body in space, one must know about the main axes which are three in number, Longitudinal, Transversal and Dorsoventral. The body can rotate about any of these three axes. The rotation about dorsoventral axis is Rolling, about transversal is Pitching and about dorsoventral is Yawing. Beside these rotations, the body can also move linearly; forward-backward on longitudinal axis, left-right on transversal axis and up-down on longitudinal axis.

Orientation behaviour takes into account a large number of factors which can be grouped under the following headings;

1. *Resources:* This includes all factors that promote well being.
2. *Stress Sources:* All the factors that are detrimental to the above i.e. well being.

Orientation can also be defined as spatial (with reference to space) adjustment of animals towards or in response to various stimuli. The orienting components of behaviour patterns are quite variable in comparison to the stereotyped form of the fixed action pattern (FAP). Orientation in space depends upon external stimuli which not only triggers it but also directs it during the performance. The difference between the two behaviour components i.e. orientation and FAP is best

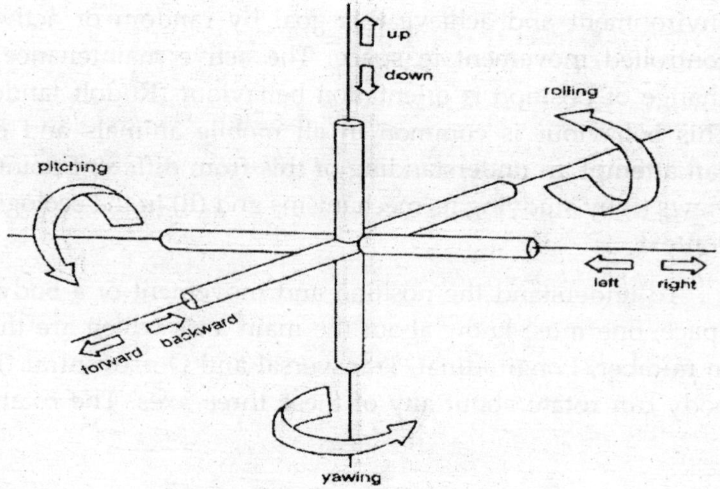

Body axes and associated movements.

explained by the example of retrieving the egg back into the nest by rolling movement in greylag goose, which is a ground nesting bird. This task involves two distinct categories of movements; 1) stretching of the neck, placing the underside of beak over the egg and pulling back the neck 2) performing lateral balancing movements of neck to prevent the egg from slipping away. If one removes the egg (i.e. external stimuli) while the bird is still pulling it back, the animal continues the retrieving movements but the lateral balancing movements cease. Egg rolling, then is a fixed action pattern that once initiated, continues whereas, balancing movements of the body are orienting responses whose presence and course depends on the external stimulus.

Here it may be seen that FAF and orienting components were going on simultaneously; but sometimes both components occur one after the other. A frog when noticing a fly will first turn towards it and line it up with its body axis and only after this movement comes the fixed action pattern i.e. the flicking out of the tongue and catching the pray.

All kinds of orientations can be broadly put under two categories; Kineses and Taxes.

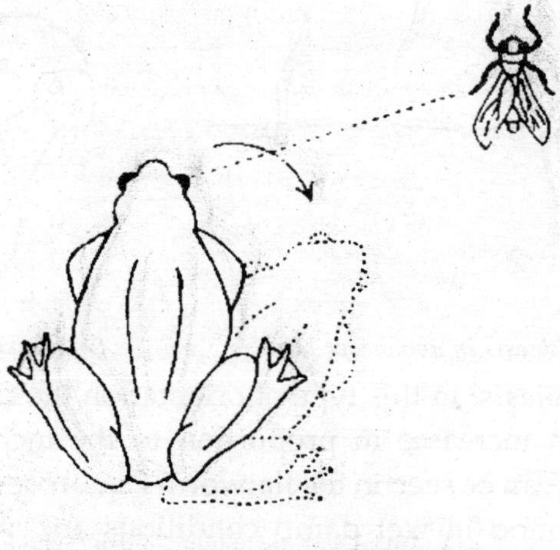

Orienting component and FAP in Frog.

Kineses

The simplest form of spatial orientation is called kineses. This type of orientation is unlearned and the movement of animal in response to stimuli is undirected. The response of the animal is proportional to the intensity of stimulus but it is independent of the spatiality of the stimuli. The kineses can be further distinguished into two types.

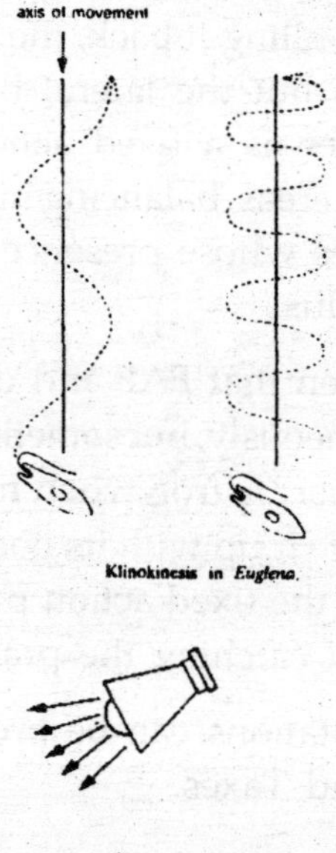

Klinokinesis in *Euglena.*

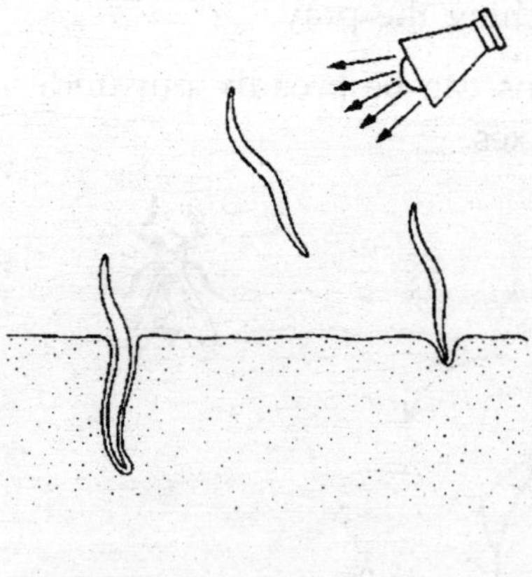

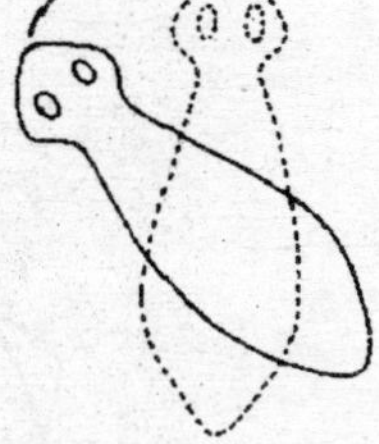

Orthokinesis in ammocoet larva. *Tropotaxis in planaria.*

Klinokinesis: In this type of orientation the rate of change of direction increases in proportion to the increase in light intensity. It can be seen in the flatworm Dendrocoelum lacteum which is found in wet damp conditions and which prefers darker places. If it is kept in diffused or dim light, it turns occasionally but when the light intensity is increased, the rate of turning of Dendrocoelum also increases. This increase helps the animal to locate a darker place which it prefers. The protozoans usually swim in a corkscrew manner and when in favourable location, the random deviations from the line of movement are increased keeping the animal in favourable environment longer. This can also be seen in higher crustacea and grazing mammals.

Orthokinesis: Orthokinesis is the orienting mechanism where speed of locomotion is related to the intensity of stimulation. The example of ammocoet larva of lamprey Lampetra planeri can be cited in this respect. These larvae are found buried in the bottom of water bodies like lakes and ponds, with their head pointing below. If the light intensity is increased and they are exposed to it, they exhibit a very active phase of swimming with their heads pointing below. This active movement helps them burrow again and move away from light.

Taxes : Taxes are those orienting mechanisms which occur with reference to the direction of stimulus. In this type of orienting mechanism, the spatiality of stimuli is of utmost importance. Interesting experiments have been carried out with Amphipodia (an order of crustacea) called Beach fleas. It has been observed that if these Beachfleas are carried away from the sea to a dry spot, they head straight back to water. These fleas were placed in closed glass tubes and kept away from sea on dry beach. It was noted that fleas always gathered in that part of the glass tube which was closest to sea. It has been established that along with moisture it was also the position of sun which was utilized by fleas to orient themselves towards sea. These are a variety of taxes depending upon the relationship of stimuli and the receptors.

Tropotaxis: When a stimulus is received simultaneously by two or more receptors of an animal., it can then compare and steer a course towards or away from the stimulus. Planaria, reacts with positive tropotaxis response to the water current carrying scent of food. The animal aligns itself in water current in such a way that both sides of its three cornered head are simultaneously and equally stimulated and then it moves upstream. Another example of tropotaxis is illustrated by the pill wood louse which has an active and inactive phase in its life. When it has fed enough, it moves away from the light (negative phototaxis) under the stones and fallen trees where it leads an inactive life. When the need of food arises again it comes out in the open (i.e. towards light; positive phototaxis) and enters into active phase of its life. The photoreceptors of this animal play a role in controlling its phases.

In tropotaxis, the animal uses paired receptors to achieve a balance in reception of the stimulus. If one receptor is destroyed then the animal is unable to reach this balance and would continuously keep on trying to achieve the balance resulting in typical movements like moving in circles e.g. the grayling butterfly escapes from the enemy by flying towards the sun. If it is blinded in one eye, it would keep on flying in circles.

It has been found out that Door beetles use their right eye for orientation by the solar compass in the forenoon and their left eye in the after noon. Biological significance of this type of orientation is still not known.

Telotaxis: The orienting mechanisms which keep a certain stimulus at a fixed place on a receptor, mostly the eye, are called telotaxis. When we see a sprinter doing a 100 meter dash at an athletic meet, we turn our eyes to get the image of the athlete on fovia centralis, the area of retina where the vision is sharpest. From then onwards even though we may turn our eye, head or body to follow the movement the image is kept locked on that specific location on retina. So inspite of the image being fixed at a place, the movement of sprinter is perceived and the motion appreciated by our brain. This appreciation comes through the motor organization involved in moving the eye, head or body. All the messages coming from these different regions are integrated into continuous perception of the movement.

In telotaxis, simultaneous comparison of the stimuli is not involved. It can be called as a goal directed orientation. Hermit crabs have two well developed eyes and when it is presented with two light sources, one dim and other bright, it always moves towards the brighter one. Male grayling butterflies, after being blinded in one eye would still fly towards the passing females showing the goal directed orientation.

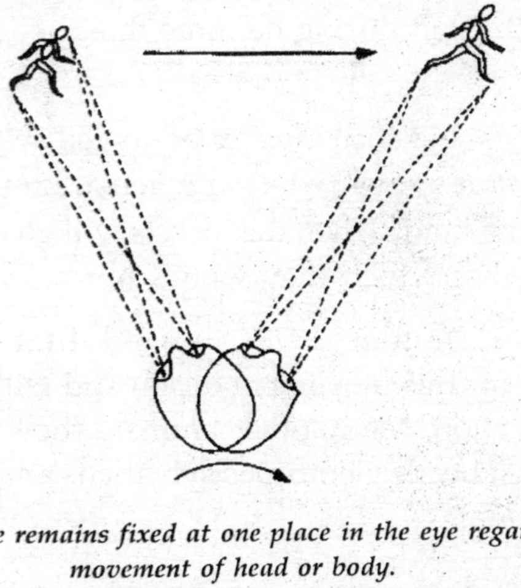

Telotaxis : image remains fixed at one place in the eye regardless of the movement of head or body.

Menotaxis: The orienting mechanisms which help the animal to maintain a constant direction by steering a course at constant angle to the incidence of the stimulus are called menotaxis (Greek word meno = I remain). In this type of orientation generally the principle of negative feedback is involved, where the output of regulating cycle acts as the input. One of the common example of this type of orienting behaviour is the light compass reaction of the insects. The insects use sun for guidance in homing. The ants' Lasius niger use sun as a reference for homing. If these ants are confined to a dark box for few hours while in the middle of their homeward journey, after their release they would maintain the same angle to the sun as the one when they were kept in the box. They would not compensate for the movement of sun during these hours which changed it's angle. This would then lead them to a different path then the one taking them home.

An ant taken far away from its hill (or path of movement) will return along a straight line, that is when the ant is displaced in open and is allowed to see the sun but if it is put in a box and then released it will make a slight directional error, the error in the angle of direction corresponding to the change in position of the sun during the time that the insect was in the box.

Ants, bees and locusts may be fooled with a mirror that reflects the sun's rays (while the actual sun is cut off by a shield of some kind). Then the insects will go off in a direction opposite to that which they would normally take.

However, Jander 1975 showed that ants can also compensate for this movement of sun and will continue in the original direction. Many other animals show this ability like Velia currens, beetle Geotrupes sylvaticus and honeybee Apis mellifera.

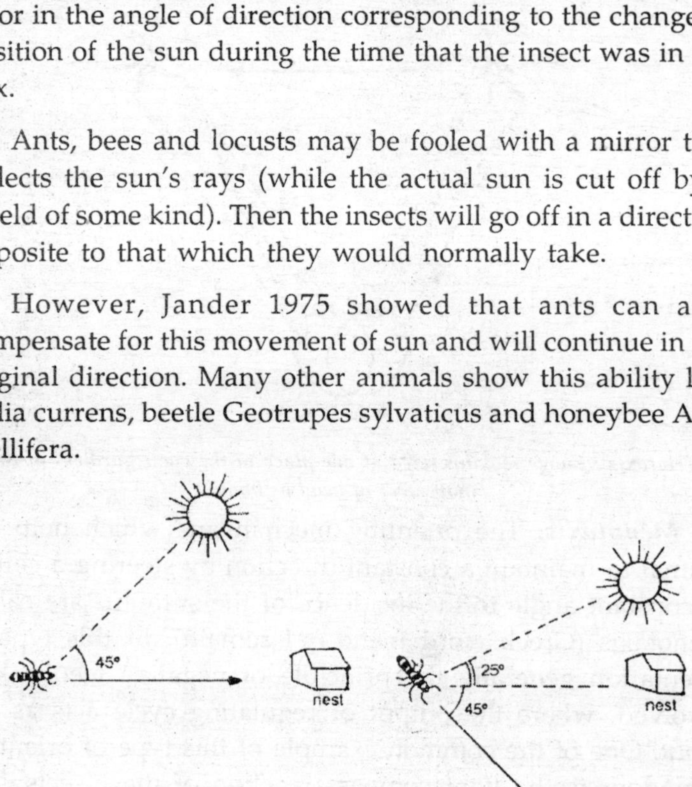

Menotaxis in ants.

However the central nervous system may, at times, make it difficult to interpret the regulating cycles in terms of negative feedback. This is so because the choice of reference value may be directed by the central nervous system any time. The importance of negative feedback can be demonstrated well when the requirements for its operation are lacking. Insects may face dangers if they get close to a light, therefore the requirement would be that light be at a great distance so no harm is caused to the insect. If the light is at a great distance, then it's rays will be parallel and would subtend an angle of 90 or more to the receptors (eyes) of the insect. If the insect chooses a reference with less value then this, then it would end at the source of light which may be hazardous. This phenomenon can be seen at street lamps in the night where in absence of any other light like sunlight, the street lamp light is taken as a reference and the insects ultimately collect around them in large numbers and are preyed upon by the predators like bats, etc.

Klinotaxis: When successive comparison of stimulus intensity are taken as reference for orientation, it is called klinotaxis. This can be seen in animals when they are confronted with environments which have; for example, gradients of chemical stimulation.

Mnemotaxis: Interestingly mnemotaxis was postulated as a possibility by Kuhn in 1919. But his critics argued against it and he succumbed to their pressure and retracted his theory as no proper example could be found then. However he has been vindicated now as many examples have been discovered. Mnemotaxis is a form of orientation based on memory. The water shrew (Neomys fodiens) exhibits this memory resident orientation when it follows the same path everytime. If its memory map is disturbed by changing a part of its path, it will stop and start exploring the surroundings, it will then retrace its steps till it locates a known land mark and then will resume the earlier course and would try to overcome the difficulty posed by the changed path. It can also be seen in children reciting poetry, who repeat the earlier lines if they forget any one line in the middle of recitation.

Negative-positive Orientation

A number of orienting mechanisms can be related to their effect in turning an animal in a direction which is determined by the external stimuli (Kuhn, 1919). Following types of orientations can be grouped under this category. If the animal moves towards the stimulus it is called positive and it moves away, then negative.

(1) *Hydrotaxis* : response to water

(2) *Thermotaxis* : response to temperature

(3) *Phototaxis* : response to light

(4) *Thigmotaxis* : response to touch

(5) *Chemotaxis* : response to chemicals.

(6) *Galvanotaxis* : response to electrical field

(7) *Geotaxis* : response to gravity

(8) *Rheotaxis* : response to water currents

(9) *Geomagnetaxis* : response to earth's magnetic field

Many migratory fishes orient and navigate with the help of odour, Silver Salmon of far East were caught in North-America, In half of the fishes the nose cavity was plugged with cotton, they were then released. It was observed that fishes without cotton plugs could follow the correct path. Plugged nose fishes were lost in the way. Bees and many other insects are very sensitive to odours and utilize them in orientation. For instance, the male gypsy moth responds to an aromatic substance secreted by the glands of the females in a concentration of a few molecules per cubic metre. This enables the males to locate females at distances up to 6-8 kilometres.

We favour darkness to have a good sleep and the larva of the house fly looks for a dark place to pupate. These orienting responses can be called as negative phototaxes as the stimulus (light) is avoided here. A grain beetle calendra granari has the tendency to climb in a vertical plane showing negative geotaxis (Birukow 1964) and if planaria is kept in a test tube of water at constant temperature and light, it aggregates at the bottom of the tube showing positive geotaxis.

Orientation mechanisms can also be grouped according to their function (Schone, 1965).

(1) *Orientation in Space:* the stimulus is used as a reference to maintain the normal position of the animal. The animal does not seek such stimuli as goals. The examples of this type of orientation are; Maintaining the normal body position against the gravitational attraction. Directional control of aquatic animals.

Control of body by optical and tactile perception.

Orientation related to light.

Compass and sun compass orientation.

Orientation with land marks.

Electrical and magnetic orientation.

Orientation in air and water currents.

(2) Stability of posture and movement by mechanical sense organs like semicircular canals and with eyes.

(3) Object orientation, where the stimulus is the goal such as; Optical target orientation (Predation)

Echolocation

Tactile location

Chemical orientation.

Orientation in ecological context can be thought of as a form of adaptive behaviour to the patterns of existence of factors like food, predation and physiographic conditions. The ability of organisms to minimize their distance from resources and maximize it from stress-sources is called orientation fitness. Orientation requires spatial informations and sources have to be there to cater to it. This information can be grouped into four classes; (i) Immediate environmental information; (ii) Learned information; (iii) Innate information; (iv) Random information.

The immediate environmental information is about the directions and distances through the sensory system. A predator would turn in the direction of its prey after seeing it and then approach it to capture or, kill it.

Learned information comes into play along with the immediate information. After the sensory information is received, the animal has many options for the orientational response that he can perform and he has to select one from this. The selection of proper response for the given situation is directed by the information already available to the animal through the memory or instinct. This information is either learned or inherited. A kangaroo, immediately after its birth, crawls only in one direction that would take it to its mother's pouch and not in any other direction, a young calf will direct its head between the horizontal and vertical axis, where it would find udders.

Random information may be a result of ignorance. An organism would locate a resource or stress source through trial and error or random movements and would then plan its next orientational response.

The orientation is very relevant to the spatial nature of the environment and the orientation behaviour in this context can be categorized into six types;

(1) Positional orientation

(2) Object orientation

(3) Strato orientation

(4) Zonal orientation

(5) Topographic orientation

(6) Geographic orientation

Positional Orientation

Most of the mobile animals maintain their position in space through positional orientation. This is done with reference to the gravity and the substrate. Large animals maintain an upright position whereas lighter animals rely more on the position of the substrate than gravity like a fly walking on the roof. In the pigeons less specialized gravity sense organs are located in the intestine along with the main sense organs in the head to maintain the position in the flight.

Many insects like locusts and dragon fly use dorsal light orientation to maintain flight position. Flies have special structure, the halters, which are stick like organs that oscillate and any aberration is detected by the small sensory organs at the base of these halters. Many aquatic animals rely on the direction of light to maintain their position along with the gravity. Larger aquatic animals prefer to keep their longitudinal body axis horizontal while microscopic animals like daphnia, nauplius larva, etc. keep their longitudinal axis vertical.

The positional orientation also helps to ward off displacement specially for swimming and flying animals which can be displaced by currents of water or air. The positional orientation also provides a means of ensuring placement of locomotory organ on the substrate. A kind of camouflage is also provided by this type of orientation, like aquatic animals have dark backs and light bellies and hence are difficult to locate against background when observed from below or above.

Object Orientation

The ability to find and approach the resources and avoid stress sources is object orientation. Typically the object orientation is two phased; search and approach. Initially the searching begins in a larger area and is called ranging which is largely a straight movement with occasional turns. If some source is detected then the search is in smaller area and is called local search which now involves mainly turns with little straight movement. If it leads to a resource then it is approached and if to a stress source then it is avoided.

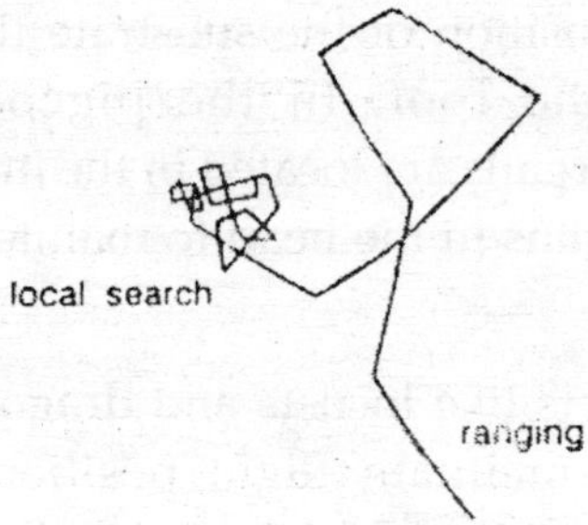

Pattern of movement in object orientation

Strato Orientation

Movement across the vertical layers of a habitat is strato orientation. The importance of vertical movement is more for the smaller forms as it is easier for them with their small weight and also because the smaller animals can not move fast and it becomes taxing to move great distances horizontally to encounter variations in physical factors whereas less vertical movement produces same changes. The sensory input for strato orientation is very easy because of the earth's gravity as reference and which can be sensed by all animals. Light may also be used as reference. The best example of strato orientation is the vertical migration of the aquatic zooplankton. Every day scores of these animals move up for the night and down for the day. This movement is guided by the direction and intensity of light and the gravity. The internal biological clock of the animals also play active role in this. The apparent advantage of this type of behaviour is the avoidance of predation in the day and the utlisation of abundance of food in the upper layer of water in the night.

Zonal Orientation

Factors like variations in the ground level and degree of forestation lead to zonal orientation. The clam Donax lives on marine beaches where wave action provides plenty of organic material as food. The tides shift this habitat back and forth. In order to remain in this zone, the clams dig themselves out of the sand and move with the outgoing tide and dig in again at proper place to get optimum supply of food. Talitrus saltator is a crustacean which prefers beaches with certain level of humidity, it goes down the slope on dry sand and up the slope on wet sand to find such a place. This simple type of zonal orientation response is many times supplemented by compass orientation where some directional or celestial clues are utilized. Zonal orientation is also important to avoid stress sources. A frog would jump into water if some predator approaches. Similarly a rodent would jump into its burrow in such a situation.

Topographic Orientation

When an animal retraces its path to the point where it started from, it is a form of topographical orientation. It is a complex form of orientation and requires many components like distance orientation, directional orientation, and orientation by land marks as well as learned information. In some animals it is only the last component which is involved in topographical orientation like some spiders and crabs. They some how seem to know the length of its outward journey and at the end of it, turn 180° and travel the same distance. However Bees, ants and wasps use all the components in their topographical orientation. The bee measures and reproduces the distance with reference to the energy spent to make the movement. They use compass orientation to maintain proper course to their nest or feeding sites. They also use landmarks and this comes handy specially when the sky is overcast.

Geographical Orientation

Many animals which move great distances require some kind of geographic orientation. These movements or migrations as they are called, are related to the seasonal changes. Animal species like birds, bats, fishes, insects, turtles and whales show this behavioural form. The locusts Schistocerca gregaria and Locusta migratoria form huge swarms and let themselves be carried away by the wind currents over a vast area. Many butterflies like Pieris brassicae and Pieris rapae fly towards the sun in fall and away from it in spring in England. The migrating bird use a kind of vector geographical orientation to move over great distances for nesting or wintering.

Besides these types of orientation behaviour there are special types of orientation responses seen in animals. Two such distinct and special orientation bahaviour are the echolocation in bats and the language of honeybee.

Echolocation : For a long time it was a matter of wonder and puzzlement as to how bats are able to move about and catch the prey in the darkness. This curiosity led Lazzro Spallanzani to experiment with bats. He blindfolded them and was amazed to see that they could still avoid obstacles. He tried removing various senses like touch, smell, taste, etc., but could not understand the mechanism behind this ability to see in the dark. Finally after years of labour, by many workers who toiled in this field, it emerged that hearing plays an important part in this. There was also the common belief that bats are blind but it is not correct. It is only that bats are nocturnal animals and hence acuity of vision is not that important in situations of low light. In fact fruit bats would not even fly on a moonless dark night, though they have better night vision. However most of the bats do fly in the night, dark or otherwise, and they are able to negotiate obstacles with ease because of a special type of orientation behaviour, the Echolocation. Here the bats emit high frequency or ultrasound and are then able to catch and analyse it's echo. Echolocation is also seen in some other animals like the oil bird, Steatomis caripensis and dolphins. However most of the research on echolocation has been done in bats.

The bats emit the ultrasound which is quite suitable for the purpose of echolocation. Its short wavelength makes it easy to bounce back from small, objects and it can give a better directional bearing than long waves. The bats utilize the echo to gain the information about the obstacles in the way or the prey in two ways; it can find out the time it takes for the echo to come back and it can detect the changes in the frequency of the returning sound.

The Vespertilionid bats emit ultrasound in short pulses with a wide band width. These characteristics make it ideal for measuring the range. Each pulse is of a very short duration in milliseconds and the band width may be like 40 kHz to 80 kHz (1Hz is one cycle in one second). The pattern of emission of sound pulses varies with the activity of bats. In normal cruising flight, the pulses are larger and have a wide band width whereas during hunting as the bat approaches a prey, the pulses became shorter, more frequent and have narrower band widths. Bats also use the Doppler effect which is the shift in frequency if the source of sound is approached or moved away. The range of ultrasound usually produced by bats is 20-100 kHz. Bats are capable of achieving remarkable degree of precision with echolocation avoiding wires as thin as 0.1 mm. The ultrasound is produced through their mouth and nose and its auditory system is exceptionally developed to interpret the echoes.

Language of Honeybees : A very remarkable animal showing a highly developed pattern of orientation is the honeybee (genus Apis). This animal feeds on nectar of flowers which may be at quite a distance from its hive. It is really amazing that the bee is able to convey to its hive mates the information about the food source complete in respect of distance and direction precisely. Karl von Frisch was the pioneer who devoted a considerable part of his career in deciphering the mechanisms by which the bees are able to communicate with its hivemates. In order to establish this fact that bees pass on this information, a very simple experiment can be performed.

If a dish containing sugar solution is kept near a hive, bees may not visit it for some time but once a bee visits it, very soon many more follow suit. Somehow the first bee conveyed the information about the sugar solution to other bees. Von Frisch made glass enclosures in which a hive could be kept and bees observed. He also used coloured markings on the bees. In summary, when a bee discovers a food source near the hive it passes on the information to other bees by performing a special dance the "Rundtanz" or the "Round dance" It turns in circle, alternately to the left and right on the face of the honey comb. After a short while the other bees start following this dancing bee and soon they fly off to look for the food and on return they also perform the dance thereby increasing the number of foragers (workers).

The first bee also carries the scent of the flower it visited so the followers seek the flower with the same scent. There is a relationship in the duration and liveliness of the dance and the richness of the food source. The richer the source, the more lively and longer the dance and if it is so, more strongly the other bees are attracted. When this large number of bees have exhausted the nectar, they gradually slow down and ultimately stop dancing altogether signalling the end of food supply from a particular source and a new source is then searched.

When the food source is located further than 250 feet, this round dance is replaced by another type of dance, the "Schwanzeltanz" or the "Tail wagging dance". The liveliness of the dance indicates the richness of the source and the scent points to the specific flower but in addition the information regarding the direction and the distance is also conveyed precisely. In this type of dance the bee runs in a straight path for a short distance on the face of the hive, wagging its abdomen from side to side and once on the top of the run, it returns back to the starting point in a semicircle and then starts on the straight run once again but would now make the semicircle on the opposite side. This sequence is repeated many times. The tempo of the dance indicates the distance of the food source from the hive. As the distance increases, the number of complete dance pattern in a given time decreases e.g. a bee

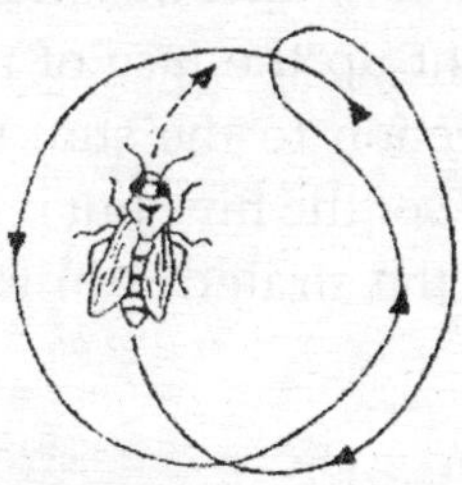

Round dance of honeybee.

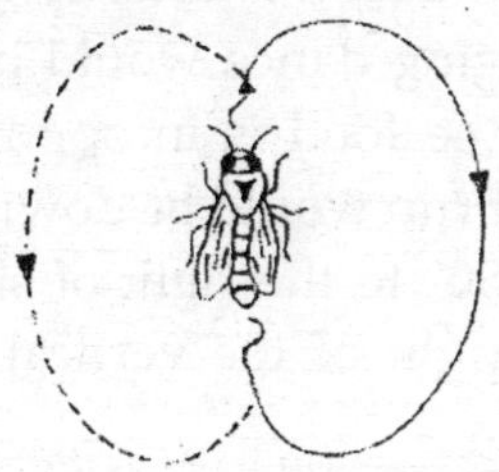

Tail wagging dance of honeybee.

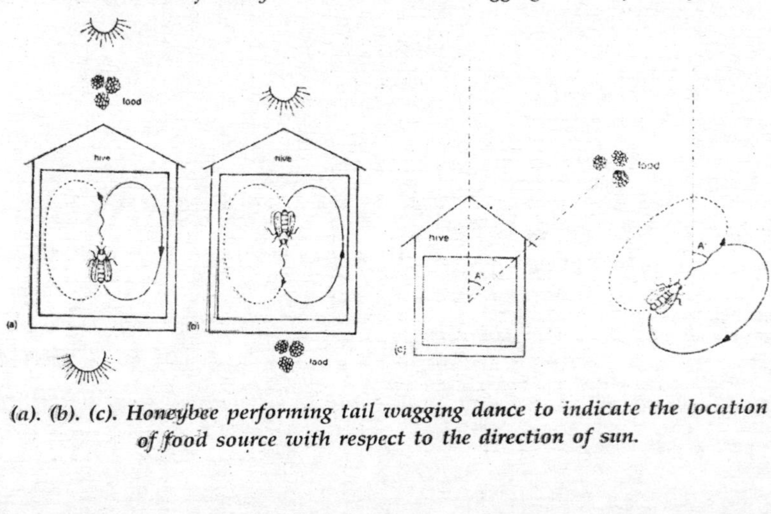

(a). (b). (c). Honeybee performing tail wagging dance to indicate the location of food source with respect to the direction of sun.

may complete 15 runs in 30 seconds for a distance of 1000 feet but may do only 11 in this time (30 seconds) if the food is 2000 feet away. However when the bees which follow this dance move out to look for the food they not only know the distance but also the direction as well. This information about the direction is conveyed taking into account the position of sun as reference. The bee while searching for food remembers the position of the sun and when she comes back and performs the dance she orients herself in such a manner that on the straight run of her dance, she keeps the sun on the same side and same angle. This is possible only when the dance is performed on the horizontal surface and most of the time it is performed on the vertical face of the hive which is also dark. In such a case the angle of its straight run to the vertical, points to the food source. The vertical represents the direction of the sun. If the food is directly in line of sun, then the straight run of wagging dance would be straight up the face of the hive, and if the food is in opposite direction to the sun, then the straight run would be down the face of the hive but if the food is, say 30° to the right of sun then the straight run would be 30° to right of the vertical.

9

The Hormones

Endocrine system consists of many ductless or endocrine glands which secrete hormones. Hormones reach all over the body through blood and act as chemical messengers. Hormones are protenaceous chemical substance produced by various parts of body or by neurons called neuro secretory cells. Here we will consider how the endocrine system acts as a behaviour regulating mechanism in both invertebrates and vertebrates through the production of hormones. Since behavioural studies are based on experiments, lets consider two examples just to understand the importance of hormones in behavioural mechanism.

Example I : If a male rat is placed with sexually mature female rat it will mount the female in matter of seconds and copulate which a castrated male rat takes longer to initiate mounting. Since the testes are a source of androgen hormones that affect reproductive behaviour, injection of synthetic androgens can replace the hormone normally and restore mount latencies to normal ievels.

The following classification will give a clear understanding of different types of effects.

Hormonal influence on behaviour

Activational effect

Organizational effect

Direct (Hor. secretion leads to a rapid response)

Indirect (require more complete sequence of stimulation and hormone secretion). Manifested during an organism development.

Activational groups.

1. Sexual attraction
2. Eclosion
3. Development Phases
4. Mating
5. Colour change
6. Aggression and sexual behaviour
7. Secondary sexual characteristics.

Example II : If we remove the prothoracic gland which secretes ecdysone (-) the hormone which controls moulting from the body of grasshopper, we will see that in an early nymphal stage the normal development sequence stops.

Dual System of Relationship between Nervous and Endocrine System : The nervous and endocrine system are feedback system that form key points of the body's mechanism. Both in vertebrates and invertebrates, these two systems show a dual system of interrelationships i.e. glands are interconnected through the circulatory system or they are closely tied to the nervous system through names connecting with the brain.

Experimental Method and their Effects : The Hormones that influence the behaviour can be divided in two parts educational and organizational effects.

Educational Effect : Its influence on the expression is an information of behavioural pattern.

Organizational Effect: It manifested during organisms development.

Techniques to Explore the Interrelationship between Hormones and Behaviour : Investigators have used several techniques to explore the interrelationship between hormones and behaviour. These includes:

(i) Removal of a particular endocrine gland to assess the absence of specific hormone on behaviour.

(ii) Hormone replacement therapy-injection of specific hormone or transplantation of a gland.

(iii) Blood transfusion to transfer the 'hormonal state' of one animal to another in order to observe the behavioural effect.

(iv) Bioassays to indirectly assess circulating hormones levels by measuring a secondary characteristic such as a skin gland that is dependent on a particular hormone.

(v) Radioimmunoassays to directly measure the circulating levels of a hormone through the use of immunological method.

(vi) Autoradiography to localize the sites at which hormone's uptake occur.

Activational Effect

(a) *Sexual Attraction :* The female cockroach and moths releases *Pheromones* which act as a sex attractants for male. Example — If the corpus allatum of female cockroach is removed surgically after moulting so the female will not produce pheromones after it become sexually mature. But if corpus allatum of other adult female is transplanted to it than it again starts producing pheromone.

(b) *Eclosion* : When adult form of an insect emerges from the pupa after metamorphosis then this process is termed 'eclosion'. This is another activational effect which is controlled hormonally. The eclosion hormone play a critical role in this process. Eclosion hormone is produced by neurosecretory cells.

Example : If the eclosion hormone is circulated into pupae that are near the end of the metamorphosis eclosion behaviour, such as abdomen movement and being spreading after emergence can be activated at any time of the deny.

(c) *Development Phase* : In adult male desert locusts sexual behaviour is exhibited when corpora allata are removed and when transplanted from other locust it restores its sexual behaviour. While similar experiments have revealed that corpora allata are not needed for sexual behaviour.

(d) *Moulting* : Many crustaceans moult periodically as they grind. Removal of both eyes talk in these animals shorten the interval between moults. If these are given the extracts of particular neurosecretory gland moulting is prevented clearly. The gland produces a moult inhibiting factor.

(e) *Colour Change* : In spring season, the short tailed measles change their colour. The metamorphose stimulating hormone (MSH) secretion increases and new brown hair replace the white coat colour during the full month. MSH secretion is inhibited by the action of another hormone melatonis secretion by the sexual gland, the hair then is not argumented and returns to white.

(f) *Aggressions and Sexual Behaviour* : Testosteron effects both on sexual and aggression behaviour. For example: when male ring doves are castrated they show decreased levels of aggressions courtship and copulation behaviour when treated with crystalline testosterons the normal levels of behaviour are restored.

(g) *Secondary Sexual Characteristics* : *Example*—Male cats which sprays urine probably as a marked behaviour, often cease to spray after their testes are removed.

Organisational Effect

These can be understand by following examples : *Example*: The organizational effect of hormone on female behaviour have been conducted on guinea pig and rhesus monkey. When the female progeny were treated with androgens, they exhibited frequent change in their external genitalia and were masculinised. They developed male secondary sexual characters.

Endocrine Environment Behaviour Interaction : Some activational effects involve complex interaction among behaviour, hormones and specific environmental stimuli. These are discussed in detail. This inter-relationship has been illustrated by taking an example of reproductive sequence in ring doves.

(a) *Reproductive Sequence in Ring Doves* : This is the sequence in which at each stage the internal state of each kind interacts with external variable to produce the observed behaviour pattern. These consist of—

(i) The hormonal state of both male and female dove except feedback loops.

(ii) The behaviour of each member of the pair that stimulates changes in the hormonal levels and behaviour of its mate.

(iii) Environmental eves, such as nests and eggs that influence hormonal and behavioural changes in both.

Daniel Lehrenan and his colleagues used three experimental groups to determine whether the presence of a mate or the nesting material affects incubation behaviour of female ring doves.

These are :

(i) Control for males housed alone.

(ii) Females housed with a mate only.

(iii) Females housed with a male and nesting material.

They assessed the results of these pairing in terms of the percentage of test females in each group that exhibited incubation behaviour when presented with a nest containing eggs.

Control females never incubated eggs. By days 6, 7 and 8 in the second group those housed with only a mate incubated eggs by the day. Hundred percent of females caged with both a male and nesting material incubated test eggs.

Thus, they concluded that presence of both a male and nesting material is necessary for complete incubation behaviour in a female.

(b) *Migratory Behaviour in Birds* : In birds, migratory behaviour is also due to the result of interaction between the environment and the hormones.

Reowan suggested that several factors influence migration. Increasing temperature, increasing day length, increasing ambition and many other environmental changes which affect the animal through pituitary activity and perhaps the nervous system directly.

Mechanism and Ontogeny of Hormonal Effects

A. *Stimuli and Mechanisms* : The mechanism by which hormones affect behaviour is very complex, so it is very necessary to know the action of hormonal effect upon behaviour and its stimulation and release. Beach (1948) separates four theoretical possibilities. Hormones may influence behaviour through their effects upon

(i) The whole organism (for example, general activity level)

(ii) Morphogenetic structures employed in specific response patterns.

(iii) Peripheral receptor mechanism.

(iv) Integrative functions of C.N.S. :

  (a) Control of this development of nervous organization.

  (b) Control of periodic growth.

  (c) Control of sensitivity.

B. *Ontogenetic Development:* There are so many examples which prove that some interaction of hormonal action and behavioural development establishes a behavioural organization which then persists in the absence of hormonal influences.

Such as if a baby which is injected with testosterone exhibits complete mating responses including covering, treading, etc.

Some Recent Development : Some recent research developed the delicate and complex interrelationship, between hormones, the nervous system and behaviour.

(i) D.S. Lehrman (1964) has recently made three discoveries about the role of prolactin in parental behaviour in doves.

(I) Prolactin alone does not elicit parental behaviour, but requires oestrogenic priming.

(II) Hormone can act directly on structure involved in behaviour without any vagus effects on the CNS.

(III) The young birds provide the tactile stimuli that elicit regurgitation from the crop.

(ii) Hinde (1958) explored the hormonal bases and external stimulus condition and controlling nest formation.

(iii) Another interesting approach to hormones and behaviour has been explained by the work of R.P. Michael regarding sexuality on the verge of nymphomania.

Conclusion : Hormones influenced approximately all behaviour in some way or the other. Hormonal flow is elicited by a very broad range of physical and social external stimuli. Hormone may act directly upon the central neural mechanism and also act upon the sensory reception.

10

Hormones and Behaviour

Nervous system and hormones both are very important underlying mechanisms to carry out behavioural activities. The effects of nervous system are quick and short lived while those of hormones have a slower onset and last for a longer time. Hormones are usually described in relation with metabolic processes. Hormones are secreted by special ductless glands of body, each hormone passes into the blood stream to produce a specific effect on some other organ called the target organ in the body. Hormones are chemical messengers or they are like instructions on sheets of paper which are carried around in the body for smooth functioning of the body physiology.

There are clear evidences that some hormones have a strong influence on some behavioural patterns; they compel the animals to break their groups, they make them more colourful and attractive albeit territorial too, hormones make animals restless and motivate them to migrate, hormones help animals in courtship, mating, nest building and feeding young ones, hormones influence growth and development of animals, hormones control the dominance hierarchy; in all, they have their influence directly or indirectly on most of the behavioural activities.

In 1849, A.A. Berthold observed the relation between gonads and sexual behaviour. Since the time of Berthold's observations a full research field has developed which is known as "Behavioural Endocrinology" or "Psychoneuroendocrinology" or "Ethoendocrinology". The basic research started with gonads and sexual behaviour. It has been known since long that gonads produce sex hormones. When the animal is young and gonads are not fully developed the courtship and mating behaviour is not seen. Higher vertebrates show a cyclic change in gonadal structure, secreting maximum sex hormones in some season and minimum in another, the volume of testes in the Jackdaws fluctuates between 4 and 1200 mm^3 over a period of one year, and in European blackbird from 1 to 400 mm^3. In some vertebrates until gonads develop fully the animal species live together peacefully in herds; after the arrival of breeding season (when the gonads are fully matured, secreting sex hormones to their maximum level) they leave the herds and become aggressive, form their own territories; the males of many species begin to court females which leads to courtship rituals and mating behaviour. Once this is over and the females have conceived, the gonads start reducing in size, lowering the hormone level and thus minimizing the mating behaviour and subsequently other hormones from pituitary come into action and rearing of young ones starts. After all these activities, when the sex hormone levels fall, the animals gather once again into peaceful herds.

Hormones can control behaviour directly or indirectly. There are hormones which have direct control or influence on the secondary sexual characters, colour of plumage and skin, courtship and subsequently mating (e.g. testosterone, oestrogen), on aggression, fleeing or fighting (e.g. adrenalin), on nest building, egg laying and parental care (e.g. oxytocin, prolactin). But hormone like angiotensin (produced from kidney) first causes rise in blood pressure, later stimulates urine production, thirst, and defecation-all these activities are included in behaviour.

Influence of hormones on behaviour may either be slow and prolonged, or quick acting and short lived, the luteinizing hormone (LH) stimulates the ovary to produce oestrogens and progesterone, which are responsible for the development of the mammary glands in females at puberty. During pregnancy there is a rise in the level of prolactin, and this stimulates milk secretion. When the newly born suckles, the mechanical stimulation induces the hypothalamus to release oxytocin from the posterior pituitary lobe, this in turn triggers milk production.

Just as the hormones affect behaviour, in turn, behaviour has its effect on hormone secretion; the females of small mammals like Rabbit and Cat start ovulating only after they have mated, the stimulus of thrusting of penis is needed to start production of luteinizing hormone from pituitary and that causes ovulation.

The hormones controlling behaviour could be categorized as : (i) Primary or direct producers of behaviour, (ii) Secondary or indirect producers of behaviour.

The glands secreting the primary producers of behaviour are gonads, placenta, adrenal glands, thyroid and the pituitary. Hormones secreted by the parathyroid, pancreas and gastro-intestinal mucosa do not have obvious direct influence on behavioural acts and could be designated as secondary or, indirect producers of behaviour. The influence of hormones on the central nervous system and consequently behaviour comes from an experiment conducted on cats who were given injections of radioactively marked oestrogen, which produced sexual behaviour after six days, the in-between observations revealed that marked oestrogen was located within few hours of injection in some part of the brain -from where it gradually disappeared. It was assumed that oestrogen first induced the formation of a messenger substance, and a sufficient amount of this intermediate substance could evoke sexual behaviour.

Various hormonal systems have influence over behavioural patterns and they are described below but before that, it must be understood that the distinction is not so clear, the hormonal systems influence one another depending largely on feedback mechanism and their functions are interdependent, interwoven.

Hormones of Gonads

The primary hormones secreted by the male gonads i.e. testes are androgens e.g. testosterone. This hormone influences the development of the male reproductive tract, sexual behaviour and secondary sex characteristics like big sized mane in lions and baboons; antlers, horns in deers and antelopes; plumage and colouration in birds. The ovaries in females produce both oestrogen and progestins. Oestrogens function in the development and maintenance of the female reproductive tract, in the development of secondary sex characteristics, in stimulating the mammary glands to secrete milk and in regulating sexual behaviour. Progestins are important in preparing the uterus for the implantation of fertilized ova and in maintaining pregnancy.

Hormones of Adrenal Gland

The medulla and cortex of the adrenal gland secrete different hormones having different functions. The medulla secretes epinephrine and norepinephrine, both function in emergency situations like catching the prey or running away from predator. Epinephrine increases pulse rate and blood pressure, to cope up with situations like fleeing from danger, and mating. Norepinephrine functions as a transmitter in the synapses, their disfunction can lead to delayed or excessive reactions and subsequently it will show in behaviour also. Adrenal hormones are useful in emotional and physical stress e.g. rescuing fledgelings from raptors and snakes, and escaping harsh climatic conditions. The adrenal cortex secretes at least 28 different steroid hormones which function in metabolism and resistance to infection and also give a feeling of general well being. Any small abnormality in metabolism of carbohydrates or maintenance of mineral levels can lead to big impairments and abnormal individuals and subsequently abnormal behaviour, for example, if in some animal salt metabolism is disturbed, it would spend hours licking salt or drinking excess of water. The adrenal cortex also secretes both male and female gonadocorticoids, these are mainly androgens and small quantity of oestrogen. The adrenogenital syndrome usually refers to a group of enzyme deficiencies that block the synthesis of gonadocorticoids, for compensation the anterior pituitary secretes more ACTH and hence increased amounts of androgen as well, thereby causing masculinisation, a female suffering with this syndrome would develop beard, deep voice, hairy body, male sexual desire and male like behaviour.

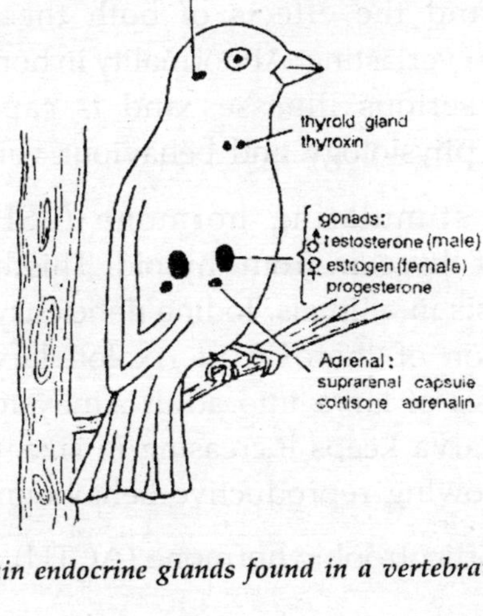

The main endocrine glands found in a vertebrate.

Hormones of Pituitary

The pituitary gland can be divided into three lobes, anterior, intermediate and posterior. The hormones produced by anterior pituitary influence the production of hormones produced by other endocrine glands, all of which control behaviour in some or the other form, therefore, anterior pituitary has indirect control over many Behavioural patterns. The somato trophic hormone (STH) or growth hormone stimulates growth and protein anabolism, any abnormality will lead to gross physiological or morphological malfunctions and later, it will show its manifestations on behaviour.

Any abnormality in the production of hormones can lead to abnormal individuals and abnormal behaviours, may it be from the category of primary or secondary producers of behaviour. Disorders of the endocrine system can involve hyposecretion (underproduction) or hypersecretion (overproduction) of hormones. If growth hormone is hyposecreted during the growth years, the normal height is not attained, this condition is called pituitary dwarfism; opposite to it hypersecretion will result in giantism in which excess of height is attained. Hyposecretion of thyroid hormones during growth years results in cretinism which causes dwarfism and mental retardation and the effects of both these symptoms on behaviour are everlasting. Abnormality in hormone production can lead to serious illnesses and is capable of altering morphology, physiology and behaviour very adversely.

Thyroid stimulating hormone (TSH) controls the production of thyroxin from thyroid. This hormone controls metamorphosis in animals. Iodine deficiency in water causes less production of thyroxin in Axolotl larva and it delays metamorphosis of larva into adult Ambystoma salamander. The Axolotl larva keeps increasing in size, develops gonads and starts showing reproductive behaviour.

Adrenocorticotrophic hormone (ACTH) stimulates cortex

of the adrenal gland, which is particularly important during emotional and physical stress. Lactogenic hormone or prolactin has direct effect on mammary glands and milk production; deficiency of this hormone leads to lack of milk production by mother, starvation in infant, and early weaning, such a dyad (mother-infant) will show unusual behaviour.

Gonado-trophic hormones (GTH) are of two types in both male and female; the follicle stimulating hormone (FSH) and the luteinising hormone (LH). In females FSH stimulates maturation of ovaries which in turn produce oestrogen and LH promotes ovulation and formation of corpus luteum which secretes progesterone whereas, in males, FSH stimulates maturation of testes and spermatozoa and the LH promotes production of testosterone from testes.

The posterior lobe of pituitary and hypothalamus form a single functional unit which controls several activities in the body. Oxytocin and vasopressin (anti-diuretic hormone : ADH) are the main hormones produced by the posterior lobe. Oxytocin promotes contraction of the uterine muscles and squeezing milk into the larger ducts behind the nipple, studies on parturition and mother-infant relationship have made an important contribution in the behavioural endocrinology. The role of morphologically tiny but functionally mighty hypothalamus in controlling behaviour has been phenomenal. Antidiuretic hormone stimulates the reabsorption of water from the glomerular filtrate. Absorption of water or excretion of water, increases or decreases the osmotic pressure of blood, such blood when passes over the osmoreceptors in the hypothalamus controls drinking behaviour in animals, an impairment can cause permanent refusal of drinking water or can force an animal to drink excessively.

The middle lobe is the smallest part of pituitary. It produces a hormone which is associated with the growth and development of melanocytes which give the skin its colour.

There are many animals which change skin colour during breeding season e.g. male sticklebacks start glowing in passion as the breeding time approaches, they look like neon lights but chameleons assume vibrant colours to attract females, and sometimes to scare away predators.

Survey of Hormonal Effects on Different Behavioural Patterns

Sexual Behaviour : The role of hormones in controlling sexual behaviour has been most extensively studied. In an experiment, it was observed that a capon (a cock whose testes are removed) could neither crow nor show any courtship behaviour due to lack of male hormone. On the other hand, female duck whose ovary was removed ceased to perform the typical movement patterns of inviting the partner to mate. All these lost characters reappeared if castrated (removal of testes and/or ovaries) animals were given the missing sex hormones by artificial means. In male parakeets, the sex hormones regulate their loud singing and their courtship behaviour.

Male rats copulate successfully only when they have produced enough testosterone. A number of vertebrate females will perform male sexual behaviour if treated with androgen. Hens which receive a sufficient amount of testosterone will perform male courtship movements and even crow like males. Several investigators have observed that after gonadectomy, i.e. removal of testes or ovaries, mating behaviour is eliminated in male frogs and in certain lizards. To some extent, mate recognition is also influenced by sex hormones. Tavolga (1950) observed that castrated gobid fish become nondiscriminating in sexual behaviour and a male fish starts courting a male only; the sexual behaviour is not shown by young ones. Eversole (1941) studied that injection of testosterone into immature male guppies produces normal mature sexual behaviour. The sexual behaviour is not manifested in young animals, because of their immature gonads and insufficient hormone production. If a chick is treated with testosterone he may be seen mounting

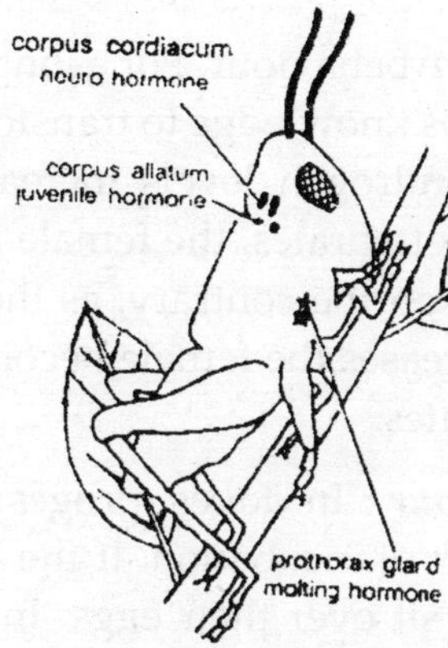

Endocrine glands in insects.

small objects in the case. In one grasshopper species females allow copulation only when under the influence of a hormone secreted from corpora allata without this hormone they would send away all males. The male desert locusts also do not elicit sexual behaviour if their corpora allata have been surgically removed. In male cockroaches the copulation movements are dependent upon a hormone produced by corpora cardiaca.

Aggressive Behaviour : There are still other areas of social behaviour more or less under the influence of hormone action. Aggressive behaviour is affected by a considerable variety of hormones from both the pituitary gonadal system and the pituitary adrenal system. In Red belted weaver bird, lutenising hormone from the pituitary gland has been demonstrated to play a role in the control of aggressive behaviour. Since in most species the males are far more aggressive than the females, it may be well assumed that testosterone plays a significant part in the aggressive behaviour.

Even among males, those that achieve high ranks in their groups generally have higher level of testosterone than those at the bottom of the hierarchy. Studies on dangerous criminals in USA have revealed that they had higher levels of testosterone. Castration in male domestic animal leads to a considerable reduction in aggressive behaviour. For a long time people have taken advantage of this knowledge to transform wild bulls into tame oxen. Higher androgen levels increase motivation for aggression in many vertebrates, the female hormone does not have the same effect, on the contrary, as the female hormone level in the blood increases the female becomes more peaceful and acceptable to mates.

Maternal Behaviour : In doves, progesterone secretion is important for the start of incubation. If the levels are low, the females will cease to sit over their eggs. In hen too prolactin initiates incubation behaviour in few birds like phalaropes and dotterel, it has been found that females secrete more testosterone than males hence it is the female who is brightly coloured, who fights with other females, courts the dull coloured male and, having laid eggs in the nest which the male has built, leaves him to incubate and raise the young. For the release of appropriate behaviour the hormones should be present in proper amount and ratio, nest building behaviour in female rabbits is triggered if the ratio between oestrogen and progesterones is 1: 4000, if this ratio is changed (1: 400) the female will cease to build brood nests. Bock (1928) and Baggerman (1957) reported complete absence of nest building behaviour following castration in the three spined stickleback males. Paradise fish will build their typical bubble nests only when the pituitary hormone prolactin is present. Doves feed their young on "Milk" produced as a secretion from the walls of the crop and for the milk to be produced, prolactin from the pituitary must circulate in the blood. Prolactin is essential for the production of crop milk in pigeons as well.

Female mice show a graded degree of responsiveness to their young which is related to hormone level. The newly born young are given maximum attention and care, but as they become stronger or as they grow, the mother mouse leaves them alone for more and more time and begins to wean (accustom to food other than its mother's milk: the young ones are known as weanlings), this slow reduction in maternal responsiveness is correlated with the slow fall in prolactin secretion during the period of weaning. There are few exceptions where this kind of direct correlation between brood care and hormonal level is not evident. Many brood care behaviours seem to occur quite independently of the hormones. Noirot (1964) showed that among certain animals virgin females and even males would show a similar level of maternal responses. Therefore, there remain certain questions which warrant further investigations.

Miscellaneous : The influence of hormones is not restricted to specific behaviour patterns such as those of sex, brood care and aggression. In some cases hormones regulate other general activities also, such as testosterone regulates the dominance hierarchy in social animals, e.g. when a low ranking pigeon was treated with a male sex hormone, it soon rose to a higher status in dominance hierarchy and with sufficient doses it even rose to the highest rank in the order. Similar results were obtained with chicken, mice and chimpanzees. Male stickleback performs its typical movements of territorial defence, nest building and zig-zag or nuptial dance only when its gonads produce sufficient amount of male hormones. Male and female gerbils secrete pheromones for marking their territory, which are actually hormone dependent pheromones, castration (removal of gonads) leads to decline in scent marking in both males and females. In certain fishes, treatment with testosterone and thyroxin leads to a general restlessness and an urge to migrate.

Three spined stickleback is known to migrate from salt water to fresh water in the spring season under the influence of TSH released by the anterior pituitary; artificial injection of thyroxin in these fish causes them to migrate in any season other than spring. Bird migration is partly dependent upon thyroid hormone. Gonadal hormones also activate migratory urge in birds (Rowan, 1931). Food hoarding is carried out with the help of hormones, hoarding by male gerbil is increased by castration and decreased by testosterone treatment. Thiessen and Yahn (1992) worked with the mongolian gerbil in which males mark their territory with a secretion from a gland on their ventral surface; they observed that castration led to regression of the gland and the disappearance of marking behaviour as well. In higher vertebrates, even the, learning ability is dependent on certain hormones. Although experiments have produced contradictory results, yet some studies have clearly indicated that learning in rats is facilitated by hormones from pituitary. When this gland was removed, animals subjected to training experiments showed poor performance. When they were treated with ACTH, their performance rose back to normal.

Among insects if the influence of juvenile hormone from corpora allata is excessive then juvenile characters will persist in the adult (metathetely or neoteny) or the influence of the juvenile hormone may be depressed, which would cause early development of adult characters in the larva (Prothetely or paedogenesis). The example of metathetely comes from Gryllus campestris where the adult fails to develop wings, absence of wings is a feature of larva and this persists in the adult. Prothetely is seen in locust's 4th instar which becomes sexually mature.

In Aphids, winged (alatae) and wingless (apterae) forms occur, due to hormonal imbalance. Caste system in social insects is controlled by pheromones which are under the influence of hormones. Through juvenile hormone only, many grasshopper species exhibit green brown polymorphism, In locust, hormone from corpora cardiaca controls darkening of body. Stick insect gets black in night and light brown during the day time, pigment granules move up and down respectively; hormones from corpora cardiaca and neurosecretory cells of suboesophageal ganglia control colour change in stick insect.

Triman and Riddifor (1974) classified hormones which affect behaviour as releasers and modifiers. According to them releaser hormones are those that trigger a response within a short time after entering the haemolymph. Modifiers affect the responsiveness of the nervous system so that the response of the insect to a given stimulus is altered.

Calling by Hyalophora is triggered by releaser hormone. Immature female Gomphocerus (grasshopper) will respond positively to male after getting juvenile hormone. Sexual receptivity is often associated with the pheromones, which in turn are controlled by the hormones of corpora allata. In locust, the median neurosecretory cells have a direct effect on sexual behaviour. The ovaries of the rabbit flea mature if the insect feeds on a pregnant rabbit, it is believed that the hormones of female rabbit help maturation.

There are instances, where the performance of certain behaviour by one animal influences hormone production in another animal and consequently behaviour of either or both. It is known that nerve cells have neurotransmitters that help in transfer of electrical impulses through synapses, some nerve cells operate neurohormonally, producing messenger substances or releasing factors, which in turn either release or block the production of hormones. These cells are able to transform nerve impulses into hormonal messages, in this way, the song of a male parakeet can stimulate the formation of gonadotropins in the female parakeet to an extent that she produces mature eggs. In canaries too, females will produce increased amounts of follicle stimulating hormone in response to male courtship behaviour. Similarly, among most of the birds sight and feeling of eggs during incubation stimulates prolactin production in them. In the bristleworm, Platynereis, a brain hormone motivates the animal to leave its tube like burrow at the bottom of the shallow seawater, swim up and swarm into open water.

Methods of Study

Ethoendocrinological research employs following methods to study the role of hormones in controlling behaviour:

(1) Correlation method

(2) Castration and replacement method

(3) Cannulation method

(4) Hormone uptake method

(5) Cross transfusion method

(6) Antihormone method

(7) Hormone assay method

Correlation Method : In the field of science basically all methods are correlation methods, an experiment is performed, the results are obtained, and a correlation is derived, but here correlation method means a very basic method, in which gross correlations are derived between specific hormones and behaviours. In fact, correlation method forms the basis of all other methods; in earlier years, when sophisticated instruments and techniques were not available this was the only method which gave information for future methods.

This was the easiest method for studying hormones and behaviour. In this case, one simply searches fluctuations or differences in behaviour which can be due to parallel hormonal fluctuations, e.g. testosterone levels were found to be higher in criminals, the correlation is that higher levels of male hormone are responsible for aggression; or non receptive female cats when given injections of oestrogen show sexual behaviour, inference is that oestrogen controls female sexual behaviour. This particular method is considered basic and crude and with new and more sophisticated techniques available, use of correlation technique has become obsolete.

Castration and Replacement : More direct evidence of a hormonal regulation of behaviour can be studied through use of the method of castration (oophorectomy: removal of ovary; orchiectomy; removal of testes) and replacement. With castration one removes the source of a hormone. If a behavioural pattern is hormone dependent, it should show a decrease with decrease in hormone. If one replaces the hormone, behaviour should reappear. Instead of operative castration, it is now possible to castrate chemically by using anti-androgen (Cyproterone), or by exposing testes to highly concentrated x-rays. Oophorectomized ducks cannot perform species specific courtship movements to invite males, but can resume all the lost activities when given female sex hormones, similarly, orchiectomised mallard drakes fail to court, mount and mate with ducks, testosterone injection would help the male in resuming its normal sexual behaviour.

Though hormones have an extremely pronounced effect on courtship and mating, the other behavioural acts are no exception. The males of mongolian gerbils scentmark their territories with a secretion produced from a gland situated on their ventral side, orchiectomy led to hypoactivity of scent gland and disappearance of marking behaviour. Testosterone injections restored the marking behaviour. Removal of thyroid at an early age would retard the growth or prevent metamorphosis, replacement of hormones may initiate the normal activity.

Cannulation : Hormones have been found to affect behaviour through effects on numerous specific brain structures. To study this particular correlation, metal cannula is inserted in the exact place with the help of brain maps and the hormone is poured through it to study the behaviour.

Barfield (1971) cannulated testosterone into the brains of castrated male ring doves, he found that preoptic area of the hypothalamus was responsible for sexual and aggressive behaviour in these birds.

Implantation of testosterone through the needles into the

diencephalon of male rats increased aggressive behaviour and if the same hormone was placed on preoptic area of hypothalamus it produced copulatory behaviour.

Hormone Uptake Method : Autoradiography, a procedure in which hormones are labelled with radioactive substance, can be used to determine the brain regions in which hormones are concentrated. If labelled hormones are injected into the animal, their location in the brain can be detected later by using radio sensitive plates over the sections of brain.

Cross Transfusion : Tokel and Rosen Blatt (1972) developed a technique whereby blood could be cross transfused between two freely moving rats. For example, when blood was cross transfused between mother females and virgin females, it was found that the blood from mother female was effective in inducing maternal behaviour in virgins.

Antihormone Method : Various drugs are available that have been shown to function as anti-hormones. Both anti-androgen and anti-oestrogen drugs have been used in this type of studies. When an antihormone is given to an animal it blocks the functioning of a particular hormone and simultaneously the animal shows the absence of a particular behaviour.

Hormone Assay Method : One of the most promising developments in the study of hormone action is the use of refined chemical assay methods to measure the hormone content of the blood directly. Direct hormone assay helps to associate changes in the body's normal hormone secretion with changes in behaviour. Hormone assay work has already yielded some fascinating results.

Mechanisms of Hormone Action

Hormones affect behaviour by altering one or more of the following:

(1) Sensory perceptual functions,

(2) Neural activity, and

(3) The effector mechanism.

Effects of Hormones on Sensory Perceptual Mechanism : Numerous studies have demonstrated that both the ability to detect stimuli and preference for particular stimuli are affected by hormones e.g. the ability of female rats to detect odours varies with varying levels of oestrogen and progesterone. In an experiment two sets of male rats were taken, one set was castrated and other one was not, all of them were blindfolded, the non castrated males could locate the females by their odour whereas, castrated males could not.

Effect of Hormones on Neural Mechanism : There is now abundant evidence that hormones affect behaviour by direct action on specialized neurons in the central nervous system. It is possible to implant hollow needles (canula) into the brains of many animals without interfering with their freedom. Brain atlases of several species are available, with the help of which the tip of a needle can be placed exactly at a desired spot. The exact position can be checked later by sectioning the brain after behavioural tests are over. Two basic techniques have been used. (i) hormones in solution are introduced into neural tissue via canula (ii) the tip of the needle before insertion is coated with the hormone. Harris and Michael investigated the role of oestrogen in the sexual behaviour of female cats. When in heat or oestrous, the female cat shows a very characteristic posture towards an approaching male; she elevates the rump, the tail is deflected to one side and the hind legs make shaky movements. A castrated female whose ovary is removed will not show this oestrous behaviour. Harris and Michael placed a needle coated with slowly dissolving oestrogen in ventromedial hypothalamus of this castrated female cat, and it immediately showed strong oestrous behaviour. In other experiments, the implantation of testosterone in the anterior hypothalamus caused castrated male doves to show courtship behaviour and aggression. Progesterone implanted in the same place suppressed these behaviour patterns and caused an increased tendency to incubate. Same hormone can act on different parts of the brain and can bring about different behaviours too. Testosterone crystals when placed on lateral diencephalon in rats elicit aggressive behaviour but if the crystals are placed on preoptic nucleus of hypothalamus, then copulatory behaviour would be evoked. Besides specific influence on central nervous system, some hormones increase the receptiveness of large area of brain, hormone progesterone has such an effect on female rats, it was noted that animals treated with this hormone responded more easily not only to sexual stimuli but to a large number of other stimuli.

Effect of Hormones on Effectors : Hormones may influence external structures, for example testosterone induces changes in the skin; the surface of rat's penis becomes more sensitive to touch and transmits stronger messages to the brain as the testosterone level rises and as a result brain responds to this "exaggeration" and releases sexual behaviour. Similarly higher levels of female hormones make the nerves of genitalia more sensitive, so that she is more stimulated by the males thrusting penis.

There are a number of ways, in which hormones can exercise control over behaviour. Under natural conditions, any one hormone may operate simultaneously in a variety of ways or sometimes many hormones may influence just a single reaction. F.A. Beach, one of the leading scientists in this field made a remark that behavioural activities e.g. courting and mating or fighting and fleeing, etc. involve complex chain of reactions controlled collectively by neural and hormonal systems of the animal; there is no behavioural activity which can be brought about by just one hormone.

11

The Pheromones

Communication or transfer of information among animals is essential for their survival. Communication is a general term for signals exchanged between animals. Many animals have been studied for their unique and remarkable ways of communication. The bird songs have been heard by everyone; in the stillness of night cricket mating calls are easily audible; the humpback whales have been identified as world's greatest singers and composers; the bees hum and perform circle and waggle dance to communicate complex messages; the bats and dolphins use ultrasonic sounds; the monkeys and apes have now been extensively studied for language. The animals can communicate by making gestures or facial expressions and by postures for example, a dominant rhesus will walk confidently, sprawl, carry the tail up. The animals communicate mating willingness by exhibiting feathers, genitalia or sniffing each other. Animals communicate by screeching, growling, barking, whooping, chirping, shrieking and moaning; animals may pass on messages through touch, e.g. chimpanzees embrace on meeting, male-female of titi monkeys twine their tails while sitting on a branch, the gorilla mothers fondle their babies almost like human beings, the ants will tap their antennae on each other; while mating snakes will entangle, the scorpions will hold chelicerae of each other and perform a dance. Almost all animals in the animal kingdom communicate through chemicals produced by glands within the body. Hediger (1944) first reported the significance of odour in communication, this specialized communication includes smell or scent and pheromones.

The chemical messages can be received via taste buds, nasal epithelium or through specialized structures found on the body. Taste (gustation), is treated more as sensation, they have limited use in communication, therefore, shall not be dealt in present chapter. Smell is also by and large considered a sensation perceived through nasal epithelium but, it forms part of communication. There is subtle difference between smell and pheromones (Table). By definition "Pheromone is a chemical secreted by an animal that influences the behaviour of other animals of the same species". Among invertebrates the chemicals or pheromones, which influence behaviour are produced and perceived by the specially modified structures, whereas, in vertebrates the chemicals, with reference to behaviour are produced by special glands but are always received by nasal epithelium, with which smell is associated.

Generalized Comparison between Smell and Pheromones

Smell	*Pheromones*
1. Chemical information received by nasal epithelium by which odours are identified This word is generally used for vertebrates and relates to a specialised sense i.e. sense of smell.	1. Chemical information produced and perceived by special structures in vertebrates. Among vertebrates pheromones are produced by special glands but are received by nasal epithelium; pheromones are also sometimes designated as scent or odour.

Contd...

Smell	*Pheromones*
2. The origin or production of smell can be from outside source viz. food, flower, organic matter, perfume and body.	2. The origin or production is from special glands or along with saliva, urine and faeces.
3. The smell can be identified as pleasant, unpleasant, and neutral by conscious brain.	3. The pheromone may or may not be identified by conscious brain.
4. The smell may or may not have direct effect on animal's behaviour.	4. Pheromones have direct effect on animal's behaviour, one who produces is 'called sender and the one who perceives is called the receiver.
5. Production and perception may not be under conscious control.	5. Production of many pheromones is under conscious control.

Exchange of information through pheromones or through scent is considered a primitive form of communication, because as animals evolved they depended more and more on other elaborate and much specific visual, auditory and vocal forms of communication, still the animals can communicate complicated messages through chemicals, they can mark and identify their territories, pathways, conspecifics, rivals, defeated animals, pups, and sexual partners through them. When chemicals are used for communication among conspecifics (individuals of the same species) they are called pheromones; if they are used to communicate with the members of other species (which is rare) they are designated as kalromones; if only the sender is benefited they are called allomones.

Earlier, these chemicals were referred to as "ectohormones", but since 1959 they have been called pheromones, unlike true hormones which are secreted internally by ductless glands and which regulate the organism's own physiology, pheromones are produced by ectodermal glands and are secreted externally to regulate the behaviour of conspecifics.

The term pheromone was originally applied to sex attractants of insects but with the accumulation of information the term has been broadened to include various kinds of chemicals released into the environment and functioning differently in almost all the major groups of animals. Among most vertebrates pheromones are used extensively in social and sex life. Among social insects pheromones play a key role in caste system and maintenance of colony. Pheromones have found important place in human life, because after exploring the secret world of scent communication, we have been able to come up with a harmless way of controlling pests. There are rising protests against pesticides destroying wildlife and endangering the health of man himself.

The battle against harmful insects would be much less costly and more efficient, and the problem of environmental contamination by toxic materials would be vastly reduced if the insect activities are controlled by pheromones, for example use of artificially synthesized sex attractants or repellants may eliminate cockroaches and termites from the house. It has also been of use in captive breeding, and breed improvement may it be poultry, horses, cattle, goats, dogs or pigs. A sow in heat recognizes a boar by the scent in his saliva. To artificially inseminate the sow, a person has to sit on the sow to apply pressure on her back (which is normally applied by the mounting boar) the person then sprays artificially synthesized pheromone of boar, and after that the sow allows the insertion of spirette, and semen through which artificial insemination is achieved.

Study of pheromones has given much information to

perfume industry, Some of the popular perfumes contain animal scents. The Importance of scent or smell is realized when familiar land marks become almost invisible and we try to find our way using smell as cues, familiar smells can come from trees, or a restaurant or a bakery. The importance of smell is also realized when we come across pleasant or obnoxious odours. Man's sense of smell is not very well developed, it is in fact so poor that when we watch a dog sniffing along a street we cannot imagine the rich world of smell in which he lives. The significance of scent is realized when we read in news papers how dogs identify criminals, drugs and bombs. Dogs have about 200 million olfactory receptors in nose, that is 20 times more than the number in human beings.

The role of body scent in human sexual behaviour has attracted much attention. Ellis as early as 1905 reported olfactosexual behaviour in human beings. In 1971 Maclintoek maintained notes on the menstrual cycles of 135 girls of a hostel and observed Lee Boot effect i.e. synchronization in onset of their menstrual dates, the reason assigned was body smell. It has also been reported that a female living with a male has relatively shorter menstrual cycle, because the body smell of male has an effect on female's reproductive physiology.

Female rhesus secretes a pheromone, copulin, from its vagina which stimulates mounting and mating in rhesus male, similar copulins have been identified in human female vaginal secretions also. Identification of human pheromone from male sweat has been carried out by Dr. George Dodd of Warwick University in U.K., it is identified as K-androstenol (related to sex hormone) and its principal function is in attracting the females, during sexual interaction.

Such informations are helping perfume industries to come out with good albeit expensive perfumes. A perfume called

Andron which contains a trace of the attractant pheromone androstenol sells in the market for $275 for an ounce. Pheromones have significant role in maintaining mother child bond. It is a common observation that a young human infant is most comfortable and quiet with its mother, this is because they start recognizing their mothers by smell when they are six days old; by sixth week the baby start orienting its head towards the mother's breast and that is again guided by the mother's scent. In western countries young mothers are advised not to use any other perfume because that counteracts with mother's natural body scent, which may subsequently affect mother-infant bond adversely.

In 18th century, the discovery of pheromones was done by chance, a French biologist, Bonnet discovered the long trails of ants. Bonnet placed a colony of ants at one end of a table and at the other end he placed a small heap of sugar. He observed the ants coming across the table to the pile, picking up sugar and going back with it. The ants moved to and fro along a definite track, none of them diverting from it Bonnet wondered why the ants followed such a definite path ? He rubbed his finger across the path of ants running towards the sugar; and saw that the ants stopped and searched about when they reached the rubbed place, waving their antennae in the air and tapping them on the ground.

There was a gradual crowding of ants on both sides of the line-some coming from nest, some from the food, until a few adventurous ants ran on to the place where Bonnet had rubbed. Ants from both sides met, recognised each other and continued their journey once again. Bonnet thus assumed that it was actually a trail of chemicals which the ants could sense and follow.

Another observation was made by Jean Henri Fabre a French naturalist in 19th century, who had an endless curiosity about the lives and habits of insects. Fabre one·day placed a female moth in a cage near a window and to his surprise within 15 hours about 60 males of the same species came out of the woods and collected around it. Fabre began a series of experiments to try to discover what had attracted the males? When he placed the females in a tightly closed transparent container where the female could be seen, no male appeared. On the other hand, a container made up of wire mesh occupied by the female moth was strongly attractive to the males, in this situation male moths could not see the females but could smell, therefore, Fabre concluded presence of scent, or pheromones in female moths.

Fabre further observed that these moths had pheromone sensitise structures located in their antennae and that they could detect the female scent in fantastically small amounts. Pheromones are produced in very small amounts in microns or millimicrons but are enough for use over a long time. E.O. Wilson of Harvard University calculated that just one milligram of the trail pheromone of the leaf cutter ant is enough to lead a small colony three times around the world.

Extraction of pheromones has been a tedious method. Robert T. Yamomoto identified the sex attractants of the cockroach Periplenata Americana by isolating it from a vapour emitted by female cockroach. The author ran a stream of air through large cans containing thousands of female for a period of eight months trapping the vapour by freezing it with dry ice. From the condensate he could obtain only 12.2 mg of the pure attractant, this pheromone was a volatile yellow coloured substance with a strong floral odour. Male cockroaches responded to the odour with intense excitement, raising their wings and making mating attempts.

This pheromone was identified as 2, 2-Dimethyl 3-isopropylidene, cyclopropyl propionate. Another team of chemists at the Institute of Chemistry in Munich isolated the sex attractants of the female silk moth, Bombyx mori. They extracted the abdomens of five million females to obtain a tiny amount (12mg) of the pure pheromone which they named as Bombaykol. The other attractants which have been isolated are Gyplure (gypsy moth), Civetone (civet), Muskone, (musk deer), Honeybee queen substance (bee).

The identification and extraction of pheromones has moved with a slow pace. Till 1976 only six mammalian pheromones could be identified but, with the advent of knowledge and technology it has been possible to identify the chemical compositions of a few more pheromones produced by different animals. Skunk, a carnivore found from Canada to northern Mexico produces a very foul smelling secretion (allomone) from its anal gland. The secretion has three sulphur compounds methyl crotyl sulphide, crotyl and isopentyl mercaptan and its main function is to repel predators and human beings.

The musk deer and civet cat emit muscopyridine. The other acids which form active ingredients of pheromones are volatile carboxylic acids (red fox), unsaturated lactone (black tailed deer), phenylacetic acid (mongolian gerbil), isovaleric acid (male proghorn-a ruminant), mixture of acetic acid, propionic acid, isobutyric acid n-butyric acid and isovaleric acid (female rhesus). Brahmchay (1981) worked extensively on a pheromone produced by tigers which he called Tigeramine. He noticed it as a milky thick fluid with a strong smell passed out along with urine in tigers. He further investigated that all male Tigers produce pheromones to mark their respective territories, he collected various samples and found out that Phenylethylamine was the main and common constituent among pheromones of different tigers; the chromatographic studies revealed that other amines differed from tiger to tiger which makes it typical of that individual and that helps them identifying their own territory from others.

Effect of Pheromones

A pheromone affects behaviour in two ways (i) It stimulates the central nervous system (CNS) of the recipient and produces an immediate change in its behaviour, it is known as releaser effect. (ii) If a pheromone alters a set of physiological reactions in the recipient which in turn prepares the animal for a particular behavioural pattern to be released later by an appropriate stimulus, is known as primer effect.

Suppose a pheromone is released by an animal (a), and it is perceived by animal (b), the pheromone acts on CNS of (b) and this releases the behaviour immediately in animal (b). This kind of quick action is called the releaser effect. Crush one snail by the side of a river or pond and the crushed cells emit a fear pheromone which sends other snails immediately back into water; a mother rat licks her nipples so that her blind pups can follow the scent of her saliva to the milk, wash the nipples, and the pups are lost; termites discovering a breach in their mound emit an alarm scent that brings masons to work, who immediately start repairing the crack. Pheromones that produce a releaser effect are widely spread in the animal kingdom. Sex attractants in insects constitute a large number of this category, trails of ants are also consequences of releaser's effect; similarly you crush a bee and within no time you will be attacked by hundreds of them.

On the other hand, in primer effect a pheromone is released by animal (a), it is perceived by animal (b), the pheromone acts on CNS of (b), which brings about changes in physiological reactions and prepares CNS of (b) for a particular behaviour, which is then released by an appropriate stimulus. It is a delayed effect of pheromone and is unlike releaser effect.

There are fewer but interesting examples to explain primer effect of pheromones. In 1955 mammalian endocrinologists, Van der Lee and L.M. Boot, discovered several unexpected effects on the female mouse. They observed that when many female mice were left in a cage to live together their oestrous cycles were synchronised.

They further observed that if female mice were kept away from males for a long time, their oestrous cycle stopped completely. Further, scent or pheromone of a male mouse could initiate the oestrous cycle of a female mouse. These scientists came to the conclusion that pheromones secreted by male mice are perceived by female mice which gradually affect the reproductive physiology of female mice and consequently its reproductive behaviour. They further investigated that when female mice were kept in groups of four, there was an increase in the number of spontaneous pseudopregnancies, which could be prevented by removing olfactory bulbs or by isolating the females.

They called it Lee-Boot effect. A still more surprising primer effect was studied by Helen Bruce (1951) of the National Institute for Medical Research in London. She observed that the odour of a strange male mouse blocked the pregnancy of a newly conceived female mouse. The odour of the original male left pregnancy undisturbed. While the odour of the strange male suppressed the secretion of the hormone prolactin so that normal oestrous was restored, this is known as Helen Bruce effect. In this, physical contact between male-female is not necessary, the Bruce effect could be seen by placing the female in an empty box that formerly housed a strange male. However, female is sensitive to the presence of a strange male only upto a maximum of five days after the first copulation; after six days Bruce effect does not occur.

Another scientist Whitten (1959) found that females that have been grouped together, do not mate as quickly when left with male as do females that have been isolated, this is known as the Whitten effect, this effect can be reversed by introducing a male into the all female group. The other examples of primer effect come from social insects. Honeybees, termites and ants have been most widely studied for this purpose. Among social insects, members of the same colony share a common pheromone which differs from that of the other colonies within the same species. The queen bee preserves her monarchy by exuding pheromone that inhibits worker bees from making new royal chambers of queen cells. Termite caste production is regulated by the number of individuals of the caste that are already present e.g. the presence of a pair of reproductive termites inhibits the production of other reproductive termites and hence the size of the colony is controlled by a pheromone released from the anus, all these examples involve delayed primer effect.

Pheromones in Invertebrates

Pheromones play an integral part in the secret lives of invertebrates as already evident from the earlier pages of this chapter. The invertebrates have special scent producing glands and structures to receive the chemical messages.

Production and Perception

Glands Producing Pheromones: Pheromones are produced by definite glands. The location and structure of these glands on the body varies in different animals. The insects have been most extensively studied for their pheromones and structures related to their production and perception.

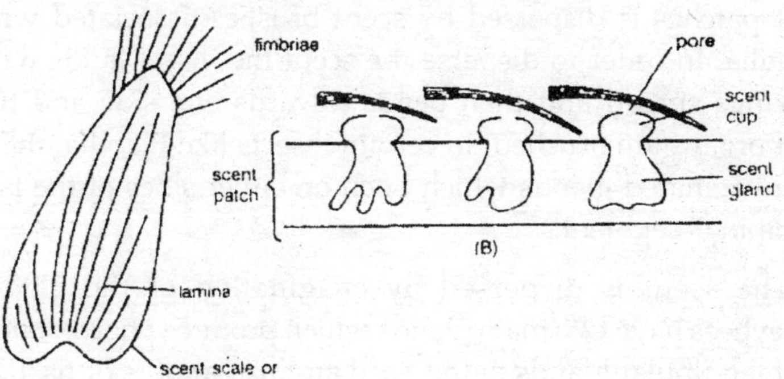

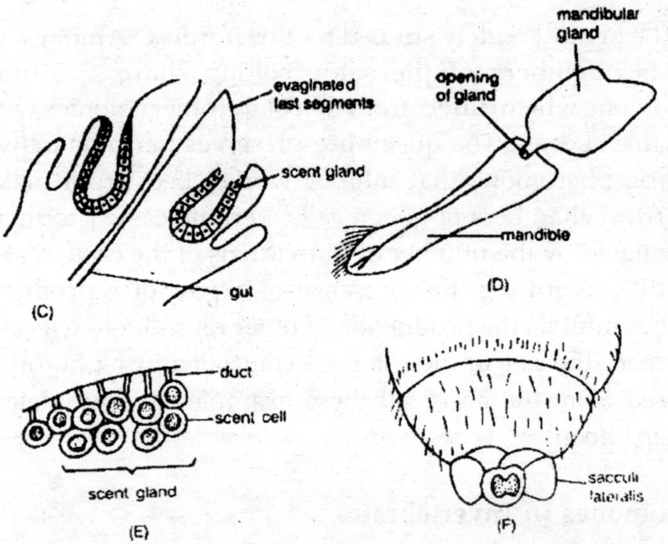

Scent producing glands.

Male Lepidopterans (butterflies and moths) are often able to produce scent from glands which are commonly associated with scales, these scales are known as androconia and are located on wings, these scent scales have an elongated form and terminate in a row of processes of fimbriae. The pheromones produced by these structures evaporate through these fimbriae. Males of Amauris have a small scent patch on each of the hind wings. These patches contain highly modified structures called scent cups, each scent cup has a median pore and below each cup is found a scent gland. Scent from these scent patches is dispersed by scent brushes associated with genitalia. In order to disperse the scent the insect lands with its wings spread and then bends towards one side and the scent organs are brushed. In certain insects like Plaodia, there are invaginated glands which open on either sides of the last abdominal segment.

The scent is dispersed by evagination of the gland. Honeybees have two main glands which produce pheromones. The mandibular glands in the head and nassanoff's or nasnov (named after the Russian who first described it in 1882) gland in the abdomen. Mandibular gland is a sac like structure, whose duct opens at the base of the mandible. These glands are well developed in the queen and workers, but generally reduced in the drones. Nassanoff's gland is found below the inter segmental membrane (arthrodial membrane) between 6th and 7th abdominal segments, which is exposed by bending the tip of abdomen. This gland is made up of a number of large cells which open to exterior by small ducts. Female silkworm moth emits a sex pheromone from a pair of sacs, called sacculi lateralis that are found on the last abdominal segment. Ants have many pheromone producing glands viz.: mandibular gland and its reservoir, labial gland, metapleural gland, pavan's gland, dufours gland, anal gland and its reservoir.

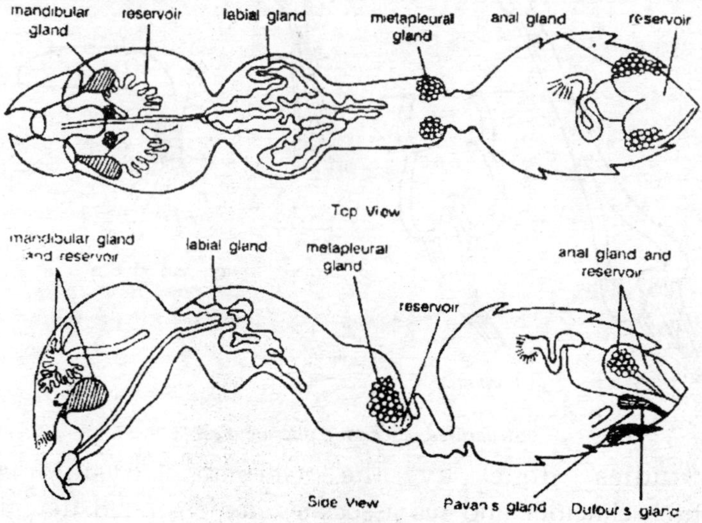

Scent producing glands in ant.

Structures Perceiving Pheromones: Great Swiss entomologist Auguste Forel found towards the end of 18th century various structures on an ant's antennae and on other insects which seemed to be particularly well adapted for receiving chemical substances. These structures were flat in some insects, cup shaped in others while in some they were in the form of short stubby hair. All these structures were supplied by nerves. The pheromones released by a conspecific are perceived by these specialised structures and the messages are sent to central nervous system. Despite many variations, all chemically sensitive structures share a basic plan : the receptor cells and nerves, the receptor cells are positioned in such a way that stimulus - substance molecules from the surrounding medium can easily reach the special areas of cell membrane. Some animals will have projecting antennae specially designed to receive pheromones.

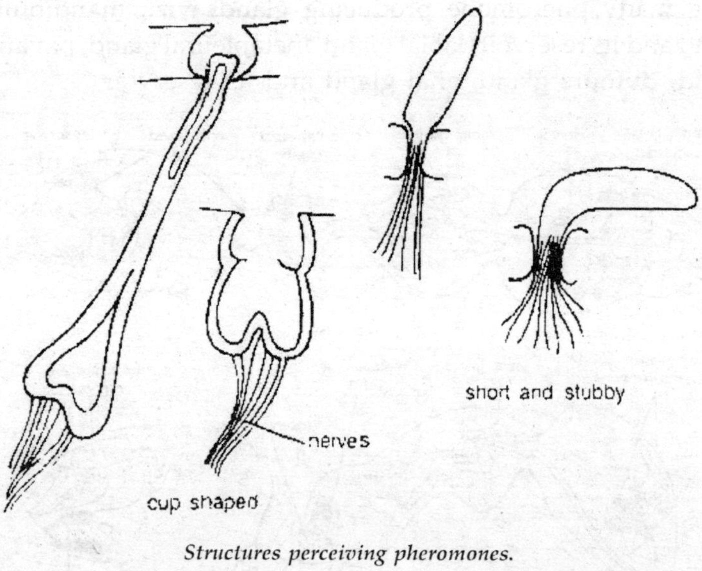

Structures perceiving pheromones.

Studies done by the German physiologist, Dietrichschneider and his associates demonstrated that the male silkworm moth receives the stimulus of pheromones by specialized structures located on their antennae which act as chemoreceptors. They further discovered two specialized sensory structures on the surface of the antenna - sensillae coeloconicae which were pit like and short rod like sensillae basiconicae.

Functions of Pheromones

Pheromones perform a variety of functions among different animals. The main functions are mentioned below.

Pheromones as Sex Attractants

Pheromones are employed by a large number of insects in bringing the sexes together for mating. These pheromones are known as sex attractants.

Pheromones of Females: Pheromones in female insects are usually released by exposing the glands by movements of abdomen. Normally, scent is released at particular times of a day, depending upon the nocturnal or diurnal nature of animal. The effect of scent is to excite the male and to promote take off towards her. Eventually, as the male approaches the female, there is continuous increase in the concentration of the chemical attractant and this serves as a guide for the remaining distance. How far and in which direction the pheromone spreads, depends largely on the wind direction.

The queen bee attracts drones by a pheromone, the principal component of which is 9 oxodecenoic acid produced in the mandibular gland. The power of this pheromone is almost unbelievable. As mentioned earlier, the queen bee chiefly produces pheromones from two glands viz mandibular and nasnoffs', she rules over a colony of 6000-8000 individuals for 5 to 7 years, it is assumed that the variety of odours she can pass around to make those 8000 individuals work, is controlled by producing varied concentrations of pheromones from just two glands. She decides the number in a caste, she prevents further construction of royal chamber, she passes orders to all the workers who execute work tirelessly. Most acute sense of chemicals exhibited in nature is that of the male Emperor Moth Endia Pavonia which according to experiments conducted in Germany in 1961, can detect the sex attractant of the female as far as 13 kms. This scent has been identified as one of the higher alcohols ($C_{16}H_{21}OH$) of which the female carries less than 0.001 of a milligram. Among many vertebrates the readiness of a female for mating is perceived by male through pheromones produced from vagina therefore, sniffing the genitalia is a common sight in ready-to-mate animals, e.g. pheromone copulin is produced from the vagina of rhesus female the male rhesus would inspect the hindquarters of a female in heat before copulating. Appearance of male dogs from neighbourhood around a pet bitch is the most common example of a pheromone produced from vagina to attract males.

Pheromones of Males: There are only few instances of male producing sex attractants. This is true of the male beetle Harpobittacus. In this insect, after the male has caught the prey and started to feed, two vesicles are everted from between the posterior abdominal segments. These vesicles are expanded and contracted and the scent is released which excites female for mating. On her arrival male presents her the remains of prey, and while female is busy eating it, the male Harpobittacus copulates.

Pheromones of Social Insects

The pheromones of social insects fall roughly into two categories according to their function

(1) Those concerned with communication between conspecifics and
(2) Those concerned with the maintenance of colony structure or caste system.

Communication: Many ant species lay scent trails by which they are able to find their way about. Edward O. Wilson (1958) did some experiments with fire ants to study the role of pheromones in communication, he observed that when a worker finds food it returns to the nest laying a trail. On encountering a fellow worker it rushes towards it and may climb on it, apparently bringing the trail substance to its notice. The trail initially consists of a series of scent spots produced by a worker ant touching the ground with its abdomen as it runs along.

Subsequently, if the trail is used by many ants the spots which they produce may merge into a continuous streak. Wilson observed that the trail of fire ants consisted of a substance secreted in minute amounts from the edge of the abdomen by dufour's gland; the substance leaves the ant's body through a pointed thin sting, which is touched to the ground and leaves the pheromone much like a moving pen or it is drawn out just like tooth-paste which has been squeezed from the tube. Wilson extracted pheromone from the dufour's glands of freshly killed workers and performed several experiments, to conclude that fire ants have an elaborate system of communication through chemicals. Wilson developed a colony of fire ants into a glass dish which had one opening and with the help of extracted pheromone he created an artificial trail with a stick. He observed that the workers were attracted from the nest. Workers followed the artificial trail in close formation. Ultimately they gathered in the confusion at its terminus because there was no food. He proposed that the trails can be of different types, one leading to food and other may just be a routine passage or it can be a trail of emergency exits.

If an ant is attacked in a colony it secretes alarm pheromone from mandibular or anal gland which attracts other members of the colony that quickly reach for the rescue of conspecific. The alarm pheromone among social insects have three main functions (i) to alert the colony (ii) to release aggression and (iii) to mark the target to be attacked. If you are stung by a bee and you reflexly kill that bee the dead stinger might invite hundreds of bees; there are alarm and distress pheromones that are released from stinging or dead bees respectively which triggers a mass attack from the hive. All pheromones are volatile, however, the extent varies. For example, distress pheromones are very highly volatile because if this was not the case then colonies of animals might remain in an almost continually disturbed state. However, pheromones for marking trails of movement or territory are much less volatile.

Worker bees are attracted to each other by a scent from nasanoff's gland. This scent is dispersed by depressing the tip of the abdomen so that the gland is exposed and sometimes the wings are vibrated too, creating an air current for better dispersion. The scent is released in various situations viz., at the time of feeding, during formation of new hive, to mark a source of water, to mark hive for recognition and for maintaining cohesion of swarm during flight.

Some insects use the pheromones to trap the prey. The bola spider tricks male moths that are searching for mates, to attract them, the spider produces a scent like that made by a female moth, as a male moth comes near, spider kills the moth and eats it.

Maintenance of Colony Structure: The queen bee produces pheromones mainly from two glands which play a part in controlling the social structure of her colony. If a queen is removed from a colony her absence is perceived immediately by the workers who become restless, immediately they start building emergency queen cells or royal chambers, by enlarging and reshaping the existing worker cells and larvae within them are fed Royal jelly instead of Bee's bread, these larvae then grow into queens. The queen which comes out first of all locates sister queens with the help of pheromones and sting them to death before they even emerge. In the presence of queen this behaviour of workers of building new royal chambers is kept in check by a pheromone produced by her mandibular gland. Like honeybees, the queen of certain Vespine wasps who are also social but live in smaller colonies, produce pheromones that maintain the stability of the colony. The differentiation and the number of a particular caste in termites is also regulated by a series of pheromones. Replacement of a queen termite is produced within a week from its absence. In termites also, more than one queen is produced and as soon as the first queen comes out, the workers eat away the remaining developing queens.

Scent Marking in Vertebrates

Many vertebrates produce odours, scents or pheromones to carry out many important functions. House shrew, striped hyena and black tailed deer produce scent under stress, other mammals mark territories, mates, pups, and their habitat with scent. The scent is released in many ways viz, along with urine, faeces and saliva, they are also produced through special glands.

Use of Urine and Faeces : Urine or faeces of many animals have pheromones or scent through which they can communicate with conspecifics. The most common use of urine and faeces is to identify or scent mark core area (an area, used extensively and defended actively), home range (an area around core area, actively used but not defended actively) or territory (a bigger area around home range used or exploited but not defended). Some animals use urine and faeces for identifying pathways, resting grounds, feeding grounds and sleeping sites. Some animals urinate or defecate on rivals, opponents, defeated conspecifics, etc.

The most interesting and strange example comes from mammals. The male hippopotamus mark their pathway between aquatic resting place and grassy feeding ground, by the deposition of dung at places along the trail, usually close to some conspicuous objects. As hippopotamus defecates while walking on the trail it keeps moving its tail rapidly from side to side, so that the faeces are splashed out and get deposited all over the vegetation above ground level and thus, places it at nose height to make it more noticeable; rhinos mark their pathway by demarcating at prominent land marks and depositing piles of faeces so that it can be both smelt and seen. Tigers select specific places and mark their territories by their droppings. Main land serow of family Bovidae, found in northern India has fixed pathways of movements, initially these pathways are used visually and later the animal marks traditional spots by depositing its droppings there, so that it can just follow the scent and others are kept away. Polecat (a weasel like animal) and civet deposit urine and faeces at particular places along their trail too. The male giant rats and skunks have their own style of marking, they place their droppings above ground level by standing on fore limbs and hindlimbs supported against some solid object as they defecate and urinate, this is done to mark the area at nose level for easy detection. Dogs, coyote and foxes use just the urine to mark their territories, the scent left with urine indicates minute details like individual identity, male or female, dominant or subdominant, adult or subadult and also how long ago that area was marked.

Slow loris found in south east Asia are small sized primates, they occupy small territories which they mark by their urine. Capuchin a new world monkey, found in southern Central America lives in large groups among the dense foliage of forest trees, each group has its own home range in which the members move about. The adult members soak their feet and hands in their urine known as "urine washing" and smear it on the foliage. Urine washing is also seen in slender loris. It is a small sized animal found in India. They sprinkle urine on tree trunks to mark their territory. Similarly, common squirrel monkey of northern South America, for demarcating its own territory, leaves a scent by soaking its body specially, tail in urine. Proghorn, an antelope found in Canada, USA, Mexico and sambhar a common deer in India live in herds, during mating season the adult males become territorial and mark their areas with droppings and urine.

In greater bush baby the habit is further peculiar, while urinating one hand is held cupped in front of the urine which is rubbed on tree branches. In both cases the scent marking is done to identify the home or area where they spend maximum time. Some lemurs (non - human primates found in Madagascar) mark their territory by smearing their faeces on branches, quite like painting the house.

Lions, tigers leopards, and all other big and small male cats mark their home range with pheromone which is passed with urine. The cats select some suitable objects, back up to it, extend the back legs, raise the tail, as the urine is passed, the tip of the penis is curved backward SO that the urine is given out as a fine jet between the hind legs. Bears use two different ways, they first mark a tree by scratching and chewing the bark, and in addition, they frequently urinate on such trees to consolidate the marking. All three species of hyena, striped, brown and spotted mark the areas of their movement with urine and faeces, they also have special anal glands for the production of pheromones for the same purpose. Use of pheromones through urine and faeces had its own limitations, that the animal has virtually no control over their production and release and this could have been one of the reasons which led to development of special glands to produce pheromones.

Use of Special Glands : Many species are not dependent on urine or faeces but have evolved special glands producing the scent, these have the advantage that their use can be restricted to necessity, that the animals could use pheromones as and when needed or production of pheromones remain under control, to prevent its automatic, unwanted release with urine and faeces.

Anal Gland: The commonest scent producing glands are anal glands. The usual procedure among animals in applying secretions is to evert the anal glands and rub it on the ground or on object to be marked. Marking of this type occurs mainly in Monotremes, Marsupials and in few Placentals.

The male beaver is tenacious about keeping strange beavers out of his neighbourhood, he virtually erects "No Trespassing" signs by pheromones on dams, lodges and feeding trails. He makes mounds by scooping up mud from his pond and on them he squats to deposit a smear from his anal scent gland. Spiny ant eater everts the cloaca and rubs it on the ground leaving a smelly mark. The brush tailed opossum and sugar glider have two types of cloacal glands with the help of which they mark their nests which is the most precious area for them. In the yellow footed marsupial mouse cloacal secretions are rabbed on twigs and branches around their small territory. The anal gland of black and red tamarin produces a strong smelling secretion which keeps others away..

Male tasmanian devil has scent glands around its anus which it rubs on objects like stones, branches, ground and grass to protect his female and young ones. In most mammals marking the territory is the job of the male but in few selected ones like, ring tailed lemur which are small sized, long tailed primates found in Madagascar, both male and female have scent glands in their fore arms and anus, both rub their glands to mark their territory, however, males have additional scent glands, under chin and by the penis which produce pheromones to attract females. Verreaux's sifakas also found in Madagascar mark territories with urine and with secretions of males' throat glands. Brown hare of Europe, normally lives solitary and marks its territory with strong smelling secretions produced from glands situated in its anal region, around its face and inside its cheeks. These secretions are also important in its mating season to attract the mate.

Salivary Gland: Saliva is a secretion available to all mammals, which has been used as a marking agent by bears, dogs, pigs and a number of other animals. Most marsupials deposit saliva on twigs by chewing them and hence marking the foliage of their territory. Lesser hedgehog first salivates on the object to be marked and then scratches it. A female rat licks its nipples to scent mark them so that its blind pups can reach them. A female wildebeest will lick its baby to scent mark it through its saliva for recognition.

Miscellaneous: There are many other glands via which the scent marking can be done. Roe deer of order Artiodactyla found in Europe and Northern Asia live in small groups, the stag marks its territory with secretions from glands on the forehead, anal and metacarpal glands and urine. Roe deers, black bucks and sambar found in India rub facial glands found just below the eyes against trees to mark them. Red brocket, found in Central and South America are small sized solitary deer, during breeding season, males mark territories by strong smelling substance produced from a gland situated at the base of the antlers, the females are attracted to this smell.

Thompson's gazelles found in Africa mark their territory by depositing scent on long twigs and grass, this scent is a tar like substance which is produced by glands situated just below their eyes. Bulls of elephants have temporal glands which produce a scent to indicate that they are musth i.e. ready to mate and looking for a cow elephant in breeding season. The dominant male reindeer have scent glands between their hind toes, which help them to leave scent trails for the remaining members of the herd. Ground squirrel and American porcupine mark their mates and territories by rubbing with sides of the head.

In marmot, the scent glands occur in the area between eye and ear. The secretions from these glands are used to mark their territory and mates. In Australian hopping mouse the scent glands are found below the chin which it uses to mark pups. In the camel the glands are situated on the neck which are rubbed against the object to be marked and plays an important role in mating. Black tailed deer has 5 different scent producing glands, located at various places, producing different pheromones to carry out different functions.

Functions of Scent in Vertebrates

Not everything is known about pheromones and their secret functions. Much remains unexplored, decoding of chemical messages has been a difficult task. More than often, the pheromones have been used as marking agents, and for mating purposes in most of the mammals. The knowledge about their utility in other functions remains obscure.

Roth-Kolar observed that for animals, any new object was a source of unease until it had been made safe by marking. It gives them a sense of belonging. In one instance, a sleeping box of a fox was removed for cleaning and returned after being thoroughly washed, the animal found it unfamiliar because all its body scent was washed out too. The animal stayed out all night and was found dead in the morning due to cold. The familiar smell gives a feeling of safety and sense of belonging by receiving its own body scent from its dwellings, it gives them a feeling that "this area belongs to me and I am safe here".

The scent increases the self confidence of the animal. Kuhme (1963) recorded that the African elephant, when facing a rival, during the breeding season will sometime turn his trunk round and smell his own temporal gland. This appeared to increase the animal's confidence and make him more likely to attack.

Scent plays an important role in sexual behaviour, the males mark their territories and ward, off the rivals which is most commonly seen in deer and antelopes or the rivals do not tresspass marked territory as seen in tigers and other cats. This behaviour can be easily observed among dogs, frequent urination by dogs, on all possible landmarks is a way of marking objects within their territory. Scent brings two sexes together in most of the vertebrates.

Many animals mark their trail just to leave a record of their comings and goings to be read by themselves and subsequent visitors as in hippo and rhino. These sign posts are regularly visited and inspected by neighbours and thus each animal can keep a check on his fellows without actually seeing them, a newcomer in the area will at once be detected and may be challenged. By marking the pathway animal does not loose path and conserves energy by not finding the way afresh every time it moves.

Scent is used for marking the territory, the territory is first acquired by fighting, then the territory is loaded with animal's scent indicating that no more fights are required for this particular territory. Sometimes males also mark their mates with scent e.g. male rabbit marks his female with his chin gland during mating. The male rabbit also marks the youngsters of his own group with his chin gland secretions. In certain animals, marking is often associated with a fight, hamsters mark a defeated rival and the marked animal subsequently retreats. Mice and rats similarly mark a defeated opponent with urine. A wildebeest sniffs her newborn calf to form a permanent olfactory bond with her offspring. The calf also sniffs mother for her odour. Many mammals, such as bighorn sheep, dogs often sniff and smell at each other and, when they do so, they can probably know the sex, the group it belongs to and its readiness for mating.

Role of Chemicals

Those chemicals which bring about coordination in the same species, are called pheromones. These are generally known as carriers of excitation, and the term itself means chemical communicator substance on the other hand endocrinology stresses the integration of individual through internal release of chemical messengers. But, now according to the literature, there are individuals which liberate materials externally to initiate the developmental or behavioural responses in the recipients. According to Karlsow and Luscher (1959), the term pheromones include such external secretion that serves to integrate conspecific individual of a group. According to Devenyort *(1955) "every symbiotic relationship including animals of different species may be influenced by similar animals."* In 1961 Parks and Bruce coined a new term known as exocrinology for these studies.

Different animals understand their own signals to communicate ideas. However, for this they also depend on chemical signals. The hunting dog reach their prey by smelling the earth. The female cat exhibit sexual play and give noise to such males which are not visible but are only heard. These are a few examples where we find chemicals as a means of communication. These types of chemical which spread or carries the information are known as semi-chemicals. These may be of two types :

Intra-specific : It is the communication occurring within the same species (pheromones)

Inter-specific : It is the communication occurring between the different species.

When the chemicals are used as inter-specific and it helps to the producers than it is named as Allomones, and when they are helpful to the receivers they are named as Kairomones.

Initially, the term pheromone was applied to sex attractants of insects but now it involves several kinds of agents released into the environment and their functions in all major groups to integrate members of the population. These could be absorbed by body surface, ingested or perceived by olfaction. They provoke specific behavioural, developmental or reproductive responses which from the ecology and survival of the species point of view are quite significant.

The chemical that stimulates gametogenesis are generally known as Termones, those which are responsible for coming together are referred as Gamones. Tyler, in his egg-sperm experiment used both these terms. During fertilization, the activation and agglutination of the sperms is brought about by the fertilization which diffuses from the egg. Although it is obvious that both termones and gamones are diffusion ectohormones. Since, in early literature, on the subject chemicals used in communication were usually referred to as 'Ectohormones' but since 1959 the term pheromone has been widely adopted.

Differences between Pheromones and Hormones : It has been observed sometimes that endocrine product may be active both with an individual (hormone) and between individuals of a colony (pheromones). According to Lischer 1963, the hormones which are secreted from the corpora allata of termites is dual in function. Still there are some differences between the two.

Hormone	*Pheromone*
1. It is the internal secretory product.	1. It is external secretory product.
2. It acts within the individual.	2. It acts between the individuals of colony or population.
3. It is not species specific.	3. It is species specific.

According to Barth (1962), while studying cockroach and Martom in guinea pig the production of pheromones by exocrine glands may be under endocrine control. Strauss and Pochi (1963) stated that the endogenic substance regulates the activity of human sebaceous glands.

Mode of Action

1. *Olfaction :* Sex attractants-insects and mammalian reproductive pheromones.
2. *Ingestion :* Queen substances secreted by mandibular gland of queen (honeybee) is aketodecanoic acid. It is ingested by the workers and act to inhibit the development of their ovaries.
3. *Contact or Absorption :* In lucosid spiders, the male's cutaneous gland secretes chemical substances which is absorbed by the skin of female and induces sexual stimulation.

Classification of Pheromones : Action pathway is the main basis of classification of pheromones.

1. Pheromones exert their influence by directly affecting the central nervous system of the recipient to produce reversible and rapid changes in behaviour. It is said to be releaser effect. For example —sex attractants, alarm pheromone and aggression pheromone.
2. Pheromones affect the central nervous system and exocrine (mostly pituitary gland and initiate a chain of physiological adjustments which eventually modifies the behavioural patterns. It is said to be primer effect and is quiet slow in reaction. For example — sex pheromones of invertebrates, queen substances, locust substances.

Behaviour of an organism is influenced by pheromones, either directly or indirectly. If a pheromone stimulates the recipient's central nervous system to produce an immediate change in behaviour then it is said to have a releaser effect. If it alters a set of long term physiological conditions so that the recipient's behaviour can subsequently be influenced by specific accessory stimuli, then it is said to have a primer effect.

Primer Pheromones : Those pheromones which have primer effect cause important physiological changes without an immediate accompanying behavioural response. L.M. Boot and S. Vander Lee the mammalian endocrinologist gave Lee Boot effect in the year 1955. According to this effect — "Female mouse placed in groups of four, show an increase in the percentage of pseudopregnencies". By removing the olfactory bulb of the mice or by housing the mice separably the normal pattern of reproduction can be restored. When more and more female mice are forced to live together their oestrous cycles become highly irregular and in most of the mice the cycle stops for long period. W.K. Whitten gave the Whitten effect according to which the odour of male mouse can initiate and synchronize the oestrous cycles of female mice. The odour of male also reduces the frequency of reproductive abnormalities arising when female mice are forced to live under crowded condition. The pheromone which is responsible for this effect is present in male urine. Castration eliminates this pheromone from the urine while testosterone therapy in castrated of either sex causes it to appear. The urinary factor for this reason is said to be the product of an androgen dependent tissue.

According to Miss Helen Bruce (National Institute for Medical Research London) odour of a single male mouse will block the pregnancy of a newly-impregnated female mouse. The odour of the original stud male of course leaves pregnancy undisturbed. Although the mode of action of mouse reproductive pheromone is not well-known but the evidences suggest that the odour of strange male suppress the secretion of hormone prolactin with the result corpus luteum (ovarian) fails to develop and normal oestrous is restored. This pheromone present in bladder urine free from accessory gland secretions but its identity and source remains unknown. In adult, due to the castration, pheromone is completely disappeared and never appears if castration is accomplished before puberty. In an ovariectomized female, a series of androgen injections are given then it acquires pregnancy blocking.

For about four days after coitus, mice remain vulnurable to the pregnancy blocking hormones, whereas the zygotes are developing in the oviducts. The blastocysts are moved into uteri but implantation is possible because the ovaries do not produce the hormone required to build up the bind of an endometrium essential for nidation. With the end of the pregnancy the female returns promptly to estrus, ovulates and again becomes sexually receptive. The stimulation provided by hypothalamus leads to the decrease in amount of prolactin secretion by the anterior pituitary, accompanied by failure of ovarian corpora allata to produce progesterone. Returning of estrus would include production of two other hormones from pituitary, FSH and LH in the following sequence.

Stimulated olfactory epithelium	→	Ordinary neurons
of CNS (Neurohumors)	→	Neurosecretary
Cell of hypothalamus (3 Neuro hormones as releasing factor)	→	3 Anterior
Pituitary gonadotrophins	→	Ovaries
(Progesteron and Oestrogen)	→	Responses to accessory sex organs and behavioural changes.

Releaser pheromone : Those pheromones which cause simple release effect, sex attractant constitute a large and important category.

Sex Attractant Pheromone

Pheromones are employed by a large number of insects bringing the sexes together for mating. These pheromones are called sex attractants. At present, two mammalian and four insect sex attractant pheromones are recognised. They are wide spread amongst the members of order Lepidoptera and are also known to occur in some of order Dictyoptera, Coleoptera, Hymenoptera and some other orders (Jacobson, 1972). In most of the cases, the pheromone is produced by the female to attract the male, while less frequently a male pheromone attract the female, or both sexes may be lured by the odour.

The exact role of sex attractant has never been rigorously established experimentally in living animals. In fact mammal seem to employ musk like compounds alone or in combination with other substances to serve functions to mark home ranges to assist in territorial defence and to identify the sexes.

There are four insect sex attractants, whose nature and role is well-known.

Around 12 mg of esters of Bombaykol is extracted from 250,000 female silk worm moths by Adolf F. J. Butenandt and Karlson et al. (Max plant institute of Biochemistry in Munich)

20 mg of Gyplure was obtained from 500,000 female Gypsy moths by Martin Jacobson, Morton Beroza and William zones (U.S. Department of Agriculture)

Bombaykol and Gyplure obtained by extracts of insects and separation is done by chromatographic method.

Another techniques were applied on American cockroach by T. lamamoto (U.S. Deptt. of Agriculture). He put virgin females in metal canes house and air was continuously drawn through the canes and passed through chilled containers to condense any vaporized materials. As a result, he obtained 12.2 mg of materials from 10,000 females, by applying the above technique.

Insect attractant pheromones possess a great power of diffusion. If we consider both volatility and diffusion then it could be estimated that the threshold concentration is no more than a few hundred molecules per cubic cm. If .01 micro gram of Gyplure (approximate of one female) is distributed with maximum efficiency it excite more than billions male moths.

Iseschwinck (Munich university) did some experiment to study how the male moth gets the source of tinting ? He observed that when male moths are activated by the pheromones they simply fly up wind and thus inevitably more towards female. If by accident they pass out of active zone they either abandon the search of fly about at random until they approach the female, there is a slight increase in concentration of pheromone and this serves as a guide for remaining distance.

Aphrodisiac Phe'omone : This term is used with respect to pheromones which facilitate copulation after the two sexes have come together. In Danaidae the males are visually attracted to the females and fly in pursuit. When a male Danaus gillippus overtakes a female he extrudes his hair pencils and dusts her with pheromone transfer particles. These carry the pheromone transfer particle. These carry the particles which stimulate via the olfactory sensilla of the antennae and induce the female to land. The male may then hover over the female, continuing to dust her with pheromone particle, until, finally he lands and copulates.

In female cockroaches, sex attractant pheromones are produced from one or more anal, sternal and tergal glands. In Periplaneta Americana the oocyte development is controlled by copora allata so the concomitant endocrine control of pheromone production provides a mating signal during correct period i.e., when ootheca is being formed and extruded. With Pycnocelus surenemensis also similar results were obtained.

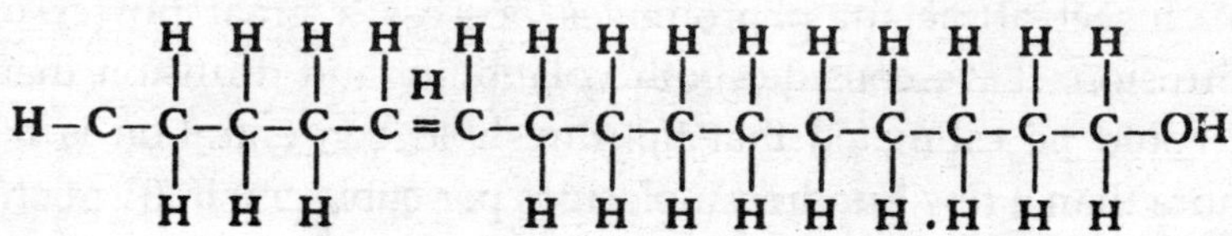

Bombay Kol (Silk Worm Moth) $C_{15}H_{30}O$

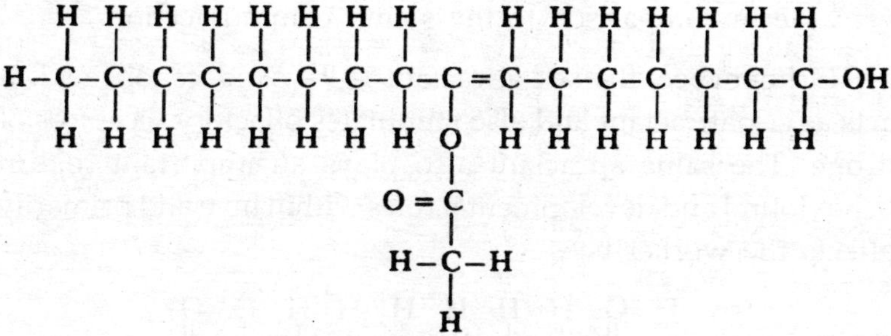

Gyplure (Gypsy moth) $C_{18}H_{35}O_3$

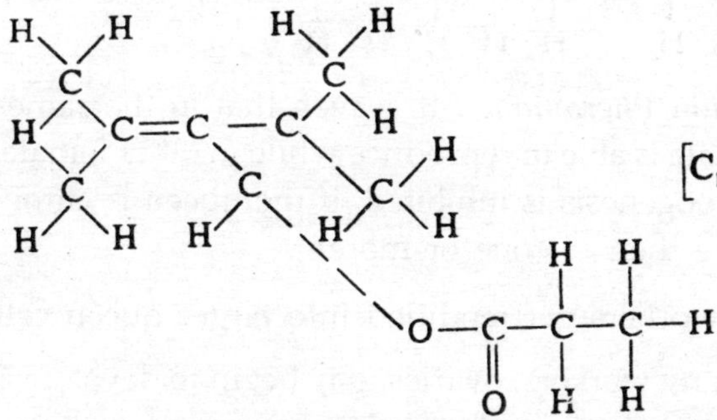

$[C_{11}H_{35}O_2]$

2,2-Dimethyl 3-Isopropylidencycloropyle propinate (American Cockroach)

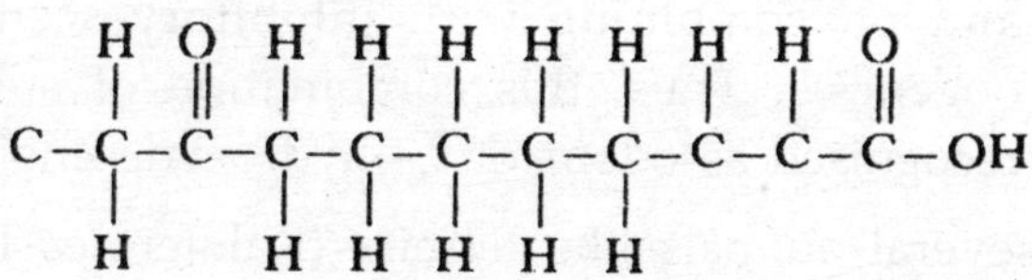

The adult endocrine system in Bombyx mori, Antheraea pernyle (the oak silkworm) Galleria mellonella (wax moth), etc. controls neither sex pheromone production nor oocyte development. So it is obvious that only those insects which have regularly repeated reproductive cycles, with intermittent periods when mating is possible, will bear an endocrine mechanism coordinating both events.

In most of the orthopterans including both male and female, a stribulatory song is produced which aids in auditory communications process. Although the chemical signal is more versatile in comparison to the sound communication.

In honeybee, the sex attractant is 9—oxodecenoic acid. It acts as sex attractant and also stimulates olfactory receptors of drone. The same attractant also plays an important role in behavioural and developmental roles within hive and primarily affects the worker bees.

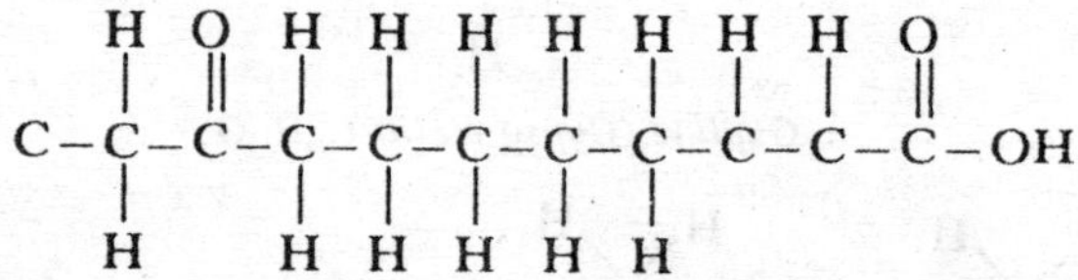

Maturation Pheromone : It is seen that in the same time only one female is able to reproduce while in other females i.e., workers the oogenesis is inhibited. If the queen is eliminated from the hive then — (one or more)

(i) The workers get modified into larger queen cell.

(ii) In many workers, ovaries may begin to develop. There is thus behavioural and developmental responses by the worker, to the disappearance of their queen. Queen substance is mainly made up of 9 —oxodecenoic acid and on combining with inhibitory scent inhibits oogenesis. Thus, this combination of substance is recognised as odour of 9—hydroxydecenoic acid.

In several animals like termite (Kalotermes flavicollis), desert locust (Schistocerca gregaria), etc. maturation pheromone activity could be observed.

Sources of Pheromones : Pheromone producing animals like bees, wasps, ants, termits, etc. are provided with reservoir.

In all kinds of ants, a well-developed exocrine glandular system is present. All those prominent glands whose functions were earlier considered as mystery by entomologists have now been recognised as source of pheromones.

Trail Substances : Trail substance is secreted by the Dufour's gland via the sting of insect. It is a limited guide for the worker ants. It helps as a guide to worker in search of nest or food. Trail substance is a volatile substance and the range of its activity is about two minutes. The artificial trail can be made by using the extract of Dufour's gland of freshly killed worker.

Alarm Substance : They are produced when a social insect is threatened in any way. The response varies according to the species, but commonly includes attraction of other workers or soldiers and the adoption of aggressive postures with the head raised, and jaws wide apart. To the human nose, the alarm substance is mild or pleasant, but to the ant they represent an urgent toxin that can propel a colony into violent and instant action. Sometimes the pheromone also functions as a defensive substance. This is the case with the formic acid produced by Formica, but in most cases, defensive substances and pheromones are quite different. The alarm pheromones of the bee Trigona are also produced by the mandibular glands. At least some of the chemicals which are used in trail laying also function in producing alarm, the nature of the response depends upon the situation.

Pheromones Among Human : It is quiet difficult to detect the primer pheromones since they can affect the endocrine system without producing overt specific behavioural responses. Striking sexual differences have been observed in the ability of humans to small certain substances.

Some Examples of Pheromones — Generally, the pheromone constituents are derivatives of fatty acid. They are conveyed from individual to individual by several ways. In olfactory, active substance air is required as medium. The alarm substance released from the skin of an injured fish requires water as medium. Orally active pheromone requires hormones like factors among colonial insects, termites substances conveyed by faeces.

Following four alarm substances are now analysed.

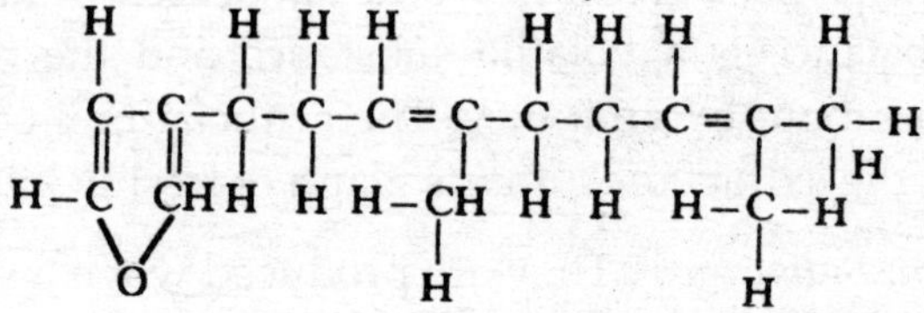

Dendrolasin (Lasius Fuliginosus) $C_{15}H_{23}O$

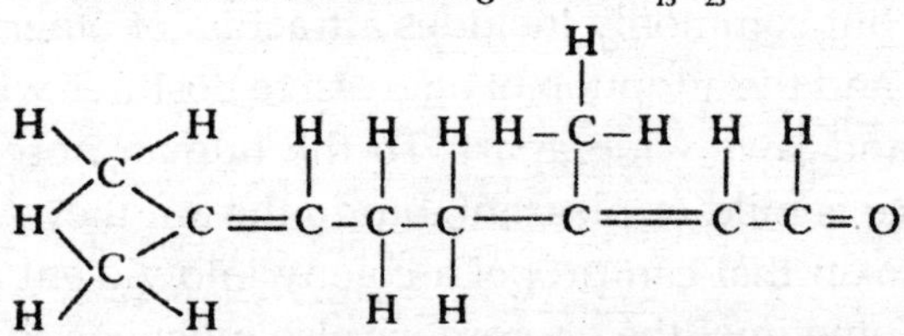

Citral [Atla Sexdens] $C_{10}H_{16}O$

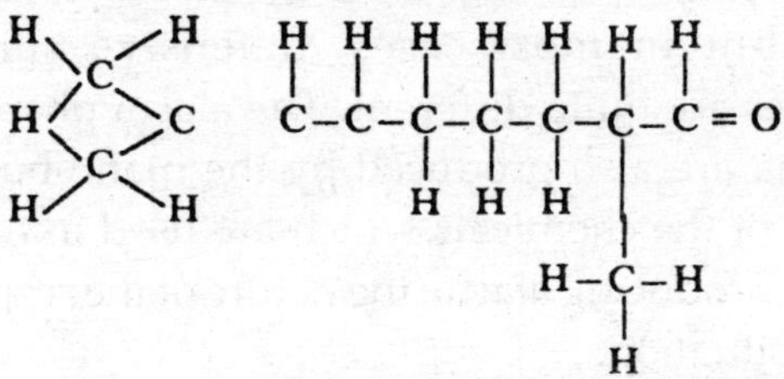

Citronellal [Acanthomyops Claviger] $C_{10}H_{14}O$

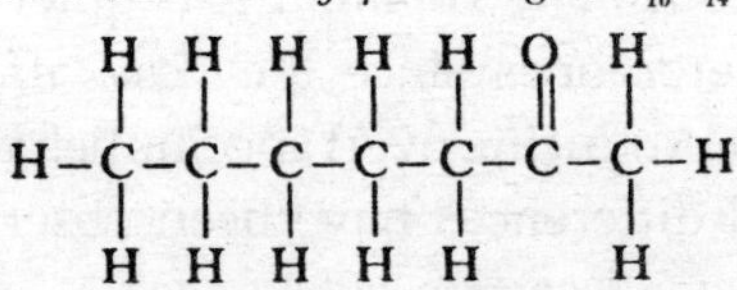

2. Heptanone (Iridomyrmex Pruinosus) $C_7H_{14}O$

According to J. Le Magnen (French), the odour of exaltolide perceives clearly by sexually mature at the time of ovulation. Males and young girls were found to be relatively insensitive but a male subject becomes more sensitive following an injection of oestrogen.

Exaltolide is used commercially as a perfume fixative. Lemagen also reported that the ability of his subjects to detect the odour of certain steroids paralleled that of their ability to smell exallolide.

Showing Some Examples of Pheromones Among Insects

Species	*Manner of effect*	*Source*	*Function*	*Chemistry*
Bombyx raori (silk worm moth)	Olfactory "	Lateral gland at the tip of abdomen of female.	Attract male of same species releases male sex behaviour	$C_9H_{30}O$
Porthetria dispar	"	"	—	$C_{16}H_{34}O_2$ 10-acetoxy 1-hydroxy cis 7-hexadecone
Periplaneta Americana (cockroach)	"	Pouch like organ in female	Attract male of same species releases male sex behaviour	?
Belostoma indica (tropical bug)	"	Special ducts in male.	Sex attractant ?	2-Hexenol acetate
Formic rufa (ant)	"	—	Marking scent to establish trails	Formic acid
Pogonomyrmex badius (ant)	"	Mandibular gland	"Schrecepst off" communicates alarm.	?
Apis mellifera (honeybee)	Oral	Head gland of queen bee.	Queen bee substance inhibits ovarian development in workers and prevents development of additional queen.	$C_{10}H_{16}O_3$ 9-keto, 2-decenoic acid
Kalotermcs flavitollis (termites)	"	Head of functional female.	Inhibits development of supplementary reproductives possibly through	?

These observations hardly represent a case for the existence of human pheromones.

Adverse Effects of Pheromones

Pheromones may act to advertise the presence of an insect to another individual of the same species. They may also advertise the insect to potential parasites and perdators and a few cases are known of such insects which respond to host pheromones. The clerid beetle, Thahasimus dubuis is a predator of Dendroctomus. It locates its host by responding to the pheromone frontalin, so that large number of predators arrive at the same time as the Dendroctonus. This response also brings the two sexes close together in order to facilitate mating.

12

Biological Clocks

You must have all experienced that you tend to wake up almost at the same time everyday in the morning and you start feeling sleepy at certain time as the night falls or why do we so often get up before the alarm is due to sound? Because there is a built in clock inside your body, which regulates sleep-wake timings. It has now been established that all organisms possess an innate biological clock (Pittern drigh, 1960).

Rhythms of various kinds have been in evidence in the biological world for quite sometime. It was during Great Alexander's march to India that one of his officers Androsthenes noted a daily leaf movement among the plants of Tamarind. About 300 years before Aristotle, men first wrote about plants which raise their leaves to the sky during the day and fold at night. Charles Darwin noted rhythms in many animals and wrote in 1880 that the periodicity among plants and animals was to a certain extent inherited. In animals rhythms were observed less as compared to plants because an animal's daily routine is more complex than that of a plant. Rhythmic events occur every wherein the environment, the sun rises and sets forming days and nights, the moon moves through its phases, the tides get high and low regularly, constituting lunar and tidal rhythms respectively.

The orbit around the sun and rotation of earth creates seasons. All the living organisms are subjected to these rhythms, and they exert great influence on them. Animals must regulate their activities in concurrence with these cycles in order to avoid adverse conditions and to get enough food and shelter. Now it is Well known that daily rhythms form the basic feature of living organisms - from Protozoa to Primates. The Paramecium will congregate for reproduction only at certain times of the day, animals requiring warm temperatures, will become active during the middle of the day. The birds sing mostly in the mornings and evenings; owls, hyenas, jackals, and porcupines are active at night, whereas deers, monkeys and apes are active during the day time. People have always wondered on the sense of time exhibited by birds, bees, worms and plants in their day to day behaviour. A built in clock makes the owls, rats and cockroaches sleep during the day, bats come out of their roost to forage in the evenings, the whole day they remain upside down hanging by the branches; parrots, mynas, squirrels, house sparrows would sleep by night, the bean plants fold their leaves during the night and open in the day. Cold blooded animals remain active in the day time during winters and during mornings, evenings in summers; during extreme cold or heat they hide at appropriate places. You must have noticed that your pets specially dogs and cats lose more hair during summers, in winters their fur coat gets thicker. Domestic sheeps have an annual cycle of wool production, rodents, house lizards, squirrels exhibit cycles of hibernation and activity.

Their body weight is maximum when they go into hibernation because they have to depend on reserve food material and the weight is minimum when they wake up.

Birds and other animals migrate at particular times in an year. Behaviour exhibits a marked periodicity in its pattern in performance, and this periodicity involves complex interactions between nervous and endocrine systems and outside world.

In earlier times the most outstanding work on biological clocks was done by Erwin Bunning, in 1936 he proposed that organisms have an inbuilt and nearly automatic rhythm which repeats itself at about 24 hours interval and these rhythms are used to measure time.

Biological clocks are also known as internal clocks or circadian rhythms these rhythms are the internal mechanisms that provide a means of measuring length of time to the living beings. These rhythms allow the animals to adapt their behaviour in the best possible way to the course of a day's changing event. By setting the internal clock, the plants clock proves itself when its petals open at dawn and close at dusk. The common example around us is of a flower called as office flower or morning glory, or Portulacca which opens when the sun comes up and closes when sun sets. The birds leave for their migration at a predetermined time to arrive at the destination at a suitable time. A dormouse cannot withstand extreme of cold weather, hence it goes into hibernation but comes out for food every three weeks, that too always near sunset. A human being's clock becomes evident after we fly to another continent, through changing time zones; we yawn through lunch, go to sleep when it is time for evening tea, wake up at two in the middle of the night and fall off to sleep at 8.00 am, but after few days the biological clock resets according to the new environment, the time taken in resetting the clock may vary from person to person.

It has been a subject of interest whether periodicity in an animal is determined by external environmental stimuli (exogenous) or controlled through physiological factors within the animal itself (endogenous). Both the stimuli are important.

Their intimate relationship was brought to notice as early as 1960 by Lorenz, neither exogenous nor endogenous stimuli are solely responsible in controlling behaviour through biological clocks, it is the intricate interaction between the two together. Initially, to set the clock and run it, external stimuli like day length, lunar phases, tides, and temperature play the most important role but after some time the clock is maintained and run by internal stimuli controlled by body physiology. Experiments have shown that many animals maintained their biorhythms even when the relevant external stimuli were either removed artificially or when organisms were displaced from their habitat and placed in conditions of constant light intensity, temperature, pressure, humidity, and chemical composition, they continued to exhibit nearly the same rhythm of behaviour they showed in their natural habitat. In other words, animals possess internal clock that controls the rhythm in behaviour, and keep it going with the help of internal stimuli while external stimuli serve only to set and reset this clock periodically; such external environmental stimuli are called Zeitgebers or Entrianers because they entrain internal clocks and the process is called entrainment, an individual's clock can be re-entrained if placed in a new time zone, as happens with rapid, long distance trams-continental travel.

During evolution, animals have acquired a variety of endogenous rhythms, their periods are matched with rhythmic events in the environment. Some rhythms are matched to 24 hour cycle of light and dark (circadian), other to the 12.4 or 24.8 hour tide cycle (circatidal), the 29 day lunar cycle (circalunar) the yearly seasons (circa-annular) or the time between successive spring low waters (14.7 days; semilunar or circasyzygic). Prefixing word circa (which means about) with all kinds of biological rhythms is very necessary because all these internal clocks are only approximately matched e.g. 24 hour cycle or rhythm does not mean 24 hours by wrist watch, it means about 24 hour or circadian.

Circannual Clocks

The activity of animals or plants, influenced by the seasons occurring once in an year remain under the control of circannual clocks. People who have been interested in gardening know that some flowers bloom only in summers while others like pansy, calendula, candituft and carnation bloom during winters because flowering in these plants is influenced by day length, temperature and humidity. Migration among birds and other animals to escape harsh climatic conditions and in search of food is a popular example. Migratory birds undertaking long distance migrations every year, show cyclic changes in their body weight, gonad size, plumage and colouration.

The larvae of many insects viz parasitic wasp, flesh fly show two phases every year (i) active (when the insect executes work actively) and (ii) inactive (diapause: a period of arrested growth and development in insects which is under the control of endocrine system), it is an adaptation to avoid adverse condition, but does not automatically end with the termination of the adverse conditions, it is genetically determined and can be broken by an appropriate environmental change or by a biological clock. Dormancy state in which warm blooded animals reduce their metabolic activities to a minimum level during unfavourable conditions so as to survive until conditions improve is also a cycle pattern, e.g. ruffed grouse do not migrate, instead, they bury themselves under the snow during the winter to lead a dormant life; several other birds, viz. small passerines also bury themselves in similar way. Humming birds move into caves from high Andes to escape below freezing temperature in winters.

This cyclic movement from Andes to caves and back remains under the control of circannual clock. Long term torpor a state of inactivity, similar to dormancy, into which some warm blooded animals (birds and mammals) enter in order to avoid adverse conditions and to conserve energy, it can be either long term or short term, is also influenced by circannual rhythms (Bartholomew, 1972). All the behaviours are tied to a clock so perfectly and the condition are so predictable that animals go in dormancy in anticipation. Hibernation and aestivation are also states of inactivity into which some cold blooded animals like frogs, toads, lizards and snakes enter in order to avoid cold (hibernation) and heat (aestivation). The timely escape from harsh weather conditions is achieved by circannual clock.

Monarch butterflies of northern states of USA and Canada show circa-annular rhythmicity in their annual migration to and from Mexico. From USA, monarchs migrate in millions every year in September covering about 3,200 km, they pass their winter in state of semi-hibernation on trees and in March they fly back. There is a small jungle near San Fransisco, known all over the world for its butterfly trees, because these trees remain covered by monarch butterflies from October to April. American thirteen lined ground squirrel generally begins to hibernate in October and emerges from its sleep during the following April. This annual rhythm of hibernation could be seen even when external conditions were kept same for the squirrel in laboratory conditions. Almost everyone has witnessed that the common Indian house lizard disappears during winters. Fish, bird, and whale migration, movements of thousands of wildebeests every year in Africa and annual reproductive cycles in large vertebrates are some of the common examples of circannual rhythms. It is true that activities of many animals are synchronized with exogenous stimuli but what happens when animals are kept in unchanging conditions?

In an experiment millow warblers were subjected to constant twelve hour day for a period of fifteen months. These birds moulted twice in the year as they would have normally done and showed migratory restlessness at the usual time of the year (just like other free living' conspecifics). It must be noted that even under such constant experimental conditions, the periodicity was maintained by warblers for three years, after that it -got' disturbed and a readjusted itself according to the changed stimuli.

Circatidal Clocks

The biological rhythm synchronized with the low and high tides (the alternate rise and fall of the sea due to gravitational pull exerted by moon) in the sea are called circatidal clocks, the organisms living in the intertidal zone of the seashore are alternately submerged in water and exposed to air, due to this, pressure, salinity, food supply, temperature and predation risks change too, therefore, animals inhabiting tidal area show behavioural periodicity associated with the tides e.g. some of the Planarians and Polychaetes which make burrows in shallow seawater or in tidal areas show vertical movement or movement out of their burrow as the high tide approaches; bivalve Molluscs increase filteration rates when high tide brings in more food; the beautiful column like Coelentrate, sea anemone increases the rate of body expansion and contraction during high tide, waving its tentacles actively to trap more and more food. It is noted that some animals get more active when the high tide approaches and some go deep into their burrows and will come out only after the tide has receded, it can lead to hide and seek between prey and predator, if the predators are coming out of the burrow it is beneficial for the possible prey to go deep down into the burrow. Significant work has been done by the marine biologists of California, on circatidal rhythms.

It was found out in one of the experiments, that sand beach hopper dig and make a burrow during the low tide and come up to swim at high tide, contrary to this, odd looking fiddler crabs leave their burrow when the tide has receded in search of getting food which the high tide must have brought and go back to the burrow when its time for high tide to flood the shore. At the same Californian coast, hundreds of small silver fish Grunion are brought to the coast alongwith a high tide. During their breeding season, the fish are washed on to the moist sand, where they spawn and then flip back toward water.

The time sense of laying eggs is so precise that their eggs hatch in fifteen days when the next high tide arrives, as the wave lashes against the sandy beach the baby fish are freed from the egg pods and start their journey into sea. A small crab, Synchellidium found in California inhabits the tidal zone, when the tide is low, these crabs bury themselves deep inside the wet sand, and during a high tide, when there is water, they come up to get food. Once this movement gets synchronized using exogenous stimuli, in this case, high tide, it was observed that crabs came out regularly in anticipation even in the absence of water which was stopped by putting barriers. There are air-breathing crabs which live on the seashores, they predict the arrival of high tide and start building a shelter in anticipation against the incoming water and predatory fishes, the soldier crab of tropical Asia starts forming a shallow crater, then pushes pellets of wet, muddy sand up on either side until it has built a roof over its head, the crab then burrows deeper, loosening more sand and plastering it overhead to form a sturdy roof, the shelter so built provides life giving inside, all those activities of this crab are intimately linked with tides and are controlled by circatidal clocks.

Circasynodic Clocks

Biological clocks which are synchronized with the moon phases are called circalunar or circasynodic clocks. The fantasized stories about warewolves have been known to many, that some persons change into a wolf and behave like animal on full moon nights, though this is truly a fiction, but in nature there are many animals specially marine invertebrates who transform into entirely different forms depending upon the phases of moon and perform particular behaviours, e.g. a marine Polychaete Platynereis dumerii is a long worm like creature, which lives in a burrow, on the, lateral sides of its body are flap 'like projections called parapodia, during their breeding season this seaworm becomes brilliantly colourful, gets bigger in size, its parapodia, eyes and tentacles also get enlarged, and it assumes a sexual form called heteronereis, which swarm actively at seashores during the full moon, perform nuptial dance and spawn. In the beginning platynereis and heteronereis were considered as two different species, because they looked so different latter it was realised that they are the non-sexual and sexual forms of one species.

The time bound activities of animals have been known from time immemorial. People of Fiji islands relish eating marine worms called palolo (Eunice viridis), which is a deep sea creature extremely difficult to catch but on the first day of the 1st quarter of the moon in November, the palolo worms swim to the ocean surface in large numbers, they swarm and spawn. Just before sunrise people get into their boats, over the coral reefs where the worms collect, it is a bonanza for people of Fiji who strain worms with large nets. Some terrestrial animals also show circalunar rhythms. Pit building activity in antlions (Myrmeleon obscurus) also reaches a peak around full moon.

Circasyzygic Clocks

The biological rhythms which are synchronized with the fortnightly cycle of spring tide (high tide occurring a day or two after the new or the full moon) and neap tide (low tide which occurs in the middle of the second and fourth quarters of the moon) are known as semilunar clocks e.g. pariwinkle, Littorina rudis shows a marked 14.5 day periodicity in its locomotory activity; the species lives high up on the shore and is only covered by the high water of spring tides when it comes out of its burrow to move around and feed. Aplysia, a mollusc also shows distinct 15 day periodicity in egg laying behaviour.

Circadian Clocks

Operating on an approximately 24 hour basis, they are the most studied and best understood biological clocks and are found in almost all the major taxonomic groups of animal kingdom. In their simplest form they are reflected in the alternating periods of activity and sleep which correlates with the light/dark cycle. In this cycle some animals remain most active at sunrise and sunset times, such animals are called crepuscular, there are animals who are active during the nights (nocturnal) but most animals are active during the day time, called diurnal. In an experiment, a flying squirrel was taught to turn a wheel in the day time and rest during the night, after that the animal was kept in continuous darkness. The observations showed the remarkable precision in the onset and cessation of wheel turning even in the complete absence of the environmental cues to which it was related.

Circadian or daily rhythm is best illustrated by the activity of bees. In an experiment bees were trained to visit a food dish at a certain time of the day, say between 12.00 to 2.00 in the afternoon when the sun shines to its maximum and temperature is also high; large number of bees learned to go to food dish between this time and they kept visiting the food dish roughly between this time even when there was an overcast, and the temperature was also lower; the bees worked according to an inbuilt clock. Some intertidal crabs exhibit cyclic variation in their shell colour. They are lighter during the day time and get darker as the night falls, which prevents their detection from predators. The internal clock of bees did not correspond exactly to 24 hours, it was little shorter than 24 hours. Therefore, daily biological clocks are called circadian rhythms (i.e. approximate rhythms). Welly (1962) noted that snow buntings spend part of the day under snow during extreme cold. Bats, rodents show daily torpor under extremely unfavourable conditions of climate or food. Black capped chickadees can decrease their body temperature as much as 7°C during the night, thus reducing their energy demands by 60% as the temperature falls in winters; nocturnal moths start shivering to generate heat. Pupfish become more active in the water having 42°C temperature. Some animals show age dependent rhythmicity. Circadian rhythms may change with the change of place, they may also change with age, e.g. young wood chucks remain active during the evenings only, while their adults, remain active almost throughout the day. In human beings also the sleeping cycle gets shorter with age. Young badgers remain most active during the early hours to the middle of the day, as they grow older, their activity time is shifted from evening to night (Eibl Eibesfeldt, 1950). Similarly differences in the daily activity rhythms of two species occupying the same habitat play important function in reducing their competition, for example fly catching birds and bats never come in direct competition for the same food because birds are diurnal and bats are nocturnal.

Not only behaviour, the biological clocks also synchronize physiological activities with exogenous factors e.g. the respiration rate of animals is not constant throughout the day, even a small, slowly wriggling creature like earthworm shows a diurnal rhythm in oxygen consumption, peak of oxygen consumption in a twenty four hour period in Lampito mauritii a species of earthworm found in India is associated with its peak activity time (Kulkarni, 1984), the animal was more active in early mornings when there was more oxygen in the environment. Patra (1984) studied circadian rhythm in oxygen consumption in three air breathing fishes anabas, clarias and channa, and observed a clear rhythm in their metabolism, these fishes were most active during the morning hours, when they moved maximum in search of food and from afternoon onwards they became inactive, or remained submerged in water, in other words these fishes respired more air in mornings and depended on dissolved oxygen afternoon onwards, this rhythmicity is closely interrelated with the diurnal fluctuation of dissolved oxygen. The pure physiological or biochemical reactions within the body of living organisms show rhythmicity. Vijaylakshmi (1984) studied synchronised shifting of liver glycogen and ATPase activity under varied light regimen in Albino rats. She maintained rats under normal 12 hours light and 12 hours darkness, then in reverse order i.e. first 12 hours dark and later 12 hours light. She maintained a set of rats under constant light and another set under constant darkness. In the last two situations rhythmic trend in glycogen, glucose and energy linked enzyme ATPase level showed entrainment. Her study indicated the interdependence of the metabolic and related enzyme activity under varied photoperiodism.

By setting the biological clock animals learn to predict environmental events, for example, a nocturnal animal will stop hunting and begin its journey back to den to reach there before the day breaks, by doing so it avoids its own detection by other potential predators, the other obvious example of prediciting ecological changes is seen in the time of onset of breeding, the animals have timed their breeding time in such a way that the young ones are born when it is time for maximum food abundance.

Animals show remarkable time sense; in Panama, pile at ed tinamou is a fascinating bird, it looks like a tiny ostrich, this bird sings every three hours, day and night, in Panama it is called three hour bird. Honeybees have been known to possess marvellous time memory, Auguste Foral, a Swiss naturalist, made interesting observations that every morning when he had his breakfast on the terrace of his house, honeybees came to his table to feed on fruits and jams, their time of arrival was always same every day.

In some cases synchronization in rhythmicity of certain plants and animals has also been noted. A German scientist, Von Butt el Reepen observed that the bees stayed in his farm around the flowers until about the middle of the day and then left; he further found out that the flower retained same fragrance all through the day, so the bees were not attracted to scent but to nectar which was produced till the afternoon. He gave a new word Zeitgedachtnis (time sense) to express his idea of animals having internal cocks.

Translocation experiments have helped scientists to study the relationship between exogenous and endogenous stimuli. In the eighteenth century French astronomer De Mairan, observed leaf movements in certain plant species with relation to light intensity, he transferred this plant to a dark cellar, to his amazement leaf movements continued for a very long time even in the total darkness. This simple experiment led him to conclude that leaf movements which deemed totally dependent on light intensity were also controlled by an endogenous mechanism which gets them going.

An interesting translocation experiment was carried out in 1960 by an Indian scientist Sirohi together with two US professors Hammer and Hoshizaki. These three scientists went to Antarctica and carried along with them few hamsters, bean plants, fruit flies and fungi to study rhythmicity in their behaviour. These organisms were placed on a turn table which rotated in such a way that it nullified the rotation of earth which is minimum at poles, the organisms were kept in constant darkness and at constant temperature. The plants' and animals showed their usual rhythmic behaviour at the pole and there was no influence of earth's rotation either. Hamsters kept their 12 hour sleep wake schedule, bean plants showed 12 hour schedule, the eggs of fruit flies hatched every 24 hours and fungi in petri dish showed light and dark rings of growth every 24 hours. These experiments indicated that once the clocks are set they can go on and on without getting disturbed;

but this would not be a permanent feature, depending upon the animal type, species and its behaviour, the clocks will get reset. Circadian rhythms which persists under constant conditions are considered as evidence for an endogenous control mechanism.

As mentioned briefly earlier in this chapter the most obviously recognizable daily rhythm in human beings is reflected in "Jet lag" which happens when we cross quickly from one time zone into another sitting in a boeing. Our bodies keep on working according to the time schedule of the zone we have left, and the "clock" will not be "reset" for five to seven days, e.g. there is almost twelve hours difference between New Delhi and New York, when it is 10 am in Delhi, it's 10 pm in New York (of the previous day), after undertaking this journey the day and night are reversed, and the one who travels, would be wide awake in the middle of the night and would feel sleepy during the day. However, this effect is temporary, till the biological clock is reset or entrained. Brown (1970) demonstrated that animals and humans can generate periods other than 24 hours while using a 24 hours clock, for example, a man can get up very late than his usual time on weekends, while still maintaining his usual sleep wake clock for week days, this is called autophasing. Human body shows clear circadian rhythm in many of its physiological functions too. The body temperature remains high at day time and goes down by night, the blood is less acidic in day time, the adrenals secrete smaller amounts of corticosteroids by night, and the kidneys excrete less urine during night. Many other activities demonstrating rhythmic nature are pulse rate, blood pressure and oxygen uptake in lungs. The intestinal cells when grown in a culture, reveal that they follow the same rhythm of growth, maturation and juice production as the other intestinal cells still within the body forming part of the whole organism.

Understanding of the biological clocks has enabled man to find a variety of its applications, the hunters know when are the animals going to be up and active; the fisherman knows the exact time to harvest marine animals specially the crabs, grunion, palolo and other worms, farmer know what fruit plants will grow in which season, or what are summer and winter flowers and vegetables. There are other more direct useful applications to human beings e.g. a lady in New York, USA was unable to sleep until about 3 am and then would feel sleepy during office hours, she had to rely on sleeping pills, it was discovered that her biological clock was totally out of harmony with most of us and she was unable to keep the same schedule as we do. She couldn't sleep earlier than 3 am, the doctors asked her to be awake till 6 am, by then she was very sleepy; two days after she was asked to sleep not before 9 am, like this, in a month's time she reached 9 pm, this experiment worked well on her and she started sleeping from 9 pm to 6 am, by shifting the bed time, doctors helped her to reset her biological clock.

The internal clocks, may it be daily, tidal, or lunar deviate little from the periodicity of respective environmental rhythm these internal clocks go either fast or slow. Hence, they have to be reset or re entrained from time to time. The exogenous stimuli are important to start/set and reset the internal clock and once it is done, the clock functions automatically by internal stimuli e.g. initially a crab living in a burrow on a seashore learns to time the tide, it leaves its deep down burrow to come upto the surface when it is filled with high tide water but gradually it learns to leave the burrow well in advance so that, by the time it reaches the surface, the tide also hits the shore. This simply means that high tide is an external stimulus or an exogenous entrainer to set the circatidal rhythm in crab and once the clock is set the internal stimuli anticipate the time of the tide and animal reaches the surface in anticipation. Similarly, in many animals length of the day acts as an important exogenous entrainer, it is useful initially to start the clock and later on, to reset the clock; once set and started the internal clocks can function independently of length of the day, e.g. most of the birds will start chirping at 5.30 am when sun rises and would settle down on trees by 7.30 pm when the sunsets in summers, even if there is an overcast and one can hardly see rising sun the birds would start fluttering almost at 5.30 am.

There are questions which remain unanswered regarding biological clocks like, is the property of having internal clocks transmitted through genes, or it is learned at an early stage of development through conspecifics? The single-celled primitive forms responsible for so much of the sea's phosphorescence in the tropics have a 24 hour rhythm to their luminosity. They are bright for 12 hours, and then relatively subdued for the remaining 12 hours of the day. At first, it was presumed that such an action was controlled by sunlight or by temperature but hundreds of generations of these marine algae were kept in constant darkness and they still followed their original schedule of 12 hours bright and 12 hours dim; from such an experiment it would seem that the periodicity is inherited at least in lower animals. Another experiment conducted by K. Hoffman (1977) supported this fording, he took three groups of unhatched lizard eggs. These were incubated in controlled conditions of light viz 24 hour day (12 hrs. light - 12 hrs. dark); 18 hours day (9 hrs. light 9 hrs. dark) and 36 hour day (18 hrs. light - 18 hrs. dark). After hatching, young lizards were placed under similar conditions; and members of all three groups showed a circadian activity very close to 24 hour day indicating the innate or genetic control over biological clocks even in higher animals. Aschoff (1963) wrote that in most of the animals studied for rhythms, the clocks seemed to be innate or inherited. Animals which were reared entirely under controlled unchanging conditions showed the presence of an internal clock which remained arythmic and on the slightest exposure to environmental stimuli became rhythmic. For example, drosophila emerge from their pupal cases at dawn and start their 24 hour rhythm. If their eggs are kept under constant lab conditions they will hatch at any time of the day; in an experiment drosophila were kept under constant conditions, for 25 generations, with each generation hatching randomly throughout the day, in one of the generations a light was flashed on eggs, all these eggs hatched 24 hours later, only the single stimulus was enough to synchronize the clocks in this generation of drosophila.

Length of the day or light hours can influence important activities like metamorphosis, which are necessary for survival among animals. The larvae of Sarcophaga and Nasonia develop into pupae and then adults after 14 hours exposure to light. If this period is shortened then the pupa enters into a period of dormancy or diapause for sometime, till it is exposed to desirable day length; on the other hand pupae of many insects hatch during the cooler parts of the day (mornings or evenings), to avoid heat, because their skin is wet and soft, and if it dries up quickly the young insect may even die, therefore, the cooler hours of the day, when the humidity is high and external temperature is low, are favoured so that the skin dries up gradually to become hard.

Biological clock of endoparasites is set according to the daily rhythm of their host animal and the vector. For example, lervae of Wuchereria bancroftii which live in the lungs of human beings, move to peripheral blood vessels below the skin, from where they are sucked by the mosquito, the vector. In Africa, the larvae are transmitted by a nocturnally active mosquito, therefore, the larvae move towards the peripheral blood vessels at night, but in Fiji Islands the carrier mosquito is diurnally active, hence here the Wuchereria larvae travel to the peripheral blood vessels during the day time. A simple organism like Plasmodium has a highly accurate biological clock. It has precise timings of its recurrent attacks, which are generally multiples of 24 hours. How does this parasite, which lives in total darkness, manage precise timings ? It was found out, that if the host is made to sleep by day and wake up by night, i.e. if the host's normal 24 hour rhythm is inverted or reversed the biological clock of malarial parasite is also inverted, indicating close relation between the two rhythms.

Two main theories about biological clock mechanism have existed (Smith, 1973). One is "The pendulum theory" which suggests that an organism is an independent oscillator, with its own intrinsic timing equalling one day. An ordinary mechanical clock can be an example of pendulum theory. A certain timing has been built into it and that is the speed with which it will tick. The second theory is "Relaxation oscillation theory" which suggests that the organism possesses no such timing but acquires a rhythmic timing from rhythmic and cyclic geophysical events going on all the time. The human heart is a near example. Its speed can be adjusted by its environment, quickened by some hormones, slowed down by others,, accelerated by activity of body and slowed' down by rest. This second theory has been more widely accepted for explaining the functioning of biological clock.

Are biological clocks regulated by biochemical reactions. It is known that temperature greatly affects the rates of chemical reaction, 10°C increase can double the rate of many biochemical reactions, will it also double up the speed of an internal clock? No, it does not happen, even the cold blooded animals whose body temperature fluctuates can run their biological clocks accurately. An animal can have many internal clocks, which are perhaps controlled by a master clock.

The various internal clocks may function independently or in synchronization. Clunio (mosquito) is found near sea and its habitat is almost always flooded with water except for twice a month i.e. soon after full moon and also when it is new moon at the time of spring low tide. Only during these two dry times the female lays eggs in the ground. For adult female, the life is very short after it hatches from pupa, soon she must find a male to mate and lay eggs in the ground. The time of hatching, mating and egg laying has to be matched perfectly with the rhythm of the tide. The males, (have wings) emerge a little earlier than the females (wingless) which hatch earlier than the arrival of spring low-tide, lay eggs when the tide reaches its lowest level they die. All this happens within a few hours and only once in every fifteen days. In this case two internal clocks : one clock to determine the right time of hatching and another clock to determine the right time of low tide are functioning in synchronization.

Where is the biological clock located ? Among lower animals it has been demonstrated by performing careful surgery specially on cockroaches and crickets that these insects can lose their daily rhythm if the connection between Lobi option (part of the brain) and rest of the brain are either chemically blocked or cut. But, in higher animals like rats by destroying certain part of the Nucleus Suprachiasmaticus in diencephalon, the circadian rhythm can be eliminated. In most higher vertebrates it is the pineal body which controls the circadian rhythm. The functional organisation underlying the circadian rhythmicity is, however, largely unknown, there are definitely many more evidences of presence of biological clock than physiological evidences explaining the rhythmicity. It was observed by Kulkarni and his team in 1984 that in Hirudo birmanica excision of brain disturbed the rhythimicity in oxygen consumption, and when such leeches were given injections of brain homogenates the rhythmicity was resumed. In Crustaceans the circadian rhythms could be altered by removing the eyestalks. It is possible that the substance responsible for rhythmicity may be a proteinaceous neurohormone released through axons (Arechiga, 1977).

The Periodicity

All the physical factors including light, temperature, with their daily progression makes itself felt in all but it is to be expected that in deep water, soil or cave communities, majority of population will exhibit periodicities that are either directly or indirectly related with the changes occurring during the 24 hours day-night period.

The term diel (day) periodicity refers to all those events which occur at intervals of 24 hours or less. Community periodicity or cycle rhythm results when whole groups of organism exhibit synchronous activity patterns in the day-night cycle. For example some are active only during the night time i.e. nocturnal, others during the day i.e., diurnal and still others only during twilight periods *i.e.,* corpuscular. Singing of bird normally signals the arrival of dawn and butterflies, dragonflies and bees become conspicuous. At dusk, light fades and diurnal animals retires while the nocturnal animals become active.

In both lakes and oceans the vertical 'migration' of zooplankton organisms occurs regularly. It is the striking example of diel periodicity in aquatic habitats. Copepods, cladocerans, larval forms, etc. which make up the vast floating life in the open waters generally move upwards or towards the surface at night and downward during the day-light hours (Fig.). Whereas in this case light is clearly the controlling factor where, these diel movements are complex and their physiological mechanisms have not yet been fully elucidated. Generally different stages of the same species and each species responds in a different manner so that the crowding of all organisms is avoided into the same region. Although stratification is much more marked at mid-day in comparison to mid-night.

Seasonal Periodicities

Seasonal periodicities are likewise nearly universal in communities and are generally responsible for almost complete change in the community structure during the annual cycle.

In regulating the structure and function of a community, the role of temperature, photoperiod, wet and dry seasons, etc. have been fully established.

Temperature interacts with day length in the temperature zone. Thus Leopold and Jones (1947) found that variation from year to year in the time of flowering of plant and the arrival of migratory birds is greater .in early spring when temperatures are critical than in late spring, even though experiments show that the length of day is the controlling factor in many of the cases. Thus in all the insects and herbaceous plants disappear, the summer resident birds return south and winter visitors arrive. The tide rise on the ocean shore and falls about 50 minutes late each day and affects the life on the edge of the sea. The movement of the earth relative to the sun and the moon are underlying these rythmicities. The alternation of day and night occurs by the earth's rotation on its axis, and the tilt of earth's axis, along with its annual revolution around the sun results in the producing seasons. Orcadian Rhythms (Circa = about; diel - day).

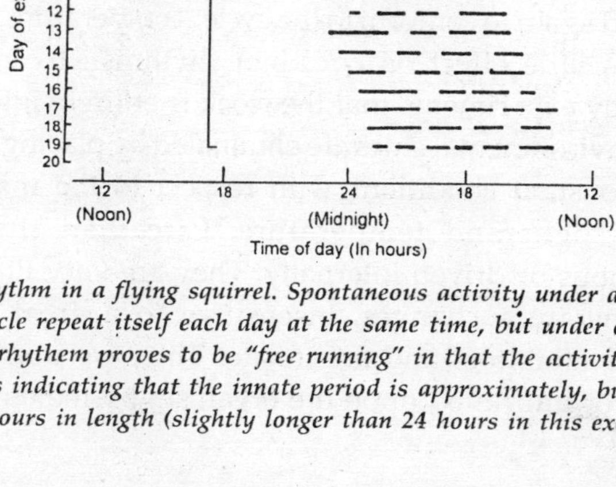

Circadian rhythm in a flying squirrel. Spontaneous activity under a regular light-dark cycle repeat itself each day at the same time, but under constant darkness the rhythem proves to be "free running" in that the activity period drifts, thus indicating that the innate period is approximately, but not exactly 24 hours in length (slightly longer than 24 hours in this example).

The persistent periodicity regulated by the biological clock that couples environmental and physiological rhythm, is called circadian rhythm. It may also be defined as the ability to time and repeat functions at about 24 hours intervals even in the absence of the conspicuous diurnal clue's such as light.

Life is evolved under the influence of daily and seasonal environmental changes. So it is natural that plants and animals would have some rhythm or pattern to their lives that would synchronize them with fluctuations in the environment.

A nocturnal animal like flying squirrel (Glaucomys) or white-footed mouse (Feromyscus) remains in its nest during the day-light hours and is active at night. The rhythmic activity is exactly 24 hours in length and is entrained or 'locked into' the daily light dark cycle generated by the rotation of the earth on its axis. If the animal is brought into the laboratory and placed under constant darkness, the daily rhythm continues, but the onset of activity will likely shift slightly each day, indicating a persistent or "free running" period a little shorter or a little longer than 24 hours. In the Fig., the period in complete darkness is shown to be about 20 minutes longer than 24 hours so that after 20 days the animal is leaving its nest for its "nightly" round six hours later than at the start of the experiment.

When the light dark cycle is re-established, the activity is entrained again to conform to the cycle. Temperature fluctuation produces little effect on circadian rhythms. It can never be proved by experiments that the clock is entirely internal since not all environmental clues are eliminated by placing the animal under constant conditions with respect to the major factors such as light and temperature. Circadian rhythms are endogenous or driven internally. They are very little affected by temperature changes, insensitive to a great variety of chemical inhibitors and are innate. They are innate and not learned or imprinted upon the organism by the environment.

The Tides

Tide is one of the most pronounced periodicities affecting the biota. Tides are produced in the ocean by sun and moon. Since, the sun and its planets exerts an attraction on each other, but the sun, being the largest exerts the most powerful force (Newton's law of universal gravitation). The earth and the moon also exerts a force of attraction on the same basis. The moon is much closer to earth than the sun and it exerts a force twice as great as that of the sun. Ocean waters exerts two types of bulges by the gravitational pull of the sun and the moon. Such bulges, tides usually consists of a regular rhythmic rise and fall of water twice a day over a range of several feet, but in shallow waters under special conditions, this sometimes amount to as much as fifty feet. The changes received at the shore line covering and uncovering the tidal zone, directly affect the biota by covering them with water, on by exposing them to air and sunlight.

At the same time, tide occurs on opposite side of the earth. Intervals between successive rises of the tide are about 12 hours, 25½ minutes, or between daily rises about 24 hours, 51 minutes. Because this is the same interval as the average daily passage of the moon across the meridian, it is almost certain that the moon is the principal regulator of tides.

Moon day is 51 minutes longer in comparison to a sun day. In the relationship of moon and sun tides there is constant change, at new and full moon, the attraction of sun and moon on tide reinforce each other, and cause the tides to rise higher at high water and fall lower at low water, without changing the mean sea level. These are known as spring tides. At first and last quarters of the moon, the forces oppose each other and the daily fluctuations are much smaller due to neutralizing effect of the force. These are called as neap tides.

Due to the immensely greater distance from the earth, the tide-raising force of the sun is only about 5/11th of that of the moon. When the sun and moon are pulling together, they tend to pull the water of the ocean toward themselves. As a result water is drawn away from the tides of the earth at 90° meridian where the combined pull is most effective and thus produces low tides on both sides of the earth at right angles to the direction of sun and moon.

The tide gradually decline in height, as the sun and moon change relative positions with respect to the earth. When they are pulling at right angles to each other, the sun neutralizes 5/11 th, of the moon pull and the tides are correspondingly reduced to neap tide. In the fig. both moon and sun tides are drawn but this does not happen actually rather this is only theoretical. The theoretical sun tide merely results in lower high tides and higher low tides. Theoretical figures indicate 11 + 5 for height of spring tides and 11 — 5 for height of neap tides, with intermediate fluctuations between the two extremes.

According to the theoretical observations the tides should occur when the moon is on meridian, but because the water cannot follow as fast as the apparent movement of the moon, there is a definite lag of tide behind the moon. In addition, the masses interfere with the ocean tides and cause great irregularity both in the tidal lag and in the height of rise and fall. Whereas the ordinary tide of ocean is usually about 2-5 ft., this figure is often highly multiplied against the shoreline. All of these factors and other minor items like biometric pressure, latitude, on shore and off shore winds, contour of shore, depth of water, natural periods of oscillations must be integrated before predictions of tides can be made at any particular place.

Lunar Orbit

Movement of moon in its orbit around the earth, causes additional periodicities. Its solid surface not being self luminous is a source of reflected sunlight, some of which reaches the earth as moonlight. Although there is constant reflection from the moon but still its reflection does not reach the earth when it is between the earth and the sun.

Lunar periodicity depends upon the lunar orbit of about twenty-eight days in which the moonlight grows to full light and fades into darkness again, requiring about two weeks for each of these two phases. In the daytime the moonlight is of little significance to organisms but at night the rise and fall of moonlight is important and effects them in several ways.

In comparison to daily and seasonal rhythms, the adaptations to moonlight periodicities are quite less common. Some of the most striking lunar responses are found in marine palolo worms and in the 'grunion fish', in which the behaviour patterns are correlated not only with moon phases, but also with seasons. The pacific palolo in Southern hemisphere rises to the surface to spawn in great numbers when the moon is full during October and November, but the Atlantic palolo in the northern hemisphere spawns at the surface about the last quarter of the moon in July. The grunion rhythm is much more complicated. The small fish of the species Leurethes tenius come to the southern California coast to spawn from March to July each year. The fish ride the high waves up the beath until they are left stranded on the sand during the nights immediately following high tide of the full or dark moon. Spawning takes place in the sand first below high tide mark and the eggs are burried in the sand. Two weeks later, most of the eggs are ready for hatching and are washed out of the sand by the next high tide.

Biological Clock

Photoperiod or the day-length period is one of the most dependable cues by which organism time their activities in temperate zones. In contrast to most other seasonal factors, day-length is always the same for a given season and locality. In the annual cycle the amplitude increases with increasing latitude, thus providing latitudinal as well as seasonal cues. Photoperiod has been shown to be the timer or trigger that sets off physiological sequences which bring about growth and flowering of many plants, molting, fat deposition, migration and breeding in birds and mammals, and the onset of diapause in insects. Photoperiodicity is coupled with what is now widely known as the organism's biological clock to create a timing mechanism of great versatility. In brief, organisms bears a physiological mechanism, for measuring time which is known as the biological clock. Among rhythms, particularly the cercadian rythm are the basic manifestations of biological clock.

Day-length serves as a sensory receptor like an eye in animals or a special pigment in plant leaves which in turn, activates one or more hormone and enzyme to bring physiological or behavioural response. It is not known just where in this sequence time is measured actually. Although there is wide morphological divergence in higher plants and animals but the linkage with environment photoperiodicity is quite similar. Similar to plants, animals may also respond to either short or long days. In most of the photoperiod sensitive organisms the timing can be altered by experimental or artificial manipulation of the photoperiod. It is observed that artificial light can bring brooktrout into breeding condition four months early. In migratory birds there is a period of several months after the fall migration when the birds are refractory to photoperiod stimulation. The short days of fall are apparently necessary to 'reset' the biological clock, as it were, and prepare the endocrine system for a response to long days. After late December an artificial increase in day length will bring on the sequence of molting, fat deposition, migratory restlessness and gonad enlargement, normally occurs in the spring. Among birds the physiology of this response in now farely well known, but it is uncertain whether the fall migration is brought about by the direct stimulus of short days or is timed by the biological clock that was set by the long photoperiods of the spring.

In certain insects the photoperiodism is noteworthy because it provides a sort of 'birth control'. 'Brain' is stimulated by long days of late spring and early summer to produce a neurohormone that brings on the production of a diapause or resting egg that will not hatch until next spring no matter how favourable are temperature, food and other conditions. Thus population growth is halted before, rather than after, the food supply becomes critical.

It has also been observed that the number of underground nitrogen fixing root nodules on legumes is controlled by photoperiod acting through the leaves of the plant. Since nitrogen fixing bacteria in the nodules requires food energy manufactured by the leaves of plant to perform their work. The abundant light, chlorophyll and maximum coordination between the plant and its microbial partner is thus enhanced by the photoperiod regulator.

The process of the biological clock has not been resolved, but two theories have been proposed:

1. *The Endogenous tinier Hypothesis* : according to this hypothesis 'clock' is an internal device that can measure time without environmental clues, and
2. *The External Timing Hypothesis* : according to this hypothesis the external clock is timed by external signals from the environment.

Zeitgebars

Many organisms exhibit circadian and circannual periodicities that appear to be closely adjusted to patterns of daylength, temperature fluctuation, or other environmental patterns. The mechanism whereby the period of a rhythm occurs, repetitively and coincides approximately with the presence of some external stimulus is called entrainment. All the cues that provide information to animals about periodicity of environmental variables are called Zeitgebars ("time givers").

Zeitgebars are the entraining agents defined as those cyclic environmental cues that can entrain free-running, endogenous pacemakers. Zeitgebars can influence rhythms by effecting both the phase and the frequency.

Considering the example, when the flying squirrel is provided with a daily cycle of light and dark, its activity rhythm generally conforms to the We cycle the onset of activity occurs at about the same time each evening. The squirrel is thus said to be entrained to the daily light-dark cycle. For most endothermic vertebrates thus far tested, the tight-dark cycle is the critical cue for entrainment. Further, in most terrestrial organisms, the daily light-dark cycle is the Zeitgebar. This is not surprising, since the daily light cycle is usually the most consistent environmental cue in the terrestrial habitat. For animals in the intertidal zone, the web and flow of the tides may be a prime zeitgebars. Ectotherms such as lizards and insects that are unable fully to control their own body temperatures may use temperature or light cues as zeitgebars. When entrained, they exhibit increased activity under warmer conditions and decreased activity during the cooler portions of the daily cycles..

Bibliography

Alcock, John : *Animal Behavior : An Evolutionary Approach*, Sinauer, 1998.

Altmann, S. A. and Altmann, J. : *Baboon Ecology : African Field Research*, Oxford, New York, 2000.

Asdell, S. A. : *Patterns of Mammalian Reproduction*, Comstock, New York, 1946.

Aslam, M. S. : *Genetics-Behavioural and Molecular*, Campus Books, New Delhi, 2003.

Aubrey, Manning : *An Introduction to Animal Behaviour*, ELBS Publication, New York, 2001.

Bahr, G. F. : *Chromosomes and Chromatin Structure*, Academic Press, New York, 2001.

Barnett, S. A. : *Study in Behaviour*, Science Paper Backs, London, 2001.

Bendall, J. R. : *Muscles, Molecules and Movement*, Heinemann Educational Books, London 1969.

Benjamin, Gutfreud : *Enzymes : Physical Principals*, John Wiley, London, 1972.

Bernhard, : *The Structure and Function of Enzymes*, Routledge, New York, 1968.

Bhaskar, Harsh Vardhan : *Animal Ecology*, Campus Books, New Delhi, 2004.

Bogert, L. J. and Calloway D.H. : *Nutrition and Physical Fitness*, Saunders, London, 1973.

Bolhuis, J. J. and Hogan, J.A. : *The Development of Animal Behaviour : A Reader*, Blackwell, 1999.

Boulenger, E. G. : *An Introduction to Animal Behaviour*, Discovery Publishing, New Delhi, 2003.

———— : *Animal Mysteries*, Discovery Publishing, New Delhi, 1993.

Boyd, J. D. and Hamilton, W.J. : *The Human Placenta*, Cambridge, London, 1978.

Buffon's, B. : *Natural History of Birds, Fishes, Insects and Reptiles*, Discovery Publishing, New Delhi, 2003.

Carpenter, K. : *Dinosaur Eggs and Babies*, Cambridge, New York, 1996.

Castillo, J. : *Progress in Biophysics and Biophysical Chemistry*, Prentice Hall, London, 1956.

Chauvin, Remy : *Ethology : The Biological Study of Animal Behaviour*, International Universities Press, New York, 1999.

Damron W. Stephen : *Introduction to Animal Science : Global, Biological, Social and Industry Perspectives*, Prentice Hall, London, 2000.

David, M. : *The Marvels of Animal Behaviour*, National Geographic Society, Washington, 2000.

David, McFarland : *Animal Behaviour : Psychobiology, Ethology and Evolution*, ELBS Publication, New York, 2000.

Dickerson and Geis : *The Structure and Action of Proteins*, Academic Press, New York, 1973.

Ernster and Drahota : *Mitochondrial Structure and Function*, Academic Press, New York, 1969.

Filley, G. F. : *Acid-Base and Gas Regulation*, Lea & Febiger, New York, 1971.

Finn, Frank : *Classification of Birds*, Discovery Publishing, New Delhi, 1994.

Florey, E. : *An Introduction to General and Comparative Animal,* Oxford University Press, New York, 1989.

Freeman, W. H. : *Animal Societies and Evolution,* A Scientific American Publication, New York, 2001.

Gene and Reynolds : *Water-soluble Vitamins,* John Wiley, London, 1963.

Gorbman, A. and Bern, H.A. : *A Text Book of Comparative Endocrinology,* John Wiley and Sons, New York, 1962.

Greenberg, Rodwell : *Amino Acid Metabolism,* Cambridge University Press, Cambridge, 1964.

Gyorgy, Pearson : *The Vitamins,* Academic Press, New York, 1967.

Houssay, Frederic : *Industries of Animals,* Discovery Publishing, New Delhi, 1999.

Huxley, H. E. : *The Mechanism of Muscular Conduction,* Spring Harbor, Harbor, 1972.

James, F. : *Animal Social Behaviour,* Wittenberger Duxbury Press, Boston, 1999.

Jencks, : *Catalysis in Chemistry and Enzymology,* McGraw Hill, New York, 1961.

Jennifer, Owen : *Feeding Strategy,* Oxford, New York, 2002.

John, Alcock : *Animal Behaviour: An Evolutionary Approach,* Sinaver Associates, New York, 2002.

Johnson, Christopher B., Franz, Chlodwig : *Breeding Research on Aromatic and Medical Plants,* The Howorth Herbal, London, 2002.

Johnston, R. G. : *Introduction to Sheep Farming,* Granada, 1983.

Jorge, Peter and Ewert Verlag : *Neuroethology : An Introduction to the Neuropathy Sociological and Fundamentals of Behaviour,* Oxford, New York, 2001.

Kamra, D. N. and Pathak, N. N. : *Nutritional Microbiology of Farm Animals,* Vikash Publishing House, New Delhi, 1995.

Kay, Ian : *Introduction to Animal Physiology,* BIOS, London, 1998.

Khanna, D. R. : *Animal Distribution,* Discovery Publishing, New Delhi, 2004.

Konrad Lorenz : *King Solomon's Ring,* MacMillan, London, 2001.

Konrad, Z. and Lorenz, Verlag : *Messengers to the Brain : Our Fantastic, Five Senses,* National Geographic Society Publication, Washington, 2001.

Konrad, Z. and Lorenz, Verlag : *The Foundations of Ethology,* New York, 1999.

Krishnan, M. : *The Hand Book of India's Wildlife,* Oxford, New York, 2002.

Kumar, Ashok : *Animal Physiology,* Discovery Publishing, New Delhi, 2003.

———— : *Aviculture,* Discovery Publishing, New Delhi, 2003.

———— : *Movement and Locomotion in Animals,* Discovery Publishing, New Delhi, 2003.

Kumar, Bipin : *Histology,* Campus Books, New Delhi, 2001.

Kung, S. D. and Yang, S.F. : *Discoveries in Plant Biology,* MacMillan, London, 2000.

Linville and Kelly : *A Text Book in General Zoology,* Discovery Publishing, New Delhi, 1993.

Lister, Harry : *An Introduction to Zoology,* Discovery Publishing, New Delhi, 1999.

Lubibiecki, A. S. : *Large-Scale Mammalian Cell Culture Technology,* Dekker, 1990.

Maheswari, S. K. : *A Textbook of Life Science,* Dominant Publishers, New Delhi, 2004.

Mary, E. Torrence and Issacson : *Microbial Food Safety in Animal Agriculture,* Blackwell, London, 2003.

McCance and Widdowson : *The Composition of Food,* Her Majesty's Office, London, 1960.

Merler, E. : *Immunoglobulins : Biological Aspects and Clinical Uses,* Natural Academy of Sciences, Washington, 1970.

Michael, A. M. : *Irrigation : Theory and Practice,* Vikash Publishing House, New Delhi, 1997.

Morgan, S. J. : *Animal Cell Culture : Introduction to Biotechniques,* BIOS, London, 1993.

Nama, Shinde and Jadhav : *The Nematodes of Indian Mammals,* Dominant Publishers, New Delhi, 2004.

Openheimer, J. H. and Others : *Nuclear Receptors and the Initiation of Thyroid Jormone Action,* Little Brown and Company, Boston, 1976.

Pathak, Nityanand : *Textbook of Feed Processing Technology,* Vikash Publishing House, New Delhi, 1997.

Price, M. R. S. : *Animal Re-Introduction: The Arabian Oryx in Oman,* Cambridge, 1989.

Prosser, Brown, : *Comparative Animal Physiology,* Saunder, Philadelphia, 1961.

Quirk, R. P. : *Chemistry of Muscular Contraction,* Academic Press, New York, 1970.

Ranjhan, S. K. : *Animal Nutrition and Feeding Practices,* Vikash Publishing House, New Delhi, 1995.

——— : *Animal Nutrition in the Tropics,* Vikash Publishing House, New Delhi, 1998.

——— : *Textbook on Buffalo Production,* Vikash Publishing House, New Delhi, 1996.

——— : *Textbook on Quality Meat Production,* Vikash Publishing House, New Delhi, 1999.

Rao, D. Bhaskara : *Animal Kingdom,* Discovery Publishing, New Delhi, 1995.

Robert, Burton : *Animal Behaviour,* Time Life International, Nederland, 1999.

Roger, Tory : *The Birds : Life Nature Library,* Peterson, London, 1999.

Ronald, H. Bailey : *The Role of the Brain*, Time-Life Books, New York, 2001.

———— : *The Secret World of Animals*, A National Geographic Society Publication, Washington, 2002.

Saharia, V. B. : *Wildlife in India*, Natraj Publishers, Dehradun, 1999.

Samuel, K. Wasser : *Social Behaviour of Female Vertebrates*,Academic Press, New York, 2002.

Samuel, Israel and Toby, Sinclair : *Indian Wildlife*, Hans Johannes Hoefer, Singapore, 2001.

Sandhu, G. S. : *A Textbook of Fish and Fisheries*, Dominant Publishers, New Delhi, 2005.

———— : *Analytic Genetics of Animal Development*, Campus Books, New Delhi, 2004.

———— : *An Introduction to Birds*, Campus Books, New Delhi, 2003.

Sarkar, Amira : *Development of Animal Behaviour*, Discovery Publishing, New Delhi, 2003.

———— : *Fundamentals of Animals Behaviours*, Discovery Publishing, New Delhi, 2003.

———— : *Social Behaviour in Animals*, Discovery Publishing, New Delhi, 2003.

Schmidt, Nelson K. : *Animal Behaviour*, Prentice Hall, New Jersey, 1970.

Sela, M. : *New Perspectives in Biology*, American Elsevier, New York, 1963.

Shreeve, W. W. : *Physiological Chemistry of Carbohydrates in Mammals*, Saunders, New York, 1974.

Singh, Gurdarshan, Bhaskar, H. : *An Introduction to Fishes*, Campus Books, New Delhi, 2003.

Spater, P. J. B. : *An Introduction to Ethology*, Cambridge University Press, New York, 1999.

Tikader, B. K. : *Threatened Animals of India*, Zoological Survey of India Publication, Calcutta, 1999.

Tortora and Anagnostrakes : *Principles of Anatony and Physiology*, Harper and Row Publishers, New York, 2002.

Tullts, L. P. : *Blood Cells and Plasma Proteins : their State in Nature*, Academic Press, New York, 1953.

Underwood, E. J. : *Trace Elements in Human and Animal Nutrition*, Academic Press, New York, 1971.

Van, Grvisen : *Wildlife of India*, Guide Book Company Ltd., New York, 2002.

Wadsworth, Cecie Starr : *Basic Concepts in Biology*, Wadsworth, New York, 1997.

Whelan : *Control of Glycogen Metabolism*, Academic Press, New York, 1968.

Whitfield, Dr. P. : *Longman World Guide to Mammals : A Visual Who's Who of the Mammal Kingdom*, Longman, London, 1985.

Williams, C. A. and Chase, M. W. : *Methods of Animal Behaviour*, Academic Press, New York, 1968.

Williams, S. R. : *Animal Behaviour*, MacMillan, London, 1973.

Yadav, Manju : *Applied Entomology*, Discovery Publishing, New Delhi, 2003.

Yadav, P. R. : *Environmental Biodiversity*, Discovery Publishing, New Delhi, 2003.

Yapp, W. B. : *An Introduction to Animal Behaviour*, Clarendon Press, New York, 1960.

Zhukov, M. F. : *Organic Chemistry of Biological Compounds*, Englewood Cliffs, Prentice Hall, 1971.

Zumdahl, S. S. : *An Introduction to General and Comparative Animal Behaviour*, Oxford University Press, New York, 1981.

ENCYCLOPAEDIA OF ANIMAL BEHAVIOUR

ENCYCLOPAEDIA
OF
ANIMAL BEHAVIOUR

(Vol. 2)

Dr. A. Tyagi

ANMOL PUBLICATIONS PVT. LTD.
NEW DELHI - 110 002 (INDIA)

ANMOL PUBLICATIONS PVT. LTD.
H.O.: 4374/4B, Ansari Road, Darya Ganj,
New Delhi-110 002 (India)
Ph.: 23261597, 23278000

B.O.: No. 1015, Ist Main Road, BSK IIIrd Stage
IIIrd Phase, IIIrd Block,
Bangalore - 560 085 (India)
Visit us at: www.anmolpublications.com

Encyclopaedia of Animal Behaviour

First Published, 2006

ISBN 81-261-2675-2 (Set)

PRINTED IN INDIA

Published by J.L. Kumar for Anmol Publications Pvt. Ltd., New Delhi - 110 002 and Printed at ARORA Offset Press, Delhi.

Contents

Preface

Behaviour could be most simply defined as movement or a way one conducts oneself. Animal behaviour is, obviously, the way in which animals-non human-lead their life, such as, feeding, migration to different parts in different seasons, sexual life, gregariousness etc.

Different factors need to be studied to have right knowledge about animal behaviour. One of them is the concept of "Natural Selection" propounded by Darwin. He observes that species compete with each other for survival.

Apart from animals, there also occurs a peaceful contest among birds. For example, the severest rivalry between the males of many species to attract their female counterpart is one of the patterns of animal behaviour. It also includes a genetic component similar to the morphological and psychological traits.

The present multi-volume encyclopaedia is a painstaking research work on animal behaviour. It would certainly prove to be an asset for psychologists and those who seek thorough knowledge on the subject.

Editor

1

Nervous System

Living organisms have a unique property of receiving stimuli from environment and respond to changes accordingly. The information received by the organism is coordinated through the nervous system so that it can act or behave effectively for its survival under any circumstances. All our activities whether we are solving one of the many problems of everyday life, watching a TV programme, jogging or playing cards, or even sleeping, we are completely dependent upon this complex arrangement of neurons in the brain and spinal cord.

The brain is like a most remarkable computer with an unmatchable filing system of billions of bits of information, endowed with software that sorts and arranges innumerable words in an almost infinite variety of combinations. It is the constant monitor of a flood of sensory impressions from the outer world, the principal generator of drives and emotions, and the operator of such vital functions as breathing, heart beat and muscle movements.

For more than 150 years scientists have been probing the brain in an effort to find out what part of it does what, when, why and how. It has still remained the most amazing part of the body. Brain has not been unique to human beings, even primitive creatures like earthworm have a nerve ring in the anterior segments crudely approximating a brain. In certain ways, some animals appear superior to human beings; migratory fishes and birds perform astonishing feats of navigation. Dogs and cats have far better sense of smell and hearing than our own. What distinguishes the human brain from other brains is the relative size of the cerebral cortex, as compared to other animals, only in human beings is the cerebral cortex so large in relation to body size.

To understand the control of brain over behaviour, study of human brain would be the most fascinating journey through the bizarre world of neurons. Exploration of human brain has not been an easy task, there are many unsolved mysteries and much has to be done. Inquisitiveness of our own mind has led to interesting findings with relation to brain and behaviour.

What is most remarkable about human brain is that it can carry out many physical activities at one time and keep the mind busy by thinking at the same time, for example a teacher can be lecturing in the class, drawing diagrams on the board, and her brain will be figuring out what she has to say next, recalling from her mental filing system, controlling movement of her limbs, maintaining body posture, making gestures while speaking and simultaneously registering the reactions of students.

Human brain is considered as the most developed; among all living organisms it is not the weight of the brain which is taken into consideration for assigning it the category of "most developed" but it is the ratio between the rest of the body with brain *e.g.* the brain of the elephant is huge, weighing an average of 13 pounds, for every pound of it an elephant has almost a thousand pounds of body (in human beings the ratio is 1 to 50 and average weight of the brain is only 3 1/2 lbs.) so most of the elephant brain is devoted to coordinating its massive body.

The ability to learn and remember is one of the brain's most astonishing capabilities. For many centuries people thought that heart was the principal organ controlling human body and activity. Later, Greek philosopher Plato correlated ambition to heart, reason to the head and hunger to stomach. Aristotle thought that the brain was merely a device meant to keep the heart and mind cool. History's first recorded mention of the brain is in Egyptian papyrus (1700 BC) that head injury could result in lameness and blindness. Sir Frances Galton (1822-1911), cousin of Darwin, wrote a book *"Enquiries into Human Faculty"* which was published in 1823.

The landmarks in the history of study of brain and behaviour have been Franz Joseph, Gall, Paul Broca, Eduard Hitzig and Gustav Fritsch. Behaviour is so much varied that brains might be expected to come in a variety of structures, sizes and shapes. They do not, except for some dependence on body weight - women's brain average 10% smaller than men's because on an average women weigh less. All brains are much alike in form and chemical make-up. The brains of notable individuals - Einstein, Lenin, Napoleon - have been examined with great interest after death, and they did not appear to be much different from any one else.

The mammalian nervous system includes central (brain and spinal cord), peripheral and autonomic nervous systems along with sense organs.

Neuron is the fundamental, functional and structural unit of the nervous system. The human brain consists of approximately 15 billion nerve cells. These billions of neurons form a complicated but accurate network all over the body. It can be compared with the telephone network a metropolitan city has. A telephone cable has many copper wires in it, which are well insulated by rubber coating. Similarly, neurons have axons insulated by Schwann cells. Even myelin sheath has its own importance in behaviour. A child can start crawling, walking and talking but the bowel and bladder control comes only after the myelin sheath has wrapped the nerves.

Neurons also do not touch one another, they are separated by unbelievable narrow gaps - 500 times thinner than the finest hair. Although central nervous system plays the most important role in controlling behaviour, yet the functions of other parts of nervous system can not be ignored, they are equally important. Performance of complex behavioural repertoire is a consequence of the entire nervous system and not just any particular part.

Reflexes and Behaviour

In all vertebrates, the spinal cord has two main functions. integration of reflex actions and conduction of nerve impulses to and from the brain, establishing connection between peripheral and central nervous system. A reflex arc is the simplest but complete functional unit of the nervous system capable of detecting change and causing a response, it is a means for automatic and rapid adjustment to maintain homeostasis without conscious effort or consulting brain. Automatic responses carried out at the level of spinal cord and without any conscious efforts are part of the every day existence of most living organisms. They are basically of three types reflexes which involve skin are superficial reflexes, those which depend on impulses from muscles are called myotatic reflexes. Inspite of the basic complexity of many behavioural acts controlled by central nervous system, the responses to simple stimuli are often automatic and generally invariable.

You tickle the side of dog and it would show scratching reflex, if a cat is dropped upside down from a height, it will land correctly on" all fours due to righting reflex, chickens with their heads chopped off have been known to run about for a short time because the stepping motion is basically reflexive unlike this smooth and effortless finger movements of a sitar player which seem automatic are not reflexes, they have been developed by training, such learned automatic responses are called acquired automatism.

A wide variety of movements in vertebrates are performed as reflex actions. A strong light held in front of the eye makes it shut. If our hand comes in contact with a hot object or is pricked, we draw the hand away, a sudden loud noise startles us reflexly. Some vital functions of the important organs of the body like secretion of juices from gastric glands, heart rate and respiration are also controlled reflexly. In mammals nervous system, there are reflex arcs corresponding to the series of thirty three spinal nerves, these reflex arcs are not separated but are interconnected by longitudinal ascending and descending tracts. These tracts ultimately connect them with the brain.

Because of connections between the different reflex arcs, complicated motor patterns are possibly carried out *e.g.* walking involves many reflex actions simultaneously or in a series. To a certain extent we have some control over our reflex actions *e.g.* as soon as a hot object is touched the hand reflexly moves away but consciously one can keep one's hand in contact with the hot object also; these activities are possible because of connection between reflex arc and brain through ascending and descending tracts.

If an animal whose brain has been separated from the spinal cord, is laid on its side on an uncomfortable surface, the limbs will make movements designed to return the body to an upright position, but they will be uncoordinated and therefore, ineffective, because of the lack of cerebral involvement. In this case the spinal cord has received sensory information from the skin and initiates impulses to the muscles to correct the posture but fails to do so as it is not in contact with chief controller *i.e.* the brain. This indicates that the spinal cord is involved in carrying out muscle movements reflexly, but requires the brain to coordinate it.

The spinal cord alone does not contain all centres for reflex actions. They are present in the gray matter of the brain as well. Many reflexes such as salivation, suckling, pupil reflex, respiratory reflexes are mediated by the brain.

Mammalian brain has three primary vesicles viz Prosencephalon (forebrain), Mesencephalon (midbrain) and Rhombencephalon (hindbrain) and five secondary vesicles : Diencephalon, Telencephalon, Mesencephalon, Myelencephalon and Metencephalon, these vesicles give rise to different parts of the brain.

(1) Prosencephalon (forebrain)	Diencephalon – Thalamus and hypothalamus Telencephalon – Cerebrum (cortex)
(2) Mesencephalon (Midbrain)	Mesencephalon – Midbrain
(3) Rhombencephalon (Hindbrain)	Myelencephalon – Medulla oblongata Metencephalon – Cerebellum and Pons

According to neuroscientist Paul MacLean, human brain is actually not one but three brains put together during the course of evolution. The first is reptilian brain because it resembles in some ways the total brain possessed by reptiles, it is mainly devoted to basic life support such as breathing, heartbeat, muscle movement, eating, drinking, mating, parental behaviour, fight and fleeing.

The second part is limbic system, which controls complex emotional responses like snarl, growl, shriek, or fear, teaching young, do tricks for reward, wag tails, appeasement, look submissive, are some behaviours which are linked to mammals but are not well developed in reptiles. Limbic system in human beings helps maintaining emotional equilibrium. The third brain, is composed of folded cerebral cortex, where reside the

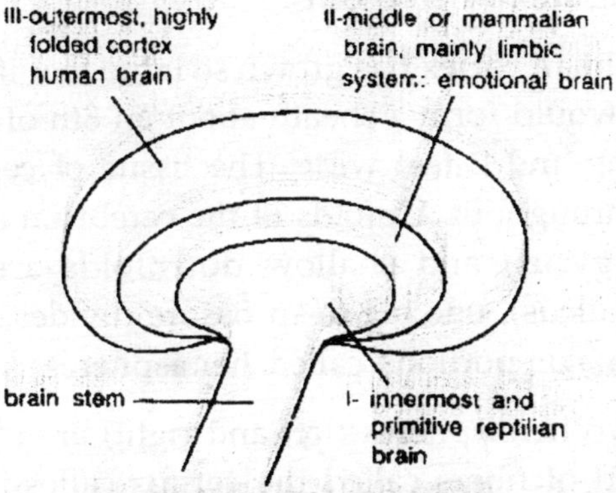

Fig. MacLean's three brain concept.

powers of speech and written language, observing, analyzing, planning ahead and imagining what may happen in the future which are exclusively human faculties. All three brains collectively make human brain very unique.

Forebrain (Prosencephalon)

Cerebrum (Cerebral cortex) : The brain's master analyzer (Telencephalon), supported on the brain stem and forming the bulk of the brain is cerebrum. It is the most important part of the brain. Thoughts originate, things are memorized and learning takes place here. It has motor and sensory cortex and some part of it is known as association cortex. You hear flutter of wings, your eyes turn and look around, you see a beautiful bird, the sound and sight are associated in association cortex and it flashes the information "house sparrow" or a "parrot". You smell something good being cooked in the kitchen, and you know what exactly it is. Different informations are associated in the cortex.

Cerebrum is a highly folded structure of pale white mass of tissue covered by an extremely thin layer (2-4mm) of gray matter called cerebral cortex. The gray matter consists chiefly of neuron cell bodies while white matter is made up mostly of axons called white cortex. This folding creates more surface for housing more nerve cells.

The human cortex has grown so large that if it is stretched out flat it would form a sheath about an 8th of an inch thick, a yard long and 2 feet wide. The tissue of cerebrum is not uniform throughout. Upfolds of the cerebrum are called gyri (singular gyrus) and shallow downfolds are called sulci (singular sulcus), one big deep fissure divides, the cerebrum into two major portions called hemispheres.

The two hemispheres (left and right) are connected by a broad band of fibers called the corpus callosum. At several other places deep fissures divide each of the two hemispheres into four lobes: frontal, parietal, occipital and temporal, the lobes are named after the bones that cover them. The central sulcus separates the frontal lobe from the parietal lobe, the precentral gyrus is the landmark for the primary motor area, postcentral gyrus distinguishes -general sensory area, the lateral cerebral sulcus separates the frontal lobe from the temporal lobe and lastly parieao-occipital sulcus separates the parietal lobe from the occipital lobe.

The four lobes have areas that specialize is certain functions, but recent research stresses the interdependence of brain parts for execution of functions.

Frontal Lobe

It is the largest lobe but least understood, much of its area is till today considered as silent area whose function is not known. The basic step in behaviour is the decision to move a muscle. The frontal lobe contains more than a million unusually large nerve cell bodies, most of them pyramid shaped; descending from them, are million axons which snake

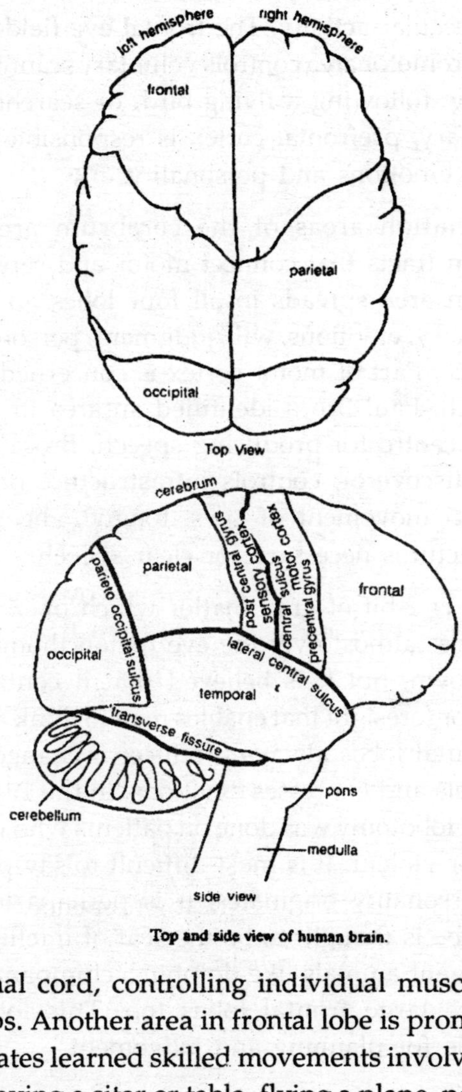

Top and side view of human brain.

down the spinal cord, controlling individual muscles in the trunk and limbs. Another area in frontal lobe is promoter area which coordinates learned skilled movements involving many muscles *e.g.* playing a sitar or tabla, flying a plane, performing acrobatics and dance, whereas, primary motor cortex controls general movements of all the muscles.

Scientists have mapped the motor cortex with astonishing precision. The map shows exactly which part of cortex controls which muscular activity. The frontal eye field area which is a part of promotor area controls voluntary scanning movements of eyes *e.g.* following a flying bird, or searching for a word in dictionary; prefrontal cortex is responsible for reasoning, planning, emotions and personality.

Association areas of the cerebrum are made up of association tracts that connect motor and sensory areas, the association area spreads in all four lobes and is concerned with memory, emotions, will, judgment, personality traits and intelligence. Part of motor cortex is concerned with memory and speech. Paul Broca identified an area in the left frontal lobe as a centre for producing speech, Broca's area (named after its discoverer; controls infrastructure or mechanics of speech *viz.* movement of jaws, tongue, cheeks, larynx and other structures necessary for clear speech.

There is a bit of gray matter which overhangs from the frontal lobes almost over the eye region though it's function is not known, but it is believed that it contains centres of intention or foresight that enables man to think ahead. Damage to the frontal lobes always produces a change in the way a person feels and expresses emotions. In the 1940's and 1950s, prefrontal lobotomy was done on patients who were extremely agitated or violent. It is most difficult to say physiologically, where personality originates; it is perhaps in frontal lobe. Frontal lobe is thought to be the seat of intelligence, most of the intelligent animals like dolphins, chimpanzees, dogs and seals have large frontal lobes too. This lobe is possibly responsible for planning and judgement.

Parietal Lobe

This lobe of the cerebrum is involved in a wide variety of general sensory functions, hence is called sensory cortex. The sensation of heat, cold, pain, touch, pressure and position of body are carried out by the parietal lobe. At one point on elbow, four different types of nerve endings are bundled up, touch, pain, heat and cold. A sharp blow there elicits a very peculiar sensation because centres for all those feelings are excited simultaneously in parietal lobe. This lobe keeps track of the movable parts of the body. This part of the cortex is also known as somesthetic area (for body sensation). Parietal lobes, which contain the primary receiving areas for the skin senses and our sense of bodily position, if damaged produces deficits in the sense of touch. A person with extensive parietal lobe damage extending beyond the primary sensory area also has unusual difficulty with spatial organization and has a distorted perception of his own body image.

Occipital Lobe

This lobe forms the extreme posterior end of cerebrum, a very special sense *i.e.* vision is centered here. When we see something, nerve endings of retina of the eye get stimulated, then the nerve impulses travel through optic nerve and reach occipital lobe, where the image is identified and gives a meaning to the consciousness and the eyes only act as receptor organs. It is responsible for shape, size, pattern and colour recognition.

Temporal Lobe

Another special sense, hearing (audition) is interpreted in this lobe of cerebrum. The seats for sense of smell (olfaction) and sense of taste (gustation) are also found in this lobe. A German neurologist, Karl Wernick, discovered another centre for speech and language which is called Wernick's area which is located in this lobe in the left hemisphere. Wernick's area has the dictionary of all those words which we learn; it is a store house of vocabulary, it is a word retriever and selector. Injury to speech area results in aphasia (without speech or an inability to speak) and agraphia (inability to write), word deafness (inability to understand spoken words) and word blindness *i.e.* an inability to understand written words.

The inter dependence of lobe functions can be demonstrated by the work shared by Broca's and Wernick's areas. If we have to repeat word we have just heard, the sequence of activity is; from the ears the sensation of a heard word travels by neurons to the auditory cortex on the temporal lobe, the word is understood by Wernick's area, also on the temporal lobe, then it goes to Broca's area on the frontal lobe, which sends messages to the motor cortex, at the rear of the frontal lobe which moves necessary muscles and you speak a word.

One individual encloses almost two brains in the form of two hemispheres which are connected together by corpus callosum. Two hemispheres though similar structurally, are different functionally, these hemispheres rest above common pons and medulla. The left cerebral hemisphere controls the right side of the body and the right hemisphere controls left side of the body *i.e.* all motor activities and sensations of the right side of the body go to left hemisphere and vice-versa. It is generally presumed that the left hemisphere does all the planning, along with controlling right hand touch, mathematical skills, language, science, writing and logic. Whereas, the right hemisphere controls the management, left hand touch, music and art appreciation, dance, sculpture, perception, and fantasy; basically, left hemisphere is more scientific and right hemisphere is artistic.

Both speech centres are in left hemisphere, whereas, singing centre is in right; a person who stammers in talking often sings very well; though, the two hemispheres are so different functionally, still they work in unison, because they keep exchanging information through corpus callosum. Experiments on "split-brain" (where corpus callosum is cut to discontinue exchange of information between two hemispheres) led to many interesting findings, Weber, Mgers and Sherry (1970, 1974) performed some of the classic experiments, they found out that left parietal lobe monitors body awareness of right side and right parietal lobe of left side. In a pathological condition called "parietal lobe syndrome", one of the two halves of the lobe is affected. If the right half of the lobe suffers, awareness of the left side of the body is lost, the victim, lacking reports from an entire half of his body, tends to forget it is there and neglects it, often, he presents the image of a split person - half well dressed, nicely shaven, neat and clean and half unkempt, unshaved and dirty.

Even after so much experimentation and exploration, there are areas in the human cortex which still remain unexplored and mysterious, perhaps human faculties like thoughts, creativity, and love lie somewhere in those unknown areas.

Basal Ganglia

Embedded within the substance of each cerebral hemisphere, lies a cluster of nuclei known as Basal Ganglia or Basal Nuclei or Cerebral Nuclei. The basal ganglia consists of head and tail of Caudata nucleus and the lentiform nucleus. Amygdaloid nucleus also forms the part of basal ganglia which is interconnected by many fibres to cerebral cortex, thalamus and hypothalamus. Caudata nucleus controls subconscious movements of the skeletal muscles, such as swinging the arms while walking, this is the part of the

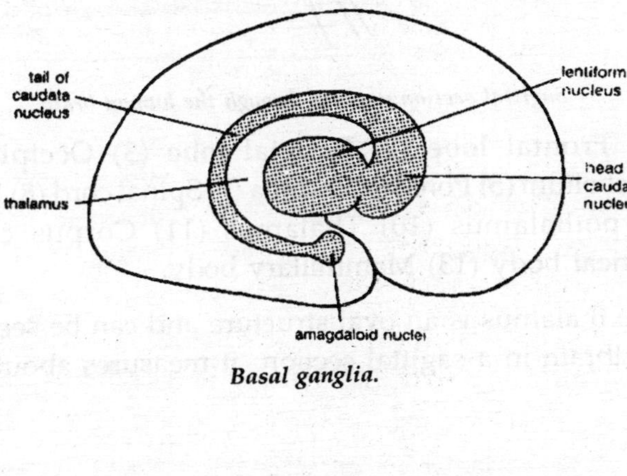

Basal ganglia.

brain which when destroyed causes paralysis. Though the muscular activities of the body are under the control of motor cortex yet the coordination of body movements depends to a large extent upon this cluster of nuclei. It is this area of brain that is known to be affected in a disease; called parkinsonism in which the nerve cells of the basal ganglia fire too rapidly resulting in muscle tremors, this continuous firing is caused by an insufficiency of an inhibitory neurotransmitter called dopamine, therefore, person suffering with parkinsonism shows uncontrollable jerky movements of limbs and head.

Diencephalon

Thalamus

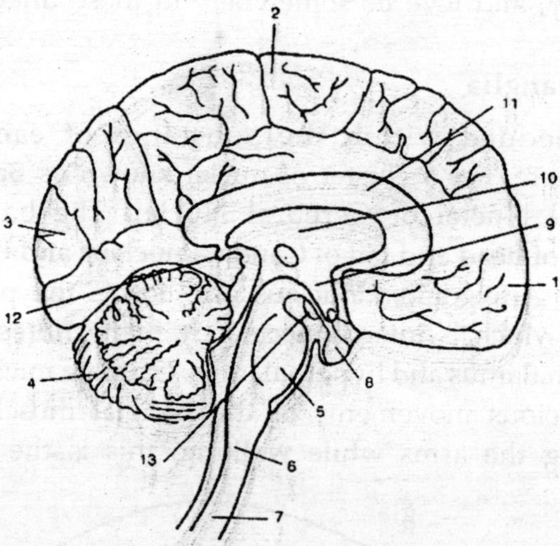

Sagittal section passing through the human brain.

(1) Frontal lobe (2) Parietal lobe (3) Occipital lobe (4) Cerebellum (5) Pons (6) Medulla (7) Spinal cord (8) Pituitary (9) Hypothalamus (10) Thalamus (11) Corpus callosum (12) Pineal body (13) Mammillary body.

The thalamus is an oval structure and can be seen above the midbrain in a sagittal section it measures about 3cm in length and constitutes four fifth of the diencephalon. It chiefly constitutes gray matter which is divided into three masses-anterior, medial and lateral within each mass are nuclei that perform different functions. The medial geniculate nucleus receives sensory impulses of audition, lateral geniculate nucleus is for vision and the ventral posterior nucleus is for general sensation and taste, other nuclei are centres for synapses in the somatic motor system, these include the ventral lateral nucleus for voluntary motor actions and ventral anterior nucleus for arousal. It also functions as interpretation centre for pain, temperature, touch and pressure. It also forms part of reticular formation, the lateral thalamus is concerned with certain emotions and memory also.

Its primary function is reception and sorting out of sensory information everything that touches a human body, from the softest touch of cotton to the sharpest edge of knife, sends nerve signals to the thalamus which acts as traffic director for most sensations, it monitors and sorts sensory signals, then sends them on to appropriate parts of the brain *e.g.* it relays eye impulses to visual cortex and sound signals to hearing centre in temporal lobe. Therefore, thalamus is like a central switchboard between the spinal cord and the brain.

Hypothalamus

The hypothalamus is located in the middle of the brain above the pituitary gland and below the thalamus, and is surrounded by two cerebral hemispheres. Although a small structure, yet it is composed of a relatively large number of nuclei. Although differentiation of the hypothalamic nuclei is not precise, yet it is possible to identify certain nuclei; within a given nucleus there may be several kinds of cells that can only be differentiated histologically. The localization of function, with a few exceptions, is not specific to the individual nuclei; certain functions tend to overlap, therefore, the functions are assigned to regions rather than specific nuclei.

Hypothalamus is a small but functionally mighty area of the brain. It weighs little more than a quarter of an ounce yet it performs many functions of greatest importance both for survival and for enjoyment of life. It works as a link between the psyche (mind) and soma (body), it links the nervous system with endocrine system. It is difficult indeed, to think of any function of the body that is not dependent directly or indirectly upon the hypothalamus. It organizes the metabolism of the body, seeing to it that the animal takes in the right amounts of fluid and food, it is essential for making the animal feel (motivation, drive), hungry, thirsty, ready to mate, and aggressive.

It brings about sleep and hibernation. Among those that hibernate, it is the essential part of the brain for controlling the temperature in warm blooded animals. Rhythmic movements of breathing are arranged at the level of spinal cord but the hypothalamus can alter the basic rhythm depending on the need *e.g.* the rate and depth of breathing specially when the animal is organizing acts of behaviour, such as running away, mating, fighting and sleeping. Hypothalamus has many vital functions to perform which are necessary for survival.

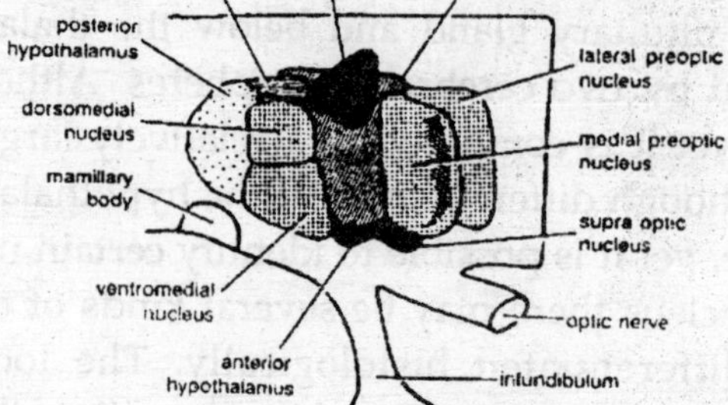

Important nuclei of hypothalamus.

Role of Hypothalamus in Feeding and Drinking : All parts of central nervous system make a contribution to such important activities as eating and drinking, yet hypothalamus is apparently the main centre organizing these acts. It controls the intake and output of water and food. Groups of neurons in the hypothalamus when stimulated make the animal go around sniffing everything to find out if it is edible or not. The animal eats voraciously when a strong current is passed through a particular region of hypothalamus. To test feeding areas in hypothalamus, experiments with male rats were conducted, these rats had electrodes implanted in lateral hypothalamus and were put in cages containing both nice smelling food and females in heat.

The males immediately showed interest in the females but as the current was passed to stimulate neurons in lateral hypothalamus, the male rats left the females and started eating. It has been known since long that tumors or lesions in particular areas of hypothalamus in man and animals lead to obesity, reason for this obesity was constant stimulation of eating centre which caused overeating and defective food intake and led to "Hypothalamic hyperphagia syndrome". Delgado and Anand (1953) observed that the lateral hypothalamus had the feeding centre and if this centre was stimulated it caused hyperphagia (excessive eating); on the other hand, bilateral lesion (destruction) in lateral hypothalamus caused complete cessation of food intake - a condition known as aphagia and caused "Lateral hypothalamic syndrome".

Under normal conditions, an animal must stop eating, when it has taken enough food, this is also brought about by neurons arranged in the hypothalamus, Anand and Brobeck observed that a centre in ventro-medial hypothalamus had reverse role, when it was stimulated, aphagia was caused and its destruction brought about hyperphagia. They named this centre as the satiety centre. Artificial stimulation of this centre makes the animal stop eating - no matter how much hungry that animal would be. Hence, two separate centres were located in hypothalamus. Lesion in pituitary gland are also often associated with weight gain or obesity, and it is identified as "Frohlich's syndrome" but it was observed by Smith in 1930 that, as long as hypothalamus remained unharmed, even removal of pituitary in rats did not produce obesity. Later in 1941 Hethrington showed that bilateral destruction of the ventro-medial hypothalamic nuclei (satiety centre) alone produced obesity, this was caused by voracious over- eating and was called "Ventromedial hypothalamic syndrome".

Hunger is actually a sensation of weakness and restlessness accompanied with powerful gastric constrictions or "hungerpangs". There are three basic theories to explain the involvement of hypothalamus with hunger and feeding among animals: Glucostat theory is based on the level of glucose in blood and can be explained as follows : Low blood sugar stimulates feeding centre in lateral hypothalamus, which initiates feeding, that causes rise in blood sugar which in turn stimulates satiety centre in ventromedial hypothalamus, the animal stops feeding; after few hours when the blood sugar level falls the entire cycle is repeated.

Thermostat theory is based on the body temperature. This suggests that hypothalamic centres for feeding and satiety are sensitive to body heat or temperature, when animal is hungry its body temperature is lowered which stimulates feeding centre, after eating, the absorption and metabolism of food generates heat that stimulates satiety centre. The other theory the Brobeck's abdominal stretch receptor theory suggests that as the stomach fills, stretch receptors are stimulated, which send inhibitory impulses to the feeding centre and stimulate satiety centre which in turn causes the animal to stop eating and when the stomach is empty accelerating impulses are transmitted to feeding centre.

Eating gives a satisfaction and pleasure to the animal. The neurons for "pleasure of eating" are near the feeding centre; in one of the experiments, rats with electrodes implanted among these neurons went on stimulating their own "pleasure centre for eating." The rats appeared to be comfortable even without eating, direct stimulation of these neurons satisfied them. Some regions of hypothalamus are responsible for hoarding food, these can not be the same neurons that make it eat, because when these neurons are stimulated, the animal stops eating and starts hoarding, therefore, in hypothalamus there are different centres for feeding, satiety and hoarding.

Involvement of hypothalamus in water metabolism or in drinking behaviour has also been well established. The antidiuretic hormone of the posterior pituitary, the hormone that promotes water conservation by facilitating reabsorption of water from the kidney, is under the control of the supraoptic nucleus of the anterior hypothalamus. This is based on findings that lesions destroying the supraoptic nucleus bilaterally, abolish normal water metabolism (Ranson and Magoum, 1939) whereas, stimulation of supraoptic nucleus produces antidiuresis in animals (Harris, 1947).

There is yet another region in hypothalamus which controls water intake, Anderson (1953, 1955) found that electrical stimulation of a nucleus in the lateral hypothalamus (a point very close to feeding centre), produces excessive drinking (hyperdipsia) in goats; he prepared a number of goats with fine hollow needles penetrating into the lateral hypothalamus, he allowed them to drink as much as they could and then introduced concentrated salt solution that caused the goats to drink frantically. These animals, although previously satiated with water even drank salt water after stimulation with saline, which they would not have normally touched even if extremely thirsty; destruction of this area resulted in permanent refusal to drink (adipsia). Dr. Anderson noted that a rat with a lateral hypothalamic lesion was disinterested in water and even developed aversion to it. If water was placed in its mouth it did not swallow and let the water run out. Neural centres in hypothalamus causing the animal to feel thirsty are under the control of osmoreceptors, these receptors are sensitive to osmotic pressure, temperature and salinity of blood.

Therefore, we feel thirsty when we have just finished our major meals because at that time the salt content of blood increases, which stimulates the drinking centre or osmoreceptors; and we also feel thirsty when our body gets heated in the sun or after running because by these actions the blood temperature rises, which in turn stimulates drinking centre of hypothalamus. Centres for feeding and drinking are set very closely near each other in lateral hypothalamus, and perhaps their circuits overlap but they function separately because of the difference in neurotransmitters they have, the drinking centre has noradrenalin at the synaptic ends while feeding centre has acetylcholine.

Broadly speaking, drinking is regulated in following steps:

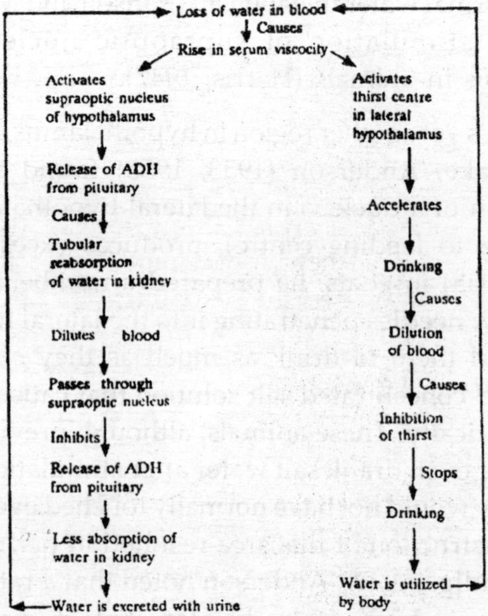

Role of Hypothalamus in Reproductive Behaviour : The hypothalamus controls reproductive behaviour either through pituitary or directly. The pituitary gland secretes several hormones concerned with regulation of sexual behaviour and reproductive physiology. The gonadotrophic hormone stimulates the development of ovaries and the secretion of estrogen in the female; it triggers spermatogenesis and secretion of androgen in the male. Other hormones of pituitary like luteinizing and lactogenic are involved in the regulation of maternal activities; and since hypothalamus influences pituitary activities indirectly it exerts regulatory influence on sexual and maternal behaviour.

The nuclei of the hypothalamus organizing sexual behaviour are situated in anterior hypothalamus. Fisher and co-workers (1938) observed that female cats with anterior hypothalamic lesions did not mate, this observation was further confirmed subsequently in female guinea pigs by Brookhard (1940). He observed that lesions in this area impaired sexual behaviour and when electrical stimulation was provided through implanted electrodes the sexual behaviour was evoked. As soon as the current was stopped, the sexual behaviour immediately stopped. In other experiment, rats had 15 to 20 ejaculations per hour, when anterior hypothalamus was electrically stimulated while having none during similar period without such stimulation (Vaughan and Fisher, 1962).

In another experiment, lesion in medial part of the anterior hypothalamus impaired the sexual activity of both male and female rats and lesion in ventral portion of anterior hypothalamus in guinea pigs abolished normal mating behaviour (Day et al., 1940). Using neurophysiological technique Harris (1958) produced estrous behaviour in female cats by introducing very small amounts of estrogen into the anterior hypothalamus through a canula. Evidence of presence of estrogen sensitive neurons in the hypothalamus was obtained histologically by means of autoradiography (Michael,

1961-1962) in which the hormone was labelled with radioactive substance and their uptake was noted on light sensitive plates.

Green in 1954 implanted electrodes in hypothalamus in female rabbits and recorded altered neural activity during courtship and mating. The same kind of electrical activity was observed from the anterior hypothalamus of estrous cats during their vaginal stimulation. In another study, it was recorded in male rabbits that sexual activity like intercourse created a great deal of activity among the neurons of anterior hypothalamus and preoptic area. All the above mentioned experiments indicated that anterior hypothalamus has the centre for sexual behaviour and this centre regulates behaviour directly or through pituitary.

Role of Hypothalamus in Reward and Punishment : It has been found that, in rats, there are at least two different pleasure centres-one is associated with satiety from eating and the other is associated with sexual pleasure.

There are groups of nuclei in posterior hypothalamus which are known as reward and punishment centres. These centres were discovered by Olds as early as 1953 in MacGill University in Montreal. He implanted electrodes in the pleasure centre of a rat, and connected them to a circuit incorporating a lever which the rat itself could press with its paws; whenever the rat pressed it, it excited its own pleasure centre; the rat found it extremely pleasing and pressed it 5000 times in an hour without any pause till he fell with exhaustion. Olds kept the same rat in a cage and starved him, later, Olds offered the rat nice odorous food, but still this rat preferred pressing the lever to get the pleasure, rats pressed the lever without any pause, though tired still he dragged the body back to lever for more and more pleasure, the rat did almost anything to get at the lever, even hopped across a painful electrified grid. Similarly, in another experiment monkeys kept awake for 48 hours stimulating the centre for sexual pleasure.

It has been shown that stimulation of another specific cluster of nuclei in posterior hypothalamus causes the animal to show all the signs of pain, displeasure and punishment (discomfort). Prolonged stimulation of such areas causes the animal to become severely sick and sometimes this may even lead to death. Stimulation in the pain and punishment centres can frequently inhibit the reward and pleasure. Administration of tranquilizers inhibit both reward and punishment centres, thereby greatly decreasing the effective reactivity of the animal.

One thing is important that pleasure centres must have certain level of excitement at any given time, eating is an obvious example; pleasure is caused by eating when one needs food and is hungry, on the other hand if one is forced to eat more and more, eating would evoke nausea, disgust and vomiting.

Role of Hypothalamus in Fighting and Fleeing : When Hess (1953) explored the hypothalamus of the cat, he discovered that stimulation of certain nuclei in lateral and posterior hypothalamus caused aggression, attack, fight or submission. Hess further found that when he stimulated these neurons in hypothalamus with a minimal current the animal just showed a state of vigilance, as the current was increased the reaction became more intense, finally when the intensity of current was increased more and more at the same locus in hypothalamus the animal showed an actual attack. The cat jumped with rage on experimenter and when the current was switched off, the animal quickly calmed down.

The extensive investigation of Hess as early as 1954, using the implanted electrodes technique revealed that lateral and posterior hypothalamus have centres for fighting and fleeing. Destruction of these regions produces unresponsiveness in animals (Ranson, 1959). However, certain nuclei in anterior and ventro medial hypothalamus also have control over fight and fleeing in cats and fowls.

Role of Hypothalamus in Sleeping and Waking : Nearly all animals sleep. Fish go through periods of inactivity. The dolphins submerge when asleep. Snakes and lizards sleep with their eyes open. Cows, sheep and goats sleep very little. Antelopes have even shorter periods of sleep. Sleep-wake phenomenon has been of great importance and has drawn attention of various workers since early. A Russian surgeon reported a very interesting case of a soldier on whom he had operated during war time. A metal fragment had entered the patient's skull and had lodged in the anterior hypothalamus. When the surgeon operating under local anaesthesia tried to pull the piece of metal out, the patient immediately fell off to sleep.

The surgeon got upset thinking that the patient had fainted. After sometime the patient woke up and when questioned, said he had an irresistible urge to sleep. The surgeon derived from the incidence that anterior hypothalamus had the centre for sleep. The involvement of hypothalamus and adjacent regions in the regulation of "sleep-wake"' was first observed by Von Economo as early as 1929 in a patient suffering with sleeping sickness. Later in mid fifties Hess re-established the fact by performing extensive experimentation that there is a waking centre in the posterior hypothalamus.

In the pathological condition of epidemic encephalitis the patient goes into a state of coma or prolonged sleep. This is due to inhibition in activity of posterior hypothalamus. The sleep centre is in the anterior hypothalamus, if by accident this centre gets destroyed the animal would never go to sleep and contrary to this, the stimulation of this centre would cause prolonged sleep. Ranson (1959) dropped the neural activity in this particular nuclei in monkeys, this caused 4-8 days continuous sleep in monkeys and even after the experiment, monkeys showed drowsiness for several months.

Hypothalamus and Other Activities : Hypothalamus is the seat for emotions *i.e.* various components of emotional reactions are organized here. This centre is situated in the posterior hypothalamus and controls fear, anxiety, pleasure, rage, irritation and attack. Lesion in the centre can make a person like a vegetable who is difficult to arouse emotionally.

Lesions in medial preoptic area of rat produce severe deficits in maternal behaviour, such females neither build nests, nor retrieve (get possession) pups, they do not even nurse them (Numan, 1974), whereas, stimulation of the same area can produce maternal behaviour.

Hypothalamus along with frontal lobes regulates the predatory behaviour in many animals. Wasman and Flymn (1962) stimulated cat's hypothalamus and noticed their predatory behaviour.

Midbrain (Mesencephalon)

The midbrain or mesencephalon extends from the pons to the lower portions of the diencephalon. It is about 2.5 cm in length. The ventral portion of the midbrain contains a pair of fiber bundles known as cerebral peduncles. These fibers convey impulses from the cerebral cortex to the pons and spinal cord. They also contain sensory fibers that pass from the spinal cord to the thalamus. The dorsal portion is called tectum which contains reflex centres for movements of eyeballs and head in response to visual and other stimuli and movements of head and trunk in response to auditory stimuli.

The mid brain serves as a motor relay station for fibers passing from cerebrum to cerebellum and spinal cord.

Hindbrain (Metencephalon and Myelencephalon)

Pons Varolii : Pons lies just superior to medulla and it means bridge. It measures about 2.5 cm in length. It is generally made up of white matter with few nuclei scattered. The fibers of pons belong to motor and sensory tracts. It connects the medulla and cerebrum with the mesencephalon and also joins the two cerebral hemispheres. 5th crannal nerve (trigeminal) is rooted in the pons while, abducens, facial and auditory are seen emerging at the junction between pons and medulla.

Cerebellum

It is the second largest portion of the brain, it is found below the occipital lobe separated by transverse fissure and very close to the pons. The gyri and sulci of the cerebellum are narrow and numerous. Functionally, the cerebellum is concerned with co-ordination of the actions of muscles or motor activities *e.g.* touching the tip of your nose with finger, suggested that cerebellum is like a recording machine that tapes all those muscular actions that are complex and habitually used. Initially the brain has to use its higher faculties to decide the sequence but after few trials, the sequence is taped or memorized by cerebellum which can be played back on command without involving inch of motor cortex. From playing a piano to threading the needle are all functions of cerebellum.

Precise body movements depend upon computer action of cerebellum which receives impulses from the cerebrum and skeletal muscles. Cerebellum evaluates these impulses and sorts them out and relays the information to those muscles that are required to bring about a given action. Thus with the aid of cerebellum's precision a cricket player times his stroke and returns the ball across the boundary, if a precise coordination of muscles was not there then this cricket player would have hit the ball right in the hands of a fielder for a catch out.

The maintenance of posture *i.e.* body equilibrium depends largely on cerebellum. Damage to the cerebellum also affects the ability to walk, victims of cerebellar disorders, lurch (roll suddenly to one side) so much as if they were drunk. Damage to cerebellum cause erratic skeletal movements or Ataxia (uncoordinated movements) such a person when blindfolded will not be able to touch the tip of his nose with a finger.

Medulla

Anatomically the medulla oblongata can be considered to be simply the enlarged upper end of the spinal cord. It provides conduction pathway for motor and sensory impulses. Functionally, it contains the nuclei of several cranial nerves and it also has vital reflex centres concerned with the control of breathing (medullary rhythmicity area), heartrate, (cardiac centre), diameter of blood vessels (vasomotor centre), swallowing, coughing vomiting, sneezing and hiccuping. A cluster of nuclei from the reticular formation, the sleep-wake mechanism in our body is also housed in medulla.

Limbic System

Limbic system includes different parts and nuclei of cortex and surrounding tissue, it is nested deep in the temporal lobe. They function singularly or collectively to generate feelings that strongly influence our behaviour. The limbic system promotes us to do what is most likely to keep us alive and well, to run from what is dangerous and to fight for what we don't like.

The major role of limbic system is to recognize emotional upsets from equilibrium and help to reestablish a state of emotional equilibrium or feelings which could be pleasant or unpleasant, collectively known as hedonic feelings. Pleasant feelings are like seeing a person you love, listening to a good song, going for picnics, love, joy, laughter and contentment and unpleasant feelings include attending a funeral, having a picnic cancelled, fear, grief, hate and resentment. The brain copes with emotional upsets with the help of limbic system. The limbic system also helps to prepare the body for action.

We all know the cold moist palms, racing heart and butterflies in our stomach when we are excited or nervous. The limbic system and hypothalamus cope up with these feelings and bring back the normal emotional level.

In one way limbic system proposes and the cortex disposes. The septum pellucidum is associated with pleasure. A rat having an electrode implanted in this region, would stimulate it through a lever in preference to food and mate. It would keep stimulating pleasure centre till the point of exhaustion; it would cross a painful electric grid to reach the lever. The amygdaloid body is associated with aggression. A patient having a small tumor on amygdaloid body used to get uncontrollable anger, he killed many people during the attack, an electrode implanted in amygdaloid region could inhibit the neural activity and cure the person. Ralph was a college student and used to get bouts of aggression. He was suffering with epilepsy and during the attack neural activities were recorded from amygdaloid region.

An electrode was implanted in this region and Ralph was provided with a cordless device; which he carried in his pocket and by pressing a button on that device he kept the attack in check. It has been observed that removal of amygdaloid nucleus can turn aggressive animals docile. The cingulate gyrus and hippocampus are associated with memory, the thalamus and mammillary body appear to act as organizers. Mammillary body also acts as relay station between the limbic system and rest of the brain. In human beings, electrical stimulation to amygdala, hippocampus and septum pellucidum have led to rage, agitation, anxiety, elation, excitement, colourful vision (hallucination), sexual interest and relaxation. The body of fornix and dorsal fornix link parts of the limbic system with each other and with other parts of the brain.

In recent years it has been learned that hypothalamus and limbic system are particularly concerned with pleasant and painful feelings. Electrical stimulation of particular regions of limbic system pleases or satisfies the animal, whereas, electrical stimulation of other regions cause extreme pain, fear, defence and escape reaction. Almost everything that we do depends on reward and punishment. If we are doing something that is rewarding we continue to do it but if it is punishing, we cease to do it. Therefore, the reward and punishment centres which are present in limbic system undoubtedly constitute one of the most important of all the controllers of our body activity.

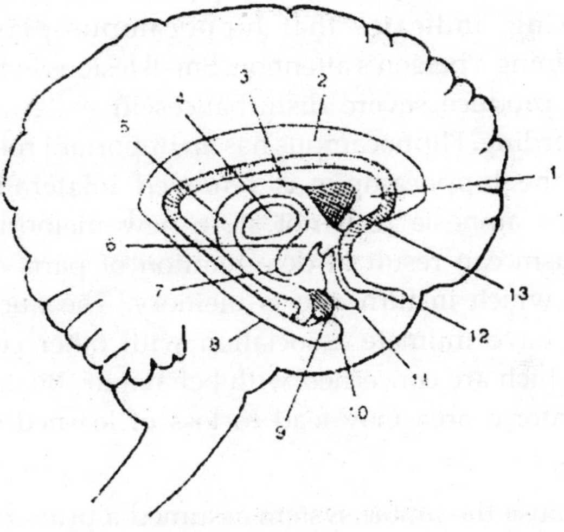

Limbic system.

(1) Corpus callosum (2) Cingulate gyrus (3) Body of fornix (4) Thalamus (5) Dorsal fornix (6) Column of fornix (7) Mammillary body (8) Hippocampus (9) Amagdaloid nucleus (10) Uncus (11) Optic nerve (12) Olfactory bulb (13) Septum peliucidum.

Amygdala stimulation also causes raising the head or bending the body, circling movements, licking, chewing and swallowing in cats or at other times stops all movements of an animal, freezing the animal in its present postural state, a phenomenon called the arrest reaction. As stated earlier stimulation of amygdala mainly produces aggressive behaviour, while their destruction causes violent animals to become passive, to exhibit no fear of object that formerly produced fear *e.g.* if amygdaloid region is stimulated in cat's brain, it will stop responding fearfully towards a dog, instead it may attack the dog. In addition, stimulation of certain amygdaloid nuclei can cause sexual activities also.

Besides this, stimulation of another group of neurons in the limbic system called hippocampus in conscious man causes immediate loss of contact with person with whom he might be talking, indicates that hippocampus plays a role in determining a person's attention. Small lesions in hippocampus of rats produce severe disturbances in maternal behaviour and boarding. Hippocampus has an important role in learning. When the hippocampus is removed bilaterally, a person develops amnesia (can not store new memories). Chronic alcoholism can result in deterioration of parts of the limbic system which in turn affects memory. The nuclei of limbic system have intimate association with other centres of the brain which are concerned with behaviour. Bilateral injury to the pyriform area may lead to loss of learned responses to odours.

Because the limbic system assumed a primary function in emotions such as pain, pleasure, anger, rage, fear, sorrow, sexual feelings, docility, and affection, it is sometimes called the "emotional-brain".

Reticular Activating System (RAS) or Reticular Formation (RF) : This refers to a particular region of the brain which includes parts of medulla oblongata, pons and midbrain. It was once a mysterious part which has recently received a great deal of attention. It is located in the central part of the brain stem, no bigger than our little finger. It extends from medulla towards the thalamus and is actually a column of gray matter.

If neural activity in the RAS drops below a certain level

the system puts us to sleep. Coma or a state of unconsciousness from which even strongest stimuli can not wake us up develops when there is a damage in brainstem. Drugs that make us sleepy depress synaptic conduction through RAS.

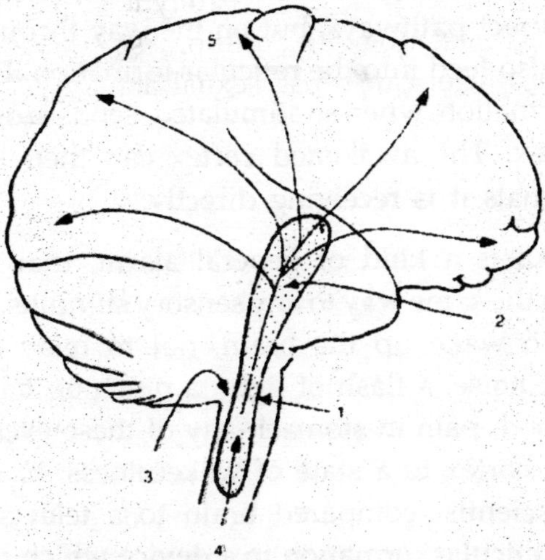

Reticular formation.

(1) Reticular activating-inhibiting system; (2) Visual impulses; (3) Auditory impulses; (4) Other sensory impulses; (5) Messages to cerebrum.

Investigators have discovered that this bit of nerve tissue is a far more important structure than anyone can think. Without it, an individual is reduced to a helpless, senseless, paralyzed mass of protoplasm.

The actual seat of the power to think, to perceive and to respond to a stimulus lies in the cortex of the brain, but the cortex cannot perceive or think unless it is "awake". The RAS is like the starter in an automobile, starts the brain engine running, and keep the individual in a conscious state. It was not until the end of 1940 that two physiologists Magous and Moruzzi discovered this fact, they were exploring the mystery of the wakefulness by means of electrodes implanted in the brain stem of a cat. They found that stimulation of this area with a small electric current would awaken a drowsing cat. It was later discovered that all the main sensory nerves in the body have brush like branches which stream into the reticular formation. Sensory signals from all parts of the body go to the cortex by direct pathways, but on the way through the brain stem they also feed into the reticular formation. Evidently the reticular formation, when so stimulated, sends arousing signals to the cortex. The awakened cortex can then interpret the sensory signals it is receiving directly.

The RAS is a kind of general alarm, that is to say, it responds in the same way to any sensory stimulus. Its response is simply to wake up the brain, not to relay any specific message. A noise, a flash of light, a pinch on hand, smell of burning wood, pain in stomach-any of these excites the RAS to alert the cortex to a state of wakefulness. E. Asratyan an American scientist compared brain to a television set and compared reticular formation to a device which regulates the loudness of the sound and the contrast and brightness of image. The reticular formation concentrates the attention of cortex on the most important stimulus. For example, in a moment of danger the pedestrian sees only the car coming towards him.

The RAS becomes selective in its sensitivity, *e.g.* mother may be instantly awakened by the faintest cry of her child, whereas, a father may sleep but would be aroused by smell of smoke or burning wire. Similarly, a city dweller may sleep peacefully inspite of the traffic noise while his visitor from the village spends a sleepless night.

2

The Communication

Acts of communication tend to be fixed in forms, to act as releasers for other actions, and to develop very similarly in all members of a species. Perhaps ideas such as that of the fixed action pattern or the innate releasing mechanism would have come less easily to mind if ethologists had been more interested in grooming or feeding behaviour than in aggressive or courtship display. But then they had every reason to be enthusiastic about studying the later, because the signal that animals produce in communication are often remarkably striking. Discovering just what they are saying to each other is one of the most fascinating fields of ethology.

Whenever we see an animal that is brightly coloured or boldly patterned, or one that is making a loud and obvious noise, we can be pretty certain that communication of some sort is taking place. Standing out from the back ground is dangerous and there must be some benefit to doing so which offsets the risk involved. Most often the communication is between members of the same species, and the risk come when predators cue in on this as a means of finding a meal. Sometimes, however, different species communicate with one another and here it is often prey animals that forsake their camouflage and communicate directly with predators that might eat them. These different sorts of relationship between predators and their prey make a useful starting point as they illustrate some basic principles of animal communication.

How Signals Convey Information ? : Normally animals are as cryptic and silent as they can be and it is amazing the accuracy of camouflage that some of them have achieved. Selection quickly removes those that fail to comply. The slightest rustle on a woodland is enough for a barn owl to home in with deadly accuracy on the mouse that made the sound.

As with other behaviour patterns, then communication has costs as well as advantages. The animal that advertises for a mate could well end up being eaten instead.

In Frog : (Physalaemus pustulosus) from Panama, the males call in choruses to attract mates. There is, however, a problem that fringe-lipped bats home in on the calls and can catch and feed on the frog with great success. The frogs cannot obtain mates unless they call, so they must do so, but the intense selection pressure that the bats provide had made them minimise their risk. On all but the darkest nights, they spot approaching bats and while they often ignore the smaller species their chorus will shut down the minute a fringe lipped one appears above their pond.

European Songbirds produce another good example of communicating when a predator is around. At the same time, as minimising the risk of being captured a high-pitched and thin 'seet' alarm call is produced when they spot a hawk Interestingly the call is almost identical in all the species that produce it. Two factors have probably contributed to this. First there is no need for the call to incorporate information about which species the caller is provided that its mate or young gather the information that there is a hawk about and so know that they should beware.

Thus there is no reason for the calls to be different. Secondly, there is good reason for the calls to be the same. This particular sound has characteristics which make it especially difficult to locate, so all species seem to have homed is on the same form to reduce the risk of capture as much as possible. Animals usually locate sounds by detecting differences between their ears for example if a sound is louder in one ear than the other, it obviously comes from the side of the head where it is most intense. However, with thin, high pitched sounds like the 'seet' call, such differences are slight and locating the caller becomes very difficult for a predator. Certainly, to a human, the call is almost ventriloquial and it is very hard when hearing it to decide where the caller is perched.

These examples illustrate how pressure from predators may mould the form that animal signals take, assuming that most animals avoid communicating with predators as far as possible. But what about cases where signals are thought to be produced specifically to communicate with predators ? Very often the information that these provide is misleading. Such signals are usually referred to as deceitful, without of course implying that the animals producing them are conscious of lying in any way. They make such signals simply because selection favours the transmission of false information.

The plover that drags its wings along the ground as if injured and so distracts a fox from the area of its nest is communicating with the fox, but the information it is providing is false. Its wing is not really broken and once it has lured the fox far enough away for its chicks to be safe, it will fly strongly off. Likewise, the eyed hawk moth that flicks its forewings forwards to reveal two large eyespots deceives the bird about to eat it by mimicking a large animal with its eyes spaced wide apart. Insects which taste nasty or having stings often advertise their distastefulness by bright colouration so that predators soon learn that they are to be avoided. In this case both predator and prey gain. The prey escapes damage and the predator avoids an unpalatable meal. The black and yellow stripes of wasps and bees warn birds of their wings hover flies are harmless but similarly striped and they may thus avoid capture without the need to be dangerous in any way.

Prey animals can thus warn predators to keep clear, or deceive them into doing so under false pretences. Predators can behave in a similar way, in this case appearing to be harmless when they are actually lethal. The polar bear is white so that it blends in with the snow and can creep up on its prey unseen. An American buzzard, the zone-tailed hawk which preys upon small mammals also merges with its background, but it does so by being just like a vulture in silhouette. Prey animals pay little attention to vultures as they eat only carrion, so the hawk gains from this deception by getting close enough to pounce without the prey running off. The green colour and rather leaf like limbs of praying mantis probably also helps them to sheik up on prey unseen. These examples are of predators being camouflaged.

However, predators also may produce signals which mislead their prey. An especially nasty example is that of firefly photuris. Male fireflies produce a signal of light flashes in a pattern typical of their species, and females respond with another pattern which is also species specific, thereby enticing the male to approach and mate. The female photuris, however, is a predator, and when she detects the flashes made by the male of another species she responds with the appropriate female version. The unfortunate male flies in to mate and meets a sticky end instead.

These examples illustrate the fact that predators and prey may either fit in with their backgrounds and so avoid communicating or may actually communicate with each other, though in this case the information they provide is often false. That communication does not always involve information which is useful to the animal receiving it is not true, however, just of predators and prey. Much of the communication between members of the same species can be interpreted in similar terms.

Information and Manipulation

In the past, ethologists often used to refer to communication between animals of the same species as involving the sharing of information for their mutual benefit. Indeed, much communication may be just this. For example, when potential mates meet it is advantageous to both of them then they communicate to each other accurately, information about the species to which they belong, for if they are not members of the same species both of them will waste their breeding effort. However, animals will produce signals when this is to their advantage, regardless of whether others benefit from receiving them.

Exactly the same happens within a species as we might expect from the idea that animals are essentially selfish, behaving so as to maximise their own inclusive fitness. So, while the information that passes from one to another when a signal is transmitted may be useful to the one that receives it, this will not necessarily be the case, and there are some very good examples where it is better to think of animals as manipulating each other rather than sharing information. One of these is in the mating behaviour of the bluegill sunfish. Most males of this species do not mature until are seven or eight years old, when they set up territories to which they attract females for mating. The females are smaller and look quiet different. Sometimes two of them may spawn simultaneously with one of these males, sperms and eggs being shed into the water as they swim around together on his territory.

Deceit enters into this scheme of things because some males mature at only 2-3 year old, when much smaller than the normal. These just look like females and join in the mating of spawning pairs. The territorial male does not drive such a rival off as he cannot tell that he is not a second mate, but actually he is a male and he shares in the fertilization of eggs that the female produces. Of course, all males cannot mature early, or there would be no territorial ones to attack females in the first place. Thus this is an example of a mixed evolutionary stable strategy with a balance between the two possible modes of behaviour.

Another nice example is in the calling of male toads studied by Nick Davies and Tim Halliday. Male toads cling on to ripe females in a posture called amplexus which ensures that they fertilize the eggs as they emerge. Males often flight to take up this position and the male already in amplexus kicks and croaks to repel boarders. Larger males generally win in such encounters, as one might expect from their greater strength. Being larger, however, they are also able to produce deeper sounds and Davies and Halliday showed that these too had an effect. They silenced males in amplexus by fitting elastic band round their arms and through their mouth. They then saw whether other males attempted to displace them.

As expected, medium-sized males spent more time attempting to displace small males than large ones. However, if the deep croaks of a large male were played from a loud speaker, a small male in amplexus was harassed much less by his larger rival. Clearly, toads can use calls to assess the size of others and will be less likely to fight with those whose calls are deeper. Here again then, the information transmitted is accurate and the animal receiving it act on it appropriately.

To summarise communication involves the transmission of information, in the form of a signal from one individual to another. As it is said that this information may be true or false, the important point being that producing the signal should be of benefit to the animal that does so. Sometimes deception may be possible and the signaller may gain from it, but this will not be possible if the animal receiving the signal can detect that the signal is misleading. Deception tends, therefore, to be rather rare, for the more the animals practice it, the more it will be best for the recipient to assume that the information it is receiving is false. If almost every plover which appears to have a broken wing is really feigning injury, then a fox discovering one would gain most by not following the adult but searching the area for a nest or chicks.

Messages and Meanings

A useful way of analysing communication was devised by the American ethologist W. John Smith. He pointed out that it could be looked at either from the point of view of the animal that sent the signal, in which case one would ask what the message encoded in it was, or from the point of view of recipients, to see what its meaning was for them.

If we turn back to the singing bird, what is the message incorporated in the song and what meanings may it have for those that are listening ? The most obvious point is that, as every bird-watcher knows, song is usually clearly distinct between species, so that it includes information as to the species to which the singer belongs. Further more, songs tends only to be produced by male birds in breeding conditions, they usually only sing when they have a territory and in some cases they stop singing as soon as they obtain a mate. In such a species, the basic message of song may be 'I am an unmatched adult male sedge warbler on my territory and in breeding condition.'

In many animals, signals are, like song, primarily concerned with reproduction, attracting mates and repelling rivals, being their main role in communication. However, especially in species that live in social groups, there are many other messages that it may be useful to convey. Animals that move around in groups often have obvious patterns on them which can be seen some distance off and enable them to maintain contact with their companions. Call notes can serve a similar function, especially where visibility is poor. These are simple sounds and patterns, and they tend not to vary much. Their messages may simply be 'I am here' though usually they also communicate with the signaller's species, and there are cases where, despite thin simplicity, they vary enough to indicate individual identity.

The 'seet' call of small birds is most reliably produced by the appearance of a hawk but, in the breeding season, birds with young may also call in this way when alarmed by a dog or a human near to their nest. Most animal signals are like this, the message they convey is a generalised one, rather than being precise like a word in our language. A famous exception is the alarm calling of vervet monkeys. These animals have several different calls which they produce when alarmed and three of these are specific to particular sorts of predators a snake call, a leopard call and an eagle call. Here the information in the call is very precise. The calling animal might as well be shouting snake, leopard or eagle, provided that its listeners were able to recognise the meaning of those words as we can do.

To understand messages, we have to study the particular situation in which an animal produces a signal, in other words, try to discover what causes it to do so. In the case of meanings, it is the recipient of the signal that we must study. How does it respond when the signal is received ? This response may vary from one animal to another. In the case of a contact call, for example, other members of the flock may approach so that the group keeps together, animals from other flocks may keep their distance, predators may be attracted with the prospect of a meal, and animals of other species are likely to ignore it as irrelevant to their interests. The same signal can thus have many different meanings, depending on the listener and on the exact context in which it is received. To stress a point made earlier, the critical factor is that, for a signal to be produced, the advantages to the signaller must outweigh the disadvantages. Otherwise, there is no reason why evolution would favour signalling.

There are thus many different messages that animal signals may convey, though it is unusual for these to be very precise. Most often they provide rather general information about the motivational state of the signaller, though their features can also incorporate more specific information such as the species or identity of the caller. The number of distinct signals that a species uses is not often large. What makes the signalling system flexible and varied is that the same signal may respond to it in different ways depending on the context in which it is received.

Form of Signals

The signal is the physical form in which the transmission from one to the other takes place. The message is what an animal encodes in a signal it sends, the meaning is what another makes of it. Signal is not necessarily related in a very clear way to the message that it conveys. However, certain types of signal are better adapted to certain functions and it is illuminating to consider the characteristics that makes them so.

Most animal communicate by using sights, sound and smells, though at close quarters the sense of touch and taste may come into play. Occasionally, species may use a 'sixth sense', like for example the electric sense of some fish; these can detect each other's discharge and so can signal using changes in the patterning of their electricity production. Such bizarre senses are rare, however, and so the three main senses are those with which we are best to be concerned.

Vision

Changes in posture and colour are the main ways that animals communicate through the visual channel. Vision is the most important sense used by predators that hunt by day, for it is the best for locating, and homing in on prey. This means that it is not the ideal sense for other animals to use when advertising for mates. Visual signals also do not travel round corners, so are not good for communicating in a dense environment such as a forest and the distance over which they can be transmitted in limited by the size of the signaller, small animals cannot be seen in long way off.

For all these reasons, the main use of visual signals is in private, short-range communication such as that between mates or between rivals disputing at the territorial boundary. When not in use they tend to be hidden so that the animal does not attract predators, most of the bright colours on birds are on their undersides, butterflies fold their wings so that their most colourful surfaces are hidden, and many fish have the capacity to change their colour so that they are less conspicuous.

The small pale grey fish which is now hiding in the weeds from a passing pike is hardly recognisable as the jet-black ten-spined stickle-back which was chasing off an intruder a few minutes ago. It is interesting that the most brightly coloured animals are found amongst the insects, fish and birds. These, through their capacity to fly or swim, can move in three dimensions and so have extra scope for escaping from predators.

Audition

Sound is not as private a channel of communication as is sight for noises travel out from a source in all directions and cannot easily be limited, so that they are only transmitted to one individual. Sounds can also pass round corners and they move rapidly through the environment, though admittedly less so than visual signals travelling to the speed of light. Using sound is thus a good way of advertising, provided the animal producing the signal is large enough to generate sufficient noise.

In addition, sound communication has the merit that a great deal of information can be transmitted very rapidly. The pattern of frequency in time which codes this information can be changed with great speed so that one signal can follow quickly on the heels of another to a much greater extent than would be possible with visual signals. Sound communication is therefore ideal not only for advertising over long range but also for other signalling, such as in our language, where a great deal of information is required to be transmitted very rapidly.

Communication by sound provides some nice examples where the exact form of the signal appears to be related to its function. Producing an alarm call in a form the predator cannot easily localise in a case we considered earlier. By contrast with this, advertising animals often produce broad frequency band sounds which are easy to locate and towards which others can orientate without difficulty.

However, some sounds travel better than others through different environments, and the habitat in which a species lives may also influence the form of the signal it is best for it to produce. In woods, for instance, sounds echo off trees so that the pattern in time of a series of sounds becomes distorted. By contrast, pure tones can pass relatively uninfluenced through such an environment. As one might expect from this, bird species living in the tropical forest of central America use tones in their songs more than do the open country species living nearby.

Olfaction

At first sight, the production of olfactory signals, or pheromones as they are usually called, might seem to have very few advantages. Smells diffuse only slowly through the environment, their speed and direction of travel being highly wind-dependent, and they can carry very little information for after one smell is released time must elapse for it to disperse before another signal can be employed. However, there are situations where pheromones are ideal. A small animal such as moth could not be seen or heard from more than 100 metres or so away, no matter how brightly coloured or noisy it was. Yet the male of some moth species can detect the pheromone produced by the female several kilometres away. The chemical involved is a small molecule so that it diffuses rapidly, yet large enough that its structure can be species-specific; the male need only sense a few molecules to start moving upwind to where the female waits for him.

Another merit of pheromones lies in their very persistence, they can continue to signal even when the animal is not there. Many mammals have different scent glands which they use for various purposes. Territorial species often mark the boundaries of thin range as a signal to others that the area is occupied. They may do this with special glands, or with scents in their urine or faeces, and a wealth of information can be contained in such signals. The potential intruder may glean, for example, not only where the territorial boundary is, but which individual is occupying it, what reproductive state he is in and even, by sensing how fresh the mark is, how long ago it was that the owner of the territory last went by.

Thus, though olfaction might seem to us to poor sister of the sense as far as communication is concerned, its very persistence and slowness of dissipation make it adapted for uses of its own for which the other senses would be of little use.

Animal Communication as Language : Most animal communication is very different from human language, as the messages it transmits are not at all precise and word like. They are much more akin to the signals we communicate to each other with our faces, the yawns, smiles, coughs and frowns which indicate to other people about their feelings. Language is different from this in many different ways. But the difference is not so great as might at first appear, for some animals are capable of transmitting detailed and accurate information just as we are. It is interesting that one of the best examples is a humble insect rather than an animal more closely related to ourselves. The dance of the honeybees, first described by Karl Von Frisch and for long questioned because it seemed so extraordinary, is certainly one of the most remarkable forms of communication to have been discovered.

On a sunny summer day, worker bees fly out from their live in search of the pollen and nectar on which the colony feeds. If a worker finds a good food source within about 100 metres or so of the hive she will return and perform a round dance on vertical surface of the honey comb. Her movements attract other workers and they crowd around her sensing the smell of the flower on which she has been finding. They will than fly out and search for this smell in the neighbourhood of the live, thus homing in on the food supply that she has found.

The round dance does little more than stimulate other bees to take an interest in the dancer, but the information in the waggle dance, which workers returning from more distant sites perform, is much more detailed. In this dance, the bee performs of eight, looping over the surface of the comb, just in one direction than in the other and waggling her abdomen as she goes. What Yon Frisch discovered was that this pattern informs other bees of both the direction and distance of the source of food that the forager has discovered. Direction is indicated by the orientation of the dance upon the surface of the comb. If the food is directly towards the sun, the bee will make the middle run of its figure of eight vertically up the hive, and if in the opposite direction it will go straight down.

In other words, the angle between the sun and the source of food as the bee emerges from the hive is the same as that between vertical and the direction in which it dances when back inside the hive. A remarkable feature of this system is that workers need neither see the dancer nor the sun to be recruited. It is dark inside the hive so they can only sense the direction the dancer takes by touch. Outside the live the sun may be hidden behind cloud but the bees will take up the correct direction as long as a small patch of blue sky is visible. This is because they are sensitive to the pattern of polarised light in the sky and can tell where the sun is from this.

3

The Migration

Migration in fish is a mysterious event of nature which has not yet been solved completely. This phenomenon is characterised by regular periodic movements like homing. According to Cohen (1970) about 8,000 fresh water species, 12,000 marine and 120 diadromous species (moving between fresh water and sea water) show regular migration. Approximately 70% of total migrants are found in tropical (warm) waters. Fish migration has been known since very early times but its knowledge was restricted to a few fishermen. In recent times, modern equipments have helped in tracing the migratory pathways, yet it is not as commonly observed as bird migration. Nearly all fish move in shoals (groups of fish with similar size and biological conditions; also known as schools) generally with no leader. Shoals have a definite shape which is hydrodynamically suited for movement. Simple movements are different from migration.

Definition of Migration

The term 'migration' has been derived from a Latin word and it means 'travelling'. It has been defined in various ways by different workers. Thompson (1942) has defined migration as "a shift in what may be called the centre of gravity of the population." Nikolsky (1963) regards migration as "an adaptation towards increasing the abundance of a species". It can also be defined as a periodic and directed travel due to changes in environmental conditions. Migration differs from dispersal, local movements and seasonal movements as migrants always return to the region from which they had migrated.

Homing

Homing is a complex, instinctively and physiologically dictated behaviour of migratory fish which compels the fish to return to its original locality, after undertaking migratory journeys; this instinct is seen most prominently in the 'Salmon'. Fish inhabiting smaller water bodies such as lakes, ponds and pools also have this remarkable instinct and if displaced, they can return to their home territories. Gerking (1959) has defined homing as "the return to a place formerly occupied, instead of going to other equally probable new places."

Types of Migratory Fish

According to Myers (1949) migratory fish can be classified into three main categories :

(1) Diadromous fish (2) Potamodromous fishes (3) Oceanodromous fish

Diadromous Fish: These fish migrate between fresh water and sea water. These can further be sub-divided into three types -

(a) *Anadromous Fish* : These fish migrate from sea to fresh waters for breeding (spawning) purpose *e.g.* Salmon, Striped bass, Shad, Lampreys and Hilsa.

(b) *Catadromous Fish* : These fish migrate from fresh water

to the sea, chiefly for spawning *e.g.* European eel, American eel.

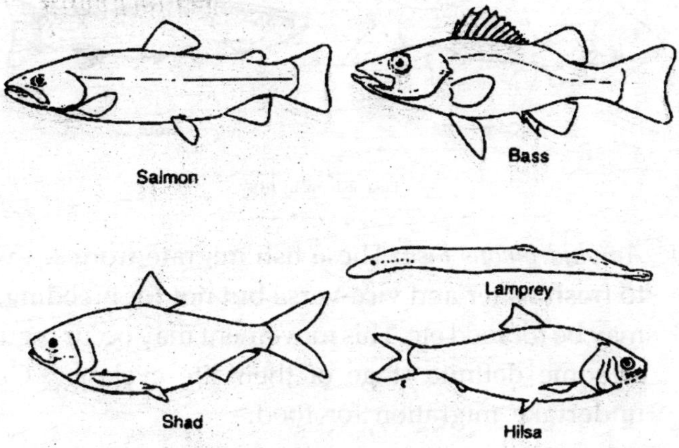

Anadromous fish.

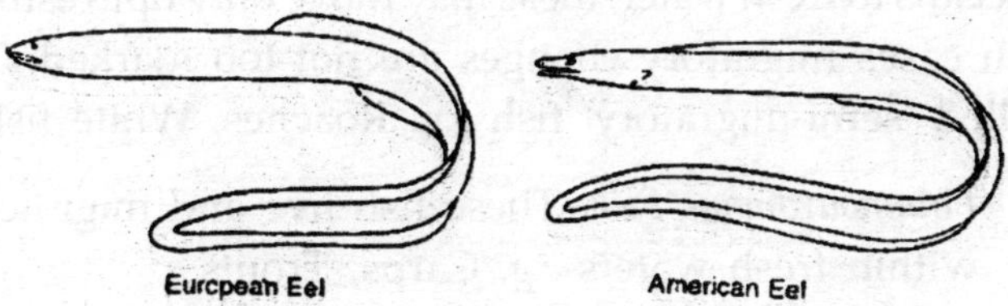

Cotadromous fish.

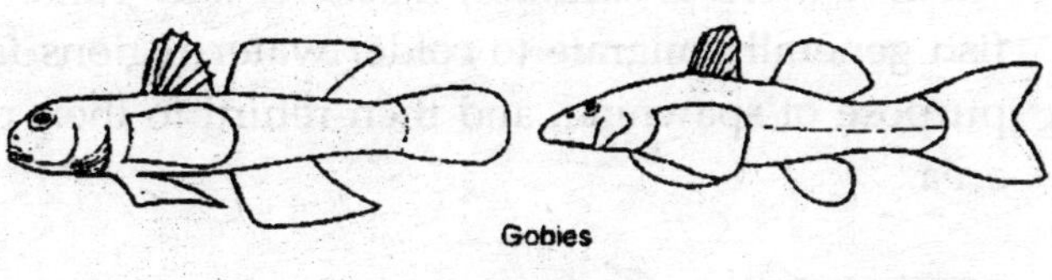

Amphidromous fish.

Semi-migratory fish.

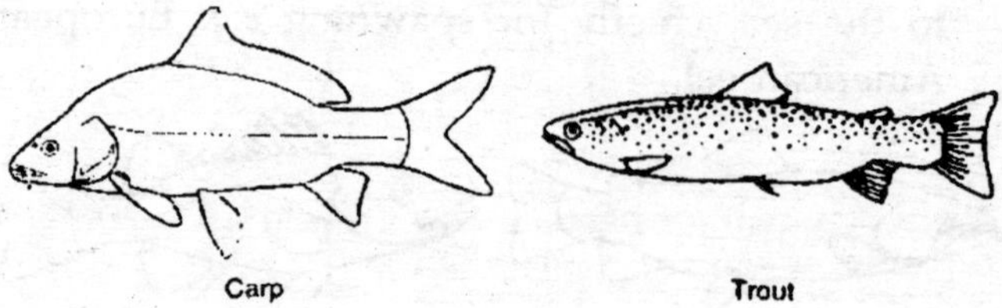

Potamodromous fish

(c) *Amphidromous Fish:* These fish migrate from sea water to fresh water and vice-versa but not for breeding, this may be for food etc. This movement may occur regularly at some definite stage of their life cycle *e.g.* Gobies undertake migration for food.

Some fish, however, do not show any significant movement from oceans to fresh water; these may move only upto estuaries. In such cases migratory changes are not too marked. These are called 'semi-migratory' fish *e.g.* Roaches, White fishes.

2. *Potamodromous Fish:* These fish live and migrate only within fresh waters *e.g.* Carps, Trouts.
3. *Oceanodromous Fish:* These fish which live and migrate only within the sea are called oceanodromous fish *e.g.* Atlantic herring, sardines, mackerel and Tuna. These fish generally migrate to colder water regions for the purpose of spawning, and then return to their parent area.

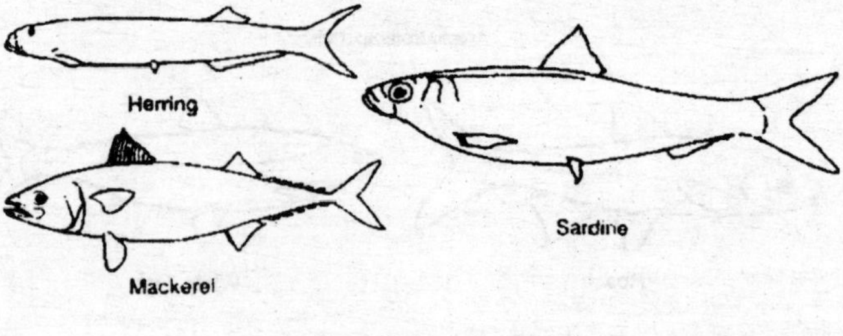

Oceanodromous fish

In some migratory species like Salmon, the young ones do not show migration, whereas in others like-the Shad, both adults and the young ones show migration.

Migratory movements are of two types (i) along with the water currents, called Denatant or against the water current, called contranatant. Some species may partly resort to both means. Movement in fish can be caused by.

Drift: When the fish are carried passively by the water currents. This type of movement is most commonly employed by larvae and rarely by adults.

Random Locomotory Movement: This may lead to a haphazard or random divergence of the species or it may lead to aggregation of the species specially if there are differences in environmental fields such as light, temperature etc.

Orientated Locomotory Movement: When movement is caused by a particular stimulus leading to migration of the species either towards or away from the source. Speed is an important factor in migrating fish.

Speed of the Fish

The speed at which different species of fish move during migration is influenced by a number of physiological and environmental factors.

Maximum Speed: The maximum speed at which a fish can move is ten times its body length per second (body length × 10/sec). This speed can, however, not he maintained by the fish for more than one minute; they have to slow down to regain stamina and may again attain the maximum speed.

Maximum Sustainable Speed: This speed can be sustained by the fish over long periods of time. It is three times body length per second (body length × 3/sec) *e.g.* in Herring of 25 cm size, maximum sustainable speed will be 25 × 3 = 75 cm/sec. in a cod of 80 cm this speed will be 80 × 3 = 240 cm/sec.

Spectacular migrants like the Salmon, Cod and Eel breed in one area but grow up and feed in another area. Interestingly, the distance between the feeding and spawning grounds may be over 700 miles.

Causes of Migration

Some scientists regard migration in fresh water species to arise basically as a result of insufficient food supply in rivers, whereas, in marine species it is due to more favourable conditions and better protection for developing eggs in rivers than in the open sea.

According to Heape (1961) fish mainly migrate for three primary reasons, thereby giving rise to

(1) Gametic migration (Spawning/breeding migration)

(2) Elemental migration (Feeding migration)

(3) Climatic migration (Wintering migration) Later Myers (1963) added another category

(4) Osmoregulatory or Protective migration (for water and mineral balance).

All the above mentioned types of migrations may occur in separate, well defined and widely separated areas or any two types may occur in the same area or a restricted area *e.g.* White fish migrate between feeding and spawning grounds but wintering migration is not seen, whereas in Gudgeons feeding and spawning occurs in the same area and only wintering migration is noted.

All four types of migrations, form a link in a single cycle which may be repeated once or several times, depending on the biological conditions of each species.

Each type of migration is being described in details in the following pages.

Gametic or Spawning Migration : According to Vasnetsov (1953) spawning migration is undertaken by a fish to ensure better survival and proper development of their eggs and larvae. Anadromous migration is more commonly noted in fish while catadromous migration is relatively uncommon as seen in Eels.

Just prior to migration, most of the up-river or contranatant (against the water current) migrants either stop feeding completely or their food intake is drastically reduced. As a result, the energy requirement for up river movement is provided by accumulated nutrients, chiefly in the form of fat deposits of the body. It is estimated that a male and female Chum salmon spend 25,810 cals. and 28,390 cals. respectively during migration.

Amongst Indian fish, Hilsa ilisha, which is found in the Bay of Bengal, provides an explicit example of anadromous migration. During breeding season it ascends the Ganges river and is seen as far off as Allahabad, traversing almost half the width of the continent.

Migration in Salmon (Anadromous Fish) : Most typical example of anadromous migration comes from Salmon. There are nine species of Salmon belonging to two genera, Salmo (S. Salar, S. Gairdneiri and S. Trutta) and Oncorhyncus (O. Tschawytscha, Chinook or King Salmon: O. Nerka, Sockeye Salmon; O. Kisutch, Coho Salmon; O. Keta, Chum Salmon, O. Gorbuscha, Pink Salmon; O. Masu, Masu Salmon). Homing instinct is strongest among Salmons and generation after generation they return to the same river or its tributary (branches of a river) where their ancestors had spawned. The Chinook or King Salmon (O. Tschawytscha) provides a legendary example of this homing instinct. They are born in a small stream and as smolts (young Salmon at the stage when it migrates from fresh water to sea) they migrate to the Pacific Ocean.

After spending five years in the sea; they return without fail to the same home shore and the stream of their birth for spawning. Salmons may travel anywhere between 700 kms to 4000 kms during such migrations. Pacific Salmon (genus Oncorhynchus) die soon after spawning, thus they spawn once in their life time but amazingly their young ones inherently know their way back to the sea. It would therefore, be interesting to know the life cycle of Salmon in general. Their eggs are laid and fertilized in fresh water streams during winter; the fertilized eggs hatch into small Salmons towards spring.

A large amount of yolk is retained by the Salmons on emerging from the eggs: this stage is called the 'alevir stage'. After further growth, yolk is reabsorbed and a size of 3-5 cm is attained. The larva is now called a 'fry' which grows further to form a 'fingerling' and remains in the river for an year or more. During this period the young fish continues to feed and grow and transforms into the 'Smolt' (stage after attaining a certain amount of maturity). Smolts stop feeding and then start their journey towards the sea, Soma species may migrate soon after hatching (*e.g.* Pink Salmon, O. Gorbuscha) while others take as long as 7 yrs. (S. Salar) before starting migration.

Salmon undergo many changes before undertaking migration. Due to cessation in feeding, changes in the size of the intestinal tract and intestinal enzymes are brought about. Some biochemical changes have also been noted in the flesh of the fish, along with changes in osmoregulation. Morphological changes result in peculiar sexual characters such as a drawn out snout and lower jaw, development of a hook on the lower jaw and enlargement of front teeth of the upper jaw in the male Chinook Salmon. These physiological and morphological changes probably provide the stimulus for the start of a migratory journey extending over thousands of miles.

The two species of Atlantic eels (catadromous fish) *i.e.* European eel (Anguilla anguilla) and American eel (Anguilla rostrata) do not differ markedly from each other. The spawning ground and spawning time of these species was discovered by Schmidt (1967). It was found that these fish spawn near the

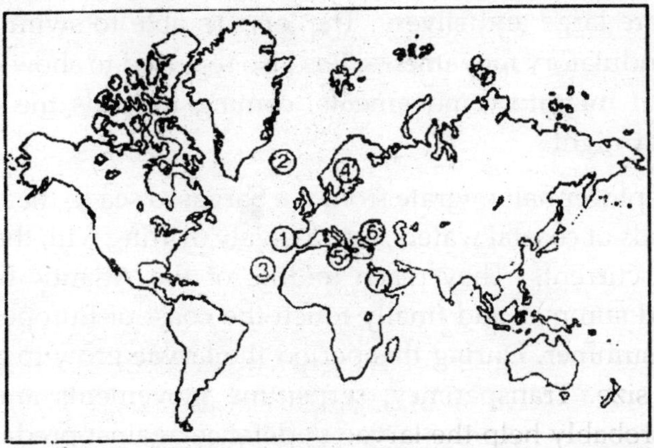

Map showing (1) Murman coast of North Africa (2) Iceland (3) Azores, Sea of Azor, Madeira, Canary Island (4) Baltic coast (5) Mediterranean coasts (5) Black sea (7) Red sea

'Sargasso Sea' in the Atlantic Ocean, south-east of Bermuda. The European eel is found towards the eastern region of the sea while the American eel is in the western parts.

Migration in European eel (Anguilla Anguilla) : The European eels are found towards the eastern region of the Atlantic ocean and in inland waters of the countries near the shores of Europe, some of them are - Murman coast of N. Africa, Iceland, the Azores, Madeira and Canary Islands, the Baltic and Mediterranean countries, the Black sea, sea of Azor and the Red. Spawning occurs during spring season and eggs are laid at a depth of 500-700 mts. with the temperature ranging between 10-12°C. On hatching, the larvae ascend to a depth of about 200 tufts. where the temperature is warmer (around 20°C). The larvae are called 'protocephaline larvae' these larvae these larvae retain a yolk sac. Protocephaline larva changes to 'Leptocephalus' after reabsorption of its yolk content. Leptocephali are confined to a depth of about 100 mts. and have a flattened leaf-like, glassy-transparent body measuring about 5 cm. The skeleton remains unossified at this stage, gut has a straight tubular structure, teeth are long and pointed and eyes are large and silvery. The larva is able to swim by eel-like undulatory movements. It is also reported to show diurnal vertical migratory movements, coming towards the surface only at night.

Leptocephali migrate from the Sargasso sea to the nursery grounds of coastal waters, by passively drifting with the warm water currents. They reach middle of the Atlantic by their second summer and finally touch the coast of Europe by the third summer. During this period the larvae grow to attain a large size. Transparency, serpentine movements and large size probably help the larvae in defence against predators on their long migratory route. During autumn and winter of the third year, Leptocephali metamorphose to form 'mini eels' or 'glass eels' called 'Elvers'.

During metamorphosis the larva stops feeding, its body becomes cylindrical with a decrease in length and the needle like teeth are replaced by new ones. Elvers are initially translucent but gradually develop pigmentation on the body. Elvers may remain in coastal waters for 1-2 years, blocking the mouth of the rivers till they are strong enough to swim up the rivers. They now measure about 15-20 cm and are as thick as a pencil.

The male fish prefer brackish water and therefore, stay in the estuarine region whereas, females prefer fresh water and are seen to ascend upriver. In the stronger fish larval teeth are lost and intestine shortens, as a result the anal aperture moves forward. Pigmentation gradually increases, transforming the eels into yellow, cylindrical fishes. These fish spend 8-10 years for feeding and growing; they become sexually mature before starting the spawning migration to the sea.

Prominent nuptial changes noted in the eels during maturation are a change in colour from yellow to silvery, enlargement of eyes, pectoral fins become pointed, shrinking of the alimentary canal and enlargement and maturation of reproductive organs. These changes trigger a movement of over 3,000 miles in search of more favourable conditions of salinity, depth and warmer temperature required by the gametes.

Migration in American eel: Interestingly, like the European eel the American eel (A. Rostrata) is also catadromous and spawns in the 'Sargasso sea'. It is similar to the European eel in many biological aspects as well. The larvae called 'Leptocephali' drift to the American coast by water currents. In about two years they reach coastal waters (nursery ground) to change into `Elvers' or 'mini eels' before ascending the rivers. On reaching parental habitat they feed and grow till sexual maturity is attained. The larvae of American eels, therefore, grow and metamorphose at an earlier age and smaller size in comparison to their European counter parts.

The two species of eels *viz.* European and American have been regarded as co-phenotypes due to similarities in their life cycles and in their spawning ground, Tucker (1959) is of the opinion that European eels are derived from the American eels. According to him the 'Leptocephali' migrate eastwards or westwards from the spawning grounds, and eventually become the European or American eels respectively.

Migration in Eels

Fresh water eels are mainly restricted to countries where the oceanic currents circulate from the spawning areas of the eels. They are altogether absent from countries on the western part of U.S., South America and Canada. They are found only on the Northern coast in African continent. Japanese eel (Anguilla japonica), another important species is found in coastal waters of Japan and China. They migrate for spawning

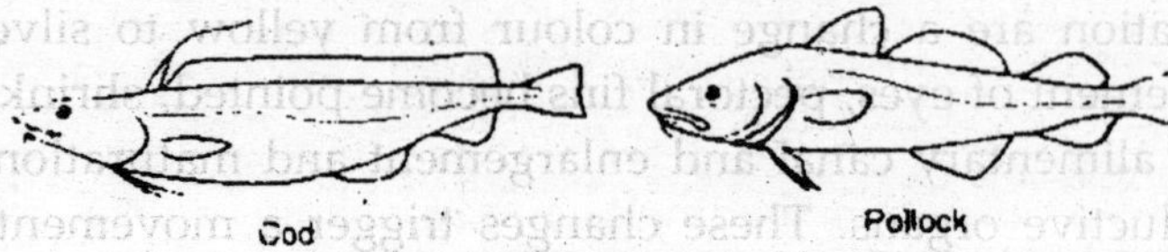

Fish showing migration within the same area.

to the North Pacific. Australian eels inhabit coastal waters of Australia, Tasmania and New Zealand; these fish spawn in South Pacific waters.

The reason for long-distance complex migrations undertaken by European and American eels are yet unresolved. Scientists believe that (i) the inland lakes, where these eels spawned during the mesozoic era, have now shifted due to continental drift and occupy the present position of the 'Sargasso Sea'. The development of eels therefore stretches over a span of a few years probably owing to the great distance between the spawning and feeding grounds. (ii) Another cause of this long distance travel is suggested to be the search for warmer temperature and higher salinity required for spawning. (iii) Vasnelsov (1953) has suggested 'safety' to be the reason of this migration as central Atlantic is a safe spawning ground due to very few predators of eggs and larvae.

Common Migrants

Some marine and fresh water species show spawning migrations within the same area. Biochemical and physiological changes are less marked in these fish as they continue to feed during migrations e.g, Cod and Pollock.

Atlanto-scandinavian Herring: The Herring (Clupea harengus) is found in the Norwegian sea where partial cyclonic circulation is seen because of mixing of warm and cold waters coming to this point from different directions. The fish spawns off the south-west coast of Norway during spring. Eggs are demersal (do not float but sink) but larvae are pelagic (upper surfaces of sea) and these larvae drift with the northward currents. The young fish spend their first two years in coastal (near land) waters whereas, the older and mature fish are fully oceanic (away from land). After spawning the fish move towards their summer feeding area between the warm Atlantic and cold Arctic waters. In Autumn, the Herring move in a Southwesterly direction along the western boundary of the Norwegian sea and subsequently change to a southeasterly direction in the East Icelandic current. In winter, they concentrate in the north of Faroes at a depth of 300-400 m in relatively cold (3-4°C) water. In March, the ripe Herring (ready to spawn) reach coastal spawning grounds to complete their annual migration, covering over 3000 kms.

Arcto-Norwegian Cod: The Cod (Gadus morhua) is widely distributed in the Atlantic, with separate stocks at Georges

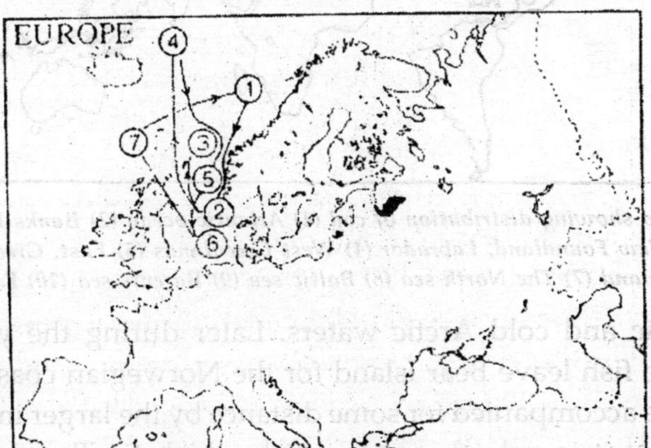

Map showing seasonal movements of Herring (1) found in Norwegian sea (2) Spawns off the south west coast of Norway (3) eggs drift northward. (4) Parent Herrings move between warm Atlantic and cold Arctic waters. (5) Then they move south westerly along the western coast of Norwegian sea (6) Then move to south easterly in east Icelandic current.(7) In winter, they concentrate in the north of Farces then back to Norwegian sea.

Bank, New Foundland, Labrador, West Greenland, East Greenland, Iceland, Faeroerne, the North sea, Baltic and the Barents sea.

The summer feeding area of the Cod in Barents sea is an extensive shelf, bounded by 200 mts. depth contours. Important places of the region are Bear Island, Murmansk coast, Spitsbergen, Hope Island and North cape. Distribution of cod recedes in autumn but during winter dense concentrations are noted on the bottom near Bear island in association with warm

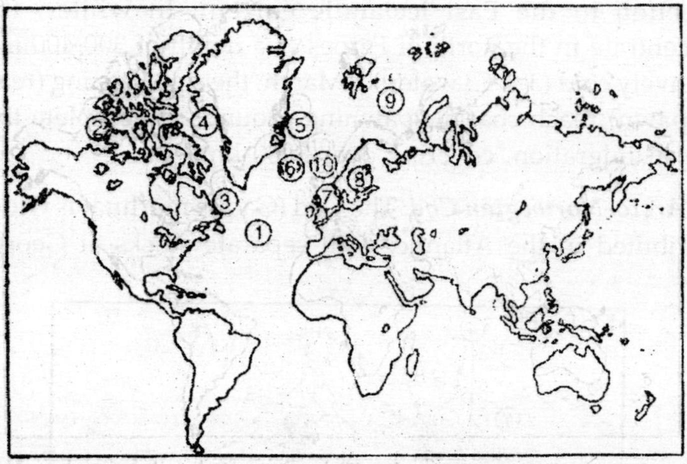

Map showing distribution of cod (1) Atlantic ocean (2) Banks Island (3) New Foundland, Labrador (4) West Greenlands (5) East. Greenlands (6) Iceland (7) The North sea (8) Baltic sea (9) Barents sea (10) Faeroeme.

Atlantic and cold Arctic waters. Later during the year, the mature fish leave Bear Island for the Norwegian coast. These may be accompanied for some distance by the larger immature fish, but it is only the mature fish which finally reach their spawning grounds within the Lofoten Islands, around February/March. After spawning in spring and early summer, the adults and larval fish return to their northerly feeding area, Pelagic eggs and larvae are carried to the north by coastal currents. The migratory path travelled by the fish covers a distance of approximately 700 miles.

Alaskan Pollock

The Alaskan Pollock (Theragra Chalcogramma) is a shelf fish, like the Cod. Its distribution in the North Pacific Parallels that of the Cod in North Atlantic. Stocks of Pollock abound in the sea of Japan, Sea of Okhotsk, Bering sea, American coastal waters as far south as California and in Alaskan coastal waters, off Vancouver Island.

The Pollock of North-east: Bering sea shows remarkable similarity with Arcto-Norwegian cod in its annual migration, probably owing to similarity between the hydrography of the two areas. The migration and distribution of both stocks is apparently related to two warm water currents *i.e.* the North-Atlantic-west Spitsbergen current for Cod and the Bering slope current for Pollock. Their summer feeding areas are spread across an extensive shelf and are located north to the spring spawning grounds. Both stocks winter in deep, relatively warm water.

Elemental of Feeding Migration : Feeding migrations are brought about mainly due to shortage of suitable food supply in the wintering and/or spawning grounds. Food requirement by the body, after exhaustive spawning or wintering conditions, probably provides the stimulus for feeding migrations, so that the fish have an access to food resources of different areas. Besides enjoying better food facilities, the individuals have a better chance of survival due to faster growth rate which helps juveniles to escape predation. Better fed individuals may grow to large size and also mature earlier and be able to produce more eggs which is for betterment and survival of the species. As a result, the life cycles, migratory patterns as well as their numbers are closely related to the production cycle and hydrographic conditions of the particular region. Commercially important marine fish, which show this kind of migration are the Salmons, Cods, Herrings, Tuna, important fresh water species are Grass-carp, Chinese roach, Gudgeons etc.

Besides regular horizontal feeding migration some species of marine fish (Mackerel, Swordfish) and fresh water fish (White fishes) show 'vertical migrations' which are correlated with the movements of their food organisms. Mackerels are reported to rise towards the surface and sink along with movement of planktons on which they feed. Sword fish simultaneously move along with Sardines which are their

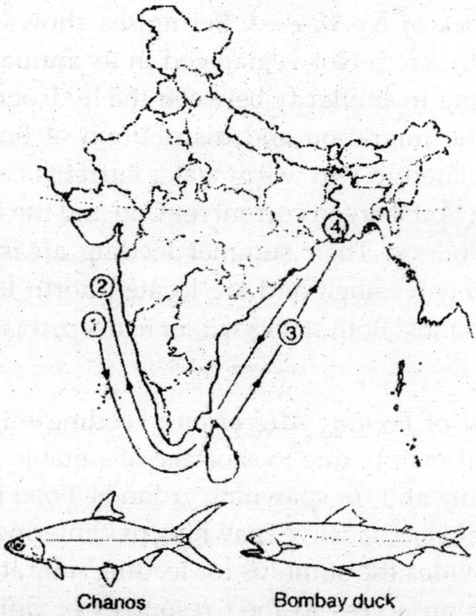

Map showing migratory routes of Chanos and Bombay duck. These fish move from (1) coastal waters of Arabian sea, off the coast of Bombay (2) migrate to Bay of Bengal (3) sometimes enter estuaries of Ganga (4) come back to Bay of Bengal and (1) return to Arabian sea.

simultaneously move along with Sardines which are their food species. Diurnal migrations are observed in Dace, Gudgeons and Razor fish, in pursuit of rising amphipods and dipterous larvae.

Among Indian species showing feeding migration are Chanos which is common in coastal waters of the Arabian sea. Harpodon (Bombay duck) is also abundant in the Arabian sea, off the coast of Bombay and migrates for feeding up to the Bay of Bengal after encircling the entire peninsular region. It migrates further upstream and has also been located in the estuaries of river Ganges.

Climatic or Wintering Migration : Wintering migrations are initiated in fish due to inactive physical conditions and low BMR (Basic Metabolic rate) either after feeding or before spawning. This depends both on the condition of the fish and environmental changes. These fish, therefore, need to move to areas with more favourable abiotic conditions which provide better protection. This migration is undertaken by fishes which are 'prepared' implying that their body can tide over the winter conditions. This is mainly achieved by hormonal and other physiological changes accompanying development of gonads. Since feeding stops, the nutritional demands of the body, although low, are met by accumulated food contents, deposited mainly as fat reserves. This ensures successful wintering. Wintering migration is commonly seen in migratory fish *e.g.* Sturgeons, Atlantic Salmon and in some semi-migratory fish like Roaches and Perches. Among fresh water species Grass carp is known to move to the wintering grounds.

Osmoregulatory or Protective Migration : Spawning, feeding and wintering migrations can all be regarded as protective migrations as they ensure further life of the fish. Yet, some mass movements may arise due to sudden unfavourable biological or environmental conditions like a stormy weather. These migrations are not cyclical.

Method of Study

Several methods are commonly employed to study the path of migration and speed at which fish move. Since fish move in water, external markings can easily be washed off and it is very difficult to follow them visually. The method to be opted varies with the species. Methods employed for different types of fish include :

Marking and Tagging: Putting specific marks by removing the scales or attaching coloured (numbered) tags on the body of the fish before they migrate and their recapture from distant areas.

Fisheries Statistics: By noting the number of fish recorded in different areas at specific times, inference from catch can be noted by the type of fish in the net at a specific time period and place.

Direct Observation: By observing shoals from land, ship or aircraft.

Echo Ranging Techniques: This is considered the most modern and accurate technique. Principally it is similar to echo-location in bats. Ultrasonic sound, is emitted from an equipment housed in a small ship. The time between transmission and return of echo from the shoals indicates the presence of the shoals as well as the depth at which they are present.

Of these, the methods most commonly and successfully used are - (1) Marking and tagging (2) Echo-ranging

Marking and Tagging

This methods includes the following techniques—

(1) Physical marking

(2) Internal marking

(3) External marking (this further has two popular varieties. viz (i) Gilbert marking (ii) Peterson marking.)

(4) Lea hydrostatic tagging

(5) Numbered bait hook tagging

(6) Plastic flag tagging

1. *Physical Marking:* Only small to medium sized fish can be marked physically, because they can be caught and

handled easily. It is generally done by clipping the fins in an orderly design or removing certain scales in an orderly fashion. However, these are not very useful for longtime study as these markings get obliterated with growth of the fish.

2. *Internal Marking:* These are usually metallic markers, in the form of strips which are introduced within the body cavity by specialized marking guns. This mark is commonly employed for Herrings. Internal markers can be detected by special electromagnetic detectors which prevent the fish being caught in canning factories.
3. *External Marking:* These are attached to the surface of the fish body either on gill covers (Gilbert mark) or at the base of the pectoral fin (Peterson mark). These usually consist of two celluloid discs attached to a wire by which they are attached to the body. These are commonly used for plaice and flat fishes.
4. *Lea hydrostatic Tagging*: These are generally used for long distance migrants like cods. It consists of a small celluloid cylinder which remains hermetically sealed. Within this cylinder printed instructions are kept for the finders, regarding informations pertaining to the fish and address of contact. Other tags in common use are the Gilbert tag and internal tags for Clupeoids.
5. *Numbered Bait Hook Tagging:* These are commonly used for Red fish and other deep sea forms which may die when brought to the surface for marking. These fish are, therefore, marked by numbered bait hooks which the fish seizes along with the bait.
6. *Plastic Flag Tagging:* These are in general used for Cod, Haddock and Whiting. Plastic flag like structures are attached to the back of the fish by braided nylon thread.

Echo-ranging Technique

This broadly includes two techniques.

1. *Acoustic Transponding Compass System:* These are used for continuous monitoring of the position, depth and direction of movement of a fish. A small radio transmitter is fitted inside the body of the fish which transmits a pulse. There are special receivers which can pick up this pulse. This reference pulse is then read through a digital decoder. Plaice, fitted with compass are released into the open sea for recording their tracks. The position of transmitting fishes is found out and plotted on maps.
2. *Ultrasonic Transmitter System*: These can be fitted into the body cavity of fish. This method is used extensively to study migration in Salmon. These transmitters send out ultrasonic sound waves with the help of which nocturnal movements can be found out.

Factors Influencing Migration

A fish may be influenced by a number of physical, chemical and biological factors during its course of migration. Some of the common physical factors are quality of water, depth, oceanic currents, temperature, photoperiod and intensity of light. Chemical factors chiefly include the hydrogen ion concentrations, dissolved oxygen, salinity and types of organic substances. Common biological factors influencing migration are sexual maturity and endocrine system, social response, response to predators and competitors and the biological clocks.

Several laboratory and field studies have been conducted to unravel the mystery by which a fish can select its course and subsequently maintain its direction during long migratory movements. Yet, very little convincing evidence is available. Some theories which have been experimented and debated at length are discussed in the following pages. These relate largely to studies on pacific salmon since it has greater commercial value in International markets. Many hypothesis have been proposed to explain this phenomenon, some of these are given below:

1. *Olfactory Hypothesis:* According to this hypothesis, a fish may recognize its surroundings and path of migration by detecting the specific odour from a stream or water body. This odour may depend on the bottom topography *i.e.* soil, plants and springs at the bottom from which the water body acquires its organic quality. Scientists have also termed this as the 'stream factor' and existence of such a factor has been substantiated by laboratory studies. Experiments by Fagurlund and co-workers (1963) further confirmed this view when they observed a positive response in Sockzye salmon exposed to 'home stream water' in the laboratory. These tests have revealed that a specific stream odour can be detected and recognized selectively by fish from several different individual odours even after long periods of non-exposure. Interference with olfaction causes inability of the fish to retrace its original path and discriminate between streams, thus confirming the view that olfaction contributes effectively to this behaviour of fish.

A.D. Hasler (1983) has associated the olfactory hypothesis with 'Imprinting' and regards it as the sensory basis for homing. He has suggested that the fish (specially Salmon) may learn to recognize the organic quality of water in the early days of its life cycle and this may become imprinted for the rest of its life. Interestingly, experiments have revealed that imprinting can take place even at a later stage (smolt) upto even the 4th year. Accordingly "a fish smell's its way home as would a fox hound". The 'paleo cortex' is, therefore, a dominant part of fish brain as it receives impulses from the olfactory tissue. This region is much less significant in non-migratory fishes as well as humans.

The foregoing account then raises the question of whether the factor responsible for a response to olfaction during migration, is genetic and is inherited due to generations of isolation or it is due to environmental imprinting of olfactory cortex which is seen throughout adulthood. Experiments with decoy odours (Donaldson and Allen, 1957) support the conditional response concept. Eggs were hatched in a stream away from the ancestral stream and the 'marked' smolts then migrated to sea. When the adult salmons returned to spawn they, however, ascended the adopted stream rather than the ancestral stream. Some probable explanations for the role that olfaction plays during movement from river mouth to natal stream by Salmon smolts which linger in estuarine waters for several weeks before ascending the rivers are -

(i) by identifying specific odour molecules of the river, (ii) by identifying a combination of odours at river mouth, (iii) by detecting the organic nature of inland water, (iv) by identifying currents and sound vibrations.

Olfaction does not appear to have a bearing in open sea migrations, yet it has been suggested that airborne odour of the sea could also be a guiding factor in salmons.

2. *Internal Biological Clocks*: The oceanic phase of salmon migration is more intriguing since landmarks and olfactory, visual, auditory or tactile features do not seem to be of much use in the open sea.

Laboratory experiments on salmons show that the fish is able to adjust its course or orientation through internal biological clocks. These adjust to seasons (through latitudes) as well as to the time of the day (through longitudes) according to appearance of light.

3. *Sun-compass Mechanism:* Distribution studies by

scientists show that open-water migrations in salmon may be directed by a sun-compass mechanism. These may be directed either by the altitude of the sun or by changing diurnal azimuth of the sun or by a combination of the two. To swim in a single direction, the fish probably 'calculates' appropriate horizontal angles to the sun during different times of the day. This, however, does not explain travel by height.

4. *Ground Water Seepage Hypothesis:* This hypothesis was given by Harden Jones (1980). According to this hypothesis, the assembly areas for spawning or otherwise are identified by marine fish with the help of chemicals that enter the sea by ground water seepage. This hypothesis could, therefore, link some anadromous species with their fully marine counterparts which spawn on or above the continental shelf and show homing to areas with ground water seepage. Nursery and spawning areas can also be linked by their geological structure and ground waters.

A report on Lofoten cod fishery states that "the idea very generally prevailing among fisherman is that cod fish seeks certain places on the bottom where there are said to be springs of fresh water which they drink in order to bring the roe to maturity."

5. *Pheromones:* It has been suggested that pheromones play an important role in identifying the spawning area. They may deter further spawning in a particular area after a sufficient number of eggs have already been laid.

6. *Celestial Reference Points:* The use of celestial points in maintaining a steady course in mid waters was suggested to be useful, especially in nocturnal travel. Nocturnal direction may also be assisted by gravity, magnetic field, oceanic currents etc. although no

conclusive evidence is available. These factors may be useful on moonless or starless nights.

7. *Tidal Transport:* This view was suggested by Dr. Arnold for Plaice of south Bight of the North sea. Accordingly to him this mode of migration saves energy and problems of direction are eliminated if the right tide is caught at the right time.

The seaward migration of salmon smolts was earlier regarded as a passive movement of drifting with downstream currents but now evidence indicates that this type of movement too is oriented in response to environmental factors.

8. *Gradient Theory:* This is based on the salmon's presumed ability to follow gradients of physio-chemical characteristics of water, mainly temperature and carbon-di-oxide. It has been shown that salmons swim in the direction of coolest waters and that high CO_2 tension repels migrating salmons.

Certain objections have, however, been raised since neither of these factors accounts for stream selectivity of a salmon. Secondly, it seems unlikely that the fish may experience sufficiently steep gradients to bring about migration and thirdly changes in physio-chemical properties must occur often enough to prevent adaptation of the fish to that stimulus.

The sensory capacity of salmon and other migratory fishes is, therefore, well established but it is difficult to pin point whether a number of senses contribute to their success or a single factor contributes at a specific period during the life cycle. The importance of age and physiological condition of the fish as determinants of the manner and time of response has been high lighted by Hoar.

4

The Motivation

It is a common analysis that the same stimulus given to the same animal at different times does not always evoke the same response, the change which the animal developed from inside. Everytime we invoke an intervening variable. As in any action beyond a simple reflex there is a change in the internal patterns of an organism which are known as drives or motives which serve the organism directly towards same specific goal. In other words, "The persistent behaviour which appears to be dominant for a time being by relatively limited group of stimuli acting upon the organisms." So in these cases, the stimulus we give and the response we get out affects the relationship between the organisms. The variables are of three types — (i) Independent variable (ii) Dependent variable (iii) Intervening variable.

To understand these variables we take an example of a rat. If a rat is presented with water, he drinks more but if he does not drink then there arises circumstances in which it becomes necessary to postulate a thirst drive. This drive is related firstly to the period of water deprivation which we termed as independent variable and secondly the amount of water which the animal drinks, we termed as dependent variable and if the postulated thirst drive arrives then we termed it as intervening variable (Tolman, 1932).

The variables which induce behaviours have specific effects on a group of behaviour patterns. Such variables are termed as motivational variables. These variables are related to the hours of food deprivation, sex hormone concentration and stimuli. We cannot observe these variables directly. Physiological studies of the brain plays an important role in providing information but the images are so complex one that we know next to nothing about the real nature of such variables. However, some people who have worked with natural conditions of animals were successful in recognizing the necessity to invoke intervening variables in behaviour. According to A. Manning, at this stage of our knowledge is beneficial to define them carefully.

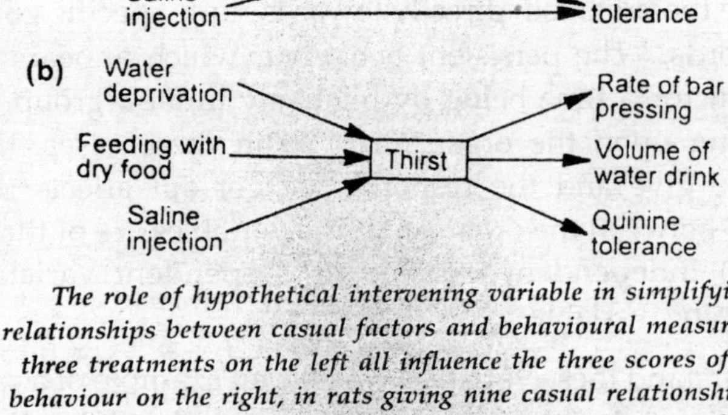

The role of hypothetical intervening variable in simplifying the relationships between casual factors and behavioural measures. (a) The three treatments on the left all influence the three scores of drinking behaviour on the right, in rats giving nine casual relationships. (b) By introducing 'thirst', the number of casual relationships is reduced to six.

More recently, the study of motivation has gained scientific status, as the psychologists are trying to understand it. The study of motivation is now studied as the study of conditions that aroused sustain and directs the behaviour.

Motivation is something in which a person or an animal moves from his activities and searches for his goal which he needs at that time. For *e.g.* a hungry animal or a man will turn from other activities and move in his environment until he finds and gets food. This behaviour is also similar when he is thirsty. So the stimulation related with such tissues are called as the simplest kind of motivation and motivated act is complete when the goal is reached and the action is called the Consummatory response.

In a complex behaviour of animal, it is easy to find that it is not just one stimulus which is effective or ineffective but a whole range of stimuli are related functionally to one another. Therefore an animal's threshold of response to all stimuli associated with food and feeding behaviour will rise and fall together. As a result of this, many workers especially ethologists have regarded motivational change as highly specific to make the animal respond in a particular way.

The result of such tendency, the animal achieves a particular goal. So we can say that Goal oriented behaviour is also known as "Motivated Behaviour". Goal oriented behaviour is better understood by taking a simple example of dog. A dog is kept without food for many hours and moves about restlessly. So if the smell or sight of food or any other stimulus such as sight of food bowl, sound of a knife being sharpened which the dog learnt to associate with food, is given then the dog becomes highly responsive towards it. So we can say that a 'drive' is striving towards some goal. This is known as goal oriented behaviour as the dog oriented towards food.

Drive concepts are useful when they lead to an economy in the number of relationships between independent and dependent variables, which must be specified. These are profitable when a positive correlation is possible between dependent variables and when it is of primary interest but drive concepts are harmful if they are used to explain diverse characteristics of behaviour. In ourselves, specific motivational states are co-related with strong feelings or emotions but we cannot measure the animal's feeling or emotions in a same way. There are various physiological changes in the body which are associated with emotion which the animal shows. For example, mouth becomes dry, sweating starts, heart beat accelerates and so on.

These changes aroused when there is a secretion of adrenalin into the blood stream but we have to only identify the motivation with reference to a function that shows reason to us. For example, if one animal attacks another animal we describe this behaviour to aggressive motivation and if it eats we describe it to feeding motivation and so on. So at last we can say that not all the behaviour in an animal are easily identified whether it is a emotion or a motivation.

Goal Oriented Behaviour

"Goal is a condition or situation which brings the behaviour in question to an end." For example having fed the dog, it ceases to search the food and having completed building its nest the digger wasp philanthus ceases digging and so on. As this condition is sometimes satisfactory but as we see in other situations apart from the achievement of the normal goal, brings activities to an end. For *e.g.* the digger wasp stops building its nest as darkness falls. So we can say that there are problems in using the concept of a 'goal' with animals.

In other cases, the definition of goal involves some kind of mental image of the desired condition. For *e.g.* a rat can learn to go to one part of a maze to get water and other part for food. When deprived of water it moves off and makes the appropriate turns to get the water. So this concept is said to be 'congnitive map' *i.e.* the goal which guides the rat movements towards the water. After studying different types of motivational states we can describe three stages in an animal's behaviour.

1. *A Phase of Searching for the Goal* : This phase is also called as 'appetitive behaviour'. It is difficult to observe the nature of appetitive behaviour unless and until we observe the actual goal. Animals searching for food, water or a mate may behave very similarly but in each case the stimuli required to bring the search to an end are different. So the observations are irrelevant to the goal may cause the animal to change its behaviour. For example —a hungry animal first of all has to search for its food and therefore its behaviour patterns may be many or diverse. This is particularly clear with predatory animals which actively hunt their prey.
2. *Behaviour Oriented Around the Goal Once it is Found* : It is seen or observed that once the goal is found, the animal's behaviour changes. This is a measure of the consummatory acts. For example — eating is under the consummatary act of feeding behaviour, copulation to sexual behaviour and drinking to thirst and so on. So there are series of responses directed at the goal are fixed action patterns.
3. *Quiescent Phase Following the Achievement of the Goal* : Quiescence is a stage when the animal is no longer responsive to stimuli from the goal and shows no further appetitive behaviour. This type of state is related to only one type of behaviour in which the animal may actively continue some other goal. In some cases, the responsiveness again builds up slowly between the threshold to goal stimuli but these stages are not universally applicable. For example, it is difficult to identify the nest building behaviour of lack bird as it may begin by searching for large twigs and ends with a fine grass and hairs in which a series of 'sub

goals' are included such as nest foundation, sides, cup formation etc. along the way, each with its own appetitive behaviour and consummatory act.

Explanations of Motivations : The psychologists explain human drives to be one sided and over simplified. Many behaviourists rejected the idea that the behaviour of men were controlled as in inherited behaviour patterns of birds. But it is the external stimuli which plays an important role in controlling the behaviour of men.

Other theorists believed that the internal sensory stimulation control the contraction of the stomach in hunger. The first approach that tells what motivates behaviour was given by K.S. Lashley (1938), who regarded motivation as being controlled by the responses of the control nervous system to a complex variety of external and internal stimuli, including the changes in the chemical composition of the blood stream. Lashley indicated that motivated behaviour does not consist of a simple chain of stimulus — response sequence but the condition within the individual alter the responses that given stimuli elicit from one movement to another.

The motivated behaviour does not depend upon a single stimulus but on a complex pattern of stimuli as a result of which the motivation varies greatly on different situations and among different organisms. After analysing these concepts, physiologists and psychologists are able to learn about the role of central nervous system in controlling motivation. Motivation and learning are interrelated because motivation cannot be demonstrated in the absence of learning just as we study the influence of heredity and environment on development.

Effects of Motivation on Behaviour : In human beings, motivation influences behaviour in two ways :

(1) Motives determine the direction of behaviour and therefore change the organism's relation to its surroundings. When sensitivity to certain stimuli is

increased it causes the external stimulus pattern to win over competiting ones. If the external stimulus are not present at that time, motives cause the individual to search until he achieved his goal.

For example : In the meal time, we stop everything what we are doing and go in search of food.

(2) Motives make energy available for the activity required. The motives or drives direct the organism towards some goal which will satisfy a type of need, it can also work as a general energizer. (Champion 1957)

In 1958, Hebb suggested that the same stimulus may arouse the organism to action and also perform a cue formation, *i.e.* it indicates the direction of activity. For example—the smell of cooking not only arouses the organism to search for food but also provides cues as to where the food may be found. So we can say that the stomach concentrations may act as arousal pattern and the olfactory stimuli act as cue functions which help in determining the direction of food-seeking activity. The arousal component of motivation can be measured by the level of activity of organs controlled by central nervous system.

Patterns of Motivation : When a goal is achieved, we cannot say that the motive is satisfied because the demand of others become more dominant. As hunger is the most incited drive when you wake up in the morning but after having your breakfast you are motivated by the need you need at that time. But as it is observed that with the achievement of one goal it lead directly to the formation of new ones.

Different motives acting within the organism are not compulsarily suitable but may be at odds with one another. For example — hungry person may abstain from eating if the only food is available to him is something that he has been taught to avoid for religious reasons.

The internal conditions which direct the individual's responses in the environment are of two types —(1) Biological drives which include water, food, oxygen, sleep, warmth or coolness. (2) Psychological motives which include social approval, self-esteem, security and knowledge.

Biological Drives : Biological drives are the unavoidable drives which motivate the behaviour of an individual in different directions that lead to the required changes in internal environment. For example — the hunger drive motivates the organism to search for food and then ingestion of food which is important for the maintenance of metabolism. If these biological drives are stayed for a long time then it results in improper health, disease or even death. These biological drives are associated with our needs such as food, water, oxygen, rest and sleep which are necessary for the maintenance of moderate temperature for the satisfaction of sexual urges, pain, bladder tensions etc.

The biological drives are inherited structures which are present in every human being and as well as in the lower animals. These drives influence an individual's behaviour from the time of birth and even during foetal life, the requirements are satisfied by the mother's body and in an adult it is satisfied through learning.

Measurement of Biological Drives : Under experimental condition, it is clear that man's knowledge including biological drives has come from study of the behaviour of animals. In any field of science the basic motive is the measurement. Physiologists and psychologists have discovered many ways of measuring the strength of biological drives by varying the intensity of drive stimulation. The stimuli which are varied are internal as well as external For example water deprivation or amount of food comes under internal and amount of electrical current applied to the skin comes under external stimuli. The biological drives can be measured by the following ways—

(i) gross motor activity; (2) rate of responding, (3) amount of obstruction given to prevent the animal from reaching the specific goal; (4) force of response; (5) speed of learning; (6) action of autonomic nervous system; (7) attention given by the animal when there are two or more goals.

As discussed above, the biological drives include food, water, air, sex, pain. So we study these drives one by one as follows :

1. *The Hunger Drive:* In all the drives, hunger is most important because the internal condition developed in this drive is quite complex.

Hunger Stimulus :

> "Hunger is referred to as a mass of sensation coming from the stomach." Many psychologists and physiologists worked together to understand the hunger mechanism and performed different types of observations on the behaviour when stomach is empty.

In 1912, Canon and Washburn conducted an experiment to show stomach contractions.

Experiment : The subject (animal) swallows a rubber sack, like a balloon which is attached to a long rubber tube. The sack is adjusted to the upper end of the subject's stomach and inflate until the sack should come in contact with the walls of stomach. The free end of the tube was connected to the recording device. After observation, they come to a conclusion that the recording device made a graphic record if there is any change in the balloon.

But newer evidences suggest that the balloon technique provides the misleading results. But some investigators found that with the presence of balloon in the stomach cause contractions instead of recording them only.

Current experiments have shown that the hunger exists in the absence of stomach contractions. Therefore, we can say that there is a difference between the sensations arising from stomach contractions and hunger drive itself. Because hunger drive says that "one should eat" and hunger sensation says that "when one should eat."

Some findings suggest that hunger drive is chemical in nature. For example if the blood transfused from a starving dog to a recently fed dog, stomach contraction starts. But if the transfusion of blood is from recently fed dog to a starving dog, stomach contraction stops. This experiment shows that blood sugar level acts directly on the brain centres that can excite the major nerves present in the stomach that causes strong stomach contractions (Janawitz and Ivy, 1949). To support this hypothesis, Goodner and Russel (1965) suggest that the blood sugar controls the hunger. They found that an injection of insulin when induced in the blood, developed hunger and stomach contractions.

The Mechanism of Hunger Satisfaction : The mechanism of hunger satisfaction involved "metering of food intake" an experiment conducted by Janowitz and Grossmann (1949) to show the metering of food intake.

Sham Feeding Experiment : In this experiment, the animals are surgically operated so that the food entering the mouth is chewed and swallowed through oesophagus but never reaches to the stomach. Animals fed in this way do stop eating but only after the food intake is much greater than it would have been reached the stomach.

In another experiment, conducted by Epstein and Teitelbaum, (1962), it suggests that a rat can regulate food and water intake in the absence of taste, smell, tactile sensory input from the mouth or oesophagus. For *e.g.* they placed food directly in the stomach of rats through oral-oesophageal by pass and eliminates taste, smell and tactile stimulation of the mouth.

After doing this they conclude that the post ingestional factors are enough to control the onset of feeding.

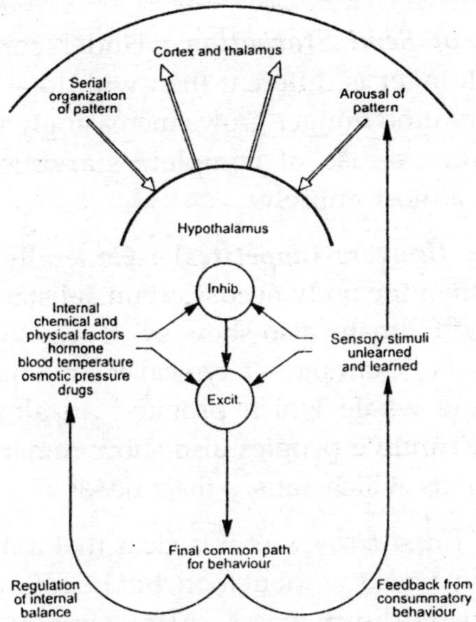

Schematic diagram of the physiological factors controlling the excitatory and inhibitory hypothalamic mechanisms that govern motivated behaviour.

So at last we conclude that the intake is normally regulated by both taste stimuli and post-ingestional factors.

Hunger and Hypothalamus : Hunger is controlled by the hypothalamus, as hypothalamus is in direct contact with the cerebro-spinal fluid and also contain blood vessels therefore these centres play a key role in controlling feeding behaviour. Other nearby centres are also been found to be important in controlling feeding behaviour, and if the hypothalamic centres are destroyed then their functions can be taken over by other structures and the animal can be kept active by forced feeding for a sufficient length of time. (Teitelbaun and Stellar, 1954)

Hunger and Sensitivity : When a person is hungry, no doubt he can recognize pictures of food more readily. The hungry organism is more sensitive as compared to the normal one. For example — hungry rats show greater reaction when changes occurs in light and sound as compared to the rats which are not hungry.

Effects of Semi Starvation : Under condition of semi starvation hunger is different than with total fasting. In case of semi-starvation hunger grows increasingly until the need is satisfied but in case of complete starvation, the hunger disappears almost entirely.

Specific Hungers (appetites) : Generally, hunger drive develops when the body needs certain substances to make up a deficiency. Humans also show evidences to satisfy specific body needs. For example —a typical eskimo meal includes the skin of white whale which provides an almost exclusively meat diet. Primitive peoples also show remarkable ability in choosing foods which satisfy their needs.

2. *The Thirst Drive :* As it is clear that a man can live for many weeks without food, but he can survive only few days without water. After several days, hunger disappears entirely but without water for long periods sensation of thirst makes him mad.

Effects of the Thirst Drive : In the lower animals hunger and thirst cause qualitative difference in behaviour. In an experiment it is found that the thirsty rats learn to find a reward of water more quickly than hungry one learns to find food when the rewards are in the same location. So, the behaviour is more stereotyped than that of hungry ones.

How Thirst Operates : To maintain a constant water content thirst drive controls the intake of water into the body. Thirst is developed by which factors are partially understood by physiologists and psychologists. When there is dryness of mouth and throat tissues thirst arises. This condition can be relieved by washing the mouth with water. Concentration of salt and other chemical substances, total amount of fluids are responsible in determining the thirst. Hypothalamus is also responsible in activating the thirst drive. According to Anderson and McCann (1955) injection containing high concentration of salt into the hypothalamus causes excessive drinking, suggesting that hypothalamus cells respond directly to the level of salt in the blood.

Thirst is the only factor that helps in maintaining the proper hydration of the tissues, just like the output of urine is important in maintaining a balance. This releasing of water is controlled by the hypothalamus as the axons of certain nerve cells send their stalk to the pituitary gland causing release of a hormone into the blood stream. On reaching the kidneys, this hormone causes them to retain water. But in case of dehydration, hypothalamic cells are stimulated, the hormone is released, the amount of water secreted by the kidney is reduced and balance is obtained. So water balance is important and helps in maintaining life.

(3) *Air Hunger* : For the survival of life, we need air in the form of oxygen. For *e.g.* if oxygen has been replaced by a gas called methane then the individual loses his consciousness without any feeling of air hunger to give him warning. Another *e.g.* if carbon-di-oxide collects in the lungs then we feel immediate discomfort and intense pain of our breathing and we are not in a position to get rid off the carbon-dioxide and draw in air. The effective stimulus to the mechanism of air hunger is the presence of excess CO_2 rather than the deficiency of O_2. The continuous supply of O_2 to the body especially to the brain cannot be neglected. As oxygen consumed by nervous system is in very small amount but if this supply is deficient actual neural damage can result.

On mountain tops or on high altitude flying where O_2 starvation in the atmosphere appears without excess CO_2 a sort of intoxication results. The person loses control of himself and may shout, fight or burst into tears.

(4) *The Sexual Drive :* The sexual drive after hunger is important for social living. As in our society there are no restrictions upon the food taking behaviour but sexual expression is governed by both law and social conventions. Because our social structures limit an individual's sexual behaviour as a result of which sex urge is more persistent. Sexual activity is necessary to the survival of race but it is not essential in keeping an individual alive as other biological drives.

Scientists are still trying to find the exact location of the internal controls of the sexual drive. But observations shows that the sexual drive is influenced by the presence or absence of hormones produced by testicles in males and ovaries in the females.

The Sexual Drive in the Male : Sexual drive in male is controlled by (1) androgens secreted by testicles and (2) hormones, secreted from the pituitary. The hormones influence the sexual development that stimulates the adrenal cortex which secretes hormones that are necessary for sexual maturation.

For Example : According to Green, Clemente and Degroot (1957), sexually immature male cat will show male sexual behaviour when given male hormone testosterone and show female sexual behaviour when given female hormone stilbestrol. This inter-reversible change cannot be produced in sexual mature cats.

The sexual urge in human beings attains maximum height in the late teens and decline in the early twenties. This gradual falling off of the sexual urge is due to the decrease in the quantity of androgens secreted. The levels of nervous system in man play a part in determining sexual behaviour. For example — in man whose spinal cords are cut, an erection of penis can occur spontaneously but on the other hand, man whose nervous system is intact can drive much pleasure in sex even without erection.

Sexual Drive in Female : The primary sex glands of a female are ovaries which secretes estrogens (influence sexual urge) and progesterone (important in pregnancy). Estrogen secretion into the blood is greatest each month before and after ovulation. In the adult human female, ovulation occurs about half way between two menstrual periods. Many experts believe that when the egg is in the oviduct or uterus, it is the only time a woman can become pregnant. In lower animals, the influence of estrogens on the female sex urge is most commonly seen. At, ovulation, when the blood stream becomes enriched with estrogens, the female loses its control and becomes highly aggressive in her sexual behaviour. This behaviour is known as 'estrus heat' and it is considered as a symbol that the female is in a condition of readiness of pregnancy.

In our society, the inadequate amount of sex education is being a cause of sexual maladjustments in women. Girls as they aré grown taught that, they should deny sexual feeling as a result of which many women find it difficult after marriage and take their husbands as uninterested. This condition is often originated by lack of understanding or consideration. In some case, due to lack of sex education, fear of sex can make a woman psychologically frigid but biologically healthy.

Sex as a Learned Appetite : In both male and female, biological process play only a part in sexual desire but hormones affect the sexual desire much greater. The women have greatest desire immediately before and after menstrual cycle than during the fertile period. According to Harpp (1964), human sexual behaviour is a learned appetite rather than an innate drive. According to this theory, sexual appetite grows with experience, for example —in a society a child is restrained from sexual activity but after puberty he is encouraged to show interest in members of opposite sex. This is a biological phenomenon.

Another example as taken, a dating couple holds hand and sit close together, the pleasurable feelings which they experience reinforce their hand holding behaviour. But the original hand holding motive is often unisexual in nature until and unless kissing or embracing are included. This action may arouse true sexual point and bring a desire to repeat the powerful excitement. So, it is important to note that sexual activities are associated with sexual desires. Thus, by reading about love or seeing erotic movie scenes can arouse sexual feelings, as they could not have done during childhood.

(5) *Pain as a Drive* : Pain is not properly considered as a drive but it is said to be a drive. For example — if you touch a hot iron accidentally, you will draw back your hand. This is a simple involuntary response but suppose that after withdrawing your hand the burn continues then you put saliva on it. Then we can say that pain is serving as a drive. Pain serves as a drive when it is prolonged. Pain has important biological significance in those situations which produce pain as well as injury. Therefore, to avoid pain we also tend to avoid being injured. So, pain drive is the important motivating force in the history of mankind.

Biological Drives as Homeostatic Mechanism : Biological drives are regulatory drives that are responsible in maintaining the physiological equilibrium of the individual though their intensities vary. The process in which an individual attains maximum heights to maintain the constancy of its normal internal environment is termed as homeostasis.

Biological drives which are developed in physiological conditions have disturbed the individuals, equilibrium. When internal state is disturbed, conditions are originated that motivates the organism and forced to seeking activities. Many homoeostatic activities are internal and automatic. Along with this they also help in maintaining constant body temperature and balance of oxygen and carbondioxide in the blood stream.

According to Stagner (1915), biological needs are temporary and can never be permanent as in man. The organism reacts to very small physiological changes just to satisfy its needs. Animals built nests and store food for winters. Man also not only learned to eat food before hunger pangs begin but has developed means of agriculture, food preservation, storage to get an adequate food supply all the time. Thus homeostasis involves automatic maintenance of chemical conditions of the body in the response to specific stimuli. It involves quick effort of an organism to establish a physical surrounding as well as social environment that is as constant as possible.

Even at the physiological level also, homeostasis does not stand for all types of behaviour. To maintain the body, an organism may occasionally behave for example — human babies, if given a chance, may eat enough salt to kill themselves. Sometimes an organism strives for goals that have no significance for adaptation etc. According to Davis (1958) homeostasis is a valuable concept but it is not sufficient to tell the entire story.

***Physiological Basis of Motivation* :** Grossman (1967) provides an introduction to the entire field of physiological psychology. Hypothalamus part of the brain has been proved to be a major importance in the control of motivation. Brain is smaller than the last joint of the little finger but is of primary importance and it is not wrong to say that any function related to the body directly or indirectly depends on the hypothalamus.

Pituitary gland, which controls the whole hormonal system develops from the fusion of a down growth of the embryonic hypothalamus with an up growth from the roof of the embryonic mouth cavity. So, from the behavioural point of view we can say that it is the most significant feature of the hypothalamus.

• Hypothalamus helps in measuring the changes in the

metabolism and to set in motion activities which will rectify them. So it serves as a homeostatic control system also. Hypothalamus is also responsible for the arousal of motivated behaviour (excitatory mechanism) and also in the reduction of motivated behaviour, (inhibitory mechanism). For example, in case of feeding the destruction of ventromedial regions of the hypothalamus on both sides results in vast increase in eating, a doubling or tripling of food intake. Therefore, ventromedial areas act as a part of inhibitory mechanisms and the bilateral destruction of the lateral hypothalamus will cause an animal to starve to death acts as excitatory mechanism.

These conditions originated when electrodes are implanted into these regions of the brain and their tips are electrically activated. So, at last we can say that electrical stimulation of the lateral hypothalamus produces increased eating and stimulation of the ventromedial hypothalamus causes reduction in eating. Similarly hypothalamic mechanisms work in the case of thirst, sexual behaviour, emotional behaviour, sleep and maternal behaviour and these mechanisms explain that the action drive is based on excitatory hypothalamic mechanism and satisfaction is based on inhibitory hypothalmic mechanism.

Hypothalamus is essential for the initiation of motivational states but cannot be proved that hypothalamus alone is essential for their control. It involves many other areas of the brain for its control.

Hormones and Motivation : There is an inter-relationship between hormones and behaviour because in many ways, the endocrine glands which produce the hormones share the common functions of communication both within the animal and between it and the outside world. The hormones provide a chemical message system which is as old as nervous system. In the animal kingdom, neurosecretory cells are present in nervous system which are modified neurons and transfer their special chemicals down through axons into the blood stream. When these cells combine and form clusters, they form glands such as corpus cardiacum in insects which has close connection with both nervous system and blood stream. The vertebrate pituitary gland originated from the fusion of neural and epithelial tissue which remain connected to the hypothalamus. The pituitary regulated by the secretions of endocrine glands which in turn regulated by the nervous system. The system of control helps the hormonal response to be matched with the environmental changes. For example control of breeding seasons in mammals and birds.

Hormones in Mating Behaviour : Hormones play an important role in the mating behaviour. The development of several types of behaviour involves both hormonal action and experimental factors in the phenomenon of mating behaviour.

(i) *Hormonal Action :* Gonadal hormones are of fundamental importance in the control of sexual behaviour in most adult vertebrates. For example —if the young are treated with gonadal hormones their behaviour resembles with males and if pregnant guinea pigs are given testosterone propionate (hormone) then the genitalia of the female offsprings are just like that of males.

(ii) *Experimental factors :* The role of experimental factors is often underestimated even then they play an important role in the development of sexual behaviour. Most of the experimental factors are social in nature. For example in birds, males reared in social isolation may fail to mate with females because they show fear than group-reared controls. But in rats it is evident that social experience on sexual behaviour is contradictory. Beach (1942a, 1958a) found that if the male rats reared in isolation were successful in mating with females than males reared with females. But according to Harper (1968) social experience is not necessary in the development of sexual behaviour.

Hormones are regarded as the prime motivating factors in animal behaviour as ethologists have suggested that they act directly on brain control centres to increase the drive. For *e.g.* an experiment conducted by Blum and Fiedler (1965). They describe the injections of prolactin (pituitary hormone) will cause isolated males of the fish Crenilabrus ocellatus for fanning movements. If the injected crenilabrus male will be devoid of all the normal external stimuli for fanning such as CO_2 in water, fertilized eggs, nest, etc. then the fanning is dependent on the dose of prolactin. By this example, we can say that the hormones are not only for motivating behaviour but also affects the other behaviour.

Roles of Hormones in Behaviour : Hormones are the chemical which produces rapid growth changes in the target organs. For example, in female mammals estrogen causes growth of uterus, growth of mammary ducts and changes in the epithelium of vagina. In some species it is also responsible for shedding of ventral feathers and increased vascularization of the ventral skin to form a brood patch.

Hypothalamus also plays an important role not only in initiating courtship and copulation but also the pattern of reproductive behaviour. Hormones also affect behaviour and behaviour in turn affects the level and type of hormone secretion. For example — female mammals, cats and rabbits will not ovulate unless they mate and from the pituitary gland the stimulus of copulation needed luteinizing hormone which helps in ovulation. Another examples of interactions between hormones and behaviour is doves and pigeons.

As it is clear that a female pigeon do not normally lay eggs if kept alone but as soon as male is introduced, the female begin to lay eggs. So, it means that the secretion of pituitary is influenced by male itself and male is enough to cause ovulation. So, at last we can say that the hormones and the behaviour are interrelated.

Concept of Motivation

The study of animal behaviour is still in its infancy, nevertheless, considerable research has been done in laboratory and in wild, from which certain basic concepts have emerged.

When you read ethology, you will come across new terms, ideas and concepts. It is very essential for a student of ethology to get acquainted with these terms before proceeding further.

Konrad Lorenz and Niko Tinbergen formulated a number of concepts on which the study of animal behaviour has been based.

A hypothetical state of the individual organism that arouses a goal directed activity is motivation. Mood (as called by Oskar Heinroth), drive, urge or readiness of an animal to perform a certain behaviour, is motivation. Each instinctive behaviour has a certain motivational level at all times. This value declines when the act is performed, then rises again. At different times the animals respond in different ways to the same stimulus, for example: a hungry dog would get up, sniff here and there, look around for food, at this time motivational level or the urge for eating food is at the highest level. When it finds goal *i.e.* food it starts eating and after eating a full meal, the motivational level goes down to its minimum level. This dog will not eat for next few hours.

By this time, the motivational level rises gradually and once again after a lapse of few hours when the level rises to its maximum, the dog would start looking for the goal *i.e.* food once again and all these steps would be repeated. Thus, dog would respond to food differently when he is hungry or full. This is true not just with food, but almost all instinctive behaviours (determined by genes) are regulated by motivation. There seems to be an urge or internal drive to elicit a particular behaviour at particular time. Motivation to perform a particular behaviour is deduced from the strength and frequency of the behaviour itself, which is called motivational analysis. The majority of behaviour patterns can be assigned to four larger

High motivational level	⟶	Appetitive or searching behaviour	⟶	Search for goal
↑				↓
Motivational level starts rising after a lapse of time				Locating the goal or orienting around the goal
↑				↓
Quiescent period showing refractory behaviour	⟵	Motivational level falls to minimum	⟵	Achieving the goal through consummatory behaviour or Fixed Action Pattern

Schematic explanation of motivation.

Motivational Systems: feeding, reproduction, fight, and flee (escape). Within the large motivational systems there are many subunits of behaviour that are more or less connected with each other. For example, subunits of reproduction include not only sexual behaviour but also nest building and care of young and these behaviour patterns have their own drive mechanisms.

The psychologists have divided a drive into three phases

(1) Search for goal (appetitive or searching behaviour)

(2) Orientation around the goal and achieving the goal (consummatory behaviour).

(3) Quiescent period (refractory behaviour).

The biological significance of drive mechanism or motivation is to remind the animal about when a behaviour should occur and to prepare an animal to perform a particular behaviour.

In an experiment, rats were trained to run for water through a maze (a zig zag path), it was observed that they ran through the maze much faster and drank more when they were kept thirsty as compared to those rats which had water ad-libitum (*i.e.* available whenever they wanted).

Concept of Fixed Action Patterns (FAP) : The behavioural patterns, which are basically determined by genes and exhibited automatically by any animal without having seen or learnt from, conspecifics or performance of a behaviour even when the animal has been raised in isolation, is known as FAP When a wasp supplies its young with caterpillars or a pregnant mouse builds a nest, they are not guided by conspecifics. Among animals there are many instances of elaborate behaviour which are found throughout a species and are not acquired through previous experience or learning. These acts are Fixed Action Patterns and are, also known as consummatory action, innate, instinctive, inborn or inherent behaviour.

The behaviour of each animal is adapted to its surroundings and is perfectly attuned to the requirements of its living and non-living environment, the information to adjust to this can enter an organism in two different ways: by means of the genes which can be stored in the genome, this represents a "species memory" that is passed on from one generation to another generation. The other information is acquired by an organism and is not transferred through the genes.

If it is derived from the genes, the behaviour is referred to as inborn or innate; if it is based on individual experience, it is called acquired or learned. The terms inherited adaptations and acquired adaptations are sometimes used to emphasize the source. By definition FAP is species specific, stereotype sequence of co-ordinated motor actions. The animals can perform them without having previous experience and without having seen another species member (conspecifics) do it.

There are innumerable number of behavioural acts in the

entire animal kingdom which are transferred through genes, these acts are already previewed or preprogrammed in the central nervous system through the genetic material. The elaborate courtship displays in most of the animals mating, feeding patterns, nest building, parental care, vocalization; songs of crickets, birds and whales; wing cleaning in insects and birds, territoriality and aggression in sticklebacks and Europian robin males, lek formation in sage grouse, construction of web by spider and of pendulum nest by weaver bird all fall into this category.

The most common example which comes across is of dancing peacock. The peacocks start dancing as they see peahens during the breeding season, they do not learn dancing by watching conspecifics. Similarly, it is worth noting that weaver bird never learns to construct the nest, it just knows, and the bees do not learn to build their hive, this property is acquired genetically. A tailor bird would bring two leaves together and literally stitch them with the help of long, flexible grass, then cushion it with soft grass and lay her eggs; these birds when kept separately from their conspecifics could still perform this behaviour although they were given no opportunity to watch their older conspecifics to learn it, this vital information was passed on to them through their genes. In human beings crawling, smiling, crying are not learned, they are the examples of Fixed Action Pattern.

Instinctive behaviour has evolved gradually along with the other morphological features through natural selection, to enable the animal to fit steadily and gradually in the environment in the best way. Instincts such as hunger, sex, fight, fear and even pleasure have been wired into brain circuits during the course of evolution to ensure physical and psychological well being of the species.

To distinguish between innate and learned behaviour, experiments are conducted on animals reared in isolation by keeping them individually, away from others, from as early an age as possible, when they mature, their responses to a variety of stimuli are tested and compared with animals reared normally. Many fishes and birds perform normal feeding, sexual and aggressive behavioural patterns even when reared in isolation. The chaffinche develops its complete song during its first 12-13 months; if it is reared in a sound proof room until the expiry of 13 months, it can not sing normally. The song of this bird, is of about the right length, and has the right number of notes but the melody is not recognizable, because for that it has to live with its conspecifics, indicating that length and notes of songs are instinctive but melody is learnt. In human beings also vocalization *i.e.* production of sound is inherent but we have to learn the language.

Experimentally isolated animals never have any experience with the particular situation but they will still perform certain behavioural pattern in appropriate situations; some squirrels were reared in isolation and were never given nuts (one of their favourite foods) to handle or to eat, but when such squirrels were given plenty of nuts they ate them and attempted to bury extra nuts in the ground, typically like all other squirrels. Therefore, this behaviour of squirrel is a Fixed Action Pattern, animals that are grown under the conditions of isolation for the study purposes are called Kaspar Hausers. The Fixed Action Patterns show following properties:

Constancy or Stereotype : Instincts are inherited and are similar in all individuals of a species *i.e.* all digger wasps and weaver birds build their nests in the same way, roosters of same species court their females in generally the same way. Instincts involve complex, and often highly rigid patterns of behaviour, in which numerous muscles, muscle groups, organs and systems function in an extremely coordinated fashion.

FAP are found to be the same in all members of a species (stereotype) and can not be influenced in performance by external stimuli. A classic example of FAP is provided in the study of egg rolling in greylag geese (plural). These birds build shallow nests in the ground, the greylag goose (singular) uses a stereotype set of behaviour to bring back eggs that roll out of the nest.

The movements are typical of this species, no matter in which environment or circumstances it may live. It need not watch conspecifics doing it, this action is performed automatically. Sight of an egg outside the nest acts as stimulus (which is known as sign or key stimulus) and that elicits egg rolling behaviour.

Resistant to Phylogenetic Changes

FAP are extremely conservative in evolution of a species. It has been assumed that once they have appeared in a species, they are resistant to phylogenetic changes occurring through evolution. It can be studied by comparing the behaviour of long tailed and short tailed monkeys. When a long tailed monkey runs along a branch, its tail moves from side to side thus achieving balance, which is a FAP. Short tailed monkeys stopped using their tail as a structure for balancing yet perform similar type of movements of tail. Probably short tailed monkeys evolved from long tailed monkeys, where tail had an important function of balancing and FAP, once evolved, persisted even though the relevant morphological structure had disappeared.

Need Sign Stimulus for Release : A characteristic feature of FAP is that they are elicited by simple but specific stimuli, which are known as sign stimuli.

Innate complex behaviour patterns can as a rule show two components that are clearly distinguishable; one is stereotyped *i.e.* consummatory action (as explained earlier to you as FAP) while the other is made up of a longer, more variable sequence of movements that lead to the consummatory action; this second component is called appetitive (as named by Wallace Craig) or searching behaviour. By definition, it is a behaviour for a situation that contains a stimulus that could release the consummatory behaviour. Eckhard Hess (1962) pointed out that appetitive behaviour may be characterized in two ways: (i) a motor pattern (usually locomotion) (ii) an orientation towards the goal or stimuli to which the animal is receptive *i.e.* smell or sight of food, water and mate.

Most of the FAT (consummatory action) are initiated by searching or appetitive behaviour. FAP are Invariant whereas, appetitive behaviour is variable, *i.e.* appetitive behaviour is terminated by -the appearance of food and then starts FAP (stereotype eating pattern). A dog may spend hours in search of a mate during mating season, which shall cease after copulation. Here the act of mating is an example of FAP, and searching for goal *i.e.* (mate) is appetitive behaviour.

Both the actions are dependent on the level of motivation for that particular behaviour. When a dog is hungry, the motivational level/urge/drive increases for eating *i.e.* it starts showing searching behaviour, after locating the food, it starts indulging in a consummatory action (eats food), by this time the motivational level of hunger starts going down and when the animal has eaten its full meal, the motivational level for that particular behaviour gets to almost zero and that ;tops both consummatory and appetitive behaviours.

Transfer of behaviour through genes has great significance in the survival of species. Firstly, there are behavioural acts like feeding, drinking, mating which form essential part in an organism's life; they save time and energy if such vital traits are transferred through genes. Secondly, some organisms have a short life span, and they never get enough time and opportunity to watch conspecifics, in such an eventuality, their life processes depend only on inherited traits.

According to Lorenz, when animals perform FAP it seems that as if the animals were releasing some kind of tension. *e.g.* a ground squirrel in captivity when presented with excess of food may retire to a comer of its cage and go through the motions pretending to bury the food on the iron floor apparently satisfied by the act, even though the food remains in full view.

Similarly, a hen canary in breeding season, deprived of nesting material, has been seen to fly repeatedly from one part of the cage to the nest site, holding one of her own feathers in her beak and on reaching the nest site, it goes through the movements of weaving the material into the nest, even though the feathers were still attached to her body. A female rat deprived of nesting material has been seen repeatedly carrying her own tail from one corner of the cage to the nest site, going through the process of building the nest; such a release of FAP is called in Vacuo.

Concept of Sign or Key Stimulus or Releasers : In general, releasers are those parts of the surrounding that trigger instinctive acts. Isolated characteristics of an object, such as colour or shape, that evoke FAP or the simple and specific stimuli that cause the release of FAP are termed as sign stimulus (SS). It was Konard Lorenz who first realized that specific stimuli (which he called as key stimulus) are needed to release the FAP He said that a stimulus acts as a key which unlocks and opens a particular fixed action pattern. Sticklebacks are small fresh water fishes; in the breeding season, the males develop a red colouration on the underside of their bodies and become highly territorial. They select a site, form a tubular nest and defend it actively by fighting and chasing an intruder male. Niko Tinbergen in 1952 performed some classical experiments to understand the intriguing mechanism behind sign stimulus. He used some models to study the specificity of sign stimulus.

Model A was very realistic but did not have a red coloured underside, whereas, models B and C were of odd shapes but with red bellies. Models with red undersides acted as sign stimuli for releasing attack behaviour in other male sticklebacks. Therefore, it was concluded that the red belly is the critical or specific sign stimulus responsible for releasing territorial defence in male sticklebacks. Similarly, the exposure of red breasted feathers of the male European robin in the territory of another male robin is a sign stimulus for aggressive behaviour. Significance and specificity of sign stimulus is reflected in many other examples : A herring gull's chick pecks instinctively at a red spot on its parents bill and that acts as a sign stimulus which induces the parent to regurgitate food called crop milk.

The chicks of song birds which are born blind and helpless respond to the jerk on their nest. In nature, when the parents arrive on the nest it creates movements, which acts as a sign stimulus for the chicks, they turn, their neck in the direction of movement and open up their mouths the colouration inside or just the sight of gape acts as sign stimulus for the parents to regurgitate food.

A hen does not rescue a chick that it can see struggling under a glass cover. She will come to rescue chicks only if she hears the distress cries of the chick, that means, only such calls can release the rescuing behaviour in mother and not the sight of helpless chicks. Similarly, Turkey hens render parental care only after hearing the calls, but not to the appearance and movement of chicks, even an unnatural model will be brooded by the hen if chick calls can be heard from a small loudspeaker mounted within.

Male flicker has a black patch under its beak resembling a moustache. If a moustache is printed on female, on her return to nest the male attacks her as though she was a rival male. Therefore, the black spot is a sign stimulus for the release of attacking behaviour among flickers. Cuban finches react aggressively during breeding season towards any model having a yellow mark that imitates the neck band of a male. An interesting experiment was done using a card board model to study the specificity of a stimulus, this model was flown over young gallinaveous birds like pheasants, partridges, turkey and domestic fowl. In one direction the model resembled a hawk or a predatory bird and it elicited an escape response in birds, whereas, in the opposite direction, it resembled a duck or a crane and hence did not receive an escape response.

Organisms use a variety of stimuli to release different behaviour patterns. Impressive results have been obtained in experiments with the grayling butterfly. During the breeding season the male of this species performs a pursuit flight or courtship flight that invites passing female to land near him, after a short courtship ritual they copulate. By the use of models that varied in colour, form, brightness, movement and

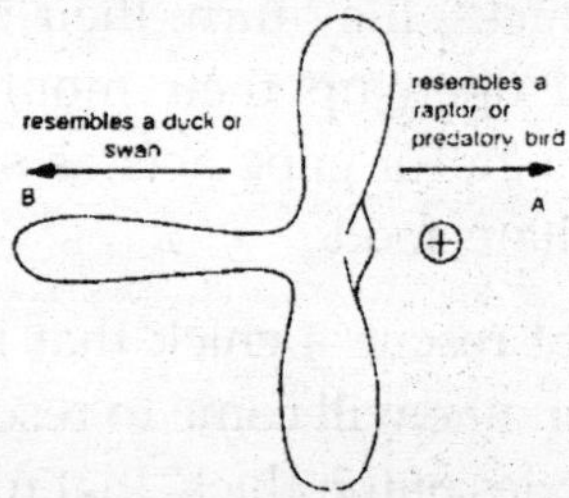

Release of escape response in gallinaveous chicks. Flown in direction (A) released escape response but when flown in direction (B) did not release escape response.

contrast with background it was determined that the first three cues were almost completely ineffective, and the important stimuli were movement and contrast.

Sometimes, the sign stimulus is not just one but a combination of many. In gulls, begging behaviour in hatchlings is stimulated by (i) the shape of the bill (long, pointed objects were preferred over short and stubby), (ii) the colour of the patch (red preferred over all other colours), (iii) the contrast of the patch with its background (more the better) (iv) a moving beak model attracted many more pecks than stationary one and (v) a beak that pointed down was more effective than one oriented in another direction.

It has been found that fighting behaviour in the male cichlid fish is elicited by five stimuli

(1) Silvery blueness of body

(2) Dark margins

(3) Broadness of animal

(4) Parallel orientation to opponent and

(5) Tail beating.

Any one of these stimuli was capable of eliciting hostile behaviour and that any two would produce about twice the reaction of one. Therefore, in many instances response to releasers appears to be additive, and is called law of heterogeneous summation.

These stimuli have been compared to a key which unlocks a behaviour pattern in the animal's neural system. The ethologists have developed a concept which assumes the existence of an inborn neural mechanism that responds in a specific way to the releasers.

Each organism, at any moment, is confronted by a wealth of information from the environment but only a small fraction is biologically significant for the animal. One of the most important task for the organism is the selection of those stimuli that should be followed by a reaction. Two organ systems are available for this sorting out or filtering of information - the sense organs and central nervous system, these are called peripheral and central stimulus filtering respectively and both are helpful in selecting relevant stimuli.

Ethologists have been able to construct models which are more effective in releasing FAP than the original stimuli. Exaggerated form of stimulus which releases a particular behaviour pattern more effectively than the appropriate natural stimulus is called super normal stimulus. This phenomenon was first discovered in laboratory conditions, *e.g.* a bird, oyster catcher prefers eggs larger than normal and will try to sit on eggs too large to incubate.

Other ground nesting birds *e.g.* killdear, herring gulls or greylag geese when presented with eggs or egg models that are more conspicuous or larger than their own eggs, retrieve those exaggerated forms into the nest and incubate in preference to their own eggs.

The existence of supernormal stimuli shows very clearly how strong the selection pressures are to make stimuli even more important and conspicuous. There are some supernormal stimuli that occur in the natural situation. Some birds *e.g.* cuckoos lay their eggs in someone else's nest. This is known as brood parasitism. The nestlings of cuckoo possess a gape marking in their mouth that is more prominently coloured than the gape marking of the host species young. The stimulus of gape marking of cuckoo is so effective that the response of the foster parents is more towards them and hence they feed them more.

In our day to day life, the TV advertisements or other marketing strategies are taking advantage of many different kinds of supernormalizing devices/stimuli. For every product a supernormal end product is predicted to entice the customer, be it a toothpaste, soap, shampoo or cooking oil.

Concept of Innate Releasing Mechanism (IRM) : The special neurosensory mechanism that releases the behaviour in response to specific sign stimuli is called the Innate Releasing Mechanism, in other words there exists an inborn neural centre called IRM that responds to a sign stimulus. The term and concept were proposed by Niko Tinbergen (1951), it was called

Angeborener Auslosemechanismus (AAM) by Konard Lorenz IRM or AAM are essential for the perception and processing of relevant sign stimulus. IRM is situated at the hypothesized locus in central nervous system (CNS) on which sign stimulus acts to release FAP. A specific sign stimulus acts on a specific locus; a lock and key analogy is often used to clarify the relationship of the sign stimulus and IRM. For each FAP there is separate IRM, which is like a gate to a reservoir of energy used to bring about FAP.

Concept of Action Specific Energy (ASE) : It is also known as Specific Action Potential (SAP) or Action Specific Potential (ASP). Lorenz in 1950, advanced the concept of ASE- which could not gain wide acceptance, probably due to lack of neurological evidence to support it. Lorenz postulated that ASE, or energy for a specific action is constantly being produced in the animal's central nervous system. The ASE is that energy which is required for carrying out a particular FAP. It is being produced continuously in an animal's CNS, but it is held in check by some inhibitory mechanism until the appropriate sign stimulus releases this energy and subsequently FAP; each FAP is thought to have its own ASE. The concepts mentioned so far, function in an organised sequence: *i.e.* appropriate sign stimulus is received through sense organs and sent to the hypothesized locus IRM in central nervous system, which releases ASE to elicit FAP. Lorenz in (1950) proposed a model known as "Psycho-hydraulic model" or "Flush toilet model" to explain the correlation between SS, IRM, ASE and FAP.

Sign stimulus (weight) acts on IRM (spring and valve) which in turn allows ASE to flow from reservoir to carry out FAP; gradually increasing pressure of ASE as well as sign stimulus act in the same direction to release the FAP. The higher the ASE levels the smaller the weights (stimulus) required. In the absence of sign stimulus, ASE is accumulated and eventually its own pressure may push open the valve to release FAP, this kind of release is known as "in-vacuo"

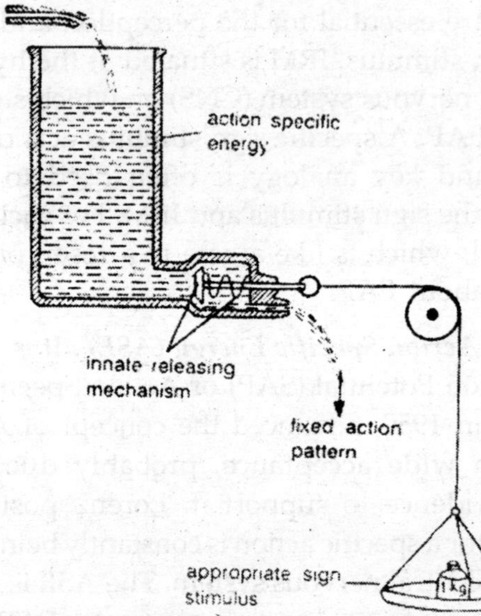

Psycho-hydraulic model of K. Lorenz (1950).

release. Once the reservoir is empty, the behaviour can no longer be elicited for some time, no matter how strong the stimulus is, but gradually the ASE would get accumulated again and would be ready to elicit behaviour.

Concept of Physiological Basis

After Lorenz, ethology began to expand into the neurophysiological field. Von Holsta, Hass, Hinde, Pavlov, Harlow, Skinner are the prominent names in this field. Neuroethology and ethoendocrinology are specialized branches of ethology. For a long time, an animal's behaviour was seen exclusively as a reaction *i.e.* as a response to external or internal sensory stimulation.

The simplest form of reactive behaviour is the reflex. Initially it was considered that many behaviour patterns are brought about by long and complex chain of reflexes. Not until the middle of 1930s did Erich Von Holst (1932) and

Konard Lorenz (1935) independently discover, on the basis of experiments and observations, that behaviour is not just a reaction to external stimuli but is also based on internal physiological conditions and spontaneous reactions controlled by central nervous, hormonal and muscular systems. This discovery made it possible to separate innate and learned behaviour patterns, conceptually from reflexes. There is one basic difference between reflex and innate behaviour, in a reflex the response to a stimulus is generally identical from one presentation to the next, *i.e.* under the same conditions and the same stimulus the response always occurs in exactly the same way, whereas in an innate behaviour pattern it can vary.

Classification of behaviours is based on the assumption that the complexity of behaviour increases with the complexity of the underlying neural mechanism. The major phyla of animals differ in the complexity of their nervous systems, correspondingly, these groups differ in their ability to carry out elaborate and difficult tasks, to solve complicated problems and to cope up with the changes in environment. Behaviour is actually a result of co-ordinated muscle activity and it is very well known to you that muscles do not contract unless stimulated by nerves. It has been rightly said that "behaviour is the symphony of muscle contractions, guided by messages from the central nervous and hormonal systems.

Behaviour and physiology are inseparable, the animal senses its external world through sense organs, which in turn send nerve impulses to brain, where it is interpreted and message is sent to respective muscles to behave accordingly. The internal urges are also sensed by special structures *e.g.* long hours of hunger and low level of glucose in the blood - causes excitation of feeding centre in hypothalamus, this causes hunger, animal shows appetitive behaviour, after finding the food the consummatory actions are performed. It is only the last part which can be seen and designated as behaviour, but it is whole lot of body physiology responsible to make these movements possible.

All the sense organs form an integral part of animal's behaviour, because it is through these organs that the organisms receive external stimuli from environment and keep adjusting to changes accordingly. The brain perhaps plays the most important role in bringing out a behaviour.

Concept of Learning or Experience

"It is the process by which an activity originates or is changed through reacting to an encountered situation" or "any change in behaviour as a consequence of experience" is learning. All animal forms from protozoans to primates learn. You have already known that some behavioural acts are inborn, that is the animals can just perform them automatically without having any previous experience and are called FAP The other is learned behaviour, which the animals can perform only after seeing the conspecifics performing it. That makes two major categories of behaviour- innate and learned and there is an adaptive significance attached to both these categories. Certain behaviours are essential to survive viz feeding, drinking, sex, maternal, aggression and escape, etc. hence their basic components are transferred genetically.

The other major advantage of having innate behaviour is for those animals who either live solitary or have a short span of life, the first ones do not want to see conspecifics and the second ones do not get enough chance to see the conspecifics. A female digger wasp emerges from her underground nest in spring, her parents are already dead, she finds a male and mates, then digs a hole on the ground and forms a complex nest of many cells by performing complicated behavioural patterns; then she goes out for hunting, kills caterpillars, brings them to the nest, lays eggs and seals the opening of nest, this is completed in few weeks time, the female leaves and dies.

The wasp has a short life span, where she does not get the time to see her parents; if she had to learn every act by watching others she could probably never achieve this tight schedule. For such organisms genetically programmed behaviours are most important.

All the complicated behavioural acts are genetically controlled, and this does not mean that they do not learn anything at all, they do. Because if all the behavioural acts were innate which are stereotype, repetitive, rigid and invariant, how would an animal adjust to day to day variations it has to face, in the habitat it lives in or adjust to the ever changing environment ? Therefore, it is very necessary to have the ability to modify its behaviour through experience or learning according to factors like habitat, geography and climate. In contrast to digger wasps situation, the tiger cubs do not have the pressure of short life span. Born quite helpless, the cub is sheltered and fed by its mother until it can move around, the cub is gradually introduced to hunting and solid food. The cub has constant opportunities to watch and learn from its parents and other members of the group.

The digger wasp, which must predominantly rely on preprogrammed inborn behaviour and tiger cub, that can learn in leisure, represent two different categories of behaviour. Instinct and learning both have adaptive significance, the former by selection operating during the history of a species, the latter during the history of an individual. Snapping at food is a routine in toads or it's a FAP, when they are mature enough they snap up flies, mealworms and other moving creatures. If a millipede is placed in toad's enclosure, the toad will snap at it automatically, as a consequence of FAP.

The millipede gives out nauseating substance, whereupon the toad will spit out the prey from its mouth. Later, this toad will always refuse to ingest a millipede species even though if it is hungry, showing that its behaviour has changed, the animal has learnt to avoid obnoxious prey, it has learnt to discriminate between edible and inedible. Similarly, chicks peck indiscriminately on all objects which they see on the ground as a FAP but as they grow, they learn by experience to peck at useful food items only and ignore the rest.

Though most of the insects live by their instincts yet they are capable of rapid learning. In a classical experiment, Niko Tinbergen studied learning in digger wasps Philanthus triangulum. He found out that a female digger wasp can learn the appearance of her nest entrance by encircling just one flight above it. Tinbergen kept few pine cones in a circle surrounding the nest of wasp as physical cues (clues) When wasp came out, it took a circling flight, and went away. Just after that Niko Tinbergen displaced those pine cones. On return wasp searched the nest in the centre of the displaced ring. The wasp had learnt the position of its nest by taking a flight around the pine cones.

In another experiment observations were made on adult songthrust, a bird which feeds on snails, it picks up a snail, carries it to a rock, where it holds it by its feet and smashes it against the rock by means of rhythmic sideways swinging motion of the head. This is achieved by a long period of learning in the young birds by seeing their adults doing it. If a song thrust is reared in isolation, it will not be able to perform the above mentioned behaviour.

It is almost impossible to differentiate between innate and learned behaviours in an adult animal without performing experiments in isolation. Probably the most spectacular instance of learning has been found among island populations of the Japanese macaques (Macaca fuscata). On one particular island a two year old female, Imo, started washing; sweet potatoes - by holding them in the water with one hand and cleaning sand away with the other. Within few years, 90 percent members of her group except the very young and very old, had learnt this habit. Something more interesting soon followed, within three years of her first learned task, the same female - Imo invented an interesting method for separating wheat grains from sand. She collected the mixture of sand and grains in her palms and carried it to the sea, threw it into the water, let the sand sink, and strained off the floating grains. This pattern spread through the group as well.

Learning in animals can be of two types :

1. *Flexible:* Which includes habituation, trial and error, classical, and instrumental. As the name suggests this kind of learning is flexible to further changes and experiences.
2. *Restricted:* This includes a very special type of learning called imprinting, once it is established, it is comparatively resistant to further changes. It plays an important role in the development of many behaviours in almost all higher animals.

Concept of Imprinting

The animals learn through their own experience, through observation and imitation throughout their lives, but there are some phases in an animal's life when it is likely to be sensitive to some typical learning processes. Many animals and even human beings pass through such a period when certain learning experiences are far more' significant than at other times. This means that if an individual encounters the same kind of experience outside this period, perhaps it may not learn it at all. K. Lorenz (1935) was the first one to notice this type of learning and he coined the term - "imprinting" for such a process. Lorenz has mentioned many interesting incidents in his book "King Solomon's ring". He was a great lover of animals, and had many animals as pets in his farm house. He was especially fond of greylag geese (duck); with which he conducted classical experiments of imprinting for which he is world renowned. K. Lorenz had unique ways of studying animal's behaviour.

There is an interesting incident - while he was studying jackdaws (crow like); he used to take them out from their nests to mark them with aluminium rings. After few days, jackdaws started making wild noises just at the sight of Prof. Lorenz. He got rather worried and thought how to stop birds from developing a permanent shyness for him, so he disguised himself. He used to wear a Devil's costume while closely observing or handling jackdaws, which was a gorgeous black furry dress with a mask covering the whole head, complete with horns and tongue and a long devil's tail. He picked up a newly hatched jackdaw from the nest and hand reared it, it got imprinted on Konrad Lorenz, as the breeding time approached, it courted Konrad Lorenz, pecked at his teeth and fed him with minced worms mixed with jackdaw saliva, in normal course the jackdaw would do this with a mate.

In early life of many animals a period starts as soon as they are hatched or born. This period varies from days, weeks, months or years depending upon the species. During this time, also known as the critical period, they learn to identify or recognize their parents by following them, according to Schultz it is known as filial imprinting, if following is carried out uninterrupted, this knowledge is utilized by the animals later in their adult life to recognize mates known as sexual imprinting. Imprinting has profound influence on animal's social behaviour.

Though it sounds strange that how a thing which is learnt so early in the life can influence adult behaviour? But it is true, that there is a critical period in early childhood when an individual can learn certain things, once this period is over, those particular behavioural patterns cannot be learnt. Konrad Lorenz observed that newly hatched precocial birds, (which start walking immediately after hatching *e.g.* chicks and ducklings) follow the first moving object they encounter, - usually they are their parents but with incubator reared birds it is likely to be a human being. If ducklings begin their walk by following a man, they will continue to follow a man in preference to adult ducks throughout their life. Moreover, they behave as if man belongs to the same species. These birds may even court a man taking him as a mate.

As mentioned earlier, that in wild or in nature the newborns see their parents, and learn to follow them or learn to recognize their mates through them. Sometimes if the parents leave or die, the newly borns have been seen following individuals of other species which they saw around during the critical time. Mallard ducks raised by parents of another species followed their surrogate mothers. Calves of buffaloes and bosons have been observed following horses, and zebras followed a car.

As adults these animals preferred courting them also. Five year old turkey raised by hand, preferred courting human beings even though they had been placed with members of their own species. Zoo authorities know that hand reared animals preferred courting their caretakers and such animals had poor reproduction when they were allowed to remain with their conspecifics. In lab condition, the animals got imprinted to match boxes, balloons, wooden blocks, toys and many other unnatural objects. The experiments are carried out in a circular runway which has a long rotating arm to which are attached different models which move slowly along the circular path.

The birds can recognize the shape, size and colour of the model. If they are imprinted with a wooden block they will not follow a toy duck. Further, according to E. Hess (1964) the more difficulty a duckling has in following a model, the stronger it will imprint on that model. Also a moving and noisy model is more effective. Socialization and an animal's ability to become a part of a social group are largely determined by early experiences. Human children may become shy loners if between 2 to 6 years they are not part of some social group *i.e.* a school. H.F. Harlow described "Deprived Syndrome" in young rhesus infants. The syndrome is expressed by general restlessness, aggressive reaction, extensive apathy and lack of exploratory behaviour, as adults most of these animals were unable to mount and copulate. According to Lorenz imprinting is a unique form of learning because of the following factors:

(1) It takes place only during a brief sensitive period early in the life

(2) It has great stability, often lasting for the rest of the animal's life

(3) It influences the animal's adult social and sexual behaviour

Concept of Behavioural Genetics

Let us now look at some of the evidences which suggest that patterns of behaviour can be transmitted from parents to offspring via genes. Actually it should have been a part of FAP, but since it now makes a separate and specialized field in ethology and explains issues based on experimental evidences therefore, it needs a special mention. In those species in which parent upbringing is absent, any sort of complex behaviour that is spontaneously expressed is very likely to be genetic. A male cupid fly will catch a prey, form a cocoon around it and present it to female before copulation, female cupid fly will open up the cocoon and eat the prey, in the meantime male will copulate. All this has to be done within the few weeks of its adult life. Such complex behaviours would hardly be accomplished by learning, it is assumed that the entire sequence of behavioural steps followed by cupid flies is coded in the fly's genes.

The example of the hygienic honeybee is a particularly dramatic instance of the genetic control of behaviour.

Certain honeybee strains are "hygienic", if their larva or pupa dies inside a cell in the hive, the workers uncap the seal and remove the dead animal, whereas in "unhygienic" strains this behaviour does not occur. Among themselves, both these strains breed true to their type. If they are mated experimentally, hybrids between hygienic and unhygienic strains are all unhygienic. If these hybrids are mated to hygienic bees, four types of progeny result.

Cross Between Hygienic and Unhygienic Strains of Honeybees.

UU = Do not uncap the cell
RR = Do not remove the dead larvae   Unhygienic UURR
uu = uncap the cell
r r = Remove the dead larvae   Hygienic uurr

UURR × uu rr
||
UuRr
×
uu rr

Gametes	UR	Ur	uR	ur
ur UuRr	Uurr	uuRr	uurr	
	(1)	(2)	(3)	(4)

(1) Those which neither uncap nor remove the dead body *i.e.* unhygienic.

(2) Those which do not uncap, but if the experimenter does so, remove the dead bodies.

(3) Those which uncap cells containing dead larvae, but do not remove the dead body.

(4) Those which uncap and also remove the dead body *i.e.* hygienic

The fascinating point is that what seems on first site to be a complex behavioural pattern can be understood on the basis of action of just two genes. An interesting cross breeding experiment between two species of love birds was conducted by William C. Dilger (1962) of Cornell uriversity. One of these species had the peculiar habit of cutting up strips of leaf or bark for its nest and then tucking them in the feathers of its rump, and in that way carrying them to the nest site. The other species used to carry its nesting material in its bill.

The hybrids between these two species carried material both ways. They started off by carrying most of the nesting material tucked in the rump feathers, but this was less efficient than transportation by bill, it kept falling out, and they subsequently learned to use their bills more and more. The roosters of different species crow in different postures, ring necked pheasant crows by stretching its neck and pointing upward, domestic rooster looks down and a hybrid of both species assumes intermediate position. Variability in species and populations is as large for behaviour characters also as it is for structural characters.

Concept of Evolution of Behaviour

Has behaviour evolved? It most certainly has, unfortunately, fossils do not behave, and so we have no direct knowledge of the course that behavioural evolution has followed in the past. Yet there is a great deal of indirect evidence which allows us to say something about the way behaviour has evolved. As in any other evolutionary process, we can assume that new behavioural developments did not take place by sudden leaps and bounds, but through small changes and stages of adaptations. In animals today, we see the combination of structural adaptation and refined behavioural patterns developed through many generations, which enables every individual to survive. For example, certain animals, like the screech owl and sphinx moth have developed behavioural traits that provide each of these with two distinct lines of defence against predator.

Both these species are nocturnal and rely primarily on camouflage to protect them while roosting during the day. They stay motionless and if at all detected by a predator, the screech owl bluffs by swelling, snapping and flapping, trying to scare away the predator. Sphinx moth hag a more subtle second line of defence, it flicks open its wings to expose eyes like spots which often startle attackers, allowing the moth to escape. This simple act is a forceful example of behaviour developing in step with the evolution of body markings. Caterpillars, on which eye spots have evolved as part of their defence, have also developed behavioural displays that show their markings to best advantage. Many larvae have eye spots concealed in the folds of their skin and only reveal them when threatened. Some other animals have them underneath and display them by turning their bodies.

Many animals gain protection by imitating physical structure and behaviour similar to objects regarded as inedible by predators. A twig caterpillar has evolved the physical dimensions and markings of a real twig, unless it remains as still as a part of the tree, a predator will not be fooled and the insect will be eaten, candid fish in the Amazon basin, resemble dead, dry leaves and can float motionless for hours to escape the notice of larger fishes and predatory birds, similarly some grass hoppers resemble fresh grass, Camouflaged mantis have their wings like plant leaves.

Studies of evolution often involve comparisons between closely related species. It is not easy to comprehend how something as complicated as the social life of honey bees may have evolved ? By studying the behaviour of many species of been Karl Von Frish was able to suggest a likely course of evolution from solitary ancestor, through various intermediates, to the social structure of hive. There is one species of bees whose individuals live solitary, each bee makes a burrow, constructs few cells in a tree trunk, lays eggs singly and puts honey in each cell, seals all the cells and leaves.

In another species, the young hatch while their mother is still laying eggs, some will help the mother before she leaves. In bumble bees, the young ones which hatch to help mother remain small and less fertile and finally comes the species where mother lays eggs, young come out that remain sterile and all of them live together with mother to form a complex social group.

It is widely accepted that all behavioural patterns that are learned or preprogrammed are for the individual's benefit, in other words, any behaviour performed by a particular animal must be directly useful to it. In terms of evolution all individuals behave to increase their own "genetic fitness". It is measured in the currency of offspring and is given in terms of the number of its offspring expected to survive to the reproductive age. In other words, it is a measure of the number of copies of genes an animal contributes towards the hereditary composition of the generation. The animal behaves to increase its own fitness irrespective of other individuals of the same group or clan. Here "fitness" does not mean the physical fitness, an animal could be physically very fit, but if it does not reproduce and leave its genes through progeny in the population it will remain an evolutionary flop.

That means an individual should behave in such a way so as to maximize its reproductive success. That way all individuals should be selfish, they should be concerned about themselves only. Then why should animals live in social groups? What is the process by which they form bonds between members of a family, leading to mutual support and cooperation? Could we not argue that each individual according to the rules governing evolutionary process, should be concerned primarily with its own reproduction. Why things like Altruism (sacrifice) and cooperation evolved in social animals?

Altruism is, when an animal facilitates the reproductive

chance of a conspecific at the expense of its own, altruism may include anything from sharing food when food is scarce to defending others at the risk of the animal's own life.

Why some of the bees give their lives in a suicide attack while proctecting sister bees of the hive? Squirrels, deer, and some monkeys do risk their lives by drawing attention of the hawk and big predatory cats while giving out alarm calls to others. There must, therefore, be circumstances under which alruistic behaviour can be advantageous. The clue to the origin of most of altruistic behaviour in the animal kingdom lies in such behaviour which are being directed toward one's blood relatives. Aunts of elephant calves and lion cubs protect and nurse them.

The animals often tend to be altruistic towards other animals which share in common, a large fraction of their hereditary units or genes by virtue of common ancestry. For example, a person might be willing to throw away his own life to save the lives of two full brothers but he may not be willing to do so to save the lives of three strange persons. Altruistic behaviour, could be explained on the basis of "kin-selection" in which any individual would sacrifice its own reproductive success for the sake of its kins or relatives who share some proportion of its genes. In other words, the same genes must be passed on or propagated, either through individual or through a kin.

The help is not rendered to related animals alone. In the troops of the African olive baboon, quite unrelated males may get together to help each other when predators attack. Baboon males are big in size and often when they get together they may kill a leopard. This kind of cooperation held at the time of need is like "you scratch my back and I will scratch yours". Or "you help me this time and I will help you whenever you need me", therefore, safe guarding their own genes.

Organisms without any social tendencies are very rare. We know of few animals that avoid all contact with others of their species, yet they have to come together in order to breed. Oyster bank, swarm of grasshopper, schools of herrings, flocks of starlings, the brood colonies of gulls, pack of wolves, herds of antelopes, pride of lion, and group of monkeys are few examples of social tendencies. The non-social behaviours may have just provided roots for the growth of higher social life.

5

The Navigation

Bird Migration

Animals cannot spread their roots in one place and synthesize their own food like plants. They must move about from one place to another in search of food. Birds with their fantastic ability to fly and with their unparalleled mobility, have explored almost the entire globe searching for food, territory, nesting sites and other necessities for survival. Because of their high metabolism and an equally high energy consumption, birds must have abundant and unfailing sources of rich food. It commonly happens that in regions with seasonal climatic changes, ecological changes occur which may require the birds to move away if they have to survive, In colder countries the deep chill of winter closes lakes and streams, and snow fall covers the land and its vegetation compelling the birds to move to warmer regions.

In countries around equator, summers or dry season may reduce the plants, insects, worms and other food sources, the temperature gets very high and to escape that the birds must emigrate to some more favourable place. The movement from one place to another may be permanent, but the true migration is when the animals return to their original place.

Disappearance of bird flocks and change in plumage during a certain season was noticed very early. Aristotle thought swallow and several other birds hibernated in holes during winter, he introduced the idea of transmutation and he wrote that European robin changed into European redstart in summer, he also noticed some kind of movement and wrote that many birds flew to other lands but perhaps didn't know why ? It was Frederick the II, a Roman Emperor about 2000 years ago, who could be called as the first serious student of bird migration, which became evident from his writings.

The ancient Greeks and Romans were aware of this ability of birds to return to their homes and they used the birds to carry messages. Julius Caeser's armies communicated through pigeons, so did the armies of the ancient Persians and Egyptians. Pigeons were used in both world wars, infact, there is a monument of the carrier pigeon in Brussels in respect to their remarkable contribution in war time and also for their homing ability. Migration has evolved among many animal groups. Frogs, newts, seaturtles, buffaloes, zebra, wildebeests, sea lions, whales, butterflies, crabs, locust, ladybirds along with a great variety of birds and fishes form some of the most common examples. The distance travelled varies from species to species. The crab Ericheir sinensis who can only crawl slowly, travels 500km, locust Schistocerca gregaria can fly and travel about 5000km, however, the longest distance is covered by arctic tern, which travels from Arctic to Antarctic, covering an astonishing distance of about 17,600 km.

As evident, many animals migrate but birds migrate large distances and that too very regularly, because of their flying ability. It must be clearly understood that not all species of birds migrate, there are birds which remain all through the year at one place and are called residents *e.g.* cotton teals and whistling teals at Keoladeo National Park, Bharatpur.

Definition of Migration

The word migration, derived from the Latin word migre (to go from one place to another) means any travelling. Any cyclical movement with respect to food and weather that occurs during the life history of an animal at definite intervals, and always includes a return trip from where it began, is migration.

Different workers have defined it differently

(1) It is the movement of animals from one place to another.

(2) It is seasonal movement from one place to another.

(3) It is a response of an animal population to changes in environmental conditions.

(4) It is the movement of the centre of gravity of a whole population from one place to another.

(5) Migration is a two way journey to the breeding ground in spring and to the winter quarters in autumn. {March-April (Spring); May-July (Summer); August-October (Autumn); November (fall); December–February (Winter).}

(6) Regular, extensive, seasonal movements of birds between their breeding regions and their wintering regions.

(7) Periodic to and fro movements between nesting and winter quarters.

(8) It is an adjustment to unfavourable conditions during part of the year made by certain species of birds in some areas.

(9) It is the movement of animals from one place to another,

preferably confined to rhythmic movements, which involve a journey from one place to another.

(10) According to Marten Ducane "Many species in the colder regions of the world seek, on the approach of winter, other lands, where more congenial conditions prevail and suitable food in sufficient quantity may be found. The return of spring, however, sees them once again on wings, making their way back to their old breeding grounds and even, as a rule, seeking the same nesting sites. Broadly speaking, the autumn migration as a whole may be said to be a journey from the arctic and temperate zones of northern hemisphere to countries lying in a more or less southern direction to the equator," or it is a movement from west to east. In northern Europe, there is a good deal of migration taking place in west to east direction.

(11) An instinctive regular two way movement of part or all of an animal population to and from a given area, usually along well defined routes. It is closely linked to the cycle of the seasons and is triggered off by seasonal factors such as increasing and decreasing day lengths in spring and autumn. Animals migrate to escape harsh climatic conditions (extreme cold and heat) and to get adequate food.

The word migration is different from such terms as "mass movement", "emigration" "immigration" or "dispersal". In migration, animals from one place move to other but always return to the same place. However, in other types of movements animals do not return to the same place. According to Salim Ali, world renowned ornithologist from India, the true migrants show "pendulum - swing" movement. A migrating bird does not travel to just any area in the spring and the fall. It has been demonstrated many times that a bird often returns to the same small area each spring to carry out its nesting activities, and that it also tends to return to the same wintering area each year.

As a rule, the more rigorous the climate, (especially in the extreme temperature may it be heat or cold) the greater is the percentage of birds that migrate. The bird migration is predominantly from north hemisphere to south hemisphere. In Canada, the percentage of migrants is higher than in the United States because Canada is further north and is cooler than USA, fewer Mexican birds migrate, because it is warmer than USA and practically no birds migrate in the rain forest of South America. Very few southern hemisphere birds are strong migrants. In southern Africa there are only 20 species that move toward the equator to spend the winter, but not a single one migrates as far north as Europe, USA or Canada.

Types of Migration

It is just wonderful to note that birds can fly as high as a boeing jet airplane while going from New Delhi to New York; that the birds can fly with a speed of 80 km/h by which an average car moves on the road from Jaipur to New Delhi; that the birds like ducks and geese which are adapted to water and swimming are few of the strongest fliers. If there were Olympic games for animals, birds would probably get most of the gold medals for their remarkable performance in various feats. The birds undertake many different types of migrations.

There are six types of migration

(1) Daily/Local

(2) Seasonal

(3) Cyclic

(4) Latitudinal

(5) Altitudinal

(6) Longitudinal

Daily / Local Migration

Many birds make daily movements from their nest in response to environmental forces such as light, darkness, temperature, humidity and food availability. Birds may make daily migrations from their resting sites to feeding areas. The house sparrow does not undertake long migration, but it moves out everyday from its permanent colony for feeding. Disappearance of paradise flycatcher, golden oriole and pitta at regular intervals is also due to local migration.

Seasonal Migration

Migrations of birds in response to seasonal changes are common. In tropical and subtropical areas, they occur in the beginning or end of the warm season. In temperate areas, the movements are triggered by onset of winters. The birds which rest in the temperate zones, go to warmer areas for the winter and then return to the breeding area the following spring. Such migrations of the birds are known in Europe, north America and Asia. This sort of migrations are best known in northern hemisphere, where there are tremendous changes in the environment. In most of the ducks, males and juvenile birds migrate short distances northward for moulting leaving behind females and young birds in the breeding ground, it is known as moult migration.

Cyclic Migration

Some migrations of birds are seasonal but do not occur at regular intervals. The cyclic migration of the snowy owl in United States in winter occurs in three to five years.

Latitudinal Migration

It is the migration from north to south or south to north. It is the most common migration birds undertake, birds migrate from north to south in the fall and in the opposite direction the following spring. Every year 5000 million or more birds fly 3200 km south to the grass lands, forests and lakes of central and southern Africa. The distance travelled may be several thousand miles or it may be only a few hundred miles. The longest latitudinal migrations are performed by several species of shorebirds.

The maximum distance (about 17,600km) between the summer and winter homes of any species is that of the arctic tern. This species migrates from Canada to Antarctica, crosses the Atlantic ocean both times during its migration flights. The American golden plover escapes winter of America by migrating to Argentina, similarly some birds leave Siberia to go to Himalayas. About 1/3 birds of Europe spend their winters in Africa. Few land and sea birds from African and South American countries reverse the process and they cross the equator to go to the northern lands and oceans. Wilsons petrel leaves its nesting grounds at Antarctica and go to Newfoundland in Canada. Greylag geese, pintails, common teals, gadwall and shovellers migrate from Palaearctic region or Eurasia to India.

Altitudinal Migration

Species that live in mountains may adjust to changing seasons (generally during winters) by moving downward to warmer and lower altitudes where more food and shelter are also available. Coots of Andes mountains in Argentina move downwards, mountain quails, which nest at elevations upto 9500 feet in central California mountains, migrate downward to areas below 5000 feet in the fall and in the summers they go back to the hills, violet green swallows of UK and willows in Siberia are also vertical migrants. Species of birds, albeit few, show the reverse movement, they migrate to higher elevations in winters, *e.g.* blue grouse, in USA moves upward in the northern Rockey mountains from low altitudinal pine forests.

Longitudinal Migration

Movement of birds from east to west or vice-versa is known as longitudinal migration *e.g.* evening grossbeaks that nest in northern Michigan spend the winter in New England, California gulls that breed in Utah migrate westward to winter on the Pacific coast, starlings move from east Europe or Asia to the Atlantic coast.

Migratory Status

Birds are classified into four major categories according to their migratory status.

(1) Permanent residents

(2) Summer migrants or Summer visitors

(3) Winter migrants or Winter visitors

(4) Transient visitors or Passage migrants

Permanent Residents: These species are found in a particular area throughout the year and do not migrate from one place to another. *e.g.* cotton teal, spotbill duck, whistling teal, barheaded goose, mallards, comb duck are all permanent residents of Keoladeo National Park, Bharatpur.

Summer Migrants: The birds which move to their breeding grounds in spring (March–April), rear their young ones during the summer (May-July), return to wintering grounds in the fall (November) and pass their winters (Dec–Feb) there, are called summer migrants.

Winter Migrants: These include those species which move from their breeding grounds to spend their winter in a more suitable place, where food is also in plenty, *e.g.* greylag goose, pintail, common teal, gadwal, shoveller (from Palaeartic region or Eurasia), common pochard (France) and European starling, are the winter migrants visiting Keoladeo National Park at Bharatpur.

Transient Visitors or Passage Migrants: Those species, which pass through a locality usually twice a year *i.e.* once in spring when they are going to their breeding grounds and again in fall on their way back to their wintering grounds, *e.g.* gargany teals migrate from Mangolia or Siberia to Bharatpur to go to South India which is their actual breeding ground, while returning they again stop over at Bharatpur. Within India, a large number of species start their migration from north - east states, stop over at Bhopal and then go to Peninsular India and Srilanka.

Duration and Distance

The birds may travel from 500 km to 17,000 km either at a stretch or by making stopovers depending upon the distance taking few hours to days to months. There is much variation in the distance covered during migration.

The birds which cover short distances, *e.g.* - vesper sparrow, blue bird and woodcock migrate from southern Canada to north America covering only few hundred kilometers. On the other hand several hundreds of kilometres may be covered by other birds at a stretch. About 100 species of long distance migrants are known. In north America, the night hawk and barn swallow have the longest migrational routes. They cover a distance of about 11,200km. Some birds, such as snipes, sand pipers travel a distance of more than 12,800km. The longest and the most remarkable of all migrations is that of arctic tern

Variation in the Extent of Migration

Species	*Breeding area*	*Wintering area*	*Distance apart in miles*	*Kms*
1. Arctic tern	Canada/Arctic	Antarctica	11000	17600
2. Pectoral sandpiper	Arctic	Argentina	9000	14400
3. Bobolink	Canada	South America	8000	12800
4. American Golden plover	Arctic Tundra	Argentina	10000	16000
5. Willow warbler	Siberia	Africa	8000	12800
6. Sooty shearwater	New Zealand	North Atlantic	6000-9000	9600-14400

Contd...

Species	*Breeding area*	*Wintering area*	*Distance apart in miles*	*Kms*
7. Eastern golden	West Alaska,	Hawaii, Indonesia	6000-8000	9600-12800
plover	North, Siberia	Australia, India	6000-8000	9600-12800
8. Great shearwater	Tristana Cunha	North Europe	6000-8000	9600-12800
9. Buff breasted	Arctic Canada	Argentina and	6000-8000	9600-12800
	sandpiper		Uruguvay	
10. Arctic warbler	North Europe	S. E. Asia	4000-7000	6400-11200
		and North Siberia		
11. White stork	Europe	South Africa	4000-7000	6400-8000
12. Blue Checked	N.India,	East Africa	4000-5000	6400-8000
	bee-eater	West China		
13. Scarlet grossbeak	N. E. Europe	India, S. E.	3000-6000	4800-9600
			Asia	
14. Ruft	Europe, Siberia	S. Africa, India,	3000-6000	4800-9600
			Sri Lanka	
15. Black and white	India	East and South	3000-5000	4800-8000
	cuckoo		East	
16. Snow goose	Alaska	California Asia	2000-3000	3200-4800
17. Long tailed	New Zealand	Africa Fiji	2000	3200
cuckoo				
18. Pintail	Alaska, California	South America	1000-4000	1600-6400
			Hawaii	

its breeding range is circumpolar. The distance covered is 17,600km. The black headed bunting travels about 6,500kms from south-eastern Europe to Bharatpur with a speed of 40 to 48 kmph. Majority of the birds migrate at night, usually these are small birds, they rest and forage during day time, waterfowl and shore birds also fly at night hours, black poll warbler make non-stop, overwater flights for more than 3680 km lasting about 86 hours. The golden plover flies non stop from Alaska to south America, covering 4000 kilometres. Greater shearwater fly across 12,800 kilometres from north Atlantic to south Atlantic. Blue geese fly 2720 kms in 60 hours, Most migratory birds can fly about 800 kms non stop.

Altitude : Formerly, it was thought that birds while migrating travelled very high and with a very fast speed. Timothy C. William noted-the altitude attained by migratory birds from his ship. However, recent knowledge, obtained by telescope, radar and radio telemetry has pointed out a great variety. Some birds fly at sea level, some fly very close to the height of Mount Everest. Most birds fly less than 7,400 ft. above sea level. In the Himalayas, migrants fly between 800-1,600 ft. and night migrants fly still lower. Strong fliers, such as ducks and geese, fly 4800 ft. to 8640 ft. above sea level. In the Himalaya, migratory birds have been found at a height of 17,280 ft. Similarly blackpoll warbler, godwits and sand pipers, fly at 21,000, 20,000 and 13,000 ft. respectively. Blackpoll can fly where it is very cold, and at oxygen starved altitude of 21,000 ft. Himalayan expeditions found griffon vultures and lammergeier at 22,400 ft. and Himalayan geese at an unbelievable height of 29,500 ft.

Speed : Most birds can compete with an efficient automobile, like a car they can also cover 90 km in an hour. One famous example is that of a bird called manx shearwater which was taken by plane to USA from its burrow in Europe, the bird was back on its nest 12.5 days after its release in USA, ten hours before the mail arrived giving details of its release.

Flying speeds of birds have been timed on many occasions. The speed of the migratory birds is measured using doppler radar, it was found out that the speed ranged from 32 to 64 kmph in small song birds, while in larger birds, such as cranes speed varies from 40 to 96 kmph. Smaller perching birds can fly 32-59 kmph whereas, falcons, ducks, and geese fly at the speed of 77-96 kmph, humming birds fly 32 kmph, and sandpipers fly 96 kmph.

Causes of Migration

The evolutionary origin of migration is extremely complex and incompletely understood. Why birds started migrating? It is assumed that birds originated in northern hemisphere and lived there till late Pleistocene when glaciation began; to avoid cold, birds flew to southern hemisphere but returned to original home when ice receded.

Therefore, bird migration, at least in the northern hemisphere, was initiated by the effects of Ice Age. It is assumed that prior to the coming of the great glaciers to the polar regions, birds lived the year round in the northern hemisphere. Though forced to retreat with the advance of the glaciers, they continued to return to nest in the summer because of an innate attachment to their homeland. The other theory is that the birds originating in warm climates probably spread northwards in search of food, many found it in northern land masses and oceans but were forced to go back to their south bound homes when winters arrived. It is also believed that intraspecific and interspecific competition for food, territory, nest sites, have also been important factors initiating migration.

Before migrating, a bird must get ready with energy requirements for prolonged flight. It accomplishes this by eating amounts of food in excess of its daily needs and thereby storing energy in the form of subcutaneous fat. The excess of fat deposits in clusters, which appear as balls called fuel tanks, the black poll warbler becomes almost double of its weight just before it under takes migration. At the same time, the bird must become predisposed to migration by developing a condition commonly called migratory- restlessness or Zugunruhe. This condition is directly under the influence of pituitary gland, whose activity at this time is stimulated by the day lengths. When the bird has attained the necessary physiological and behavioural conditions, an outside stimulus is required to trigger migration. The stimulus is a combination of climatic factors, such as a change in temperature, change in direction and velocity of wind, the onset of cold, and the day lengths.

Apparently photoperiodism *i.e.* the effect of length of day, triggers bird migration, the day length affects pituitary and pineal body in the brain which in turn prepare birds for migration, the migratory restlessness is initiated from here.

Advantages of Migration

(1) Securing a better climate for living by avoiding cold and stormy weather.

(2) If all the birds that migrate should remain in their winter quarters the year round, then they and non-migratory birds would make great demands on the food supply, and many birds might have difficulty in nourishing their young. However, by alternately exploiting two different habitats for food, more birds are able to exist.

(3) Geographic wandering, or change in habitat provides greater variety in bird's diets, and perhaps more nourishing diets, and more chances for adaptive radiations.

(4) The long summer days provide birds with longer working hours to gather more food to feed young ones.

(5) In their breeding grounds the migratory birds arrive in large numbers hence the pressure of predation is divided among a greater number of eggs and young, and this results in greater individual survival.

(6) Parasites and infections of micro-organisms are reduced to minimum in far northern habitats due to long cold winters.

(7) It provides certain genetic and evolutionary benefits. A long distant migrant, is subjected to different kinds of natural selections in its breeding grounds, its winter quarters and while on journey between the two habitats. Certainly, migratory birds are subjected to greater ecological diversity and therefore, have a greater range of adaptability than resident birds.

(8) Migration promotes the geographic dispersal of birds. Dispersal may isolate small populations of a species

from each other, it tends to increase the rate of evolution in that species because genetic variation or mutation makes a greater impression on a small population than on a large one.

(9) Last but not the least bird migration provides golden opportunity to human beings for watching birds from far off lands. Keoladeo National park is considered one of the best parks in the world for watching birds.

Methods of Study

Birds migrate day and night, in summers, in winters. They fly above the snow covered mountains, rivers, oceans, they may fly very high with a great speed. This makes the study of migration extremely difficult. Geese, ducks, gulls and shore birds travel day and night, but this is not the case with many other birds: hawks, eagles, falcons, storks migrate only during day, while nearly all passerine birds migrate during the night. Birds are excellent weather sensors and can sense favourable winds. Broad winged birds like eagles, storks and cranes use thermal current for soaring by which they spend less energy for flight whereas, for small winged birds such as passarine birds it is advantageous if they fly during night hours. Two explanations have been advanced for the development of nocturnal migration:

(1) Movement by night offers birds, which normally live in thick vegetational cover and rarely take long flights away from it, the protection of darkness against their day predators.

(2) It gives the birds opportunity of using all the daylight hours for feeding, thereby enabling them to build up sufficient energy resources for sustained long distance flights.

There are six field methods of studying migration:

(1) By observing movements across the face of the moon.

The principal equipment needed is a telescope. The birds crossing the face of moon are counted.

(2) By radio-tracking, the equipment needed includes a transmitter and receiver. When a bird cannot be followed visually, it may be equipped with a miniaturized radio transmitter which is either put around the neck or stuck with an adhesive to the skin of the back between the wings. The transmitter is then monitored by a receiver and the movement of the bird is plotted on a map, the method is known as radiotelemetry or biotelemetry.

(3) By radar, this method also requires implantation of transmitters, which sends out signals, these signals are caught by radar.

(4) By tape recording, the bird calls through a tape recorder, with the help of a parabolic receiver, microphone, and amplifier. This is used to estimate the number of migrating birds by tabulating the call notes heard at night. The method is also known as chip counting.

(5) By collecting and studying dead migratory birds killed at television towers or by airport ceilometers or other man-made high towers and poles interfering with migration.

(6) The last and most popular method of studying bird migration is by banding or ringing. Large scale banding has enabled ornithologists to plot migratory paths or routes of many bird species. There are many schemes for banding migratory birds throughout the world. All it involves is the use of coloured metal or plastic leg bands, systematically numbered and with the address of the banding agency to which the recoverer of the banded bird may report. The birds are first caught or trapped alive in large number by using mist nest or boomnet, this net is thrown into air by explosives and

spread out by metal weights which soars over a flock of birds. A wide variety of markers have been devised for studying large migratory birds to enable their recognition from a distance. Most such markers consist of plastic tags of different flashy, striking colours hanged around the bird's neck.

Various combinations of coloured plastic bands enable identification of individual birds visually without having to recapture them. By devising a coding system *e.g.* green band on the left leg of one bird, green band on the right leg of another, green and yellow bands on the left leg of still another can denote different geographical areas, seasons and species. Sometimes tags made up of coloured plastic are attached on the birds bill by very thin nylon thread passing through the nostrils called nasal tags. Birds may even be "feather-marked" by colouring their feathers with different colours, when it is desirable to mark birds which can not, for one reason or another, be captured. It is sometimes possible to spray fast colour with the help of a water pistol, however, this method is difficult to follow.

The bird banding laboratory at Lawrel and Maryland, in USA have maintained more than thirty million banding records and add a million new ones every year. In India, at Keoladeo National Park (formerly called as Ghana 'Bird Sanctuary), Bharatpur, extensive research on migratory birds is in progress specially on their ecology, behaviour, and migratory routes. The work is primarily being conducted by Bombay Natural History Society (BNHS) in collaboration with State forest department. More than 100000 birds have been ringed during past 25 years.

In one of the studies Spanish and Turkistan sparrows ringed at Bharatpur were recovered in Kazakhastan. By using this technique it has been possible to establish the migratory routes of birds, *e.g.* the main migratory routes into India run down the valleys of the Indus in the west and the Brahmputra in the east. The ringing or banding at Bharatpur has established that European swallows return not only to the same locality but even to the same place for nest building purposes year after year, covering a distance of 6000 miles. This is the case with other true migrants also. Salim Ali observed that a ringed grey wagtail returned from Himalayan breeding grounds to a particular lawn in Greater Bombay year after year on almost the exact date in September covering a distance of about 2000km.

BNHS is the only authorised institution to carry out bird banding studies in India. A total of 360 species of birds have been recorded from Keoladeo National Park and a great majority of them are migratory. Most birds migrate from Himalayas, Laddakh, Tibet, Central Asia, Siberia and Eastern Europe. Some of the most commonly seen migratory species are greylag goose, pintail, common teal, gadwall, shovellers, tufted duck, crested pochard, common pochard, mallards, wigeon, common shelduck, marble teal. In Bharatpur, winter migration of birds starts in the August with the arrival of garganey teals and ends in December by the arrival of Siberian cranes which is a fascinating feature in this park, because it is a very rare bird and not more than 200 are left in nature, they come to the park covering 4600km. The return migration of most species starts in March.

Navigation

Most of the birds are remarkable path finders. The knottiest problem in ornithology is how a bird finds its way home? Year in, year out, millions of migrating birds successfully travel hundreds and even thousands of kilometers, back and forth between their breeding and their wintering grounds. To do this successfully, a bird must know where it is, it must know the direction of its goal, and it must be able to maintain its course or navigate in that direction. Banded long distance migrants have returned not only to the same restricted wintering area, but also to the same pinpoint nesting site. Many individually marked birds have been experimentally displaced thousands of kilometers into completely strange areas, and yet have returned to their home nest sites, often with astonishing speed.

For example a Puffinus puffinus carried by airplane to Boston, over 5100 kilometers from its breeding island off Wales, was back in its nesting burrow 13 days later. Homing ability is widespread in the animal kingdom., but it is perhaps best developed in birds. This recognition may be hereditary, or more likely, acquired early in life, perhaps through imprinting by watching their conspecifics. It has been proposed that birds may have an, inborn, builtin innate or instinctive programme to migrate, meaning that birds just know inherently that this is the time to migrate. Where to migrate and how to migrate, that they do not have to learn by watching conspecifics, *e.g.* there is a migratory bird bronze cuckoo found in New Zealand, its young was allowed to rear by foster non- migratory parents and 'as the young bronze cuckoo matured, it migrated 4000 km over the open sea.

In few species of fishes and birds the young ones have been seen to take the lead in return journeys, therefore, it may be true for just few, otherwise it is believed that most of the birds learn to carry out this amazing task. In an experiment indigo buntings were reared in isolation, these birds when released as adults, could not attain accuracy of orientation suggesting, that they depended on experience or learning from conspecifics.

For human beings, solar and stellar navigation still remain very complex procedures requiring sextant, chronometer, compass, charts, and mathematical tables. Do birds have special builtin instruments, what are they and where are they located in their small bodies, still remains unknown. How do birds determine their way and position themselves correctly remains a mystery for man. Daytime migrants use their 'vision to steer by the sun and land marks, some migrants see ultraviolet and polarized light and hear low frequency sounds that can emerge from distant surf; night fliers take compass cues from moon and star positions, migrants can sense earth's magnetic field, probably in combination with gravity. It is almost established that birds use some or all of these abilities in complex interrelated ways that vary with species, age, experience, location, weather and season.

A builtin compass in animal's brain helps in finding ways, it is equivalent to our ideas of north, south, east and west. These compasses use magnetic field of earth and position of sun. The magnetic compass has been demonstrated in a number of vertebrate animals, it is assumed that animals have the ability to perceive the earth's magnetic field and thereby decide directions. Sun compass orientation has been extensively studied in crustaceans, spiders, insects, fishes, amphibians and birds. Using an internal compass to localize geographic direction offers new possibilities of orientation, the animal can remember in which direction it has it's home and in which direction it is moving and reaches back home by returning on a compass bearing which reverses this direction; such a strategy is known as "route reversal" and is applied by many arthropods.

Do birds have specially modified sense for navigation? Animals do have specially modified senses which they use for different purposes *e.g.* butterflies have exceedingly acute sense of taste; the blood hounds and moths have acute sense of smell, bees can see ultraviolet rays and can discriminate the plane of polarized light, a blind folded pit viper can perceive body heat of a small mammal by special pits found between the eyes and nostrils; echo-location in bats is well known but few know that two species of birds *viz.* oilbird and cave swiftlet can also avoid obstacles in the dark using the same techniques as bats; dogs have extraordinary sense of smell; dolphins can hear ultrasonic sound. Among birds the theories involving special sense perception is the retracement theory, which says that a bird stores in its memory kinesthetic and semi-circular canal impressions of every twist and turn of an outgoing journey, and recalls them, in reverse order, for guidance on the return trip. Many experiments, conducted by displacing the birds either by anesthetizing them or by putting them on constantly moving circular turn tables during their outgoing trip, showed no differences in homing ability. Moreover, this theory could not explain the correct orientation achieved by young-unaccompanied, inexperienced birds on their return trips.

Another hypothesis which says that homing birds have an ability to sense the earth's magnetic field and thereby acquire a compass like sense of direction has been widely accepted, German investigators Wolfgang and Roswitha Wiltschko, kept European robins in a featureless circular cage and blocked out the sky so that the birds couldn't see the landscape, sun, moon or the stars, the birds in a state of migratory restlessness faced the proper migratory direction. The investigators then surrounded the cage with Helmholtz coils, which produce a magnetic field when an electric current passes through them. When current in the coils altered the normal magnetic field, the birds changed their direction too.

In another interesting experiment, William T. Keeton of Cornell University used homing pigeons, he put little bar magnets on the backs of some pigeons to disturb their magnetic fields and put non-magnetic brass bars on controls. Released in sunny weather, both groups reached home equally well, but released under overcast skies, only the brass birds or controls flew normally, those with magnets, lost their way perhaps they were not able to use their sun compasses and received false information from magnetic compass and hence became confused. They concluded that the birds used two compasses simultaneously, sun and magnetic, to orient themselves correctly in time and space.

Scientists have recently found that some bacteria synthesize magnetite, a compound of iron and oxygen, and use it to orient themselves according to the earth's magnetic field, magnetite has been found in the brain of pigeons and abdomen of bees, do these animals also rely on magnetite for orientation, still remains unanswered.

Another hypothesis, proposed by Ising (1946) suggests that, because of their motions over the surface of the earth, birds can add to or subtract from the effects of the earth's centrifugal force, and that they are able to feel the gravitational differences. Another force assumed as a possible clue to orientation - is the coriolis force. This is the lateral force generated in moving bodies that crosses lines of latitude on the earth's surface, the force that causes the clockwise deflections of winds in the northern hemisphere, but the use of this force has not been convincingly demonstrated in birds. Yeagley (1947, 1951) suggested that birds perhaps respond both to terrestrial magnetism and to coriolis force.

A sense of direction, or compass sense, is worthless unless a bird also possesses a sense of position - that is, know where it is with reference to its goal. It needs, a compass, a chart, and knowledge of its map location. It would be strange indeed if birds did not use their superior eyesight to aid them in navigation.

In some species, visual clues may be all that are required to guide birds on their migration travels. Birds following similar route every year only need to remember the different and prominent visual landmarks over which they fly *viz.* shores, islands, mountains, rivers, valleys, deserts and forests. Birds flying in large groups or flocks orient themselves better than those birds which fly alone because of the possibilities of following navigational resources, information from other members and eliminating individual errors and avoiding deviations from the correct path.

In an experiment marine birds - gannet, were displaced 342 kilometers south west from their nests on Bonaventure island and released deep inland in Maine, they were observed by airplane to perform more or less random movements until they encountered the familiar Atlantic Coast line, after which they made a more rapid and direct return home indicating their topographic memory or visual orientation. Is it vision which helps? Keeton and Walcott put opaque lenses on homing pigeons, so that the animals could not see, but the pigeons could still reach the correct destination, that means, visual orientation is an added facility along with other navigational skills.

How do night migrants find their way? Do they inspect the landscape by day and memorize it? Or they can see at night because even on moonless nights, the earth is by no means featureless, rocks, trees, fields, rivers, and hills are clearly visible by starlight specially to dark adapted eyes. How do birds orient when they fly over sea? By judging wind and wave direction, by identifying islands, reefs, cloud formations over oceanic islands and over cold waters. They can even sense temperature and humidity differences in the air; colour, and turbidity differences in the great surface currents of the sea water, the presence or absence of organisms visible from the air such as dolphins, flying fishes, jelly fishes, and whales.

Though, these do not make really a very informative list of traffic signs, but they may, nevertheless, be extremely useful clues to some who have lost their ways. One difficulty in the use of land marks for navigation is that they change with the seasons, if the changes are gradual over a period of few years the birds adapt themselves accordingly but an early snow fall can completely change the appearance of the land and a chilled storm can freeze the lakes and rivers.

One of the most dependable navigation a migrating bird could use in orienting itself is by the sun - in the same way as sailors use it in their navigation. Experiments have shown that some birds definitely use sun in maintaining their migratory sense of direction. Main work has been done by Kramer and his students (1957, 1961), they placed migrating sterlings in a cage exposed to sun and observed that the bird fluttered on their perch facing towards the normal migration direction as long as the sun was visible but on cloudy days their activity was disoriented, when the apparent position of the sun was artificially shifted by using mirrors, the orientation of the birds also shifted correspondingly. It was further explored that the birds responded to the azimuth (arc of the sky) or compass bearing of the sun along with the altitude *i.e.* its height above the horizon. This type of orientation is called sun azimuth orientation.

Since birds are able to determine compass directions in the day time with the help of the sun's position, it will be logical to assume that night migrants may be able to use the stars in the same way. Night fliers take bearings from moon and stars. In an experiment S.T. Emlen of Cornell University, one of the leading scientist in avian research, kept indigo buntings in conical cages in a planetarium. The conical cages had ink pad at the bottom. In one situation when the stars were visible the indigo buntings made tracks to the north east side of the cage which was its spring course or route for migration, in another situation the stars were switched off, and the bird kept flying aimlessly. Emlen concluded that indigo buntings take bearing or direction from stars.

German scientists Franz and Eleonore Saver, put European warblers in a cage with a glass top so they could see only night sky but nothing else. The birds oriented in the right migratory direction as long as they could see night sky and stars.

The other popular belief about how do birds find their way are that they have acute sense of hearing. Human ears will not pick up sound waves under ten to twenty cycles/sec but birds can detect much lower levels therefore, while flying high they may hear a thunderstorm approaching, tides lashing at seashores and even rhythmic pulse of the ionosphere, birds may use this sense to gather information and orient themselves correctly.

Some birds can detect very minute differences in barometric pressure, even a difference of pressure between the floor and ceiling of a room can be detected by birds. Italian scientist Floriano Papi thought that pigeons use their sense of smell to locate home, and in one of his experiments he blocked the nostrils of pigeon which showed poor homing behaviour as compared to the ones which had open nostrils.

It has now been accepted that for navigation, birds use a variety of methods singly or in combination to effectively migrate long distances.

6

The Learning

One of the most characteristic attributes of animals is learning. The modification of behaviour in response to experiences. The most important feature of learning is that it is adaptive. A learned animal responds in such a way that its survival and reproductive success get improved. It is observed that different animals, even of the same species, if are exposed to different environments, learn different things. Learning is demonstrated when animals in two groups, one given an experience denied to the other, develop different behaviour patterns. These experiments, sometimes reveal that experiences do modify behaviour, but at the other times they show that experience is not essential. For example, in most of the young birds, before they leave their nest, start "practice flying". They stand tall and flap their wings vigorously as if they were testing them for the first take off.

But according to the experiments, it is concluded that birds can develop the ability to fly purely by maturation. An experiment was performed in which pigeons were reared in a narrow tube in order to prevent the movement of their wings. They were not able to practice flights. On the other hand, control group of birds were allowed to practice, flight every day. When the controls had learned to fly, the experimental pigeons were released from their tubes. The results were surprising, since the experimental pigeons flew just as well as the control birds.

Hilgard (1956) said that learning comes when the probability of certain behaviour patterns in specific stimulus situation has been changed as a result of previous encounters or other similar stimulus situations.

According to Hinde (1970), learning are those changes which cannot be understood in terms of maturational growth process in the sensory adaptations or nervous system fatigue.

Wallace (1973) defined learning as the adaptive changes in behaviour that result from the individual experience and probably associated with physical changes in the central nervous system. May be it functions as innate pattern or but ultimately may be separable from these patterns.

Forms of Learning

Many different procedures have been used by animal behaviourist to study learning. According to Tinbergen (1963), a description of the several forms of learning is analogous to the observation and description that provides the starting point for the study of all animal behaviour. Most of the distinctions which are to be made among different "forms" of learning are based on differences in the procedures used in the study of learning. There may be several kinds of processes included under different forms of learning.

Habitation

Whenever a stimulus which elicits the response is presented repeatedly, then there is probability that amplitude of a response decreases. It is known as habituation. Thompson and Spencer (1966) in an excellent discussion of the phenomenon of habituation listed nine characteristics of habituation that are manifested in many systems. Few of these could be considered. If, after repeated stimulus presentation and habituation of the response, the stimulus is no longer presented, the response tends once again to be elicited by the stimulus.

The longer one withholds the stimulus within limits, the greater is the likelihood that the response originally elicited by the stimulus will again be elicited by it, this mechanism is known as spontaneous recovery. Time requirement may be a few minutes to several days, depending upon the conditions. If a series of habituation training trials and spontaneous trials are repeatedly given then the habituation occurs progressively and more rapidly. This is called potentiation of habituation. Habituation of a response to a given stimulus shows generalization to other similar stimuli (stimulus generalization). Thus if an animal habituates a response to a tone of a given frequency, it is likely that responses to tones of frequencies close to the training stimulus will also show habituation.

The more dissimilar the test tone from the habituation tone, the less would stimulus generalization be apparent. Finally whenever any other stimulus is presented, it results in a sudden recovery of the habituated response *i.e.,* dishabituation. To consider true habituation, it is essential that the response decrement must clearly be the result of changes in the central nervous system rather than that of adaptation in the fatigue in the receptors or in the sensory system.

Whenever we apply a puff of air to the gills of a horseshoe crab, it responds with a movement of its telson or tail. In the year 1975 Lahue, Kokkinidis and Corning studied the characteristics of habituation of telson reflex, dominating the movements of muscles electrophysiologically. They observed a clear habituation of the telson to repeated presentations of the air-puff stimulus. When a short squirt of saline from a syringe was aimed at the site of habituation, it resulted in dishabituation. After an interval of one hour without stimulus presentation (spontaneous recovery) the initial levels of responsivity were normal.

Potentiation of the habituation was observed after about two hours. When the stimuli were applied to other regions, then no generalization of the habituation were observed. Logan (1975) observed habituation of contraction responses of sea anemones to water stream stimulation. In 1974 Gilpin studied the habituation and retention of habituation of the response to air puffs in both decerebrate and normal earthworms.

Classical Conditioning

It is one of the simplest form of conditioning. It is so called because it was discovered by Pavlov, the father of condition. He performed a classical experiment, he lightly restrained a dog in a harness and blew meat powder repeatedly into its mouth. He recorded the amount it salivated accurately. Then he associated the sound of a bell with the meat powder and repeated this procedures many times at successive intervals. Although, the bell did not at first elicit salivation but after repeated pairings with meat, it happened so. Pavlov, in describing this experiment, called the salivation to the bell a conditioned reflex (CR), the bell a conditioned stimulus (CS), the salivation to the meat an unconditioned reflex (UCR) and meat itself an unconditioned stimulus (UCS).

This same experiment is been repeated several times on different animals and with many different stimuli and responses. But in all the best way to describe classical conditioning is as a process in which a previously neutral stimulus (CS = bell) is enabled to elicit a response (CR = salivation) that it never elicited before training.

Pavlov in his experiments observed that in between conditioned stimulus and unconditioned stimulus, the time relation was critical. If the unconditioned stimulus preceded the conditioned stimulation, there was very little, if any, conditioning, and if the unconditioned stimulus by more than about a second, conditioning become more and more difficult to establish instrumental conditioning in which the sensory stimuli involved become more numerous and more complex and in which the animal is given a greater and greater measure of choice in the stimuli it uses. Generally, learning becomes more difficult and more easily disrupted by brain injury as we move from simple to the complex learning tasks. By considering these points in mind, it becomes interesting to compare the learning ability of several animals that represents different levels of the phylogenetic scale and thus different levels of development of the nervous system.

Operant Conditioning or Instrumental Learning : Instrumental or operant conditioning includes a wide variety of procedures. In every instance, the animal learns to associate its behaviour with the consequences of that behaviour *i.e.,* the sequence of events is dependent upon the behaviour of the animal. Generally, some sort of punishment or reward is involved, like a rat trained to press a lever in a skinner box, the task performed by the animal may be relatively simple. In this example also, pressing is rewarded with water or food.

Mechanism of Learning

Learning abilities of an animal are easy to describe in comparison to understanding the mechanism. Evidences suggest that the mechanism by which new information is first stored in the brain are different from mechanisms of long term storage. Immediately prior to the injury, brain concussions produce amnesia (forgettings) for events. This is the reason that normally people are not able to recall that what had just happened before accidents. Short term memory may be based on continuously circulating nervous impulses. A rat or a hamster is fully trained or learned a simple maze. In this maze, there is a T-shaped run way and the rat had to make correct solution by taking a right turn at the junction. But if an electric shock is given to its brain within five minutes of its training, then it does not remember anything regarding its training. If the shock is given upto an hour after the learning experience, slight amnesia effects are still detectable. Although, these electric shocks, convulsions, concussion or deep anaesthesia does not disturb the long-term memory.

Since a long time, the investigators are trying to determine whether specific memories are stored in particular cells or group of cells in the brain. Experiments have included stimulating specific parts of the brain to determine what behaviour is evoked and making lesions or cuts in the brain to determine which memories are destroyed by particular cuts or lesions. According to the work done on rats, it is observed that only limited regions of the brain are essential for retension of some memories but that particular small groups of cells are not at all responsible for specific memories.

To search the sites of specific memories, some simple animals were studied which have fewer cells in their nervous system. The most suitable animal has been the marine gastropod mollusc, Aplysia. It is commonly called sea hare. Similar to other molluscs, the venous system of Aplysia bears several distinct ganglia that innervate different body structures. The cerebral ganglia innervates the anterior tentacles, mouth, eye and rhinophores. The buccal ganglia innervates the pharynx, salivary gland, oesophagus, crop, gizzard and muscles that control the protraction and retraction of the odontophore. The pedal ganglia supply the foot, head, parapodia, penis sheath, caudal part of the penis and penis retractor muscles. The abdominal ganglion controls a variety of respiratory, circulatory, excretory and reproductive functions. It also controls movement of the external organs of the mantle. Aplysia habituate to a number of stimuli although it is not a spectacular learner. The abdominal ganglion also control the habituation of the defensive withdrawal of the siphon and gill.

Normally, withdrawal of both the organs is brought about by the electric shocks. But if 10 stimuli are given at short (30-second) intervals, the animal stops withdrawing its siphon and gills and does not resume doing so for several hours. Four such training sessions, each separated by approximately 1½ hour produce habituation that lasts upto 3 weeks. In this learning several neural changes are involved. One in a reduction in transmission of messages from the receptors. One of this is the decrease in the transmission of messages from the receptors in the siphon skin and the motor neurons that cause withdrawal. According to Eric Kandel and his associates (Columbia University) just a few neurons decrease the amount of neurotransmitters they release. Conversely when withdrawal is enhanced by following shocks with other negative reinforcement, a few neurons release an amine that acts on the terminals of sensory nervous release an amine that acts on the terminals of sensory nervous to increase the release of transmitters. Although for the withdrawl of both these organs several muscles are involved, and just a few neurons are required to change for learning.

Phylogeny of Learning

To study the evolution of learning, we cannot retrace the steps, only a small survey of the learning capabilities and capacities of organisms from various animal phyla should provide some insight into the possible generality of any laws of learning and a sampling of the differences in learning ability across various phyla. Four points should be taken into consideration:

1. There has been a disproportionate examination of learning across the various phyla. For more studies have been conducted on learning processes in invertebrates than in vertebrates and even within vertebrates much of the attention is been focussed on mammals.
2. There are valid ongoing disagreements among scientists regarding what constitute proper criteria for demonstrating several types of learning and concerning the evaluation and interpretation of research results.
3. To explore the physiological and behavioural aspects of learning in most of the animals, some more refined techniques and objectives, bias-free methods are required.
4. We measure performance and not a genetically determined ability or mechanism. Several factors like test design can influence performance.

With the considerations of these problems in mind, we can survey the known information in a number of animal phyla regarding learning.

Protozoa and Coelenterata

Among protozoans, learning capability is still a debated question. For example, in Amoeba and Paramecium, it has been observed that these are capable of habituation to noxious sensory stimulus such as strong light or mechanical shock. On giving repeated stimulations, their response grows weaker and weaker until in few cases they totally become unresponsive. However, regarding this work two criticism have been given :

1. None of these experiments has satisfactorily demonstrated that such diminished responses lasted long enough to be anything more than adaptations to sensory stimuli.

2. The noxious stimulation may have temporarily injured the organism so that they were made capable of responding with each successive stimulation.

Several attempts have been made to demonstrate few forms of associate conditioning or learning to get around these objections. But every claim has again met with cogent criticisms. There is one experiment which serves to illustrate the point. First of all a sterile platinum wire was lowered into the centre of a dish of paramecia. No special reaction was observed. Next the wire was 'baited' with bacteria, and the paramecia responded by congregating around the wire, clinging to it and feeding. After then many such presentations of baited wire, it was sterilized and dipped into the same spot and the paramecium responded by congregating around and clung to it. This was claimed to be a well controlled demonstration of learning of a new response to the sterile wire. However, a simple control demonstrated that the training was unnecessary.

In this control experiment, bacteria were dropped into the dish and the paramecia congregated and fed. Then the sterile platinum wire was lowered for the first time into the same spot, and the paramecia clung to it. Careful investigation showed that the congregation around the spot, the wire contacted was due to residue of bacteria bait. The increased clinging to the wire resulted from the increased acidity. The bacteria contributed to the medium, since further controls exhibited that in Paramecia, clinging was a junction of acidity.

There is also ample evidence for the habituation response for coelenterates, but for association learning there is very little evidence. Investigators generally assumed that coelenterates exhibited only certain involuntary, stereotyped reflexes, in the early years of this century. We now know that there are endogenous neural rhythms in many coelenterates and that, rather than being totally "passive", many of them are able of spontaneous active responses in the course of interacting with their environment (Rushforth 1973). According to Ross (1965), one possible demonstration of a form of association learning is based on the fact that sea anemones (genus stomphia) respond to chemostimulation from certain starfish for example Dermasterias imbicata. They respond by stretching their bodies, detaching from the substrate and swimming away.

Soon the animal lands and gets reattatched to the substratum. The chemostimulation is paired with gentle pressure applied near the base of the anemone. The application of the pressure stimulus alone leads to some reduction in the "swimming" response, with repeated trials. Ross considered it as conditioned inhibition. The stimulus, however, may have induced some related effect which lead in the movement of anemone.

Platyhelminthes

There is a lot of controversy on the research conducted to learn planarians. These flatworms are interesting because a nervous system with bilateral symmetry and a primitive brain appears only in the platyhelminthes.

A classical conditioning technique was used in this case. When the worms were gliding along in a trough of water, a light was turned on, followed two seconds later by electric shock which caused the worms to contract longitudinally. The worms contracted to the light alone over 90 percent of the time after 150 trials. Followed by this the worms were cut in half and allowed four weeks for regeneration. Both the regenerated head and tail section showed a high degree of retention of what had been learned earlier. Even after these regenerated worms were cut and a second period of regeneration occured, there was retention of the conditional response. Similarly in this case also learning is not confined to the interior portion of the nervous system, but effects throughout.

Annelida

At the level of worms, where a bilaterally symmetrical, synaptic nervous system is already developed, the first unequivocal evidence for learning is found. Here we observe clear instances of habituation and associate learning. In one experiment, earthworms were trained to go to one arm of a T-maze leading to a dark, moist chamber and to avoid the other arm which led to electric shock and irritating salt solution. In this simple learning of a position habit is required about 200 trails on the average to reach the criterion of 90 percent correct responses. Interestingly enough, the worms were able to retain what they had learned after removal of the first five segments of body bearing the cephalic ganglion and untrained worms were able to learn after removal of the head ganglion. Apparently, all those neural changes which are involved in learning can take place in the ganglia of the lower segments of the body.

Molluscs

Some learning of a simple T-maze by snails has been demonstrated among the molluscs. But in these the experiment was not highly successful, because all snails did not learn and some which did rarely reached high levels of consistency in their performance. In case of octopus, much better learning has been demonstrated. It bears a well-developed eye and an elaborate brain. In these studies for example it was shown that the octopus could readily learn to discriminate the presence of a white card signifying electric shock. At first, the octopus was trained to come forward from its rocky nest to seize a crab that was lowered into the far end of the aquarium by a thread.

Then on half the trials, a white card was lowered with the crab, and in these trials, the octopus was driven away from the crab by strong electric shock. The octopus began to inhibit its approach to the card after 12 trials and by 24 trials consistently remained in its nest when the card was present and consistently came out to feed when the crab was presented alone.

The octopus is able to learn more difficult problems where two different cards were used, a small one concerned with feeding and a large one related with electric shock. Discriminations of various shapes have also been demonstrated in similar manner. However, it is interesting that the octopus was rather poor in problems requiring it to 'detour' around a barrier. Thus when a glass partition was lowered between the crab and an octopus, the octopus persisted in swimming straight into the glass and failed to swim around it through an open space.

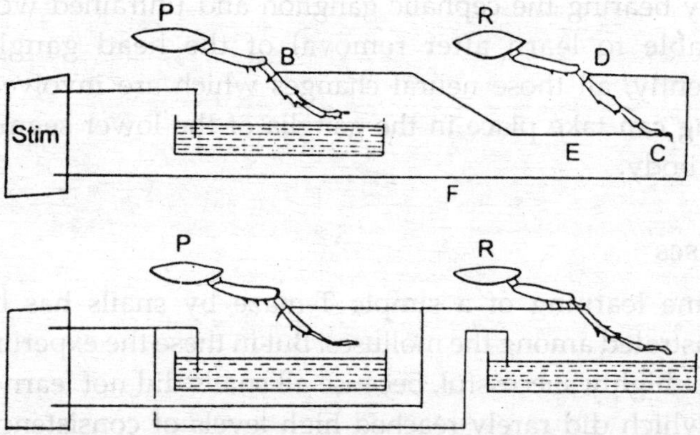

(b) Test

Experimental arrangement for the conditioning and testing of insect avoidance learning (a) During training the two animals are arranged so that both animal P and animal R receive shocks when P lowers its leg beyond a critical profit. (b) in the retest situation, the animals are connected in parallel so that either receives electric shock separately if its leg is lowered. When trained and tested in this situation, animal P typically demonstrates passive avoidance learning, as indicated by fewer instances of the lowering of its leg than of animal R, the "yoked control."

Arthropoda

There are a number of arthropodan species in which habituation has been demonstrated including horseshoe crabs

(Lahue et al. 1975), crayfish (Krasne, 1969), and the pupae of grain beetles (Hollis, 1963). Nelson in 1971 reported a well-controlled study of classical conditioning of proboscis extension in blowflies. The UCS was an application of a sugar solution to the mouthparts of the fly, with the CS being the application of water or a saline solution to the feet. A lot of attention is been paid to the study of instrumental learning in arthropods.

The studies of maze learning in ants conducted by Schneirla (1946) are classics in comparative psychology. In larva, the effects of maze-learning experiences have been shown to survive through metamorphosis in adult grain beetles (Borsellino : Pierantoni and Schieti-Lavazza, 1970). Bermant and Gary (1966) and Wenner and Johnson (1966) demonstrated conditioning of responses related to food getting in honeybees. Horridge (1962, 1965) studied the acquisition of avoidance responses in insects, with or without their heads, by using a rather clever procedure. During training, the legs of two cockroaches (P and R) are connected in series to an electrical stimulator.

The electrical circuit is completed and both animals receive shock whenever animal P lowers its leg into a saline solution, thus completing the electrical circuit. Animal P comes to hold up its leg with training to avoid shock. Animal R is a "yoked control" that receives the identical number and temporal patterning of shocks, but they are delivered in such a way as to not be contingent on its behaviour. When the animals are shocked independently for lowering their legs, learning is demonstrated as a difference between animals P and R in a test situation. In comparison to animal R, animal P experiences fewer shocks in the test situation. In this test situation both intact and headless locusts and cockroaches have displaying avoidance in learning. (Horridge, 1962,1965; Disterhoft, 1972)

Learning in Vertebrates

Several learning studies have been conducted on species from most vertebrate classes. Since for these studies rat and

Rhesus monkey have been used in large number. Rats could be trained in an operant conditioning apparatus to press a lever to receive food. They could also be trained in an apparatus with two levers (designated L for left and R for right) to press the levers alternately LR or RL, for a food reward can rats be conditioned to press the levers in a LLRR or RRLL sequence? Travis-Niedeffer, Niedeffer and Davis (1982) tested this question by conditioning rats first on the single lever task, either R or L followed by the single alternation task, RL or LR. They then rewarded only double alternation performances.

The rats learned this task at a level exceeding chance expectations, and their performance improved over days. It was necessary however in some instances, to permit the rats to give extra responses to the first level before pressing the second lever. Thus LLRR was rewarded the same as LLRR. When the investigations attempted to condition the rats to a sequence LLRRLLRR or RRLLRRLL (double alternation with a fixed ratio schedule of two repetitions), no rats could successfully perform at better than a chance level. The results are significant because of the new information provided regarding the capacity of the rat to associate a series of responses required to obtain the reward and because previous attempts to condition double alternation tasks in rats had failed.

There is an apparatus called Wisconsin General Test Apparatus (WGTA) which is used in many of the studies of learning behaviour in macaques (Meyer, Treichler, 1965 for the history of this and related techniques). One type of learning problem studied with this apparatus is the delayed response problem (Fletcher 1965). The initial procedure starts with the test tray out of reach but in full view of the test animal. Food is placed in one food well and then two identical objects are placed over the wells. The test tray is moved closer to the monkey after a prescribed delay, which then responds by lifting one object or the other. The entire procedure includes four stages baiting phase, covering phase, delay phase and response phase. When the monkey picks up one of the object, a trail ends uncovering either the correct food well containing a reward, or the empty well. With the help of this procedure, several variables can be investigated with other animals such as cats and raccoons with little modification in the apparatus. Among the variables that have been manipulated are length of the delay phase, the nature and size of the reward.

The nature and the similarity or dissimilarity of the objects used to cover the food wells, and whether the animal is permitted to watch the test tray during the delay phase or, instead has an opaque screen lowered in front of the tray (Meyer and Harlow 1952). In all it is concluded that Rhesus monkeys are capable of learning basic discriminations in this procedures, with delays of upto 30 seconds or more. More food reward leads to better performances and imposition of the opaque screen during the delay phase increases error rates upto 50 percent.

There is an interesting and striking finding included in these studies is that the behaviours and performances of individual monkeys differ markedly. Generally monkeys that exhibit hyperactivity in test situation and those that are more easily distracted during the delay phase exhibit lower levels of performance. In studying the learning behaviour, we should clearly consider the importance of individual difference in performance, regardless the task being performed or the species being tested.

Neural Mechanism of Hearing

If in the phylogenetic series, we were to attribute in improvement of learning ability to one thing, then it would be the evolution of the central nervous system. Several investigators sought to answer two major questions firstly where does learning takes place in the nervous system ? and secondly what is the nature of the change ? Although both the questions are interrelated, but still we shall take up each of these questions separately.

Locus of learning question has been mainly approached by the technique of experimentally destroying parts of the nervous system. According to Pavlov, the cortex is essential for conditioning, but studies have shown that simple conditioning is possible in the dog after its cortex has been removed. Usually this type of decorticate dog gives emotional and generalized responses and is greatly deficient in sensory capacity, but it can be successfully trained in the classical conditioning technique using shock as the UCS.

Lashley performed an experiment on rats, on just the visual area in the back of the cortex and tested the animal for visual learning and retention. He used a simultaneous discrimination situation in the jumping stand where the animal had to choose the correct one of two doors containing visual stimuli. When he used black and white doors, those rats which were devoid of visual cortex learned the discrimination almost normally. If they had learned to discriminate black from white before the lesion of the visual cortex, however, they lost the habit postoperatively and had to learn it all over again.

When pattern discrimination was used and a choice was involved between a triangle and a circle, then it turned out that the operated animals could never learn. Apparently in this case, they lost the capacity of detail vision, while in the brightness discrimination, capacity was impaired and only memory was affected. However, actually it was studied that in brightness discrimination case, it was not memory that was affected, but rather it was a loss of sensory capacity required to respond to spatially separated black and white doors.

In other words, Lashley concluded that cortex operates on a mass action principle in learning and memory so that the larger the lesion, the poorer the ability. He also found that it hardly matters where the lesion was in the cortex, a lesion of a given size have a given effect whether it was in the visual area in the back of the brain or the somatic sensory and motor areas in front of the brain. Lashley formulated a principle of equipotentiality on the basis of his findings. According to this principle, all parts of the cortex are equal in their contribution to learning and memory.

Another approach to the understanding of brain mechanism underlying learning is through the use of electrical recording methods in which it is possible to trace changes in the electrical activity of many parts of the brain during learning. In these experiments, the animal has electrodes chronically implanted in its brain, and changes in pattern of electrical activity are noted as the animal is trained. Here, the striking thing is that the changes occurring in many parts of the brain, cortically and subcortically within the sensory systems and outside of them as well.

It may be that the electrical method is so sensitive in recording changes that go on learning that it cannot isolate the essential form the non-essential parts of the brain for learning and memory. Or it may be that learning, or different faces of learning, occur in many places of the brain at once, and therefore no one part is completely essential for its retention or formation.

In human beings, lesions of the temporal cortex on both side of the brain may result in a similar defect. These patients can learn something simple and retain it for about 15 minutes to 1 hour, but after the time they forget completely and may not even remember having learned. Yet their life long memories are left undisturbed. When studied combinedly, this finding suggests two part process memory *i.e.*, an initial, vulnerable perhaps physiological process lasting fifteen minute to an hour and later, invulnerable, perhaps anatomical process, providing permanent basis of memory.

Many mysteries remain in our quest for the physiological basis of simple learning. We know that learning in such a property in which all animals possess a synaptic nervous system. The major question is whether the superior learning of mammals and especially primates is due to the development of superior neural mechanism for learning and memory or to their greatly increased sensory and motor capacities of both. The development of superior neural mechanisms must be an important factor, for the capacities for complex learning, problem solution and reasoning emerge with the evolution of the central nervous system.

7

The Aggression

Aggression may be defined as an 'agonistic behaviour which is a system of behaviour patterns that has the common function of adjustment to situations of conflict among conspecifics.' The term agonistic refers to all aspects of conflicts like submission, chase, threat and physical combat. Aggression is a complex behaviour with many functions and many causes and it may include even predatory behaviour.

Aggression may be expressed in two ways — (i) Social use of space and (ii) Dominance hierarchy.

Aggression can also be listed in the following manner :

1. *Territorial:* excluding others from a particular physical space.
2. *Predatory:* act of predation, possibly including cannibalism.
3. *Anti-predatory:* defensive attack by prey on predator.
4. *Dominance :* the behaviour of a conspecific or central as a result of previous encounter.

5. *Moralistic* : in humans, it is enforced conformity to group standards generally based on ideology and religion.
6. *Sexual:* in order to obtain and retain mates, males generally use threats and physical punishment.
7. *Parent offspring:* disciplinary action of parent against offspring.
8. *Weaning:* access of milk is restricted to the offspring.

Aggression and Competition : Mostly the agonistic behaviour includes competition for some source. Competition may be of two types — scamble and contest.

Scamble means to get victory, win for *e.g.* if we throw any piece of food in between two squirrel monkey, both of them will go after it and the one catches it first, takes it away. Contest means to compromise for *e.g.* if any food material is thrown in between the two rhesus monkeys one would probably gravies and face away, and other would then calmly work and take the food.

Types of Aggressive Behaviour

Territoriality or Social Use of Space : When an animal or organism spends most of time in a particular area, it becomes habitual to it and then that area is its home range. Within the home range, the heaviest use portion is called the core area. This may include a nest, sleeping trees, water sources or feeding trees. The minimum distance kept between an animal itself and other members of their own species is known as its individual distance. Territory is an area occupied by a group or an animal and defended by overt aggression or advertisement.

Size and Boundaries of Territory : Size of the territory depends on the size and diet of the animal as well as the particular resource the animal is defending. A large number of vertebrates are found defending the source of food.

Most of the time, the territorial animals are found patrolling the boundaries of their space, visiting scent posts, singing and making other activities.

Territorial Model : In order to predict territorial behaviour of an animal, a model was constructed by Carpenter and MacMillian in 1976. They argue that in order for territorial behaviour to occur, the basic cost of living (E) + the added cost of defending a territory (T), must be less than the yield to the individual. If not, territorial + the extra yield gained by the reduced competition of territorial *i.e.* fraction of b of productivity P, in all it is E+T aP+bP. All the required resources by an animal are not found in all territories, *e.g.* a mating system Lek involves a peculiar type of territory Lek. In this case, the resource of the organism defends is the place for mating. On the lek the males do little or no feeding. They almost spend all their time and energy in patrolling the boundaries displaying to other males, and attempting to attract females into their area. We can observe animals founded territories for defending the special, for mating place, nest building, feeding or play etc.

Dominance : The animal or organisms identify each other by odour or by earlier encounters which beat them previously and which ones they were able to dominance. This establish dominance hierarchies which needs learning and recognition based on previous encounters among the individuals.

Dominance Hierarchies : Schjelderup-Ebbe in 1922 first of all gave the idea of dominance and worked out the peck order in domestic checkers. Dominance hierarchies are most oftenly seen in arthropods and vertebrates that live in semipermanent or permanent social groups. Hierarchic means the ranking of individuals for different objects like assessment of water, favourable breeding place etc.

According to Scott (1966) if an animal controls the behaviour of another animal it is dominant. In another sense according to Rowett (1974) a prediction is being made about the outcome of future competitive interactions. For example, if four or five mice which are strangers to each other are put together, several outcomes are possible. Most likely a despot will take over and all the subordinates will be more or less equal. Another possibility is a linear hierarchy where A dominates B, B dominates C and so on *i.e.*, $A \rightarrow B \rightarrow C \rightarrow D$. Sometimes triangular relationship is formed like.

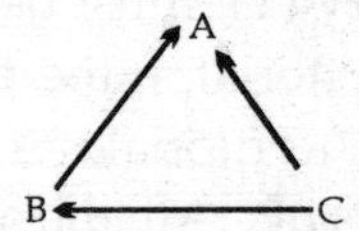

Coalitions may affect dominance in other species like

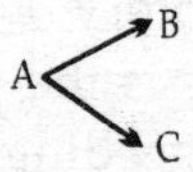

but $C + B \rightarrow A$

Whenever socially living organisms engage in context competition, dominance hierarchies seem to be a common result. For example—dominance hierarchy in Rhesus monkeys.

Dominance in Females : According to Sade (1967) and Maxdew (1968), position in the hierarchy is determined by the others rank. Adult females are linearly ranked with daughters and sons generally ranked below their mothers. In these families, dominance manifests itself subtly but constantly at a water or food source or increases to resting spots with the dominants supplanting subordinates.

Dominance in Males : Among males, dominance is more complex because males have the natal group immediately after puberty whereas in the natal group they rank below their mothers. Dominance relationships are formed quickly as the males move in new groups. Their duration of stay in the new group is not at all influenced by the size and previous fighting experiences. According to Duckenar and Nessey (1973), their rank is positively co-related with seniority those in the group the longest, rank the highest.

Advantages of Dominance

(a) When the group living birds and mammals are observed, it is seen that the dominant animals are well-fed and healthy.

(b) Christian and Davis (1864) have reviewed data for mammals showing that low ranking individuals, frequently losers in fight have higher levels of adrenal cotical hormones in comparison to the dominants. These hormones elevate blood sugar and prepare the animal for flight or fight.

(c) Higher rank is generally directly correlated with reproductive success, as was noted by Le Boeuf in 1974 while studying on elephant seals.

(d) Several famous treatments have emphasized the importance of dominance and aggression in keeping the species fit, since in the fierce battles, the survivors are most vigorous and therefore by passing on those traits they improve the species.

Factors Affecting Aggression : Two factors are there which affect aggression : (i) Internal and (ii) External.

Internal factors include limbic system, hormones, and genetic control. External factors include learning and experience, pain and frustration, lenophobia, crowding breeding and feeding.

Internal Factors in Aggression

Limbic System : Limbic system includes the brain structures involved in emotion. The hypothalamus is involved in escape and defense behaviour. Various workers have found that different sites of brain are responsible for different types of aggression. For *e.g.* in cats, electrical stimulations of the ventromedial nucleons of the hypothalamus produces hissing, attacking with claws and growling.

Hormones : Hormones and neurosecretions interact with the limbic system. Epinephrine are concerned with physiological arousal. Some other substances like serotonin, and dopamine may also act as a neurotransmitters and affect aggressiveness. Some early work done on the chickens demonstrated that they become more aggressive by the testosterone hormone which also increase their dominance rank. Other than testosterone reproductive hormones also affect aggression. In some species of birds, the leutenizing hormone increases aggression. According to Davist (1963) estrogen lowers aggression in some species and raises it in other.

Genetic Control : Genetic control includes synthesis of nervous structure, neuro secretions and other compounds. Agonistic behaviour is shaped by natural selection, as in any other behaviour. Aggression level changes significantly by the artificial selection within few generations.

External Factors in Aggression

Learning and Experience : According to several researchers, it is believed that some factors outside the animal triggers the response. Previous experience can produce semipermanent changes in the expression of agonistic behaviour and animal can be conditioned to lose or win. Scott in 1976 emphasized the role of learning and experience in the development of aggression. Although he acknowledges the genetic and physiological bases of aggression, he emphasizes the importance of early experience and learning in the development of aggression.

Pain and Frustration : According to Ulrich, Hutchinson and Azrin in 1965, pain includes noxious stimuli like wind noises, foot shocks, tail finches and intense heat which cause different lab species to attack a wide variety of objects from conspecific cage mates to tenis balls.

Different researchers have also explored the frustration aggression convection in humans. In some previous experiments, children were allowed to view a room full of toys but denied them to access to it, but when finally they entered the room, smashed the toys frequently and fought more than did a control group that was permitted access to the room. But now frustration is the only cause of aggression.

Xenophobia, Crowding, Breeding and Feeding

Xenophobiais another strong stimulus for aggression. Xenophobia refers to the presence of a stranger. Whenever a strange animal is introduced into a establish social group, a violent reaction of xenophobia is seen. (Southwich et al 1974). Crowding may result in increase of interaction but does not always increase aggression. Aggression is more powerfully effected by overall group size, presence of strangers, or restricted access to resources in comparison to by reducing the available space for existing social group.

Breeding—Aggressive components also include courtship and cooperation. At the time of mating, some of the aggression stems from the fact that males tend to court females rather discriminately, generally making advances when the females are not sexually receptive.

According to some evidences, feeding in groups of fishes intragroups agonistic behaviour. According to Albrecht (1966) predatory and aggressive behaviour are functionally different but motivationally linked together.

Restraint of Aggression

1. *Evolutionary Model :* The argument regarding species preservation has become unpopular now because it

implies that natural selection operates at species or group level. It has been previously pointed out by Tinbergin in 1951 that literalized contests would reduce the risk of injury to the aggression and that threats, which consume less energy than physical combat are advantageous to both groups and individuals.

2. *Displays* : Display, that communicate an animal's intentions and mood inhibit attack by another. Scott (1966) Lorenz (1966) and others argued that these types of behaviour are essential to avoid killing and wounding on large scale.

3. *Social Control and Disorganization* : The idea of social control of aggression has been developed by Scott (1975). When the agonistic behaviour in aichlid fish was observed, it showed that when strange fish of the same species are continually introduced into the group, fighting remains at the peak. If group membership is stablized, then the fighting level declines to a relatively low level.

   Social disorganisation appears to be one cause of aggression. Whenever a high ranking male or female Rhesus monkey leaves the group or dies, aggression may increase within the group.

4. *Relevance for Humans* : Among animals, control of aggression may have special relevance for humans. According to Lorenz, humans also become violent, because we have few harmless outlets for aggression. Lorenz says that if aggression is inevitable then we must find harmless ways to vent it. This cathartic approach suggests the importance of ritualized tournaments, for example, sports events in which participants and spectators alike can work out their hostilities.

Conclusion **: In all, the above concept could be concluded by saying that maintenance of a position or territory in a dominance hierarchy requires constant reinforcement by threats, sings or other displays. Individual recognition and memory of the results of the previous experiences are always involved. It has also been suggested that aggression could be controlled by the modifications of the social environment through early experience and avoidance of social disorganization.**

8

Mating System

The sexual selection is based upon the struggle for mating. Sexual selection implies presumptive choice on her part of the most desirable male to be the father of her progeny. Biological fitness provides an ability to contribute to gene pool of the next generation and subsequent generation to reproduce. An animal that successfully copes with its problems of obtaining food, water and shelter but fails to reproduce still may have fitness of zero. Reproductive behaviour is of obvious importance to fitness and lies very close to many definition of species. Reproductive behaviour possesses a set of related questions about mating and rearing of offspring. Before conception, there may be a struggle among males and a characteristic sequence of behaviour pattern in which the sexes court each other. The term courtship is used by the ethologist refers to behavioural interaction for the fertilization. In some species, such as stickle back it lasts a few minutes while in others may last for month but in some species it does not even exist.

Courtship ensures that animals mate with other individual of the same species, sex and condition.

Choosing a Member of the Right Species : Animals do not mate with different species because it is adaptive in animals. Mating between members of different species are very rare. It is proved by several examples, in a sample of 725 copulating pair of two species of cicada, which are distinguishable but similar in appearance of human eye, only seven contained members of two species while rest were mating of the same species. Hybrid individuals are inferior or weaker to those produced by mating of members of the same species because they are sterile. The best-known example of a sterile-hybrid is the mule. Natural selection will favour animals that produce normal healthy offspring by choosing to mate with members of their own species. There have been many investigation factors for the insurement that they mate with right species. In case of crickets there are approximately 3000 species of cricket in the world. Male cricket, produce a song for attracting the female of same species. The songs are produced by males when they rub their wings together and are heard by the female through the ears on the front legs.

The all importance of song in the species recognition of cricket can be shown by another experiment. Female cricket can be trained to walk along a "y-maze" on which they have to turn to the left or right. If for example the loudspeaker on the left were playing the song of the female's own species, while the one on the right played the song of some other species, the female turns to the left, the side of the song can be reversed to control for any bias the female may have for turning in one direction or the other.

Sexual Selection

Individual can be successful by the quality of probable offspring. Thus, it becomes important who its mate will be. Darwin (1871) introduced the concept of sexual selection. Sexual selection can be divided into two types : one in which members of one sex choose certain mates of the other sex (intersexual selection), and a second in which individual of one sex compete among themselves for access to the other sex (intrasexual selection)

Intersexual Selection : In intersexual selection, individual of one sex shows that they are worthy of an investment so the member of other sex choose among them. Fisher (1958) used birds as an example. Bowerbirds are unique in that they build a structure whose only function is to attract mates. The male satin bowerbird (Ptilonorhynchus violaceus) of Australia and New Guinea station itself alone in the forest, clears a space weaves the bower from twigs, and decorates it with brightly coloured objects. Gilliard (1963, 1969) reasoned that bowers function in place of showy plumage, and noted that those species with the fanciest bowers had drabest plumage.

Intrasexual Selection : It is a competition (usually males) within one sex. The winner gaining access to the opposite sex. This competition may take place in several animals such as deer (family Cervidae) and antelope (family Bovidae). The winner gain dominance and do most of the mating.

Females may incite competition among males and thus may gain some control over the choice of mate. For example, female elephant seals (Mirounga angusiirostris) vocalized loudly whenever a male attempt to copulate. This behaviour attracts other males and tests the dominance of male attempting to mate. In some other cases a male will sometimes attack a mated pair, displacing the previous male and mating with the female.

Among insects such as moths and butterflies, two kind of sperms are produced. One is eupyrene that fertilizes the eggs and the second type apyrene contain no nuclear material. Silberglud, Shepherd and Dickinson (1984) suggest that apyrene are the result of sperm competition, possibly displacing eupyrene sperm from first male or delaying remating by the female. Following conception, male-male competition may take a different form. In mice, the "Bruce effect" operate early in pregnancy, a strange male causes the female to abort and become receptive. Among langeur monkeys (Presbytis entellus) strange male may take over a group, driving out the resident male. The new male may then kill the young sired by the previous male (Hardy 1977). Females who have lost their young soon become sexually receptive.

Timing or Seasonally : Animals breed in a particular time or season. So these animals are known as seasonal breeders. In lower animals, timing and seasonality is less understood. But in higher animals, it is very complicated affair. In fishes, birds and mammals of the northern temperate zone, reproduction begins in spring. The first phase is migration and is done by all individual at approximately the same time. In some cases, the timing process is not very accurate. There may be a considerable difference between the male and female of a pair. It has been found in pigeons and in other animals. When a male and a female are kept in adjoining cages so that they can see and even touch each other, but are prevented from copulating the persistent courtship of the male will finally induce the female to lay eggs. These eggs are infertile.

Sex Ratio

Suppose there are 20 females for every male in a population. Every male has 20 times the expected reproductive success of a female and therefore a parent whose children are exclusively sons can expect to have almost 20 times the number of grand children produced by a parent with mainly female offspring. A female biassed sex ratio is therefore not evolutionarily stable because a gene which causes parents to bias the sex ratio of their offspring towards males would rapidly spread a greater proportion of males than the initial 1 in 20. But now imagine the converse, if males are 20 times as common as females, a parent producing only daughter will be at an advantage. Since one sperm fertilizes each egg. Only one in every 20 males can contribute genes to any individual offspring and females therefore have twenty times the average reproductive success of a male. So a male biassed sex ratio is not stable either. The conclusion is that the rare sex always has an advantage and parent which concentrate on producing offspring of the rare sex will therefore be favoured by selection. Only when the sex ratio is exactly 1 : 1 will the expected success of a male and a female be equal and the populations label. This means that parents should have invested equally in sons and daughters.

(a) *Local Mate Competition :* We can understand it by following examples. Suppose there are two sons and they have only one chance to mate and that they compete for the same female. Only one of them can be successful in mating so according to their mother's point of view one of them is wasted. Therefore, the mother should bias her ratio of investment towards daughters.

The exact degree of bias predicted by Fishers theory depends on the degree of local. mate competition. Extreme competition is to be expected in species with limited powers of dispersal, and therefore, it tends to be associated with inbreeding. The best sex ratio in this instance is to produce just enough sons to fertilize the daughters, since any other mates will be wasted. The crucial difference between this and the earlier argument for a 1: 1 sex ratio is that here the ratio of males to female in the rest of the population does not matter.

(b) *Local Resource Competition or Enhancement:* Anne Clark found that the South African prosimian Galago crassiccudatus has a male biassed investment ratio among its offspring. As with most mammals, female Galago disperse less far than males, and often end up competing both with their mother and with each other for rich sources of food such as gum and fruit trees in the mother's home range. This local resources competition among females reduces their value as offspring in the extreme case only one daughter might be able to survive on the food available near home and so investment in other daughters would be wasted.

(c) *Maternal Condition :* Male red deer always compete for the females by prolonged roaring and antler wrestling contest. In these contests, it is an advantage for a male to be big and size depends among other things on how well-fed the male was as a youngster, which in turn depends on his mother's ability to compete for good source of food and hence produce a plentiful supply of milk. Now if a mother knew that her sons would be highly successful harem holders, it would pay her to invest heavily in sons would not grow up to be big and strong would be better to have daughters. Since a daughter's future reproductive success does not depend so much on her mother's milk. Exactly, this pattern has been found in red deer.

Ardent Males : During mates, male often fight and ritualized and have evolved weapons for fighting. Male may dispute over direct access to females. The most intense fights in many species occur when females are ready to be fertilized and once a male finds a female he often guards.

In some invertebrates, the male cements up the female's genital opening after copulation to prevent other males from fertilizing her. The male of moniliformes dubius parasitic acanthocephalan worm in the intestine of rats, produces a chastity belt of this kind but in addition to sealing up the female after copulation. The male sometimes copulate with rival males and applies cement to their genital region to prevent them from mating again. A male xylocoris may inject his sperm into a rival male. The sperms then swim and wait to be passed on to a female.

The other example of the acrane method of male-male competition found among invertebrate is the use of anti-aphrodisiac smells. Larry Gulbert noticed that female Heliconius erato butterflies always smell peculiar after they have mated. He was able to show experimentally that the scent does not come from the female herself but is deposited by the male at the end of mating. Gilbert also found that the scent discourages other males from mating with the female.

Reluctant Females : Females in the great majority of species are the chief resources of zygote. Not very surprisingly, females are the whole chooser than males during courtship, by which female get or obtain good genes for their offspring.

(a) *Good Resources and Parental Ability* : For the betterment of the young ones, male often defend breeding territories containing resources which play a crucial role in the survival of a female's egg or young. For example, male North American bullfrogs defend territories in ponds and small lakes where female come to lay their eggs. Some territories are much better for survival of eggs than other and these are the ones which females prefer. The bullfrogs also show that female choice and male-male competition may go hand in hand.

Food is a resource which often limits a female's capacity to make eggs and during courtship females may choose whether or not to mate with a male on the basis of the ability to provide food. In some birds and insects, for example, males may provide food for the females during courtship which makes a significant contribution to her eggs. Female hanging flies (Hylobitacus apicalis) will mate with a male only if he provides a large insect for her to eat during copulation and the more eggs he fertilizes. In birds the male usually helps to feed the young and courtship feeding may play the additional role of indicating to the female how good the male is in bringing food for the young.

Good Gene

Good gene are ones which increase the ability of her offspring to survive, compete and reproduce. Linda partridge experimented on female fruitflies (Drosophila) and allowed them to mate freely with a population of males or forced each female to mate with a randomly choosen partner. The offspring of the "choice" and "no choice" females were then tested for their competitive ability by rearing the larvae in bottles with a fixed number of standard competitors. Partridge concluded that the offspring of the choice group did slightly but consistently better than those of no choice females in the larvae competition experiments. This experiment suggests that females are able to increase the survival of their offspring by choosing good genes in their mates but it must be born in mind that the results could also be in part explained by intrasexual competition in the choice experiment the males that mated may have superior competition against other male.

Male Investment : In many birds, some amphibians and arthropods both male and female invest about equally in the eggs or young by feeding guarding or brooding. Sometimes the usual sex roles are completely reversed so that males do the investing and female competing. The idea about sexual conflict and sexual selection can still be applied in modified form to species with equal or primarily male investment. When both sexes care equally for the offspring, for example, courtship may involve assessment and choice by males as well as by females. Males of species with internal fertilization can never be absolutely sure that they have fathered the children of their mate.

Sexual Conflict

Recall the view of the origin of anisogamy as the primaeval example of sexual conflict. The conflict was one about mating decision. Today, the conflict is not only with respect to mating decision but also in the contexts of parental investment, multiple mating and infanticide.

(a) *Mating Decision* : As we already know that female tends to be more choosier than males. An extreme manifestation of this conflict is enforced copulation as exemplified by scorpionflies (Panorpa spp). Male scorpionflies usually acquire a mate by presenting her with a nuptial gift in the form of a special salivary secretion or a dead insect. The females feed on the gift during copulation and turns the food into eggs. However, sometimes a male enforces copulation : he grasps her with a special abdominal organ enforced copulation appears to be a case of sexual conflict. The female loses because she obtains no food while the male benefits because he avoids the risky business of finding a nuptial gift.

(b) *Parental Investment* : It is sufficient to note that in species with investment beyond the gamete stage, each sex might be expected to exploit the other by reducing its own share in the investment. The outcome of this sexual conflict may depend on practical considerations such as which sex is the first to be in a position to desert the other which has the possibility of deserting the female immediately after fertilization and leaving her with the eggs or young to care for it.

(c) *Infanticide* : Male lion may slaughter the cubs in a pride shortly after they take over as group leaders. This behaviour increases male reproductive success and clearly decrease female success. This seems to be a case of sexual conflict in which the males have won but it is perhaps surprising that females have not evolved counter adaptation. For example, they killed their young ones to recoup as much as possible of their losses.

(d) *Multiple Mating* : When a female mates with more than

one male termed as multiple mating. Multiple mating is always advantageous to the male. For example, when two dungflies males struggle for a female the female is sometimes drowned in cowdung due to the fighting males on top of her.

Mating System : Despite the argument based upon relative parental investment by the sexes in animals polygamy should be the normal mating system, whereas monogamy is common in some groups for example in 90% of birds. In many birds parental investment is same for male as well as females. Young altricial birds in nests are very vulnerable to predation and survive best if they can develop rapidly. If the female alone can seldom rear two young, it is advantageous for males to help with parental care.

Emlen and Oring (1977) summarise the essential conditions for the occurence of polygamy in terms of the relative economics of defending one or several males. This depends upon the spatial distribution of resources, the sex ratio and the time at which mates become available. If female exhibit synchrony in their sexual receptivity, polygamy is more difficult. An intermediate between polygamy and polyandry in which female lay a second batch of eggs which the male incubates or guards but where the male also mate with other females was rapid multiple clutch polygamy. This is common in fishes and rare in birds. Polyandry has not been reported in mammals (Eisenberg 1966) but Jenni (1974) describes it in several species in birds.

The females of American Jacana (Jacanaspinosa) defend large territories in march areas which include the smaller territories of several males. The males perform all incubation and parental care but they defend their territories against other males. Emlen and Oring also described harem polyandry and female access polyandry in addition to resource defend polyandry.

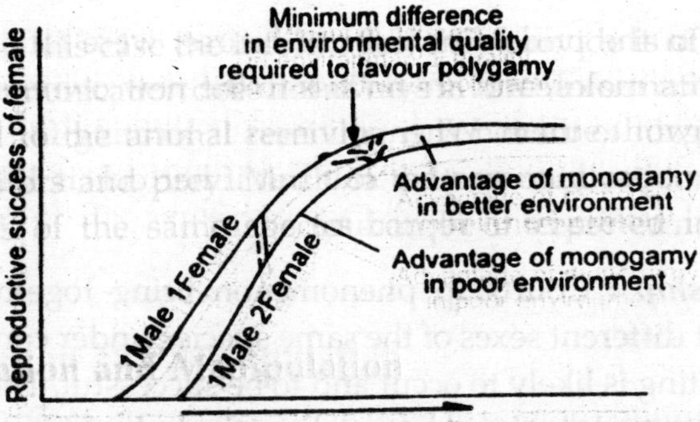

Graphical representation of reproductive success and territory

The Struggle of Males and Females : Due to the reason that the males reproductive success is probably limited by the number of females they can attract and defend from other males rather than their sperm supply natural selection will favour any property in a male that enables him to mate with more females. The most obvious form of competition for females is straight-forward fighting. In toads for instance, the males sit on the back of the females for a few days before the female lays her eggs. Other males try to dislodge the sitting males from the females by pulling them off. In experiments larger males are much better at dislodging smaller males from the females than vice versa.

There is an advantage in being large in males, and large males are more likely to mate with females than are small males.

In species that fight, a male can increase his effective strength by forming a coalition with another male, and ganging up on competitors. In lions, a group of two or three male lions try to take over harems of females by forcibly erecting the existing male owners. Bigger coalitions of males are more successful in taking over harems. However, physical fighting is not the only means by which males compete with each other. When a coalition of male lions successfully takes over a harem of females, the jut thing the males do is to kill all the young lion cubs in the pride. The cubs of course, were fathered by the previous males, and are no loss to the new owners. There is even a gain to them. When a lioness is locating for her cub she will not produce another cub but if her cub is killed she soon becomes ready to reproduce again.

Courtship : Courtship phenomenon bring together two animals of different sexes of the same species under conditions where mating is likely to occur and to be successful. The initial problem is to locate a potential mate and obviously the conspecific must be of the opposite sex if reproduction is to be successful. Courtship generally entails a complex sequence of interacting signals, that function to ensure that an animal mates with an appropriate partner. Timing is an important part of successful reproduction; both male and female must be in an appropriate physiological condition for mating, although mating system varies greatly from species to species.

In arthropods, initial detection occurs via olfaction for example silk moths, Bombyx mori. These are noted for the extreme sensitivity of the male to a sex attractant, bombykol, produced by the female. There are several other combination of behavioural patterns like wing vibrations, leg vibration, wing seamaphoring, circling and licking.

In fishes various species display diverse courtship patterns. In the tropical aquarium fishes, behaviour has been studied very carefully. Blackchin mouthbreeders, (Tilapia melanotheron), display four pure courtship patterns — patterns that occur in virtually no other contexts. These are the nod *i.e.*, rapid titling down; quiver *i.e.*, a pattern of head-shaking; nib *i.e.* biting the substrate and skim *i.e.*, remaining motionless near the nest (Barlow and Green 1970). In the courtship of guppies the most prominent display is the adoption of "sigmoid' or S-shaped posture by the male. In amphibians the male bullfrogs adopt territories from where they emit their familiar loud choruses to attract the females.

In reptiles the crocodiles also emit loud roars. Courtship in most species of snakes is based on tactile stimulation of the female and olfactory stimulation of the male. The so-called dramatic courtship dances in which two snakes become closely intertwined have usually proved to be male-male interactions which may be aggressive in nature (Porter 1972).

In birds there are many dramatic examples of courtship behaviour. The complex displays of grubes, gulls, ducks, herons and other birds have been favoured subjects of study by ethologists. In green herons male set up territories. The full forward display functions to drive away other males from a resident male's territory. The calls of the male attract the female. Initially the females are repelled with full forward displays, but later on they persist and courtship is eventually initiated with the appearance of the snap display and the stretch display. The two birds fly about the territory after pairing and occasionally exhibit the more intense flap-flight display. According of Meyerriecks (1960) until copulation displays patterns become more contact oriented.

In most of the birds, their song functions to repel males from territories and to attract females.

Among mammals Ewer (1968) surveyed some courtship patterns and according to him olfaction plays a large role in the regulation of courtship in most of the species. Patterns of anogenital investigations and urine testing are common. Female mammals often solicit mountings, sometimes by approaching a male, nuzzling or licking him and often by darting from him.

In bottle-nosed dolphins, courtship entails vocalizations, mouthing of the partner's genitalia, rubbing of bodies, displaying of the white underside, chasing, head butting, leaping. (Puente and Dewsbury, 1976)

Mating : Actual mating occurs as a result of consummation of courtship activity where the fertilization of ova is accomplished.

In most of the species of arthropods, sperm are 'packaged' in a spermatophore, which is a kind of bag or sac containing the sperm. Many aquatic species have external fertilization for *e.g.* Limulus, it hooks itself on behind the female and remains there until the eggs are released. Most of the terrestrial forms exhibit internal fertilization.

In fishes fertilization may be either external or internal. In anabantid fish like bettas and gourames, a bubble nest is prepared and spawning generally occurs beneath it. Eggs and sperms are released simultaneously and egg are fertilized as they float up to the nest.

In amphibians, many species of salamanders exhibit internal fertilization through use of a spermatophore, although parthenogenesis and external fertilization are also of common occurence. Most frogs and toads have external fertilization, eggs are fertilized as they are laid. Exceptions include tailed frogs, which have an intromittent organ, and African toads in which cloacal glands of the male and female are brought together. Rabb in 1973 stated that if male clasps another male or in clasping a female that has finished spawning, a release call by the recepient results in release by the clasping male.

In reptiles the fertilization is internal through the use of intromittent organs.

In birds the copulations generally show little variabilities. It most of the species penis is lacking, thus the transference of sperm occurs from cloacal glands of the male to that of the female.

In mammals a lot of research on copulation is being done using laboratory rats Rattus norvegicus. Its copulatory behaviour is described in detail. Whenever a receptive female is introduced into a cage containing a vigorous male, a predictable sequence of events occurs. After initial courtship behaviour like sniffing, chasing, etc., the male persues the running female. As the male mounts the female from behind, the female adopts a stereotyped posture, called "Lordosis" The head and hindquarters get elevated as a result back is produced into a concave arch, tail remains deflected to the side. Two kinds of copulations occur.

In some called 'intromissions' the male inserts its penis into the vagina of female for a period of 1/4 second before the male displays a rapid dismount. Although prior to insertion male may exhibit extravaginal pelvic thrusting, but there is no intravaginal thrusting. After about ten such short intromissions, the male displays 'ejaculation'. During ejaculation the male shows a kind of convulsive thrusting and clasps the female for a few seconds before he dismounts. A complete group of intromissions ending with an ejaculation is known as "series". Without prior intromissions there is no ejaculation and sperm transfer ejaculation is followed by a temporary cessation of sexual activity. After about five minutes, activity resumes and a second series occurs. Rats normally completes about seven ejaculatory series before they may be considered satiated, according to a criterion of half an hour without intromission (Beach and Jordan, 1956).

9

Feeding Behaviour

In the whole animal kingdom, feeding behaviour is incredibly varied and entirely very complex. In those animals which feed almost continuously, there is a complete spectrum of temporal for example certain filter feeding invertebrates to others, like some parasitic insects and carnivorous vertebrates that feed rarely, at intervals of many hours, days or weeks. In some animals, when there is no feeding at all, their life cycle exhibits phases, such as in the adult of some insects and in the male fur seal while he is ashore defending his territory. Most of the animals which feed at relatively short intervals measured in minutes or a few hours exists somewhere in between these extremes. Rats, mice, birds, fishes are flies and included in most of the examples and we can restrict our discussion to those animals which are restricted to any of the above animals which is more familiar to the conditions.

In rats, feeding occurs at an average interval of about 2 to 4 hours, and this interval could be changed resulting into wide variations and individual difference by training. Nevertheless, a cyclic pattern of one kind or another persists, with statistically predictable intervals between feeding bouts. Among Catties feeding occurs with a regular pattern of distinct bouts in one complete day. In Dolichonyx oryzivorus, the bololink, a small passerine bird feeding occurs with a rhythmical pattern.

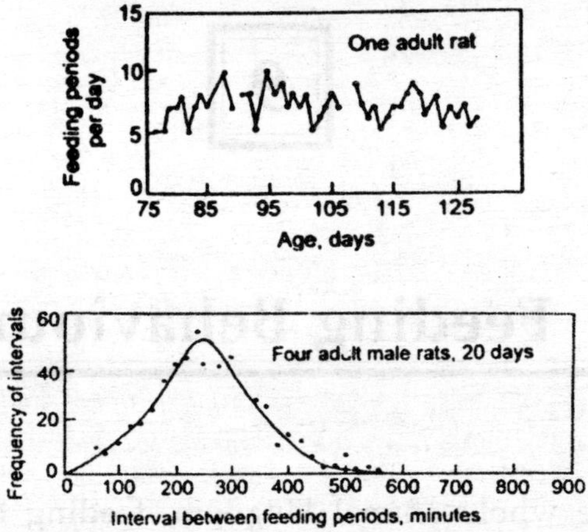

Rhythmical expression of the feeling behaviour recurring feeding in different ways.

By expanding the time scale and observing the behaviour exhibiting one cycle of feeding activity, more details of the pattern could be revealed. Generally, locomotion is the prelude to a bout of eating either in free exploration of the environment or moving in such a location where food is found habitually. This appetitive phase is followed by the behaviour associated with selecting, preparing and ingesting food *i.e.*, the consumatory phase. There are a variety of other activities like resting sleeping, drinking, social encounters and locomotion which occupy the intervals occurring in the above phase. Cattle spend most of its time in ruminating, *i.e.*, regurgitating food, chewing it and swallowing it again in between bouts of feeding.

Feeding Strategies

On the basis of the primary feeding strategies, animals can be grouped into the following classes:

(I) *Autotrophs* : It includes those organisms which synthesize the essential organic compounds from inorganic substances. Thus these organisms feed independently. Autotrophs, not only include plants but euglena, volvox and few bacteria are also said to feed independently.

Depending upon the way of producing energy, organism could be classified into two groups:

(i) Photosynthetic autotrophs (ii) Chemosynthetic autotrophs.

Photosynthetic Autotrophs : Those organisms which convert solar energy of sun into chemical energy by a process of photosynthesis are known as photosynthetic autotrophs. For example, plants, euglena, volvox. They are able to synthesize their own amino acids, fats, sugars etc. from inorganic substances like CO_2, nitrates etc.

Chemosynthetic Autotrophs : Those organisms which manufacture their all protoplasmic requirements from inorganic substance through a process called chemosynthesis are called chemosynthetic autotrophs. For example sulphur and iron bacteria (also called chemotrophs) which use the energy from oxidation of simple inorganic chemicals like iron and sulphur.

(II) *Heterotrophs* : It includes animals and some non-green plants which can neither utilize solar energy nor can synthesize organic substances from inorganic material. (CO_2, nitrate etc.) Heterotrophs need organic substances for their life.

There are several ways by which heterotrophs obtain their food from plants and animals:

(i) *Holozoic Nutrition* : In this type of nutrition organism obtain energy by oxidising pre-existing organic compound of plants or animals. There are a number of ways by which they feed, like:

Herbivores : feeding upon plants and vegetation for example, cow, horse, rabbit, buffalo etc.

Carnivores : feeding upon other animals or on flesh, for example, lion, tiger, dogs etc.

Omnivores : Feeding upon any type of food directly upon both plants and animals, for example, human beings, flies, cockroaches etc.

Cannibulus : feeding upon its own species. Although it is not a common method of taking food but still it is usually seen to be exhibited by fishes. It is under the direct control of nature.

Detritus : feeding upon organic matters present in soil or mud. For example, earthworm.

Voraceous : feeding upon crustaceans and larvae. For example, gambusia fish, gastropods etc.

Predators : feeding upon lower species especially on invertebrates, for example, birds like vultures, cranes etc.

Insectivores : feeding upon insects, for example, frog, bat, lizards etc.

Ant Eaters : feeding specially upon ants due to adaptations, for example, echidna (spiny ant eater), manis (scaly ant eater) etc.

Food Robbers : feeding upon fully or semi-digested food material of the host from its alimentary canal, for example, ascaris, taenia etc.

Saprozoic : deriving food from dead decaying matter. They are also called scavengers. For example, euglena, paramecium.

(ii) *Saprophytic :* Those organisms which take dissolved decaying organic material from their environment are called saprophytic. For example, fungi, amoeba, euglena etc.

(iii) *Osmotrophic* : Those organism which feed upon predigested food material from the host by diffusion through their body wall are called osmotrophic. For example, trypanosoma, taenia solium. etc.

(iv) *Sanguivores* : Those organisms which feed upon blood of other animals are called Sanguivores. In these organism an anticoagulin like heparin, hirudin, lamphredin etc. is secreted in order to prevent clotting of blood in their sucking apparatus. For example, leeches, lampreys, mosquitoes etc.

(v) *Parasitic* : Those organisms which feed at the expense of living organisms which entertain them as "host" normally without any fatal results. For example, ascaris, liver fluke, head louse, tapeworm etc.

(vi) *Symbionts* : In this two organisms exhibit a sort of mutual association where both of them are benefited.

Food Preferences : No two species living together at the same time and place eat exactly the same staple food. All animals are selective in choosing food in their natural habit, some more so than others. According to Lack (1954) and Mayr (1963) this is the competitive principle. Thus selectivity of food is a matter of major biological importance, not only because it has a bearing effect on nutrition but also because of its relationship to problems of interspecific competition. According to Hafez and Schein (1962), when animals are given a choice they show preferences. "Cattle exhibit preferences not only for certain plant species but for the same species at different stages of growth and even for the various parts of an individual plant and for individual plants within a species". Among the seed eating birds (Morris 1955, Wood-Gush 1955, Kear 1962), rats (Barnett 1963) and primates (Harlow and Meyer 1952, Berkson 1962), preferences are equally clear.

Feeding in Group-living Herbivores : The animals which live in huge groups which together make up some 90% of the total weight of mammals living on the serengeti includes buffalo, zebra, wildbeest, topi, and Thomson's gazelle. These five in the just sight appear to live on the same species of grass, herbs and small bushes, however, the appearance is illusory. When the stomach contents of all of these except buffalo was studied by Bell and his colleagues, it was found that each species was living on a different part of the vegetation. These different parts of the vegetation differ in their food qualities lower down these are succulent, nutritious leaves, higher up are the harder stems Highly nutritious fruits are also found distributed sparsely. Bell and his colleagues found that only the Thomson's gazelle eats much of these nutritious fruits.

The other three species differ in proportions of lower leaves and higher stems that they eat. The zebra eats mostly stem material, the wildbeest depends on the leaves and the topi is intermediate.

In order to understand the different feeding preferences, we can consider two things firstly to study their digestive system and secondly their body sizes. The digestive system can be differentiated into the non-ruminants (the zebras) and the ruminants (wildbeest, topi, and gazelle which are like the cow). Non-ruminants are not able to extract much energy from the hard part of the plant; however, this is more than made up for by the fact that food passes much more quickly through their guts.

Thus, when there is only a short supply of poor quality food, the wildbeest, topi and gazelle enjoy an advantage. They are ruminants and have special structures in their stomaches (the rumen), containing special micro-organisms which can break down the hard parts of the plants. Food passes only slowly through the ruminant's guts because ruminating, digesting the hard parts, takes time. The ruminant continually regurgitates food from its stomach to its mouth to chew it up further (that is, what a cow is doing when 'chewing cud'). Only when it has been chewed up almost to a liquid can the food pass through the rumen, and on through the gut. Larger particles cannot pass through until they have been chewed down to size.

Therefore, when food is in short supply, a ruminant can last longer than a non-ruminant because it can extract more energy out of the same food. The difference can partially explain the eating habits of the serengeti herbivores. The zebra chooses area where there is more low quality food. It migrates first to unexploited areas and chomps the abundant low quality stems before moving on. It is a fast in/fast out feeder, relying on a high throughout of incompletely digested food. By the time the wildbeest (and other ruminants) arrive, the grazing and trampling of the zebras will have worn the vegetation down. As the ruminants then set to work they eat down to the lower, leafier parts of the vegetation. All of which fits in with the differences of stomach contents with which we began.

Body size includes the other parts of the explanation. Larger animals require more food than smaller animals, but smaller animals have a higher metabolic rate. Smaller animals can, therefore, live where there is less food, provided that it is of high energy content. That is why the smallest of the herbivores, Thomson's gazelle, lives on fruit, which is very nutritious but too thin on the ground to support a larger animal. By contrast, the large zebra lives on the masses of low quality stem material.

The differences in feeding preferences lead, in turn, to differences in migratory habits. We have seen that wild beest follow, in their migration, capricuous pattern of local rainfall. The other species do likewise, but when an area is fuelled, by rain, for exploitation, the mammals migrate towards it in an orderly pattern. The larger, less fastidious feeders, the zebras, move in first, the choosier, smaller wildebeest come later and the smallest species of all, Thomson's gazelle, arrive last. The later species depend on the preparations of the earlier, for the action of the zebra fits the vegetation for the stomachs of the wildbeest and gazelle. So if we are to understand the feeding habits of the species, we must consider it in relation to the whole ecology of the species, and its relation to other species. Behaviour is an inseparable part of a whole system, made up of body size, gut morphology and the habits of associated species.

Social Carnivores

Among the mammalian carnivores, one member of the family Felidae and several members of the family Canidae have evolved complex social behaviour that is related to cooperative capture of prey. All of these species have been the subjects of intensive behavioural and ecological study in their natural habitats.

Lions : The lion (Panithero leo) lives in closed social unit and is most abundant in the grasslands and open woodlands of Africa. Schaller has studied the behaviour and ecology of this species, its competitors and its prey.

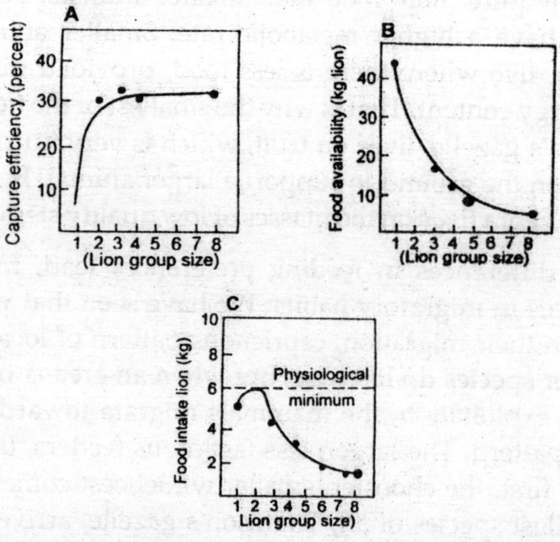

Capture efficiency, food availability, and estimated food intake as functions of lion group size for Thomson's gazelle prey.

Lions gain from cooperation in several ways. First, by hunting in groups, they increase the resource spectrum; specifically, they add to their diet two species buffalo (Syncerus caffer) and giraffe (Giraffa camelopardalis) that an individual lion could never attack alone. Second, cooperative hunts are at least twice as successful as solitary ones. Third a group can consume a captured prey more fully than can an individual. Fourth, the group can drive other predators and scavengers from the food Schaller found that plain dwelling prides kill less than half their food, relying on other predators, such as hyenas (Crocuta crocuta), to make kills for them. Although lions hunt in groups, the amount of cooperation is not extensive. When the lions encounter a herd of prey, such as zebra (Equus burchelli), the pride female fan out, sometimes encircling the herd. Lions are not endurance runners and rely on a stalk and rush tactic. Although lions are more effective when hunting upwind. Schaller found no evidence that they do so more often than would be expected by chance.

Once a kill has been made, the males, which are larger than the females but which do little hunting, may drive the females from the kill. The cubs are the last to get food. Schaller believes that lion populations are regulated directly by food supply through starvation of cubs. The reproductive rate remains constant, but the mortality rate of the young can be very high, exceeding 80% when food is scarce. Not in all cub, mortality is the result of starvation; strange males that move into a group may kill cubs. Caraco and Wolf (1975) used Schaller data to relate ecological factors to lion foraging group size. When hunting small prey, such as Thomson's gazelles (Gazella thomsoni), two lions are more than twice as efficient as one, but three or more do no better than two. Since the available food for each lion per kill decreases with increasing group size, we would expect foraging groups of about two. Caraco and Wolf went on to show only pairs of lions can take in the minimum daily amount of food if they feed exclusively on

Thomson's gazelles; Schaller reported that the mean foraging-group size ranged from 1-5 to 2, close to expected. In hunting larger prey such as zebras and wildebeest (Connochaetes taurinus), a foraging group size of 2 was again the most efficient, but groups of 1 to 4 could still attain their minimum daily requirement. However, Schaller found larger than expected foraging groups of about 4 to 7 Caraco and Wolf were forced to suggest that factors other than prey size, such as the higher reproductive success documented for larger prides (lionesses share in the feeding of cubs), could influence group size.

African Wild Dogs : These are unlikely big game predator weighing only about 18 kg (40 lbs), typically catches prey weighing as much as 250 kg (Schaller 1972). The dogs live in mixed-sex packs that average ten adults and, when there are young in a den, range out from it to hunt. The pack is tightly organised. Of particular interest here is the prehunt ceremony, in which dogs draw back their lips to expose their teeth, nibble and lick each other's mouths, and run winning from pack member to pack member.

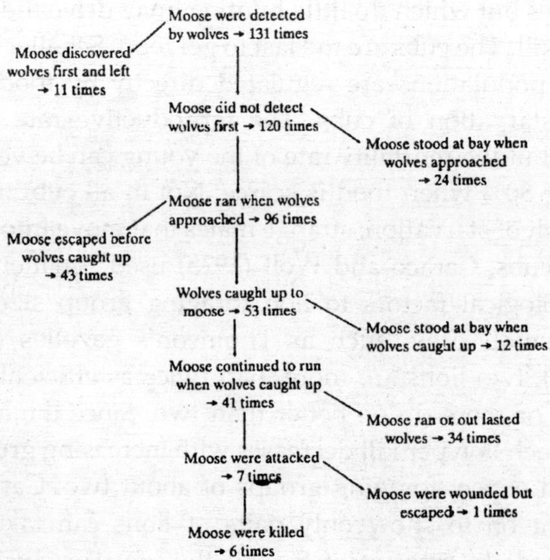

The greeting seems to represent ritualized food begging. Setting out on a hunt, the dogs travel at a trot and fan loosely over the terrain; certain adults consistently lead some packs, but in other packs there seem to be no leader. Schaller (1972) describes the hunt in his field notes :

Moose-Wolf Interactions

For African wild dogs, the hunting success is extremely high; say about 90% of all hunts result in the capture of at least one prey item.

Wolves (Canis lupus) : It is a social carnivore of northern temperate and arctic regions. It typically preys on deer (Odocoileus spp.), moose (Alces americana), buffalo (Bison bison), sheep (Ovis spp.), Caribou (Rangifer tarandus) and elk (Cerous canadensis). Pack size varies greatly (Mech 1970), but most have fewer than eight members. Wolves hunt by travelling over their territory after consuming their previous kill. Most often they use scent; the sea animals stop when they detect the odor of prey. The group may stand nose to nose with tails wagging for a few seconds, and then all follow the leaders directly towards the prey. They may also use chance encounter and tracking. If they detect the prey at some distance, they proceed with restraint and stalk sometimes to within 10 metres of the prey. When the prey detects the wolves, it may stand its ground or flee. Once the prey is in flight, the wolves rush. Wolves must get close to their prey during the stalk and rush on they will quickly give up; if they cannot make an attack, they may chase the prey, usually for less than half a mile. Some earlier studies described wolves as coursers, but Mech has concluded that they are mainly rushers and stalkers.

The musk ox (Ovibos moschatus), one of the prey of wolf, defends itself in an unusual way. Adults form a circle, facing outward and protecting the young inside. There is little evidence that wolves can break this defense, in areas where musk oxen are present, wolves eat mainly other prey or capture an occasional post-reproductive individual.

When hunting moose, which are solitary, wolves rarely attack a prime individual (between one and six years old). Most studies of wolf predation have indicated that they are highly selective of young, old, and sick prey. Although earlier reports described the main killing tactic as hamstringing, wolves actually avoid the hooves and direct bites at the rump, flanks, shoulders, neck and nose.

Anti Predator Behaviour

As we know that most of the species are preyed upon by animals of at least some other species. Thus in survival and reproduction, avoidance of predation plays an important role.

There are certain defensive mechanism which are operated throughout much of the life of an individual but others are used only when a predator is detected or when a predator attacks the individual. Edmunds (1947) refers to two types of mechanism, primary and secondary. Following Kruuk (1972) Edmunds define primary defence mechanism as those which operate regardless of whether or not there is a predator in the vicinity. The behavioural components of such mechanism may occur before any predator is detected. The mechanism include : (1) hiding in holes; (2) the use of crypsis; (3) mimicking inedible objects; (4) exhibiting a warning of danger to predators; (5) mimicking individuals in category; (6) timing activities so as to minimise the chance of detection by a predator; (7) remaining in a situation where any predator attack is likely to be unsuccessful because of possibilities for secondary defence; (8) maintaining vigilance so as to maximise the chance of detecting the advent of a predator.

All of these mechanism decreases the chance that a predator attack will occur, many of them have a considerable influence upon habitat selection. For some species which are very vulnerable to predation, the avoidance of predators limits the distribution of the animals even more than do factors related to body maintenance and feeding. The selection of a suitable habitat may be accomplished by the recognition of a factor which is not in itself advantageous but which is an indicator of the presence of advantageous features, such as those which make possible feeding or predator avoidance. For example, a bird, such as a nightjar, which is camouflaged when amongst dead wood might fly over open downland towards a dark green mass of woodland because suitable hiding places are likely to be found around the edge of the wood.

Edmunds defined secondary defence mechanisms as those which operate during an encounter with a predator. The encounter may just involve the supposed or actual detection of a predator by the prey or it may involve an attack by a predator. Some secondary defences are entirely passive. The position spined sea urchin, diadema or the thick-skinned rhinoceros are adequately protected from attack. No behavioural specialisation is needed. Active forms of secondary defence include: (1) exaggerating primary defence, *e.g.* a camouflaged animal remaining motionless; (2) withdrawal to a safe retreat; (3) flight; (4) use of a display which deters attack; (5) feigning death; (6) behaviour which deflects the attack to the least vulnerable part of the body, to an inanimate object, or to other individual; and (7) retaliation.

Mostly, animals use both primary and secondary defence mechanism but very often different predators use different mechanism. In encounter with a predator, the prey animal may be able to use several alternative anti-predator behaviours. Decisions must be taken about, for example, whether to remain motionless, to flee, to display, or to fight. If one method is tried, without successful avoidance of attack, others may be tried. There are many possible strategies which involve alternative courses of action depending upon the characteristics and behaviour of the predator.

Concealment

Many prey species protect themselves from predators by remaining in protected locations such as burrows, crevices, or huts. Cryptic colouration which functions to enable animals to blend in with the background, can be found in virtually all animal taxa. There are so many extreme examples found in insects which may resemble leaves, twigs or even bird droppings. Often a particular behavioural pattern is associated with protective colouration such that the animal orients itself in a particular relationship to its environment, usually remaining motionless.

Camouflage

A phenomenon in which a species evolves to resemble its background is called camouflage. The nochid moth's defence is to seek escape in active flight. The opposite defence is to sit deed still and try to be invisible. Camouflage is of course an adaptation of appearance and colouration. The world is a patch-work of different colours : the animal is only camouflaged if it settles in the right place. Consider the European grasshopper Acrida turrita, it comes in green and yellow form. In nature the green form lives in green places and the yellow form in yellow and brown places. In a simple experiment, the German ethologist S. Ergene gave yellow and green grasshoppers and choice between yellow and green background. The green grasshooper fittingly tended to go and settle on the green backgrounds, and the yellow grasshoppers on the yellow.

Warning Colouration and Mimicry

There are some animals which are provided with poisonous or sickening substances in order to protect themselves from being eaten up by the predators. Some such animals make their own poisons; others takes them from sources in their environment. The wings of the monarch butterfly, for example contain powerful heart stopping poisons called cardiac glycosides. The monarch eats the poisons as a caterpillar, when its food plant is milkweed which contains cardiac glycosides. The caterpillar is not harmed by the poisons, it just stores them, and they are then retained by the adult.

For the behavioural problem of defence by poison we must turn from the prey to the predator. If the defence is the work, the predator must learn not to eat poisonous animals, because natural selection will not favour a trait, by which an animal, after it is dead, makes its attacker sick. If the trait is to evolve, it must ensure survival. The tactic used is to enable predators to learn to recognize sick-making prey, in which case they will avoid them. When a bird eats a monarch, as an adult or caterpillar, it will be violently sick within minutes, an experience it would learn not to repeat. The bird's problem then is to distinguish sickening from edible prey. Now, the colours of the monarch are bright gold, and ethologist suspect that the bright colouration evolved to make the monarch more memorable to birds. It is therefore called warning colouration. Experiments have shown that predators learn to avoid sickening prey, J.V.Z. Brower, offered the monarch butterfly as food to the Florida scrub jay. On the first meal of monarch the jays were violently sick, but after only a few trials they had learned not to eat monarchs but they continued to eat other, tasty food. What has not yet been conclusively demonstrated is that bird learn to avoid brightly coloured sickening prey more-quickly than equivalent duller coloured prey. Few ethologists would doubt, however, that such an experiment would be successful.

Freezing

It is an another strategy of prey in response to predators. The presence of protective or cryptic colouration is often associated with this behaviour, as in the spotted white-tailed dur fawn (Odocoileus virgianus). Some animals carry this strategy further and feign death. For example the opossum

(Didelphis virginiana), if harassed by a predator such as a dog, remains motionless on its back. The physiological correlation of this behaviour are not completely understood. Hognosed snake (Heterodon patyrhinos) exhibit similar behaviour, although they may precede it by threat behaviour in which the hognosed snake resembles a cobra. Freezing and feigning death may work because many predators are seen to respond only to the moving prey. For example, wolves will not attack prey that stand motionless (Mech 1970). Accompanied with this, rapid and agile locomotion provides the best and probably the most common means of escaping from predators.

Social Antipredator Behaviour

Confusion Effect : One of the main benefits of sociality is getting protection while living in groups. Living in flocks, herds or schools increases the chance of spotting predators. When approached or attacked by predators, groups usually become more compact. Fish schools, however usually form a vacuole around a predator. Tightly packed prey make a difficult target for the predator which must select an individual toward which it can aim its attack. The predator tries to isolate an individual from the group, any individual that strays from the group or is in any way different from the rest is likely to be attacked. For example, when experiments presented hawks with set of ten mice, the hawks preferred the oddly coloured mouse. (Muller 1971).

A primary means of protection is the confusion effect, described by Miller (1922). A variety of predators that attack swarming prey exhibit a lower prey capture rate, a longer period of hesitation, and more irrelevant behaviour than when they attack solitary prey. Factors that enhance the confusion effect are swarm number, swarm density, and the uniformity in appearance of swarm numbers, (Milinski 1979). The mechanism producing the confusion effect is not understood, but a predator seems to be more easily frightened when it feeds in a dense swarm of prey; perhaps the predator finds it harder to detect other animals that might prey on it. In laboratory experiments with three spined sticklebacks (Asterosteus aculeatus) feeding on water fleas (Daphnia), fish with experience in feeding on dense swarms fed more efficiently when tested with dense swarms than did those experienced only in feeding in low-density populations (Milinski 1979).

Detection : We might expect to find groups of mixed species if the combination of different sensory capabilities is advantageous. Baboons and ungulates are frequently together on the African plains. The arboreal habits and keen vision of baboons and the well developed olfactory system of ungulates probably increase the distance at which prey can be detected. It is clear that these different species recognize and respond to each other's alarm calls.

Social living increases the likelihood that predators will be detected and allows group members to spend more time in other activities. Individuals in large, dense colonies prairie drops (Cynomys spp.) spend less time in alert postures than do those in small colonies (Hoogland, 1979). Once predators are detected, alarm calls alert neighbours to danger. Marler (1955) has argued that in some birds the alarm call is difficult to locate because it is a high-pitched note in a narrow frequency range. The alarm-calling ground squirrels studied by Sherman (1977) emitted calls that were easily located, and calling squirrels were more likely to be caught than non-callers. Close relatives of the caller were likely to be close by and thus could benefit from this seemly altruistic behaviour.

Development of Anti-predator Behaviour : A baboon which is attacked by a leopard but which survives, learns something about the speed and attack methods of leopards. Future encounters with leopards will be different as a consequence of the experience. Field studies of the improvement in techniques of anti-predator behaviour are difficult, but many laboratory studies of avoidance conditioning mimic certain aspects of this type of behavioural change. Suppose that a rat in a cage learns to avoid shock by pressing a set of levers in a particular sequence when a certain combination of stimuli is presented. The rat may be using abilities, which would also be of use if faced with successive exposures to a predator. The effects of experience are often easiest to study in developing animals since some aspects of previous experience can be controlled.

In species where parental care is prolonged, the most effective anti-predator behaviour which young animals can show is often to recognise and maintain contact with the parents and the place where they are put by parents. The over-riding importance of finding the parent leads to some apparently anomalous responses to predators. Lorenz (1935) describes how newly hatched precocial birds, such as goslings, show no avoidance behaviour to a variety of moving objects including, for example, man, a goose predator.

A system which results in the young animal showing antipredator behaviour to any novel change in its surroundings will operate when a real predator appears. Responsiveness to potential predators by young animals does not depend solely upon degree of novelty. It is also affected by the nature of the sensory input which results from the presence of the predator. Bodily damage, a looming object, or a loud noise will elicit more extreme anti-predator behaviour than the sight of small objects the sound of quieter noises, etc. (Hebb, 1946).

The existence of perceptual analysers in the brain which enable an individual to recognise dangerous events, for example, a looming object (Ewen, 1971). Some of these analysers are known to develop in a different way according to specific visual experience by others probably depend on general environmental factors such as light and metabolic stall at key times during embryological development. The average individual, therefore, would have some other predator detection mechanism whether or not there had been previous visual experience of the predator. An elaborate and specific possibility is a detector for flying bird of prey Lorenz (1939) and Goethe (1940) reported that young precocial birds did not respond to flying geese but showed a flight response when a hawk or falcon overhead.

Experiments by Schleidt (1961) showed that the results could have been due to habituation to the geese but not to hawks. Further work on any mallard duckling by Green, Oreen (1966) and by Mueller and Parker (1980), however, indicates that the hawk and goose silhouette can be distinguished and do elicit different reactions. The reactions were not clear anti-predator reactions in either case. The linking of the perceptual recognition to the command which initiates such reactions, therefore, may require experience which the experimental subjects of these experiments does not have.

10

Wildlife

The concept of live and let live, conservation and preservation of plants and animals has been an age old practice in India. Three hundred years before Christ, Chanakya wrote in Arthashastra that it is forbidden to cut trees for fire, destroy leaves, burn wood, kill animals, remove their skins and collect bones, because this would disrupt normal lives of animals and subsequently the delicate balance of nature. Later, Ashoka the great, further realized the significance of animals and banned hunting in the third century BC and got the names of all those protected birds and animals engraved on pillars. The first conservation law in India was enacted by Emperor Ashoka in his fifth pillar Edict, that animals such as bats, monkeys, rhinos, porcupines and tree squirrels were to be strictly preserved and that "forests must not be burned".

Hinduism and Buddhism have been very closely associated with protection of animals and trees. India has been a paradise of animals. Due to religious beliefs, mythology and iolklores an average Indian has been a conservationist. Significance of air, water and fire was realized way back and they were given the status of Gods, people, still worship them as "Pavan, Varun and Agni Devtas" who use antelope, crocodile and goat as their "Vahanas". By doing so, natural resources and the associated animals both were rendered protection.

It was an intelligent and unique way of checking their indiscriminate use and killing. Many other animals also found protection due to the belief that they are associated with Gods, viz elephant and rat with Ganesh, snake with Shiva and Vishnu, hans with Brahma, swan with Saraswati, lion with Durga and cow with Krishna. Animals are given much importance in our mythology. Hanuman and his extraordinary place in Ramayana is very well known. Incarnation of Vishnu as a fish (Matsya), tortoise (Kurma) and once as a boar (Varaha) signifies the role of animals in saving the world. Water has been an important natural resource, therefore, river goddesses - Ganga and Yamuna are worshiped too, their points of origin are treated in high esteem for pilgrimage.

The religious connotation attached to the tree of Peepal, Ber, and Kela is known to all. These beliefs have led to conservation of flora and fauna of India and perhaps due to these reasons India has remained one of the richest countries of the world in its variety of wild-life. There are about 15000 plant species, 500 Mammals, 1230 Birds, 220 Snakes, 150 Lizards, 30 Turtles, 3 Crocodiles, 142 Amphibians, 105 Freshwater fishes (Mojupuria, 1986). Inspite of such biodiversity in modern India only elephants and tigers have caught national and world awareness.

Although we have 5% of all known living organisms of the world in India, but we have only 2% of world's total land mass, which is heavily occupied by 16% of the world's human population, therefore, the biodiversity we have may not remain like this forever because one billion people and further enormous human growth in India may sweep away the forests which were about 40'% of the total land in 1947 and now remain only 19% against the desired 33%.

Geology : It is generally recognized now that peninsular India was once part of the supercontinent Gondwanaland, which included Africa, Antarctica, Madagascar, Australia and South America. Evidence of the link includes the remains of a primitive lizard the Lystrosaurus, found in South Africa, Antarctica, Bengal and China. As Gondwanaland broke up, India became separated and drifted across the globe on its molten crust, to crash slowly but with titanic force into Eurasia, 40 to 60 million years ago. It extinguished the ancient Tethys Sea whose bed has been thrust upto 4000 meters in Tibet (along the suture line of the Indian and Eurasian plates, just south of the Yarlung Zangbo river). Fossilized sea shells such as ammonites can be seen here embedded in the sandstone and limestone rocks. These are now 800 kms from the nearest sea and 4 kms above the present sea level. The Indian plate has continued to thrust under Eurasia and the Himalayas which have risen in several great upheavals, are still rising at a rate of about a centimeter a year and northwards at 5 cm a year. The great height reached by the range in the late Pleistocene created a weather barrier which is responsible for India's monsoon climate.

India, the sub-continent, perhaps the only country in the world, has all the five representative zoogeographic zones on the planet, namely those of the tundra, the alpine, the savanna, the Ethiopian and the mangrove; in north the Himalayas, the greatest mountain systems in the world, in higher reaches tundra like regions of perpetual snow and ice; foothills wet with thick forest cover. Moving down the map, are the vast open spaces of the Gangetic plains and the Deccan plateau-the peninsular India of grass lands and tropical deciduous forests, dry most of the year, dependent on seasonal showers. In west are the Desert and Saline wastes of Kutch, and zones while in east and southern part are coastal regions of heavy rainfall and dense tropical rain forests. Such diverse geographical features produce equally diverse and varied living forms. The number of bird, species recorded in Indian subcontinent is almost twice that in Europe or North America.

Indian forests give shelter to many exclusive and rare animals, some of the important ones are : the Kashmir stag or the Hangul (Cervus elaphus Hangul) (related to the European Red deer) at the Dachigam National Park, One horned Indian rhinoceros (Rhinoceros unicomis) at the Kaziranga National Park, Thamin or the Brow antlered deer (Cervus eldi) at the Keibul Lamjao National park, Manipur, elusive Golden langoor (Presbytis geei) at Manas in the Bhutan segment close to Assam, Swamp deer or the Barasingha (Cervus duvauceli) at the Kanha National Park and Dudhwa, Snow leopard (Panthera uncia) in Laddakh and Sikkim, Gharials (Gavialis gavialis) at the National Chambal Sanctuary (M.P.), endangered Great Indian bustard (Ardeotis nigriceps) at the Karera Sanctuary (M.P.), Hispid hare (Caprolagus hispidus) at Manas, the acrobat Hoolock gibbon (Hylobates hoolock) at Kaziranga, Asian elephant (Elephas maximus) at Manas, Corbett National Park (U.P.), Periyar Wild-life Sanctuary (Kerala) and Mudumalai (Tamil Nadu).

Olive Ridley sea turtles (Lepidochelys olivacea) in Sunderbans (West Bengal), Wild ass (Equus hemionus khur) at the Dhrangadhra Sanctuary (Little Rann of Kutch). In the family of big cats, India surpasses the wild world of Africa with its royal, elegant Tiger (Panthera tigris), the Common leopard, the Snow leopard and the Clouded leopard. Majestic Asiatic lion (Panthera leo) at Sasan Gir (Gujarat), Gaur or the Indian bison (Bos garus) at Bandipur (Karanataka) and Kanha, Bengal florican (Eupodotis bengalensis) at the Jaldapra Sanctuary (West Bengal), Dhole or the Wild dog (Cuon alpinus) at Kanha, Periyar and Melghat (Maharashtra) and Sariska (Rajasthan), shy Lion tailed macaque (Macaca silenus) and the Nilgiri langoor (Presbytis johnii) in Silent Valley (Kerala), big eyed Slow loris (Nycticebus coucang) in South India, Chital

(Axis axis), Chinkara (Gazella gazella), Nilgai (Boselaphus tragocamelus), Sambhar (Cervus unicolor), Blackbuck (Antilope cervicapra), Wild boar (sus scrofa), and Pigmy hog (Sus salvanius) also form part of India's exlusive fauna.

Historical Perspective

How did a situation of dwindling Wild-life come to pass in this land which has an ancient culture where the Elements and Animals are referred as gods, in Vedic scriptures, going back to 200/300 BC. In India, where it has been a great tradition of respect for nature and all life forms since time immemorial when saints and sages contemplated on the mysteries of life and destiny of man and wrote the sacred texts, vedas and then Buddha and his message of ahimsa, love and harmony with all living creatures, how could then animals and trees vanish?

It was in medieval times especially during the reign of the Mughals that hunting of animals became an organised affair, a kingly pastime and passion. Mughals themselves were not merely hunters but naturalists in their own right, Babar, Akbar, and Jehangir were keen observers of Indian birds, beasts, fruits and flowers. Jehangir miniatures of mansur are extraordinary pictorial studies of the flora and fauna. As the Mughal empire slowly declined, stepped in the English who continued hunting traditions but also began scientific and systematic documentation of India's flora and fauna ' although texts on medicinal uses of plants already existed (Ayurveda). To mobilize resources for the war, forest were cut, being enormous resources of revenue, and when the wars were over, tight arms and weapons became readily available to a large segment of society hitherto unarmed and for whom killing wild animals became a possibility denied so far (a privilege as opposed to a royal pastime).

Meanwhile with advances in modern medicine and rapid industrialisation, urban centres started growing in sheer numbers; rising populations put pressure on the land; forests were cleared to exploit the wood and fuel the fires; grasslands disappeared as fodder for cattle and where once forests grew and herds of deer roamed free, it now came under the plow of the farmer for more food. With their dwellings snatched, animals had nowhere to run and this brought many animals to extinction.

Deteriorated State of India's Wild-life : Red Data Book (this book actually comprises of long directories compiled by IUCN (Internatonal Union for Conservation of Nature and Natural resources) where rare and/or endangered species are listed along with relevant information from all over the world}; lists 103 animals from India alone. Like Tiger, the elephant is also fighting its last battle to survive along with many other animals. Satellite imageries have shown that the forest cover has shrunk to a bare minimum of 7 per cent of the total land which is also constantly under population pressure causing adverse effects on plant and animal forms.

In India, an estimated 13,000 species of mammals and birds were known in 17th century but now more than 130 are already gone for ever; some of the extinct animal species are Mountain quail, Pink headed duck, Lesser one horned rhino, and Jerdon's courser (this bird was rediscovered in 1986 at Cuddapah in Andhra Pradesh) and the streamlined spotted feline cheetah which once ranged freely across India's vast grasslands. Their abundance in India could be judged by the fact that Akbar had a stable of over 1,000 cheetahs in his army. The number of official endangered species has risen from 13 in 1952 to 240 today. Brow antlered deer, Hispid hare, Lion tailed macaque and Pygmy hog are considered very rare and are near extinction. The world is losing one life form every year. Extinction of animals and plants can be considered perfectly natural, after all, the dianosaurs who had once ruled the earth for millions of years also became extinct; it is true that extinction is a natural process, which can not be prevented but when we talk about endangered species in the twentieth century it means unnatural extinction due to manmade factors, which can be prevented.

India's Awakening to the Wildlife and Environment : After independence a serious concern for the forest and Wildlife heritage of India was recognised in the Indian constitution. The founding fathers of our constitution realised the need for environmental ethics and in Article 51 A inter-alia of the Indian constitution, it was written "It shall be the duty of every citizen of India to protect and improve the natural environment, including forests, lakes, rivers and Wild-life and to have compassion for living creatures." In the section of Directive Principles of State Policy Article 48 laid down the following duties for the State and the People : "the State shall endeavour to protect and improve the environment and to safeguard the forests and the Wild-life of the country."

As a new republic slowly began to stand on its feet, lot of damage to environment, forest and Wild-life started taking place, some due to huge technological projects and plants which were set up sprawling over vast areas; rivers dammed for irrigation and power resulted in inundation of large areas destroying animal and plant life. After independence, with rapid decline in Wild-life, the Government of India instituted the central board for Wild-life in 1949 later it was renamed as the Indian Board of Wild-life (IBWL) in 1952 to safeguard and monitor Wild-life. In 1952 an Animal Welfare Board was also constituted and the areas, for National Parks and Sanctuaries (National Park : area dedicated by statute (legislation) for all times to come, to conserve the natural or historical objects of National significance and to conserve Wild-life therein, in such a manner, and by such means, as well as leave them unimpaired for enjoyment of future generations, with such modifications as local conditions may demand. Sanctuary :

area constituted by competent authority in which killing, hunting, shooting, or capturing of any species of animals is prohibited except by or under the control of the highest authority in the department responsible for the management of the area. Protected area : area where special protection is granted to Wild-life on the verge of extinction to reestablish them; protection to Wild-life attracted to water impounded in river valley projects; protection to Wild-life around large towns.

A Reserve forest is an area in which Wild-life is protected under the forest laws, Prime Minister is always the chairperson of IBWL This Board has State Wild-life Advisory Board under its wings, scientists, naturalists and citizens of eminence are also included in board which advices Government of India on all major policy issues related to Wild-life. In the States, Forest Departments are responsible for Wild-life conservation and management. The Chief Wild-life Warden is the authority under whom Wild-life Wardens and Field Directors work to look after the Sanctuaries, National Parks and Zoos.

The main achievements of IBWL have been:

(1) Conservation of Wild-life through legislation and other measures.

(2) Establishment of National Parks, Sanctuaries and Zoological gardens.

(3) Promoting public interest and education in Wild-life.

(4) Import and export policy of Wild-life and its products.

(5) Formulation of a National Environmental Conservation Policy and revision of the National Forest Policy.

It was an alarming situation in the decade of the 60s when forests were being cut recklessly and Wild-life was vanishing, few species were already gone forever and other animals were either threatened or facing extinction, The depletion continued as did hunting and the loss of habitat, deforestation continued, contrived by rapacious timber traders plundered the forests to satisfy industry's insatiable demands that grew unchecked spewing out its wastes into our rivers, lakes, land and air which started environmental pollutions. It was the Tiger, symbol of India's Wild-life, that drew National and International attention which raised a cry of, concern from the turn of the century to 1970, a survey revealed the Tiger had declined from an estimated figure of 40000 to barely 2100-1800 in the late 1960s. Meanwhile, in the 70's an International concern on the global ecology started. Government of India also responded and Project tiger was launched to save this dying magnificent stripped feline, 10 reserves were set up for the Tiger.

Turning point came in 1969 when International Union for Conservation of Nature and Natural Resources (IUCN) held its convocation in New Delhi. Then Prime Minister Mrs. Indira Gandhi gave the inaugural address, strongly endorsing conservation and in 1970 India outlawed the shooting of Tigers.

A comprehensive central legislation was enacted in 1972 called the Wild-life (Protection) Act, for providing special legal protection to our Wild-life and to the endangered species of fauna and flora, by this Act State Governments are empowered to declare an appropriate area a Sanctuary or a National Park for the purpose of protecting, propagating or developing Wild-life therein, or its environment. This act was adopted by all the states of the Indian Union; the act provided for

(1) Provision to include plants in protected categories

(2) control trade in Wild-life products and

(3) creation of biospheres

At the same time International hue and cry was raised over an alarming situation and the Tiger became a focus of

International attention. In 1973 an International movement, initiated by World Wild-life Foundation, was started to protect the Tiger. India has since then slowly but steadily improved Wild-life conservation and management. Hangul Deer of Kashmir, Barasingha of Kanha and Tigers have been brought back from brink because of collective efforts of Governmental and Non-Governmental Wild-life management machinery.

The awareness has consistently gone up in large policy making bodies. There were only 80 National Parks and Sanctuaries in 1965, but now there are 70 National Parks and 385 Sanctuaries which cover approximately 90,000 sq.km, which is 12% of the forested area and 3% of the total geographical area. Today, most people wonder why should we spend so much money to save animals in a poor country like India, Forests and Wild-life therein must be preserved to preserve biological diversity, and gene pools, and also to save the delicate ecological balance between all the living organisms. Small disruptions may cause irreversible, slowly noticeable but ever lasting large effect even on the existence of supreme organism *i.e.* human being. The answer is not clearly understood by a commner that animals are the sign of healthy forest and environment, the same environment we live in.

All animals can not have same status with reference to conservation therefore, International Union for the Conservation of nature and Natural resources (IUCN) has categorized plants and animals into six types according to their distribution, population, abundance, habitat and potential value of the species.

1. *Endangered:* Taxa in danger of extinction and whose survival is unlikely if the casual factors continue operating. These are taxa whose numbers have been reduced to a critical level or whose habitats have been so drastically reduced that they are deemed to be in

immediate danger of extinction; *e.g.* Lion tailed macaque, Crocodile, Blue whale, Musk deer, Rhino.

2. *Vulnerable:* Taxa likely to move into the endangered category in the near future if the casual factors continue operating. These are taxa of which most or all of the populations are decreasing because of over exploitation, extensive destruction of habitat. The taxa with population that are still abundant but are under threat from serious adverse factors throughout their range. *e.g.* Crab eating macaque, Golden langoor, Blood pheasant, Chinkara.

3. *Rare:* The taxa with small populations in the world that are not at present endangered or vulnerable, but are at risk. These taxa are usually localized within restricted geographical areas or habitats or are thinly scattered over a more extensive range. *e.g.* Indian egg eating snake, Olive Ridley turtle, Indian Pied Hornbill, Crestless Himalayan porcupine, Golden cat.

4. *Threatened:* Taxa which are in one of the categories of Endangered, Vulnerable and Rare.

5. *Out of Danger:* Taxa formerly included in one of the above categories, but which are now considered relatively secure because of effective conservation.

6. *Intermediate:* Taxa that are suspected of belonging to one of the first three categories, but for which insufficient information is currently available. *e.g.* Gangetic Dolphin, Pallas's Cat, Binturong, Hog-badger, Indian Pangolin.

Many animals are surviving well because they have been granted total protection by the Wild-life (Protection) Act, 1972 which has seven chapters followed by five schedules. Schedule I has the following list of rare and endangered species which are totally protected throughout the country, live or dead or part thereof. The list below also gives a brief idea of great biodiversity we have in India.

Schedule I

PART I : MAMMALS

1. Andarnan Wild Pig (Sus andamanensis)

   IA Bharal (Ovis nahura)

   B Binturong (Arctictis binturong)

2. Blackbuck (Antelope cervicapra)

3. Browantlered Deer or Thamin (Cenus eldi)

   3A Capped Langoor (Presbytis pileatus)

4. Caracal (Felis carcal)

   4A Cetecean species

5. Cheetah (Acinonys jubatus)

   5A Chinese Pangolin (Manis pentadactyla)

   5B Chinkara or Indian Cazelle (Gazella gazella bennerti)

6. Clouded leopard (Neofelis nebuleosa)

   6A Crabeating Macaque (Macacairus tubrosa)

   6B Desert Cat (Felis libyca)

7. Dugong (Dugong dugon)

8. Fishing Cat (Fells viverrina)

   8A Fourhorned Antelope (Tetracerus quadricornis)

   8B Flying Squirrels (all species of the genera Hylopetes, Petaurista, Belomys and Eupetaurus)

   8D Gangetic Dolphin (Platanista gangetica)

   8E Bison/Gaur (Bos gaurus)

9. Golden Cat (Fells temmincki)

10. Golden Langoor (Presbytis geei)

10A Giant squirrel (Ratufa macrovro)

10B Himalayan Ibex (Capra ibex)

10C Himalayan Tahr (Hernitragus jemlahicus)

11. Hispid Hate (Caprolagus hispidus)

11A Hob-Badger (Arctonyx collaris)

12. Hoolock (Hylobates hoolock)

12A

12B Indian Elephant (Elephas maximus)

13. Indian Lion (Panthera Leo persica)

14. Indian Wild Ass (Equus hemionus khur)

15. Indian Wolf (Canis lupus pallipes)

16. Kashmir Stag (Cerus elephus hanglu)

16A Leaf Monkey (Presbytis phayrei)

16B Leopard or Panther (Panthera pardus)

17. Leopard Cat (Fells bengalensis)

18. Lesser or Red Panda (Ailurus fulgens)

19. Liontailed Macaque (Macaca silenus)

20. Slender Loris (Loris tardigradus)

20A Little Indian Porpoise (Necomeris phocaenoides)

21. Lynx (Fells lyns isabellinus)

22. Malabar Civet (Vverra megaspila)

22A Malayan Sun Bear (Helarctos malayanus)

23. Marbled Cat (Fells marmorata)

24. Markbor (Capra falconeri)

24A Mouse-Deer (Tragulus meminna)

25. Musk Deer (Moschus moschiferus)

25A Nilgiri Langoor (Presbytis johnii)

25B Nilgiri Tahr (Hmitragus hylocrius)

26. Nayan (Ovis ammon hodgsoni)
27. Pallas's Cat (Fells manul)
28. Pangolin (Manis crassicaudata)
29. Pygmy Hog (Sus salvanius)

    29A Retel (Mellivora capensis)
30. Rhinoceros (Rhinoceros unicornis)
31. Rustyspotted Cat (Feiis rubiginosa)

    31A Serow (Capricornis sumatraensis)

    31B Otter (Anoyx cinerea)

    31C Sloth Bear (Melursus ursinus)
32. Slow Loris (Nycticebus coucang)

    32A Small Travancore Flying Squirrel (Petinomys fuscocapillus)
33. Snow Leopard (Panthera uncia)

    33A Snubfin Dolphin (Orcacella brevirostris)
34. Spotted Linsang (Priondon pardicolor)
35. Swamp Deer (All subspecies of Cervus duvauceli)
36. Takin or Mishmi Takin (budorcas taxicolour)

    36A Tibetain Antelope of Chiru (Pantholops hodgsoai)

    36B Tibetan Fox (Vulpes ferrilatus)
37. Tibetan Gazelle (Procapra picticaudata)
38. Tibetan Wild Ass (Asinus hemionus kiang)
39. Tiger (Panthera tigris)
40. Urial or Shapu (Ovis vignei)
41. Wild Buffalo (Bubalus bubalis)

    41A Wild Yak (Bos grunniens)

    41B Tibetan Wolf (Canis lupus chanco)

Part II : Amphibians And Reptiles

1. Agra Monitor Lizard (Varanus griseus)

   1A Audithia Turtle (Pelochelys bibroni)

   1B Barred, Oval or Yellow Monitor Lizard (Varanus flavescens)

   1C Crocodiles (both the Estuarine and Marsh Crocodiles, Crocodylus porosus and Crocodylus palustris)

   1D Batagur terrapin (baska)

   1E Eastern Hill terrapin (Melanochelys tricarinata)

2. Gharial (Gavialis ganeticus)
3. Ganges Sofshelled Turtle (Trionyx ganeticus)

   3A Golden Gecko (caloductyloides aureaus)

4. Green Sea Turtle (Chelonia mydas)
5. Hawksbill Turtle (Eretmochelys imbricata)
6. Himalayan Newt or Salamandar (Tylototriton verrucosus)
7. Indian Egg-eating Snake (Elachistodon westermanni)
8. Indian Flapshelled Turtle (Lissemys punctata punctata)
9. Indian Tent Turtle (Kachuga tecta tecta)
10. Large Bengal Monitor Lizard (Varanus bengalensis)
11. Leathery Turtle (Dermochelys coriacea)
12. Logger Head Turtle (Caretta carena)
13. Olive-back Logger Head Turtle (Lepidochelys olivacea)
14. Peacock marked or Softshelled Turtle (Trionyx hurum)

    14A Pythons (Python sp.)

    14B Sail terrapin (Kachvuga kachuga)

14C Spotted back terrapin (Geockmys hamilloni)

14C Water Lizard (Laranus salvator)

15. Water Lizard (Varanus salvatar)

Part III: Birds

1. Andaman Teal (Anal gibberifrons albogularis)

IA Assam Bamboo Partridge (Barnbusicola fytchii)

1B Bazas (Aviceda jerdoni and Aviceda leuphotes)

IC Bengal Florian (Eupodotis bengalensis)

1D Black-necked Crane (Gras nigricollis)

I.e. Blood Pheasants (lthaginis cruentus tibetanus, Ithaginis cruentus kuseri)

2. Cheer pheasant (Catreus wallichji)

2A Eastern White Stork (Ciconia ciconia boyciana)

2B Forest Spotted Owlet (Athene blewitti)

2C Frogmouth (Genus Batrachostonius)

3. Great Indian Bustard (Ardeotis nigriceps)

4. Great Pied Hombill (Buceros bicornis)

4A Hawks (Accipitridae)

4B Hooded Crane (Grus rnonacho)

4C Hornbills (Ptiloleamus tickelli austeni, Aceros nipalensis, Rhyticeros undulatus ticehursti)

4D Houbara Bustard (Chamydotis undulata)

4E Hume's Bartailed Pheasant (Syrmaticus humiae)

4F Indian Pied Hornbill (Anthracoceros malabaricus)

5. Jerdon's Courser (Cursorius bitorquantus)

6. Lammergeier (Gypaetus barbatus)

7. Large Falcons (Falco peregrinus, Falco biarmicus and Falco chicquera)

7A Large Whistling 'real (Anatidaee)

7B Lesser Florican (Sypheotides indica)

7C Monal Pheasants (Lophophorus imperianus, L. sclateri)

8. Moutain Quail (Ophrysia superciliosa)
9. Narcondam Hombill (Rhyticeros undulatus narcondami)
10. Nicobar Megapode (Megapodius freycinet)

10A Nicobar Pigeon (Caloeruts nicobarica)

10B Osprey or Fish-eating Eagle (Pandion haliaetus)

10C Peacok Pheasant (Polyplectron bicalcaratum)

11. Peafowl (Pavo cristatus)
12. Pink-headed Duck (Rhodonessa caryophyllacea)
13. Sclater's Monal (Lophophorus sclateri)
14. Siberian White Crane (Gras leucogeranus)

14A

14B Tibetan Snow cock (Tetraogallus tibetanus)

15. Tragopan pheasants (Tragopan melanocephalus, Tragopan blythii, Tragopan satyra, Tragopan ternminckii)
16. White-bellied Sea Eagle (Haliaeetur leucogaster)
17. White-eared Pheasant (Crosspotilon crossoptilon)

17A White Spoonbill (Platalea leucorodia)

18. White-winged Wood Duck (Cairina scutulata)

Organisations

A number of National and International organizations are actively involved in Wild-life conservation.

(1) Bombay Natural History Society.

(2) Wild-life Institute of India.

(3) Wild-life Preservation Society of India.

(4) World Wild-life Foundation in India.

(5) Zoological survey of India.

Bombay Natural History Society : The society was founded in 1883 by seven residents of Bombay. In 1886 they started publishing a journal, which has continued till date, Society has increased by many folds and has been actively engaged in collecting information and specimens of fauna, and flora all through the sub-continent, it has played an extremely important role in drawing public attention to the need of wild-life conservation, it has a museum in Bombay which exhibits skins of rare animals, stuffed birds, reptiles and other animals. The Society has published some excellent books viz : The Book of Indian Birds (Salim Ali), The Book of Indian Animals (S.H. Prater), The Book of Indian Reptiles (J.C. Daniel), Some Beautiful Indian Climbers and Shrubs (Bor and Raizada) and Encyclopaedia of Indian Natural History (R.E. Hawkins). Society has been carrying out various extremely useful research projects, it is affiliated to the University of Bombay and guides post graduate research leading to M.Sc and Ph.D degrees. The Society is represented on the Indian Board of Wild-life and Wild-life Advisory Boards of many States.

Wild-life Institute of India (WII) : The institute was set up in 1982, by Ministry of Environment and Forests, Government of India. WII's aim is to develop Wild-life science and promote its application in the field according to our economic and socio-culture background, its main objectives are

(1) Train biologists and managers for protected area management and Wild-life research.

(2) Conduct and coordinate applied Wild-life research, and evolve techniques relevant to the Indian situation.

(3) Give training to education and extension workers to acquire skills in eliciting public support for Wild-life conservation.

(4) Provide consultancy services in conservation matters to government and private agencies.

(5) Create a data base leading to a National Wild-life Information System.

(6) Provide conservation orientation courses.

Research in Wild-life is a major activity of Wll, their projects cover different ecological and geographic regions of the country, projects on Snow leopard, Nilgiri langoor, Grizzled giant squirrel, Freshwater turtles, Mugger crocodiles and Rhino have already gathered useful information; ongoing studies cover high altitude, outer Himalaya's terai, moist and dry peninsular forests, western ghats, arid, desert and river ecosystems. Institute's faculty and field researchers regularly publish scientific papers in journals, four manuals have been brought out and WII publishes a quarterly newsletter.

Wild-life Preservation Society of India, Dehradun : It was founded in 1958, its main objectives are

(1) Promote interest in conservation through Journal Cheetal, Monographs, Films and Bulletins etc.

(2) Impart knowledge about conservation.

(3) Promote Wild-life tourism.

(4) Assist in forming Wild-life Protection Act.

(5) Help Wild-life administrators in maintenance and protection of National Parks and Sanctuaries.

International World Wild-life Foundation (WWF) : This was formed in 1961 and was launched in India in 1969. Its headquarter is in Switzerland. It takes up conservation projects all over the world. WWF works in close collaboration with IUCN, and is supported by United Nations, UNESCO, UNDP,

UNEP and FAO. WWF supports a number of projects in India and its most successful is "Project Tiger" which is the single largest conservation project of its own kind in the world.

UNEP	:	United Nations Environment Programme
FAO	:	Food and Agriculture Organisation
UNESCO	:	United Nations Education, Scientific and Cultural Organization
UNDP	:	United Nations Development Programme

The other WWF projects are Gir Ecological Research Project (1968), Project Hangul (1970), Project Himalayan Musk Deer. Its other major contribution is in "Conservation Education Programmes".

Zoological Survey of India (ZSI) : It was established in 1916 to promote survey, exploration and research leading to the advancement of faunal resources with the following objectives:

(1) Exploration and survey of faunal resources

(2) Taxonomic studies

(3) Status survey of endangered species

(4) Publication of results through 251 journals and of Fauna of India

(5) Maintenance and development of National Zoological collections and museums

(6) Environmental impact studies

The head office is located in Calcutta with 36 other offices located all over India. Recently ZSI has laid special emphasis to programmes relating to environment, ecology, and Ethnozoology. Intensive surveys of Silent Valley (Kerala), Damodar Valley Hydel project, Lesser cats of eastern India, Lion tailed macaque, Nilgiri langurs, and deep water fishes have added remarkable information.

Wild-life Trade

Animals and animal products have been of great use to human beings; association of animals viz elephants, horses, cows, goats, buffaloes, dogs, cats, fishes, pigs, poultry with human beings and use of their products like meat, milk, fur, skin and bones have been as old as civilisation itself. For economical benefit or amusement or recreation, use of animals and animal product is acceptable to some extent for those animals which breed well and are found in large numbers. Ironically animal products of wild and rare animals have also been of great use, without realization that the indiscriminate killing and use of rare wild animals will wipe them off from the surface of this earth. These animals need to be protected and conserved.

There are various purposes for which the wild animals and their products are being used, the Wildlife trade has been very lucrative and specially for those who want big money in short time. There was little or no control on Wild-life trade in India till 1952. Indian Board of Wild-life formed the laws and the policies to prevent uncontrolled wild-life trade but it was only by 1972, when - All India Wild-life Legislation in the form of the Indian Wild-life (Protection) Act came into existence in an effective way to prevent hunting and trade. The Wild-life trade is still going on, it seems that enforcement of laws has only curbed the activities.

Ivory Trade

Ivory or the white gold as it is popularly known as, has been the biggest trade among Wild-life products. Ivory has been in demand due to its whiteness and easy to carve the finest possible motifs on it. Use of ivory in jewellery, statues, tables, chairs and chess boards has become a fashionable thing. Fashion becomes a collective need created by a handful of designers, of which common persons become victims. Anybody who claims to be somebody special, has to have ivory or something rare unavailable to common person. Ivory became the most sought because it fulfilled all these qualities. Man is the biggest threat to elephants.

In nineteenth centuary, according to figures compiled by Natural History Museum in Paris, England alone imported annually about 550,000 tons of ivory. By 1900 between 60,000 and 70,000 elephants were slaughtered every year to satisfy the demands of the European market.

The market for ivory was unlimited, but the source of ivory was not. Elephants were massacred in China, Africa, Far East Asia and India to meet the world requirement. Indian Ivory carving and handicrafts are famous all over the world. Capture and poaching of tuskers (elephants with tusks) for ivory has taken a heavy toll in north-eastern and southern India.

Once Periyar (Kerala) was famous for its tuskers, today not many of them are left. There are about 7,500 ivory craftman in India, after a ban by Government of India in 1972 on Wild-life products, today trade in ivory is controlled, there are registered Wild-life product dealers who procure license or take permission from Government of India to pursue this trade, the buyers of ivory goods have to carry photocopy of that permission to get clearance at major world airports. Controls have not succeeded in eliminating misuse, fraud and poaching. Because products bring in lots of money hence illegal trade remains in operation.

Bones, teeth, horns of many other wild animals have also been used *e.g.* teeth of Whales and Walrus tusks are used as decorative pieces and for carving. Dikdik are killed to get horns which are sold for use as buttons and as good luck charms, Venice, is one of the centres of this business. Bone jewellery has also been common among tribal people of various countries.

Decoration

Display of animal trophies and stuffed animals, specially tigers, lions, cheetah bear and leopards was a royal hobby, palaces of kings of India and abroad still retain and display them. Leopard and other small cats are killed, stuffed and displayed as decorative pieces in the spacious rooms of rich and wealthy persons. Hands and feet of bears, apes, are used as decorative pieces and as ash trays. On Isles des Saintes in Gaudeloupe there were once many Iguanas, the high tourist demand for stuffed Iguanas as souvenirs has pushed these harmless lizards to near extinction. In south America the young crocodiles are killed, stuffed and sold as souvenirs.

The elephant legs are cut, stuffed and are used for sitting. Feathers of big colourful birds specially peacocks, decorate many houses. Polynesians use unusual tail feathers of colourful birds to decorate their hair. Antlers and horns of Sambhar, cheetal, swamp deers, wild buffaloes are displayed in houses; people also make handles of cutlery, umbrellas, revolvers using bones and horns of rare animals to make them unique and expensive.

Medicinal Property

Rhino horn is believed to cure headaches, colds and increase sexual powers, it would be interesting to note that it consists mainly of Keratin for which there is no need to kill a rhino, a person can obtain this by eating hair and chewing his or her own nails. It is a strong but wrong notion that the horn of the rhino is a powerful aphrodisiac, this legend is based on the phallic shape of the horn and also on the fact that the mating act in Rhino lasts for 1 to 2 hours.

The Rhino horn was used as a drinking glass by royals because it was believed that it would split if poison was added to drink. People in South America, far east Asia and south India eat brains and meat of rare breeds of monkey with an assumption that it cures insane people and give vitality. The Tiger bones are also believed to carry aphrodisiac property if eaten in powdered form, Tiger nails worn around the neck are believed to give boundless courage to the wearer. The extensive use of Rhesus, Chimpanzee and other non-human primates in bio-medical field such as in Vaccine, Contraception, Cosmetics, Space research and in Biological warfare has been very well known.

Perfume Industry

This industry thrives on scent collected from animals like Musk, Civet and many other such animals. To get just a few milligrams of scent, glands are extracted from hundreds of animals. The musk is used as an important ingredient in colognes soaps, hair tonics, body lotions and other cosmetics. In south India, Peacocks were killed by local people to eat meat and to take out oil for some medicinal value. Tiger fat, Rhino's urine, Uromastix oil are used as cure for rheumatoid arthritis, gout, purification of blood and for increasing sexual powers.

Dress Material

Skins and furs are important items to make coats, shoes, bags, gloves, belts, watch straps and caps from animals like mink, panther, fox, Sambhar, jungle cats and reptiles. The reptilian skins are the symbols of status among the ladies of high aristocratic societies, the atrocity is that the furriers insist on furs in excellent condition. There are different ways poachers use to kill animals and gather products. Tribals use bows and poisoned arrows to kill animals but sophisticated poachers use metal traps or rope noose because it is a silent method albeit more painful. Use of sophisticated weapons like automatic guns and helicopters for large scale gathering of animals and collection of their products is also adopted by rich poachers.

The animals are killed in a particularly horrifying way *e.g.* when a Leopard is trapped, it is tied, and then pressed by a plate from top or dorsal side so that it spreads on all four limbs called eagle spread, then a red hot iron bar is inserted into the Leopard's anus and pushed recklessly here and there to destroy some vital organs causing death of the animal but leaving its fur uncut, intact and shinning for the buyer to get a dress stitched out of it. Asian Crocodiles are killed for skin to make shoes and purses. These animals are skinned alive because the skin comes off more easily without stretching and cracking. The reptilian skins remain in demand because of the natural designs and patterns over them which can be tanned in different colours.

Pet and Amusements

Through Government trade statistics and previous studies India was known as one of the major bird exporting countries in the world. In India, birds are trapped primarily in the northern states of Bihar and Uttar Pradesh and also from Orissa, Madhya Pradesh, Maharashtra and Assam. The birds are caught and killed for meat and feathers to make fancy sleeping bags, pillows, jackets, greeting cards and decorative pieces. But the best trade is export of live birds. The total number of birds exported from India between 1970-76 was nearly 13 million, an average of about 1,850,000 birds per year. USA, UK, Japan, Italy, France and Belgium were the chief importers.

Of 1,160 species of birds 289 species were exported. Their value ranged from few to hundreds of dollars *e.g.* Ring necked Parakeet (Psittacula krameri) and Sarus crane (Gras antigone) were sold in UK for $ 27.00 and $ 700.00 respectively. Initially there was no check on bird trade but in 1976 CITES (Convention on International Trade on Endangered Species) imposed ban on export of eight rare and endangered species viz Hill Mynahas (Gracula religiosa.G.intermedia, G. peninsularia,

G. indica), Green Avadavat (Amandava Formosa), Finn's Baya Weaver (Pioceus megarhynchus), Hill blue flycatcher (Cyomis banyumas) and Malabar Parakeet (Psittacula columboider). Export of rare birds can cause total disappearance of that species and on the other hand illegal entry of exotic animals can be disastrous. Outbreak of new castle disease was carried by such birds to domestic fowl in California in 1971 and that costed 56 million dollars to control.

Export of live and rare animals for the purpose of display in the zoo has been of common occurrence as zoo are a common man's window to the animal's world. But it involves handling and capturing of wild animals and their transportation over long distances. Keeping a pet also seems to be man's inherent tendency. Parrots, parakeets, koalas, snakes, lizards, apes, monkeys are some of the pets owned by many persons. Hunting is also a source of amusement for human beings. Rare animals like deer, partridge, peacock, bear, shark, octopus, turtle, and fur bearing animals are killed for eating their meat and are considered a delicacy. Almost all fur bearing animals are stuffed and displayed. Sea turtle shells are used to make jewellery and wall hangings, oil from Whales and other rare sea animals is used for cosmetics, their meat is also eaten, fats are used for soup and skin is used for making leather goods.

Wild-life is big business; Wild-life products are sold at very high rates in International markets. Purses, coats, belts, shoes of Crocodile, Snake and Lizard skins are sold for exuberant amounts. Exotic birds can get big amounts *e.g.* $ 10,000 are paid for a Hyacinth macaw (Brazil). Wallets of Crocodile hide were sold in 1981 for £ 250 a piece and today they can fetch $ 560 for a single purse, a fur coat made from Lynx can be sold in the market for $ 1,00,000 at fashionable shops in developed countries. Tiger skin can sell for $10,000-15,000. Rhino horn is sold at the rate of $ 450 an ounce or Rs.1,50,000/horn. A pair of Golden shoulder parakeet, Brown parakeet, and Naretha Blue bonnet parakeet could cost $ 12,000, 8,000 and 10,000 respectively. Musk extract from musk deer, which is used in perfumery industry is sold for $ 40,000 to $ 60,000 per kilogram. Ivory may go for $ 150 a kilo.

In India about 400-600 tigers are thought to be killed every year. In the world, there is a great demand of parts of tigers specially in China. The Chinese system of medicine uses all the dried and powdered parts of body of a tiger. For example :

Organ	*Usage*
1. Nose	for epilepsy and convulsions
2. Teeth	for rabies, asthma and sore on penis
3. Whiskers	for toothaches
4. Blood	for strong body and mind
5. Gall bladder	for healthy eyes.
6. Stomach	for stomach aches and better digestion
7. Tail	for skin diseases
8. Testes	for tuberculosis
9. Skin	for mental illness.
10. Fat	for vomiting, fissures
11. Flesh	for nausea, malaria, vitality
12. Hair	for driving away pests
13. Brain	for laziness
14. Eyeballs	for epilepsy, nervousness, convulsions

A poacher gets US $ 5000 for a tiger skeleton and upto US $ 300 per kg of tiger bones.

Status of The Tiger, 1994

Sub-species	*Minimum*	*Maximum*
Bengal (Indian) Tiger		
Total	3250	4700
India	2750	3750

Contd...

Sub-species	*Minimum*	*Maximum*
	Status of The Tiger, 1994	
Bhutan	300	460
Nepal	150	250
Caspian Tiger	Extinct in 1970s	
Siberian Tiger		
Total	150	200
China	Species present	
Russia	150	200
Javan Tiger	Extinct in 1980s	
South China Tiger		
China	30	80
Bali Tiger	Extinct in 1940s	
Sumatran Tiger		
Indonesia	600	650
Indo-chinese Tiger		
Total	1050	1750
Cambodia	100	200
Lao PDR	species present	
Malaysia	600	650
Myanmar	species present	
Thailand	150	600
Vietnam	200	300
Species Total	5080	7380

There are many countries which have put a ban on export of wild animals and their products. Export of primates, lizards and snakes has been totally banned in India. In USA the Lacey Act of 1935 was one of the first steps towards discouraging business in Wild-life products, this act prohibited even the importation into U.S.A. of any Wild-life acquired illegally from another country. Later in USA, in 1969, Endangered Species Conservation Act totally banned import of any animal which appeared on the list of endangered species. The first serious effort to control all Wild-life sales took place in Washington DC in 1973 when 80 nations met to draft the Convention on International Trade in Endangered Species (CITES) of Wild Fauna and Flora. It divided Wild-life into three appendix.

Appendix I: Dangerously threatened with extinction and totally banned from any trade *e.g.* Rhino and Tigers, Green sea turtle.

Appendix II: Potentially threatened but sale could take place; if shipment are accompanied by authorizing documents from the country of export *e.g.* American alligator.

Appendix III: Not endangered in the world at large, but at some countries their number is abundant, at some it is a rare animal *e.g.* the Scarlet macaw can be shipped from Panama, but Costa Rica bans its export because it is rare in that country.

Another big organization called TRAFFIC (Trade Records Analysis of Fauna and Flora in Commerce), monitors the movement of animal products around the world its offices are in London, Washington, Nairobi, Frankfurt and Tokyo. TRAFFIC keeps all the data about the animals and animal products being moved around the world. In India, to control Wild-life trade, four offices were established at Delhi, Bombay, Calcutta and Madras, the export of Wild-life products is possible only after getting clearance from the Wild-life regional offices, under the rules vide section 39-49 in the Chapter V of the Wild-life (protection) Act 1972.

Problems Regarding Wild-life Conservation and Management : In the past geological times, extinctions have taken place due to earth's upheavals, volcanic eruptions, ice age, periods of rains and draught, desertification, forest fires, but simultaneously new environmental conditions resulted in the evolution and speciation. This has been nature's own course. All the species are important and integral part of ecosystem and all the species of plants and animals are equally vital, our own species is subjected to evolutionary processes and hence the more we learn about the life processes and biology of animals, the better able we will be in managing our own evolution and biology. It is also true that we can not live without other living beings because we are all part of the unity called life on earth, if any part is destroyed the other parts will be harmed too. In recent times, man has been the sole and biggest causative factor for polluting and disrupting life on the entire globe and causing extinction of other living beings.

The newly added factor is militancy and terrorism, control of many national parks has gone into the hands of armed militants, for example Nagarjunasagar, Indravati, Kanha, Palamau, Duowa, Corbett, Manas and Kaziranga National parks are infested with militants. Management of such parks is getting difficult.

Man has been the biggest cause of environmental degradation. For years, in the name of development, fuels have been burnt, chemicals harmful to atmosphere and living beings have been dumped literally on land and in water and spread in air. The accidental oil spills in the oceans and burning of oil wells have played havoc to pollute earth's environment. Trees have been cut to the point of denuding the earth. "Lungs of the World" - the Amazon rain forest is being cleared at an alarming rate, as a result we are confronted with a phenomenon of global warming, and green house effect caused by excessive carbon-di-oxide accumulation and depletion of ozone layer in the atmosphere. Various official and non-official agencies have been crying out loudly to save the forest and Wild-life therein, but little notice was taken about it. Now, when the situation has gone worse even for human beings, environmental issue is gradually getting into the top priority and hence it was the subject of the largest ever Earth summit meeting held in Rio De Janeiro, Brazil in June, 1992.

The major causes affecting Wild-life and environment adversely have been extensive use of wood in making furniture, doors, windows and packing cartons, all segments of Indian society peasants, tribals, rural people, slum dwellers and industrialists-depend greatly on the produce of forests as the source of fuel, fodder, construction and industrial raw material. In India over, a lakh of trees are cut every year just to make wooden cases to pack fruits. Further, due to deforestation, industrialization, pollution, and ignorance, forests are vanishing at a rate of some 17 million hectares per year. Hydro-electric projects have proved disastrous to forest, because they need clearing of ground, construction of dam, settlement of labour and development of town.

By turning down the Silent Valley hydro electric project the conservationists could save one of the world's best biodiversity reserve. When India got its independence 33% of the land was covered with forest, by mid-eighties it came down by 10%. Agriculture, settlement, house constructions and rehabilitation of refugees have also been some of the reasons for India's depleting forest and Wild-life.

Nearly all the tiger reserves in our country are constantly struggling with the local population of humans and cattles. The population in India of human beings has increased by 30 crores and of live stock by 10 crores in the past two decades.

Villagers living inside or at the periphery of National Parks get fodder and wood from the Park, their live stock gives a tough competition in grazing to wild animals. It has been very difficult to explain it to a common villager that why is it necessary to save a Tiger and not his live stock or his child from the beast, why is it necessary to save trees in a forest by disallowing him to cut fire wood and let him sleep hungry. Initially it was thought that villages pose the toughest problem in Wild-life conservation but lately line of thought has changed because people realized that tribals and villagers have been living in the forests from time immemorial, they used to maintain the balance because their lives depended on forest and Wild-life. Many animals and habitats have existed because of villagers alone. *e.g.* Bishnois of Rajasthan worship Black bucks and give them protection, similarly Kokre-Bellur villagers live with Storks and Pelicans in their village, they provide total protection and consider them auspiciuos. In the Chipko movement in Garhwal, initiated by Sundarlal Bahuguna the villagers prevented felling of trees from wood contractors by simply hugging the tree trunks, this movement received International attention.

The Tribal people in India used to be saviours of forest and Wild-life, today 48 million people live within or around the forests because these two things have been their means of survival they know how to balance between the supply and demand, villages have been rightly called as Ecopeople; the problems began when villagers started exploiting it commercially for big merchants. In present context a well managed interaction between Government personnel and these people can be of immense benefit, good planning, and execution can save India's forests and Wild-life.

Our nature conservation policies need to be far more sensitive to the needs of the communities living in and around protected areas, these people must be fully involved in the management of Parks and Sanctuaries, assured employment and a share of tourist revenues, and adequately and promptly compensated for crop damage and man slaughter by wild animals, by creating community forest and pasture lands under the joint management of local people and forest officials, remarkable success has already been achieved in West Bengal by Village Forest Protection Committees and by Arabari Forest Range of Medinipur district. It is suggested by Gadgil and Guha, (1992 that for the social and ecological benefit following four steps should be taken into consideration.

(1) Maintain soil and water conditions.

(2) Maintain biological/genetic diversity.

(3) Provide fuel, fodder and other materials for local communities.

(4) Provide non-wood forest produce such as tendu leaves, sal seeds and pine resin, for local and National economy.

Management of habitat by ways of establishing more National Parks, Sanctuaries, Reserves and Zoological gardens can be one of the method to provide protection to Wild-life.

Breeding the near extinction animals in captivity and their reintroduction in their natural habitat has also proved to be an excellent albeit difficult and expensive way of conserving wild animals. Pere David's deer found in China was extinct in wild it was bred in captivity and then reintroduced in their natural habitat, Golden lion tamarin, an extremely attractive, small sized non-human primate used to be found in the jungles of Brazil, got extinct in wild, it was then successfully bred in captivity by a team of workers of Smithsonian Institution, in USA and was reintroduced in its natural habitat.

The urban population has to learn about preserving healthy environment for themselves and for remaining living organisms. Most people think that conservation plans are useless and in a poor country like India the priority must be given to people, not to wild animals. People must be educated via newspaper, television and radio about the significance of each animals in the delicate ecological balance, that all plants and animals are parts of a single string if it breaks it will have effect on every one. Audio-visual has the best impact on masses these days. Naturalists, Wild-life photographers and filmmakers have an important role to play in carrying the message of conservation to the common people. Making general public aware of environment, protection of earth and creature therein specially human beings themselves in a popular way by showing movies could catch the attention of a commoner very fast. Education at school level about delicate ecological balance, significance of each trophic level, conservation of gene pool, maintenance of gene diversity is very essential.

First making the public aware and then involving them at root level will greatly benefit. Advertising companies can also contribute substantially. The Bureau of Indian standards has decided to put the "Echo-mark" on soaps, detergents, textiles, aerosoles, food additives and cosmetics along with many other products to indicate that this product is not hazardous to environment while it is manufactured and after its use.

The packings of cosmetics have already started bearing the note that the product has not been tried on animal. Or no animal has been killed to test the product. This is catching up and more and more people are getting aware of it, similarly, use of recycled paper, and less use of plastic bags is being popularized through media. It is assumed that people will use more and more of such products to save environment, forest animals and subsequently themselves.

Protection of Wild-life does not mean just preservation, it actually means conservation. Each given area has certain carrying capacity, meaning, it has fixed space and food to sustain certain number of animals, now if all the animals in that area are strictly preserved, their number is likely to exceed the carrying capacity, the animals may in turn suffer lack of food and struggle for space.

In USA, UK and Australia the population of game animals is kept in check by licensed hunting. In India any kind of hunting is totally banned. Total protection to Nilgai has caused significant increase in their population, in Buxar (Bihar) they roam around freely and destroy cultivated lands; hunting of few/animals in a restricted way may help conservation.

What we also need to do is to save the global environment and life therein by making

(1) A shift from fossil fuels to efficient solar based energy systems.

(2) New transportation networks and city designs that reduces automobile use.

(3) Redistribution of land and wealth.

(4) Control on population growth.

(5) More funds for Wild-life projects.

(6) Better coordination between various agencies involved in conservation work (Gadgil and Guha, 1992).

The wider concept which is coming up now is of having Biosphere Reserves, which has been evolved under UNESCO's Man and Biosphere (MAB) programme. It has three main objectives

(1) Conserve diversity and integrity of biotic communities of plants and animals to safe guard the genetic diversity.

(2) Provide areas for ecological and environmental research.

(3) Provide facilities for education and training.

Research and Training

Forest departments have been training their officers in the field of Wild-life - a Directorate of Wild-life Environmental Research and Education has been established at the Forest Research Institute, Dehradun, which is conducting a post-graduate diploma course in Wild-life management for professional foresters. There are short term courses for foresters run by Indian Forest College and Forest Research Institute, Dehradun.

Promotion of Wild-life studies and research in the Universities has also been taken up by University Grants Commission (UGC), New Delhi to promote education in Indian Universities; UGC recommends introduction of Wild-life study as a subject at the B.Sc level and M.Sc level. UGC funds projects pertaining to Wild-life studies.

There are only three institutions in India which offer full courses in Wild-life:

AVC College, Mannampandal, Mayiladuthurai, Bharathidasan University, Tiruvar, Tamil Nadu started M.Sc. Wild-life Biology course in the year 1980. It was the first college in India to offer a degree course in this field. The study of Wild-life Biology comes under the Department of Zoology.

Wild-life Institute of India (WII) Dehradun, Uttaranchal was set up in 1982, under the Ministry of Environment and Forests of the Government of India. Initially, it promoted research and conservation activities. It started its education programme in 1987 with the introduction of M.Sc. course in Wild-life biology. One of the primary responsibility of Institute is to train inservice personnel of the State Forest Department in the Wild-life management via a months post graduate diploma or 3 months certificate course. For this, Institute conducts short term and long term courses.

Centre of Wild-life and Ornithology, Aligarh Muslim University, Aligarh, was established in 1986 as a department of post graduate studies and research. It has the distinction of being the first University devoted to Wild-life studies in the Indian University system. It offers M.Sc. M.Phil. and Ph.D. courses in Wild-life Biology.

Salim Ali Centre for Ornithology and Natural History, Coimbatore, has started recently in 1991. Presently it is devoted to research in pure and applied Ornithology and Natural History.

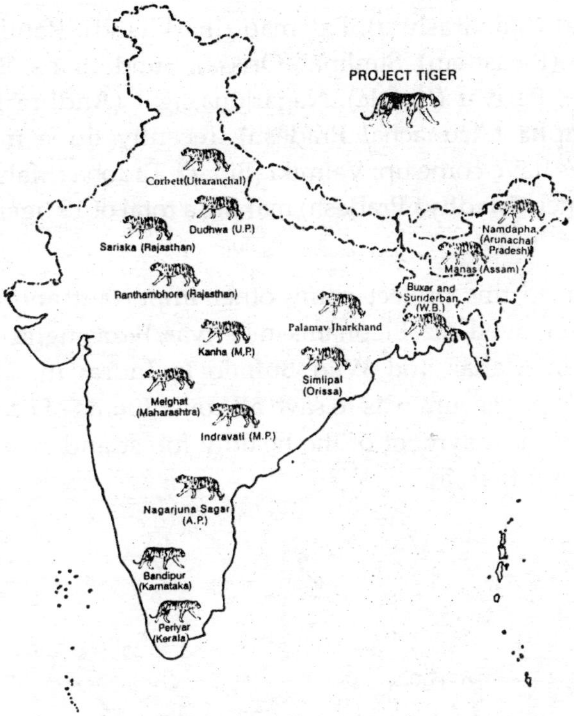

Locations of Project Tiger in India.

Major Projects

Project Tiger : The Tigers (Panthera tigris) had reached the brink of extinction, hunting was one of the major reasons, along with habitat destruction due to agriculture, timber business, industrialization and fire wood. In 1972 tiger was declared as National animal in place of lion. An International movement, initiated by World Wild-life Foundation, supported by then Prime Minister Mrs. Indira Gandhi was started in 1973 in Reserves to save Tiger with an expected population of 268 Tigers. By 1981 their population rose to 757 within seven years in 1988 the estimated number went up to 4000. Presently there are 16 Tiger reserves : Corbett Uttaranchal, Dudhwa (U.P.), Bandipur (Karnataka), Kanha, Indravati (M.P.), Manas (Assam),

Melghat (Maharashtra), Palamau (Jharkhand), Ranthambhor, Sariska (Rajasthan), Simlipal (Orissa), Sunderbans, Buxar (W. Bengal), Periyar (Kerala), Nagarjunasagar (Andhra Pradesh), Namdapha (Arunachal Pradesh), recently three more tiger reserves have come up: Valmiki (Bihar), Tadoba (Maharashtra) and Pench (Madhya Pradesh) making a total of 19 tiger reserves in all, in India.

Due to this project many other important animals have also been saved, *viz.* Elephants in Periyar, Barasingha in Kanha, Rhino in Manas and Wild Buffalo in Indravati. The Tiger project's prime aim was to save all constituents of eco-system. The Tiger is a symbol of the healthy forest and all the living organisms therein.

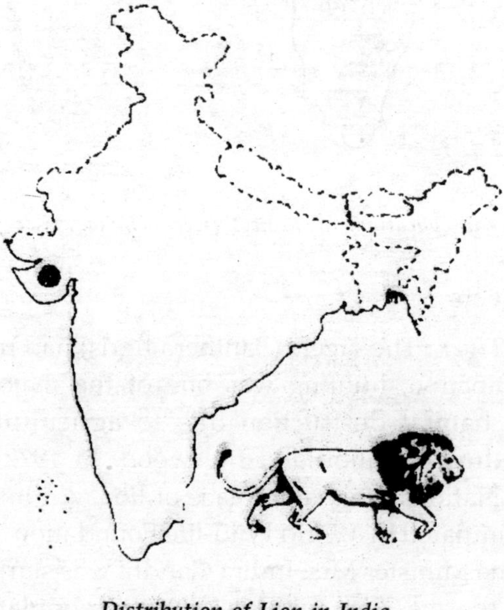

Distribution of Lion in India

Project Lion : Lions (Panthera leo) were once found over the whole of northern and central India but it is now restricted to the Gir Forest in Kathiawar, Gujrat. The project to save them was started in 1972. In 1947 their population was 180. A core area of 258.7 sq.km was declared as National Park. By 1979 their population increased to 205. Initially the main cause for depletion in their population was hunting, but now it has been grazing because thousand of cattle are left inside the Gir Forest during monsoon months, they not only compete for fodder with wild animals, they pollute the water and spread diseases. This causes reduction in herbivore.population and subsequently affects the Lion.

Due to scarcity of natural food, Lions prey upon livestock, which creates direct conflict between cattle owners and park managers. Due to such reasons, the Lion population has become unstable and the project has not been a success like Project Tiger.

Project Crocodile : Government of India started this project in collaboration with FAO and UNDP on recommendation of Dr. H.R. Bustard in 1975. India has only three species of Crocodiles: Gavialis gangeticus (Gharial), Crocodylus palustris (Mugger) and Crocodylus porosus (Salt water Crocodile). All three were declared endangered in India in 1971. Large scale hunting for their skins has been one of the biggest reasons for their disappearance, fisherman may also kill them due to difficulty in fishing. The first Crocodile breeding farm was set up in Orissa, later on breeding farms were started in many other States.

The main aims of the project were to (i) locate the best Crocodile areas (ii) collect eggs soon after they are laid in nest (iii) transporting them to a breeding/hatching station (iv) rear the hatchlings (v) release them back into the open water when they are ready to lead an independent life. Between 1976 to 1980 about 2700 Gharials, 600 Saltwater Crocodiles and 900 Muggers have been hatched at different rearing stations and a total of 324 Gharials, 168 Salt water Crocodiles and 218 Muggers have been released into the open water. Today Crocodile breeding is managed successfully in following

Sanctuaries : Bhiterkanika, Hadgarh and Satkosia Gorge (Orissa), Katernia Ghat (U.P.), Krishna, Lanjumaduga, Manjira, Papikonda (W. Bengal) and National Chambal Sanctuary spread over M.P., U.P. and Rajasthan. The biggest non-governmental Crocodile farm is managed by R. Whitakar in Madras.

Project Hangul : Hangul (Cervus Hangul) is found in Jammu, Kashmir and Himachal Pradesh. Its concentration is in Dachigam Sanctuary. This animal has become rare mainly due to human interference in the form of wood cutting, poaching, grazing etc. From an estimated number of 3000 in 1940, it was reduced to 170 by 1970. In this very year Hangul project was launched by Government of India with the help of IUCN/WWF. Because of effective management their population increased from 320 in 1978 to 347 in 1980 (Holloway, 1970). Besides this, it is estimated that only about 30 more Hanguls have survived outside Dachigam (Saharia, 1981).

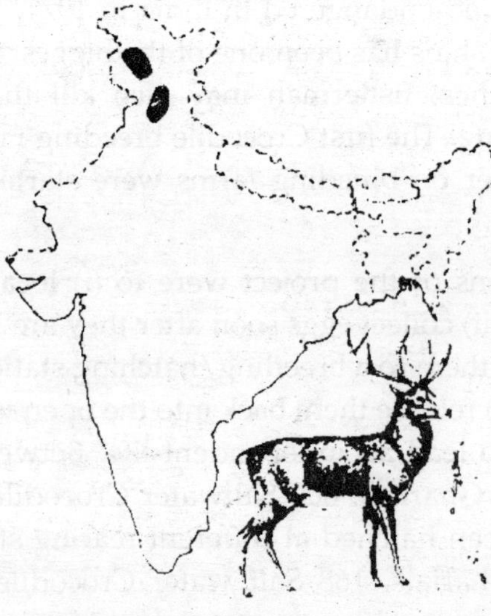

Distribution of Kashmir stag in India.

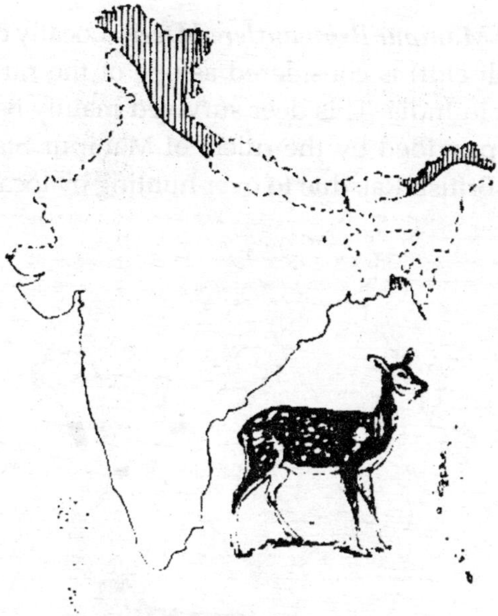

Distribution of Musk deer in India.

Project Musk Deer : Musk deer (Moschus moschiferus) was once found from Pakistan, through North India, Nepal to Bhutan and Burma and also into parts of Tibet and China. Today, in India, it is found in small population in Jammu-Kashmir, parts of Himachal Pradesh and Uttar Pradesh. A unique feature of this deer is the presence of a musk gland in the male, thick, brown secretion of this gland "musk" is used in preparing medicines, and perfumes.

It is sold at a very high rate in International markets (40,000-59,000 Rs. per kg). Habitat destruction has been the main cause for their decline and their population is now restricted to some isolated, protected areas in India. As part of the threatened deer programme of the IUCN and WWF was launched in 1974. The project was started in Kedarnath Sanctuary (U.P.) where enclosures have also been constructed at certain places to carry out breeding in captivity for the purpose of reintroduction into the wild.

Project Manipur Browantlered Deer : Locally called Sangai (Cervus eldi eldi) is considered as one of the rarest mammal found only in India. This deer survived mainly because of the protection provided by the rulers of Manipur State, but later during the British Raj, due to over hunting by local people and

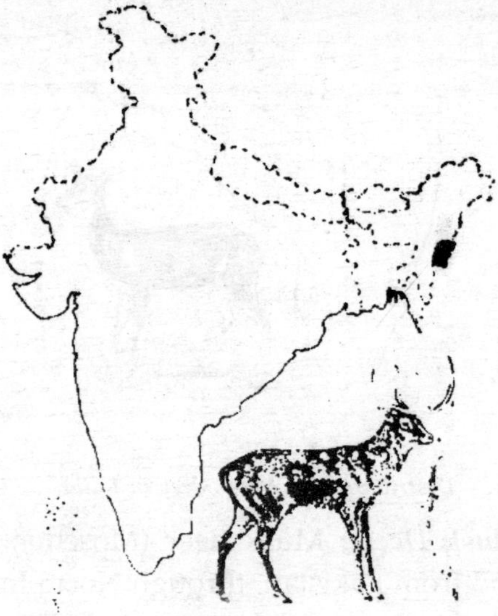

Distribution of Manipur browanilered deer in India.

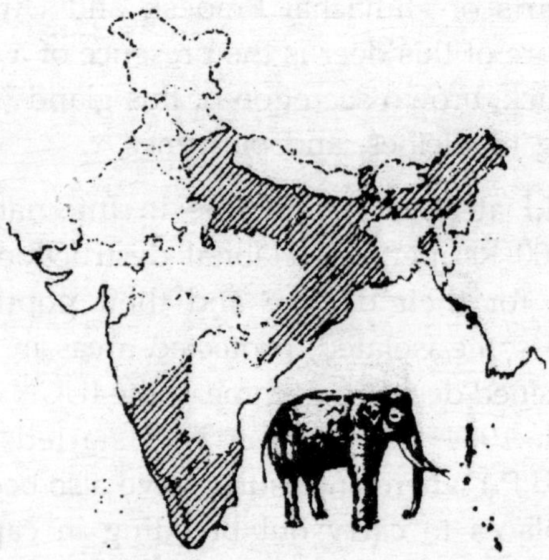

Distribution of Elephant in India.

army personnel it was declared extinct in 1950. Later, it was located once again in a small area called the Keibul Lamjao (Manipur) which was declared as Sanctuary in 1954, and in 1977 was given the status of a National Park. Their survival in this area is because of the floating swamp through which this deer manages to move far better than man and cattle. The National Park covers an area of 35 sq. km only. From an estimated number of 100 deer in 1960, it was reduced to 18 in 1977 which increased to 23 in 1980's. However, more numbers (49) have been preserved in captivity.

Project Elephant : The project on Elephants (Elephas maximus), launched recently by the Ministry of Environment and Forests provides basis for restoring the lost and degraded habitats of this largest land animal. This would include creation of migration corridors, establishment of a data base of their population dynamics. Today there are about 18,000 to 22,000 elephants in India. Elephants in India are found naturally in the states of Assam, Arunachal Pradesh, Meghalaya, Tripura, West Bengal, Jharkhand, Orissa, Uttar Pradesh, Karnataka, Kerala and Tamil Nadu.

Project Lion Tailed Macaque : Macaca silenus is the zoological name of a highly endangered species of a monkey (non-human primate) found only in India that too in a very small geographical region of western ghats of Kerala. As the name suggests this monkey has a tail resembling to that of a lion, but more than that it has a mane around the head like a lion. The colour of the mane is brownish gray, whereas, colour of the body is dark black. The size of the body is not bigger than a rhesus (the red faced monkey - Macaca mulatta), but in overall appearance it is bigger than rhesus because of its thicker fur coat. From head to body the males measure about 51-61 cms; the tail alone measures approximately 46 cms. Like other monkeys lion tailed macaque also has bare face, palms and soles of gray colour. Their young ones have

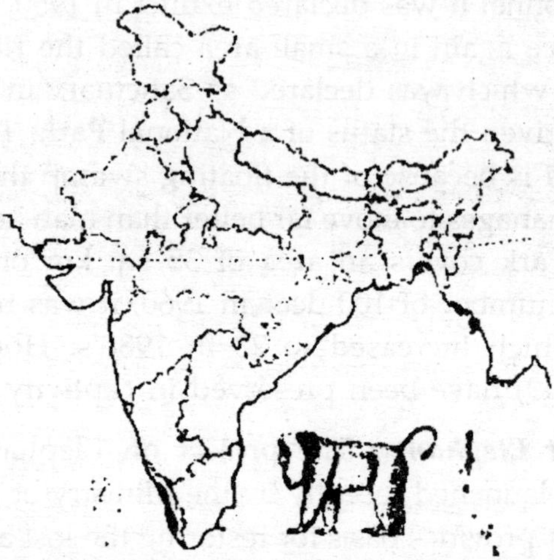

Distribution of Lion tailed macaque.

pink face, palms and soles. These animals are very shy in nature, they prefer living from human beings. They live on high canopy of trees of wet evergreen forests. Because of this reason very little scientific studies have been possible on them. These animals have become a great concern of governmental and non-governmental agencies related with nature. Many projects have been launched to save them and their habitat.

After executing much research only little information has been gathered about their behaviour so far. Lion tailed macaques feed on fruits, insects, frogs and sometime fungus. The total population of lion tailed macaques is only 735 to 750 in the entire state of Kerala. In Western Ghats its range is restricted to Nilgiri, Anamalai, Cardamom hills, Periyar National park and silent valley. Actually lion tailed macaques came to lime light when the controversy over the silent valley hydro-electric project was raging the country. These monkeys became one of the key animals in getting this controversial project ultimately dropped.

Few research projects which have been carried out on lion tailed macaques (LTM) have thrown some light on their activities. According to these studies LTM are arboreal, diurnal. Their sighting is difficult because of their shy nature, slow movements through thick foliage of tall trees, Their black fur coat is further difficult to locale in dimly lit surroundings. These animals can hardly be located by untrained eyes.

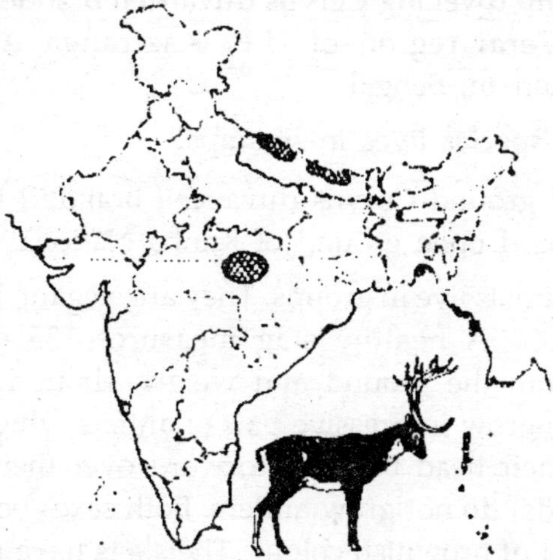

Distribution of Swamp dear

LTM live in groups, their group size can be any where between 4 to 30. Each group has 1-3 adult males, remaining members are adult females with or without new borns, subadults and juveniles. These animals move in a small home range of about 2 sq. km. Females in estrus do not show any conspicuous swelling of sexual skin at hind quarters (which is typically seen in other macaques). The mating takes place in trees, extensive grooming and embracing has been observed in consorting or mating couples.

Very little is known about territoriality, dominance hierarchy and mother infant relationship.

Many important research projects are going on to protect LTM in their narrow geographical range.

Project Swamp Deer **:** Swamp deer are commonly known as Barasingha, because of their large beautiful set of antlers. They are found only in India and are declared as highly endangered species. There are two sub-species of Barasingha which are found in very small areas of India -.

(1) Swamp dwelling Cervus duvauceli branderi found in the Terai region of U.P. Kaziranga Assam and Sunderban, Bengal.

This sub-species lives in swamps.

(2) Hard ground Corvus duvauceli branderi is found in the hard open ground of Kanha National Park, M.P.

These animals live in groups. They are elegant, large sized, beautiful deer. A healthy stag measures 135 cms at the shoulders from the ground and weighs about 175 kg. The males (stags) grow impressive pair of antlers which typically grow over their head bending forward over their face. The females (hinds) do not grow antlers. Both sexes bear shining, thick fur coat of brownish colour. The stags have mave (long hair on the underside of neck) and are darker than hinds. Light spots on the body are seen in young ones, the adults do not have such spots.

In Terai, swamp deer live in marshy land and do not like to come out of water. Opposite to this, in Kanha National Park the sub-species Cervus duvauceli branderi inhabit grass covered open dry fields.

Swam deer live in groups of 20 - 80 individuals. They are active in early mornings and late evenings. They rest in day time and nights. They are alert animals and give warning signals to each other when a predator approaches them. More than eyes and ears they rely on sense of smell, which is very acute in swamp deer.

Few decades ago, these deer were found widely distributed in India. Swamp deer are not found any where else in the world. Their population decreased consistently and today it is very alarmingly low. Many projects have been started to protect these highly endangered deer from getting extinct. Human interference has been the major cause for driving these elegant animals to small protected forests of Tarai, Kaziranga and Kanha. Success has been achieved by putting large enclosures around their populations to render them perfect protection from human and predators. Extensive studies are being carried out on Swamp deer by the research personnale of Wild-life Institute of India, state forest departments and other non-governmental organisations. Behavioural observations on swamp deer in these protected regions have revealed important information of them. Some are as follows:

(1) Barasingha females become sexually mature in 36 months and males in 18 months after their birth.

(2) Swamp deer breed in winters, this is known as their rutting time.

(3) The antlers of males drop after the breeding is over. Sometimes these dropped antlers are eaten by pregnant females to supplement calcium in their diet.

(4) Adult dominant males male with 'ready' females. Wallowing (loud, deep throated, long calls) is a peculiar activity of adult male Barasingha during and after mating.

(5) After mating these adult dominant males stay for few days' with females then leave to join an all male group.

(6) Related females who form the core of the group stay back.

(7) The gestation period is of 245 days.

(8) The babies are born very weak, they gain strength in 2-3 months, but like to remain hidden till they are 4-5 month old.

(9) The males who had left the females after rutting also shed their antlers.

(10) When the summers start the stags start growing their antlers, as the months pass by the antlers keep growing but remain covered by a soft skin called 'valvet'.

(11) After the rainy season in the month of Sept-Oct the stags start rubbing their antlers against trees, tearing and removing the velvet from their anthers. During this process they also sharpen the edges of antlers. Because of the availability of such informations it has been possible to control their reproductive behaviour and in turn it has been possible to have more young ones and juveniles. The population of swamp deer is not declining any more.

Wild Life of Rajasthan

When we think about Rajasthan few things make lightening appearance in our mind-dry, desert, thorns, little vegetation, scarcity of food and water. In the name of animals people perhaps can just think of camel. Above mentioned things definitely form an integral part of this desert state of India but that's not all. Rajasthan does not mean desert alone, it includes hills of aravalli, slopes of vindhyas, maze of ravines, Chambal river, many lakes, marsh lands, vast lands and of course above all-extremely hot and arid Thar desert. In spite of harsh climatic conditions, Rajasthan supports a variety of flora and fauna.

The great Indian desert or Thar desert spreads to Gujrat and Rajasthan states. The vegetation in desert regions of Rajasthan is mainly of Rohira (Tecoma undulata), Khejri (Prosopis sinreria), Thor (Euphorbia cadvca), Aak (Calotropis procera). Representatives from most of the insect orders are found in Rajasthan. There are velvet mites, grass hoppers, variety of beetles, butterflies and months. 43 species of reptiles including snakes, tortoise, varanus, uromastix, skinks and crocodiles are found in this desert state of India.

Avifauna of the state is perhaps the richest in the country. The world renowned heronry (a place where herons breed; generally referred to a bird sanctuary) Keoladeo National Park is located in Bharatpur, Rajasthan. This park will be described in details further in this chapter. It is necessary to high light, here that even outside the famous keoladev there exists a rich avifauna in certain other regions of Rajasthan. One only needs a keen eyesight and a bit of interest in the nature. Thousands of Greater Flamingo and Lesser Flamingo visit Rann of Kutch, Cranes, Herones, Ducks, King fishers, Woodpeckers, Shrikes, Drongoes, Larks, Babblers are regular visitors of various ponds and lakes of Rajasthan such as Chhaparwada, Jalmahal, Ramgarh, Jaisamand, Ramsagar and Silisesh (detailed ahead).

Among other birds, the most important is the Great Indian Bustard, (Ardeotis nigriceps) which are inhabitants of desert. This bird has been declared as 'The state bird'. Rajasthan also has Sand goose, Quails, Parakeets, Ring doves, Pigeons, Bee eaters and of course a moderate population of Peafowls (Pave cristatus - our National Bird). Among mammals Rajasthan is famous all over the world for tigers. Ranthambhore and Sariska National Parks are two of the very best-tiger reserves we have in India.

Rajasthan has large populations of Black Bucks (an antelope) which enjoys religious protection of Bishnoi community of Rajasthan. Chinkara, Neelgai, Sambhar, Chital roam around freely in the protected forests of the state. Small population of highly endangered Wild Ass is found in Rann of Kutch. Gerbils (rodents), Hares, Rabbits, Hedgehogs, Porcupines, Wolves, Jackals, Hyenas and Wild Boars are also found in this dry state of camels. Rajasthan state has three national parks, many sanctuaries, reserves and protected forests, some of them are described in this chapter.

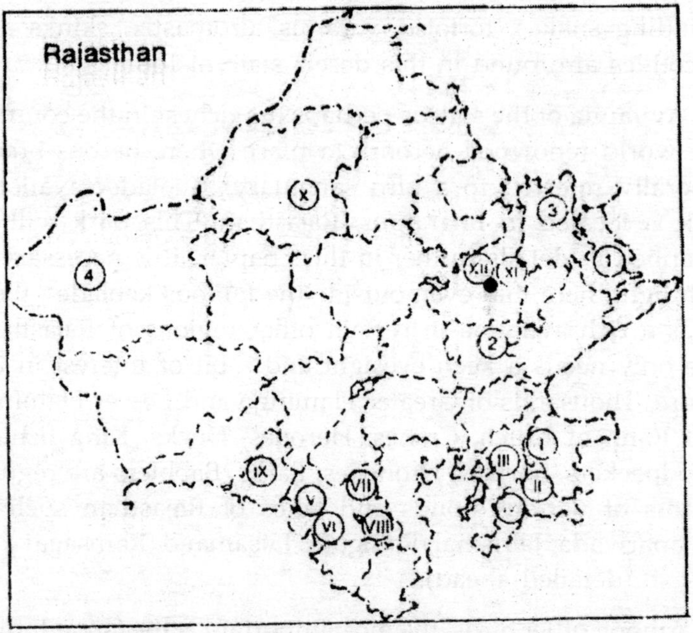

Outline map of Rajasthan indication locations of National Parks and important sanctuaries.

(1) Kaeladeo National Park, Bharatpur, (2) Ranthambhore National Park, Sawai Madhopur, (3) Sariska National Park, Alwar, (4) Desert Biosphere Reserve, Jaisalmer, (i) Shergarh Sanctuary, Baran, (ii) Ramgarh Sanctuary, Bundi, (iii) Darrah Sanctuary, Kota, (iv) National gavial Sanctuary, Kota, (v) Jaisamand Sanctuary, Udaipur, (vi) Kumbhalgarh Sanctuary, Udaipur, (vii) Sajjangarh Sanctuary, Udaipur, (viii) Sitamata Sanctuary, Chittor, (ix) Mount Abu Sanctuary, Mount Abu, (x) Talchapar Sanctuary, Bikaner, (xi) Jamva Ramgarh Sanctuary, Jaipur, (xii) Nahargarh Sanctuary, Jaipur.

1. *Keoladeo National Park, Bharatpur:* Earlier it was known as Bharatpur bird sanctuary or Ghana bird sanctuary. In 1982 it was declared as a Keoladeo National Park (NP). In 1985 the NP was honoured by UNESCO with a title of 'World Heritage.' It is a world renowned park and is considered one of the finest and richest heronries. This NP spreads around a shallow fresh water marsh surrounded by grasses, herbs, scrubs and mixed deciduous trees. The total area of NP is about 29 sq.

km, this park is 370 meters above sea level. The park has:

(i) Resident birds which can be seen all through the year.

(ii) Migratory birds which visit the park temporarily.

Over 350 species of birds find a refuge in the shallow waters of this park. The lakes of the park get their water from rains (average rainfall is 26 inches per year) and also through rivers Gambhir and Banganga. The NP is almost full with Babul trees (Acacia nilotica). The lakes have Tortoise, Otters, Frogs, Water snakes, Fishes, Leeches, Snails and also many Insects.

Besides a large variety of birds which have been already mentioned in this book there are Storks, Egrets, Spoonbills, Cormorants, Plovers, Sandpipers, Snipes, Warblers, Pipits, Vultures, Eagles and Cranes. Pride of the park is the Siberian Crane.

Bombay Natural History Society (BNHS) has done wonderful indepth studies on various aspects of Keoladeo National Park along with the state Forest department. BNHS has listed in their publications the presence of Pythons, Hyenas, Wildboars, Black Bucks, Mongoose, Wildcats, Toddy cat (Common palm civet), Jackals, Foxes, Porcupines, Hares, Sambhar and also Neelgai.

2. *Ranthambore National Park:* It is a world famous tiger reserve. This is located amidst the Aravali and Vindhya mountains range. An attractive lake adds beauty to the whole park. The area of the park is 390 sq. km and it lies 214 meters above sea level.

Ranthambore is famous for its tiger population (43 to 54) and there is plenty of prey population of Sambhar, Chital, Chinkara, Neelgai, Wildboar and Langoors. There are others predators too such as Leopards, Jungle cats, Caracal, Sloth bears, Crocodiles, Jackals and Hyenas.

Ranthambore supports a moderate population of Wildboars, Indian hare, Mongoose, Monitor lizard, Foxes, Flying foxes and Snakes. There are also Storks, Spoonbills, Green Pigeons, Variety of ducks, Jacana, Fly catcher, Quails, Sandgoose, Partridges, owls and Eagles.

For better management this NP is divided into two zones.

(1) Larger and important zone with an area of 285 sq. km is also known as the core area (an area which is extensively used by wild animals) which is protected actively by State Forest Department. The government does not allow grazing, wood cutting, tresspassing and village habitations within this area.

(2) Smaller, less important zone with an area of 107 sq. km is known as the buffer area. It is an area which is shared with domestic cattles and villagers and is not used much by wild animals.

3. *Sariska National Park:* Previously known as sariska tiger reserve. This park has an area of approximately 800 sq. km. The park spreads into the forests of Aravali hills near Alwar. This forest was the favourite hunting ground of Alwar Maharaja. In 1958 it was declared as a sanctuary, then it became a tiger reserve under the scheme of Project tiger in 1979. The status of National Park was given in the year 1982. The total area of this park is also divided into two zones (1) core (480 sq./km) and (2) Buffer (320 sq./km). Like Ranthambore, Sariska is also very famous for its tiger population. Leopards, Jungli cats, Jackals and Hyenas are also other common predators. Caracals and Wild dogs have also been sighted in this park.

Among the prey population are the Sambhar, Chital, Chinkara, Neelgai. Occasionally four horned antelopes have also been seen. There is a moderate population of Wildboars, porcupines, Bats, Chamelion, Skinks and Snakes. There are plenty of Peafowls, Partridges and other varieties of birds.

4.*Desert Sanctuary/biosphere Reserve:* Large area (3,210 sq./km) of thar desert has varied flora and fauna, but only a small area (360 sq./km) was declared as sanctuary in 1980. This sanctuary is 30 km away from Jaisalmer.

Black bucks and Chinkara are found in moderate numbers. There are many Bishnoi villages in the sanctuary area who protect these animals. Small populations of Wolf, Desert cat, Desert fox, Gerbils. Long nosed hedgehogs, Saw scaled viper, Sand lizards and Varanus are found. Desert courser, Great Indian bustard and Lesser bustard breed and nest here. The lagger falcon, Tawny eagle, Short toed eagle, Spotted eagle, Sand grouse, Sparrows, Demoiselle cranes are seen hovering in the scorching heats of that desert.

The desert sanctuary is very famous for its rich source of fossils of plants and aquatic animals.

There is a proposal to convert this area to a Biosphere Reserve to conserve the biological diversity under the Man and Biosphere Programme of UNESCO. The main objectives of the Biosphere Reserves are to - (i) conserve diversity and integrity of plants, animals and micro-organisms, (ii) promote research on ecological conservation and other environmental aspects, (iii) provide facility for education, awareness and training. Rajasthan has many sanctuaries, reserves and protected areas. Only few important sanctuaries are described in the following paragraphs.

(a) *Shergarh Sanctuary:* It is located in the distinct of Baran. There is a patch of forest covering 100 sq. km giving shelter to Sambhar, Chital, Jackal, Foxes, Hyenas, Snakes, Skinks, Chameleon and few Leopards. In 1982-83 this forest was given the status of a sanctuary.

(b) *Ramgarh Sanctuary:* Close to Shergarh, there is another forest in Bundi district, covering an area of about 300 sq. km. This forest is the abode for Leopards, Sloth bear, Mongoose, Wildcats. Wildboars, Neelgai and Chinkara. In 1987 this forest was taken under the protection of wild-life cell of State Forest department. The other commonly found animals are Snakes, Chameleon.

(c) *Darrah Sanctuary:* Near Kota in Hadoti region is an enchanting forest reserve. It sprawls in 266 sq. km area. It is a thick jungle and once upon a time people had sighted tigers in this beautiful forest which has plenty of water resources and variety of animals viz Neelgai, Sambhar, Chital, Sloth bear, Jackals, Hyenas, Langoors, Hares, Porcupines, Hedgehogs and crocodiles.

(d) *National Gavial Sanctuary:* River Chambal flows through Kota division. The whole length of the river from Rana Pratapsagar to Jamuna has been declared as a sanctuary for Gharial. Government of Rajasthan supports an extensive project for the protection and breeding of Gharial along the banks of Chambal river near kota. This unique 'flowing sanctuary' of Rajasthan has a rich variety of Fishes, Water snakes, Tortoise and Insects.

(e) *Jai Samand Sanctuary:* Udaipur division of Rajasthan state is considered the most beautiful because of its lakes, mountains, palacial buildings, finely carved temples and above all its lush green forest. The thick jungles provide shelter to wild animals. Many forest patches have been declared as sanctuaries in Udaipur division. This sanctuary includes an artificially created lake; it is 52 km away from Udaipur. In the rainy season the lake spreads in an area of 160 sq. km. It holds a sizable population of Crocodiles and Tortoises along with a variety of Fish. On the surrounding land and adjoining hills Chital, Chinkara, Wildboar and a vast range of Birds are found. Occasional sightings of Leopards are also reported from this sanctuary.

(f) *Kumbhalgarh Sanctuary:* Kumbhalgarh is a famous historic fort 120 km away from main city of Udaipur. It has famous Jain temples with awesome carvings on marble. Aravali hills and thick forest make the whole area serene and peaceful. During the rainy season the forest gets denser and greener, temporary water falls further add to the scenic beauty. The hills of Kumbhalgarh have many Sloth bears. The four horned Antelope, Wolves, Leopards, Wildboars, Porcupines, Neelgai, Chital, Chinkara, Hyenas and Jackals are found in this sanctuary.

(g) *Sajangarh Sanctuary:* Sajangarh is a tourist resort in proper city of Udaipur, it is a small palace built at the top of a hill clearly visible from all points of Udaipur city. This palace is surrounded by hillocks, covered with thick vegetation. The forested area including all these big and small hills spreading in 520 sq. km was declared as a sanctuary in 1985. This sanctuary has free ranging populations of Sambhar, Neelgai, Chital, Wildboars, Langoors and Rhesus.

(h) *Sitamata Sanctuary:* 200 km away from Udaipur city in Chittor near Pratapgarh this sanctuary is located. The forest in this region extends in 423 sq. km which also includes range of Aravali hills. This is the only sanctuary in Rajasthan State from where flying Squirrels have been reported. These Squirrels have thin flaps of muscles and skin between the fore and hind limbs. When they jump from trees they stretch their all four limbs using them for a smooth glide. Leopards, Sloth bears, Wildcats, Wildboars, Foxes, Jackals, Sambhar and Neelgai are also found in Sitamata sanctuary.

(i) *Mount Abu Sanctuary:* About 100 km from Udaipur is Mount Abu, the only hill station of Rajasthan State. It is 1220 m above sea level. In the jungles of Abu live Sloth bears, Wildboars, Jackals, Foxes, Langoors and Leopards. Speciality of this forest is the Grey jungle fowl. A big lake in the centre of the city has variety of Water birds, Crocodiles and Tortoise.

(j) *Talchhaper Sanctuary:* 200 km from Jaipur on way to Bikaner is located Talchhaper sanctuary. During the dry seasons it is characterized by flat, tree less, saline land. From distance this barren land sometimes look like a lake because of mirage. The characteristic of this sanctuary is that there are hundreds of Black Bucks. The other animals seen are Varanus, Uromastix, Partridges, snakes, and a large variety of Arthropods.

(k) *Jamva Ramgarh Sanctuary:* 30 km away from Jaipur there is a beautiful lake surrounded by hills from its three sides making it to be an excellent catchment area. The forest of this area is extensive and merges with the forests of Alwar district. Jamva Ramgarh was the hunting ground of Maharaja of Jaipur. In 1985 this forest was declared as a sanctuary covering 300 sq. km area. At one time many tigers roamed in these jungles but not any more. The vegetation is chiefly of Anogeissus pendula, A. latifolia, Cassia fistula, Acacia leucophloea, and Zizyphus spp, several types of grasses, shrubs and herbs are also found. This vegetation supports herbivores like Chital, Sambhar, Neelgai, Hyene, Wildboars, Desert cat, Jungle cat, Toddy cat, Jackals, Foxes, Mongoose, Porcupines, Hedgehogs, Cobras, Kraits, Vipers and a large variety of Molluscs and Arthropods.

The Ramgarh lake is a rich source of fish like Carps,

Catfishes, and Channa. There are Crocodiles, and Tortoise too in this lake.

Leopards, Caracal, Wolf, Varanus, Skinks, Hares, Spur fowl and Bats are also found in the forests of Jamva Ramgarh.

(*l*) *Nahargarh Sanctuary.* As evident from its name, the fort was built on top of a hill, surrounded by thick forest, where tigers roamed around freely. This jungle was also used as a hunting resort by princely state. Deforestation, construction of tarred roads, restoration of fort and other monuments, human habitation and other such factors such as tourism and led to the destruction of this densely forested area. In 1980 this jungle with an area of 70 sq. km was declared as a sanctuary. This region includes thick forest, and water sources viz Akera dam, Kukas dam, Tal sagar, Ram sagar, Maotha, Gujarghati Tal, Jaisla dam, Meena bandh, Babuka bandh and most beautiful Jalmahal. Nahargarh sanctuary has few Leopards, Neelgai, Jungle cat, Wildboars, Foxes, Desert cat, Jackals, Porcupines, Hares, Mongoose, Varanus, Snakes, Chameleon Skinks and a variety of Bats (Indian flying fox, Great Horse shoe bat, Indian false vampire, Painted bat and Indian pipisterella.

Animals in Indian Mythology

Nowhere in the world has there been a deeper understanding of the animals than in India. This is evident because down the long line of Indian history, the animals have come to be tolerated, respected and even worshipped, depending upon their significance in human life. Undoubtedly, it is the insights provided by the ancient Indian thought that has given them this special place in its culture.

The animals are personified, symbolised and celebrated in

art and literature, custom and ritual, fair and festival. Above all, animals find unique position in the mythology of this land where they are seen in the company of the gods.

In the context of India's vast religion and mythology, it is only appropriate to make a beginning with the elephant. Venerated as Lord Ganesh or Ganapati, the elephant headed god is invoked and propitiated before beginning any enterprise, function or ceremony, for he is the remover of obstacles-just as an elephant would crash his way through an impenetrable jungle. However mighty, the elephant headed god is incomplete without its mount the rat. This is not surprising at all because by its very small size and powerful set of incisors, a rat is also able to cut his way but below the earth, much like its master does, only on the ground above.

Together, they are able to surmount any barrier. Hence man worships that power in nature, personified as Ganesh, that helps him overcome problems and difficulties, much like the combination of the elephant and the rat. The little rodent is often depicted sitting at the feet of Lord Ganesha.

Riding smoothly through space, we find Indra the king of gods in the Vedic pantheon, seated on a giant elephant called 'Airavat'. As the mount of Indra, Airavat is the celestial ancestor of all elephants on earth. Just as Airavat belongs to Indra the king of gods, so earthly elephants belong to temporal kings hence symbol of royalty, Naturally then, it was a king's duty to keep pachyderms in royal stables because earthly elephants would also attract their ethereal relatives the clouds. Since Indra was also the god of Rain and Thunder, the dark and heavy water bearing clouds became synonymous with the elephants that bring in the bountiful rain.

In tropical India where monsoon is of critical importance, elephants became associated with rainfall, hence also with the fertility of crops and therefore, the general well being of the country. Symbolising agricultural prosperity, elephants later came to be associated with Lakshmi the goddess of Fortune and Wealth depicted seated on a lotus with two white elephants on either side, sprinkling holy waters.

Floating in the primeval ocean of Hindu cosmology is another animal, an enormous snake called the Ananta Nag or the Timeless Serpent on whose coils reclines Lord Vishnu personifying the force of preservation in nature. This nine hooded serpent is also called Shesha or the Residue, which is left after creation has taken place. It is extraordinary that the serpent came to symbolise Time. In Hindu mythical imagery, it is the Great Serpent Vasuki who is considered to be the king and progenitor of all the snakes that crawl upon the earth.

In one of India's classic myths the Samudra Manthan, Vasuki was used as a rope, with gods aligned on one side and demons on the other, to churn the primeval ocean so as to obtain divine nectar. But in the great churning, there was just as much poison produced which threatened the universe. Lord Shiva personifying the destructive force in nature, is said to have drunk up the deadly poison, not allowing it to go below his throat, which justly gave him the name of the Neelakantha or the Blue Necked One.

But still, one poisonous drop slipped and fell on the earth out of which were born the snakes. Symbolic of the poison on earth, the snake inevitably came to be associated with death and therefore Shiva, also called Nagaraja or the Lord of the Serpents. Entwined around the poisonous blue neck, the cobra also symbolised hypnotic spell and meditative trance of this ascetic god.

Another evident attribute was the fact that the cobra seemed to thrive on land but disappeared with equal ease under ground or water. This made it the custodian of the either worlds and the treasures buried therein. The cobra became a symbol of wealth and the guardian of treasures in the womb of the earth. Associated with vegetation and water around which it was abundantly found, the cobra acquired the status of demigods the Nagas and the Nagins that dwelled and ruled the water bodies whether they were springs, rivers or lakes The restless stirring character of the serpent became a symbol of creative energy.

In one of the systems of Yoga the Supreme Power or Cosmic Energy is represented as a coiled serpent that lies dormant in every human body at the bottom of the spinal cord. With practice and ritual the snake moves up the spinal cord with a burst of energy. This is called the awakening of the kundalini. The serpent in Indian mythology is generally held to be a benevolent creature, symbolising divine protection under its spreading hood. In India, it is not a dreaded creature to be killed when encountered with rather, to be treated with respect and deference. Of the 3 species of Cobra found in India, it is the Indian Spectacled Cobra or Naga that is seen in the company of the gods and worshiped even today during the festival of Nagpanchami.

Partly human with a pair of wings and aquiline features indicating his avian character, Garud is a serpent killer and seen as the controller of poison on the earth to maintain balance. It was Jatayu, the mythical eagle from the epic Ramayana, that swooped down on villen Ravana and put up a valiant fight till the end to rescue Sita, wife of Lord Rama, as she was being abducted. Love and affection is reserved for the cow who is considered generous, unselfish and kindhearted, almost a human. Undoubtedly, the cow is a sacred animal considered to be the gift of the Gods to the human race.

It is symbolic of the divine mother that gives milk to all human beings and brings them up as her very own offspring. In the great Indian epic the Mahabharata, we find long descriptions elucidating the cow as the sustainer of life. It is in this conception that this gentle creature is understood as Kamadhenu, the wish fulfilling mythical cow. But it is with Krishna that the cow is closely connected. Among other deeds, Krishna is said to have lifted mount Govardhan to protect his group of cows, cowherds and milkmaids. Hence to stake care of this innocent and self sacrificing animal is a matter of virtue for the Hindus who identify the act as dharma or moral duty.

Consideration of conscience aside, it was natural that in a predominantly agricultural and pastoral country like India, cow and her calves were and still are, considered to be the real wealth of people. India being basically a country where main profession was agriculture, cow was always held in highest esteem since she gives not only milk but was the best converter of green biomass which was produced in excess, compared to food grain, and was available as dung to increase the fertility of soil. Her male calves were the main source of energy in agriculture, for ploughing fields and threshing grains.

Even her hide was used for making shoes. It is not surprising that the cow is actually worshiped for the largesse and bounty that it gives without ever complaining. Similarly, the bull still remains an integral part of the Indian agricultural scene and a significant asset in the rural economy as it has been since time immemorial. Its remarkable virility and association with man in agricultural also made it a symbol of both masculinity and fertility. The bull however, best symbolises kama or the sexual impulse. Among those who have mastered the bull or controlled desire, is Shiva, also known as the Pashupati or the Lord of the beasts whose mount is the Nandi bull.

Here the bull symbolises the sex principle as Ananda or the timeless delight of Shiva arising from his activity as the creator. No Shiva temple is quite complete without an idol of the Nandi bull. Although in some instances the bull also appears as a demon, this malignant character is reserved for the buffalo that presignifies ignorance and death. Like the dark and ugly bovine, Man's ignorance and ego is projected as a monstrous buffalo Mahishasur that is killed by the Goddess Durga. The buffalo also drives terror into the heart of people when it arrives at one's doorstep as the mount of Yama or the Lord of Death.

Bounded by bhakti of Lord Rama, Hanuman, the monkey faced god was a symbol for the pure of heart. Enticing damsels of Lanka couldn't shake the resolve of Hanuman who set fire and destroyed the legendary city of Desires. Although there are several species of monkeys in India, it is only the black faced Pnesbytis entellus, also known as Langoor which is associated with Hanuman and is given special sanction.

Among the two thousand different species of birds in this country, it is undoubtedly the peacock which is uniquely Indian, besides of course being the most majestic. Perhaps no other bird has so deeply influenced Indian ethos as the peacock that is found in the company of various gods and goddesses. As a killer of serpents, the peacock is the mount of Kartikay, the god of war. Enchanted by the music of Saraswati's veena and Krishna's flute, the peacock is always depicted in their company. But the multicoloured feather of the peacock finds its place of pride adorning the headgear of Krishna. Symbolic of many qualities, peacock is the National Bird of India, protected not only by a religious sentiment but also by parliamentary statute.

But the bird belonging to the gods is the gander, said to live in the subliminal heights of the Himalayas. The Gander or the Hansa is the vehicle of Brahma, the creator of the cosmos. However, the gander best symbolises freedom or the divine essence in human body. Hansa is the homeless free wanderer between the celestial and the earthly spheres, equally at ease 'here' and 'there', not bound by either - remaining forever free. Symbolic of wisdom and intellect, the swan is also depicted at the feet of the Saraswati, goodess of learning.

There are other birds and animals like the owl, cock, crow, chakravaka, sarus and the horse, deer, lion, tiger and boar that find themselves also in the company of the gods. Indians have been traditionally preserving and respecting all the denizens of the forest, be it animal or plants. The scientifically sound teaching was given the religions sanction so that the message of conservation goes to masses, particularly those who are living in close proximity of forest. With our increasing knowledge on scientific techniques, at times acquired from overseas, and our traditional love and respect for wild-life the future appears to be brightening up for the flora and fauna of this country.

11

Wild-life Conservation

India, because of its unique geographical location and variety of climatic conditions has been harbouring and giving protection to a large number of wild animals since time immemorial. The vast expanded bountiful land area from Kashmir to Kanyakumari, the lofty Himalayas and the mighty oceans are very rich in natural wealth which has been handed over to us from our ancestors in the form of natures present 'the wild life.' The conservation of wild life in India is as old as Indian history.

The 'RISHI ASHRAMS' were our historical parks where every form of life was protected. Kalidas, in his famous work *'Abhigyan Shakuntalam'* has mentioned the touching farewell given to the heroine by the plants and animals of her father's Ashrams'. This is the height of human understanding where man is in communique with nature which is the outstanding example of wild life conservation in Indian Culture. Even in the very beginning of Indian history as early as 242 B.C. Ashoka, the then Emperor, had motivated the protection of animals and plants.

Earlier, India was big living storehouse of unlimited wild animals and it was not possible to know its real stock, only two or three centuries ago. The records obtained from archaeological and historical data show that tigers, lions, rhinoceros and elephants were common in northern and north-western India. Thus, it can be easily assumed that the fertile plains of Uttar Pradesh, Madhya Pradesh and Punjab were once very rich in natural wealth. Five thousand years ago rhinoceros existed in the plains of Indus valley in the vicinity of Mohen-jo-Daro but nowadays they are restricted to West-Bengal and Assam.

The Asiatic lion was once common in India but today they are limited to Gir Forest of Gujarat. In the beginning of this century, tigers were estimated to be 40,000 to 50,000 but according to national census of 1972 they were limited to only 1827 in number.

The causes of decline of wild life are many. During the last two centuries, the whole country was happy hunting ground of the employees of the East India Company and the British soldiers, Nawabs, Rajas and Jagirdars and the civil servants. The trigger happy movement destroyed most of the magnificient wild life of India.

In the early part of the present century, a few naturalists realised the danger of vanishing species and raised their voice for the protection of wild animals. First wild life sanctuary was established in 1900 in Madhya Pradesh, the then famous Banjar Valley of Central Provinces but the idea of wild life preserves did not appeal to the 'shikar' oriented rulers. In 1935 first National Park was established in India in the foothills of the Himalayas and was named as Hailey National Park, after Sir Hailey the then Governor of United Provinces, later as Ramganga National Park and now-a-day's Corbett National Park. Later on other National Parks were established in other parts of the country.

Immediately after World War II and during the transfer of power and partition of the country, poachers took advantages of the transition phase and started killing wild animals and damaging their habitat. Even sanctuaries and parks were not safe. The death cries of these innocent natural beauties reached the late Prime Minister Pt. Jawaharlal Nehru who issued strict orders to check the hunting and set up an enquiry committee. Thus, the Indian Board for wild life preservation was constituted in 1952. This was the beginning of the legal era for the protection of the Indian wild life, as a result of which sixty six sanctuaries and National Parks were established within five years.

At present a number or projects for the conservation of forests and wild life are underway with the aid from a number of organisations like International Union for Conservation of natural Resources, World Wild Life Fund, The convention on International Trade in threatened and Endangered species of Wild Fauna and Flora, the Smithsonian Institution and others. These projects are focussing attention on a number of species such as pygmy hog, lion, tiger, gharial, rhinoceros, hispid hare, Hangul, thamin and great Indian bustard.

Threatened Wild Animals

MAMMALS

Order-Primates

Slender Loris
(Loris Tardigradus Linn.)

It is distributed in Sri Lanka and southern part of India *viz.* Tamil Nadu, Kerala, Karnataka and Andhra Pradesh. Slender loris is a small primate with slender body. The long limbs, well developed index finger, large ears and thin appearance are the characteristic features of this loris. It is tail less and measures about 25 cm in length. Its body colour is dark grey or earthy brown with beautiful ornamental silvery hairs.

Selender loris is nocturnal in habit. During the day period it remains hidden among foliage and starts moving from tree to tree at dusk. It feeds on berries of *Lantana*, lizards, tree frogs, insects and smaller birds. The newly born young ones have been seen in the months of March, May, June and October.

Before 1925 they were quite large in number but due to heavy killing for medicinal use in eye diseases and as an experimental animal their number reduced to a great extent. Nowadays they are in the category of vulnerable due to the destruction of their habitat.

Slow Loris (Nycticebus Coucang Boddaert)

The slow loris is distributed in tropical rain forest south of the Brahmaputra river in northern India and in eastern Bangladesh and Burma to far east. It is also reported from Goalpara, north of Brahmaputra. Its body and limbs are stout with rounded head and dark brown encircled rounded eyes. Its body length is from 30-40 cm but tail is rudimentary. The body is covered with dense fur which are silvery white on the head and shoulder regions but grey on the flanks and rump. There is a brown strip on the middle of its back which terminates on the crown. Its toes are provided with flat nails except the second toe which is clawed.

The slow loris is a very seclusive nocturnal animal. It is arboreal and rarely comes to the ground. It sleeps during day period in the holes or crevices of trees and comes out during night in search of food. The main food of slow loris are fruits, young shoots, leaves and insects. Due to destruction of its habitat slow loris is in vulnerable category.

Hoolock Gibbon (Hylobates Hoolock (Harlan)

It is distributed in the dense forests of north eastern India, south and east of Brahmaputra and Loni rivers. It also inhabits in northern Burma, China and Bangladesh. It is the only ape found in India. This slender bodied tail less animal measures about 90 cm in length. Its rounded head are very small. Its arms are more than double the length of the legs.

This gibbon prefers to live in evergreen, semi-evergreen and moist deciduous forests particularly along the hill slopes. They live in a definite territory in the form of a small family of 2 to 4 members including young ones. They are very active and reach on to the hill sides tree tops at dawn but during the night it remains in the forests of valleys. The main food of the gibbon are fruits, leaves, flowers, twigs and insects. In the search of food it moves most of the times during early mornings and evenings. The male gibbon is monogamous which remains faithful and satisfied with his female partner. They mate early in the rainy season and female produces the young between the months of December and March.

The population of hoolock gibbon is getting reduced due to fast destruction of pristine forests and also due to heavy killing by the tribals for flesh. The total population of hoolock gibbon in Assam, Mizoram and Meghalaya was counted to be about 7,9000 in 1971-72 but according to the latest position their population has been reduced to be very low level.

Negroes

They are locally called as Siddis. Gir has been the home of a few families of Negroes. It is said that Nawab at Junagarh, who protected the Gir lions from being killed, once went to Africa and brought some lion cubs along with some slaves to look after them. These slaves are the ancestors of the present community of Siddis. They had settle in Jambur village long before the Nawab rule and later moves to Shirvan. They retained their cultural and physical characteristics to the present days. The Negroes are short, dark skinned, provided with projecting jaws, thick lips, broad flat nose and black hairs on the head.

Golden Langoor (Presbytis Geei, Khajuria)

It was for the first time reported in 1907 E.O. Shebbeare, an officer of Bengal Forest Department. The colour of this langoor changes in accordance with the variation of the season. During winter its colour is deep golden or light chestnut which changes to light golden or cream-white in summer. The colour of face is always black. Body length is about 50 cm while tail is about 76 cm in length. It prefers the forest with mixed deciduous tree. They are found very active in the morning and evening hours. The breeding season of this langoor is from December to February and young ones are born in the months of September to November and are almost white coloured. They are getting full protection in Manas wild life sanctuary in Assam.

Black Nilgiri Langoor (Presbytis Entellus)

They are found in the hills of western ghats down to Kanyakumari, Nilgiri, Anamalais, Piani and the adjacent ranges. They are found in a number of different habitat types like evergreen forests, semi evergreen forests, moist deciduous forests and temperate evergreen forest. It has soft glossy, black skin and pale-brown head. Its body length is about 80 cm and the tail is 75 to 90 cm long.

It feeds on leaves and fruits and some times raids orchards. They come out early in morning and evening in search of food. They live in troops of 5 to 15 which is led by a male. The youngs are born in June and sometimes in September. The young ones are reddish brown coloured up to the age of 10 months after which gradually turn dark black.

It is in great demand in fashion markets. It is commonly believed that its palatable flesh; blood and certain internal organs of the body can cure many ailments. Thus, it is the most hunted monkey.

It is heartening to know that the number of students among

75,000 to 90,000 tourists that visit Periyar annually is steadily increasing. This is only due to the extension lectures being delivered in schools around the sanctuary on the importance of wild life and its conservation which will no doubt create interest in wild-life among children and make the conservation task easier.

Lion-tailed Macaque (Macaca Silenus)

It is a black coloured monkey which is shy and retiring in nature. The body and the head of this monkey is 50-60 cm and the length of the tail is 25-28 cm. There is tuft of hairs like that of lion or cattle at its tail end which is responsible for its name. It is called as 'Goduma bandar' in Hindi and Golangulkapi or Simhapuchchha-Vanar in Sanskrit. It prefers tall trees for living. Its food consists of leaves, fruits and buds. The call of male macaque is similar to that of a man lost in the forest calling for his companions-Koihai ? The forest dwellers say that such a call is followed by a lad pigeon-like echoing sound-Coo..Coo ? It is hunted for its skin and meat which the local people think increases a man's potency.

It is an endangered species which is in the process of making periyar its permanent home. It is proposed to combine the Bandipur sanctuary of Karnataka, Madhumalini of Tamil Nadu and Wynad Sanctuary of Kerala and establish it to name Jawahar Lal Nehru National Park which would cover an area of 2,000 square km and will be meeting point of Nilgiris, Deccan plateau and Western Ghats.

Order-Pholidota

Chinese Pangolin (Manis Pentadactyla Aurita Hodgson)

It is distributed from Nepal to north eastern India, Laos, Burma and Southern China. It is a slender animal measuring 70-75 cm from head to tail. The dorsal surface of its body along with tail is covered with large overlapping scales while ventral surface is provided with somewhat curved long and blunt claws which help for digging the soil.

It is terrestrial animal which inhabits the hilly as well as forest areas and nocturnal in habit. During day period it passes its time in the self-dug burrows.

During the movement its front toes are bent under the soles. They feed on eggs and adults of termites and ants.

Chinese pangolin is Indeterminate in status. It is rarely seen due to its nocturnal habit. The tribals kill them due to the belief that its scales and flesh are of great medicinal importance as a result their population is declining day by day.

Indian Pangolin (Manis Crassicaudata Gray)

It is little larger than the Chinese Pangolin and measures 60 to 75 cm from head to body. The tail alone is about 45 cm. Its scales are larger and arranged in 11-13 longitudinal rows. Indian pangolin is widely distributed through the plains and lower slopes of hills from south of the Himalaya to the southern extremity, of India. It is also reported to inhabit Sri Lanka.

The common habit of Indian Pangolin is similar to that of the Chinese Pangolin. The young ones are fond to be present during the months of January, March and July. It produces single young one at a time but sometimes two young ones are produced at birth. The real status of Indian Pangolin is not known because of its nocturnal habit. Like Chinese Pangolin, the Indian Pangolins are also killed by tribals due to their medicinal value, therefore, its population is declining.

Order-Carnivora

Wolf (Canis Lupus Linnaeus)

It is distributed from the upper Himalaya to the desert zone, and dry open country in the peninsular India. It is also found in northern, south western and central Asia, Europe and

North America. It measures from 90-105 cm in length from head to body region and the tail is black at the tip. The characteristic features of wolf are long powerful jaws, large skull and elevated fore head. The dorsal surface of the body is greyish-reddish yellow usually with brownish tinge but sometimes mixed with black colour. The face and limbs of wolf are reddish and the ventral surface of the body is dull white.

The wolf inhabits dry open country around human habitat and small places in rocky hills. They get shelter in caves, crevices in rocks and burrows in sand. The main prey of wolf are rodent, fox, hare, antelope, sheep, goat etc. Sometimes they have been reported to lift children. Their breeding season is at the end of rains and 3 to 8 young ones are born in the month of December.

It comes under the status of rare species. Due to its habit of lifting children there is a tendency of killing wolves by the people. Bihar, Rajasthan, Gujarat, Karnataka and Maharashtra are the states where the surviving population of wolf if reported.

Himalayan Brown Bear (Ursus Arctos Isabellinus Horsfield)

It is distributed at higher elevations of the Central and North Western Himalaya. It is also reported from Bhutan, Pamirs, North Western Pakistan and Afganistan. It is one of the subspecies of the brown bear of Europe and Northern Asia. The male bear is about 175 cm in length. Its body is covered with brown coat of fur which is thick, shagy with underwool in the months of winter but shorter, thinner and darker in the months of summer. The shade of the brown fur vary from dark pale to reddish brown according to the variation in season. But some bears are provided with silvery grey with white tip to the fur.

It is commonly found on the grass slopes above the tree line close to the snow. It feeds berries, young sprouts of grass, back wheat, wild fruits, rats and insects. Sometimes it takes larger animals also as food. Breeding takes place in the months of early summer and usually two cubs are born in winter season at sheltered places.

Due to the destructive activities of human-beings in its habitat the population of Himalayan brown bear has been reduced and reached the status of endangered species.

Red Panda (Ailurus Fulgens F. Cuvier)

It has a wide range of distribution in the Himalaya from Nepal and Sikkim to Upper Burma and Southern China. It is a handsome creature with rounded head, large ears, stumpy muzzle and retractile claws. The length of its body and head is about 60 cm and that of tail is about 40 cm. It is covered with dense wooly coat of bright chestnut. The colour of the ventral side and leg is black.

It inhabits temperate forest at a wide range of altitudes from 1500-3500 M. Most of the time it lives in pair. Being an arboreal it lives in the hollow of trees, fork between two branches and rock crevices. Basically Red Panda is a nocturnal animal but looks to be actively moving during early morning and before sunset. It feeds on fruits, roots, sprout of bamboo, insects, eggs and chicks. The young ones are born in spring which are two in number.

Due to rapid destruction of its habitat Red Panda is rarest of the rare status in Indian territory. It is still being hunted for its charming skin. However, nowadays it is reported to be present in Chunthang, Lachen, Singhik, Menshithang and Lechung area of Sikkim.

Cheetah (Acinonyx Jubatus Venaticus Griffith)

It is mainly distributed in Africa and drier parts of south western Asia including India. In head and body length it measures from 90-130 cm and in tail length from 60-70 cm. It has long limb, narrow chest and retractile claws. Cheetah inhabits rough hill-dotted planes and arid areas. Its nocturnalism is less than the other members of cat family. It is known to be fastest running animal and can achieve a terrific speed of about 120 km per hour within a shorter distance of about 375 m from starting point. Its speed is helpful in catching the prey very easily. It preys on deer, black bucks, hares and birds. It can be tamed by human-beings and used for hunting the other animals.

In older days cheetah was one of the most common cat animals of India but due to the rapid destruction of its habitat, high rate of mortality of the cubs and killing by hunters its population in our country has been reduced drastically since 19th century. Complete failure of its breeding in captivity is also one of the most important factors of its declining population. Its presence in southern part of India was reported in 1940-52 but after that the presence of cheetah has not been heard. Now it is in the category of extinction.

Asiatic Lion (Panthera Leopersica)

Since time immemorial, the lion has been the symbol of prowess and righteousness. In the ATHARVA VEDA, the king has been compared to the mighty lion. Within historic time the range of the lion outside Africa extended throughout Asia Minor, Arab, Persia and India. In the beginning of the 20th century they were exterminated from Iraq. In India the extermination of this species was started region-wise, *viz.*, Bihar-1814, Saharanpur and Ludhiana 1822, Delhi-1834, Bahawalpur-1842, Bundelkhand and Eastern Vindhyas-1865, Rajasthan-1970 and Aravali-1880. Thereafter, lions were not recorded outside Gujarat. During British regime 'British Viceroy, Governors of British Provinces, Commander-in-Chief,' the rulers of princely states, state guest wanted to be invited for Shikar of lions. A British officer of the Indian cavalry killed 80 lions in a period of 3 years.

At the turn of the 19th century Lord Curzon was invited to shoot a lion but he did not accept the offer and told the Nawab to protect lions from extinction. This was the first attempt to protect the lions in British period. In 1900 the lion was declared a protected animal.

In 1956 when late Pt. Jawaharlal Nehru visited the lion reserved area, he spurred interest and advocated that proper steps should be taken to develop in into a tourist spot. In 1952, Indian Board for Wild Life put the Indian lion on the list of 13 rare faunas to be given full protection. In 1966 about 600 square kilometres area was developed as the 'Gir Lion Sanctuary' and the 6th meeting of the State Wild Life Advisory Board in 1968 recommended further to upgrade it as National Park within the sanctuary.

Indian lion is mentally alert and physically competent. It can attain the length of 2.5-2.8 metres with an average of 200-250 kilograms weight. A very good sense of vision and hearing helps lions in hunting their food. The lions in pair remain in heat period for about eight days. The lioness can give 2 or 3 cubs at a time. The life span of a lion is about 20 years. Within the onset of summer, the shortage of drinking water is common in the forests so some small cemented water tanks have been constructed in the parks.

The post-mortem reports of some of the old lions found dead in the forest reveal that their canine teeth had fallen and their stomach was empty. Apparently, they had died of starvation.

Conservation : Nowadays the lions are very well protected in Gir Sanctuary by all means. The Gujarat Government is taking interest for the well being of the lions on the lines advocated by Kautilya. Adequate common grazing ground is also provided with the progress of 'Maladharis' shifting from Gir forest region. Thus the area of National Park would be increased in phases to cover the entire sanctuary. A 464-

kilometre-long low rubbed and live-hedge wall has been constructed to protect the living area of lions in the forest.

Tiger (Panthera Tigris Linnaeus)

Tiger has been an animal of interest in the hunting history of India since 5,000 years which was known after the excavation of Mohen-jo-Daro and Harappa. Even today, when emphasis has shifted from hunting to conservation, tiger still remains the centre of attraction in the forest for conservations, wild life enthusiasts and photographers.

Tiger, a rare combination of magnanimity, colour, ferocity and awe, is a noble animal which is afraid of man. Man does not come under the common item of its menu. Perforce, it may kill man, and after killing 4-5 men, its fear of man disappears. So, man becomes an easier prey of tiger and thus, it becomes a man-eater. As per report of the Director, Tiger Project, the man-eating by the tigers of Sunderbans and Dudhawa National Park is still continuing. Twenty-three persons had been killed by the man eaters of Dudhawa and Sunderbans in 1985 only. In the month of November and December tigers mate for eight to ten days, once in 2 to 2.5 years. After the gestation period of 15-16 weeks the tigress gives birth to 2-3 cubs which are born with closed eyes, helpless and totally dependent on the mother. The hunger makes the mother sullen and irritable. She is always apprehensive that someone may cause harm to her cubs. So, a lactating mother tigress becomes a killer due to hunger and fear for cubs safety. They are excellent swimmers and drink water frequently and cool themselves in summer by bathing.

Habitat: Seventy five years ago, the Indian tiger was found almost all over the Indian subcontinent except in the alpine snow-clad mountains of the Himalayas, dry region and the extreme tropical thorn forests of western desert of Rajasthan. In the north-western part of the country, the dry Sal *(Shorea robusta)* of sub-montane forests of Kalashar and the tropical bamboo *(Bamboosa stricta)* forests of the Morni hills of Haryana, the sub-montane sat forests of Nahan in Himachal Pradesh were tiger hunting blocks in the past. The tigers also strayed out of the forests in open lands when standing crops provided cover. Now there are no tigers in these forests.

The correct population of tigers in India has been subject of controversy. According to Corbett (1955) the tiger population in 1953 was only 2,000 which reached to 4,000 in 1964 according to Gee. Pery (1964) was of the opinion that by the end of twentieth century, we may see the last of the great mammals in the wild state and tiger will be one of them.

The causes of declining population of tiger are production in the areas of forests, non availability of natural prey, tiger shooting tendency, poisoning by villagers due to its man-eating nature and the foreign trade of tiger's skin. The Indian Board for Wild Life, 1969, approved a total ban on the export of all cat skins.

Project Tiger : Due to regular declining number in tiger population in India, steps were taken by the Government to protect them and provide shelter in separate definite areas of forests resulting in the establishment of sanctuaries and National Parks. In 1982 the steering committee of 'Project Tiger', headed by late Prime Minister Mrs. Indira Gandhi, decided to increase tiger reserves and thus the number of reserves reached 15, covering an area of about 24,600 square kilometres.

They are : Manas Wild Life Sanctuary-Assam; Similipal Wild Life Sanctuary-Orissa : Palamau Wild Life Sanctuary-Bihar; Bandipur Wild Life Sanctuary- Karnataka; Ranthambor Tiger Project-Rajasthan ; Namdapha Wild Life Sanctuary-Arunachal ; Nagarjuna Sagar Shri-Shailam Sanctuary-Andhra ; Indrawati Wild Life Sanctuary-Madhya Pradesh ; Buxa Wild-life Sanctuary-West Bengal; Gir Sanctuary-Gujarat; Kanha, Shivapuri and Bandhavgadh National Park-Madhya Pradesh;

Corbett National Park-Uttar Pradesh, and Sunderbans Tiger Reserve-Assam.

In the history of wild-life conservation 1973 is a memorable year as an excellent 'Project Tiger' was started which was for the first time introduced in Corbett National Park. In 1972, their number was 1827 but now the number has increased up to 3,015. In this park, project tiger was started with 44 tigers only but in 1981 the number was 90. This park has limited space and cannot accommodate such a huge population hence many tigers were transferred to neighbouring forests of Amangadh, Aam, Pokhra, Fanto, Garjiya and others. There are other areas where tigers live which should be improved to rehabilitate them. These areas are Chilla, Khara, Moti Chur, Rajaji Sanctuary, Pilibhit jungles, Haldwani jungles, Vindhyas, Bundelkhand and Varanasi jungles.

Man-eaters of Sunderbans : The Sunderbans acquires an area of 10,000 square km across the delta of Ganga and Brahmaputra of which 6,000 square km are now in Bangladesh. There were 195 tigers in Indian and 300 in the Bangladesh region according to an estimate of 1971 which was highest number of tigers in the subcontinent.

Tigers in Sunderbans are too bold to hesitate in lifting a sleeping person. They generally attack man within their own territories in the forest and, unlike those in northern India, do not enter villages which indicates that they are forced to man eating due to the scarcity of food in the forest. The study of the area and the man-eaters revealed that the scarcity of freshwater sources in the Sunderbans compels them to take water from tidal creeks. The ingested salt affects liver and kidney of the animal. So, freshwater tanks should be constructed in the natural habitat of the tigers. Chital is their main food in Sunderbans but the chital's population is not enough to sustain them. So, they feed on wild boars, monkeys and other smaller animals and even on monitor lizards, crabs, fishes and salt water crocodiles. Sunderbans are the last shelter of the tiger, where it has been struggling for survival under adverse conditions. In Sunderbans, tigers are of weaker constitution because of inadequate nourishment, it gets from its food. The area of tiger reserves in the Sunderbans is 2,585 square km. It is being planned to develop 5 of its 15 subdivisions into a national park covering about 1,310 square km.

The World Wild life Fund has been associated with 'Project Tiger' in the Sunderbans and has provided fast jet-boats, long-range firearms, wireless sets and other useful equipments.

Factors that make the life of a tiger difficult are :

(1) Illegal encroachment by human beings in forests.

(2) Unplanned felling of trees in forests.

(3) Unplanned grazing of cattle in the reserved forest which causes shortage of fodder for wild herbivores resulting in declining their population, therefore, scarcity of natural food for tigers.

(4) Killing of herbivores by poachers for food and skins.

(5) Killing of wild animals by farmers to protect their crops.

Still there is a long way to go in preservation of these animals. Luckily the species is very cooperative due to its adaptability in varying habitats, short gestation period, 2-8 cubes per litter every alternate year, breeding almost all the year round, with one major peak period, sex ratio with feminine dominance, fast rate of growth during first three months, deep concern of mothers in rearing and schooling their cubs, maturity at the age of four years, unfixed nature of pairs and long span of life but now tiger needs protection.

The Snow Leopard or Ounce (Panthera uncia, Schreber)

This beautiful cat inhabits the Himalayas from Kashmir to Bhutan. It has a long tail of about 90 cm. This cat preys on wild goat, wild sheep, musk deer, goat, antelopes, hares, marmotes and rodents. Though it is protected species yet hunted for its fur which has very high price in international markets. In Jammu and Kashmir State the snow leopard has been protected in 170 square metre area of Dachigam Sanctuary, since 1951. The International Fur Trade Federation imposed a total ban on the trade of snow leopard fur in 1971.

The Clouded Leopard (Neofelis Nebulosa, Griffith)

As the name indicates there are cloudy marks on its body. It is provided with a long tail, stubby legs, rounded black ears and well developed upper canine. They are inhabiting the dense forests of Assam, Bhutan, Burma, Sikkim, South China and Malaya. Though it is a small cat yet it attacks big animals with the help of its powerful jaws and large canines. Its skin is very valuable and its sabre-like fangs are very popular in Malaysia as ear ornaments.

Tibetan Lynx (Felix Lynx Isabellina, Blyth)

It is distributed in Asia and Northern Europe. It also inhabits the upper Indus Valley in Gilgit, Laddakh, Tibet, Lahaul, Spiti, Pangi and Kinnaur regions of Himachal Pradesh. It preys on hares, marmotes, mice and wild goats. It has a very keen eyesight.

Himalayan Black Bear (Selenarctos Thibetanus, Go. Cuvier)

It is found in the lower mountain ranges of Kashmir, Assam and Himalayas and provided with a 'V shaped mark of yellow or white colour on its breast. In winters, it inhabits the lower valleys at a height of about 1525 metres whereas, in sunmers at about 3500 metres. It feeds on fruits, flowers, berries, nuts and honey. In autumn it is attracted towards ripening maize and buckwheat. In the spring and summer they peel off the bark of pine and deodar trees to obtain the sap and termites. They are known to be preying on sheep, hanguls, cattle and goats near the villages.

Malayan Sun Bear (Helarctos Malayanus, Rafles)

It is distributed in the hill forests of Manipur, Meghalaya and Nagaland. This jet black coloured bear is 104-140 cm in height. The main food of this animal is fruit. It holds the branches of the tree by its prehensile tail.

Otter (Lutra Lutra, Linnaeus)

It is basically aquatic animal generally feeding on fishes. But it is equally active on land and can catch other animals for its food. The fishermen of Sind and Orissa use to keep the otter and train it to drive the shoal of fishes towards net. It is found in Manas wild life sanctuary in Assam.

Wild Dog (Cuon Alpinus)

The body of wild dog is longer than that of wolf, with shorter legs and muzzle similar to that of domestic dog. It is reddish in colour having bushy and darker tail with a black tip. The method of hunting by this dog is very well planned. Two or three dogs chase the prey and the rest of the pack watches and as soon as any of perusers gets tired, the other member of the pack replaces it and when the animal drops down in exhaustion wild dogs kill it. They hunt deer, boar, sheep leopard, buffalo, goat etc.

Although wild dogs are found throughout India but its number is reducing day-by-day. During the census of 1980, 150 wild dogs were counted in the Kanha National Park, 148 in Bandipur wild life sanctuary and 80 in Palamau.

Order-Perissodactyla

Lesser One-horned Rhinoceros (Rhinoceros Sondaicus, Desmarest)

It was distributed from Assam to Bengal and south of Brahmaputra in India, and Sumatra, Java and Borneo. But nowadays it is said to be present in Java only. It is a dusky grey animal which is slightly smaller than the Great One-horned rhinoceros. The hide of the animal is divided by cracks into small and polygonal discs. Its horn measures about 28 cm in length. The horn is only present in males but if present in females it is only a low boss. It prefers to live in tree forests but has also been noticed to live in swampy areas. Its presence at higher elevations up to 2135 m has also been recorded.

Now it is extinct in India, Burma and eastern areas up to Malaysia. A few dozen of lesser one horned rhinoceros, are, however, still surviving in Western Java.

Kiang (Asinus Kiang Moorcroft)

It is distributed in Tibet and Tibetan facies of India in Laddakh and Sikkim. It is larger, more darker and dark red in colour than the Asiatic wild Ass. Absence of shoulder strips is the distinguishing character of kiang. The narrower mid dorsal strip is without white border. The hooves of kiang are larger than the horses.

It inhabits snowy, dry, cold deserts in very high plateau areas. They are mostly found in small groups of about a dozen or more than it. Due to natural calamities and destruction of habitat its population has been reduced drastically. The status of kiang is that of extinct in India.

Rhinoceros (Rhinoceros Unicornis, Linnaeus)

It is distributed throughout the low marsh lands on the banks of the Brahmaputra, the low land of the Himalayas, Tarai land along Rapti and Naraini rivers in western-central Nepal.

The population of rhinos is dense in the marshy lands of the park Kaziranga and this area of 430 square km is unable to sustain the population of 500 rhinos. Thus there is struggle for existence and they fight particularly for grazing areas. The last rhino in Uttar Pradesh was killed near Pilibhit at the beginning of this century but in Dec. 1986 one rhinos, escaped from Nepal, was killed by police men in Deoria district of Uttar Pradesh.

The Indian one horned rhinos may attain the weight of 2,000 kg. The horn found on the snout of rhino is composed of matted mass of horny fibres. The horn is not used by them as organ of offence and defence but they attack with their tusks.

The price of the horn is very high which may sell for an unbelievable price of 50,000 rupees per kilogram in far-eastern countries. Its skin was used for shield of warriors. It is estimated that about 1654 rhinos are in Kaziranga, Jaladapada and Manas National Parks.

Conservation : For the protection of rhinoceros the Bengal Govt. has passed 'Rhinoceros Act 1932' and for that 'Assam rhinoceros Bill 1954, was also accepted. In 1975 walkie-talkie sets were given to check the movement of poachers. Armed guards and patrol parties in Kaziranga and Manas now keep a round the clock vigil to check poaching activities.

'Project Rhino' : Centuries ago in Dudhawa area the rhinos once existed. Keeping this in view under this scheme a batch of six rhinos were brought to Dudhawa National Park from Kaziranga National Park for rehabilitation, out of which only two survived. A second batch of four female rhinos were procured from Nepal. Due to the proper care and keen interest taken by the authorities of this park the rhinos started flourishing in this well protected 'rhino village'.

This area was kept under constant surveillance of the authorities for flushing out the tiger menace from this village. This provides fruitful result and three calves born in early 1989 have survived raising the number of these animals from six to nine.

Tibetan wild Ass (Equus hemionus kiang, Moorcraft) (=Asinus kiang)

It is commonly known as Skyang by Laddakhis. This ass is found in Laddakh, Tibet, Northern high open plateaus. In Himachal Pradesh, they are reported from only kibber, valley of Spiti. Though it is not utilized for any purpose yet they face the bullets for the sake of enjoyment of sportsmen. Sometimes the tribes eat its meat when the stock of food is exhausted. Around 4,000 Tibetan wild asses are inhabiting Laddakh and nearby regions.

Wild Ass (Equus hemionus khur Lesson)

In Sanskrit it is called as Gaura Khara means white ass. They are nowadays found in Little Rann of Kutch (Rann of Dhrangadhra) only. This rann is dry seabed, a salt desert acquiring an area of about 2500 Km. The people of this region believe in the Vaishanava cult so are vegetarians and firm believers in non-violence. Killing apart, they do not even harass any animal or bird. It is a handsome beast, looks like a piece of white rock when viewed from a distance. They are 110-120 cm in length. They can run at a speed of about 70 Km per hour for 5-10 minutes if chased. They do not care for food and hence named 'Carefree King of Rann'. They graze in the shrubs around the periphery of Rann. They make very strong hind kick if delivered with full force. In 1948 wild asses from Dhrangadhra Rann were presented as a gift to General Cariappa, the then Commander-in-Chief of the Indian Armed Forces with the idea to use them for military purposes. Fortunately the wild ass has no natural enemy in the Rann, except man. Just for fun of hunting some people still kill the wild ass. But nowadays for the conservation of this species trees, shrubs and grasses for the grazing of wild asses are being grown in the area. According to the census of 1976, carried out by Gujarat Forest Department, 720 wild asses were counted in the Little Rann of Kutch.

The Indian Board of Wild-life had put the wild ass in the list of 13 endangered species in 1952.

Order-Artiodactyla

Swamp Deer (Cervus duvauceli Go. Cuvier)

It is distributed in Tarai and Duars of eastern and northern India from Uttar Pradesh to Assam. It is also found in Kanha National Park. It is about 180 cm in the body and head length, 23 cm in tail length and 130 cm in height at shoulder. The body colour of swamp deer is lighter than Sambhar. It is provided with branched antlers. The male deers are provided with much branched anthers. The base of antler remains unbranched for more than half of its length which further divides and subdivides into 10 to 12 forked spikes. The colour of its body is variable from brown to yellowish brown in winter and paler in summer. The neck of swamp deer is provided with the long hairs.

It inhabits in swampy and dry grasslands in Sal forest and on plane and undulating grounds. It is gregarious in nature and grazes in herds. During the breeding phase in winter they congregate in very large heads of about 50 or hundreds of individuals. The gestation period is about 180 days after which female swamp deer gives birth mostly to a single young one.

The number of swamp deer has drastically reduced and due to the destruction of its habitat and killing the species is in the status of endangered. In Kanha and Kaziranga they are in hundreds but in Dudhwa they are more than 2000.

Hangul (Cervus Elephus Hangul Wagner)

It is distributed in north Chamba, Himachal Pradesh and Dachigam. Hangul is also called as the 'Kashmir Stag' which is a large deer provided with a pair of spreading antlers. It measures about 125 cm in height at the shoulder and about 230 cm in head and body length. Antlers is provided with generally six or sometimes more times. They are light to dark brown colour with white patches on rump. The white patch of the rump extends up to the tail. The ears and chin are also whitish, whereas, limbs and sides are of pale colour. During the months of summer the body colour fades.

It inhabits in small groups of 3-15 individuals in dense forests or in the riverine areas of lower Dachigam between 1800 to 3700 m of elevations. In winter they, in the group of large herds, come to the lower elevations.

The antlers of males are shed in March and new antlers get perfection by September. The males establishes a harem. The young one is born in the months from April to May.

Due to rapid destruction of natural habitat, cattle grazing and interference of man in its living place the population of Hangul has been depleted. However, due to the 'Project Hangul' started in 1970 and its effective conservation plan the population of Hangul is increasing year by year.

Gaur (Bos Gaurus H. Smith)

It is distributed in Nepal to Arunachal Pradesh, Orissa, Bihar, West Bengal, Assam, Mizoram and in peninsular India. It is also called as Indian Bison.

It is a heavy animal provided with short limbs, an ashy forehead and white feet. It measures from 175-195 cm in height at the shoulder. The male is slightly larger than the female. Its paired horns measure 70-75 cm which are narrower in cow than those of bulls. The body of cow is dark or reddish brown but that of bull is black and almost hairless.

It inhabits in herds of 6-12 gaurs in dense tropical forests. It feeds on leaves and bark of trees, grasses, young bamboo-shoots. It moves in search of food from late evening to early morning in forest and open meadows. They mate in winter or in early spring. The cow gives birth to one calf in a litter in the months of September and October.

Due to rapid destruction of its habitat the population of Gaur is declining rapidly as a result it has come to vulnerable state. The total estimated population of gaur is about 4500 in forests of Assam, Bihar, Madhya Pradesh, Tamil Nadu, Kerala, Karnataka, West Bengal and Nepal.

Thamin Deer (Cervus Eldi, McClelland)

In Sanskrit it is known as 'Bhru-Shringa Mriga' which means that the antelers grow from near the brows. This is a very rare species of deer inhabiting the 'Keibul Lamjao National Park' which is one of the unique parks in the world which floats in the Vast Logtak Lake of North-East Indian State of Manipur. Nowadays, this species is confined to just 25 square Km of the park. In 1951 it was declared extinct species but again discovered in the same area. The Indian Board of Wild life declared it one of the 13 rare species in 1952 and recommended its preservation.

The detailed study of this rare deer is not so easy because it lives in high bush or reeds, rests during the heat of the day under cover, comes out for grazing early in the morning or late after dusk. However, one or two may be sighted after a number of attempts. The male is dark brown, whereas, female is light brown coloured. The male pursue the females just before the start of the breeding season to mate. After 236-256 days of gestation period fawns are born which are generally found to be larger than Chital and smaller than Nilgai calf.

Conservation : In 1953, Keibul Lamjao was declared a sanctuary but the position of safety did not improve. As per report of Zoological Survey of India, 30 deers were in this area in 1979. With the declaration of this sanctuary a National Park in 1977, and adoption of Wild-life Protection Act, 1972, authorities have made several efforts to have a check on the tilling of land, grazing by cattle and fishing. There is need for a biological study on which a sound management programme can be planned.

Pygmy Hog (Sus salvanius, Hodgson)

It is one of the most rare animals in India. This small pig is distributed in the Tarai areas of Eastern Himalayas covering a part of Bhutan and Western Assam. It is very difficult to trace it in high grasses. It is about 70 cm in length and 30 cm in height and is provided with short muzzle, tusks and ears. The tail is about 3 cm in length. The body is covered with scattered hairs of blackish-brown colour. It is very light-footed animal having 3 pairs of teats. It is a diurnal animal, generally found active between dawn to midmorning and from mid after noon to 3 hours after dusk. They dig their burrows usually at dusk.

The pygmy hog is an omnivorous animal and its food consists of a variety of birds, insects and monitor lizards, roots and tubers etc.

They are solitary in habit and meet the sows only during the mating season. The young are born in litter consisting of 2-6 piglets in the months of April to the middle of May.

They are generally chased by grass cutters, poachers and herdmen and thus harassed hogs reach the fields where they can not survive. In 1977, about 14% of pygmy hogs were killed by poachers. To save this endangered species a suitable environment should be prepared in sanctuaries for the breeding. The protection of this hog should be maintained on their natural habitat.

Conservation : The Indian Board for Wild life included this animal in the list of rare species in 1952 and recommended for its total protection. It has been included in the list of endangered species in the 'Red Data Book' of the International Union for Conservation of Nature (IUCN).

Barasinga (Cervus Duvauceli Duvauceli Cuvier Cervus Duvauceli Branderi Pocock)

India is the only country where barasinga is found. Two species of this beautiful deer is found of which first one is *Cervus duvauceli duvauceli* which are main attraction of Dudhawa National Park situated in Uttar Pradesh. The name barasinga means a deer with 12 horns or antlers (Branch of point) but careful observations indicate that barasinga should mean a multi-pointer deer in which points may vary from 10-14, grown on the antlers of the deer. A well developed male deer can attain about 135 cm height with a weight of about 170 kg. The average length of an antler is about 75 cm. The skin of this beautiful Indian deer is covered with wooly hairs of brown to pale colour. The males are darker than females. This species of barasinga is termed as 'Swamp deer' and found in swampy areas of Tarai-Uttar Pradesh, Assam and Sunderbans. Unlike the Sambhar, they are less nocturnal and are found grazing late in the morning and evening but they take rest during day period. They have good response to sound and strong sense of smell.

The skin is used as leather, meat is palatable and head and antlers are decoration pieces of drawing rooms.

Second species of barasinga is Branded Barasinga, *Cervus duvauceli branderi* (Peacock), which is found is Madhya Pradesh. Earlier this species was distributed from Hoshangabad district to Bilaspur and Bastar in east and even in the neighbouring state of Orissa. But nowadays the deers of this race are restricted only to Kanha National Park.

They are found feeding mainly on grasses and area grouped into different parties of different sexes and age groups. These groups are termed as 'Schools'. It also rests in day time. It feeds, breeds and seeks protection in the meadows. But the meadows near villages are constantly becoming naked by the grazing cattle, as a result of which its natural habitat has shrunk considerably. The hunting of barasinga is easier than any other deer.

Conservation : Dudhawa National Park and Sunderbans are very rich with regard to the population of swamp deer. Out of 4000 barasingas, about 2,000 are at Dudhawa National Park. Out of nine deer species found in India, five *viz.*, barasinga, hogdeer, Chital, barking deer and Sambhar, are at Dudhawa.

The Branderi barasingas are well protected in Kanha National Park which has provided a sanctuary to barasingas since 1933, and the present range of distribution shows that they are more protected here than any other place. But tigers and men have contributed to the decline of barasinga population here also. In 1969, an American wild life conservationist donated $ 3,000 for its conservation. The Indian Government had sanctioned Rs. 60,000 for their conservation in the same year. In 1970, the number of barasinga in Kanha National Park was 66; 118 in 1973 and 344 in 1979.

Chital (Axis axis axis Erxleben)

It is one of the most graceful animals of our forests which is most beautiful deer in the world. It is found in the foothill forests of Himalayas. In Assam, it is found in the districts of Goalpara, Kamrup and Darrange. They are of gregarious nature and are found in groups of about 100. Chital is main prey of tiger. Wild dogs, Jackals, leopards and hyenas are also reported to prey upon chitahs.

The chitahs are in well protected conditions in Kanha National Park.

Sambhar (Cervus Unicolor Niger Blainville)

It is the largest deer found in India. This deer is found in the foothill forests of Himalayas. The food of Sambhar includes grasses, leaves and variety of wild fruits. It grazes in the field during night and takes rest during the day in dense forests. An average height of this deer is 5 feet and the weight of male Sambhar is between 225-320 Kg. But the movement of such a bulky animal in the forest is surprisingly noiseless. Sambhars are found in the Kanha National Park in Madhya Pradesh and they are very well protected there.

Musk Deer (Moschus Moschus Moschiferus, Linnaeus)

It is distributed throughout the whole Himalayan belt as Kashmir, Laddakh, Garhwal, Himachal Pradesh, Kumaon including Sikkim, Bhutan and Nepal. It inhabits the forests adjoining alpine meadows and in alpine scrubs. This little deer is about half a metre in height and is provided with well developed upper canines or tusk projecting below the lower jaw. They are devoid of horns and only males carry musk. It can easily jump about 5 metres long distance and about 4 metres in height. They feed on a variety of leaves, grasses and field crops.

The musk has a penetrating and persistent fragrance and is used as base for high-grade perfumes, drugs and aphrodisiacs. Ten grams of musk costs about. Rs. 5,000 in the market. The musk pod of a deer may be 20-50 grams. So, one musk deer is worth between Rs. 10,000-25000. In the international market one kilogram of musk costs between Rs. 8 to 12 lacs.

Although musk is obtained only in males but hunters kill on an average of one thousand deer per year in Himalayan belt. One should know that musk can be obtained without any damage to deers. As deers attain the age of one year, the milky secretion of musk starts which gradually hardens, dries up and forms bean-sized granules. The musk can be extracted by milking technique before drying. But for the extraction of dried musk silver spatula is inserted in the external orifice of the musk pod. The milk musk is extracted twice a year in spring and autumn. The musk has been a subject of poetic thought in Hindi and Sanskrit literature.

Blackbuck (Antelope Cervicapra, Linnaeus)

It is considered as the favourite antelope of Lord Krishna, therefore, called as 'Krishna Mriga', and Krishan Sara and Saranga in Sanskrit. Kalidas had mentioned in his literature that Shakuntala kept domestic black bucks in Kanvarishi's Ashram. The saints and hermits of India are usually keeping the skin of black bucks for sitting during meditation. Mansaraja was the name of Jehangir's pet black buck. Akbar used to use trained cheetahs to hunt the black buck. In Gujarat this black buck is called as Kadiyas.

The black buck is a slender animal, provided with well developed muscles and spiral horns of about 15 cm. Its height is about 80 cm and weight about 40 Km. They are gregarious in nature and live in large herds. It can run with a speed of about 80-96 Km per hour. They drink water regularly, but during dry season they can adapt themselves to the scarcity of water. It is a hardy animal and can resist a number of diseases.

In the breeding, season best male mates with the best female. So that individuals with good physical fitness are added to the herds. After a gestation period of five to six months female gives birth to one or two fawns which can graze on new succulent shoots. The life span of black buck is about 12 years which may increase in zoos where it is protected against the hardship of wilds. The leadership of the herd is usually entrusted to an experienced and vigilant doe.

The horns of black bucks decorate the drawing rooms and are believed to possess curative properties for asthma and other diseases of thoracic region.

Conservation : In 1976, the Velavadar reserve was declared a national park in which best arrangements have been made for the survival of black bucks. Six big water holes have been constructed in this park for drinking water to black bucks.

According to the census in 1977, a total of 1676 antelopes were counted. But due to construction of road, poachers use to come to this park from Ahmedabad, Mumbai and Jaipur for hunting them. The dogs, jackals and jungle cats prey upon their fawns. So, effective arrangements must be made to protect black bucks.

Wild yak (Bos mutus, Przewalski)

It is a massive short-legged blackish-brown, bison-like animal with a dropping head. They are found only in the Changchenmo valley in northern Laddakh and some passes in east Kumaon. This animal is very important to the people of Trans-Himalayan ranges. It is said that Lord Shiva's mother-in-law, offered Shiva and Parvati a silky white coat of a yak to sit on during their marriage. Yak eats snow even in summer when streams flow. Its bulls occasionally mate with domestic yak cows, producing large offspring. The yak is used for ploughing the land. The hairs, hides, horn, flesh and hooves are used by the people of the region. In upland pastures, where horses are of no use, yaks are found to be most useful for higher altitude journey.

Macro Polo's Sheep (Ovis Ammon Polii, Blyth)

It is one of the rarest species of sheep and nowadays only few are still to be observed in Laddakh. It is provided with deeply wrinkled horns which become unnecessary burden for the ram in old age.

The Blue Sheep (Pseudois Nayaur, Hodgson)

It is connecting link between sheep and goat. It is also known as Bharat sheep. Its ram has no beard, unlike the male goat and devoid of goatish smell. They are hunted for mutton in Laddakh and Bhutan. It is most common sheep of the regions like Kumaun, Laddakh, Nepal and Sikkim.

Markhor (Capra Falconeri, Wagner)

It is called as king of wild goats. The name Markhor is

derived from two Persian words, Mar-means snake and Khor-means eater, thus snake eater. It is provided with long spiral horns of about 165 cm length. They inhabit the regions below the snow-line.

Ibex (Capra ibex, Linnaeus)

It is most handsome animal provided with long, deeply notched sweeping horns found in the western Himalayas in Kashmir, Laddakh and Baltistan. It is believed in Laddakh that in one of his previous births Lord Buddha was born as an Ibex. It is hunted for its meat, hide and fur. The finest Kashmiri shawls are made from ibex wool and the Kashmiri long boots are made from its hide.

Chiru (Pantholops Hodgsoni, Abel)

It is a handsome Tibetan antelope which is about 80 cm in height. It bears long slender horns of about 70 cm. The chiru comes down to Laddakh in summers for grazing. The wool (pasham) obtained from this animal is used for the preparation of soft and costly shawl (Shah tush) in Kashmir.

Nilgiri Tahr (Hemitragus Hylocrius, Ogilby)

It is commonly called as ibex which is distributed from the Nilgiris to the Anamalai and the Western Ghats. It is close relative of the Himalayan tahr. The tahrs are sharp-eyed and sure-footed animals and at the time of danger they disappear into dense forest. It is slightly larger than its Himalayan cousin. The height is about 100 cm at shoulder and the weight between 80-100 kg. The horn of males are longer than those of the females. The female has two teats while those of Himalayan Tahrs hale four (2 rudimentary).

The adult male is provided with white colour on the back and neck and a white spot on the knee. Tahrs are generally found in flocks of 6-12. They graze on grass-lands on mountain top and slopes early in the morning and late in the after noon.

They are very much watchful and can detect their enemies below but is not conscious of enemies lurking above. The handsome males are a temptation to the trigger happy sportsmen. The flesh of young tahrs is delicious so it is much hunted by hunters due to which its number has been reduced to a great extent. The jackals, wild dogs, panthers and tigers prey upon this wild goat.

In Periyar Wild-life Sanctuary Nilgiri tahrs are very well protected since 1975 when it was taken over by the Kerala Government and was declared as wild-life sanctuary.

Wild buffalo (Bubalus Bubalis, Linnaeus)

It is found in wild-life sanctuaries of Manas, Kaziranga, Laokhowa and Sonairupa. It is very dangerous and fearless of the Indian animals. It shows exceptional courage in protecting its young ones. Maharajas and Princes used to shoot down the male wild buffaloes just for the sake of satisfaction of their ego.

Order-Logomorpha

Hispid hare (Caprolagus Hispidus Pearson)

It is distributed in Duarsa area of West Bengal and Assam, and in Bangladesh. It measures from 45 to 50 cm in the length of head and body region and that of tail is about 5 cm. The colour of dorsal side is dark brown and that of ventral side is white except the chest which is brownish. Its hind legs are shorter but exceeds the four legs in length.

It inhabits the grass jungle in the primeval sal forest region. It is found singly or in pairs and seldom invades the cultivated fields. The main food of hispid hare are young shoots, leaves, barks and roots. Hispid hare breeds, twice or thrice in a year and the litter size is 2-5. The exact population of this animal is not known. In some parts of its habitat the population of this species is becoming rare due to destruction of its natural habitat and hunting by man for its flesh.

Order-Rodentia

Grizzled Giant Squirrel (Ratufa Macroura Dandolena Thomas and Wroughton)

It is distributed in the hill forests of Tamil Nadu and Sri Lanka. It is a large sized squirrel measuring 30-40 cm in head and body length. The tail of this squirrel is slightly longer than the total of head and body length. The dorsal part of its body is brownish, grizzled with white, whereas, the ventral parts are buffy white. The shoulder, toes and head are blackish brown. It inhabits high trees in dry deciduous and moist evergreen forests. Sometimes they have been reported coming to the ground. This squirrel is diurnal in habit. The main food of the squirrel consists of nuts, fruits and insects. One or two young ones are born in a large nest of leaves and twigs situated on the branch of tree. It is in the status of vulnerable due to the destruction of its habitat.

Order-Cetacea

Sperm Whale (Physeter Catodon Linn.)

It is distributed in the tropical and the subtropical seas. It is reported to be common in the Bay of Bengal and of Sri Lanka.

It is also called as Cachalot and specifically distinguished from all other members of the order due to its enormous square-cut head which is provided with a blow hole at its tip. The lower jaw of sperm whale is shorter than the upper jaw and provided with about 30 teeth. The male sperm whale measures 15-18 m in length, whereas, females are 8-10 m in length. The flippers are rounded and broad and dorsal fin is a round hump. The tail of sperm whale is wide, notched and horizontal. The colour of sperm whale is from grey to dark bluish grey and black.

Sperm whale moves rapidly and can make dive for a longer duration. In open seas they are observed to be in small group but they form large schools and move in search of cuttle-fish and squids for food. In northern hemisphere it breeds from March to May but in southern hemisphere breeding season is from September to December. The gestation period is from 15-16 months. Single young one is born at a time. For obtaining the sperm oil, spermaceti and ambergris from its blubber, head and intestine respectively man indiscriminately kills the sperm whale. Thus, its population has decreased and the species has reached the status of endangered ones.

Blue Whale (Balenoptera Musculus Linn.)

It is distributed throughout all the seas. It is reported from the Bay of Bengal and the Arabian sea. This whale is also called as Sulphur-bottomed whale as it is yellow coloured on the ventral side. The general body colour is slate blue with blue grey. It measures generally from 22.5 to 23.5 m, but rarely 34 m, with a body weight of about 80 tonnes. The maximum weight of 152 tonnes has also been recorded for this species. Except a few hairs on lips the body of blue whale is naked and streamlined. The head of this whale is very long. The baleen plates are characteristically jet black.

It is a migratory whale which moves to Arctic and Antarctic zones in search of planktonic Crustacea for feeding in summer. In breeding season it migrates to temperate region in winters. After every 2 to 3 years blue whale gives birth of one young at a time. Due to extensive hunting by man the number of blue whales has severally depleted and reached to the category of extinction. The hunting of blue whale is prohibited nowadays.

Common Dolphin (Delphinus Delphis Linn.)

It is commonly distributed in temperate and tropical seas which sometimes enters in fresh water. In India, it is reported from coastal water of Tamil Nadu, Andhra Pradesh and Gujarat. It is a clipper built slim animal provided with elongated snout forming the beak. The beak is separated from the fore head by a groove. The common dolphin measures about 2.5 m in length. The pectoral fins are three times long as broad and pointed. The dorsal fin is quite prominent. The colour of dorsal body surface is from black to dark grey while that of the ventral part is whitish.

It is social and very intelligent and moves in schools of large number of common dolphin. The speed of its movement is very fast like ships. The main food of common dolphin are cuttle-fishes and marine fishes. The gestation period is about 270 days. The young one is produced from winter to summer season. It is scarcely observed in Indian water where its status is indeterminate.

Gangetic Dolphin (Platanista Gangetica Lebeck)

It is distributed in Ganga, Brahmaputra and Indus rivers and their larger tributaries. This fusiform dolphin is provided with a pair of triangular pectoral flippers, rudimentary dorsal fin and a horizontal tail fluke. The head is elongated forming compressed rostrum. The rostrum is longer in female dolphin than in male. It measures from 2 to 3 m from head to body length. The colour of the whole body is blackish.

It is common in tidal waters. Though it is not gregarious yet a number of individuals are seen in the same area of the river. For breathing, it uses to come on the surface of water. The main food of the dolphin are crustaceans arid fishes. The jaw is well adapted for browning in bottom muds. The gestation period vary from 240 to 270 days. The young one is born from April to July. Due to the loss of habitat and population of tidal waters its population has declined drastically in few years. Its status is indeterminate.

Order-Sirenia

Dugongs or Sea-cow (Dugong Dugon Muller)

It is distributed in tropical waters and estuaries, from red sea and east Africa to New Guinea and Australia. In Indian limits it is found along the western coastal area, Palk strait, around the Andaman Islands and Gulf of Mannar. It measures about 2 to 3 m in length. It is provided with massive head, small mouth, small eyes and circular aperture like ear. The upper lip is projected and forms fleshy pad provided with long bristales. It is devoid of distinct neck. In males the incisors (tusks) are projected through the skin of the upper lip. The colour of sea-cow is grey, bluish grey or brownish grey.

Though the sea-cow is pelagic yet rarely ascends rivers. This sluggish sea-cow shows preference for warm and shallow coastal waters. It feeds on marine grass and algae. Its breeding season is throughout the year. The gestation period is about 325 days. The young one is carried on the back of the mother till maturity. It is important for its oil, meat and fat which are of economic value to human-beings. Due to severe persecution its number is declining day by day. Nowadays its status is of vulnerable species.

Order-Proboscidea

Elephant (Elephas Maximus)

They are found in forests of Uttar Pradesh, Bihar, Orissa, West Bengal, Assam and Karnataka. The Indian elephants are smaller but more intelligent than their African cousins. So, can be easily trained for a number of purposes like dragging and lifting timbers, riding etc.

The male elephant has large tusk but the female none. An elephant with single tusk is called one-tusker 'Ganesha' and is regarded auspicious by the forest people. Their sense of smell is very acute, whereas their eyesight and power of hearing is poor. An elephant drinks 60-80 litres of water at a time and daily intake is about 230 litres. It enjoys playing in water. After bath they have a dust bath for protection against flies. Normally they are harmless but it does not take them long to turn violent if they are disturbed.

The leader of the herd is generally an old and experienced female. The master of the group is a strong tusker. There are 3 or 4 other tuskers in the herd who do not act against the wishes of master.

They breed in the beginning of monsoon and give birth to only one calf at a time. The young elephant sleeps between the legs of its mother like human baby in his mother's arms.

'Project Elephant' : The Government has launched 'Project Elephant' in the eighth plant worth over more than Rs. 20 crore. Considerable emphasis is being given to conserve the habitat of elephants. The habitats are rapidly fragmenting leading to the destruction of the most valuable and economically important forests resulting in confrontation between man and the animal (elephant).

BIRDS

Order-Ciconiiformes

Eastern White Stork (Ciconia Ciconia Boyciana Swinhoe)

It is a winter visitor in Assam and nearby states of north eastern India, south to the Sunderbans. Its breeding area is from southern Ussuri and Amur rivers east to Korea and Japan. It is large sized bird with black and white plumage.

Its height is about 105 cm. It is provided with long neck, very long and red legs and black large bill. The sexes are alike.

It inhabits marshy grassland, drained out follows and moist ploughed field. It is very difficult to approach this bird. It likes to feed on fishes, frogs, rodents, reptiles, crustaceans and also insects. It is known for its bill-clattering sound. This bird was never a common visitor in India. After the early part of 20th century it has not been reported in India. The population of this subspecies is small in the world. Now it is in endangered category.

Order-Anseriformes

Pink Headed Duck (Rhodonessa Caryophyllacea Latham.)

It was distributed in Bihar, Bengal, Orissa, Manipur and Assam. Stragglers were sporadically recorded in winter from Uttar Pradesh, Punjab, Andhra Pradesh, Maharashtra and Tamil Nadu. It was also resident of Burma, Nepal and Tibet. It is about 60 cm in length. The neck of male is bright pink and rest of the body colour is dark brown with a pinkish buff speculum, the female adult is dull in colour. Its pink coloured area is not sharply distinguished from the rest of the dull brown body colour. The immature bird is of light brown colour with pale rosy white neck and head region.

The bird was generally met with singly or in small groups. During the months of breeding (May-July) they were in pairs. The bird was of shy and secretive nature. It was a surface feeder. The food consisted of aquatic vegetables and animals. Due to shrinkage of its habitat and hunting the population of pink headed duck started declining as early as 1878. The last authentic sight record of this bird in India is 1935 in Darbhanga (Bihar). It may be considered an extinct species in India.

Order-Falconiformes

Himalayan Golden Eagle (Aguila Chrysaetos Daphanea Severtzov)

It is distributed approximately south to the Himalaya east to Nepal, eastern Arunachal Pradesh between 1850 to 5500 meter elevation. This golden eagle is also reported from eastern Iran, Baluchistan, Afganistan to Central Asia, This large and graceful bird measures about 100 cm in length. The feathers are blackish brown to golden buff or tawny. The wings are dotted with buffy white. The coloration in both the sexes are alike.

It inhabits high mountainous tract of land. During flight it makes powerful wing-beats coupled with long glides. It attacks its prey from a height with terrific velocity. The main food of this eagle is snowcocks, pigeons, monals, chukors, marmots, foxes, hares, martens, flying squirrels and sometimes fawns of musk deer. This silent looking eagle calls like a dog shrill yelp. The breeding season is from January to March. Within the Indian limits it is becoming rare due to destruction of its habitat. Now it may be considered as endangered category.

Shahin Falcon (Falco Peregrinus Peregrinator Sundevall)

It is distributed in India, Pakistan, Sri Lanka, Burma, Bangladesh and in eastern China. It is found throughout India from Kashmir east to Arunachal Pradesh and south through the peninsula to the Southern Kerala.

This broad shouldered falcon measures from 38 to 45 cm in length. Its dorsal part is slaty, black head and provided with check-strips. The breast and throat are pinkish white while rest of the ventral parts being ferruginous cross-barred with black abdomen downwards. The size of the male is smaller than the female.

It inhabits steep rugged hills, thin jungle as well as open country area. With a few rapid wing beats followed by a glide it can attain tremendous speed. It feeds on pigeons, bats, mice, rats, partridges, parakeets, nightjars etc. It prefers to hunt soon after dawn and in the afternoon well into the dusk. Its breeding season is between March to May in the Himalaya and from January to April in the peninsular Indian hills. It is considered as the most prized falcon for the faloconery. The shahin falcons are killed by human-beings and are caught by faconers. Now it may be treated as endangered species.

Order-Galliformes

Pheasant-grouse (Tetraophasis Szechenyii Madarasz)

It is distributed in northeastern border of India, Southwestern China and eastern Tibet. It is a brownish coloured bird and measures about 65 cm in length. The female is almost similar to the males but devoid of spur. The head is provided with narrow streaks of black colour. The chin and throat are fawn coloured. The feathers of the lower back and rump are with narrow edges. The outer tail feathers are provided with black and white tips.

It inhabits the alpine zone between 3350 to 4600 m altitudes, covered with fir and rhododendron. They are found in a group of small parties of about four-six during foraging among the scrub and short grassy vegetation. It feeds on green leaves, tender shoots, bulbs and roots. In India it is a rare species.

Assam Bamboo Partridge (Bambusicola Fytchii Hopkinsoni Godwin-Austen)

It is distributed in the hills of northeastern India south of the Brahmaputra river extending into Sylhet and Chittagong Hill Tracts of Bangladesh, and Chin Hills and Burma. It measures about 35 cm in length. The mantle is grey brown, streaked and spotted with chest nut and black. The tails are provided with brown and buff bars. Both the sexes are similar in colouration.

It inhabits open scrub jungles near cultivation in the foothills and adjoining plains. It damages mixed scrubs of willow, oak and tall grasses along the bank of streams. They are generally found moving in groups of five to six birds during morning and evening hours foraging near streams. The main food of this bird is buds, grains, tender shoots, seeds, as well as ants, grasshoppers, termites, crickets etc. and grubs. Its breeding season it from March to May. Due to excessive hunting and destruction of its habitat the population of this bird has declined much and now it has become a rare bird of the country.

Peacock-pheasant (Polyplectron Bicalcaratum Linn.)

It is distributed in the hills and foothills of Northeastern India, Bangladesh, Burma, Thailand, Sikkim, Laos, Vietnam and Hainan Island. It measures about 64 cm in length. It is greyish brown with a long tail and an erect crest of hair like feathers. The mantle, wing-coverts and the tail are marked with violet greenish blue ocelli with white rims. The colour of female is almost similar to that of male but the colour of female is slightly duller. The immature male is like the adult female.

It inhabits dense evergreen forests from foothills to 1200 m elevation. It may be in pair or singly. The main food of this bird consists of seeds, grains, berries, figs, snails, grubs, etc. Its usual call is deep guttural hoo quickly repeated about 7 times. The breeding season is from March to June. The male has a spectacular courtship display. Due to rapid destruction of its habitat and random killing for flesh its population has reduced much. Nowadays it is under the category of an endangered species in India.

Indian Peafowl (Pavo Cristatus Linn.)

It is distributed in India, Bangladesh and Sri Lanka. In India, it is found from south and east of the Indus river, Jammu and Kashmir, east to Lakhimpur district in Assam and whole of the Indian Peninsula. It measures from 90 to 120 cm (excluding train) while its total length including the train is between 200 to 225 cm. It is provided with small head bearing a fan shaped crest, a white patch under the eyes and long slender neck. The sexes are dimorphic. The cock (male) is provided with metallic bronze-green train of about 200 elongated feathers. A part of the wing is chestnut in colour. The female is smaller than the male and brown coloured and devoid of the train. The immature male is like the female.

It inhabits the dense scrub jungles and deciduous forests and prefers to be near the water resource. If disturbed, it runs away and flies during crossing the river. It feeds on fruits, seeds, young shoots, lizards, insects, snakes, grubs etc. The common call is a loud and horse 'may-awe'. The breeding season is from January to October during which a cock is surrounded by five or six hens. The male has an elaborate courtship display by erecting and quivering of its trains.

Being a National bird of India it is fully protected under the Wild Life (Protection) Act of India.

Order-Gruiformes

Blacknecked Crane (Grus Nigricollis Przevalski)

It is a breeding bird of central Asia from Laddakh to Kokonor. It is a winter visitor from Vietnam, Yunnal in China, Arunachal Pradesh and Bhutan. It measures about 156 cm in height. It is a pale ashy grey crane with black head and neck. The crown and lores are bare and dull red coloured. The tail is covered by drooping black plumes. The female is smaller than male but both are similar in coloration.

It inhabits reed beds and open fallow paddy fields in the winter quarters. They breed on the shores of high altitude lake areas and swampy bogs between 4300 to 4600 m. The main food of this bird are fallen grains, roots, tubers, insects and small vertebrates. The breeding season is between May to July. It has elaborate courtship display in which both the sexes take part. It constructs the nest on the ground with vegetations. There is no exact information regarding the population-size of this bird but reports from different places indicate dwindling in its population. This crane may be considered as endangered species.

Lesser Florican or Leekh (Sypheotides Indica J.F. Miller)

It is distributed in India and southern Pakistan except in

very humid area. It measures about 45 cm in length. The male is slightly smaller than the female. The chin, centre of the throat and a broad band running down the sides of the lower neck of males are white. The upper plumage is sandy buff, vermiculated and dotted with blackish colour. The females are sandy buff in colour, dotted with blackish, and has blackish arrowhead markings on the back. In the non-breeding plumage males are like the females but provided with much white part on the wings.

It inhabits tall grass patches, scattered bushes and cultivated tracts. The main food of this bird is lizards, frogs, insects, centipedes and also fruits, herbs, grasses and shoots. The males emit a harsh frog like croack during the course of courtship display. The breeding season is between July to September. Due to heavy destruction of its habitat and hunting for its flesh, the Lesser Florican has become now rare species.

Tibetan Crane (Grus Nigricollis, Przevalski)

This beautiful and rarest bird inhabits from Laddakh to the KokoNor region of Tibet. In Laddakh it is called as 'Cha thung thung' meaning the bird with long legs. Its population is very poor and if not protected will be extinct. It is assumed that only 100 birds of this species are found in Tibet and Laddakh. Therefore, the World Wild life Fund has sought the cooperation of army to protect this species.

They migrate to the lower parts of Bhutan and Arunachal Pradesh during winter. For feeding, they are noticed to be in a group of one male and one female with young ones. Each party has its own feeding territory separately at a distance of about one kilometre.

Before dusk each party comes back to a specific field closer to the roost to have their last feed of the day and after sunset they reach their roosting ground in a marshy fallow field, a little away from the river bed.

Great Indian Bustard (Ardeotis Nigriceps, Vigors)

In Rajasthan every one is aware of the fact that the main interest of the hunters in the desert is the Great Indian Bustard. It is a handsome bird of about one metre height, 2.5 metres with wing span and about 18 Kg in weight. The female is lighter and similar in appearance to the young of Australian ostrich. The dorsal side is brown and ventral surface is white. The males are provided with black crests of 5 cm on head. A black band runs across he breast which is wider in young males. It has got larger eyes of pale yellow colour. The tail is short and brown coloured and provided with light grey and brown bands. The bird has long and powerful legs of dull yellow colour. The great Indian Bustard keeps its head and bill raised up at an angle of 45 degrees. The neck of male is thicker and whiter in colour.

It feeds on snails, centipedes, insect, lizards, small snakes, grains grasses, leaves and fruits of its choice. It relishes dung beetles during rainy season. It voraciously feeds on field mice as such very helpful to the farmers. It feeds early in the morning and late in the evening and takes rest during day time. The bustard is a good runner and heavy in takeoff, but once launched if flies very well.

The bird is polygamous and maintains a harem of several females. In mating season male courts like peacock and dances like a sarus crane. The mating calls are very specific, mixed with the roar of lion and snuffing of a camel which may be heard long before sunrise and long after dusk. The female after mating lays single egg at the selected spot undisturbed by man. After 20-28 days of incubation by mother, the hatchlings come out which çan walk immediately.

Success has been achieved in breeding the great Indian bustard at Bikaner and Jodhpur zoo which will prove to be of much importance for saving this species from extinction. The breeding centres should be established in all the zoos of Rajasthan and Gujarat.

Conservation : It is a migratory bird, declared as rare species by the Indian Board for Wild-life and was recommended for total protection in 1952. They were included in Wild Life Protection Act,1972, and in the schedule I of October 1977. The preservation of this bird is of great significance. The International Council of Bird Preservation, the International Union for Conservation of Nature and the World Wild life Fund are very serious for its preservation.

In the Thar desert of Rajasthan a desert sanctuary has been established for the preservation of rare species of flora and fauna and their environment for which 3,000 square Km has been set aside in Jaisalmer and Badmer Districts. This sanctuary would be established as a Desert National Park.

Siberian Crane (Grus Leucogeranus)

It is among the rarest and most famous birds of Asia which is more beautiful than the Indian cranes. It is also called as white crane means 'Shveta Sarasa' in Sanskrit. This bird was previously found in China, India and Pakistan. In India it is regular visitor to Prayagpur lake in Uttar Pradesh, eastern Bihar and lakes at Keoladeo Ghana in Bharatpur. The hunting of this bird in Soviet Union and Afganistan, dearth of its natural habitat in India, and disturbance by man are pushing Siberian crane to the point of extinction.

The Siberian cranes are jolly birds. In the morning they salute each other by sending out sharp and loud calls 'Turn' which is known as morning greeting. They start dancing and emit long call whenever friends from neighbouring lakes reach them. At Bharatpur Sanctuary Siberian cranes are mostly found in flocks of five or six and some live in pair. Some pairs have the young ones of 4 to 5 months as a third bird. All the young ones are as large as their parents but can be distinguished by their mottled-white and cinnamon-brown markings. They are completely dependent on parents for their food. It is notable that it remains mateless for the rest of its life after the death of its partner. Thus, they are unbelievably faithful towards their mates.

Greater Flamingo (Phoenicopterus Ruber Roseus, Pallas)

It is called as 'Hans' in Sanskrit and poets have mentioned the ability of this 'Rajhans' to separate milk from water. Undoubtedly, the Rajhans is skilled in separating palatable and unpalatable (Kshira-Nira) items from the mixture of slush and food it picks up in its bill. Flamingoes feed on worms, insects, larvae, seeds of marsh plants etc. They usually stay here up to March every year.

Two species of flamingoes are found in India *i.e.* residents of India and nomadic. The food habits of both the species are different so both coexist with harmony. They can feed during the day time or at night and after feeding, they collect in smaller groups to take a nap.

Jame's flamingo, a rare bird, was considered to be extinct in 1951 and was given its names *Phoenicoparrus jamsei* in the honour of a British Ornithologist, H. Berkeley James. Further it has been found living in icy saline water of Laguna Colorado at altitudes of 4112 metres in the Belivian Andes.

The greater flamingo is white in colour and its wing coverts are more white and red than lesser flamingo. It is about 132 cm in height, whereas, lesser flamingo is about 100 cm in height.

Flamingoes are found in large flocks and their breeding ground in India is the Great Rann of Kutch where they established a colony. The sky appears like a painting in pink and blue as thousands of flamingoes fly off together from the lake. They attain maturity at the age of about two years and mass wedlock takes place after initial courtship display. They are monogamous and their family bonds are strong.

Flamingoes arrive here every year at Vedaranyan Bird Sanctuary and the number of flamingo was about 50,000 in these lakes in 1980-81. There is abundance of food for water birds in this sanctuary lakes. So, flamingoes arc getting full protection in this Bird sanctuary.

Order-Charadriiformes

Jerdon's Courser (Cursorius Bitorquatus Blyth)

It is distributed in India, Pakistan, Bangladesh, Burma and erstwhile Indochina. It measures about 40 cm in length with a blackish brown upper plumage and white under parts. A dark stripe passes down the middle of the white rump and tail. The tail is slightly forked and wings are long and pointed. The bill is laterally compressed and knife like. The colour of the leg is bright red. The male is slightly larger but the colour of male and female is similar. It inhabits large rivers with sandbanks, estuaries and inshore coastal waters. They are found in small flocks skimming over the water. Its food consists of crustaceans and fishes. The usual time of its feeding is day out sometimes they also feed in moonlit nights. Its call is a nasal 'Kap Kap'. The breeding season is from February to April. Due to heavy population in the rivers its population declined and becoming scarce. Now this species is under the category of rare bird.

Order-Strigiformes

Great Horned Owl or Eagle-owl (Bubo Bubo Linn.)

It is distributed throughout India, Europe, Pakistan, Northern Africa, Western Asia, Bhutan, Nepal, Bangladesh and Burma. It measures about 55 cm in length and with dark brown streaks and tawny-buff mottlings. Above the head two prominent horns or ear-tufts are found. The tail is moderately long and banded. The eyes are large, yellowish and forwardly directed. The breast is provided with broad and black stripes while upper abdomen has narrow stripes. The sexes are alike.

It inhabits rocky hills, ravines, clifts and cultivated country area. It is commonly found near the bank of rivers and streames. It is a nocturnal bird but can be seen active during the day period. The main item of food are mice, rat, birds, snakes, lizards, frogs, fishes, crabs and insects. The breeding season is from November to May. It lays the eggs on bare soil. Due to gradual destruction of its habitat and shortage of food items this bird is becoming scarcer day by day. In India this bird has reached the status of endangered species.

Order-Caprimulgiformes

Ceylon Frogmouth (Batrachostomus Moniliger, Blyth)

It is distributed over the southern rain forest tracts of Western Ghats from north Kanara district of Karnataka, south to Trivandrum district of Kerala and Sri Lanka. It measures about 25 cm and of the size of myna. There is colour differentiation in both the sexes. The male bird is greyish brown, vermiculated and dotted with white, buff, black and chestnut. The head region is black spotted provided with white margin. The immatured bird is barred with brown.

It inhabits dense evergreen forest and secondary jungles. This bird is nocturnal in habit. The main item of its food are beetles, moth and grasshoppers. The breeding season is between January to April in India. Generally only one egg is laid. It is a rare bird at least in the Indian part of its distributional range.

Order-Coraciiformes

Rufousnecked Hornbill (Aceros Nipalensis Hodgson)

It is distributed in India, Bangladesh, Burma, Thailand and erstwhile Indochina. In India it is found in the sub-Himalaya from Nepal, east to Arunachal Pradesh and the hills south of the Brahmaputra river in northeastern states. It measures about 125 cm in length. The colour of two sexes differ. The colour of the head, neck and breast, in the male, are rufous while its abdomen is maroon. The bill is horn shaped, yellow coloured and provided with blackish vertical ridges on the sides of the upper mandible. The dorsal part of the body are black coloured glossed with dark green colour. The tail feathers at the terminal half are white. Around the eyes there are blue patches. The female is smaller than the male and is black in colour. The immature hornbill is provided with smaller bill which has no lateral ridges.

It inhabits evergreen and deciduous forests with tall trees. It is primarily an arboreal bird found in pairs or in flocks of about five on the branches of trees. For feeding it occasionally descends to the ground. It makes variety of noise like croaks, loud roars and cackles. The main food of this bird are berries, fruits, nutmeg, larger drupes etc. The breeding season is between April to May. Though it is very commonly found in some of the protected areas but in many of its distributional ranges this bird has become rare species in India.

Indian Pied Hornbill (Anthracoceros Malabaricus Gmelin)

It is distributed in Himalayan foothills from Punjab, east to Bhutan, northeastern states of India, Orissa, Madhya Pradesh, Andhra Pradesh, Bihar, and West Bengal. It is also found in Burma and Bangladesh. It measures about 90 cm in length. The dorsal part of the male bird is black, glossed with green and the ventral parts are white. The black tail feathers are provided with white tips. Its bill is large wax-yellow coloured with blackish patches. The bill is surmounted by a high, ridge-like casque with pointed anterior end. The males are larger than the females.

It inhabits moist deciduous and evergreen forests, often invading groves of large mango trees. This arboreal bird is found in a group of 5 to 10 hornbills. It has also been found feeding on the ground. Its call has been rendered as 'Kak-Kak, Kak-Kak'. The food items of this bird consist of fruits like berries, drupes and wild figs, and snakes, lizards, fishes, nesting birds and other small animals. The breeding season is in the months of April and May. Due to deforestation and hunting and less availability of food this bird has become rare species.

REPTILES

Order-Testudines

Leatherback Turtle (Dermochelys Coriacea Linn.)

It is distributed in the warmer and temperate water of Indian, Atlantic are Pacific oceans. It is the largest species of the living turtles of the world. The length of its carapace is between 150 to 180 cm and the average weight about 400 Kg but the maximum weight of this turtle has been recorded be 735 Kg. There is no cornified epidermal structures and jaws are devoid of flattened horny plates. The body of this turtle is almost triangular in shape. The head of this turtle is large and the neck is massive. The lower jaw provided with a median hook while the upper jaw has a pair of strong cusp. The limbs are clawless and paddle shaped. The carapace is slaty black with white spots. The ventral surface is a mixture of pink, white and black colours. The mature females are provided with the pink spots on the crown of the head.

It inhabits open seas but highly migratory in nature. The food of leather-back turtle consists of large marine invertebrates, fishes and marine plants. The pea of breeding season is during May and June but they can breed three or four times in a year. It lays 100 to 200 eggs at a time in 50 to 100 cm deep pit on a sandy beach. To confuse the predator of its eggs the female turtle makes a number of spots similar to real egg deposition spot. The incubation period of eggs varies from 60-70 days.

The status of this turtle is indeterminate but it is rare in its habitat.

Green Sea Turtle (Chelonia Mydas Linn)

It is distributed in all the tropical and subtropical oceans of the world. They have also been reported in the Mediterranean sea but do not breed there. It is very common in the area of Andaman Islands. It is largest of the hard-shelled species of marine turtle and is considered to be the most valuable reptile of the world. The carapace of this flat, heart shaped turtle is never completely deceived. Its rounded head is very small in comparison to the body. Neither its head nor legs can be pulled under the shell. Its flippers are long. The carapace of adult is about 105 cm in length and its colour is green, olive, brown, buff or black. The colour pattern of hatchlings is dark blue-black. The heaviest turtle recorded to be of 430 Kg but generally its weight is between 150 to 175 Kg.

It inhabits inshore waters less than 25 m in depth and sheltered by reefs. The simultaneous sweeping of the front flippers help in movement in water and on land both. The main item of its food are seaweeds and submerged marine algae. For the purpose of breeding and food search it migrates up to great distances. It mates in the months of May to July at sandy beaches. For egg laying the females are congregated in large numbers on sandy beaches between June to November. Single female turtle can lay 350 to 600 eggs.

The eggs are laid during early part of night in (60 cm) deep hole dug by female turtle. Just after egg laying the mother turtle fills the egg containing hole with sand and comes back to the sea. The incubation period of egg is between 40 to 50 days. The population of green sea turtle is declining due to excessive exploitation for eggs, cartilage and flesh. It serves as an important food for persons near seashores and Islands. The soup of its cartilage is regarded as delicacy. For the protection of this species some societies, are working throughout the world.

Olive Ridley Turtle (Lepidochelys Olivacea Eschscholtz)

It is distributed in the tropical waters of the Pacific, Indian and south Atlantic oceans. The length of the carapace of this flattened marine turtle is between 65 to 70 cm. The head of the turtle is large and triangular and different bones do not fuse completely until late in life of the turtle. The plastron consist of paired gular, pectoral, humeral, abdominal, femoral and scutes. The orbits are distinctly large. There are one or two claws in the limbs. The head is grey-brown, the carapace olive grey, the shield yellow white and the plastron is creamy yellow in colour.

It inhabits the coastal waters as well as the open sea. The main item of its food are shrimps, crabs, fishes, jelly-fishes, molluscs and marine algae. It breeds between September to March. The number of eggs laid per female may vary between 100 to 135 and the incubation period of eggs ranges between 50 to 60 days.

Order-Squamata

Common Indian Monitor (Varanus Bengalensis Daudin)

It is distributed throughout the Indian subcontinent. This medium sized, dark brown monitor measures from 70 to 75 cm in the head and body length. They are provided with about 100 cm long, strong and compressed tail. The scales on the crown are larger than other regions. The snout is convex terminally. They are provided with acute, long, sharp and recurved teeth. The tongue is forked, very long and protrusive.

It inhabits desert, forest, bank of river, side of nullah, tidal cheek, marshy land and sea coast. This monitor occupies hollows of trees, burrows, dense vegetation, cracks and crevices etc. It is a very good swimmer and climber, and graceful in movement. When it is caught or approached in the wild state then it can bite hard, lashing with the tail and scratching vigorously with the help of its powerful claws. It feeds on small terrestrial vertebrates, ground birds, eggs, arthropods and fishes. The breeding season is from July to September. It lays 20 to 30 eggs in a clutch in the holes which are further covered with leaves, sand and rubbish. Its population is declining alarmingly throughout the country. The main cause of its declining population is excessive exploitation of the adults for their valuable skins.

Indian Desert Monitor (Varanus Griseus Daudin)

It is distributed in India, Pakistan, Afganistan, Iran and north Africa. In India it is found in Maharashtra, Punjab, Rajasthan and Madhya Pradesh. It measures about 50 cm in the head and body length. Its tail is about 80 cm in length. In some case the body of monitor is cross-barred with brown particularly on tail. The snout is depressed at the end. The digits are elongated. The tail is whip like and rounded. The scales of crown are large than the scales of other parts.

It inhabits burrows in undulating sandy grounds with vegetation. As per need it can run with great speed and the body remains raised above the ground. It feeds on grasshoppers, crickets, locusts and other small vertebrates. The eggs are laid in sand in a clutch of 15 to 20 eggs. It is also hunted for its valuable skin, therefore, its population is declining day by day.

Order-Crocodylia

Crocodile

The literature and our knowledge about Indian crocodiles is scanty because they live under water surface while feeding. Crocodiles belong to three families *viz.*, Alligatoridae, Crocodilidae and Gavialidae. Of these, only 3 species of crocodiles found in India are *Crocodylus porosus*, the estuarine crocodile found in Sunderbans of Bengal, *C. palustris* Lesson, commonly known as the Magar or Naka and *Gavialis gangeticus* (Gmelin) or Gharial. *C. Porosus* is a salt water crocodile which inhabits coastal mangrove areas.

Gharials inhabit the river Ganga and are provided with long snouts. They were earlier distributed in Nepal, Pakistan, Bangladesh and in the rivers Indus, Brahmaputra, The Mahanadi and Brahmani. Before world war second they were in large numbers but after independence they began to disappear due to establishment of a number of developmental projects. It is dark-brown on dorsal side and pale coloured on ventral surface. The young ones are pale brown and grow darker with age. The snout is long slender. The female is known as Goh. They can attain a length of about 30 feet.

The Magar is dark brown on the dorsal surface but the ventral part is pale coloured. The average length is 10 to 12 feet but specimen of 18 feet or more may be obtained. The snout is short and broad. The top of the head is rough with ridges. Male and female can not be distinguished.

The crocodiles and alligators can move on land for few kilometres but gharial can not move fast due to heavy abdomen. To move forward the gharial has to drag its heavy body, therefore, they do not go far on land, leaving the water source. During day, gharials come out of water to take rest and sleep. In the month of February and March they prefer to bath in sun light for the whole day. During cold waves they generally live under water. They possess very sensitive ears.

The Magar feeds on almost any living or dead creature it can find in the water or on the ground near by. The young starts feeding on insects and smaller fishes. It can run at a considerable speed with the body raised well above the ground to catch its prey in the jaw and drag it quickly into the water. It can walk leisurely also as it happens during its travels in the search of rivers or bigger tanks which is done mainly in the dark. How they find their way is a mystery as none has been recorded as going in the wrong direction. They have been found attacking goats and sheeps grazing near the bank of rivers. Sometimes they even drag cows into the river. Hindu mythology has a story of a crocodile dragging an elephant drinking water on the bank of river by the fore-foot into the water until only the tip of the trunk was left when supernatural powers intervened to save the elephant. Crocodiles usually feed in the morning and evening.

The Gharial is reputed to be a fish eater but like the magar it also eats any thing dead or alive. Bony animals and human body were also observed in the stomach contents of dead crocodiles.

The crocodiles can be netted but the bigger ones tear the net and escape. In shooting crocodiles with rifles the brain and neck shots are fatal. The professional crocodile catchers drag them with the help of rope tied around the belly.

The Gharials mate during the months of January and February. The female attains adulthood after 10 years of age. The big female has been observed mounting small female which may be due to the mate's unability to serve too many females in the area. During courtship male and female move their heads up and down in water, slowly come closer and parallel to each other and then move to shallow water where male mounts the female and the pair glide in deeper water for coition.

After 32 days of gestation period female gharials look for suitable nesting ground far from the water line on the high sand bank. After egg laying, female fills the nest with sand gently pressed with her hind legs. One female can lay a maximum of 100 eggs. The gharial eggs, believed to be a cure for certain diseases, used to be eaten by tribals in the past. After incubation period of 60-80 days eggs develop into hatchlings. The hatchlings make a cracking sound from beneath the sand calling for the mother. The hatchlings are 30-40 cm in length and about 100 grams in weight. Only one per cent of hatchlings are recorded to survive in the initial struggle for existence. The Narayani and Girwa, originating in the snowy Himalayas, are the two rivers in which gharials breed in their natural habitat today.

The hides of gharials are very tough and durable and have a beautiful texture. Its hide is very costly as after being polished it takes best finish, so the demand in leather industry is very high.

Conservation : The Government of India has provided full protection to the 3 species of crocodiles under wild-life (Protection) Act 1972 and banned the trading and export of gharial hide. The protection and conservation of gharials and magars has been provided by the Govt. of India since 1975 in cooperation with the Food and Agriculture Organization and the United Nations Development Programme. A number of protection and conservation projects are under way at 22 sites in Andhra Pradesh, Orissa, Uttar Pradesh, Rajasthan, Madhya Pradesh, Kerala, Tamil Nadu and West Bengal since 1980. For the first time gharial's eggs were hatched in captivity with human efforts at Tikerpada and Kukrail near Lucknow in 1975. In 1978 the Chambal river, flowing through Uttar Pradesh, Madhya Pradesh and Rajasthan was declared The National Chambal Sanctuary' encompassing an area of 12,980 square Km. The breeding of gharials and magars, establishment of hatcheries, nursing of hatchlings, canning of flesh of big gharials, tanning of hides and manufacture of variety of leather goods can be taken up by the people of the valley on cooperative basis.

Necessity of Wild-life Conservation : It is fact that no animal or plant species is everlasting and that none has yet existed for more than a few million years or so without evolving into some different forms. No doubt that this process is usually very slow in nature but the trend of extinction- of species has been noticed to be swifting alarmingly in the recent years. The so called developmental activities of human beings-appear to be responsible for the extinction of the magnificent animals of our jungles. Along with the alarmingly decreasing population of some wild species, their habitat has either shrunk drastically or is being made unfit for their survival.

The question arises that why we are making efforts for the conservation of wild-life not only in India but throughout the world by spending a lot of money, energy and man power, when we are struggling for economic independence? It may also ask that what difference does it make if Cheetah and pink headed duck have disappeared, and the lion, tiger and rhinoceros would vanish in future. But the reality is far away from the range of these questions.

Our former Prime Minister Smt. Indira Gandhi in her message while inaugurating the 'Tiger project' on 1st April, 1973, said that "the narrow outlook of the accountant must give way to the wider vision of recreational, educational and ecological value of totally undisturbed areas' of wilderness. Is it beyond our political will and administrative ingenuity to set aside about one or two per cent of our forests in our pristine glory for this purpose ?"

One of the most important heritage of human-beings is the magnificent wild-life which is in so many ways more important than the man made wonders of the world, because if once destroyed wild-life may not be restored. The future generations could miss the majestic beauty of tiger, lordly elephant, grazing rhino, magnificent bison and the elegant swamp deer. If these wild animals become extinct then how dull and colourless would animal kingdom be? If proper conservation measures are not be taken in near future these animals may fall in the category of 'Endangered' ones.

It is a fact that the life on earth is a delicate balance. Due to the presence of chloroplasts, plants convert carbon dioxide of the air and water into organic substances. Herbivores obtain energy by feeding the plants. The carnivores derive their energy by feeding the herbivores. Lastly decomposers release this bind energy back to the nature. Thus, producers, consumers and decomposers form ideal food chains in the nature. The continuation of such type of food chains and cycles, in nature, is essential for existence of the living world. Any major disturbance at the level of plant and animal resources may result in serious disturbance in the natural balance which may threat the existence of man. Therefore, renewability of the natural resources and other life supporting systems must be maintained for endless duration. Species may be able to adopt themselves to the slowly changing environment in its natural process. The integrated approach of these considerations, in essence, is the crux of wild-life conservation.

Categories of Rare Species of Animals : International Union for Conservation of Natural Resources (IUCN) has categorised the rare animals in Endangered, Vulnerable, Rare, Threatened, Out of danger and Indeterminate. This categorisation is based on the status of present and past distribution, the rate of decline of population, position of natural habitats and biology and potential value of the species.

Endangered (E) : Species, whose number has declined to a critical level or their habitats have been destroyed drastically and reduced up to a great extent, are in danger of extinction if these casual factors continue operating.

Vulnerable (V) : The taxa likely to enter into the category of endangered species in near future if the casual conditions continue operating. These casual conditions are the declining population of species because of over-exploitation, extensive destruction of habitat or other environmental disturbances ; taxa with populations that have been seriously reduced and whose ultimate security is not yet assured ; and species still with abundant population but are under threat due to adverse conditions in the whole region.

Rare (R) : Though some species are in very small population in the world yet they are not under the category of endangered or vulnerable. But due to their population presently the taxa are at risk. Such species are commonly localized within restricted geographical regions or they are distributed in very wide range but with very thin population density.

Threatened (T) : Species which are under one of the three categories *viz.*, Endangered, vulnerable or Rare, are termed to be Threatened Species from the conservation point of view.

Out of Danger (O) : The species earlier included in one of the Endangered, vulnerable or Rare categories but presently they are in more secure condition due to the application of effective conservation programmes, come under this category. The earlier conditions of threat to their existence has been removed.

Indeterminate (I) : The species which are suspected to be in one of the threatened (Endangered, Vulnerable, Rare) categories but for which insufficient information is available, are included in this category.

Wild-life Conservation Activities

Though by now we have lost too much of our natural resources since the turn of twentieth century yet it is encouraging to note that we have gradually developed an awareness for the conservation of wild-life throughout the world. The British rulers started to feel the importance of wild-life and they enforced different acts *viz.*, the Arms Act, the Elephant Protection Act, the Rhino Protection Act etc. for the conservation of wild-life. First major step towards the conservation of wild-life was taken in free India just after independence in 1952 with setting up of a Central Board for Wild-life which was renamed as Indian Board for Wild-life-IBWL. Later on a number of State Governments constituted such boards but due to lack of proper Central control and coordination they were unable to do any thing substantial.

The Indian Board for Wild-life is now responsible to advise the Government of India regarding wild-life policy in the country. Prime Minister of India is ex-officio Chairman of this Board. The Director Zoological Survey of India, the Director Botanical Survey of India and State Chief Conservators of Forests are ex-officio members of this Board. There are many specialised wings like Zoo wing, the Bird wing and the Flora wing for giving proper advise to the IBWL. Most important achievement of IBLW is the enacting of wild-life (Protection) Act 1972. Other achievements of IBWL are the establishment of Zoological gardens, Sanctuaries, National Parks, formulation of a National Environmental Conservation Policy, revision of the national forest policy etc.

India became a member of the International Convention on Trade in Endangered Species of Wild Fauna and Flora (CITES) in 1976 to enforce restriction and proper regulation of the growing global trade in Endangered species. For proper management, the Inspector General of Forests and Director of Wild-life preservation have been designated as the Management Authority. The Management Authority is assisted by the Director Zoological Survey of India, the Director Botanical Survey of India and the Director Central Marine Fisheries Research Institute.

For the efficient execution of the planning to control the trade in endangered species, four Assistant Directors of Wild-life have been appointed at the four international air ports of the country *viz.*, Delhi, Kolkata, Chennai and Mumbai. The wild-life and its products can be exported or imported, only after proper inspection and final permission of the Assistant Directors appointed at these air ports.

In last few years a number of International Organisations like, International Union for Conservation of Nature and Natural Resources (IUCN), World Wild-life Fund (WWF), International Council for the Bird Preservation (ICBP) etc. came into existence for creating awareness and boosting up the conservation programmes of wild-life throughout the world. India is an active member of all these organisations.

For the proper management of wild-life three basic requirements *viz.*, adequate food and water, place of refuge and place of breeding in safety must be provided to wild animals.

For achieving the above requirements following practices are common :

(1) National Parks, Sanctuaries and Reserved forests are not exploited.

(2) Grazing of domestic live stocks in protected area is dangerous for the health and prosperity of wild animals, therefore, grazing by domestic animals is avoided up to a great extent.

(3) To prevent the ecological hazards caused by pesticides, cultivation near sanctuaries and reserved forests are avoided.

(4) Monoculture of plants in forests is discouraged.

(5) Water holes and river sides of the forests are properly protected from poachers.

(6) For the proper health of wild animals, dust baths and artificial salt licks are managed.

(7) Researches on threatened species of wild animals are encouraged to promote their status even by breeding them in captivity.

Agencies Working for the Conservation of Natural Resources International Union for Conservation of nature and Natural resources (IUCN)

IUCN, founded in 1948, is a non-Government International Organisation dealing with the conservation of Natural Resources. This organisation is highly efficient network of both, Government and non-Government organisations, Scientists, conservation experts dedicated for the conservation of wild-life and other natural resources.

More than 500 member organisations and 116 Government agencies of 110 countries, and most of the National and International non-Government Conservation Organisations are concerned with the IUCN. It receives financial support from its own members, from WWF, UNEP and a number of other sources. Function :

(1) To monitor the status of conservation throughout the world.

(2) To prepare plan for dealing with conservation problems.

(3) To provide assistance for implementing conservation.

(4) To coordinate between members and commission.

(5) To manage the conservation projects sponsored by the World Wild-life Fund (WWF) around the world.

(6) To manage United Nations Environmental Programme (UNEP).

World Wild-life Fund (WWF) : WWF is the greatest voluntary organisation set up in 1961. It raises funds for immediate conservation of wild-life. It also promotes public awareness with regard to conservation of wild animals. Its headquarter is at Gland, Switzerland. The Giant Panda is the symbol of WWF. HRH Prince Philip, the Duke of Edinburgh is the President of WWF.

An Executive Committee consisting of the President and

10 members of the Board of Trustees of WWF is responsible for the administration, disbursement of funds and conservation awareness activities. On the recommendation of this committee the Board decides WWF policy. The members of the Board of Trustee are of various fields like Science, Art, Medicine, Finance, Diplomacy etc. Fund creation. The fund for WWF is collected through national organisations and international headquarters in Switzerland. It raises money through following sources :

(1) Interest on an endowment fund of more than 10 million dollars contributed by 1001 persons of the world.

(2) Donations given by general public.

(3) Over one million regular members contribute to WWF.

(4) A large number of charitable foundations give regular grants to WWF.

WWF promotes its activities in collaboration with IUCN. Its activity gets cooperations from UNO, UNESCO, UNEP and FAO. The Conservation Programme Committee, consisting of members from WWF, National organisations, Trustees and Senior Staff of WWF and IUCN, reviews the conservation objectives twice a year. WWF affiliated organisations are entitled to allocate one third of its income to domestic programmes and the remaining two third are allocated to international projects through or in cooperation with WWF using the technical advice of the IUCN.

World Wild-life Fund-India: It came to existence in 1969. Its headquarter is in Mumbai and three regional offices are in Kolkata, Delhi and Chennai. WWF-India has a Board of eight Trustees. The Boards duty is to draw attention of Indian Government and State Government regarding the conservation of wild-life. It promotes wild-life conservation education programmes through educational institutions and 'Nature clubs.'

Research projects on the conservation of wild-life are

supported by WWF-India. A number of projects like Gir Ecological Research Project (1968), Project Hangul (1970), the Project Tiger (1973), the Himalayan Musk Deer Project (1979) and Lesser Cat Project (1981) etc. are being funded by WWF-India. A major project for the conservation and management of crocodiles was started in 1975 under the FAO of UNO and United Nations Development programme (UNDP). Twelve sanctuaries and several crocodile breeding centres have come up in the country.

International Council for Bird Preservation (ICBP) : ICBP, federation of national and international organisations concerned with wild birds is an oldest International Conservation Organisation founded in 1922. The founder of this Council is Dr. Jean Delacour, Dr. S.D. Ripley, leading ornithologist, was the President of the Council from 1958 to 1982. It is a charitable, non-profit making organisation consisting of national sections as basic unit. The national sections, comprised of bird conservation local societies (member organisation), promotes bird conservation in their home countries. ICBP annually reviews programmes of conservation activities and projects for the conservation of birds.

In the World Conference of ICBP, held once in four years, members of the Council (all National Sections) resolve the Council's polices regarding bird preservation. The Executive Committee works as per resolutions of the Council. But while executing the resolutions of the Council, the Executive Council is free to constitute sub-committees, specialist groups and task forces which ever is appropriate, for efficient functioning of the Council's programmes. ICBP initiated the publication of 'Red Data Book' for birds. This organisation also sanctions grants for needful and important bird conservation projects as per priorities of ICBP.

The Expert Committee on birds of the Indian Board of Wild-life is the National Section of the ICBP. All the matters regarding the preservation of Indian birds are reviewed by the Expert Committee. The National Section of ICBP formulates the policy of export and import of birds and their products. It suggests to the Government of India for the conservation of rare and endangered species of birds. A treaty concerning the birds that migrate between India and Russia has been signed for their protection.

Indian Board for Wild-life (IBWL) : For the conservation of wild-life of India, an adhoc committee under the Chairmanship of Inspector General Forest was formed in 1950 which drafted the constitution of a Central Board for Wild-life. Thus, a Central Board for Wild-life was constituted in 1952 as an Advisory body to the Government of India and further it was renamed as the Indian Board for Wild-life (IBWL). The board has a Chairman, two Vice-chairman, a member Secretary, four regional representatives, and the Director Zoological Survey of India, the Director Botanical Survey of India, and the Chief Conservator of Forests of different states as members. The Natural History Societies, Wild-life Preservation Societies and Bird Watcher's Societies are also represented in the Board. This Board has a Bird wing, a Zoo wing, and Flora wing. The Prime Minister of India is ex-officio Chairman of the Board. The meeting of the Board is supposed to be held once in a year.

The State Wild-life Advisory Boards have also been constituted in the States which is responsible for implementing the decisions of IBWL.

Functions of IBWL

(1) Board takes decisions for the management of wild-life through legislative and practical measures under which it can declare certain species as protected animal and prevents the indiscriminate killing of particular species.

(2) It promotes and sponsors the wild-life reserves (National Parks, Sanctuaries, Zoological gardens) throughout the country.

(3) It creates public interest in the conservation of wild-life.

(4) It advises central Government while making the policy regarding the export of wild-life products.

(5) It reviews the progress of wild-life conservation in the country and suggests for further improvement in conservation activity.

(6) It encourages for establishing the wild-life societies in the country and helps in their activities.

(7) The Board appoints standing Committee to watch the implementation of the recommendations of the Board.

Achievements of IBWL

(1) The most important achievement for IBWL is the constitution of the Indian Wild-life (Protection) Act in 1972 for providing legal protection to wild-life.

(2) Establishment of a number of wild-life reserves like National Parks, Wild-life Sanctuaries and Biosphere reserves.

(3) By the 42nd Amendment, in 1976, protection of forests, wild animals and birds have been included in the concurrent list in the constitution. Under the provisions of this amendment certain forests and sanctuaries can be acquired by the Government under its power of acquisition of property.

(4) The illegal trade of wild-life products has decreased because India has become a party to the CITES in 1976 under the advise of IBWL.

(5) National Environmental Conservation Policy and National Forest policy have been accepted. The National Forest Policy direct for the conservation rather exploitation of forest resources.

Indian Wild-life (Protection) Act : This act was constituted in 1972 which provides special legal protection to our wild animals, birds, reptiles, amphibians, insects, etc. and to the endangered species in particular. This act is applicable to the entire country except Jammu and Kashmir. The act consists of seven chapters *viz.*, definitions, hunting of wild animals, authorities to be appointed; formation of wild-life sanctuaries, national parks, nature reserves and closed areas; regulation of trade or commerce in wild animals, animal product and trophies; prevention and detection of offences; and miscellaneous regulations, followed by Schedule 1 to 5. These schedules change from time to time according to the status of animal.

Schedule 1 deals with the rare and endangered species protected throughout the country. Schedule 2 deals with special game species deserving more protection. Schedule 3 deals with big games, whereas, Schedule 4 concerns with the small games. Schedule 5 deals with animals destructive to crops (Vermin).

Convention of International Trade in Endangered Species of Wild Fauna and Flora (CITES) : For the prevention of over-exploitation of wild-life through international trade, a convention on International Trade in Endangered Species of wild Fauna and Flora (CITES) was signed at Washington on 3rd March, 1973 which came into existence on 1st July, 1975. The Government of India became a party to it from 18th October, 1976. The Inspector General of Forests and the Director of Wild-life Preservation are the managerial authorities in India. The managerial authorities of India are assisted by the Director, Zoological Survey of India, the Director, Botanical Survey of India and the Director, Central Merine Fisheries Research Institute. The meeting of the representatives of all the party countries are scheduled to be held after every two years. In this meeting the convention and its implementations are reviewed thoroughly every time. The first meeting of the CITES was held in Berne (Switzerland) in November, 1976 and the third in New Delhi in February, 1981.

Aims and Objectives of the Convention

The Party to CITES agree that :

(1) the wild fauna and flora are an irreplaceable part of the natural system.

(2) they are conscious of scientific, aesthetic, cultural, economic and recreational values of wild fauna and flora.

(3) state and people are and must be best protectors of their wild flora and fauna.

(4) for the protection of certain species of flora an fauna an international cooperation is essential against over-exploitation through international trade and,

(5) it is urgent to take suitable measures to this problem.

With the above objectives certain rules and regulations were formed to control the trade in endangered wild-life and its products. The convention has its text in five languages *viz.*, English, Russian, Chinese, French and Spanish and the text is deposited with the Swiss Government. The convention comprises of 25 articles.

National Parks

It is an area to conserve the natural or historical objects of national importance and to conserve wild-life therein, in such a manner, and by such means, as will leave them undisturbed for the enjoyment of future generations with the required modifications in accordance with necessary local demand and variable conditions. National Park is created by Central Legislation, therefore, it has a permanent status.

Wild-life Sanctuaries

The place or area provided with some area, wild, indigenous mammals, birds, reptiles and any other form of wild-life in good numbers are considered to be in the need of protection together with their natural environment, is declared as wild-life sanctuary. The sanctuary is created by the State Forest Department by Gazette Notification. After some time if the state forest department feels that now there is no need for the protection of wild animals in particular sanctuary, it can be abolished in similar manners.

National Parks and Wild-life Sanctuaries in India

Corbett National Park : It is situated in Uttar Pradesh and covers an area of 520 square Km. There are 50 species of mammals and 585 species of birds in this park. An excellent scheme 'Project Tiger' for the multiplication and protection of tigers is functioning successfully where tigers are in better condition than the other jungles. The Project Tiger was started with 44 tigers which reached up to 90 in 1981. This park has limited space for tigers, so, many tigers are transferred to neighbouring jungles like Aam Pakhra, Amangadh, Aam Danda, Fan to, Mohan, etc.

In 1935, after passing National Park Act, the forest reserve near Ramnagar in U.P. was declared as India's First National Park named as Hailey National Park. In 1954 it was named Ramganga National Park and in 1956-57 State Government had declared it as Corbett National Park after Jim Corbett, the renowned big game hunter and wild-life authority.

Dudhawa National Park : It is situated in the district of Lakhimpur-kheri in Uttar Pradesh. It covers an area of 490 square Km. The main attraction of this park are barasinga and tiger. India is the only country where barasingas are found. Out of 4,000 barasingas in India about 200 are at Dudhawa. According to a census made in 1979 there were 50 tigers, 41 elephants, 76 bears in the park. Out of 9 species of deer found in India, 5 species like, barasinga, hog deer, Sambhar, chital and barking deer are found here. There are 3 species of antelopes *viz,* black buck, nilgai and chowsinga. The chowsinga is the only 4 horned animal found in the world. The crocodiles and more than 400 species of birds are also inhabiting this park.

Kanha National Park : The forest of Madhya Pradesh cover an area of about 1,62,390 square Km which is 37% of total land area of M.P. Out of this forest area, 1,156 square kms have been acquired by Kanha National Park. Bandhavgadh National Park and Shivapuri National Park. In 1974, Kanha National Park was brought under 'Project Tiger' scheme and acquired an area of about 945 square Km. Nowadays it is known to be the largest National Park of the country. The main attraction of this park are the tigers and barasingas.

Kaziranga National Park : Assam in famous for its flora and fauna which attracted the naturalists and wild life lovers since long. The Kaziranga National Park and 7 other sanctuaries of Assam have about 1,900 wild elephants, 1,300 rhinoceroses, 300 tigers and 1,000 wild buffaloes. As per census of 1978 there were 773 elephants, 931 rhinos, 610 wild buffaloes in this park. Some other species of wild animals like barasinga-700, hog deer-7,032, wild boars-700, barking deer-300, tigers-50, otters-300, leopards-20 and bears-47 are also getting protection in this park.

Manas Wild Life Sanctuary : It is situated in Kamrup district of Assam and acquires an area of about 54 Km. This area includes nullahs, meadows, hillocks, water courses and water of river Manas. The Manas forest was managed into a sanctuary in 1928. This sanctuary is inhabited by tiger, rhinos, buffaloes, gours, black panthers, golden langoor, leopards, Indian ottor, elephants etc. The fishing is the main attraction for the tourists in this sanctuary.

Keibul Lamjao National Park : It is one of the unique parks in the world which floats in the vast Logatak Lake of the North- Eastern Indian State of Manipur. The rarest species of deer, *i.e.*, 'Thamin deer' is the centre of attraction in the park for wild-life lovers.

Sunderbans : The Sunderbans cover an area of 10,000 square Km across the delta of the Ganga and Brahmaputra, of which about 6,000 square Km are in Bangladesh. After a census in 1971 there were 195 tigers in India and 300 in Bangladesh part of Sunderbans which was highest number of tigers in the subcontinent. There are numerous 'Sunderi trees's *(Heritier minor)* which are of great economic importance for villagers of the area. It is also known as 'Sunder trees' and because of their large number, the area is called as Sunderbans.

The tigers of Sunderbans have developed a tendency of eating human flesh because of the scarcity of barasingas, chital, gaur and rhinos which have been wiped out by hunters. The tigers of Sunderbans also feed on wild boars, monkeys and other smaller animals even on monitor lizards, crabs fishes and salt water crocodiles.

The area of tiger reserve is 2,585 Km in Sunderbans. It is under planning to develop five of its 15 subdivisions into a National Park covering an area of about 1,310 square Km. In association with the Project Tiger of Sunderbans the World Wild-life Fund has provided wireless sets, fast jet boats, long range fire arms for the effective protection of tigers.

Gir Sanctuary : It is a dense forest of 70 Km length and 40 Km width, situated in Gujarat. This forest is divided into 9 subdivisions namely-Hiran, Shingavado, Shingavadi, Raval, Dhatervadi, Malan, Machhundari, Papatidi and Shattunji. The climate is hot and humid. The lions are the main attraction of the forest. The Gujarat Government is very keen for the protection of lion. Thus, the Gir sanctuary was extended over both vegetational zones and open area of about 140 square Km in 1973 which was upgraded into a National Park in 1975. A long live hedge wall of 468 Km has been maintained to protect the living area of lions in this forest.

Periyar Wild Life Sanctuary : The Periyar Wild Life Sanctuary is situated in the State of Kerala. The construction of a dam across the river Periyar has created a number of lake which are the centres of attraction of this sanctuary. Such lakes acquire an area of about 26 square Km. The natural beauty of the lake attracts the tourists. A number of wild animals like tigers, wild elephants, deer, wild boars, wild dogs, gaurs and Sambhars are found in this sanctuary. The best time for the tourists is from February to May when wild animals come out to take water at the lake because of change in season. There is complete ban on hunting in this area so the animals are free from fear in the sanctuary. The elephants of Periyar enjoy swimming in the lake which looks very interesting.

Bandipur Wild-life Sanctuary : It is situated in Karnataka and covers an area of 690 square Km. These are degraded forests of inferior quality teak, sandal and other species of trees. The area of this sanctuary was shooting ground of Maharaja Mysore famous as an experienced hunter, nature lover, scholar and wild-life photographer. After independence he was nominated as the Chairman of newly formed 'Wild Life Board' by the Government of India. According to the census of 1980, there were 3,000 elephants in the southern forests of which 1,400 were in Bandipur wild-life sanctuary. There were also 1,800 chitals, 800 wild boar, 140 leopards and 36 tigers in this sanctuary. The roads of Bandipur have been named after the wild animals as an expression of closeness to wild animals.

Bird Paradise Bharatpur (Keoladeo national park) : It is also known as Ghana sanctuary. In 1981 it was declared as 'Keoladeo National Park'. Earlier Maharaja uses to invite his guests for bird shooting at this spot. In free India, it was declared as protected area under the provisions of the 'Rajasthan Wild Animals and Bird Protection Act 1951'.

This park has an area of about 29 square Km of which 500 hectares of land has been transformed into a vast natural lake which is divided into smaller sectors by bunds, roads etc. The shrubs and trees growing on the bunds and in the water look very attractive. Across the rivers Banganga and Gambhir, the Kohni and Ajan dams respectively have been constructed which divert the flow of water into the lake.

About 3,00,000-3,50,000 birds of 331 species arrive at Bharatpur every year. It is a good breeding ground of about 103 species of birds. The water birds, from remote places in India, migrate here from July-October for nesting and breeding which are termed as local migrants. About 102 species of migratory birds migrate to this spot from Middle East, China, Siberia, Central Asia etc. Where winters are covered with snow.

It is because of the presence of aquatic vegetation, fish, frog, insects, molluscs and plankton in the lakes of Bharatpur, that it is regarded as a very good breeding ground particularly for the birds. The variety of fishes serves as food for the millions of birds that stay at this park during monsoon and winter.

List of Some Important National Parks in India

S.N.	*Name of National Park*	*Location District/State*	*Area (Sq.km.)*	*Wild Fauna*
1.	Kaziranga	Sibsagar and Nowgong (Assam)	430	Tiger, Sambhar, one horned rhinoceros, elephant, hoolock, gibbon, hog deer, Indian fox, leopard, sloth bear, python, wild buffalo, gaur, pangolin, partridge, grey pelican, sloth bear, florican.
2.	Gir	Junagadh, (Gujarat)	140	Asiatic lion, Sambhar, hyena, Chital, jungle cat, chinkara, mugger, four horned antelope.
3.	Marine	Jamnagar, (Gujarat)	110	Common dolphin, crustaceans, sea turtle, molluscs, fishes.
4.	Dachigam	Srinagar (Jammu and Kashmir)	141	Hangul, musk deer, brown bear, black bear, leopard, serow, pheasants.
5.	Kistwar	Srinagar (Jammu and Kashmir)	400	Leopard, red fox, wild boar, small Indian civet, flying squirrels, rhesus macaque, weasels.
6.	Bandipur	Mysore (Karnataka)	874	Tiger, elephant, wild dog, leopard, hyena, jungle cat, sloth bear, palm civet, Sambhar, Chital, jungle fowl, gaur, Malabar squirrel, partridges.

Contd...

S.N.	Name of National Park	Location District/State	Area (Sq.km.)	Wild Fauna
7.	Nagarhole	Coorg (Karnataka)	572	Tiger, elephant, leopard, Chital, Sambhar, sloth bear, gaur, jungle fowl, partridges, malabar squirrel.
8.	Periyar	Idukki (Kerala)	350	Tiger, elephant, Sambhar, leopard, wild dog, Nilgiri langoor, lion-tailed macaque, sloth bear, barking deer, mouse deer, gaur, wild boar.
9.	Bandhavgarh	Shahdol (Madhya Pradesh)	105	Tiger, leopard, Sambhar, Chital, chinkara, four horned antelope, barking deer, nilgai.
10.	Kanha	Mandla and Balaghat (Madhya Pradesh)	940	Tiger, leopard, swamp deer, Sambhar, chital, wild dog, wild boar, mouse deer, gaur, nilgai, black buck, four-horned antelope.
11.	Indravati	Bastar (Madhya Pradesh)	1258	Tiger, leopard, wild buffalo, chital, gaur.
12.	Sanjai	Sidhi and Surguja (Madhya Pradesh)	1938	Tiger, leopard, chital, Sambhar, gaur.
13.	Panna	Panna (Madhya Pradesh)	543	Tiger, leopard, chital, Sambhar, wild boar, blackbuck, peafowl, nilgai.
14.	Satpura	Hoshangabad (Madhya Pradesh)	524	Tiger, leopard, chital, Sambhar, blackbuck, four-horned antelope.
15.	Nawegaon	Bhandra (Maharashtra)	134	Tiger, leopard, Sambhar, chital, sloth bear, barking deer, nilgai, gaur, migratory birds.
16.	Pench	Nagpur (Maharashtra)	257	Tiger, wild dog, Indian fox, jungle cat, sloth bear, leopard, Sambhar, chital, gaur, blackbuck, four-horned antelope.
17.	North Simlipal	Mayurbhanj (Orissa)	303	Tiger, leopard, mouse deer, hyena, chital, small Indian civet, gaur, elephant, wild boar.
18.	Desert	Jaisalmer and Barmer (Rajasthan)	3162	Desert cat, chinkara, blackbuck, red fox, great Indian bustard, partridge, sand grouse.
19.	Keoladeo	Bharatpur (Rajasthan)	29	Sambhar, chital, jungle cat, herons, coot, stork, Siberian crane, blackbuck, nilgai, cormorant, spoonbill, teal, tern, quail, egrets.
20.	Ranthambore	Swami—Madhopur (Rajasthan)	392	Tiger, leopard, chital, Sambhar, chinkara, jungle cat, hyena, nilgai, partridge, green pigeon, red spurfowl.
21.	Sariska	Alwar (Rajasthan)	492	Tiger, leopard, hyena, chital, Sambhar, jungle cat, nilgai, chinkara, four-horned antelope, partridges.
22.	Khangchen-dzonga	North Sikkim (Sikkim)	850	Musk deer, leopard, red panda, black bear, snow leopard, clouded leopard, blue sheep,

Contd..

S.N.	Name of National Park	Location District/State	Area (Sq.km.)	Wild Fauna
				kiang, binturong, serow, goral, partridge, pheasants.
23.	Corbett	Garhwal (Uttar Pradesh)	525	Tiger, leopard, hyena, chital, Sambhar, black bear, jungle cat, hog deer, nilgai, peafowl, partridges, barking deer, gharial, mugger, python.
24.	Dudhwa	Kheri (Uttar Pradesh)	490	Tiger, leopard, Sambhar, barking deer, swamp deer, hyena, sloth bear, hog deer, nilgai, peafowl, partridges.
25.	Nanda Devi	Chamoli (Uttar Pradesh)	630	Musk deer, snow leopard, Himalayan tahr, brown bear, black bear, goral, marmot, serow, pheasants, bharal.
26.	Sunderbans	24-Parganas (West Bengal)	2585	Tiger, leopard cat, Indian fox, Rhesus macaque, jungle cat, fishing cat, chital, wild boar, herons, egrets, pelican, estuarine crocodile, python, jackal.

List of Some Important Wild-life Sanctuaries in India

S.N.	Name of National Park	Location District/State	Area (Sq.km.)	Wild Fauna
1.	Eturnagaram	Warangal (Andhra Pradesh)	803	Tiger, leopard, chinkara, chital, Indian fox, four-horned antelope, gaur, hyena, jungle cat. python.
2,	Kawal	Adilabad (Andhra Pradesh)	890	Tiger, leopard, chital, Sambhar, sloth bear, blackbuck, gaur.
3.	Nagarjuna Sagar Srisailam	Guntur, Prakasham, Kurnool, Mahboob nagar Nalgonda (Andhra Pradesh)	3568	Tiger, leopard, wolf, sloth bear, Chital, blackbuck, nilgai, mugger.
4.	Pakhal	Warangal (Andhra Pradesh)	803	Tiger, leopard, chital, chinkara, Indian fox, hyena, jungle cat, four-horned antelope, gaur, python.
5.	Pranahita	Adilabad (Andhra Pradesh)	136	Tiger, blackbuck.
6.	Itanagar	Lower subansiri (Arunachal Pradesh)	140	Tiger, leopard, Sambhar, hog deer, barking deer, elephant, gaur, sloth bear.

Contd...

S.N.	Name of National Park	Location District/State	Area (Sq.km.)	Wild Fauna
7.	Namdapha	Tirap (Arunachal Pradesh)	1907	Tiger, leopard, snow leopard, clouded leopard, musk deer, hog deer, elephant, binturong, golden cat, gaur, wild buffalo, flying squirrel.
8.	Manas Tiger Reserve	Kamrup and Goalpara (Assam)	390	Tiger, leopard, hog deer, swamp deer, rhinoceros, elephant, golden langoor, leopard cat, black bear, Indian fox, gaur, wild buffalo, pygmy hog, hispid hare, great pied hornbill, florican, python.
9.	Sonai Rupa	Darrang (Assam)	175	Tiger, leopard, Sambhar, elephant, swamp deer, rhinoceros, sloth bear, wild buffalo, gaur, hornbill, imperial pigeon.
10.	Bhimbandh	Monghyr (Bihar)	631	Tiger, leopard, chital, Sambhar, wolf, wild boar, water birds.
11.	Gautam Buddha	Gaya (Bihar)	260	Tiger, leopard, Sambhar, chital, barking deer, peafowl.
12.	Kaimur	Rohtas (Bihar)	1342	Tiger, leopard, Sambhar, barking deer, nilgai, crocodile.
13.	Palamau Tiger Reserve	Palamau (Bihar)	979	Tiger, leopard, elephant, hyena, Sambhar, palm civet, chital, barking deer, wild boar, gaur, peafowl, partridge, jungle fowl.
14.	Rajgir	Nalanda (Bihar)	36	Tiger, leopard, chital, Sambhar, barking deer.
15.	Valmikinagar	Champaran (Bihar)	148	Tiger, leopard, chital, Sambhar.
16.	Bhagwan Mahavir	Goa (Goa Daman and Diu)	240	Tiger, leopard, Sambhar, chital, mouse deer, small Indian civet, barking deer, gaur, malabar squirrel, hornbill.
17.	Cotigao	Goa (Goa Daman and Diu)	105	Hanuman langoor, slender loris, leopard, Sambhar, gaur, Malabar squirrel.
18.	Barda	Jamnagar (Gujarat)	192	Asiatic lion, chinkara, chital, python.
19.	Gir lion sanctuary	Junagadh (Gujarat)	1265	Asiatic lion, leopard, chital, Sambhar, chinkara, hyena; four-horned antelope, mugger, python.
20.	Narayan Sarovar	Kutch (Gujarat)	765	Chinkara, nilgai.
21.	Daranghati I and II	Simla (Himachal Pradesh)	168	Leopard, black bear, marten, goral, serow, pheasants.
22.	Gobind sagar	Bilaspur (Himachal Pradesh)	100	Crane, goose, duck, teal.
23.	Kanswar	Kula (Himachal Pradesh)	55	Leopard, Himalayan tahr, black bear, serow, goral, partridges, pheasants.
24.	Kugti	Chamba (Himachal Pradesh)	118	Leopard, snow leopard, musk deer, Himalayan tahr, brown and black bears, marten, goral, ibex, serow, partridge, pheasants.

Contd...

S.N.	*Name of National Park*	*Location District/State*	*Area (Sq.km.)*	*Wild Fauna*
25.	Neggu	Mandi (Himachal Pradesh)	278	Leopard, snow leopard, red fox, black bear, marten, musk deer, goral, serow, partridge, pheasants.
26.	Manali	Kulu (Himachal Pradesh)	28	Leopard, Himalayan tahr, black bear, serow, goral, partridges, pheasants.
27.	Shikari Devi	Mandi (Himachal Pradesh)	214	Leopard, snow leopard, musk deer, marten, black bear, goral, serow, pheasants.
28.	Sri Naina Devi	Bilaspur (Himachal Pradesh)	164	Leopard, snow leopard, barking deer, Sambhar, black bear, nilgai, partridge.
29.	Tundah	Chamba (Himachal Pradesh)	65	Leopard, snow leopard, Himalayan tahr, goral, serow, ibex, red fox, marten, weasel, brown and black bears.
30.	Nandini	Jammu (Jammu and Kashmir)	33	Indian grey mangoose, black bear, goral, rhesus macaque, starling small Indian civet, jungle cat, flying squirrel, paradise, fly catcher.
31.	Overa	Srinagar (Jammu and Kashmir)	60	Hanuman langoor, red fox, brown bear, rhesus macaque, black bear, leopard, chakor.
32.	Bhadra	Shimoga and Chikmagalur (Karnataka)	506	Tiger, leopard, Chital, Sambhar, sloth bear, wild boar, elephant, gaur.
33.	Biligiri Rangaswamy	Mysore (Karnataka)	325	Elephant, Sambhar, chital, barking deer, leopard, sloth bear.
34.	Dandeli	Dharwar (Karnataka)	573	Tiger, leopard, chital, Sambhar, palm civet, sloth bear, wild boar, gaur, flying squirrel.
35.	Mookambika	Shimoga (Karnataka)	255	Elephant, leopard, sambhar, wild boar.
36.	Nagu	Mysore (Karnataka)	30	Tiger, elephant, sambhar, chital, barking deer, gaur, wild boar.
37.	Settihalli	Shimoga (Karnataka)	407	Elephant, leopard, Sambhar, chital, gaur, barking deer, sloth bear, wild boar.
38.	Sharawathy Valley	Shimoga (Karnataka)	443	Tiger, leopard, elephant, Sambhar, Chital, barking deer, four-horned antelope, lion-tailed monkey, wild dog, crocodile.
39.	Tungabhadra	Shimoga (Karnataka)	224	Leopard, Sambhar, chital, blackbuck, sloth bear, four-horned antelope, florican, sand grouse.
40.	Neyyar	Thiruvananthapuram (Kerala)	128	Tiger, leopard, elephant, Sambhar, sloth bear, Chital, nilgai, mugger.
41.	Wynad	Kozhikode Kannur (Kerala)	344	Elephant, Sambhar, barking deer, chital, gaur.

Contd...

S.N.	Name of National Park	Location District/State	Area (Sq.km.)	Wild Fauna
42.	Achanakmar	Bilaspur (Madhya Pradesh)	552	Tiger, Sambhar, chital, gaur, wild boar, peafowl.
43.	Bagdara	Sidhi (Madhya Pradesh)	478	Leopard, Sambhar, chinkara, wild boar, black buck.
44.	Bori	Hosangabad (Madhya Pradesh)	518	Tiger, leopard, Sambhar, chital, barking deer, chinkara, wild boar, gaur.
45.	Chambal	Morena (Madhya Pradesh	3582	Gharial.
46.	Ghatigaon	Gwalior (Madhya Pradesh)	512	Great Indian bustard, partridges.
47.	Ken Gharial Sanctuary	Panna and Chatarpur (Madhya Pradesh)	45	Gharial.
48.	Noradehi	Sagar, Damoh and Narsimhapur (Madhya Pradesh)	1035	Tiger, leopard, chital, Sambhar, chinkara, nilgai.
49.	Ratapani	Raisen (Madhya Pradesh)	531	Tiger, leopard, Sambhar, chital, chinkara, nilgai.
50.	Pachmarhi	Hoshangabad (Madhya Pradesh)	462	Tiger, leopard, Sambhar, chital, gaur, barking deer, sloth bear.
51.	Sitanadi	Raipur (Madhya Pradesh)	553	Tiger, leopard, chital, Sambhar, gaur, wild boar, pea fowl.
52.	Dhakna Kolkaz	Amravati (Maharashtra)	382	Tiger, leopard, Sambhar, Chital, barking deer, gaur, four-horned antelope.
53.	Great Indian Bustard Sanctuary	Ahmadnagar (Maharashtra)	7818	Great Indian bustard, partridges.
54.	Radhanagari	Kolhapur (Maharashtra)	2072	Leopard, Sambhar, gaur, black buck.
55.	Dampa	Aizawal (Mizoram)	973	Tiger, leopard, Sambhar, gaur, bear, barking deer, elephant.
56.	Intangki	Kohima (Nagaland)	202	Tiger, leopard, clouded leopard, elephant, bear, gaur, barking deer, pangolin.
57.	Kothgarh	Baud-Khondmals (Orissa)	3995	Tiger, leopard, elephant, chital, sloth bear.

Contd...

S.N.	*Name of National Park*	*Location District/State*	*Area (Sq.km.)*	*Wild Fauna*
58.	Satkosia Gorge	Dhenkanal, Puri, Cuttack and Baud-Khondmals (Orissa)	795	Tiger, leopard, Sambhar, chital, sloth bear. gharial, four-horned antelope, nilgai.
59.	Simlipal	Mayurbhanj (Orissa)	2750	Tiger, leopard, elephant, chital, mouse deer, hyena, small Indian civet, flying squirrel, gaur.
60.	Darrah	Kota (Rajasthan)	201	Tiger, leopard, Sambhar, Chital, chinkara, sloth bear, wild boar, sand grouse, partridges.
61.	Kumbhalgarh	Udaipur (Rajasthan)	500	Leopard, Sambhar, red spurfowl, jungle fowl, nilgai.
62.	Mt. Abu	Sirohi (Rajasthan)	113	Leopard, Sambhar, chinkara, langoor, hyena, sloth bear, partridge, jungle fowl.
63.	National Chambal	Kota (Rajasthan)	280	Gharial.
64.	Sita Mata	Udaipur (Rajasthan)	500	Black buck, chinkara, partridge.
65.	Van Vihar	Dholpur (Rajasthan)	59	Tiger, leopard, Sambhar, Chital, chinkara, nilgai, peafowl, partridge.
66.	Anamalai	Coimbatore (Tamil Nadu)	959	Tiger, leopard, elephant, Sambhar, chital, nilgiri tahr, gaur, dog, sloth bear, nilgiri langoor, lion tailed macaque.
67.	Mudumalai	Nilgiris (Tamil Nadu)	321	Tiger, leopard, hyena, sloth bear, wild dog. jungle cat, Sambhar, chital, barking deer, mouse deer, elephant, gaur, four-horned antelope.
68.	Mundanthurai	Tirunelveli (Tamil Nadu)	567	Tiger, leopard, Sambhar, chital, barking deer. mouse deer, wild dog, lion-tailed macaque, nilgiri langoor.
69.	Nilgiri Tahr Sanctuary	Nilgiris (Tamil Nadu)	78	Nilgiri Tahr.
70.	Chila	Garhwal (Uttar Pradesh)	249	Leopard, sloth bear, Chital, Sambhar, nilgai, tiger, goral, elephant.
71.	Chandra Prabha	Varanasi (Uttar Pradesh)	78	Leopard, Sambhar, chital, sloth bear, nilgai, chinkara, wild boar, peafowl, partridge.
72.	Govind Pashuvihar	Uttar Kashi (Uttar Pradesh)	953	Sambhar, snow leopard, musk deer, brown bear, black bear, bharal, serow, goral, Himalayan tahr, pheasants.
73.	Kaimur	Mirzapur (Uttar Pradesh)	500	Tiger, leopard, Sambhar, chital, caracal, python, cobra.
74.	Keterniaghat	Bahraich (Uttar Pradesh)	400	Tiger, leopard, Sambhar, chital, black buck, sloth bear, gharial.
75.	Kedarnath	Chamoli (Uttar Pradesh)	957	Snow leopard, musk deer, brown bear, bharal, serow, goral, Himalayan tahr, pheasants.

Contd...

S.N.	Name of National Park	Location District/State	Area (Sq.km.)	Wild Fauna
76.	National Chambal	Lalitpur (Uttar Pradesh)	635	Gharial.
77.	Chapramari	Jalpaiguri (West Bengal)	09	Tiger, elephant, Sambhar, gaur, barking deer, birds.
78.	Jaldapara	Jalpaiguri (West Bengal)	115	Tiger, leopard, sloth bear, jungle cat, elephant, rhesus monkey, Sambhar, barking deer, one-horned rhinoceros, hog deer, gaur.
79.	Lothian Island	24-Parganas (West Bengal)	38	Tiger, chital, wild bear, otter, gangetic dolphin, estuarine crocodile, water birds.
80	Mahananda	Darjeeling (West Bengal)	127	Tiger, elephant, Sambhar, hog deer, barking deer, Hanuman langoor, Assamese macaque.
81.	Sajnakhali	24-Parganas (West Bengal)	362	Tiger, Chital, stork, herons, egrets, cormorant, pelican, ibis.

Fate of Wild Animals

Since time immemorial India has always had a variety of wild-life because of its varied climatic conditions and unique geographical position. The present period is the glorious phase for wild-life conservation due to awakening in the lovers of natural beauty who have raised the voices for their protection. The Central theme in all this thinking is that the wild animals have also a right to coexist. If serious consideration is not given to this aspect, within time just possible, a number of wildlives may become an article of imagination, for future generations, like dinosaurs. Today due to continuous exploitation the number of species which are on the verge of dwindling has reached up to 137 (1995) which was only 13 in 1952.

The list of species which have vanished includes 81 mammals, 38 birds and 18 reptiles including some aquatic and terrestrial animals. As per International Union for Conservation of Natural Resources 'Red Data Book' in India 16 species of mammals, 5 of birds and 3 of reptiles are in endangered list. Among endangered mammals are Asiatic lion, tiger, leopard, snow' leopard, Indian elephant, pygmy hog, marshy deer,

Indian forest ass, hangul, Manipuri broatalard deer, wild asian bison, hispid hare, lion tailed monkey, golden langoor, Indian dolphin, Malabar large striped mushk otter ; in bird species Indian bustard, Jordan curser, Chir pheasant, weston trogopan, white- feathered wild duck and in reptilian species, ghariyal, estuarine ghariyal and Khir terapin. It is true that death is a natural routine phenomena, yet the disappearance of any group of animals so fast, may be due to any reason, is a cause of worry.

Today in India of the total geographical area only 4.2% is reserved area in which there are 75 National Parks and 421 sanctuaries. A number of organisations like International Union for Conservation of Natural Resources, World Wild-life Fund, the Conservation on International Trade in threatened and endangered species of wild-life Flora and Fauna and the Smithsonian Institute, all have paid attention for the well being of a number of wild species like lions, tigers, rhinoceros, pygmy hog, gharial, hangul, hispid hare, thamin and Indian bustard.

At the same time it is equally important to be careful from those wild animals which are dangerous for human-beings living around National Parks and Sanctuaries. Due to the problem created by man eaters of Sunderbans and Dudhawa National Park, the villagers have started poisoning the tigers as a measure of retaliation which is against the objective of conservation of tigers. Thus, it is imperative to remanage the whole issue incorporating the safety measures particularly for the tigers and other dangerous wild animals. We have to take a balanced view and choose the middle path.

The educated youth have already started trading in snake's venom produced in their snake farms. There may not be any financial restrictions in setting up commercial establishments in dealing with wild-life products useful to man kind. It is hoped that the coming generation will make more use of valuable items like skin of wild animals, wool, meat, fat, musk and ivory on commercial scale. This will be only possible by proper management of wild animals.

People have started thinking that the wild-life should be reared and brought up in parks but this thinking is not correct. As a matter of fact the wild-life should live in the forest where they should be provided adequate protection. Presently the treatment that we are giving to Giri lions will make them completely tame and gradually they will loose their ferocious and rough behaviour which is the natural character. If this continues during the course of time they will loose their hunting habit and this capacity will disappear. One can never imagine such a situation. In Africa there is very good arrangement for wild animals and they are maintained in their own natural habitat. There in sanctuaries the animals occupy more privileged position than the visitors and are not disturbed however, important personality the visitor may be. In the same way here also for the conservation of the wild animals the rules must be most strictly followed.

It is a matter great regret that even today the matter of wildlives is being casually taken. The time has come when this important work concerning wild plans should be entrusted to such officials who are sincere, experienced, honest and alert to environment. At the same time, for their maintenance the need of co-workers and assistants should also be made proportionately reasonably fulfilled. If the Central and State Governments and environmentalists will show alertness and activity during symbolic wild-life week celebrations only, all these thoughts, ideas and aspirations would not achieve the desired targets for wild-life protection and conservation. In preparing any plan it must be kept in mind that it may not affect the other species of animals.

Any chapter about wild-life will be incomplete without mention of the tribals—the advises and forest dwellers living in and around the forests. Sometimes it is found that the lives of the devisees become hard, challenging and very difficult due to the interference by the wild-life conservation programmes and non-devisees. It is really a matter of great regret. Forest guards should actually cultivate friendship with the tribals and should always bear in mind that the primary needs of tribals could be fulfilled from forest products. In case of conservation of elephant, tiger, lion and ghariyal confusing rumours are spread unnecessarily that these powerful animals sometimes go out of their areas and create havoc and damage the property and life both. It must be clearly understood here if proper arrangements for these animals is made for their roaming in their surroundings, and providing them timely food in the natural way, there is no reason why they would go out and create danger for human-beings. Wild animals in one way guard and protect the forest wealth.

In the present day world when the life is full of tensions, tourist spots, national parks and sanctuaries may provide place for recreation and solace. Thus, it is believed that the interest in wild-life among general public particularly in future generations will increase day by day. The conservation of wild-life should be given a shape of constructive movement in which there is need of a programme based on scientific technology. Thus, they would, on the one hand, provide recreation to people and on the other hand safety and protection to wild animals.

It is true that all plants and animals under nature directly or indirectly depend on one another. Man is the supreme creature of this nature. Thus, it is our moral responsibility that all plants and animals should be allowed to have independent freedom of living in their own reserved area. This can become a source of inspiration for the protection and conservation of wild-life. Hence proper and planned steps should be taken by natural beauty lovers and as well as Government to develop them.

of the arrangers proving hard, challenging and very difficult due to the interference by the wild-life conservation programmes and their devotees. It is really a matter of great regret. Forest guards should actually cultivate friendship with the animals and should always bear in mind that the primary needs of tribals could be fulfilled from forest products. In case of conservation of elephant, tiger, lion and giant, in continuing numbers are spread unnecessarily that these powerful animals sometimes go out of their areas and create havoc and damage the property and the folk. It must be clearly understood here if proper arrangements for these animals is made for their roaming in their surroundings and providing them timely food in the natural way there is no reason why they would go out and create danger for human beings. Wild animals in this way guard and protect the forest wealth.

In the present day world when the life is full of tensions, tourist spots, national parks and sanctuaries may provide places for recreation and solace. Thus, it is believed that the interest in wild-life among general public particularly in future generations will increase day by day. The conservation of wild-life should be given a shape of conservation movement in which there is need of a programme based on scientific technology. Thus, this would, on the one hand, provide recreation and on the other hand safety and protection to wild animals.

It is true that all plants and animals, under natural process, directly or indirectly, are dependent on each other. Man is the supreme creature of the nature. Thus, it is our moral responsibility that all plants and animals should be allowed to live independently in freedom and in their own environment. This can be done only when all the people will extend protection and co-operation. Therefore, with this idea proper and planned steps should be taken to save natural beauty, wildlife and as well as environment for the nation.

Bibliography

Alcock, John : *Animal Behavior : An Evolutionary Approach*, Sinauer, 1998.

Altmann, S. A. and Altmann, J. : *Baboon Ecology : African Field Research*, Oxford, New York, 2000.

Asdell, S. A. : *Patterns of Mammalian Reproduction*, Comstock, New York, 1946.

Aslam, M. S. : *Genetics-Behavioural and Molecular*, Campus Books, New Delhi, 2003.

Aubrey, Manning : *An Introduction to Animal Behaviour*, ELBS Publication, New York, 2001.

Bahr, G. F. : *Chromosomes and Chromatin Structure*, Academic Press, New York, 2001.

Barnett, S. A. : *Study in Behaviour*, Science Paper Backs, London, 2001.

Bendall, J. R. : *Muscles, Molecules and Movement*, Heinemann Educational Books, London 1969.

Benjamin, Gutfreud : *Enzymes : Physical Principals*, John Wiley, London, 1972.

Bernhard, : *The Structure and Function of Enzymes*, Routledge, New York, 1968.

Bhaskar, Harsh Vardhan : *Animal Ecology*, Campus Books, New Delhi, 2004.

Bogert, L. J. and Calloway D.H. : *Nutrition and Physical Fitness,* Saunders, London, 1973.

Bolhuis, J. J. and Hogan, J.A. : *The Development of Animal Behaviour : A Reader,* Blackwell, 1999.

Boulenger, E. G. : *An Introduction to Animal Behaviour,* Discovery Publishing, New Delhi, 2003.

———— : *Animal Mysteries,* Discovery Publishing, New Delhi, 1993.

Boyd, J. D. and Hamilton, W.J. : *The Human Placenta,* Cambridge, London, 1978.

Buffon's, B. : *Natural History of Birds, Fishes, Insects and Reptiles,* Discovery Publishing, New Delhi, 2003.

Carpenter, K. : *Dinosaur Eggs and Babies,* Cambridge, New York, 1996.

Castillo, J. : *Progress in Biophysics and Biophysical Chemistry,* Prentice Hall, London, 1956.

Chauvin, Remy : *Ethology : The Biological Study of Animal Behaviour,* International Universities Press, New York, 1999.

Damron W. Stephen : *Introduction to Animal Science : Global, Biological, Social and Industry Perspectives,* Prentice Hall, London, 2000.

David, M. : *The Marvels of Animal Behaviour,* National Geographic Society, Washington, 2000.

David, McFarland : *Animal Behaviour : Psychobiology, Ethology and Evolution,* ELBS Publication, New York, 2000.

Dickerson and Geis : *The Structure and Action of Proteins,* Academic Press, New York, 1973.

Ernster and Drahota : *Mitochondrial Structure and Function,* Academic Press, New York, 1969.

Filley, G. F. : *Acid-Base and Gas Regulation,* Lea & Febiger, New York, 1971.

Finn, Frank : *Classification of Birds,* Discovery Publishing, New Delhi, 1994.

Florey, E. : *An Introduction to General and Comparative Animal,* Oxford University Press, New York, 1989.

Freeman, W. H. : *Animal Societies and Evolution,* A Scientific American Publication, New York, 2001.

Gene and Reynolds : *Water-soluble Vitamins,* John Wiley, London, 1963.

Gorbman, A. and Bern, H.A. : *A Text Book of Comparative Endocrinology,* John Wiley and Sons, New York, 1962.

Greenberg, Rodwell : *Amino Acid Metabolism,* Cambridge University Press, Cambridge, 1964.

Gyorgy, Pearson : *The Vitamins,* Academic Press, New York, 1967.

Houssay, Frederic : *Industries of Animals,* Discovery Publishing, New Delhi, 1999.

Huxley, H. E. : *The Mechanism of Muscular Conduction,* Spring Harbor, Harbor, 1972.

James, F. : *Animal Social Behaviour,* Wittenberger Duxbury Press, Boston, 1999.

Jencks, : *Catalysis in Chemistry and Enzymology,* McGraw Hill, New York, 1961.

Jennifer, Owen : *Feeding Strategy,* Oxford, New York, 2002.

John, Alcock : *Animal Behaviour: An Evolutionary Approach,* Sinaver Associates, New York, 2002.

Johnson, Christopher B., Franz, Chlodwig : *Breeding Research on Aromatic and Medical Plants,* The Howorth Herbal, London, 2002.

Johnston, R. G. : *Introduction to Sheep Farming,* Granada, 1983.

Jorge, Peter and Ewert Verlag : *Neuroethology : An Introduction to the Neuropathy Sociological and Fundamentals of Behaviour,* Oxford, New York, 2001.

Kamra, D. N. and Pathak, N. N. : *Nutritional Microbiology of Farm Animals,* Vikash Publishing House, New Delhi, 1995.

Kay, Ian : *Introduction to Animal Physiology*, BIOS, London, 1998.

Khanna, D. R. : *Animal Distribution*, Discovery Publishing, New Delhi, 2004.

Konrad Lorenz : *King Solomon's Ring*, MacMillan, London, 2001.

Konrad, Z. and Lorenz, Verlag : *Messengers to the Brain : Our Fantastic, Five Senses*, National Geographic Society Publication, Washington, 2001.

Konrad, Z. and Lorenz, Verlag : *The Foundations of Ethology*, New York, 1999.

Krishnan, M. : *The Hand Book of India's Wildlife*, Oxford, New York, 2002.

Kumar, Ashok : *Animal Physiology*, Discovery Publishing, New Delhi, 2003.

———— : *Aviculture*, Discovery Publishing, New Delhi, 2003.

———— : *Movement and Locomotion in Animals*, Discovery Publishing, New Delhi, 2003.

Kumar, Bipin : *Histology*, Campus Books, New Delhi, 2001.

Kung, S. D. and Yang, S.F. : *Discoveries in Plant Biology*, MacMillan, London, 2000.

Linville and Kelly : *A Text Book in General Zoology*, Discovery Publishing, New Delhi, 1993.

Lister, Harry : *An Introduction to Zoology*, Discovery Publishing, New Delhi, 1999.

Lubibiecki, A. S. : *Large-Scale Mammalian Cell Culture Technology*, Dekker, 1990.

Maheswari, S. K. : *A Textbook of Life Science*, Dominant Publishers, New Delhi, 2004.

Mary, E. Torrence and Issacson : *Microbial Food Safety in Animal Agriculture*, Blackwell, London, 2003.

McCance and Widdowson : *The Composition of Food*, Her Majesty's Office, London, 1960.

Merier, E. : *Immunoglobulins : Biological Aspects and Clinical Uses,* Natural Academy of Sciences, Washington, 1970.

Michael, A. M. : *Irrigation : Theory and Practice,* Vikash Publishing House, New Delhi, 1997.

Morgan, S. J. : *Animal Cell Culture : Introduction to Biotechniques,* BIOS, London, 1993.

Nama, Shinde and Jadhav : *The Nematodes of Indian Mammals,* Dominant Publishers, New Delhi, 2004.

Openheimer, J. H. and Others : *Nuclear Receptors and the Initiation of Thyroid Jormone Action,* Little Brown and Company, Boston, 1976.

Pathak, Nityanand : *Textbook of Feed Processing Technology,* Vikash Publishing House, New Delhi, 1997.

Price, M. R. S. : *Animal Re-Introduction: The Arabian Oryx in Oman,* Cambridge, 1989.

Prosser, Brown, : *Comparative Animal Physiology,* Saunder, Philadelphia, 1961.

Quirk, R. P. : *Chemistry of Muscular Contraction,* Academic Press, New York, 1970.

Ranjhan, S. K. : *Animal Nutrition and Feeding Practices,* Vikash Publishing House, New Delhi, 1995.

———— : *Animal Nutrition in the Tropics,* Vikash Publishing House, New Delhi, 1998.

———— : *Textbook on Buffalo Production,* Vikash Publishing House, New Delhi, 1996.

———— : *Textbook on Quality Meat Production,* Vikash Publishing House, New Delhi, 1999.

Rao, D. Bhaskara : *Animal Kingdom,* Discovery Publishing, New Delhi, 1995.

Robert, Burton : *Animal Behaviour,* Time Life International, Nederland, 1999.

Roger, Tory : *The Birds : Life Nature Library,* Peterson, London, 1999.

Ronald, H. Bailey : *The Role of the Brain*, Time-Life Books, New York, 2001.

———— : *The Secret World of Animals*, A National Geographic Society Publication, Washington, 2002.

Saharia, V. B. : *Wildlife in India*, Natraj Publishers, Dehradun, 1999.

Samuel, K. Wasser : *Social Behaviour of Female Vertebrates*, Academic Press, New York, 2002.

Samuel, Israel and Toby, Sinclair : *Indian Wildlife*, Hans Johannes Hoefer, Singapore, 2001.

Sandhu, G. S. : *A Textbook of Fish and Fisheries*, Dominant Publishers, New Delhi, 2005.

———— : *Analytic Genetics of Animal Development*, Campus Books, New Delhi, 2004.

———— : *An Introduction to Birds*, Campus Books, New Delhi, 2003.

Sarkar, Amira : *Development of Animal Behaviour*, Discovery Publishing, New Delhi, 2003.

———— : *Fundamentals of Animals Behaviours*, Discovery Publishing, New Delhi, 2003.

———— : *Social Behaviour in Animals*, Discovery Publishing, New Delhi, 2003.

Schmidt, Nelson K. : *Animal Behaviour*, Prentice Hall, New Jersey, 1970.

Sela, M. : *New Perspectives in Biology*, American Elsevier, New York, 1963.

Shreeve, W. W. : *Physiological Chemistry of Carbohydrates in Mammals*, Saunders, New York, 1974.

Singh, Gurdarshan, Bhaskar, H. : *An Introduction to Fishes*, Campus Books, New Delhi, 2003.

Spater, P. J. B. : *An Introduction to Ethology*, Cambridge University Press, New York, 1999.

Tikader, B. K. : *Threatened Animals of India*, Zoological Survey of India Publication, Calcutta, 1999.

Tortora and Anagnostrakes : *Principles of Anatony and Physiology*, Harper and Row Publishers, New York, 2002.

Tullts, L. P. : *Blood Cells and Plasma Proteins : their State in Nature*, Academic Press, New York, 1953.

Underwood, E. J. : *Trace Elements in Human and Animal Nutrition*, Academic Press, New York, 1971.

Van, Grvisen : *Wildlife of India*, Guide Book Company Ltd., New York, 2002.

Wadsworth, Cecie Starr : *Basic Concepts in Biology*, Wadsworth, New York, 1997.

Whelan : *Control of Glycogen Metabolism*, Academic Press, New York, 1968.

Whitfield, Dr. P. : *Longman World Guide to Mammals : A Visual Who's Who of the Mammal Kingdom*, Longman, London, 1985.

Williams, C. A. and Chase, M. W. : *Methods of Animal Behaviour*, Academic Press, New York, 1968.

Williams, S. R. : *Animal Behaviour*, MacMillan, London, 1973.

Yadav, Manju : *Applied Entomology*, Discovery Publishing, New Delhi, 2003.

Yadav, P. R. : *Environmental Biodiversity*, Discovery Publishing, New Delhi, 2003.

Yapp, W. B. : *An Introduction to Animal Behaviour*, Clarendon Press, New York, 1960.

Zhukov, M. F. : *Organic Chemistry of Biological Compounds*, Englewood Cliffs, Prentice Hall, 1971.

Zumdahl, S. S. : *An Introduction to General and Comparative Animal Behaviour*, Oxford University Press, New York, 1981.

ENCYCLOPAEDIA OF
ANIMAL BEHAVIOUR

ENCYCLOPAEDIA OF ANIMAL BEHAVIOUR

(Vol. 3)

Dr. A. Tyagi

ANMOL PUBLICATIONS PVT. LTD.
NEW DELHI - 110 002 (INDIA)

ANMOL PUBLICATIONS PVT. LTD.
H.O.: 4374/4B, Ansari Road, Darya Ganj,
New Delhi-110 002 (India)
Ph.: 23261597, 23278000
B.O.: No. 1015, Ist Main Road, BSK IIIrd Stage
IIIrd Phase, IIIrd Block,
Bangalore - 560 085 (India)
Visit us at: www.anmolpublications.com

Encyclopaedia of Animal Behaviour

First Published, 2006

ISBN 81-261-2675-2 (Set)

PRINTED IN INDIA

Published by J.L. Kumar for Anmol Publications Pvt. Ltd., New Delhi - 110 002 and Printed at ARORA Offset Press, Delhi.

Contents

Preface

Behaviour could be most simply defined as movement or a way one conducts oneself. Animal behaviour is, obviously, the way in which animals-non human-lead their life, such as, feeding, migration to different parts in different seasons, sexual life, gregariousness etc.

Different factors need to be studied to have right knowledge about animal behaviour. One of them is the concept of "Natural Selection" propounded by Darwin. He observes that species compete with each other for survival.

Apart from animals, there also occurs a peaceful contest among birds. For example, the severest rivalry between the males of many species to attract their female counterpart is one of the patterns of animal behaviour. It also includes a genetic component similar to the morphological and psychological traits.

The present multi-volume encyclopaedia is a painstaking research work on animal behaviour. It would certainly prove to be an asset for psychologists and those who seek thorough knowledge on the subject.

Editor

Preface

1

Immune System

The term *immunity* refers to the specific reactions by which the invading microorganisms and their toxic products or other substances foreign to the animal are inhibited, inactivated or destroyed. In mammals, this system has the remarkable ability to produce *antibodies* against large number of different foreign agents that may invade the body. The antibodies kill the invaders. In addition, the immune system "remembers", and a second exposure to a foreign substance produces a greater and more rapid response.

Plasma Proteins and Immune Reactions : A special group of blood plasma proteins, the *globulins,* plays a major role in the so-called immunological reactions. These include reactions resulting from *blood type* differences and reactions involving the *antigens* and *antibodies.*

An *antigen* is usually a foreign protein molecule (in some cases it is a polysaccharide) which enters the body fluids of an animal as part of an infectious agent and is capable of inducing the production of specific antibodies. An *antibody* is a molecule produced by animals in response to antigen and has the particular property of combining specifically with the antigen which induced its formation. It is a blood protein (globulin) which may make an infected animal immune to a foreign antigen. This response is called the *immune response.*. The product of an antigen-antibody reaction, which may also contain components of the complement system, is called *immune complex*.

Cells Involved in Immune Response : All the cells of the immune system are derived from pluripotent stem cells which give rise to two distinct progenitors, one for *myeloid cells* producing phagocytes and other cells, and the other for *lymphoid cells* producing *lymphocytes*. The phagocytes are of two basic kinds, monocytes and polymorphonuclear granulocytes.

The lymphocyte appears to be the key cell in the recognition of antigen and the subsequent immune response. It also appears that there are two populations of lymphocytes, the T-lymphocytes (T for thymus) as T cells differentiate initially in the thymus, and the B- lymphocytes (B for bursa), as in birds B cells differentiate in bursa of Fabricius). But in mammals B cells differentiate in foetal liver, spleen and bone marrow.

The *small lymphocyte* population comprises both B cells and T cells. Although they appear similar, T cells and B cells can be distinguished from one another since they carry different cell surface proteins which act as 'markers'. For example, human T cells, but not B cells, carray a marker which binds to sheep erythrocytes. B cells or B lymphocytes represent about 5-15% of the circulating lymphoid pool and are classically defined by the presence of endogenously produced immunoglobulins (antibody). During their development, both T and B lymphocytes acquire specific receptors for antigen which commit them to a single antigenic specificity for the rest of their life-span.

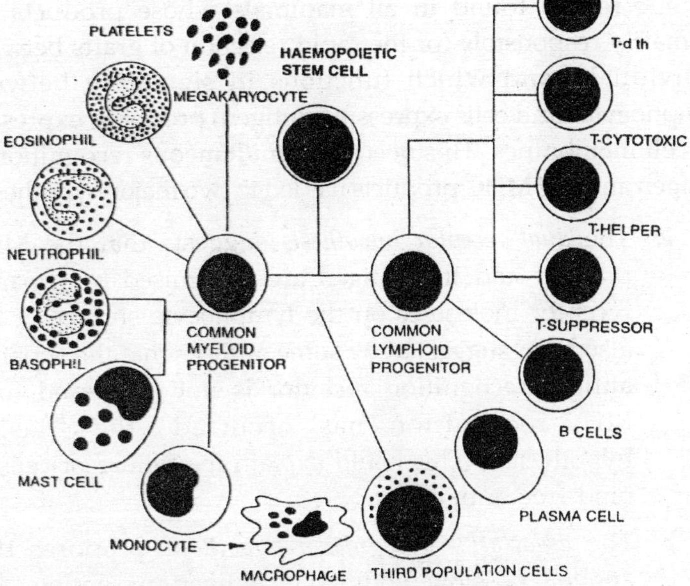

Origin of cells involved in the immune response.

Cell-mediated Immunity

The term *'cell-mediated immunity'* (CMI) or *Cellular immunity* was originally coined to describe localized reactions to intracellular pathogens, mediated by T-lymphocytes (and phagocytes) rather than by antibody. However, it is now often used in a more general sense in which antibody plays a subordinate role.

It is well known that T cells have *antigen specificity, i.e.,* they are able to recognise specific antigen. The cell-mediated immune response follows antigen presentation and activation of T cells. Certain T cells (macrophage activating cells) produce *lymphokines*. These molecules activate macrophages to enhance their phagocyte and bactericidal functions.

However, helper (Th) and cytotoxic T cells (Tc) do not bind free antigen. They appear to recognize antigen only in association with MHC (major histocompatibility complex : a genetic region found in all mammals whose products are primarily responsible for the rapid rejection of grafts between individuals, and which functions in signalling between lymphocytes and cells expressing antigen) products, expressed on cell membranes. This need for simultaneous recognition of antigen and self MHC products has led to two major hypotheses:

1. The *dual receptor hypothesis* suggests that the MHC product and the antigen are recognised by separate receptor molecules on the lymphocyte surface. It has also been suggested by some authors that the separate antigen-recognition receptor is not expressed until MHC-recognition has occurred (the 'flasher hypothesis'). This could explain the failure of cells to bind free antigen.

2. The *associative recognition hypothesis* proposes that antigen, possibly modified by antigen-presenting cells, associates with the MHC products, and is recognised by a single T cell receptor. This hypothesis has been modified to give the *altered self hypothesis* which suggests that the self-MHC on the antigen-presenting cell is somehow modified by the presence of antigen. It is proposed that a single T cell receptor recognises both antigen and altered self.

Cytotoxic T cells are activated by antigen and receive help from helper T cells. Helper cells also cooperate with B cells in the production of antibody which may arm cells (*e.g.*, K cells) carrying Fc receptors (Fc is defined as the portion of an antibody which is responsible for binding to antibody receptors on cells and the Clq component of complement). This is called *antibody-dependent cell-mediated cytotoxicity* (ADCC). Thus antibody dependent cellular cytotoxic cells also kill non-specifically, but this occurs via antibodies bound to their target cells. Natural killer (NK) cells act non-specifically, particularly against cellular targets. They may kill tumour cells and virally infected cells and play a role in regulating the immune response.

Finally, it may be mentioned that macrophages play a central role in cell-mediated immunity, because they are involved both in the initiation of responses as antigen-presenting cells, and in the effector phase as inflammatory, tumoricidal and microbicidal cells, in addition to their regulatory functions.

Cellular immunity constitutes a major defense against infections due to viruses, fungi, and a few bacteria such as the tubercle bacillus. It also helps defend against tumours.

Immunoglublins **:** In 1890, *Von Behring, Buchner,* and others discovered certain proteins when they examined the serum of persons and animals who had been vaccinated against (or who had recovered from) certain infectious diseases. These proteins, now known as *gamma globulins* and called *specific antibodies* or *immunoglobulins,* were found to be defensive only against the specific microorganisms, or their toxins, that evoked those particular antibodies.

For example, diptheria antibodies (antitoxin) "neutralize" diptheria toxin, but not the toxin for tetanus or any other kind. Specific antibodies in the blood were therefore found as the real basis of immunity to disease and gave rise to the doctrine of *humoral* (fluid) *immunity.*

It was also demonstrated by *Tiselius,* the Swedish Biochemist, that the antibody activity of blood was restricted to the gamma globulin fraction of the serum proteins. But in 1954 it was shown that the proteins related to the antibody in structure and function may also be found in other globulin fractions like the beta and even alpha. It has also been shown that all the gamma-globulins need not have the antibody action. Hence, the *term" immunoglobulins"* has been given to these proteins.

There are five distinct classes of immunoglobulins in most higher mammals, including human species. They are IgG, IgA, IgM, IgD and IgE. These differ from each other in size, charge, amino acid composition and carbohydrate content. All immunoglobulins are glycoproteins but the carbohydrate content ranges from 2—3% for IgG to 12—14% for IgM, IgD, and IgE. Within each class two immunologic types have been defined. They are K and L named after Doctor *Korngold* and *Lipari* who first distinguished them. About two thirds of the immunoglobulins of the normal individuals belong to type K or type I, and the rest to type L or Type II. Some important physical and biological properties of human immunoglobulins are shown in Table.

Structure of Immunoglobulin Molecule : Of all the immunoglobulins, the IgG fraction often accounts for 70 to 75% of the antibody pool present in normal adult human serum. As such IgG has been more extensively studied. The IgG molecule may be considered as a typical example of the basic antibody structure.

Each molecule of IgG is composed of 4 polypeptide chains—a pair of identical high molecular weight chains, called *heavy* or *H chains,* and a pair of identical low molecular weight chains called *light* or *L chains.* The H chains have a moleculer weight of 50,000 to 55,000 and contain about 450 amino acids. The L chains have a molecular weight of 20,000 to 25,000 and contain about 214 amino acids. In associating with each other, the four chains form a molecule whose quaternary structure resembles the letter "Y" (Fig.). Each arm of the Y contains a complete L chain and part of an H chain, while the leg of the Y contains the remaining parts of the H chains. The four polypeptide chains are held together by disulphide (—S—S—) linkages. There is a disulphide link near the C— terminus of each L chain, linking it to the corresponding H chain. Two additional disulphide bridges link the H chains together.

Each of the four polypeptide chains is divided into separate regions called variable and constant *domains.* There are two domains in the L chains (V_L, C_L) and four in the H chains (V_H, C_H1, C_H2, C_H3). Within each of the domains, folding of the polypeptide chain produces two parallel planes each containing several segments with folded beta structure. In each folded beta structure, neighbouring stretches of the polypeptide have opposite polarity, *i.e.*, they are antiparallel. The two parallel plans in each domain are held together by disulphide bridges and also by Van der Waals interactions between hydrophobic side chains of amino acids in each sheet.

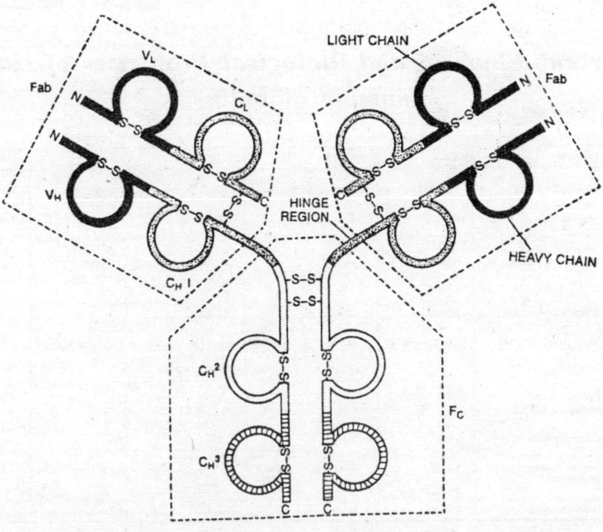

Fundamental architecture of an immunoglobulin molecule.

Enzymatic digestion of the immunoglobulin for subsequent chemical analysis cleaves the molecule in the "hinge" region and produces F_c and F_{ab} fragments. The *variable domain* (portion) of the four polypeptide chains in each immunoglobulin is believed to be responsible for the antibody specificity of the molecule *i.e.*, the sequence of amino acids in this portion of the molecule determines the spatial configurations for binding a specific antigen. There are only two types of L chains, either κ (kappa) or λ. (lambda), in IgG and also in other classes of immunoglobulins. They have different but homologous amino acid sequences. However, each class has a unique H chain-γ (gamma) in IgG, α (alpha) in IgA μ (Mu) in IgM, δ (delta) in IgD, ε (epsilon) and in IgE. Like κ and λ chains, the amino acid sequences of the H chains differ but are also homologous. Moreover, there are several subclasses of heavy chains and four are known for γ-chains, designated γ1, γ2, γ3, and γ4 to give IgG_1, IgG_2, etc. The sequences of the H chains within a subclass are also distinct but show very extensive homology in sequence. Similarly there are α_1 and α_2 chain subclasses of IgA.

Important Physical and Biological Properties of Human Immuno globulins.

S. No.	Property	IgG	IgA	IgM	IgD	IgE
1.	Molecular Weight	150,000	160,000 to 500,000	900,000	185,000	200,000
2.	Sedimentation value	6—7 S	7—11 S	18—20 S	6—7 S	8 S
3.	Heavy chains molecular weight	τ 53,000	α 64,000	μ 70,000	δ 58,000	ε 75,000
4.	Light chains	κ or λ	κ or λ	κ or λ	κ or λ	κ or λ
	molecular weight	22,000	22,000	22,000	—	22,000
5.	Molecular formula	$\lambda\kappa_2 \gamma_2$ or $\lambda_2 \gamma_2$	$(\kappa_2 \alpha_2)n$ or $(\lambda_2 \alpha_2)n$ where n = 2, 4	$(\kappa_2 \mu_2)n$ or $(\lambda_2 \mu_2)n$ where n = 5 or multiple of 5	$\kappa_2\delta_2$ or $\lambda_2 \delta_2$	$\kappa_2\varepsilon_2$ or $\lambda_2 \varepsilon_2$
6.	Percent total immunoglobulins	80	13	6	1	0.002
7.	Carbohydrate %	2.9	7.5	11.8	—	10.7
8.	Serum concentration (mg/ml)	8—16	1—4	1	0.2	0.03
9.	Half-life (days)	25—35	6—8	9—11	2—3	1—2

κ = kappa

It should be noted that in humans the IgG (but not the IgA or IgM) passes to the foetus and gives passive protection against many infectious agents for upto about 6 months of age.

Other Immunoglobulins

Besides IgG, the other major classes of immunoglobulins are IgA and IeM. The minor classes of immunoglobulins found in human plasma are designated as IgD and IgE. It should be noted that IgG, IgD and IgE are monomers. Most IgA and IgM molecules are oligomers, in which n=2 or 4 for IgA and n=5 or a multiple of 5 for IgM. Thus, IgM is a pentamer, whereas IgA may be monomer, dimer, trimer, or tetramer.

IgA : It represents 15-20% of the immunoglobulins in human serum. But it is the principal immunoglobulin in secretions such as saliva, tears and colostrum. It can also be found in the lumen of the intestine, where it is known as *copro antibody.* In secretions, it is generally adimer with an attached S-globulin of about 50,000 molecular weight (called T-piece, transport piece, or secretory piece). The secretory piece is attached to the IeA as it passes through the glandular cell producing the secretion. IgA is thought to be of special importance in the defense of membranes and mouth and vaginal cavities against infections.

IeM : It accounts for about 10% of the immunoglobulin pool. As a pentamer it has increased binding capacity for antigens and thus is extremely efficient. IgM is usually one of the first types of specific antibody to appear in the blood during an infection or after immunization with most antigens. Thus, it may have a special significance in combating a bacteremia or viremia.

IeD : It constitutes about 1% of the total immunoglobulins. It is known to be present in large quantities on the membrane of many circulating B lymphocytes. IgD molecules were first discovered in patients suffering from a disease of the lymphoid tissues known as multiple myeloma. Their main biological role is not yet described, but they are apparently present in certain intestinal fluids of normal individuals.

IeE : IeE, also called *reagins,* is a trace serum protein found on the surface membrane of basophils and mast cells. IgE may play a role in active immunity to helminthic parasites but in

Western countries is more commonly associated with immediate hypersensitivity diseases such as asthma and hayfever.

Humoral (noncellular) Immunity : Humoral means pertaining to the extracellular fluids, including the serum and lymph. B-lymphocytes are primarily responsible for the production of circulating specific antibody, *i.e., humoral immunity.* B-lymphocytes, upon contact with appropriate antigens, are transformed into plasma cells which are responsible for the production of antibody (immunoglobulin), *i.e.,* humoral immunity. Humoral immunity is a major defense against bacterial infections.

Thus, humoral immunity is due to the presence of *immunoglobilins* or humoral antibodies, which are a group of glycoproteins present in the serum and tissue fluids of all mammals. Their production is induced when the host's lymphoid system comes into contact with immunogemc foreign molecules (antigens) and they bind specifically to the antigen which induced their formation.

In order to achieve humoral immunity, each immunoglobulin molecule is bifunctional; one region of the molecule (the V domains) is concerned with binding to antigen while a different region (C domains) modiates binding of the immunoglobulin to host tissue. The binding of *antigen* to the *antibody* takes place by the formation of multiple non-covalent bonds between the antigen and amino acids of the binding site. For successful binding there must be suitable atomic groupings on opposing parts of the antigen and antibody. Further, the shape of the combining site must fit the antigen, so that several non-covalent bonds can form simultaneously. If the antigen and the combining site are complementary in this way, there will be sufficient binding energy to resist thermodynamic disruption of the bond.

The strength of a single antigen-antibody bord is termed the *antibody affinity* and it is produced by summation of the attractive and repulsive forces. A high affinity antibody implies a good fit and conversely, a low affinity antibody implies a poor fit.

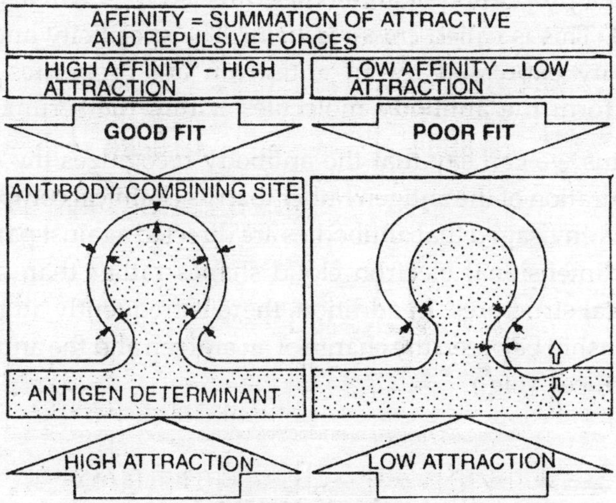

Antibody affinity.

The strength with which a multivalent antibody binds a multivalent antigen is termed *avidity* to differentiate it from the *affinity* of the bond between a single antigenic determinant and an individual combining site. Thus, the *avidity of an antibody* for its antigen is dependent on the affinities of the individual combining sites for the determinants on the antigen, but is greater than the sum of these affinities if both antigen and antibody are multivalent.

Antigen-antibody reaction shows a high level of specificity, *i.e.*, the binding sites of antibodies directed against determinants on one antigen are not complimentary to determinants of another antigen. For example, antibodies to a virus like measles will bind to the measles virus and confer humoral immunity to this disease, but will not combine with, or protect against, an unrelated virus such as polio.

The specificity of an antiserum is the result of the summation of actions of the various antibodies in the total population each reacting with a different part of the antigen molecule and even different parts of the same determinant. However, when some determinants of an antigen, A, are shared by another antigen, B, then a proportion of antibodies directed to A will also react with B. This is called *cross-reactivity*. The specificity and cross reactivity expressed by an antiserum are properties which result form the antibody molecules within the serum.

Thus we can say that the antibody recognizes the overall configuration of the antigen rather than its chemical composition and it is envisaged that antibodies are directed against particular three-dimensional electron cloud shapes rather than specific chemical structures. In addition, there is frequently an inverse relationship between the charge of an antigen and the antibodies it produces (Fig.).

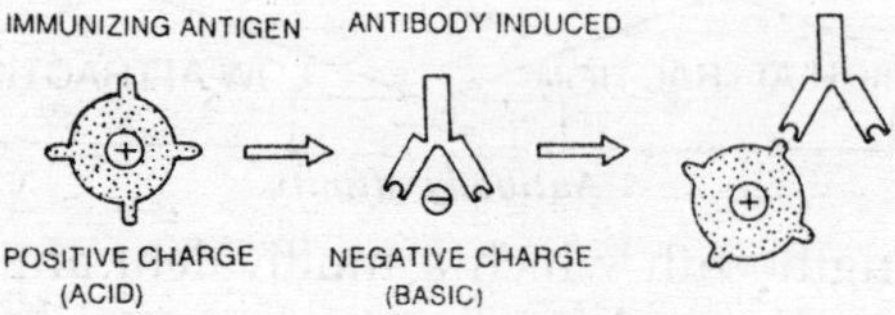

Charge specificity between antigen and antibody.

When an antigen and antibody are allowed to come to equilibrium, it may be written as follows

$$\text{Ab} + \text{Ag} \rightleftharpoons \text{AbAg}$$

Data from these systems can be analysed by application of the law of mass action to give the affinity constant, K since :

$$K = \frac{[\text{AbAg}]}{[\text{Ab}][\text{Ag}]}$$

where [AbAg] is the concentration of complexed antigen, [Ag] is the concentration of free antigen, and [Ab] is the concentraction of free binding sites at equilibrium. When half the binding sites are occupied by antigen, [Ab] = [AbAg] and it follows that, K = l/[Ag]. In other words, a high affinity antibody only requires a low antigen concentration to achieve binding of antigen to half its combining sites.

Theories of Antibody Formation : Several theories of antibody formation have been proposed. One current theory, called the *clonal selection theory*, is described. According to this theory, recognition of a substance as self or nonself depends on the following mechanism. During foetal development, the lymphoid cells in the primary lymphoid tissues differentiate so that many different cells are obtained, each having one, or perhaps more, recognition sites on its surface. After a short time, thousands of cells, each with one or more different recognition sites, are present in the cell population. As the cells migrate through the developing foetus, some of the cells encounter foetal tissue or soluble antigens which are complementary to their own recognition sites. The two combine, and the lymphoid cell dies. In this way, any lymphoid cells carrying recognition sites for "cell" are destroyed during foetal life. Cells which survive this purge can recognize and react only to nonself. They are free to locate in some lymphoid tissue and produce a *clone, i.e.,* progeny like themselves. It has been postulated that by the time the individual becomes competent immunologically, large numbers of lymphoid cells having between 10,000 and 100,000 different kinds of recognition sites are present in the tissues. It is thought that this is sufficient to recognize almost any nonself antigen to which the host may be exposed.

Thus, when an antigen enters the body and makes contact with a B-or T-lymphocyte bearing the proper recognition site, the lymphocyte is stimulated to reproduce and form a clone (clone means a family of cells or organisms having a genetically identical constitution). Its progeny (B or T) become either *plasma cells* (that secrete humoral antibodies with the same binding specificity as the original clone that differentiated in response to the specific antigen) or *lymphoblasts* (which activate cellular immunity, *i.e.,* phagocytic cells). Thus, B cells become plasma cells and T cells give rise to several types of *effector cells*. They may become (1) *killer T cells,* which eliminate foreign cells by lysis ; (2) *helper T cells,* which aid B cells to differentiate and other T cells to proliferate ; (3) *suppressor T cells,* which suppress the immune response; and (4) *amplifier T cells,* which amplify killer-T-cell response and proliferation. The more the antigenic stimulation of B and T cells, the greater the proliferation of the clones of cells possessing the homologous recognition sites.

Immunologic Memory

The clonal selection theory also accounts for *immunologic memory.* Thus, antibodies are synthesized within a few days after initial exposure to an antigen. After this time, however, reexposure to the antigen elicits a secondary, or anamnestic, response that rapidly results in production of a greater number of antibodies than those produced in the primary response. The *immunologic memory* is based on the fact that the original antigen exposure induces not only effector cells that generate humoral antibodies, but also *memory cells,* which are capable of producing further effector and memory cells after a second exposure to antigen. Unlike effector cells which have lifetimes of days, memory cells survive within the individual for its lifetime and respond rapidly to generate effector cells on any subsequent exposure to antigens.

Mechanism of Immunization : What does "making immune" mean ? It means the combining of an antibody with an invading antigen. Such a union abolishes the free mobility of the antigen and so prevents the antigen from acting in a damaging manner. The combining process is a lock-and-key reaction similar to the enzyme reaction. The molecular configuration of the antibody protein apparently is such that it "fits into" the configuration of the antigen. In other words, antigen-antibody reactions are specific and any given antigen requires a particular kind of antibody if immunity is to be developed. The specificity has been referred to as an attribute of antigens and antibodies that is a result of mutually corresponding molecular structures called determinants or recognition sites.

The structural resemblance between molecules of an antigen and molecules of the corresponding antibody may be visualized as analogous to the relationship between a mould and its replica or casting. Each is specific for the other. If the antigen molecule is small and contains only one combining site (univalent), then only one antibody molecule can combine with each such antigen molecule (though 2 antigen molecules may combine with each antibody molecule since antibody molecules are typically bivalent). Such antigen-antibody combinations are small and soluble.

The more complex the form of the antigen molecule, the more recognition sites there are for specific combination with antibody, and the larger, more stable and insoluble the combination is. For example, most protein antigens have up to about 5 different determinant groups, each of which evokes an antibody molecule specific for that site. Thus, even though such a multivalent antigen is a pure substance, the antibody produced in response to it is really a mixture of antibody molecules, each specific for a different determinant group on the antigen molecule. Saturation of all the determinant group sites of antigen with specific antibodies forms a large, stable, insoluble and precipitable colloidal complex.

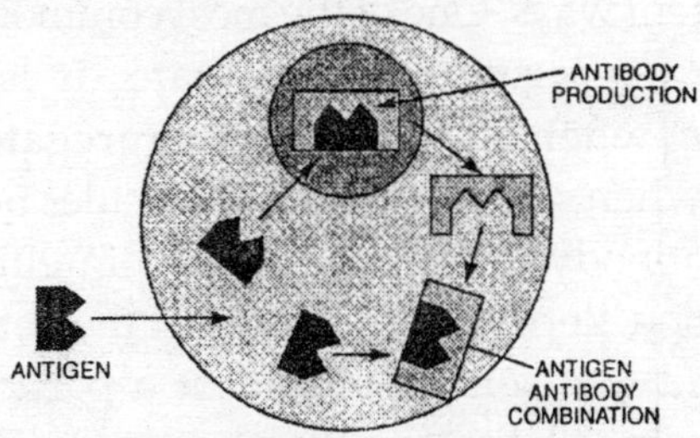

Antigen-antibody reaction.

The following stages are recognized for antigen antibody reactions. The first stage of antigen-antibody reaction is *adsorption* or complex physico-chemical combination (sometimes reversible) between antibodies and antigens. This combination may proceed quite rapidly at 37°C. The bonds between antigen and antibody are non-covalent; they are hydrogen bonds, ionic bonds, etc., and hence are not strong. The second stage of reaction is usually a *visible reaction* (precipitation, cell lysis, or other effect). This stage often develops slowly and may be demonstrated best after 12 to 18 hrs. at 4 to 6°C. However, the presence of electrolytes (NaCl, $MgCl_2$) and certain enzyme-like components of serum *(complement)* is necessary for certain types of second-stage reactions. Complement consists of a complex series of proteins many of which are proteinases. The complement system performs three vital functions : cell activation, cytolysis and opsonization.

However, animals do not normally possess preexisting antibodies for all types of antigens which might possibly invade the blood. This means that a specific type of antibody must be manufactured after a given antigen has invaded. Among the organs which produce antibodies appear to be the lymph nodes, the liver, the spleen, and the bone marrow. Among the best understood types of antibodies are : antitoxins and precipitins, agglutinins, cytolysins, immobilizing antibodies, and protective and neutralizing antibodies.

The combination of antibodies to antigens is manifested in several different ways. One of the most common is the formation of a visible aggregate or precipitate. It is usually called *agglutination* when antigens are aggregated. It is called *precipitation* when soluble antigen molecules become insoluble. The same antibody might be involved in either precipitation or agglutination, depending on whether the determinant groups were attached to a soluble colloid or a particulate substance, such as a whole cell. Frequently when bacteria have combined with antibodies, they are effectively phagocytized by leucocytes and macrophages, a phenomenon called *opsonization*. Another method of antibody-antigen reaction is the disruption, or *lysis* of cells. This is often observed with erythrocytes (resulting in a release of haemoglobin) and with certain types of bacteria and involves a rupturing of the cells. In addition to the suitable antibodies, the occurrence of opsonization and lysis requires a component of normal serum called the complement.

Active and Passive Immunization : The immunity that is conferred by a specific infection or by introducing an antigen through vaccination into the tissues is called *Active immunization.* Under these conditions the tissues of the host synthesize a relatively large amount of antibodies. *Active natural immunity* means a period of specific immunity that is frequently observed following a natural infection. By this we mean that either the individual is not susceptible to a repeat infection with the same microorganism, or if a repeat infection does occur, the illness is terminated in a much shorter time than initially and usually with less severe symptoms. This type of immunity is long-term or even life long. Vaccinations establish active artificial immunity for periods up to several years or throughout life.

Active artificial immunity depends almost solely upon the induction of specific antibodies, the major exception being when certain live attenuated vaccines are used. In the latter case, both cellular and humoral types of immunity may be elicited.

The principle of *Passive immunization* involves the use of various animals (rabbit, sheep, horse, guinea pig, etc.) that may be infected and allowed to manufacture specific antibodies against some infectious agent.

Blood serum of these animals, containing the antibodies, may then be used to protect man against the same infectious agent. Diphtheria antitoxin is an example of an immunoglobulin which confers passive immunity. Immunity here is not permanent. Thus, passive immunization involves antibodies received from another individual. There may be natural or artificial transfer of such antibodies. An example of *passive natural immunity* is the transfer of maternal antibodies via placenta or colostrum. They serve to protect the child for some months after birth. By 3 to 6 months after birth maternal antibodies have largely disappeared and the child becomes susceptible to many infectious diseases. It is therefore, advisable to begin active immunization of the child early by injections of combined antigens of diphtheria, tetanus, and pertussis, followed by the oral administration of the trivalent oral polio vaccine at 6 months.

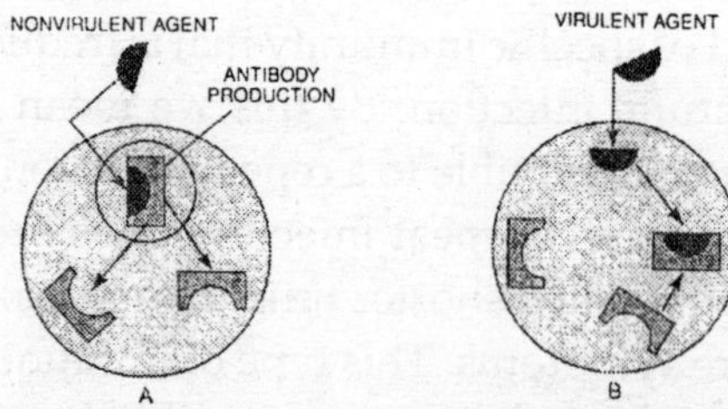

The principle of active immunization (A) antibodies formed before serious infection, (B) preformed antibodies protect against serious infection.

Immunity resulting from injections of ready-made antibodies is immediate and is called *passive artificial immunity*. Its duration is very short, 6 weeks or less. Passive artificial immunity is used in the prevention *(prophylaxis* from Gr. prophylassein = to guard against) as well as the treatment of diseases . For example, if it is known or suspected by a physician that a person has very recently been exposed, or is likely to become exposed, to certain diseases, a small quantity of serum or some derivative of serum *(e.g.,* purified gamma globulin) containing the appropriate antibodies as a prophylactic measure, is injected.

But to avoid allergic reactions, gamma globulins of human origin are commonly used for preventive purposes notably against rubella (german measles) and infectious hepatitis. Diseases against which equine serum is occasionally used for prophylaxis are diphtheria and tetanus. Effective human sera are also available for prophylaxis against rabies and pertussis.

Hypersensitivity

When an adaptive immune response occurs in an exaggerated or inappropriate form, causing tissue damage, the term *hypersensitivity* is applied. It is a characteristic of the individual and is manifested on second contact with a particular antigen.

Coombs and *Gell* (1963) have described four types of hypersensitivity reaction (type I, II, III and IV), but in practice these types may not necessarily occur in isolation from each other. It should be noted that these reactions are no more than expressions of the beneficial immune responses, acting inappropriately, and sometimes causing inflammatory reactions and tissue damage. The first three types are antibody mediated, and the fourth mediated primarily by T cells and macrophages.

Hypersensitivity—Type I : Type I, or immediate hypersensitivity occurs when an IgE response is directed against innocuous antigens, such as pollen, and the resulting release of pharmacologicai mediators (such as histamine) by IgE-sensitized mast cells produces an acute inflammatory reaction with symptoms such as asthma or rhinitis. Thus, type I hypersensitivity is characterised by *allergic reacitions* immediately following contact with antigen (allergen).

The term "allergy" was originally coined in 1906 by Von Pirquet meaning 'changed reactivity' of the host when meeting an 'agent' on a second or subsequent occasion. In recent years 'allergy' has become synonymous with Type I hypersensitivity. The immediate anaphylactic response to bee venom is a clearcut example of Type I hypersensitivity, due to the release of pharmacological mediators, including histamine, from mast cells. The reaction can produce generalized anaphylaxis and even death since the allergen is injected into the patient rather than being inhaled.

Hypersensitivity—Type II : Type II, or antibody-dependent cytotoxic hypersensitivity, occurs when antibody (IgG) binds to antigen on cells leading to phagocytosis, killer cell activity, or complement-mediated lysis.

In this form of hypersensitivity, antibody, directed against cell surface or tissue antigens, interacts with molecules of the complement pathways and a variety of effector cells (K cells, platelets, neutrophils, eosinophils, macrophage/monocyte) to bring about damage to those cells and surrounding tissues.

The mechanisms by which phagocytes damage cells in type II hypersensitivity reactions reflect their normal physiological functions in dealing with infectious pathogens.

Hypersensitivity—Type III : Type III, or immune complex mediated hypersensitivity develops when complexes are formed inlarge quantities, or cannot be cleared adequately by the reticuloendothetial system, leading to serum sickness type reactions.

Eeural diseases result from immune complex formation and may affect kidney, skin, lungs etc. For example, in viral hepatitis, the combined effects of a low grade persistent infection together with a weak antibody response leads to chronic immune complex formation with the eventual deposition of complexes in the tissues.

Hypersensitivity—Type IV : Type IV, or delayed type hypersensitivity (DTH) is manifested when antigen, for example, tubercle bacilli, trapped in a macrophage, cannot be cleared. T lymphocytes are then stimulated to elaborate lymphokines which induce inflammatory reactions and activate and attract macrophages which release mediators. Other aspects of DTH reaction are seen in graft rejection and allergic contact dermatitis.

There are a considerable number of chronic diseases in man which manifest delayed hypersensitivity, and most are due to infectious agents such as mycobacteria, protozoa and fungi. Important diseases in this respect include tuberculosis, leprosy, leishmaniasis, listeriosis, deep fungal infections (*e.g.*, blastomycosis) and helminthic infections (*e.g.*, schistosomiasis). These diseases are caused by pathogens which present a persistant chronic antigenic stimulus. The threat they pose is met by lymphocytes and macrophages. Although these diseases are liable to induce protective immunity, the protective immunity and delayed hypersensitivity do not always coincide.

AIDS

AIDS or Acquired Immune Deficiency Syndrome is a killer disease for which, at present, there is no effective medicine or vaccine. December 1 is celebrated as World AIDS Day to create public awareness regarding this disease and to stress the immediate need to improve monitoring and control the disease. Princess Diana said in an interview in 1993, "The vicious circle of fear, prejudice and ignorance has increased the spread of AIDS to an alarming level. Due to fear and prejudice, many still do not want to listen. After all, AIDS is a killer".

The first cases of AIDS were described in 1981 by *Michael Gottlied* on homosexual men living in Los Angeles, USA. By the end of 1981 epidemiological data indicated that AIDS was an infectious disease transmitted through sexual relations and blood contact. In May 1983, first description of the AIDS infectious agent, the virus LAV (Lymphadenopathic Associated Virus) was given on the basis of the work done by *Prof. Montagnier* and his team at the Pasteur Institute (Paris).

In May 1984, *Prof. Robert Gallo* (USA) and his team isolated the HTLV III (Human T cell leukemia Virus type 3) as infectious agent of AIDS. These two denominations of the same virus

(LAV and HTLV III) have now been changed into HIV I (Human Immuno deficiency Virus type 1). In 1985, first routine tests were available for detection of HIV antibodies in blood banks and blood products in North America and Europe. In March 1986, first description and isolation of a second AIDS virus, HIV II a found in West African countries by *Prof. Montagnier* and his team and a Portugese research team in Lisbon, was made. Due to the high rate of change or mutration that HIV can undergo, new subtypes of this virus are still being discovered. The latest is subtype O in West and Central Africa, taking the total number of HIV subtypes to nearly seven.

It is well known that HIV (human immuno deficiency virus), the retrovirus that causes AIDS, attacks the CD_4 T-cells of the body. These cells are the watch dogs of the human body, the cells that fight an infection. By killing these cells, the body's defence mechanisms are lowered and multiple infections lead to AIDS. While CD_4 is the doorknob to which the HIV virus attaches itself, it has now been discovered that it is the CCR-5 that actually opens the doorknob for the entry of HIV in the body. So, while people with HIV and a normal CCR-5 get full blown AIDS, people with a defective CCR-5 move very slowly towards AIDS. However, people who do not have CCR-5 do not get AIDS at all. Almost in all cases HIV spreads through intravenous drug use, gay sex and sexual activity involving multiple sex partners. In many cases mother-to-child transmission of HIV through breast feeding also occurs.

The HIV viral load tests are conducted by the polymerase chain reaction (PCR) thermocyclen which measures the amount of virus present per ml of blood. In full blown AIDS, the virus load varies between 5 lakh to 32 lakh viral particles per ml.

The first clinical results showing a certain amount of efficiency of AZT *(Azydothymidine)* in the treatment of AIDS patients came at the end of 1986. By 1988 AZT was considered a great hope for HIV-positive people, but its benefits did not last long due to drug resistant HIV strains.

The most recent research from the laboratory of Dr. David Ho at the Aaron Diamond AIDS Research Centre, New York shows that *"three-drug combination therapy"* could, over three years, flush out the HIV-1 virus from all its hiding places in the body. Referred to as anti-retroviral drugs, the therapy consists of the use of AZT, 3TC and three protease inhibitors. While AZT and 3TC work at the level of reducing the multiplicity of the virus, the protease inhibitor, sequinovir, prevents the development of new virus particles. These expensive drugs suppress HIV to the point that the virus is undetectable. Unfortunately, these therapies can cost upto 25,000 US dollars a year and require patients to follow a strict regimen with nearly 20 pills per day for the rest of their lives. These drugs also have some side effects including nausea, headache, anaemia and toxicity to liver and kidneys. However, no vaccine has been developed against HIV as the virus constantly changes itself.

The latest global statistics show an estimated 30.6 million people with HIV/AIDS infection, of whom 5.8 million have been infected in 1997. AIDS scenario in some of the developing countries is very alarming and children also have been hit by the disease. India has now more than 8 million HIV positive cases. The National AIDS Control Organization (India) now considers infected children a group large enough to be defined as a separate category. It estimates that 4% per cent of the total AIDS cases in the country are children.

AIDS has claimed its first victim in India in Bombay on June 9,1986. The Indian Council of Medical Research has purchased 2,000 Elisa kits for screening of AIDS patients. One kit could screen 88 patients and costs over 400 U.S. dollars.

Blood Groups

At least 15 different blood group systems have been recognised in human beings, each system consisting of a gene locus specifying antigens appearing on the erythrocyte surface. The five major blood group systems are : ABO, Rhesus, Kell, Duffy and MN.

The ABO system was discovered by *Landsteiner* (1900). According to him, the human blood was not the same in all individuals and could be classified into four well defined groups (A, B, AB and O). Their percentage or phenotype frequencies are about : A 42%, B 8%, AB 3% and O 47%. Blood grouping is based on the presence or absence of specific *antigens* (agglutinogens) located on the surface of red blood cells.

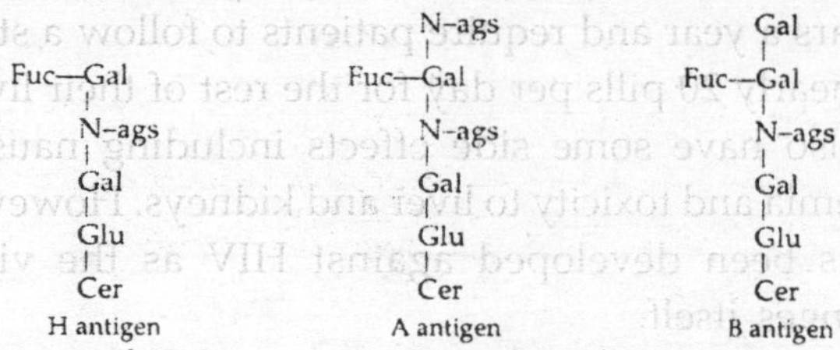

Antigens of ABO system on the surface of red blood cells. Cer, ceramide; Glu, glucose; Gal, galactose; N-ags, N-acetylgalactosamine; Fuc, fucose.

The best known and most important of antigens are the A and B antigens, which are inherited as mendelian dominants. Individuals having blood type A have the A antigen, type B have the B, type AB have both, and type O have neither. The A and B antigens are actually complex oligosaccharides that differ in their terminal sugar. On red cells they are mostly glycosphingolipids, whereas in other tissues (*e.g.*, salivary glands, pancreas, kidney, liver, lungs, testes, etc.) they are glycoproteins. However, H antigen is usually present in individuals of all blood types. An H gene codes for a fucose transferase that puts a *iucose* on the end of H antigen. Individuals who are type A have a gene which codes for a transfrase that catalyzes placement of a terminal sugar *N-acetylgalactosamine* on the H antigen. Individuals who are type B have a gene which codes for a transferase that places *galactose* as terminal sugar. Individuals who are type AB have both transferases *i.e.* having both N-acetylgalactosamine and galactose as terminal sugar. Individuals who are type O have neither, so the H antigen persists.

Antibodies against red cell antigens are called *agglutinins* and are present in plasma. Thus, type O contains antibodies *a* and *b;* type A contains antibody *b* (Anti-B); type B contains antibody *a* (Anti-A) and type AB contains no antibodies. In these groups normally no blood agglutinates itself because the corresponding antigen and antibody, as, for example, A and *a,* or B and *b* are not present together, for *a* is anti-A and *b* is anti-B. On the other hand A is compatible with *b* and B is compatible with *a.*

Blood type is decided by mixing an individual's red blood cells with antisera containing the various agglutinins on a slide and seeing whether agglutination occurs.

Principle of Blood Transfusion

The basic principle of transfusion states that if the cells of the blood to be transfused will be agglutinated by the recipients plasma, the transfusion must not be made. Thus, since the cells of type A blood are agglutinated by B plasma, A blood cannot be given to a B patient. On the other hand, the cells of O blood are not agglutinated by any plasma, therefore O blood can be given to patients with other types of blood. Accordingly, the individuals having O blood are termed *universal donors.* The plasma of AB blood does not agglutinate the cells of any blood. Hence, the individuals having this type of blood are termed *universal recipients* (Table). However, in actual practice doctors insist on giving the blood of the same type as the patient's blood to prevent any possibility of incompatibility.

Main Blood Groups with their Antigens and Antibodies and their Possible Combinations in Blood Transfusions.

S.No.	*Blood*	*Antigens on erythrocyte surface*	*Antibodies in serum*	*Can give blood to*	*Can receive blood from*
1.	AorA$_b$	A	*b* (Anti-B)	A,AB	O,A
2.	B or B$_a$	B	*a* (Anti-A	B,AB	O, B
3.	AB	A and B	none	AB	all
4.	O or Oab	none	*a, b* (Anti-A & Anti-B)	all	O

Transfusion Reactions

Transfusion of blood into a recipient who has antibodies to those red cells produces an immediate transfusion reaction. For example, when the plasma of a type A individual is mixed with type B red cells, the anti-B antibodies cause the type B red cells to clump (agglutinate). Clinical indications of transfusion reactions are fever, hypotension, lower back pain, feeling of chest compression, and nausea and vomitting.

The severity of the reaction depends on the class and amounts of the antibodies involved. Antibodies to the ABO system antigens are usually of the IgM class; they cause agglutination, complement activation and intravascular haemolysis.

Other blood groups induce IgG antibodies, and although these agglutinate the cells less well than IgM antibodies, they activate Type II hypersensitivity mechanisms and cause red cell destruction. The cell destruction may cause circulatory shock, and the released contents of erythrocytes can produce acute tubular necrosis of the kidneys.

Transfusion reactions to incompatible blood may also

develop over days or weeks as antibodies to the foreign cells are produced, resulting in anaemia and jaundice.

Rh Blood Group

The Rh antigen was first discovered by *Landsteiner* and *Weiner* (1940) in the rhesus monkey. Later on it was shown that most human beings also have this factor and are said to be *Rh-positive* (RhD^+), while only about 15% are *Kh-negative* (RhD^-) Rh-positive and Rh-negative bloods are incompatible and cannot be mixed.

Since the discovery of Rh factor, it has been established that there are no less than 6 different antigens (C or c, D or *d*, E or *e)* involved in the Rh system. The most important of which is referred to as the *antigen* D or the *Rh factor,* which is usually of serious clinical significance as it evokes a stronger reaction in an incompatible recipient than the other Rhesus antigens.

There is no antibody that normally accompanies the *Rh* antigen as is true of the antigens of ABO blood groups. If the blood of an Rh-positive person is transfused into an Rh-negative person, the Rh antigen will, however, cause the formation of antibodies. If at some subsequent time another transfusion of Rh-positive blood is made to this Rh-negative person, then the Rh antigen reacts with the antibodies, causing a serious transfusion reaction, agglutination, which may result to death.

Another serious result of Rh-incompatibility, where RhD^+ is the most commonly involved antigen, may be caused when an Rh-negative pregnant woman carries an Rh-positive child. Rh-positive erythrocytes may pass from the bloodstream of the foetus into the mother's blood where they stimulate production of IgG antibodies.

These antibodies of the mother may diffuse into the

bloodstream of the foetus and cause destruction of foetal erythrocytes, a characteristic form of anemia known as *erythroblastosis foetalis* or *haemolytic disease of the newborn* (HDNB). This results in varying degrees of injury to the foetus, depending upon the concentration of the antibodies present. The effects range all the way from a very mild anemia to more severe anemic conditions, various structural abnormalities, abnormalities of the nervous system, sometimes causing mental deficiency, to miscarriage or still birth.

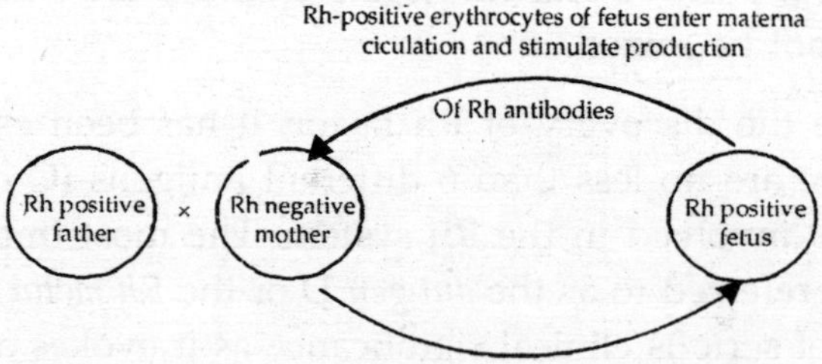

Antibodies produced by mother enter fetal circulation and cause erythroblastosis

However, for reasons as yet unexplained, the transfer of erythrocytes and antibodies across the placenta occurs only in a small percentage of cases. Thus, in only one mating out of 28 between Rh-positive males and Rh-negative females is there any sign of the effects of Rh incompatibility. The first and the second child born of parents with Rh incompatible blood groups usually escape harm. But by the time the third child is conceived there may be sufficient antibodies in the mother's blood to produce serious effects like the death of the newborn infant or abortion of the foetus.

M and N Factors : The two antigens, M and N, give rise to three blood groups called M, N and MN. Their percentage is 28%, 50% and 22% respectively. The presence of these antigens is determined by two genes that reside in the nuclear chromosomes.

Use of M and N Factors for Establishing Parentage.

If child is	*The mother*	*The father*	
		Must be	*cannot be*
M	must be M or MN	M or MN	N
N	must be N or MN	N or MN	M
MN	may be M	N or MN	M
	N	M or MN	N
	MN	M, N, or MN	

Differing from the ABO blood groups, the antigens M and N associated with the red blood corpuscles are not accompained by antibodies in the plasma. Hence, in blood transfusions these antigens are not taken into consideration. However, they are of great value in deciding the parentage of a child when this is in dispute.

2

Special Senses

Specialized receptors, or sense organs, are made up of the cells that are usually sensitive to certain types of stimuli, and will not respond to the other types. For example, the eye is specialized for sensing the light rays, but not the sound waves. Receptors are in direct contact with the environment and receive and transmit impressions of what is going on in the environment. In humans, there are special receptors for the perception of sound, light, smell, taste and touch.

The Ear

The ear is a specialized receptor for detecting sound waves in the surrounding air. The mammalian ear consists of the following three parts: (Fig.).

Outer Ear : The outer ear or external ear of higher vertebrates is made up of two parts : *(a)* the *pinna,* or skin-covered flap of elastic cartilage and muscles, and *(b)* the external auditory meatus (auditory canal). In humans, the pinna serves the functions of collecting sound waves. In many other mammals, such as the rabbit, the pinna is provided with well developed voluntary muscles and it may be moved so as to collect sound waves more effectively.

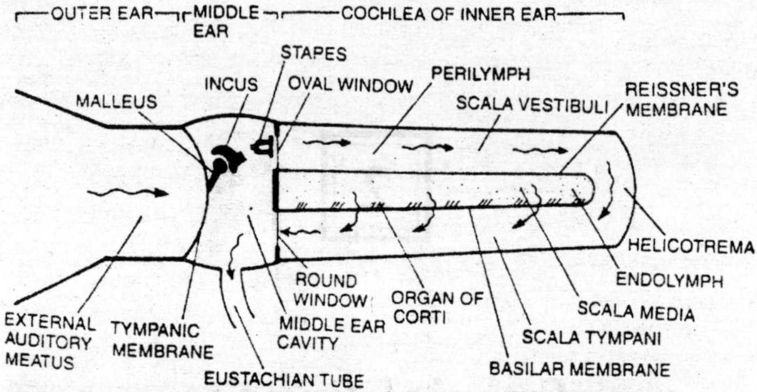

Functional anatomy of the mammalian ear. Arrows indicate the pathway of vibrations.

The auditory canal (about 2.5 cm long) extends inwards and forwards to the ear drum (tympanic membrane). It prevents hard objects from striking the tympanic membrane directly and its walls are lined with hair and wax-secreting glands as a protection against the entrance of foreign objects.

Middle Ear : The middle ear is separated from the outer ear by the *tympanic membrane,* commonly called the *ear drum.* It consists of connective tissue covered with skin on the outside and mucous membrane on the inside. The ear drum is in the shape of a shallow funnel with its apex (umbo) pointing inwards. Movements of the ear drum, set up by the pressure variations caused by the sound waves, are relayed to the three ossicles of the middle ear and thence to the inner ear. The *ossicles* are *malleus, incus* and *stapes.* They are so arranged that the handle of the malleus (hammer) is attached to the inner side of the tympanic membrane. The head of malleus is firmly bound to the incus. The long process of incus passes downwards and parallel to the handle of the malleus and articulates with the head of the stapes (stirrup). The oval base of the stapes fits into the oval window of the cochlea, being surrounded and attached to the margins of the oval window by the annular ligament. The ossicles increase the efficiency of transmission of the sound waves to the inner ear.

To increase the tension of the tympanic membrane for delicate sounds or to counteract under displacement of the membrane, a muscle *tensor tympani* is inserted on the malleus. Contraction of this muscle pulls the malleus and the tympanic membrane inward. A similar muscle *stapedius* is fastened to the stapes and tends to pull the stapes out from the oval window. Like tensor tympani it also serves a protective function and can be reflexly excited by loud sounds.

The middle ear communicates with the pharynx by means of the *eustachian tube,* which regulates the air pressure in the middle ear. Both the middle ear and the eustachian tube are lined with mucous membrane.

Inner Ear : The inner ear consists essentially of two labyrinths, one within the other. The inner one is called the membranous labyrinth which is a closed ectodermal sac filled with endolymph. The *cochlea,* the part involved in hearing, is coiled like a snail's shell, making two-and-a-half turns. The membranous labyrinth is surrounded by a bony labyrinth, which is a hollowed out part of the temporal bone and conforms to the shape and contours of the membranous labyrinth. The space between the two labyrinths is filled up with perilymph.

The cochlea is divided into three longitudinal canals that are separated from each other by two thin membranes called the Reissner's membrane and basilar membrane. These canals become progressively smaller from the base of cochlea to the apex. One of these canals is called the *scala vestibuli* (vestibular canal) whose base is closed by the oval window. The *scala tympani* (tympanic canal) has its base closed by the round window.

At the apex of the cochlea a small opening, the helicotrema, permits continuity between the tympanic canal and the vestibular canal. Between these two canals is the *scala media* (median or cochlear canal). The scala media contains the actual *organ of corti* (Fig.). The actual sensory receptors in the organ of corti are two types of *hair cells.* There are in all 3500 *inner hair cells* and 20,000 *outer hair cells.* They are innervated by nerve fibres of the cochlear nerve, a division of the eighth nerve. From the upper surface of the hair cells, which are 8-12 μ in diameter, project tiny hair or *stereocilia* 4 μ. long and 0.1 μ thick which pass through a thin dense granular *reticular lamina* and are embedded in the *tectorial membrane.* It is a spiral ribbon attached at one end to the limbus and by one surface and its outer edge to the organ of corti in the region of the Hensen's cells, which lie outside the external hair cells. The tectorial membrane is a stiff gelatinous structure composed of glycoprotein material. Between the two triangular rods of corti lies the tunnel of corti, which is filled with perilymph.

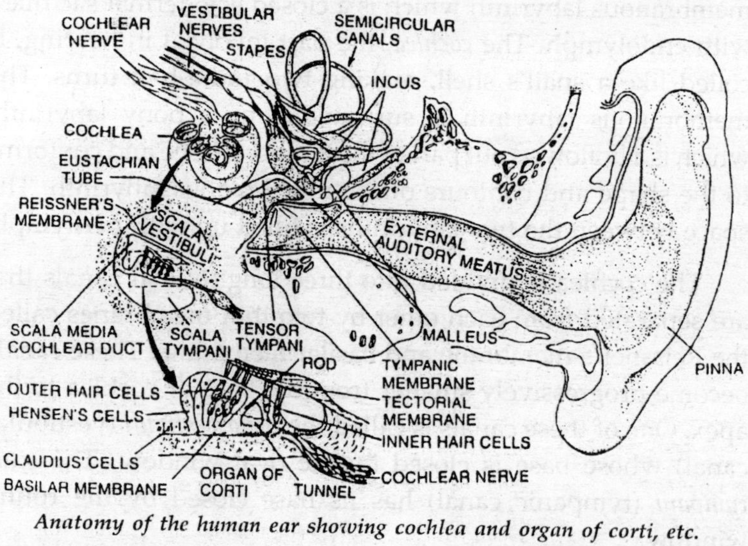

Anatomy of the human ear showing cochlea and organ of corti, etc.

Neural Pathways and Mechanism of Hearing : In hearing, sound waves are received by the external ear and transmitted through the auditory canal to the tympanic membrane which is caused to vibrate. The vibrations are conducted by the chain of ear ossicles to the oval window which transmits them to the fluid in the vestibular and tympanic canals. The vibrations of the endolymph cause the basilar membrane, with its hair cells, to vibrate so that the latter strike the tectorial membrane. This stimulation of hair cells causes them to initiate nerve impulses in the fibres of the auditory nerve, with which they are connected.

There is a spatial organization of the nerve fibres in the cochlear pathway all the way from the cochlea to the cerebral cortex. And recording of signals in the auditory tracts of the brain stem and in the auditory receptive field of the cerebral cortex shows that specific neurons are activated by specific sound frequencies. Therefore, the major method used by the nervous system to detect different sound frequencies is to determine the positions along the basilar membrane that are most stimulated. This is called *place principle* for determination of sound frequency.

According to the place principle, when sound waves strike the inner ear, the entire basilar membrane is set in vibration by a wave of displacement, which increase the amplitude from the oval window towards the apex of the cochlea. This displacement wave reaches a maximum at the region of the basilar membrane that resonates with the frequency of the incoming sound waves (*von Bekesy*, 1956). The particular hair cells in that region will be stimulated and the impulses conveyed to the fibres of the auditory nerve. Those impulses that are carried by certain fibres of the auditory nerve are interpreted by the hearing centre as particular tones. The loudness of a tone depends on the number of hair cells stimulated, whereas the quality of a tone is produced by the pattern of the hair cells stimulated by sympathetic vibration. The latter characteristic of tone enables one to distinguish between different human voices and different musical instruments even though the tones in each case be of the same pitch and loudness.

Recent findings demonstrate that the sound signals are not transmitted unchanged directly from the ear to the higher levels of the brain; instead, information from the sound signals begins to be dissected form the impulse traffic at levels as low as the cochlear nuclei. Another significant feature of the auditory pathways is that low rates of impulse firing continue even in the absence of sound, with this occuring from the cochlear nerve fibres all the way to the auditory cortex. When the basilar membrane moves towards the scala vestibuli, the impulse traffic increases; when the basilar membrane moves towards the scala tympani, the impulse traffic decreases.

Deafness

It may be defined as a state in which one is totally or partially unable to hear. The two main types of deafness are: *conductive deafness* and *nerve deafness.* The former may be due to foreign bodies, wax, otitis media which damage the drum and ossicles, or pathological fixation of the stapes in the oval window (otosclerosis). The hearing loss is fairly uniform throughout the frequency range. In patients with conductive deafness air-conducted sounds are less well perceived and bone-conducted sounds may persist longer. In normal subjects the air-conducted sound from the tuning fork is heard longer than that heard by bone conduction when the base of the vibrating fork is placed on the mastoid process (Rinne's test positive). In conductive deafness bone conduction is greater than air conduction (Rinne's test negative).

Nerve deafness may be degenerative, hereditary, toxic, acoustic trauma, meningitis, and acoustic neuroma. In nerve deafness both air-conducted and bone-conducted sounds are less well perceived (Rinne's test).

Sense of Equilibrium

Closely connected to the inner ear and forming a part of

it are two small sacs, the utriculus and sacculus, and three semicircular canals (Fig.). Like the cochlea they are filled with endolymph. They are concerned with the sense of balance and are innervated by the nonacoustic branch of the auditory nerve. The utriculus and sacculus are hollow sacs lined with sensitive hair and each contains a small otolith. The stimulations of the hair cells are interpreted in a certain way with reference to position. The three semicircular canals are at right angles to each other, one in each plane of space. They are filled with endolymph and at the opening of each canal into the utriculus there is an ampulla which contains sensory hair cells but no otoliths. The hair cells are stimulated by the movement of the endolymph. Thus, these parts of the internal ear are the principal centres of balance.

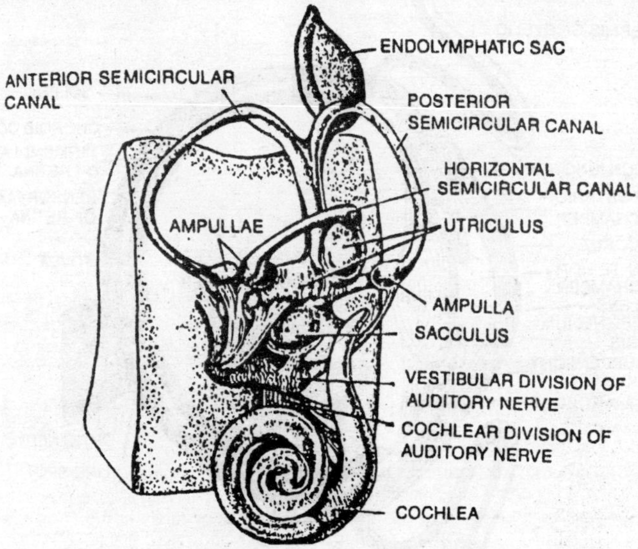

The membranous labyrinth of man.

The Eye

The human eye is optically equivalent to a photographic camera. It has a lens, a variable aperture (the pupil), and a retina that corresponds to the film. The eye wall is composed of the following three layers

1. The outer layer : The outer layer, known as the *fibrous tunic,* is complete and consists of two regions: (a) the *cornea,* which is the transparent, exposed portion of the tunic, and (b) the *sclera,* the remainder of the tunic to which the extrinsic muscles moving the eye are attached.

The oneness and continuity of the sclera and cornea are, however, more apparent than real. The sclera is a product of a mesodermal envelope that encases the embryonic eye, whereas the cornea is essentially an area of the skin of the head and thus has both ectodermal and mesodermal components. In the adult this head skin is doubled back on itself to form the underlinings of the eyelids, and its epidermal component continues over the cornea as the *conjunctiva.*

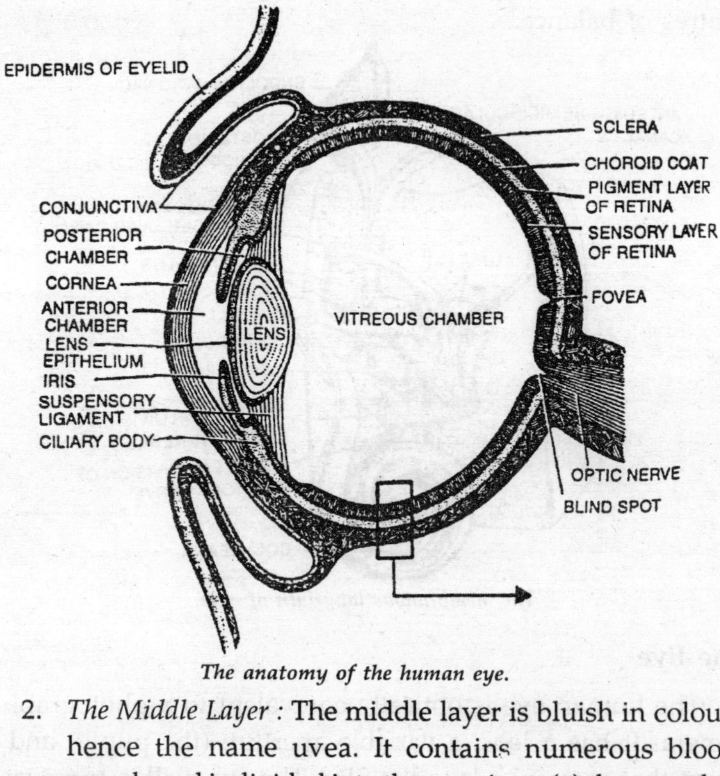

The anatomy of the human eye.

2. *The Middle Layer :* The middle layer is bluish in colour, hence the name uvea. It contains numerous blood vessels and is divided into three regions: (a) the vascular,

pigmented *choroid layer,* which is closely adherent to the sclera, (b) the *ciliary body,* which is a relatively thickened, muscular part of the *uvea* at the level of transition between the sclera and cornea, and (c) the *iris.*

From the ciliary body the uvea turns inwards and away from the sclera and joins with the margin of the innermost layer (retina) to form the *iris,* which is also equipped with muscle. The opening in the centre of the iris is the *pupil.* It can be altered in diameter by the smooth muscle of the iris.

3. *The Innermost Layer :* The innermost layer or the retina abuts the entire extent of the uvea. The part of it in contact with the choroid layer is fairly thick and is the region sensitive to light. The remainder is thin and nonsensory and contributes to the ciliary body and iris. The retina consists of the following layers (Fig.).

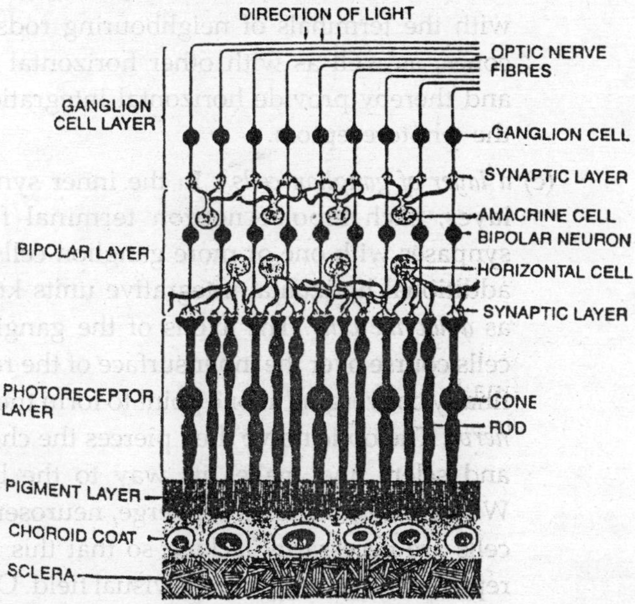

The layers of the retina.

(i) *The Outer Pigmented Layer* : It lies against the choroid coat.

(ii) *The Sensory Layer* : The inner sensory or nervous layer of the retina contains definitive light receptors and a succession of neurons in synaptic series. The cellular components of the nervous retina are, in turn, arranged in three layers. These are :

(a) *a layer of photoreceptors* consisting of long and narrow rods and shorter, thicker *cones*. In the human retina, there are 125,000,000 rods and 7,000,000 cones. The free ends of rods and cones are oriented towards the outer pigmented layer of the retina.

(b) *a layer of bipolar neurons*: Centrally the rods and cones send processes into the outer synaptic layer where they form synapses with the bipolar neurons. At this level of the bipolar layer there also lie *horizontal cells,* whose processes connect with the terminals of neighbouring rods and cones, as well as with other horizontal cells, and thereby provide horizontal integration of the photoreceptors.

(c) *a layer of ganglion cells* : In the inner synaptic layer, each bipolar neuron terminal forms synpases with one or more ganglion cells and additional horizontal integrative units known as *amacrine cells.* The axons of the ganglionic cells course over the inner surface of the retina, finally converging at one point to form the *optic nerve.* The optic nerve then pierces the choroid and sclera and makes its way to the brain. Where the optic fibres converge, neurosensory cells are necessarily lacking so that this point represents a *blind spot* in the visual field. On the other hand, there is an adjacent area where visual activity is greater than in any other region of the retina. This is *optic fibre centralis*, a highly specialized depression containing many thousands of closely packed cones, but devoid of rods. There are no blood vessels in this area and the ganglionic nerve fibres diverge so that sharply focussed light falls directly on the cones.

Thus, the arrangement of the receptors and neurons of the retina is such that the rods and cones are farthest removed from the light entering the front of the eye. In other words, except for the fovea, light must pass through the layers of ganglionic and bipolar neurons in order to reach the rods and cones. By the same token, stimuli thus set up must pass back through these layers of cells to reach the optic nerve. Because of this reason the retina of vertebrates is said to be "inverted".

The Lens

In the interior of the eye, just behind the iris lies the lens. The lens is clear and glassy in appearance and biconvex in form, with refractive index of 1.42. The lens is held in place by a *suspensory ligament* or zonule composed of fibres radiating from the ciliary body.

The lens system of the eye is composed of four refractive interfaces :

(i) the interface between air and the anterior surface of the cornea,

(ii) the interface between the posterior surface of the cornea and the aqueous humor,

(iii) the interface between the aqueous humor and the anterior surface of the crystalline lens of the eye, and

(iv) the interface between the posterior surface of the lens and the vitreous humor. The refrective index of air is 1; the cornea, 1.38; the aqueous humor, 1.33; the crystalline lens (on the average), 1.40; and the vitreous humor,

1.34. The total refractive power of the eye is about 59 diopters while the total refractive power of the crystalline lens of the eye is only 20 diopters.

Physiology of Vision

Photopigments and Photochemistry of Vision : The photopigment of rods is called *visual purple* or *rhodopsin.* This is a compound composed of a colourless protein opsin, and a yellowish carotenoid, retinaldehyde or *retinene,* a vitamin A derivative. The photopigment of cones is known as *visual violet* or *iodopsin. Wald* (1954) has shown that rhodopsin is bleached by light via intermediaries (lumirhodopsin and metarhodopsin) to the colourless protein *retinene* and *opsin.* Retinene is reduced to vitamin A by a dehydrogenase working together with $NADH_2$. The equilibrium between vitamin A and retinene lies far over towards vitamin A. If the system were allowed to come into equlibrium retinene would be completely reduced. However, retinene is trapped by opsin to form rhodopsin and is thus continuously removed. Moreover vitamin A, from the blood and from the pigment epithelium, and the continuous oxidation of NAD by the respiratory enzymes assists the vitamin A-retinene equilibrium to move over to the right. The resynthesis of rhodopsin following its breakdown by a light stimulus is responsible for the phenomenon of *dark adaptation.* The importance of vitamin A in the resynthesis and formation of rhodopsin explains how vitamin A deficiency causes night blindness (Fig.).

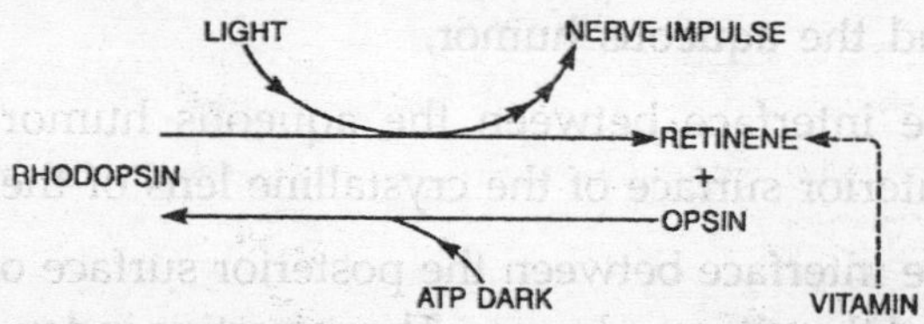

The rhodopsin cycle. The photopigment is rhodopsin in rods and iodopsin in cones. Vitamin A may replenish the supply of retinene and thus indirectly the supply of rhodopsin.

When quanta of light strike the rods or cones, they perform no photochemical work, but trigger the discharge of nerve impulses by the receptor cells. *Wald* (1954) has shown that when rhodopsin is bleached, two to three sulphydryl groups are liberated per molecule of retinene formed. He has suggested that these may initiate the optic nerve impulse either by altering the oxidation-reduction potential or by binding some metallic cation.

Dark Adaptation and Light Adaptation : It is experienced that on passing from a brightly lit room into a dark place, vision at first is very poor but improves as the eye becomes dark-adapted. Thus, *dark adaptation* is the property of the retina which enables the eye to see in the dark. When changing from a bright place to darkness, cone adaptation takes place rapidly, in about 10 minutes. Thereafter, the sensitivity shows a further increase which occurs more slowly but which is of much greater degree. This increase in sensitivity is due to dark adaptation of the rods, for it does not occur in congenitally night-blind (rod function absent) subjects. Thus, if the retina has been .exposed to bright light for a long period, complete dark adaptation may take several hours.

In dark adaptation the change in sensitivity of rods depends upon the presence of rhodopsin. As explained earlier, in light the rhodopsin is bleached and is formed only under conditions of low illumination. Thus, the lower the intensity of light, the greater the production of rhodopsin, which increases the sensitivity of rods. As vitamin A is necessary for the formation of retinene and thus indirectly for the supply of rhodopsin, vitamin A deficiency prevents dark adaptation and causes *night blindness.*

Light adaptation occurs within a few minutes when a subject re-enters the light after sitting in darkness. However, the first symptoms are those of dazzle, but the sensations, which may be initially painful, quickly pass off as the eye becomes less sensitive.

In addition to adaptation caused by changes in concentrations of rhodopsin, the eye has two other mechanisms for light and dark adaptation. The first of these is a *change in pupillary size*. This can cause adaptation of approximately 30-fold within a fraction of a second, because of changes in the amount of light allowed through the pupillary opening. The other mechanism is *neural adaptation*, involving the neurons in the successive stages of the visual chain in the retina itself and in the brain. This occurs in a fraction of a second although the degree of adaptation is only a fewfold rather than the many thousand-fold that occurs during adaptation of the photochemical system and takes many minutes to hours.

Formation of the Image

Functionally the eye resembles a photographic camera and the formation of an image on the retina depends on the following:

(i) *Accommodation* : In the eye, the distance between the lens and the retina cannot change. The ability of the eye to focus on objects at varying distances is known as *accommodation* and is due to a mechanism which allows changes of curvature of the anterior surface of the lens. The range of accommodation depends upon the ability of the lens to become more round. In the normal eye the range of accommodation is about 10 dioptres. The distance of an object at which the lens reaches its maximum refractive power is termed the *near point*. It measures about 10 cm for the normal young eye. Thus, such an eye can see objects clearly from infinity down to about 10 cm.

The shape of the lens is controlled by the ciliary body (muscle). In other words, the dioptric strength of the lens depends on the ciliary muscle. When the ciliary muscle is completely relaxed, the dioptric strength of the lens is as weak as it can become. On the other hand, when the ciliary muscle contracts as strongly as possible, the dioptic strength of the lens becomes maximal. Thus, the lens is adjusted to the distance of the object (far object, flat lens; near object, curved lens) by the ciliary muscles until the image becomes focussed on the retina.

The ability of lens to change shape progressively decreases with age. As a person grows older, the lens grows larger and thicker and becomes far less elastic, partly because of progressive denaturation of lens proteins. The loss of accommodation is called *presbyopia*, which is due to sclerosis of the lens substance which is unable to adapt itself to a more spherical shape. The power of accomodation decreases from about 14 diopters in a child to less than 2 diopters at age 50 and to about 0 at age 70.

(ii) *Convergence :* In order to perceive only one object when there is an image focussed on each retina, the images must fall on corresponding points of each retina. To accomplish this as an object is brought closer to the eyes, the eyes must converge, *i.e.,* turn inward. Such movement is brought about by the extrinsic eye muscles.

(iii) *Change in the Size of the Pupil :* The pupil, corresponding to the diaphragm of a camera, narrows or widens and thereby regulates the amount of light admitted to the eye. The pupillary aperture is controlled by the iris muscles. As these muscles extend, the pupil narrows and less light is admitted. This takes place when bright light falls on the eye. Conversely, when the light intensity is low, the iris muscles contract, thereby enlarging the pupil and permitting more light to enter. The diameter of the pupil can be changed in man from about 1.5 to 9 mm. Since the area of a circle varies with the square of its diameter and because the amount of light admitted by the pupil varies directly with its area, it can be calculated that the amount of light admitted to the eye can be varied about 36 times (9.0/1.5 = 6, 6 x 6 = 36).

The iris, by altering the aperture of the pupil, not only regulates the amount of light admitted to the eye, but also varies the *depth of focus*. It is the range of distance in which all objects are in focus. In a camera, the depth of focus is increased by decreasing the diaphragm opening. The same is true of the eye. When a lens system has great depth of focus, the retina can be considerably displaced from the focal plane or the lens strength can change considerably and still the image remains in sharp focus, whereas when a lens system has a shallow depth of focus, moving the retina only slightly away from the focal plane causes extreme blurring. The greatest possible depth of focus occurs when the pupil is extremely small.

The iris muscle is innervated by both sympathetic and the parasympathetic nervous system. The activity of the former increases the pupillary aperture, while the activity of the latter decreases it. The effective stimulus that controls the iris is the intensity of the light striking the retina. The reaction of the iris to a light directed into the eye is termed the *pupillary light reflex*. As the iris of the eye into which the light has been directed changes, so does the iris of the other eye. This is called the *consensual light reflex*.

(iv) *Impulse Interpretation* : An external object is "pictured" on the retina as a series of points, each point corresponding to a rod or a cone. Impulses from these points are transmitted into the brain according to the pattern illustrated in Fig. All fibres from the left sides of both eyes lead into the left half of the brain and all fibres from the right sides of both eyes lead into the right half of the brain. In each brain hemisphere the fibres from the eyes lead to an optic lobe which contains the visual centres. It can be shown that for each group of rods and cones there exists a corresponding group of interpreter neurons in the visual centres. In other words, the "point picture" of an object on the retina is duplicated more or less faithfully in the optic lobes by impulses from specific rods and cones to their correlated interpreter neurons. In these neurons the impulses register as vision. Each optic lobe thus receives impulses from both eyes and normally the visual interpretations of both lobes give a simple, smoothly superimposed picture of the external world.

The image of an external object is projected on the retina in an inverted position, just as an image is inverted on the film of a camera, but such images are not perceived inverted because the optic centres in the brain give visual experiences correct orientations.

It is also known that visual interpretation in the optic lobes are assisted by the retinal neurons, which carry out a considerable amount of "data processing" before impulses are sent to the brain. For example, it has been found that the retina of a frog generates four distinct sets of impulses to the brain. These four sets of impulses are neatly superimposed in the brain and appear to be interpreted as a single picture. Thus, a frog sees the world in a unique, subjective manner, and a similar conclusion holds for all kinds of light-sensitive animals.

Neural Pathways and Perception of Image : The different neuronal cell types in the retina are : *(a)* the photoreceptors themselves *(rods* and *cones), (b)* the *horizontal cells,* which transmit signals horizontally in the outer plexiform layer from the rods and cones to the bipolar cell dendrites, (c) the *bipolar cells,* which transmit signals from the rods, cones, and horizontal cells to the inner plexiform layer, where they synapse with ganglion cells and amacrine cells, (d) the *amacrine cells,* which transmit signals in two directions, either directly from bipolar cells to ganglion cells or horizontally within the inner plexiform layer between the axons of the bipolar cells, the dendrites of the ganglion cells, and/or other amacrine cells, and (e) the *pinglion cells,* which transmit output signals from the retina through the optic nerve into the brain. A sixth type of neuronal cell in the retina, is the *interplexiform cell,* which transmits signals from the inner plexiform layer to the outer plexiform layer. These signals are all inhibitory and believed to control the lateral spread of visual signals by the horizontal cells in the outer plexiform layer. Their role possibly is to help control the degree of contrast in the visual image.

The neural organization of the retina clearly shows that the visual pathway from the cones to the ganglion cells is different from the rod pathway. The retina has both an old type of vision based on rod vision and a new type of vision based on cone vision. The neurons and nerve fibres that conduct the visual signals for cone vision are considerably larger than those that conduct the visual signals for rod vision, and the signals are conducted to the brain two to five times as rapidly.

The neurotransmitters released by retinal neurons have not been entirely delineated. Both the rods and the cones release *glutamate* at their synapses with the bipolar cells, and amacrine cells secrete at least eight types of transmitter substances, including acetylcholine, dopamine, glycine, gamma-aminobutyric acid, and indolamine, all of which normally function as inhibitory transmitters. However, the transmitters of the bipolar, horizontal, and interplexiform cells are not clear, but at least some of the synapses of horizontal cells involve electrical transmission rather than chemical transmission.

The transmission of most signals occurs in the retinal neurons by *electrotonic conduction,* not by action potentials. Electrotonic conduction means direct flow of electric current (not action potentials) in the neuronal cytoplasm from the point of excitation to the output synapses. Infact, even in the rods and cones, conduction from their outer segments where the visual signals are generated to the synaptic bodies is by electrotonic conduction. That is, when hyperpolarization occurs in response to light in the outer segment, almost the same degree of hyperpolarization is conducted by direct electric current flow to the synaptic body, and no action potential is noticed. Then, when the transmitter from a rod or cone stimulates a bipolar cell or horizontal cell, once again the signal is transmitted from the input to the output by direct electric current flow, and no action potential occurs. The importance of electrotonic conduction is that it allows *graded conduction* of signal strength. Thus, for the rods and cones, the strength of the hyperpolarizing output signal is directly related to the intensity of illumination; the signal is not all-or-none as would be the case for action potential conduction.

It is probable that some of the amacrine cells provide additional lateral inhibition and further enhancement of visual contrast in the inner plexiform layer of the retina as well. It is also known that some bipolar cells depolarize when the rods and cones are excited and others hyperpolarize. Because depolarizing and hyperpolarizing bipolar cells lie immediately against each other, this provides a mechanism for separating contrast borders in the visual image even when the border lies exactly between the adjacent photoreceptors. The visual pathways can be divided roughly into an *old system* to the midbrain and base of the forebrain, and a *new system* for direct transmission into the *visual cortex*. In humans, the new system is responsible for perception of virtually all aspects of visual form, colours, and other conscious vision. In many primitive mammals, even visual form is detected by the older system, using the superior colliculus in the same manner as the visual cortex is used in mammals.

Colour Vision

The human eye can recognise about 150 different colours in the visible spectrum (400—750 $m\mu$). The *Young-Helmhotz theory* suggests that there are three types of cones, each perhaps possessing its own characteristic photosensitive substance, one type responding to red, another to green, and third to violet light. These are the primary colours.

The different colour sensations are supposed to be produced by stimulation of various combinations of these three types of receptors. When the three cone types are stimulated equally, a sensation of white light is produced. More recently, evidence has been obtained that there are three different *opsins* in the retina, each in a different cone cell. Their absorption spectra have been measured, and curves were found with peaks at wavelengths of 435, 540, 565 nanometers which are close to the wavelengths of blue, green, and red respectively. Thus, light of a specific wavelength activates one type of cone more than the others, thereby establishing a pattern of impulses different from that evoked by light of a different wavelength. These impluse patterns provide the .clue for the appreciation of the colour.

The theory suggests that there is no colour sensitivity in the rods. In order to see colour, there must be relatively high illumination. In dim light everything appears in varying shades of grey because low illumination vision utilizes mostly the rods.

Colour Blindness

Lending support to the Young-Helmhotz theory of colour vision is the clinical finding of patients who are colour blind in one or more of the primary colours. Such findings can be explained on the basis of the existence of three different types of opsins, each specific for a particular colour range. Most commonly, colour-blind people have a deficit either in the red or green range. The terms *protanope, deuteranope* and *tritanope* are used to describe the individuals who have a deficit in the red, green, and blue ranges respectively.

Abnormalities of the Vision

The following defects may be present:

(i) *Myopia* : If the eyeball is abnormally long, or if the lens in its most elongated state has increased refractive power, parallel rays of light will reach a focus in front of the retina and, therefore, the image will be blurred. Objects close to the eye will be in focus, however, because diverging rays of light from the object will fall in the retina which lies behind the principal focus of the lens. Such a refractive error is termed *myopia,* or more commonly, *nearsightedness.* Glasses with concave lenses of proper refractive power will permit the individual to focus parallel rays of light and to have the normal range of accommodation.

(ii) *Hyperopia* : If the eyeball is too short, or if the lens, in its most relaxed or non-accommodating state, has decreased refractive power, parallel rays of light will reach a virtual focus beyond the eye. This condition is termed *hyperopia,* or more popularly, *farsightedness.* However, such an eye can see distant objects clearly by accommodating.

If there is sufficient accommodation remaining, near objects may also be brought into focus. In this case the need of corrective glasses depends upon two factors : (a) the degree of accommodation required to focus parallel rays of light and *(b)* the remaining range of accommodation. A convex lens is used to correct hyperopia.

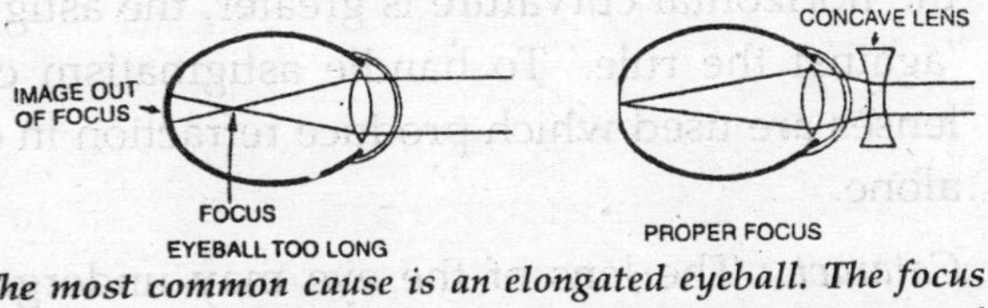

Myopia. The most common cause is an elongated eyeball. The focus point lies in front of the retina. A concave lens that causes parallel rays of light to diverge is required.

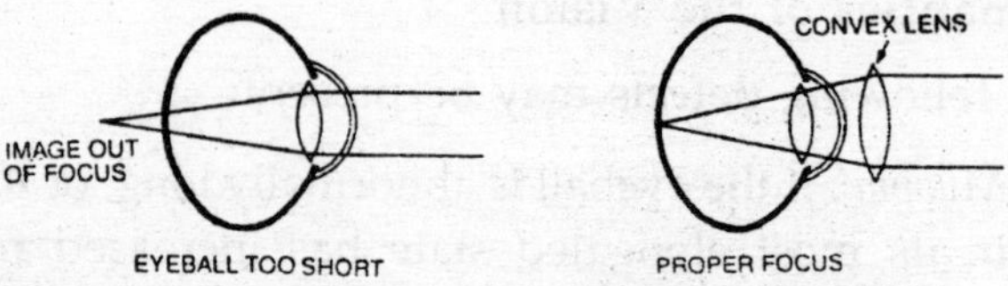

Hyperopia. The most common case is a too short eyeball. The focal point lies beyond the retina. A convex lens that causes parallel rays of light to converge is required.

(iii) *Presbyopia* : With age the lens loses its elasticity and ability to accommodate. Since this decreased range of accommodation is most commonly associated with age, it is termed *presbyopia*. The eyes can no longer accommodate for both near and far vision. Therefore, to see clearly both in distance and nearby, an older person must wear *bifocal glasses* with the upper segment normally focused for far-seeing and the lower segment focused for near-seeing (that is for reading).

(iv) *Astigmatism* : As the name suggests this is an error of vision in which the light rays are not brought to a point focus on the retina. The focus for horizontal rays differs from that for vertical rays. Hence an astigmatic subject looking at a piece of graph paper may focus on the vertical lines and may fail to focus the horizontal lines and vice versa. The defect is most commonly due to a difference in the horizontal and vertical curvatures of the cornea. Occasionally the same abnormality affects the lens. If the curvature is greater along the vertical meridian, astigmatism is said to be 'with the rule.' If the horizontal curvature is greater, the astigmatism is 'against the rule.' To handle astigmatism cylindrical lenses are used which produce refraction in one plane alone.

(v) *Cataract* : The lens of the eye may undergo changes which reduce its transparency. Ultimately the lens becomes cloudy or opaque and is termed a *cataract*. The only treatment is the removal of the opaque lens.

Since the normal lens has a refractive power of about 25 dioptres, glasses with comparable convex lens must be substituted. Because removal of lens removes all power of accommodation, at least two pairs of lenses are required.

(vi) *Strabismus* : The term strabismus means "squint", but it is used to describe a condition in which the image does not fall upon the corresponding point of the two retinas. This is generally due to an abnormality of the extrinsic eye muscles. As a result, two images are seen, a condition termed *diplopia*. It is treated by surgical alteration of the extrinsic muscles, or lenses which will bring the image to bear upon the corresponding points of the two retinas.

(vii) *Argyll Robertson Pupil* : If the pupil fails to change in size in response to light but still responds during accommodation, it is termed as Argyll Robertson Pupil. This is due to the fact that the pathway for the light reflex is different from that for pupillary responses during accommodation.

Smell and Taste

Physiologically, smell and taste are related to each other and are generally classified as *visceral senses* because of their close association with gastrointestinal function. Both smell and taste receptors are *chemoreceptors* that are stimulated by molecules in solution in mucous in the nose and saliva in the mouth. The smell receptors are distance receptors *(teleceptors)*. The smell pathways have no relay in the thalamus and there is no neocortical projection area for olfaction. The taste pathways pass up the brain stem to the thalamus and project to the post-central gyrus along with those for touch and pressure sensibility from the mouth. However, both senses are strongly tied to primitive emotional and behavioural functions of the nervous system.

Sense of Smell

Structure of Olfactory Membrane : The olfactory receptors are located in a specialized part of the nasal mucosa or the *olfactory mucous membrane..* It contains *sustentacular cells,* and *progenitor cells* for the olfactory receptors. Interspersed between these cells are *receptor cells* (olfactory cells). Each olfactory receptor is a neuron having a short thick dendrite with an expanded end called an *olfactory rod* (knob). From each rod, 6 to 12 olfactory hair or cilia project to the surface of the mucous layer. The cilia are unmyelinated processes about 2 μm long and 0.1 μm in diameter. The olfactory membrane also has small *glands of Bowman* which secrete mucous on to the surface of the olfactory membrane.

The axons of the olfactory cells pierce the cribriform plate of the ethmoid bone and enter the olfactory bulbs. In the olfactory bulbs, the axons of the olfactory cells synapse with the primary dendrites of the *mitral cells* to form the *olfactory glomeruli.* The *tufted cells* and the *periglomerular cells* also participate in the formation of glomeruli. The axons of the mitral cells pass posteriorly to the *olfactory cortex* and terminate on the apical dendrites of pyramidal cells in the olfactory cortex. In humans, the olfactory cortex is in the piriform cortex which is bilaterally stimulated by the olfactory stimuli.

Physiology of Olfaction

Mechanism for Transmission of Olfactory Signals : The cilia of the olfactory cells first of all respond to the olfactory chemical stimuli. The odorant substance, on coming in contact with the olfactory surface, first diffuses into the mucous that covers the cilia. Then it binds with a *receptor protein* protruding through the caliary membrane. This receptor protein is a long molecule that threads its way through the membrane seven times, folding inward and outward. The odorant binds with the portion of the receptor that folds to the outside. The inside of the folding receptor, on the other hand, is coupled to a so-called *G-protein,* which is a combination of three subunits (heterotrimeric). On excitation of the receptor, an alpha subunit breaks away from the G-protein and immediately activates adenyl cyclase attached to the inside of the ciliary membrane near the receptor cell body. The activated cyclase in turn converts many molecules of intracellular ATP into cAMP. Finally, this cAMP activates another nearby membrane protein, a gated sodium ion channel that allows large numbers of sodium ions to pour into the receptor cell cytoplasm. The sodium ions contribute positively to the inside of the cell membrane, thus exciting the olfactory nerve cell and transmitting action potentials into the CNS by way of an olfactory nerve.

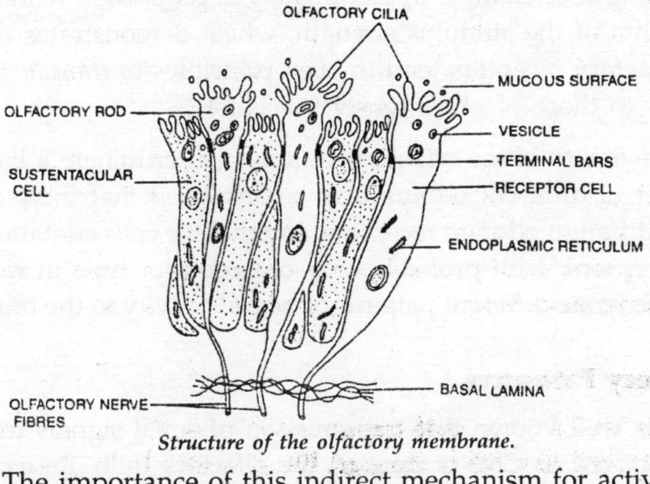

Structure of the olfactory membrane.

The importance of this indirect mechanism for activating olfactory nerves is that it greatly multiplies the excitatory effect of even the weakest odorant substance. In addition to the basic chemical mechanism which stimulates the olfactory cells, several physical factors affect the degree of stimulation. First, only volatile substances that can be sniffed into the nostrils can be smelled. Second, the stimulating substance must be at least slightly water-soluble, as that it can pass through the mucous to reach the olfactory cells. Third, it is helpful to be at least slightly lipid-soluble, because the lipid constituents of the ciliary membrane repel nonlipid-soluble odorants from the membrane receptor proteins.

The membrane potential of unstimulated olfactory cells is about -55 millivolts. At this potential olfactory cells generate continuous action potentials at a very slow rate. Most odorants cause *depolarization* of the olfactory cell membrane, decreasing the negative potential in the cell from -55 to -30 millivolts or even less. Along with this, the number of action potentials increases to about 20 per second, which is a high rate for the minute olfactory nerve fibres. However, a few odorants hyperpolarize the olfactory cell membrane, thus decreasing, instead of increasing, the nerve firing rate. The rate of olfactory nerve impulses changes approximately in proportion with the logarithm of the stimulus strength, which demonstrates that the olfactory receptors tend to obey principles of *transduction* similar to those of other sensory receptors.

The ability of the olfactory system to discriminate a large number of different odours may be answered that there are many different odorant receptors, and that the cells containing the receptors, with probably only one receptor type in most cells, generate different patterns of neural activity in the brain.

Olfactory Pathways

It is well known that transmission of smell signals from olfactory cell to CNS is through the olfactory bulb. Research suggests that different glomeruli in the olfactory bulbs respond to different odors. Therefore, it is possible that the specific glomeruli that are stimulated are the real clue to the analysis of different olfactory signals transmitted into the CNS. The olfactory tract that enters the brain at the anterior junction between the mesencephalon and cerebrum, is divided into two pathways—one passing medically into the *medial olfactory area* and the other passing laterally into the *lateral olfactory area*. The medial olfactory area is responsible for more primitive responses to olfaction, while lateral olfactory area seems to be important for learning to like or dislike certain foods. However, a third olfactory pathway (that newer pathway) that passes through the thalamus, helps especially in the conscious perception of odour.

Sense of Taste

Taste is mainly a function of *taste buds* in the mouth. The importance of taste lies in the fact that it allows a person to select the desired food. For practical analysis of taste, the receptor capabilities have been arranged into four general categories. They are *sour, salty, sweet,* and *bitter*. The sweet and salty tastes are located principally on the tip of the tongue, the sour taste on the two lateral sides of the tongue, and the bitter taste on the posterior tongue and soft palate.

Structure of Taste Buds

In humans, the taste buds are located in the mucosa of the epiglottis, palate, pharynx and on three types of papillae on the tongue—circumvallate papillae, fungiform papillae, and foliate papillae. Adults have 3000 to 10,000 taste buds.

Each taste bud is an ovoid body measuring 50-70 μm. It is made up of four types of modified epithelial cells — *basal cells,* type 1 and 2 cells, which are supporting or *sustentacular cells,* and type 3 cells, which are the gustatory receptor cells or *taste cells* that make synaptic connections to sensory nerve fibres. The basal cells arise from the epithelial cells surrounding the taste bud. They differentiate into new receptor cells, and the old receptor cells are continuously replaced in about 10 days.

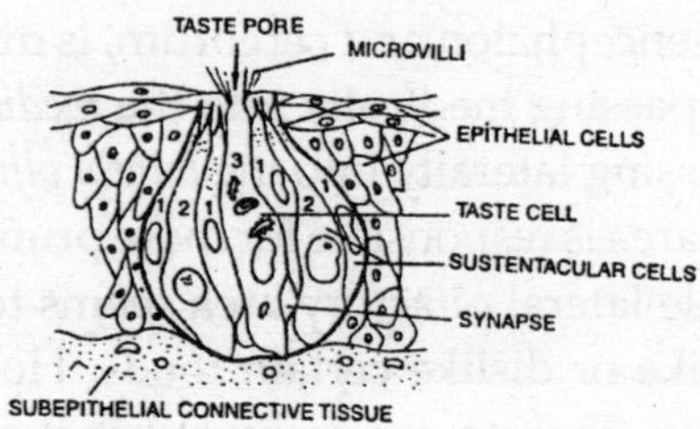

Taste bud, showing type 1,2 and 3 cells.

The outer tips of taste cells are arranged around a minute *taste pore.* From the tip of each cell, several *taste hair* (microvilli) protrude outward into the taste pore. These microvilli provide the receptor surface for taste.

There is a branching terminal network of several *taste nerve fibres* interwoven among the bodies of the taste cells. These fibres are stimulated by the taste receptor cells and some of them invaginate into folds of the taste cell membranes. Many vesicles form beneath the cell membrane near the fibres. These vesicles possibly contain a neurotransmitter substance that is released through the cell membrane to excite the nerve fibre endings in response to taste stimulation.

Physiology of Taste

Stimulation of Taste Buds : The membrane of the taste cell, like that of other sensory receptor cells, is negatively charged on the inside with respect to the outside. As soon as the taste hairs are touched by a taste substance, there is partial loss of this negative potential—that is, the taste cell is *depolarized.* The decrease in potential is approximately proportional to the logarithm of concentration of the stimulating substance. This change in potential in the taste cell is the *receptor potential* for taste.

The mechanism by which a stimulating substance reacts with the taste hairs to initiate the receptor potential is by binding the taste chemical to protein receptor molecule(s) that protrude through the taste hair membrane. This in turn opens ion channels, which allow sodium ions to enter and depolarize the cells. Then the taste substance is gradually washed away from the taste hair by the saliva, which removes the stimulus. The type of receptor protein in each taste hair determines the type of taste that will elicit responses.

Taste Pathways

Taste impulses from the anterior two thirds of the tongue pass first into the V cranial nerve, then through the chorda tympanic into the VII nerve, and finally into *tractus solitarius* in the brain stem. Taste impulses from the posterior of the tongue and from other posterior regions of the mouth are transmitted through the IX cranial nerve into the tractus solitarius. Finally, a few taste sensations are transmitted into the tractus solitarius from the base of the tongue and other parts of the pharyngeal region through the X cranial nerve. All taste fibres synapse in the nuclei of the tractus solitarius and send second-order neurons to a small area of the ventral posterior medial nucleus of the thalamus. From the thalamus, third-order neurons are transmitted to taste area in opercula of insula region. Taste does not have a separate cortical projection area in the brain but is represented in the portion of the postcentral gyrus in the parietal cortex that subserves cutaneous sensation from the face.

Taste reflexes are integrated in the brain stem to exercise control on the secretion of saliva during the ingestion of food.

Physiology of Senses

The Eye

The organ of vision consists primarily of the eye-ball sheltered within the *body orbit* and optic nerve. Associated with the eye-ball are various accessories like *extrinsic muscles, eyelids* and *tear apparatus.*

The spherical *eye-ball* consists of three coats that enclose the various refractive structures. The outer coat is the tough, fibrous *tunica fibrosa* comprising the transparent *cornea* and the opaque *sclera.* The middle coat is the vascular and pigmented *tunica vasculosa or uvea,* consisting of the *choroid,* the *ciliary body* and the *iris.* The iris ends about a circular opening, the *pupil.* The innermost coat is the *tunica internet,* or *retina.* Its nerve fibres combine posteriorly and leave the eyeball as the optic nerve. A refractive *lens,* held in position by a *suspensory ligament,* lies just behind the pupil and iris.

The vitreous body is a jelly that occupies the large space behind the lens. The space in front of the lens and vitreous body is occupied by watery aqueous humor. The eyelids are movable folds of skin that can cover the front of the eyeball. The lacrimal apparatus secretes *tears* into the *conjunctival sac,* a space between eyelids and eyeball.

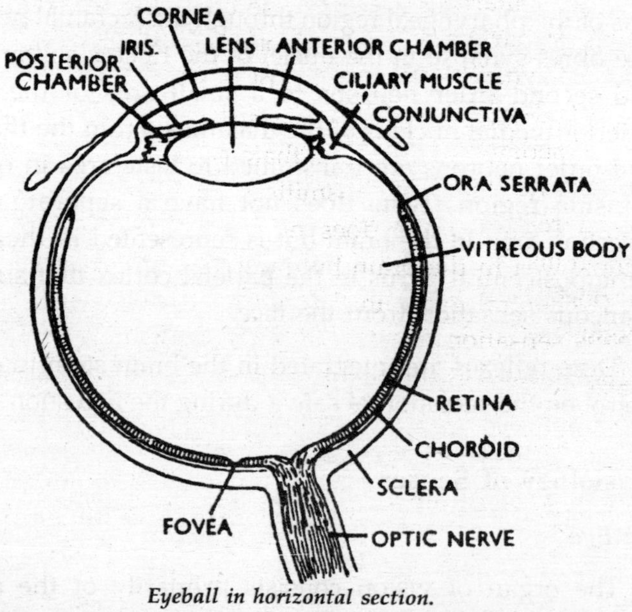

Eyeball in horizontal section.

Fibrous Tunic : The outer coat of eye-ball consists of an opaque sclera and the modified, transparent anterior cornea.

A. *Sclera* : This external coat forms the posterior 4/5 of the eyeball and is continuous anteriorly with the cornea. It ranges from 0.13 to 1.0 mm in thickness. It is fairly expansive and is commonly known as the "white" of the eye. Tendons of the eye muscles insert on this coat.

B. *Cornea* : This is the anterior transparent and non-vascular anterior continuation of the sclera which forms the bulging portion of the eye-ball. From front to back, it consists of 5 well-defined layers.

Vascular Tunic : This middle coat is primarily vascular and pigmented. It is bound closely to the retina but joins the sclera loosely. It has two circular openings : (1) the *pupil* and (2) the one about the optic nerve. From back to front, the middle coat comprises the *choroid, ciliary body* and *iris.*

A. *Choroid* : The choroid is the vascular coat of the eye and consists of four layers.

B. *Ciliary Body* : Anteriorly the choroid is thickened and forms the ciliary body which consists of ciliary processes and the ciliary muscle. The choroid thickens progressively as it nears the iris and here bears about 70 radially arranged ridges called ciliary processes. These processes anchor the fibres of the *suspensory ligament* of the lens. Besides ciliary processes the anterior end of the choroid encloses a prominent ciliary muscle, a mass of smooth muscle. The fibres of this muscle are arranged in three groups : (1) an inner *circular* group near the base of the iris ; (2) a middle *radial* group ; (3) an outer *meridional* group.

The circular fibres are responsible for relaxing the tension on the lens. This allows the lens to bulge by its own elasticity and accommodates the eye for near vision. The radial and meridional fibres are believed to focus the lens for distant vision.

C. *Iris* : This is a further continuation forward of the choroid coat beyond the ciliary body. It serves as a circular adjustable diaphragm that projects into the eye with an aperture, the *pupil,* in its centre. The anterior surface of the iris is covered by a layer of squamous epithelium continuous with that covering the posterior surface of the cornea. The posterior surface of the iris is clothed with an epithelium about 2 layers thick. This is continuous with the epithelium covering the ciliary processes and is derived from the retina. It contains a dark coloured pigment, the *melanin.* The substance of the iris consists of a spongy fibro-elastic tissue which is very vascular. Its upper layer contains many branched pigment cells *(Chromatophores)* that determine the "colour of the eye". A little pigment gives the iris a blue colour, more pigment produces hues of grey and green, while much dark pigment produces a brown iris. In Albinos the pigment as a whole is very scanty and the red colour of the blood in the vessels, combined with the colourless tissue, produces a pink iris.

The iris also contains two groups of smooth muscles, the *circular constrictor* of the pupil near the pupillary margin, and the *radiating dilator* of the pupil situated more externally.

Internal Tunic or Retina : This innermost ocular coat lines the eye-ball ending at the margin of the pupil. Its nervous elements extend only upto the posterior limit of the ciliary body where they end in a wavy line, the *ora serrata.* The extension of the retina forward to the or a serrata is entirely non-nervous. The nervous part of the retina and the optic nerve are essentially parts of the brain. The inner surface of the retina lies against the viterous humor and its outer surface is adjacent to the choroid.

The retina may be divided into ten layers ; from the outside these are : 1. Pigment layer. 2. The photoreceptor layer of rods and cones. 3. Enternal limiting membrane. 4. External nuclear layer (cell bodies of rods and cones). 5. External plexiform layer (synapses of the rods and cones). 6. Internal nuclear layer (bipolar cells). 7. Internal plexiform layer (synapses of the bipolar cells with the ganglion cells). 8. Ganglion cell layer. 9. Nerve fibre layer. 10. Internal limiting membrane.

1. The pigment layer consists of large, cuboidal epithelial cells firmly bound to the Bruch's membrane of the choroid. The pigment cells bear numerous pigmented, cytoplasmic processes that extend between the rods and cones. Each pigment cell covers about nine or ten of the rods and cones.

2. The photoreceptor layer consists of light-sensitive visual cells of two kinds, *rods* and cones, which correspond to the first neurons in an afferent sensory receptor system. These photoreceptor cells convert light into atomic motion and then into a nerve impulse. These cells are arranged vertically and parallel in a palisade fashion. The rods outnumber the cones in human retina. There are three million cones and a billion rods in a human. The rods are slender cylinders ; cones are shaped like long-necked flasks. Both possess a slender *outer segment* and a thicker *inner segment.* The outer segment of cones is much thicker and conical when seen in section, hence their name.

   (a) Rods are slender, elongated structures, typically 1 μ m in diameter and 40 μ m in length. *The outer segment of a rod is specialized for photoreception.* It contains a stack of about 1,000 discs, which are closed, flattened sacs about 160 Å thick. These membranous structures are densely packed with photoreceptor molecules. The photosensitive molecule in rods is *visual purple* or *rhodopsin,* which consists of *opsin,* a protein, and 11-cis- *retinal,* a prosthetic group. Opsin does not absorb visible light. The colour of rhodopsin and its responsiveness to light depend on the presence of 11-cis-retinal, which is a very effective chromophore. Rhodopsin is synthesized by ribosomes that are attached to the endoplasmic reticulum. A slender cilium joins the outer segment and the *inner segment.* The inner segment of the rod cell, has a large number of mitochondria which are known as the *ellipsoid.* The inner segment generates ATP at a very rapid rate and is highly active in synthesizing proteins. It is contiguous with the nucleus, which is next to the synaptic body.

The discs in the outer segment have a life of only ten days and are continually renewed. The visual pigment rhodopsin present in this segment absorbs the light energy and initiates the visual stimulus from the rod cell. These cells are used for peripheral vision and for vision at *low levels of light intensity* (twilight or scotopic vision), for example, the dim light in the interior of a darkened room.

(b) The lamellate plates present in the outer segment of a cone cell do not contain rhodopsin but another visual pigment often called *visual violet* or iodopsin. The cones are thinnest in the fovea where the visual acuity is greatest. The cones are sensitive to *bright light and colour* (photopic vision). Spectrophotometric studies have revealed that these are in fact *three types of cone cells : blue, green,* and *red-absorbing.* The effect of different wavelengths of light on the cones results in the perception of different colours.

3. The rods and cones are separated from one another by radially arranged *glial fibres* also called *Midler's fibres.*

These fibres extend through the bulk of the thickness of the retina. Lateral processes of Muller's fibres contact the bases of the rods and cones. When seen with the optical microscope, wese contacts appear as a dense line known as the external limiting membrane.

4. The external nuclear layer contains the nucleated cell bodies of rods and cones. Cone nuclei are ovoid and lie just beneath the external limiting membrane whereas rod nuclei are rounder and lie at deeper levels.
5. The external plexiform layer is the region of synapse between the fibres of the rods and cones and the dendrites of the bipolar cells. These latter cells are the second neurons in the afferent chain.
6. The internal nuclear layer contains the nuclei and cell bodies of *bipolar cells*, of a few large *horizontal cells* and of *'Muller's fibres'*.
7. The internal plexiform layer represents a second synaptic region in the neuron chain. The chief synapses are between axons of bipolar cells and dendrites of ganglion cells.
8. The ganglion cell layer contains the cell bodies and nuclei of large ganglion cells which furnish the third link of afferent neuron chains. The ganglion cells are large neurons with prominent Nissl granules.
9. The axons of ganglion cells form the nerve fibre layer. Here they converge on the papilla to leave the eye-ball as the optic nerve. The retinal elements are absent at the papilla.

   Therefore, this is an area which is insensitive to light and is known as the blind *spot.*
10. The internal limiting membrane forms the innermost layer of the retina. It arises from the expanded ends of *Muller's fibres* that are vertically arranged ependymal

cells. These latter cells are supporting elements extending between the two limiting membranes.

An area near the centre of the posterior part of the retina which lies in the direct optical axis is called the 'yellow spot' of *macula lutea.* It is nearly 2 mm in diameter and yellowish when fresh. In the centre of the macula, there is a little depression (0.5 mm wide) called the *fovea centralis.* The number of rod cells in macula is progressively decreased towards its centre, and the fovea only cones are present. The fovea is the region of greatest visual acuity and clearest vision.

Optic Nerve : The *optic nerve fibres* are non-myelinated axons of the ganglion cells of the retina. In the beginning, these fibres pass radially in the nerve-fibre layer of the retina. Near the posterior pole of the eye-ball they turn 90° and enter the optic nerve. The point of exit of the optic nerve is known as the *optic papilla, optic disc* or 'blind spot' because it is insensitive to light. Just after they leave the eye, the named fibres acquire myelin sheaths. The nerve fibres in each optic nerve total one million.

The optic nerve is an outgrowth of the central nervous system, therefore, the *meninges* (duramater, pia-arachnoid) of the brain cover the optic nerve also. A *retinal artery* and *vein* enter the optic nerve more than midway toward the eye-ball. They proceed to the eye-ball in a strand of connective tissue.

Refracting Media : *Cornea* is the strongest refracting agent of the eye. Behind the cornea, there are four other media that fill the interior of the eye-ball and co-operate in image formation. They are: *crystalline lens, ciliary zonule, vitreous body* and *aqueous humor.*

A. *Crystalline Lens :* The lens is a transparent, elastic, biconvex disc situated just behind the pupil. The lens is surrounded by a homogeneous elastic *capsule.* A single-layered *lens-epithelium* is present on the anterior surface but this layer of cells is absent from the posterior surface of the lens. Rest of the lens is made up of a fibrous *lens substance*. The component *lens fibres* are altered cells in the form of elongate bands which are united by a structureless cementing substance. The central part of the lens substance is older and called the *lens nucleus*. Rest of the lens substance is arranged in concentric *lamella*.

The posterior surface of the lens is more convex than the anterior surface. The two surfaces meet at the rounded off edge called the *equator*.

The major portion of the refraction of light at the eyes does not occur in the lens but at the corneal/air interface. The lens is capable of small changes in shape and thus serves merely to accommodate the eye for distinct vision at varying distances. The lens hardens and becomes opaque in old age due to a deposition of calcium salts. This impairment is termed *cataract* when the lens losses its power of accommodation. This condition is cured by the surgical removal of the lens after which distinct vision is still possible by using biconvex spectacles. However, for near and distant vision lenses of different focal lengths will be required.

B. *Ciliary Zonule* : Attached to the periphery of the lens capsule, both on its anterior and posterior surface is a system of fibres known as the *ciliary zonule*. As the fibres pass outwards they are inserted into the ciliary body. This radially arranged set of fibres serving the lens is a suspensory ligament. Contraction and relaxation of the ciliary muscle changes the thickness of the lens which bends the light rays entering the eye to focus them on the retina. This accommodation is produced by the ciliary muscle and *zonular fibres* of the suspensory ligament.

C. *Vitreous Body* : This is a transparent gelatinous semisolid jelly that occupies the bulk of the posterior part of the eyeball. It contains fibres, connective tissue cells and some leucocytes. After fixation the jelly appears fibrillar and a *vitreous membrane* can be seen. A *hyaloid canal* runs through the vitreous from the head of the optic nerve to the lens. This marks the site of the hyaloid artery in the embryo. After birth this vessel is not found in the eye.

D. *Aqueous Humor* : The space between the cornea and vitreous body is called the ocular *chamber*. This chamber contains *aqueous humor*, which is a clear, lymph-like fluid. The ocular chamber is incompletely divided by the iris into two parts :

(1) the *anterior chamber* lying between the cornea, iris and front of the lens, and

(2) the *posterior chamber* which is a small ring-shaped space bounded by the iris, ciliary body and vitreous body.

Accessory Organs

Eyelids : These are movable folds of skin that serve to protect the eye and shut out light. The eyelid is covered externally by skin which is modified on the inner surface as a transparent mucous membrane named the *conjunctiva*. The latter continues, by reflection, over the front of the eye-ball. The conjunctiva is a 4-5 cells thick stratified cuboidal epithelium. It is often the site of inflammations, when the blood vessels supplying it show considerable swelling. This condition is called *conjuctivities.*

The skin on the outer surface of the eyelid consists of a thin epidermis, with small dermal papillae. Sebaceous glands and small hairs are present in this skin. At the edge of the eyelid are present large hair follicles for the *eyelashes.* The latter are coarse hairs with large *sebaceous glands of zeis.* The eyelashes are replaced every 100 to 150 days. Between the hair follicles of eyelashes are large, spiraling sweat glands, the *ciliary glands of Moll* These are provided with large myoepithelial cells.

The shape of the lid is maintained by an internal fibrous plate, the *tarsus.* Numerous sebaceous *tarsalglands of Meibom are* embedded in this plate. Each gland had many alveoli opening into an axial excretory duct. These ducts open externally on the margin of the lid behind the eyelashes. Their oily secretion helps to lubricate the edges of the eyelids so that they do not sick together. It also prevents the entry of water when the eyes are closed. A thin sheet of muscle, the *orbicularis oculi muscle* acts to close the eyelids. The upper lid is raised by the *levator muscle.*

Lacrimal Gland : Each eye is provided with a lacrimal gland which grows out from the upper lateral margin of the conjunctival sac. The lacrimal gland is of the compounds, racemose type. The alveoli have tall cells and a wider lumen, fine cytoplasm of the alveolar cells contains large, pale secretion granules and fat droplets. The secretion provides the clear salty tears that flush the conjectural sac. The tears flow to the inner angle of the eye by 6-12 ducts and are collected by a *lacrimal canaliculus* in each eyelid. By way of the *lacrimal sac* and *naso-lacrimal duct,* they reach the nasal cavity.

Eye Muscles : Four *recti* and two *oblique* voluntary *ocular muscles* insert on the sclerotic coat of the eyeball. A single muscle can produce only one type of movement *i.e.* direct lateral whereas all other movements of the eye result from the combined action of at least two muscles.

Physiology of Sight

Light is reflected into the eyes by the various objects within the field of vision. White light is actually a combination of all the' colours of the visual spectrum of rainbow, that is, red, orange, yellow, green, blue, indigo and violet. This can be demonstrated by passing white light through a glass prism which refracts or bends the rays of the different colours to a greater or lesser extent. The different colours have different wavelengths, therefore when white light is passed through a prism the different colours take different paths. Red light has the longest wavelength and violet the shortest (see Fig.) This range of colour is called the spectrum of visible light. The spectrum of light is extremely broad but only a small part of it is visible to the human eye.

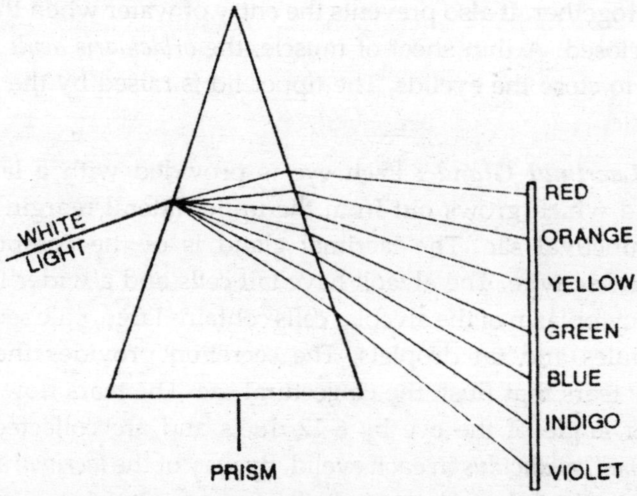

White light broken into the colours of the visible spectrum when passing through a prism.

A specific colour of the visible spectrum is perceived when only one wavelength of light is *reflected* by an object and all other wavelengths are *absorbed*. For example, when an object appears white all wavelengths are being *reflected*, when it appears *black* all wavelengths are being *absorbed* and it appears *red* when only the *red wavelength is reflected* and all *others are absorbed*. Black and white are not regarded as colours.

The achieve a clear vision light reflected from objects within the visual field is focussed on to the retinae of both eyes. Two processes are involved in producing a clear image.

1. Refraction of the light rays.
2. Accommodation of the eyes to light.

Although these may be considered as two separate processes, effective vision is dependent upon their co-ordination.

Refraction of the Light Rays

Refraction refers *to bend* or *turn aside.* When light rays pass from a medium of one density to a medium of a different density they are refracted or bent. This principle is used in the eye to focus light on the retina. Several structures which are more dense than air are involved in the refraction of light *in the eye.*

Before reaching the retina light rays pass successively through the cornea, the aqueous humour, the crystalline lens and the vitreous humour.

A lens is a transparent substance that bends rays of light and causes them to converge or diverge. Lens used in spectacles is generally of glass. The ability of a lens to bend light called *refractive power* is expressed in diopters. Focal distance is the distance between a biconvex lens and its *principal focus.* The power of a biconvex lens is expressed in plus diopters and the power of a biconcave lens is termed in minus diopters.

Accommodation of the Eyes to Light : When the eye is at rest the ciliary muscles are relaxed and the lens is thin and stretched, *i.e.,* it has a long focal length. Thus, an eye at rest is adapted for viewing distant objects. When the ciliary muscles contract, the tension in the suspensory ligaments is reduced, thus causing the lens to shrink and become thicker so that it has a short focal length. As the degree of contraction of the ciliary muscles varies, it is possible for the lens to focus on objects at varying distances from the eye. Hence, objects which are very distant *(i.e.* at the limits of visibility) or very near (about 25 cm) can be brought to a focus on the retina. The ability of the eye to bring about the necessary alterations in the focal length of the lens, with regard to the distance of the object from the eye, is known as accommodation.

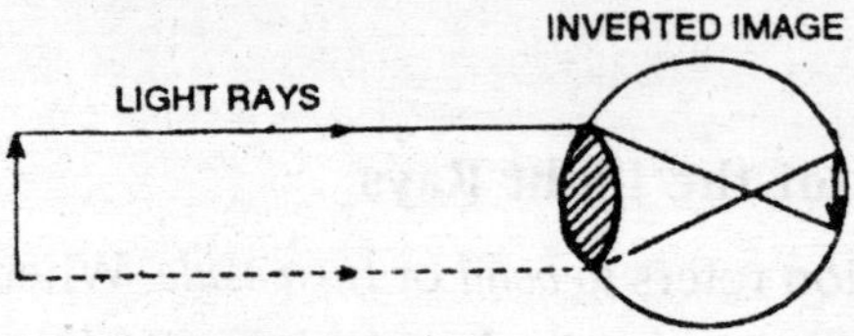

Normal sight.

The normal resting eye (emmetropic) is set for parallel rays travelling from objects 20 feet or more and does not require any accommodation. As objects move nearer to the eyes the rays of light reflected by the object become more divergent and require to be refracted more accurately.

Three factors are involved in accommodation :

A. The size of the pupils.

B. The movement of the eyeballs, called convergence.

C. The increased curvature of lens.

The Size of the Pupils : Pupil size influences accommodation by controlling the amount of light entering the eye. *In a bright light the pupils are constricted. In a dim light they are dilated.*

If the pupils were dilated in a bright light too much light may enter the eye and damage the retina. In a dim light if the pupils were constricted, insufficient light may enter the eye to stimulate the nerve endings in the retina and nothing would be seen.

The iris consists of a layer of circular and a layer of radiating smooth muscle fibres. Contraction of the circular fibres constricts the pupil, and contraction of the radiating fibres dilates it. The size of the pupil is controlled by nerves of the autonomic nervous system. *Sympathetic stimulation dilates the pupils. Parasympathetic stimulation constricts the pupils.*

The Movements of the Eyeballs—convergence : Light rays from objects enter the two eyes at different angles.

It is important that the rays entering should stimulate the corresponding areas of the two retinae. If this is not achieved the individual complains of double vision. Movement of the eyeballs is necessary to ensure this. For near objects both eyeballs move inwards so that their visual axes *converge* on the object to be seen. The nearer the object the greater the convergence. For a distant object less convergence is necessary. The movements of the eye-balls occur due to the contraction of the *extrinsic muscles* of the eyes.

There are six extrinsic muscles situated round the eyeball. Four of these are *straight* and two *oblique*. They are named as follows :

Medial rectus

Lateral rectus

Superior rectus

Inferior rectus

Superior oblique

Inferior oblique.

These muscles have their origin in the bony walls of the orbital cavity and are inserted into the sclera of the eyeball. They are responsible for moving the eye :

The medial rectus rotates the eyeball *inwards.*

The lateral rectus rotates the eyeball *outwards.*

The superior rectus rotates the eyeball *upwards.*

The inferior rectus rotates the eyeball *downwards.*

The superior oblique rotates the eyeball so that the cornea turns in a *downwards and outwards* direction.

The inferior oblique rotates the eyeball so that the cornea turns *upward and outwards.*

The movement of the eyes to look in a particular direction is under voluntary control but the co-ordination of movement of the eyes during convergence and accommodation to near or distant vision is controlled by reflex action.

Increased Curvature of Lens : The lens is a bi-convex, elastic, transparent body suspended by the suspensory ligament from the ciliary muscle. When light rays enter the eye from near objects the function of the lens is to refract these rays on to the macula lutea. When the eyes are accommodated for near vision the *ciliary muscle contracts forward* releasing its pull on the suspensory ligament. As a result there is relaxation of the capsule of the lens. Due to this the lens assumes a more convex form. The anterior surface bulges more than the posterior and the refracting power of the less is increased.

When the eyes are accommodated for distant vision, the *ciliary muscle is relaxed.* The choroid, due to its elastic property, recoils and pulls the suspensory ligament peripherally. As a result, the anterior surface of the lens becomes flattened and the image is correctly focussed on the retina.

Formation of Image on the Retina : The mammalian eye works on the same plan as a pin-hole camera. The pupil of the eye corresponds to the diaphragm of the camera, the crystalline lens to the glass lens of the camera and the retina of the eye to the film of the camera. In both the retina and film of the camera, the reaction to light is a chemical one. In the camera, the lens may be moved backward or forward; in the eye, this is accomplished by "accommodation". The inside of the camera is painted black to prevent reflected light from striking the film ; in the eye, the choroid coat, an almost black coat, accomplishes the same purpose.

Light rays from an object travel through the cornea, aqueous humour, lens and vitreous humour. In the process, they are refracted. They finally fall on the retina to form a mirror image which is upside down and smaller than the object. The rods and cones are stimulated and messages are relayed to the visual area of the cerebral cortex through the optic nerve. The messages are then interpreted by the brain cells, which invert the image, and an impression of an upright object is achieved.

Photochemical Changes : George Wald showed that when light falls on the retina *11-cis-retinal group of rhodopsin is isomerized to all-trans-retinal* through several sages. Within a few *picoseconds* of the absorption of a photon, rhodopsin is first converted into a photolytic intermediate, called batho-rhodopsin or prelumirhodopsin. In a *nanosecond,* batho-rhodopsin is converted into lumirhodopsin and then to metarhodopsin I in a *microsecond.* The transition from metarhodopsin I to metarhodopsin II takes about a *millisecond.* Metarhodopsin II is hydrolyzed in about a *minute* to yield opsin and all-trans-retinal, which diffuses away from the protein.

Rhodopsin (500 nm)

p sec ↓ Light

Bathorhodopsin (543 nm)
(Prelumirhodopsin)

n sec ↓

Lumirhodopsin (497 nm)

μ sec ↓

Metarhodopsin I (480 nm)

m sec ↓

Metarhodopsin II (380 nm)

sec ↓

Opsin + All-*trans*-retinal (380 nm)

Intermediates in the photolysis of rhodopsin. The wavelength of the absorption maximum of each species and the time constant of each transition are given alongwith.

All-trans-retinal is isomerized in the dark to 11-cis-retinal, which associates with opsin to regenerate rhodopsin :

$$\text{All-trans-retinal} \xrightleftharpoons[\text{in the dark}]{\text{Isomerization}} \text{11-cis-retinal}$$

$$\text{11-cis-retinal + Opsin} \longrightarrow \text{Rhodopsin.}$$

Finally, the ultimate precursors of 11-cis-retinal are the carotenes, which have a very broad biological distribution.

Dark Adaptation : *Visual purple* or rhodopsin is present only in the rods. It is bleached by bright light and when this occurs the rods cannot be stimulated. Rhodopsin is quickly reconstituted when the intensity of light is decreased and an adequate supply of vitamin A is available. When an individual moves from an area of bright light into a dark area, the vision at first is very poor but after sometimes the vision is improved. This is known as dark adaptation. The rate at which dark adaptation takes place is dependent upon the reconstitution of rhodopsin.

Common Defects of the Eye : The common defects of the eye are short-sightedness, long-sightedness and astigmatism. They are mainly caused by excessive eye strain and old age.

Short-Sightedness : This is also known as myopia. This condition occurs when parallel rays of light, *i.e.* rays from a distant object, are brought to a focus in front of the retina and in the vitreous chamber. The defect is the result of the eyeball being too long. The eye is said to be short-sighted and cannot, therefore, focus on distant objects clearly. This defect can be corrected by using a suitable concave spectacle lens which will diverge the light rays entering the eye to the correct extent and bring them to a focus on the retina (divergent-negative power).

Long-Sightedness : This defect maybe due to (i) the eyeball being too short (hypermetropia), *the childhood variety* or *(ii)* the lens and ciliary muscles losing their elasticity with advancing age (presbyopia), the *old age variety* (between 40 and 45 years age group). Long-sighted eyes can only focus on images of distant objects clearly. In both cases, light rays from near objects are brought to a focus behind the retina. This condition can be remedied by the use of a suitable convex spectacle less which will converge the light rays entering the eye to the correct extent and bring them to a focus on the retina (convergent-positive power).

Astigmatism : This is the most common defect of the eye. It is due to the surface of the cornea not being perfectly smooth or spherical, so that rays of light are not brought to sharp points on the retina. This defect is usually present in all eyes to a small extent. In certain cases of astigmatism vertical figures can be seen clearly but horizontal figures appear blurred. Thus, the vertical bars of a window grill will be clear while the horizontal bars will be blurred. In a number of cases, astigmatism may be remedied by the use of cylindrical lenses or by a combination of spherical and cylindrical lenses.

Binocular Vision : Binocular of Stereoscopic vision provides much more accurate assessment of the distance depth, height and width of an object. In this type of vision, each eye 'sees' a scene slightly differently. There is an overlap in the middle but the left eye sees more on the left than can be seen by the right eye and vice-versa. The images from the two eyes are fused in the cerebrum so that one image is perceived.

The Ear

The ear is the 'organ of hearing', as it is receptive to sound waves. It also responds to the effects of gravity and to movements of the head. The ear consists of three parts : (1) The external ear to receive the sound waves, (2) the middle ear to transmit the vibrations, and (3) the inner ear to receive the vibrations and to originate the nerve impulses which are perceived by the central nervous system as sound. The inner ear is also concerned with equilibrium and perception of motion.

With the exception of the pinna, the structures which form

the ear are encased within the petrous portion of the temporal bone. The ear is supplied by the *8th cranial nerve.*

The External Ear : The external ear consists of the *auricle, the external auditory maetus* and the *tympanic membrane.*

A. *Auricle (Pinna or Concha) :* It is covered by a thin layer of skin containing hairs, large sebaceous glands and small sweat glands. The pinna is shaped and supported by an internal plate of *elastic cartilage.* It is attached to the skull by small vestigial muscles. In old men large stiff hairs often develop at the dorsal edge and lobe of the pinna. It is deeply grooved and ridged. The most prominent outer ridge in humans is called the *helix.* The auricle functions as a sound-collecting appendage.

B. *External Auditory (Acoustic) Maetus :* This is a tube from the auricle to the tympanic membrane. It consists of a cartilaginous *portion* and an *osseous portion.* The outer one-half the tube is supported by elastic cartilage whereas the remainder is a tunnel through the temporal bone. The external acoustic maetus is lined with a continuation of the skin of the auricle. In the outer part of the canal the skin is thick and contains protective hairs, sebaceous glands and large coiled ceruminous glands. These are modified sweat glands of the apocrine type. They discharge a yellowish secretion which mixes with the sebaceous secretion and desquamated cells in the external auditory maetus. Drying of this mixture produces a thick waxy product, the *cerumen.* It has a protective function in that it traps foreign material, such as dust or insects, preventing them from reaching the tympanic membrane.

C. *Tympanic Membrane :* The *ear drum* is a thin semi-transparent membrane set at a strong slant at the deep end of the external maetus. It serves as a common partition between the maetus and tympanic cavity.

The membrane is about 0.1 mm thick and consists of three separate types of tissue :

1. The outer covering of *stratified epithelium* is hairless skin, continuous with that of the maetus.
2. The middle coat consists of *fibrous tissue.*
3. The inner lining consists of *cuboidal epithelium* which is continuous with that of the middle ear.

The Middle Ear : The middle ear consists of the tympanic cavity, an air-space traversed by the ear bones, their ligaments and a nerve. It lies within the petrous portion of the temporal bone. The tympanic cavity communicates with the nasopharynx by the *pharyngotympanic* (Eustachian or auditory) *tube.* The cavity is lined by a thin mucous membrane. The epithelium for the greater extent is of squamous type, but near the opening of the Eustachian tube it is columnar and ciliated. Beneath the epithelium is a thin connective tissue which is attached to the periosteum of the surrounding bone.

The medial wall of the tympanic cavity, which separates the middle ear from the inner ear bears an upper oval-shaped vestibular window *fenestra ovalis,* and a lower round cochlear window, *fenestra rotundum.* The oval window is closed by the foot plate of stapes. This window separates the tympanic cavity from the scala vestibuli of the cochlea. The round window separates the tympanic cavity from the scala tympani of the cochlea and is closed by a thin connective tissue membrane often called the *secondary tympanic membrane.* It is covered on the tympanic side by mucous membrane and on the cochlear side by a single layer of squamous cells.

The Auditory Ossicles : Within the middle ear there are three minute bones known as the *auditory ossicles.* The three bones extend across the cavity from the tympanic membrane to the fenestra ovalis like a chain. They form a series of movable joints. The auditory ossicles are : the *malleus,* the *incus* and the *stapes.*

The malleus lies in the lateral aspect of the middle ear. It is shaped like a hummer and possesses a *head, neck* and *handle.* The handle of the malleus is in contact with the medial wall of the tympanic membrane and the head forms a movable joint with the incus.

The incus is an anvil-shaped bone which has a middle Position and articulates with the other two bones. It possesses a *body* and *two processes.* The body articulates with the malleus, the long process with the stapes and the short process is fixed by fibrous tissue to the posterior wall of the cavity.

The stapes is shaped like a stirrup presenting a *head, neck, two limbs* and *a base.* The head articulates with the incus and its base fits into the fenestra ovalis.

The three ossicles are maintained in position by fine ligaments. The chain of ear-ossicles acts like a bent lever in transmitting sound vibrations from the tympanic membrane to the endolymph in cochlea.

The Pharyngo-tympanic (Eustachian) Tube : It has a bony and a cartilaginous portion. The bony wall, nearer the tympanic cavity, surrounds 1/3 of the tube, whereas the cartilaginous wall, nearer the pharynx supports 2/3 of the tube. Both portions are lined with a *mucous membrane* whose epithelium near the tympanic cavity is simple columnar and ciliated. Near the pharynx it is pseudostratified and ciliated. The lamina propria contains some *lymphnodules.* The submucosa possesses mixed *sero-mucous glands.*

The Eustachian tube serves to keep the air pressure in the middle ear at atmospheric level, thereby maintaining an equal pressure of air on either side of the ear-drum. This takes place when the tube opens during swållowing.

The Muscles : A *tensor tympani muscle* connects the malleus to the inner wall of the tympanic cavity. This muscle maintains the tone of the tympanic membrane. A *stapedius muscle,* inserted on the inner wall of the tympanic cavity, is attached to the stapes where it joins the incus. This latter muscle is concerned with the articulation of the-ossicles.

The Inner Ear : The inner ear consists of a *bony* or *osseous labyrinth* containing a watery fluid named the *perilymph.* Suspended in the perilymph is a slender *membranous labyrinth* containing a separate fluid, the *endolymph.* The bony labyrinth is hollowed out of the petrous part of the temporal bone. It comprises the *vestibule* that houses the saccule and utricle of the membranous labyrinth, the *semicircular canals* and the *cochlea.* The bony labyrinth is lined by a layer of flattened cells that covers the periosteum. The membranous labyrinth fits into the bony labyrinth, like a tube within a tube.

Structural Plane of Membranous Labyrinth : It has two divisions. One deals with the static and kinetic senses ; the other, with hearing. The first division comprises five constituent parts:

(1 to 3) Three *semicircular ducts,* each bearing a swollen *ampulla* at one end. The ducts are set in planes at right angles to each other.

(4) Both ends of the semicircular canals unite with an ellipsoidal *utricle.*

(5) The utricle joins with a spherical *saccule* by a slender duct. An *endolymphatic duct* also arises from this duct. It extends through the bone and ends in a dilatation, the *endolymphatic sac,* located just outside the duramater.

The hearing sense is served by a tube, the *cochlear duct* coiled spirally like a snail shell. Near its base, it connects with the saccule by the slender *ductus reuniens.* The cochlear duct ends blindly at the apex of the coil. Six areas of the lining epithelium of the membranous labyrinth are specialised as sensory *neuroepithelium.* Innervated by branches of the acoustic nerve, they occur in the utricle, saccule, semicircular ducts and cochlear duct.

Semicircular Ducts : In cross-section, these ducts are oval and are found eccentrically in the bony canals. On the inner side of the duct, the epithelium is columnar. On the outer side, the thin layer of the connective tissue of the wall is continuous with the strands traversing the perilymphatic space.

Muculae : Both the utricle and saccule bear a thickened, sensory area called the *macula.* It is oval in utricle and neart-shaped in saccule. Their columnar epithelium has two kinds of cells — (1) sensory hair cells and (2) supporting cells.

1. *Sensory hair Cells :* These are shaped like short flasks with a rounded base. Their nuclei lie in an upper row and the cells do not reach to the basement membrane. The free surface of these cells bears a tuft of sensory *hairs* that are nonmotile cilia.
2. *Supporting Cells :* These are slender cells whose nuclei lie at a deep level. The upper portion of each supporting cell is thin and ends at the free surface in a cuticular plate.

The surface of the macula is covered with a gelatinous *otolithic membrane.* The sensory hair tufts project into slender recesses in this gelatinous substance. The recesses are filled with watery endolymph. Towards the upper surface of the membrane are found densely packed minute crystals of calcium carbonate and protein. These particle are called *otoconia.*

The maculae in the utricle and saccule inform of stimuli arising from *linear acceleration* of the body.

Cristae : Each ampulla of the three semicircular ducts bears a prominence called *crista.* Here the epithelium is columnar and consists of two kinds of cells: (1) the supporting or *sustentacular cells* and (2) the *hair cells.* The supporting cells are narrow in the middle, wide at each end and with basal nuclei. The hair cells are flask-shaped with the nucleus in the expanded portion. They have a tuft of microvilli and one modified cilium or "hair" on the free surface. The hair cells occupy only the upper half of the epithelium.

Covering the surface of the crista is a layer of gelatinous material, the *cupola* into which project the ciliary tufts of the hair cells. The nerve fibres ramify among the hair cells. The cupola is comparable to the otolithic membrane of the macula but it lacks otoconia. The cristae in the ampullae of semicircular ducts inform about any rotational *movement* involving the head. Any degree of rotation of the head causes movement of the endolymph in the duct leading to a deflection of the cupola. As the processes of the hair cells are embedded in the cupola, any such movement will bend them and stimulate the sensory hair cell. This would initiate nerve impulses leading to the movement of the eyes, head and limbs.

Bony Cochlea : It resembles a snail shell, with a broad base and narrow apex. The osseous cochlea consists of a central axis, the *modiolus* round which a spiral bony canal makes about $2\frac{1}{2}$ turns. A bony shelf projects from the modiolus into the cavity of the shell. This flange is the *spiral osseous lamina.* From its free border a *membranous spiral lamina* or *basilar membrane* continues to the outer wall. At the line of attachment of the basilar membrane to the outer wall of the cochlea the periosteum is greatly thickened, forming the *spiral ligament*. In this way the spiral bony cochlear canal is divided into two compartments, an upper *scala vestibuli* and a lower *scala tymparni.* These are filled with perilymph and communicate with one another at the distal end of the cochlea by the *helicotrema.*

The Membranous Cochlear Duct : The membranous cochlear canal or *scala media* (lying between the scala vestibuli and the scala tympani) is found near the bony cochlear tube. It is filled with *endolymph.* In cross-section, the cochlear duct is triangular and lined throughout by ectodermal epithelium.

The upper surface, or roof, of the cochlear duct is the slanting *Reissner's membrane* (vestibular membrane). Its lower wall consists of the *bony lamina* and *basilar membrane.* The outer or lateral surface blends with thickened periosteum of the bony cochlea. This thickening is the *spiral ligament* The epithelium covering the outer wall is thickened and is very vascular. It is named the *stria vascularis.*

The epithelium resting, on the basal surface or floor of the cochlear duct (formed by bony lamina and the basilar membrane), is highly differentiated. The cells of the outer part are columnar, but those of the inner part are modified into the *spiral organ of Corti,* the receptor for hearing.

Organ of Corti : It extends throughout the two and a half turns of the cochlea. The organs of Corti has a number of special components:

1. *Pillars of Corti :* The pillars of Corti are arranged as two rows of cells — *inner* and *outer pillars.* Each pillar cell has a broad base and an expanded top or 'head'. The heads articulate together so that the diverging pillars enclose a gable-shaped space, the *inner tunnel of Corti.* A *Nuel's space* lies between the outer pillar and the nearest of the outer hair cells and phalangeal.
2. *Hair Cells :* On either side of the pillars of Corti lie the *hair cells.* These are found is a single row as *inner hair cells* next to the inner pillar, and in four rows as *outer hair cells* outside the outer pillar. Each hair cell possesses a number of microvilli ("hairs") on its outer surface. The tips of the hairs are embedded in the tectorial membrane. The hair cells are short and do not reach the basilar membrane. Their round bases are supported by being enclosed in the phalangeal cells.
3. *Phalangeal Cells (The Supporting Cells Of Dieters) :* These enclose the cell bodies of hair cells. They also send long cytoplasmic processes in between them, terminating in

an apical umbrella-like expansion. Electron microscope has revealed that the phalangeal cells also enclose afferent and efferent nerve fibres which are passing to the hair cells.

4. *Cells of Hensen :* Outside the outer row of hair cells are 7 or 8 rows of columnar supporting cells, the cells of Hensen. A space between the nearest outer hair cell and Hensens cell is the outer tunnel.

5. *The Tectorial Membrane :* It is a non-nucleated, gelatinous, cuticular membrane which is attached to the osseous spiral lamina. It projects outward peripherally, hanging free over the organ of Corti at the outer limit of the hair cells.

The organ of Corti is receptor for sound. It is responsible for the initiation of nerve impulses which are conveyed to the hearing area in the temporal-lobe of the cerebrum by the auditory nerve (8th cranial nerve). In the brain, these nerve impulses are interpreted as sound.

The *vestibular branch* of the 8th cranial nerve supplied the vestibular part of the inner ear. Its sensory nerve endings are found in the maculae and the cristae. The *cochlear branch* supplies the cochlea.

Physiology of Hearing

Every sound that is produced causes vibrations or disturbances in the atmospheric air. These vibrations travel as a succession of waves known as *sound waves.*

Sound waves travel at approximately 335 metres (1088 feet) per second. It is the function of the ear to pick up these vibrations and direct them towards the cochlea, where the organ of Corti is adapted to transmit them, as nerve impulses, by the 8th cranial nerve to the *hearing area* of the cerebral cortex where they are perceived as sound.

Function of the External Ear : The pinna is formed in such a way that the vibrations produced by any sound can be concentrated and directed through the external auditory meatus towards the tympanic membrane. The tympanic membrane vibrates in harmony with the sound waves.

The cerumen secreted by the lining of the external auditory meaturs protects the ear canal.

Function of the Tympanic Cavity or Middle Ear : The vibrations set up in the tympanic membrane are conveyed through the middle ear by a chain of three audited wall of the tympanic membrane and, as it vibrates a corresponding movement of the malleus takes place. The movements of the malleus are transmitted via the incus to the stapes. The foot plate of the stapes fits accurately into the *fenestra vestibuli* The vibrations transferred from the incus to the stapes cause the foot plate of the stapes to rock to and fro in the fenestra vestibuli. This carries the vibrations to the inner ear. The auditory ossicles can only move and transmit the vibrations of the tympanic membrane to the inner ear when air at atmospheric pressure is present in the middle ear. This air enters the middle ear from the nasal part of the pharynx through the Eustachian tube.

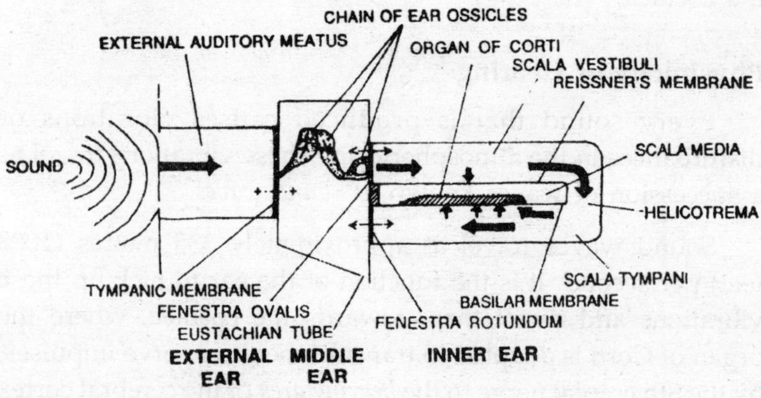

The course of sound waves in the ear.

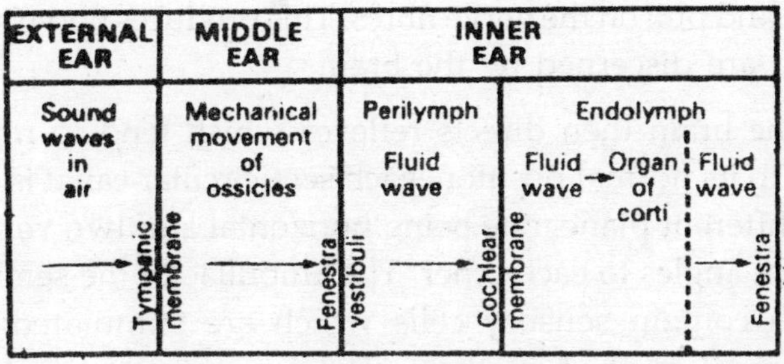

A diagrammatic representation of hte transmission of sound through human ear.

Functions of the Inner Ear : On the inner side of the fenestra vestibuli there is perilymph which surrounds the membranous labyrinth. The rocking movement of the stapes sets up wave motion in the perilymph which in turn causes vibration of the membranous labyrinth. The vibrations of the membranous labyrinth stimulate movement of the endolymph, and subsequently the special neuroepithelial cells forming the organ of Corti are stimulated. High-pitched sounds stimulate the cells at the base of the cochlea, and low pitched sounds the cells at the apex.

These nerve impulses are transmitted via the nerve fibres of the *cochlear nerve* which subsequently becomes the *8th cranial nerve.* The *vestibulocochlear nerve* transmits the impulses to various nuclei in the pons varolii and midbrain. Some of the nerve fibres pass to the hearing area of the cerebral cortex on the same side and some cross over to the corresponding area in the opposite hemisphere. It is in the hearing area of the cerebral cortex that the nerve impulses are perceived as sound.

Balance : The semicircular canals, the utricle and the sacculus are of position of head causes movement in the perilymph and endolymph. The sacculus and utriculus contain certain crystals called otoliths which lie near sensitive areas. Nerve fibres form these sensitive areas are embedded in the otoliths. Any movement of the body causes the otoliths to move and pull on the nerve fibres. This produces nerve impulses which are discerned by the brain.

The brain then directs reflexes which tend to return the body to its normal position, each semicircular canal is situated in a different plane, one being horizontal and two vertical but at right angles to each other. The ampullae of the semicircular canals contain sensory cells which are stimulated by any movement of fluid in them. This movement stimulates the nerve endings and the hair cells in the utricle, the sacculus and the ampullae. The resultant nerve impulses are then transmitted by the vestibular nerve, which joins the nerve from the cochlea of the inner ear, to form the *vertibulocochlear nerve.* The vestibular branch of the 8th cranial nerve then passes to the *cerebellum.*

3

Nervous System

All multicellular animals contain elongated nerve cells, called *neurons,* collected together into parallel bundles called *nerve trunks.* Every nerve trunk contains thousands of neurons. The human central nervous system contains about 100 billion neurons. It is clear that these control the movement of muscles (via motor nerves) and the conduction of information from the sense organs to the central nervous system (via sensory nerves).

Structure of a Neuron

Each neuron has a cell body, called the *neurocyton,* 4 to 25 microns in diameter, containing a nucleus and cytoplasm, and from this radiate out one or more highly elongated processes. One type, called an *axon,* usually extends for some distance, and it is axons that make up the bulk of the visible nerve trunks. An *axon is the process of a nerve cell that carries impulses away from it.* The axon membrane is about 50 to 100 Å thick. The axon has many of the properties of an electric cable and is specialized to conduct the brief electrical waves called nerve impulses. The axons run parallel to one another, and each is surrounded along its whole length by a series of *sheath cells* (Schwann cells), which may have *myelin (medullated)* sheath, containing alternate layers of protein and fat molecules. In vertebrates, the larger axons have thicker myelin sheaths, and the axon proper may be up to 50 microns in diameter.

In addition to axon the cell body of a neuron gives off many smaller processes, which are called *dendrites* because they are usually highly branched like trees. A *dendrite or dendron is the process that carries impulses towards the cell body.*

Cell bodies and dendrites are usually aggregated into *ganglia,* and an animal's main assemblage of ganglia is called its *central nervous system* (CNS). Within ganglia the ends of axons come into very close proximity with cell bodies and dendrites of other neurons. These places of contact are called *Synapses.* At a *Synapse* the two cell membranes are usually somewhat thickened, and the gap between the two may be as small as 200 Å and hence only visible by electron microscope. A specialized type of synapse occurs where the axon of a motor nerve ends on a muscle cell.

Types of Neurons

There are three common types of neurons : *motor, sensory* and *connector*. A sensory neuron has only a single process or fibre which divides close to the cell body into two main branches. Connector neuron consists of a cell body, one axon and numerous dendrites. Connector neurons occur exclusively in the CNS and serve as the immediate links between numerous sensory and motor neurons. The afferent or sensory neurons are located within the receptor organs or tissues, while the efferent or motor neurons are found in the spinal cord and their axons end in the effector organ, generally a muscle or gland which responds to the incoming nerve impulses. Thus, sensory neurons transmit the nerve impulses toward the CNS, whereas the motor neurons transmit them away from the CNS.

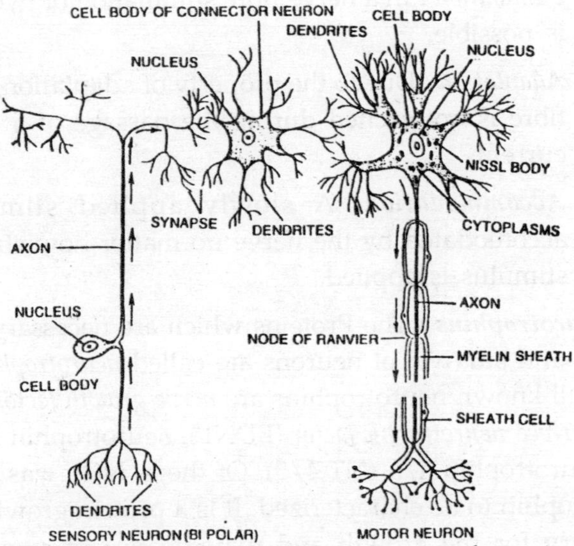

Three common types of neurons and the direction of nerve impulse transmission.

Functional Properties of the Nerve Fibres

1. *Excitability* : The nerve fibre can be stimulated by a suitable stimulus, which may be *mechanical, thermal, electrical* or *chemical* in nature.
2. *Conductivity :* The nerve impulse is conducted along the nerve fibre (normally in one direction only), and the conduction velocity depends upon the diameter of nerve fibre (the thicker fibres showing higher velocity) and temperature (in poikilotherms the velocity is lower than in nomeothermic animals).
3. *All-or-none-response* : A single nerve fibre will always give a maximum response to a suitable stimulus. If the strength or duration of the stimulus be further increased, no change in the response will occur.
4. *Refractory Period :* When a nerve fibre is once stimulated, it will not respond to a second stimulus for a brief period called the absolute refractory period.

5. *Summation* : In a nerve fibre summation of two stimuli is possible.

6. *Adaptation* : Due to the property of adaptation, a nerve fibre is not excited during the passage of a constant current.

7. *Accommodation:* A slowly applied stimulus is accomodated by the nerve no matter how strong the stimulus is applied.

Neurotrophins : The Proteins which are necessary for the growth and survival of neurons are called *neurotrophins*. The four well known neurotrophins are *nerve growth factor* (NGF), *brain-derived neurotrophic factor* (BDNF), neurotrophin 3 (NT-3) and neurotrophin 4/5 (NT-4/5). Of them, NGF was the first neurotrophin to be characterized. It is a protein growth factor necessary for the growth and maintenance of sympathetic neurons and some sensory neurons.

Neurotrophins bind to receptors (made up of two proteins that when dimerized have tyrosinkinase activity at the endings of a neuron. They are internalized and then transported to the neuronal cell body where they foster the production of proteins associated with neuronal development, growth and survival.

Physiology of the Transmission of Nerve Impulse : *The nerve impulse is that electrochemical change in the nerve fibre membrane that is aroused by a stimulus and is carried along the nerve fibre to its termination.* The transmission of the nerve impulse is a metabolic process, involving the consumption of O_2, CO_2 production, heat production, and the movement of ions. In very thin axons these impulses travel at less than one metre per second, but in large axons that activate muscles, they travel as fast as 100 metres per second.

Transmission in the Nerve Fibre : The transmission of nerve impulse has been explained by the so-called *membrane theory of nervous conduction*. According to this theory, the impulse itself is a transient set of biochemical reactions occurring mainly at the cell membrane of the nerve fibre (axon), and accompained by a small and very brief change in membrane potential (Em), known as the *action potential.* However, the action potential is not the nerve impulse itself, it is only a means of showing that a nerve impulse is passing. The cell membrane of an inactive axon has a slight accumulation of Na^+ at its outer surface. Hence, the outside of the axon is electrically positive relative to the cytoplasm. This means that at rest there is a potential difference across the membrane of about 75 mV, (inside negative).

When such a *resting potential* exists, the membrane is said to be *polarized.* But when an impulse passes a given part of the axon, the cell membrane suddenly becomes much more permeable to positively charged ions than when it was at rest. Na^+ *ions then diffuse inwards,* and K^+ *outward* during the very small fraction of a second before the membrane recovers its resting properties and by means of active transport restores the previous distribution of Na^+ and K^+ ions. The influx of Na^+ ions causes *depolarization* of the membrane and as impulse passes away, depolarized region becomes polarized again.

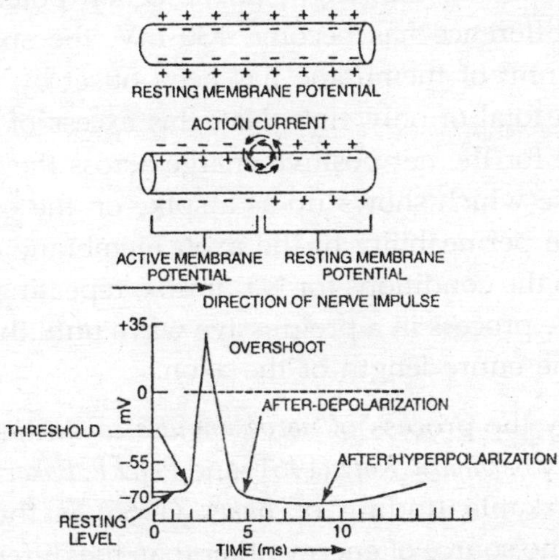

Showing the resting axon (above), the axon during passage of impulse (middle), and the action potential (below).

The electrochemical events that underlie the nerve impulse or action potential are determined largely by the axon membrane's differential permeability to Na^+ and K^+ ions and by the voltage difference across the membrane. As Na^+, ions leak through the membrane, they cancel out locally a portion of the excess negative charge inside the axon, thereby further reducing the voltage drop across the membrane. When the voltage drop has been reached to the *threshold* level, Na^+ ions enter in such numbers that they change the internal potential of the membrane from negative to positive. In other words, when potential difference of a small patch of membrane is lowered to its threshold, about -50 mV, the permeability of the membrane to Na^+ rapidly increases by as much as a hundredfold; Na^+ rushes across the membrane, in accord with the sodium gradient and attracted by the negative charge. This usually results in an *overshoot,* and the potential difference may go from -75 to +30 mV. This process "ignites" and flares up to create the nerve impulse, or the *action potential* as it is called. The most convincing evidence for the movement of ions being responsible for the action potential was provided by the *Voltage clamp technique.* At the height of the action potential, when potential difference has become +30 mV, the six K^+ deficit behind 10 nm^2 of membrane has been offset by the inward arrival of a total of only eight Na^+; this excess of two Na^+ is responsible for the net positive charge across the membrane. The impulse which shows up as a spike on the oscilloscope, changes the permeability of the axon membrane ahead of it and sets up the conditions for Na^+ influx, repeating the whole regenerative process in a progressive wave until the spike has travelled the entire length of the axon.

Recently the process of *nerve impulse conduction* has been explained by *Bernhard 'Katz* (1961) and *Peter F. Baker* (1966). The most remarkable finding of *Baker* (1966) is the fact that apparently no source of energy other than the difference in *ion concentrations* on the two sides of the membrane is needed to amplify an electrical impulse and propagate it along the axon. During the passage of an impulse in an intact nerve there is small net gain of Na^+ and a small net loss of K^+. If this continues unchecked for sometime, the nerve would lose K^+ and gain Na^+, and the *concentration gradients* on which nerve activity depends would be destroyed. The tendency is counteracted by the mechanism known as the *"Sodium pump"*, which uses *metabolic energy* in the form of ATP to extrude Na^+ ions from the axon in exchange for K^+ ions. Although an intact nerve can function for sometime without an operating sodium pump, the pump is essential for the long-term maintenance of nerve activity.

Saltatory Conduction

The impulse conduction in myelinated nerve fibres has been described by *Hoyle* (1965) as *saltatory conduction.* In nommedullated nerves the spread of depolarization is continuous along the nerve fibre membrane and conduction is a uniform process. But in medullated nerves, in which myclin provides electrical insulation, the active generation of the current is confined to the nodes of Ranvier, and the impulse behaves as though it jumped from one node to the next (saltatory conduction). Actually, it does not "skip from node to node" but spreads along the internodal regions at a finite rate by cable conduction from amplifier to amplifier. Thus, only the membrane in the nodal region becomes active and the current flows from the next nearest resting node back over the outside of the myelin and returns in the core of the fibre.

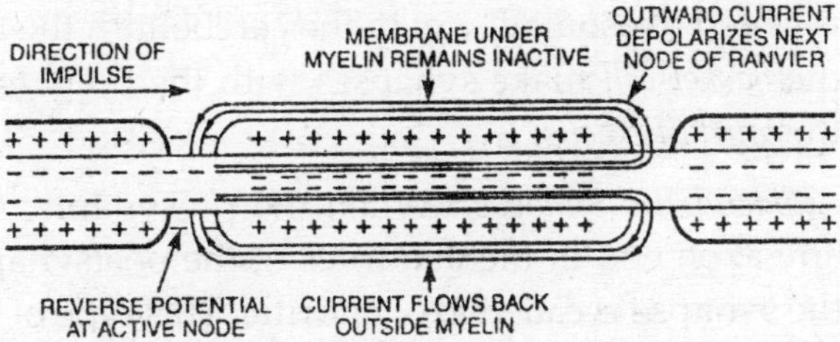

Saltatory conduction in myelinated nerve fibre.

Saltatory conduction is an evolutionary achievement of the vertebrates and possible of some higher invertebrates. This type of transmission confers two distinct advantages. Of them the first is the high conduction velocity a frog's medullated fibre of 20 μ diameter conducts at the same rate of 20m per second as a squid's giant axon of 500 μ diameter. The myelinated axons conduct up to 50 times faster than the fastest unmyelinated nerve fibre. The second advantage is one of metabolic economy. If only the nodes are depolarized, there is relatively less movement of ions across the membranes and naturally fewer demands on the "ion pumps.

The Synapse

The name "Synapse" was given by the Physiologist *Sir Charles Sherrington* (1906) to the junction point where the axon's terminal fibres make contact with another nerve cell.

The word synapse is derived from a Greek word meaning "to clasp tightly" *A synapse may be defined as the functional region where one neuron and other begins ; or synapse is the region where two neurons are functionally connected.*

Classification of Synapses : A synapse or *synaptic junction* is found in nerve-nerve and nerve-muscle interactions. According to the *nature of connection,* the synapses can be classified as :

(i) *Axosomatic Synapse :* When the presynaptic terminal of the axon ends in the cell body of the neuron, the synapse is called axosomatic *e.g.* in the cerebellum the axons of the *basket cell* make synapses with the body (soma) of the *purkinje cells.*

(ii) *Axodendritic Synapse :* When the presynaptic fibres of any axon end in the dendrites of the postsynaptic cell, the synapse is called axodendritic. This type of synapse is also present in the cerebellum where climbing fibres form axodendritic connections with the dentine of the purkinje cell.

(iii) *Axo-axonal Synpase* : When presynaptic fibres of any axon end in the axon of the postsynaptic cells, the synapse of called axo-axonal.

According to the *nature of transmission* there are two types of synapses.

(i) *Chemical Synapses* : A chemical synapse is a junction between two nerve cells. The space between them is about 20 nm. In the nervous system of vertebrates transmission at the synapse is dependent upon the secretion and release of a *chemical* (neurotransmitter) which is released from *synaptic* vesicles, after they have fused with the nerve cell membrane into the *synaptic cleft* and is bound to receptors in the Postsynaptic membrane This allows ions to flow through the receptor and causes the depolarization of the postsynaptic cell membrane.

(ii) *Electrotonic Synapse* : In electrotonic synapse, a type of *gap junction,* the space between the cells is approximately 2nm. In this case no neurotransmitter is involved. The depolarization of the presynaptic cell membrane during an action potential causes ion currents to flow through the gap junction.

In another type of electrical synapse, there is no gap between the cell membranes, which are fused to form a *tight junction.* The nerve impulse can be transmitted electrically across this tight junction often in either direction. Electrical synapses are far more frequently found in invertebrates than in vertebrates. Some vertebrates have electrotonic synapses in specialized areas ; for example, neurons controlling the electric organs of fish, the giant Mauthner cells at the base of the brain in teleost fish where transmission is electrical.

Functional Properties of Synapse

1. An impulse is allowed to pass through a synapse only *in one direction, i.e.,* from the axon of one neuron to the dendrite of the next. This is known as the *law of forward conduction.*
2. A synapse retards the passage of an impulse more than does a nerve fibre. Consequently, there is a *synoptic delay* in the passage of an impulse from presynaptic to the postsynaptic fibre.
3. It is a site where impulses are received and discharged. However, a synapse not only acts as a *relay station* but it may also act as an integrator.
4. The physiological *seat of fatigue* probably lies in the synapses. The mechanism underlying the synaptic fatigue is presumably due to exhaustion of transmitter material from *synaptic vesicles* following repeated presynaptic stimulation at a faster rate.
5. It has the characterstics of *summating* and *inhibiting* the impulses. Summation may be spatial or temporat.
6. Synapses bring about the *convergence* and *divergence* of nerve impulses.
7. It is very vulnerable to the action of many chemical compounds, some of which increase its responsiveness *(e.g., straychnine* blocks the inhibitory activity and causes reduction of inhibitory postsynaptic potential), whereas others decrease it *(e.g., anesthetics).*

Structure of Neuromuscular Synapse : A neuromuscular synapse is formed when axon fibres (motor nerve or motoneuron) come in contact with the muscle cell (skeletal muscle fibres). The axon terminates in little knobs on the surface of the muscle fibre but does not fuse with it. Within the terminals of the axon are the spherical *synaptic vesicles* (40-200 nm in diameter) containing acteylcholine, and the many mitochondria needed for the active synthetic processes occuring in the terminals. Separating the axon terminal from the muscle fibre membrane is a gap of about 40 nm. This is called *synaptic cleft* and is wide enough to permit the flow of currents generated by the action of the chemical transmitter on the *postsynaptic membrane*. This membrane has junctional folds or postsynaptic folds (postsynaptic membrane thickining). The *junctional folds* contain particles which are believed to be acetylcholine receptors. The postsynaptic region responds only to chemical stimulation or inhibition.

Summation and Inhibition : Whether a given impulse crosses the synapse at which it arrives, is determined by two important factors called *summations* and *inhibition.*

Two or more impulses arriving at a given synapse are more likely than one to initiate another impulse in cell across the synaptic gap. They may arrive over the same axon within a small fraction of a second, in which case the phenomenon is called *temporal summation* or, they reach the synapses on a given cell over two or more different axons, in which case the process is called *spatial summation. A* common type of temporal summation is observed when a sensory axon entering the spinal cord sends impulses at different frequencies to a synapse on a motor neuron activating the muscle of a leg. Spatial summation often occurs at the same type of synapse when nerve impulses reach it over several sensory axons within a fraction of a second. Temporal and spatial summation are necessary for transmission of all impulses across synapses in the central nervous system.

Sometimes nerve impulses converging on a given synapse may produce just the opposite effect to summation. In that case one set of impulses may offset the effects of another, and this phenomenon is called *inhibition.* The higher the impulse frequency in an inhibitory nerve, the less will be the transmission across the synapse. Most cell bodies of neurons or their dendrites receive many synaptic endings where axons of other neurons can either excite or inhibit. Hence each neuron is always subject to both excitatory and *inhibitory* influence. It is the balance between these two influences that determines whether a given cell will itself be stimulated to convey the nerve impulse along its own axon. However, now it is known that a given axonal ending is either excitatory or inhibitory. But this is a property of the ending and not of the *postsynaptic neuron.* It is possible that many axons branch and lead to inhibitory endings at some synapse and excitatory endings at others. According to *Sir John Eccles* (1965), a nerve cell responds to the algebraic sum of the internal voltage changes produced by excitatory and inhibitory synapses.

Biochemical Mechanism of Synaptic Transmission and Role of Neuro transmitters : The transfer of nerve impulse across a synaptic gap is called synaptic transmission. All axons end at synapses. In the synapse a nerve impulse is transmitted from one nerve cell to another, or from *presynaptic* axonal endings to either the dendrites or the cell body of the *postsynaptic* cell. It has been shown that, when an impulse reaches an axon terminal, the terminal acts like a miniature endocrine gland. It secretes minute amounts of a hormone (or neurotransmitter substance) that diffuses through the synaptic gap or cleft. Some of it eventually reaches the dendrite terminals of the adjacent neurons, and the neurotransmitter there affects a dendrite in such a way that a new impulse is initiated in it.

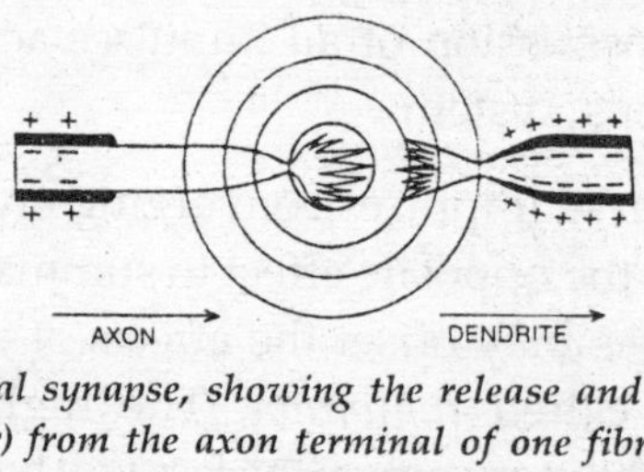

Diagram of a neural synapse, showing the release and local spreading of hormones (colour) from the axon terminal of one fibre to the dendrite terminal of another. Impulses are transmitted across the synapse by such chemical means.

In many instances the neurotransmitter is a *monoamine,* a substance the nerve cell synthesizes by making minor changes in an amino acid. Norepinephrine, epinephrine, dopamine, and serotonin are *catecholamines,* compounds that contain a *catechol nucleus* (a benzene ring with two adjacent hydroxyl groups) and an *amine* group. These monoamine neurotransmitters are all derived from the amino acid *tyrosine.* Serotonin is an indolalkylamine derived from the amino acid *tryptohphane.* They are all classified as *monoamines* because they all possess a single amine group. Sometime the neurotransmitter is the amino acid itself (example, glutamic acid). In any case, its release and reception takes milliseconds *(Bloon,* 1981).

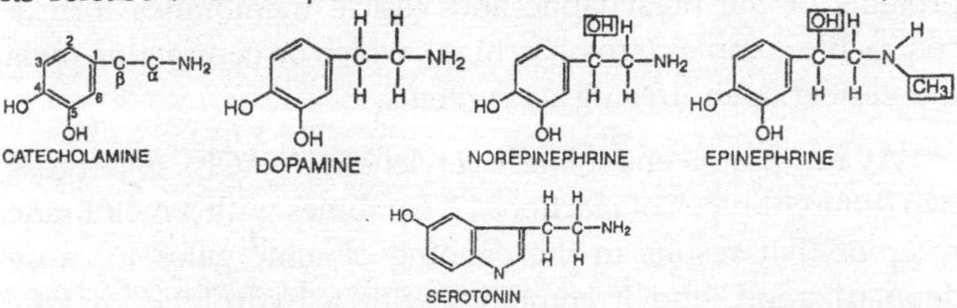

Structures of monoamines. Dopamine, norepinephrine, and epinephrine are catecholamines (compounds containing a catechol nucleus-a benezene ring with two adjacent benezene ring with two adjacent hydroxyl groups and an amine group). Serotonin is an indolamine.

In vertebrates, four neurotransmitter substances have been identified. They are *serotonin, acetylcholine, adrenaline* (epinephrine) and *nor adrenaline* (norepinephrine), and are collectively called *neurohumors,* meaning that they are produced by the nerve cells and stimulate or inhibit other cells. Adrenaline or noradrenaline is secreted by the axon terminals of the sympathetic postganglionic fibres, which are, therefore, said to be *adrenergic.* Serotonin or acetylcholine is produced by sympathetic pre-ganglionic fibres, all fibres of the parasympathetic system, and probably also by the fibres of the central nervous system. All such fibres are said to be *cholinergic.*

Gamma-aminobutyric acid (GABA) is the major inhibitory mediator in the brain where it helps in the transmission of about 20% of synapses. It is also present in the retina and is the mediator responsible for presynaptic inhibition. Gaba transaminase (GABA-T) is the enzyme that catalyzes the transmission.

Mechanism of Release and Action of Acetylcholine : The release of acetylcholine from synaptic vesicles is absolutely dependent upon the entry of Ca^{2+} which occurs when the action potential reaching the nerve terminals depolarizes the membrane sufficiently (about 30 mV). One theory holds that the transmitter is released by *exocytosis* as the vesicle fuses with the presynaptic membrane. Another theory suggests that proteins of the presynaptic and vesicle membranes form a contractile complex (crosshatching), which on contraction pulls the vesicle open, freeing its contents.

At the motor end plate, acetylcholine (ACh) undergoes two different types of reactions : it combines with a membrane receptor that results in the opening of ionic gates to cause depolarization, and it combines with a hydrolytic enzyme *acetylcholine esterase,* which rapidly inactivates it. Most evidence indicates that the ACh-receptor (AChR) is a protein consisting of 5 subunits surrounding a central channels (2a, and one each β , γ , δ) and a change in its structure occurs when ACh binds to it, resulting in the opening of ionic gates and a change in permeability.

Katz and *Miledi* (1967) estimated that there is a very high density of receptors (3×10^7) per end plate, which is enough for the 10^4 quanta of ACh liberated. There are from 12,000 to 21,000 molecules of ACh per quantam, which can easily be packed into a vesicle. Because the synaptic gap is very small, there is an excellent chance for most of the ACh molecules to colloide with a receptor site within a few milliseconds of their release and cause depolarization within the requisite time for proper transmission.

In synapses formed by cholinergic fibres, the enzyme

acetylcholine esterase (AChE) splits acetylcholine (ACh) into separate acetyl and choline fractions and so makes the hormone ineffective as shown below :

$$\underset{\text{Acetyl choline}}{CH_3CO{-}O{-}CH_2CH_2N(CH_3)_3} \underset{\substack{-H_2O \\ \textit{choline acetylase}}}{\overset{\substack{+H_2O \\ \textit{acetyl choline esterase}}}{\rightleftharpoons}} \underset{\text{acetic acid}}{CH_3COOH} + \underset{\text{choline}}{HOCH_2CH_2N(CH_3)_3}$$

The presence of this enzyme is advantageous adaptively. If it did not exist, ACh would linger in the synapse and would stimulate the dendrites repeatedly. If, therefore, two impulses arrived in a synapse in rapid succession, ACh produced by the first would still be effective when the second produced more hormone. The impulses would then merge into each other and the result would be one long drawn-out response, not two sharply distinct responses. Acetyl choline esterase prevents the merging of impulses. There is just time for ACh to stimulate the dendrites once or a few times and the hormone is destroyed immediately thereafter. An equivalent enzyme cannot be demostrated in synapses formed by adrenergic fibres, the adrenalines do have a lingering effect. This circumstance contributes to the comparative slowness and sustained quality of the responses produced by the sympathetic part of the autonomic nervous system.

Drug Modified Transmission at Synaptic Junctions : Several drugs are able to modify transmission at synaptic junctions. *Muscarine,* an alkaloid, mimics the stimulatory action of ACh on glands and smooth muscle, although it has little effect on the receptors in autonomic ganglia. These muscarinic actions of acetylcholine involving muscarinic receptors are blocked by the drug *atropine.* In sympathetic ganglia, large amounts of ACh block transmission of impulses from pre to postganglionic neurons and this action, although unaffected by atropine, is mimicked by *nicotine.* As a result, these actions of ACh involving nicotinic receptors are called nicotinic actions. Both muscarinic and nicotinic ACh receptors occur in abundance in the brain.

Inhibition of Acetylcholine

The *bacterial toxin of botulism,* the most poisonous substance known, prevents nerve endings from liberating acetylcholine and so killed by paralysis of the muscles controlling breathing.

Inhibitors of Acetykholinesterase : Acetylcholinesterase is located in the postsynaptic membrane. It is different from acetylcholine receptor and can be readily separated from it. Acetylcholinesterase does not bind *neurotoxins.* The toxic and therapeutic properties of acetylcholineasterase inhibitors are of considerable practical importance. Physostigmine, an alkaloid derived from the Calabar or ordeal bean, which was once used as an ordeal poison in witchcraft trials, inhibits acetylcholinesterase by forming a covalent intermediate that is hydrolyzed very slowly.

However, acetylcholinesterase inhibited by organic phosphates such as di-isopropyl phosphofluoridate (DIPF) can be reactivated by derivatives of hydroxylamine (NH_2OH), which is used as a designed antidote for organophosphorus poisoning.

Inhibitors of Acetylcholine Receptor : Neuromuscular transmission can also be impaired by compounds that act directly on the *acetylcholine receptor* (AChR). For instance, *Curare* (one of the active components of *curare* is *d-tubocurarine),* a plant extract that South American Indians used to poison their arrow tips, blocks excitation of the muscle by binding to AChR on the sarcolemma and produces paralysis. It is used medically as a muscle relaxant during surgery.

Recently *α-bungarotoxin,* a protein molecule purified from snake venom, has been shown to bind on to a-subunits of the AChR and thus works as a specific receptor antagonist. Since the snake venom binds very tightly to the receptor, the estimation of how many receptor sites there are, can be made by counting the number of radiolabeled α-bungarotoxin molecules. A density of $3 \times 10^4 / \mu m^2$ has been observed in a neuromuscular junction in frog. The *Cobra toxin* also acts in a similar way.

Mechanism of Action of Catecholamines : *Julius Axelrod* (1974) has described the synthesis, release, mode of action, and the regulation of the neurotransmitters that are chemically known as the catecholamines (noradrenaline, adrenaline and dopamine). The enzymes involved in the synthesis of the transmitters are made in the cell body of the sympathetic neuron and transported down the axon to the nerve endings (studded with varicosities) where the transmitters are manufactured and stored in the vesicles (Fig.).

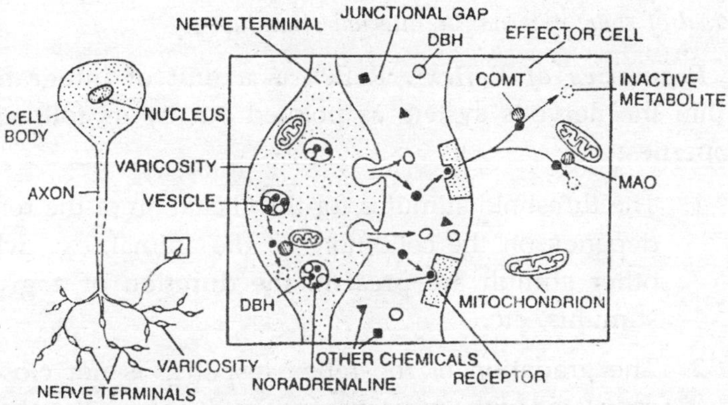

Diagram showing the release and inactivation of the neurotransmitter noradrenaline.

When the nerve is existed and depolarized, vesicles move to the cell membrane and fuse with it, releasing the neurotransmitter into the junctional gap. Once the neurotransmitter is liberated, it diffuses across the cleft between the nerve terminal and the adjacent cell. The capacity of the effector cell to respond to the transmitter then depends on the ability of its receptor to selectively recognize and combine with the neurotransmitter. When receptor and transmitter interact, a series of events is triggered that causes the effector cell to carry out its special function, including the propagation of the impulse across a synapse.

After interacting with the post junctional cell, the adrenaline is inactivated by the enzymes catechol-O-methyltransferase (COMT) and monoamine oxidase (MAO) or more generally through its recapture by the presynaptic nerve terminals. The noradrenaline that leaks out of vesicles may be inactivated by the MAO stored in the membrane of the mitochondria.

Reflex and the Reflex Arc : The entire nervous activity is based on reflexes, the functional unit of the nervous system. *Sherrington* (1906) defined reflex as the unit of reaction in nervous integration because every reflex is a complete act of integration. A reflex may be defined as *an immediate and rapid response given without our awareness by an effector organ on the arrival of some external or internal stimulus.*

Properties of Reflex : Reflex is a unit of integration within the nervous system as pointed out by its following properties:

1. The threshold stimulus for the elicitation of the reflex depends on the conditions of the animal, *e.g.*, what other stimuli are present, the duration of a given stimulus, etc.
2. The gradation of the reflex response is not closely correlated with gradation of a stimulus above threshold.
3. Reflex action is usually the result of temporal summation. A single afferent impulse does not normally elicit a response.
4. There is little correspondence between the rhythm of a repetitive and the rhythm of the reflex response.
5. Most reflex actions are followed by a state of depressed excitability, which is often of long duration.
6. In many reflex actions there is an after-discharge phenomenon, *i.e.*, a prolongation of motoneuron activity after the stimulus has ceased.

7. The integrative action of a reflex is also shown by the fact that when two or more separate reflexes are stimulated simultaneously, one reflex is completely inhibited whereas the other proceeds normally. This is particularly the case when reflex actions are antagonistic.

Types of Reflexes : Reflexes are usually classified into two, the simple or unconditioned reflex, and the *conditioned reflex. A simple reflex is an inborn, inherited or unlearned response to a stimulus.* A familiar example of a simple reflex is the *knee-jerk reflex. A conditioned reflex is the response acquired as a result of training or experience to a stimulus that originally failed to evoke the reaction.* It was first clearly demosnstrated by *Pavlov*. Under normal conditions the secretion of saliva when food is ingested is a simple reflex. *Pavlov* demonstrated on dogs that if an additional stimulus is regularly given (such as the ringing of a bell) at the time of giving the food to the animal, in time the mere ringing of the bell, without even seeing or smelling the food, would result in the secretion of saliva. Thus, a previously ineffective stimulus (the ringing of the bell in this case), by association with a stimulus (food) that evoked a response, had become effective in causing a response, the secretion of the saliva by the salivary glands. In other words, a conditioned reflex represents an inborn response which has been displaced to a new stimulus.

Reflexes are sometimes also classified as the *spinal reflex* and the *cranial reflexes.* The former are so-called because their basic neural path leads through the spinal nerves and spinal cord. Reflexes at the spinal level (involving skeletal muscle action) have the purpose of removing the animal from harmful stimuli and reflex actions are purposive and adaptive actions at the unconscious level. The cranial reflexes are executed by the brain and 12 pairs of cranial nerves. A large number of central nervous system reflexes are partly cranial, partly spinal. For example, usual stimuli may lead to muscular response in arms or legs. In such cases the sensory parts of the reflexes are cranial (via the optic nerves), and the motor portions are spinal (via the spinal nerves). Conversely, in other reflexes the sensory portions may be spinal and the motor portions may be cranial.

However, reflexes in the autonomic nervous system generally do not become conscious, irrespective of their course through the brain or the spinal cord. These reflexes regulate the functioning of body parts which are not under voluntary control, smooth muscles, heart, and most glands and other internal organs.

Reflexes may be phasic or tonic. The vertebrate knee jerk is a *phasic reflex* in which the muscle is quickly stretched and as quickly responds with a contraction to counteract the stretch. The nerve impulses that arise from the stimulus are quickly conducted to the CNS as a synchronous volley. The reflex efferent discharge from the CNS likewise passes a synchronous volley back to the muscle and causes a rapid contraction of all the fibres of the muscle.

The *tonic reflex* is characteristic, for example, of muscle activity used to maintain posture against the force of gravity. Thus, when a muscle is stretched slowly, varying numbers of stretch recepotrs are excited, only a few at first, but more as the muscle is stretched further or if the stretch is maintained. The result is an asynchornous discharge of impulses into the CNS and a resultant asynchronous discharge conducted along the motor nerve to the muscle fibres. As a result, a sustained contraction is produced in which, at any one time, only a relatively few muscle fibres are contracting.

The *myotatic reflex* depends on the activity of the stretch receptors in the muscles. Such reflexes are composed of both a phasic reflex and a static reflex. When the stretch on the muscle is brief, only the phasic reflex is seen (the best example of this is the knee jerk reflex). The peak of tension in a myotatic reflex occurs at the time of completion of the stretch movement

(= phasic reflex action). This is followed by a phase of static action represented by a plateau of gradually declining tension. The degree of tension maintained in the static phase is regulated by the degree of static stretch imposed on the muscle. The stretch reflex also has an inhibitory component, which can be demonstrated by recording the responses of an extensor muscle when its antagonistic flexor muscle is stretched. When stretched an extensor muscle shows a typical stretch reflex action. If a corresponding flexor muscle is stimulated while the extensor muscle is stretched, the stretch reflex does not occur. The stretch reflex plays an important role in walking.

The flexor or *withdrawl reflex* is the response evoked by the application of a painful or harmful stimulus to the limb of a spinal animal. This was called by *Sherrington* (1906) as *nocioceptive reflex*. The *scratch reflex* results in a motor action in which the animal scratches its skin with a foot.

A simple example of scratch reflex in frog is seen when the animal is suspended so that the lower legs hang free. The application of small amount of dilute hydrochloric acid to the thigh evokes a reflex response in which the frog raises its foot in order to remove the painful stimulus. If the toes on the stimulated side are held so they cannot be moved, the opposite leg and foot are used to remove (scratch) the stimulating agent.

Reflex Arc : The structural and functional basis of the simple reflex is called the *reflex arc* and a *reflex action is the response to a stimulus carried over a reflex arc.* The most simple reflex arc consists of only two neurons, one sensory neuron and one motor neuron. But this type of reflex arcs are hardly met within higher animals. Most of their reflex arcs consists of more than two (at least three) neurons. A typical reflex arc consists of five parts or components: a receptor, an afferent or sensory neuron, modulator, efferent or motor neuron, and effector.

In a theoretically simple reflex arc consisting of a few

neurons, the nerve impulse is normally initiated in the dendrites of the so-called sensory neuron located within the receptor organ. The impulses travel along the sensory neuron until they reach the axon endings and the synapse, through which the impulses are initiated in the second neuron in the arc, called the connector neuron. The connector neuron is always found in the CNS. The nerve impulse then proceeds along the length of the connector neuron, to its axon endings, and at another synapse initiates impulse in the third neuron, called the motor neuron. The terminal axon branches of the motor neuron end in effector organ, that is, in a muscle or gland which responds to the incoming nerve impulses.

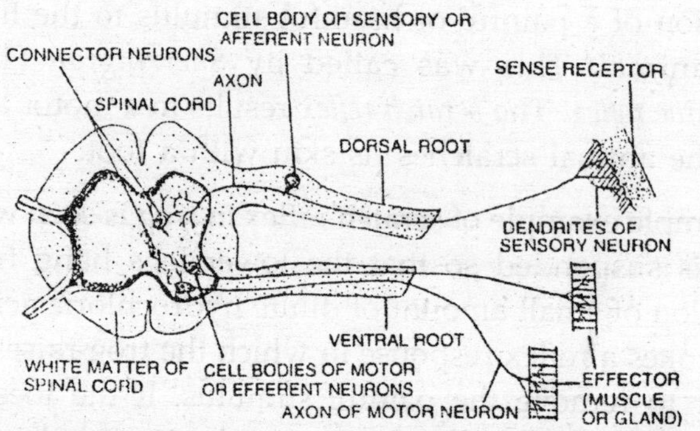

Showing the pathway of nerve impulse transmission in a reflex arc.

However, in higher animals a single stimulus may produce several simple reflexes, and their reflex actions almost always involve more than one reflex arcs. For example, should a person receive a sudden and unexpected burn on his finger tip, not only would he jerk his hand as a result of a simple reflex, but he would also respond in other ways, his behaviour and reactions involving a number of reflex arcs. This happens because a single sensory neuron may have synaptic connections with other neurons, thus accounting for the ability of a single and highly localized stimulus to elicit several responses or reflexes. Similarly, a singled motor neuron may have synapses with numerous other neurons, thus accounting for the ability of different stimuli in different parts of the body to elicit the same response. As an example an injury to any part of the body may evoke the same use of the muscles of the larynx to emit a cry of pain. Thus, most reflex actions are classified as complex and involve a number of neuronal chains. Complex reflexes involve several simple or even complex reflex pathways. Most reflex actions appear to be composed of chains of complex reflexive actions.

Reflex Centre : The term reflex centre refers to an assemblage of nerve cells (often including synapses) in the spinal cord or bixin which control the activity of a particular group of effectors, For example, the gray matter of the *medulla oblongata* contains several vital reflex centres including those that control the rate of heart beat, breathing, and constriction and dilation of blood vessels. Other reflex centres located in the medulla are those responsible for vomitting, coughing, sneezing, and swallowing, etc. The spinal cord also serves as an important reflex centre for great many reflexes. Other parts of the brain also serve as reflex centres, and in turn exercise influence on various body activities.

Nerve Physiology

In the animal kingdom, the "higher" the organism, the more important becomes the system of cells that function to coordinate its activities. Nature has developed two distinct coordinating mechanisms. One depends on the release and circulation of "chemical messengers", the hormones that are manufactured by endocrine glands that are capable of regulating the activity of cell in other parts of the body. The second mechanism, which is in general far superior and selective, depends on a specialized system of nerve cells, or neurons. whose function is to receive and to give instructions by means of electrical impulses directed over specific pathways.

The nerve cells can be divided into two functional groups

: sensory and motor. The sensory neurons collect and relay to higher centres in the nervous system the impulses that arise at the periphery of the body, *e.g.*, skin. The motor neurons carry impulses from the higher centres in brain and spinal cord to the "working" cells, stimulating muscular concentration or glandular secretion.

Functional Architecture of a Neuron : A *neuron* may be defined as a nerve cell consisting of a cell body which is drawn out into many processes. It is the morphological and functional unit of the nervous system. Neurons occur within the grey matter of the brain and spinal cord and in the ganglia of the cranial and spinal nerves and of the autonomic system.

A neuron consists of:

I. the cell body or perikaryon.
II. the processes of the nerve cell: an *axon* and *dendrons* or *dendrites.*

Cell Body or Parikaryon : The part of a complete neuron that looks like an ordinary cell is;the cell body (cyton or perikaryon). The cell membrane is a thin limiting border of condensed cytoplasm. The relatively unspecialized cytoplasm of the cyton is called neuroplasm in which *mitochondia* and *Golgi complex* are present in their typical form. Neurofibrils are firm, protoplasmic threads, embedded in the neuroplasm. They exist in both the cell body and cell processes. Nissl bodies are small masses of granular material which have an affinity for basic dyes, owing to their content of ribonucleic acid. Nissl bodies are present in the cyton and branching dendrons but absent from most axons and their conical bases. With the electron microscope, typical centriole can be seen in the cytoplasm, even though the neuron will never divide. Its role in the neuron is, therefore, very uncertain. Granules of pigment of two types can be distinguished in the neuroplasm. *Melaninis* present as dark brown or blank granules in certain cell groups. *Lipochrome* granules of a yellow or orange colour are found in the cells of autonomic ganglia, and more especially in the region of the axon hillock.

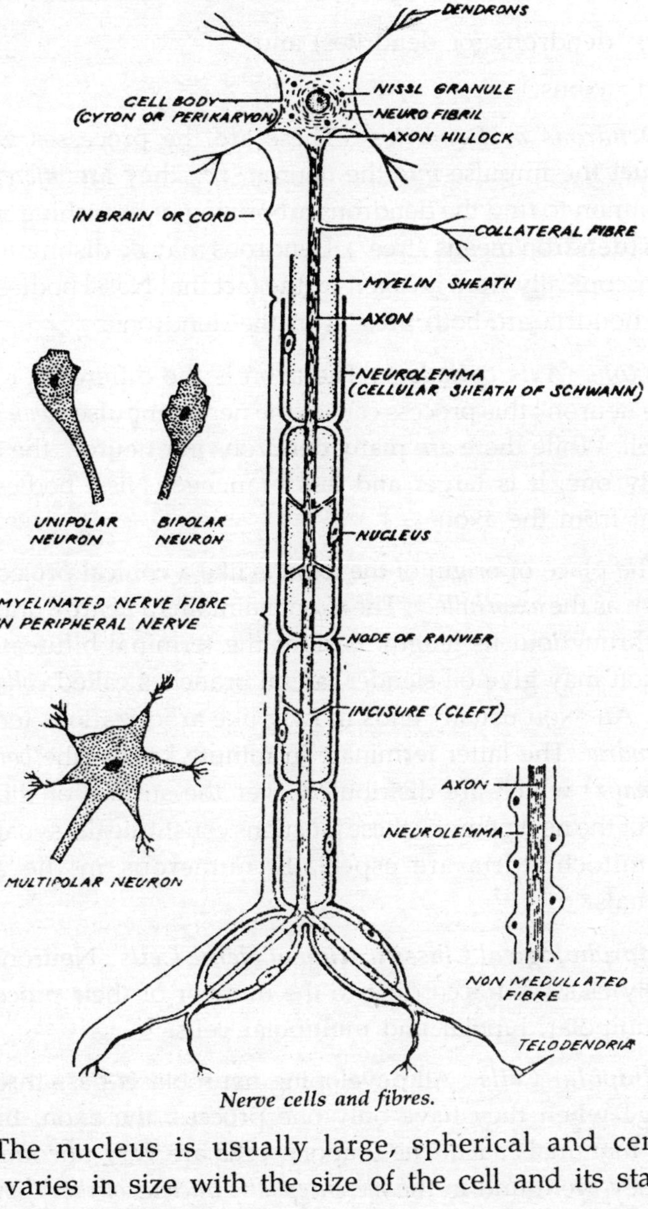

Nerve cells and fibres.

The nucleus is usually large, spherical and central, and varies in size with the size of the cell and its state of activity.

Cell Processes : Cell cytoplasm is drawn out into thread-like processes which are of two types :

(i) dendrons (or dendrites) and

(ii) axons.

Dendrons or Dendrites : These are the processes which conduct the impulse *into* the neuron, *i.e.,* they are *afferent.* It is common to find the dendrons arborizing or branching many times (dendron means "tree"). Dendrons may be distinguished microscopically from an axon by the fact that Nassl bodies and mitochondria are both present in the dendrons.

Axons (Axis cylinder) : The axon is the different process of the neuron ; this process carries the nerve impulse *away* from the cell. While there are many dendrons per neuron, the axon is only one; it is larger and less branched. Nissl bodies are absent from the axon.

The place of origin of the axon is like a conical projection, known as the *axon hillock.* The axon maintains a nearby uniform size throughout its length. Besides the terminal bifurcations, an axon may give oil slender lateral branches called *collateral fibres.* An axon usually ends in twig like arborizations termed *telodendria.* The latter terminate in minute knobs (the *boutons terminaux)* which are distributed over the surface of the cell body of the next neuron ; these junctions constitute the synapses. The mitochondria are especially numerous in the axon terminals.

Morphological Classification of Nerve Cells : Neurons are usually classified according to the number of their processes into unipolar, bipolar and multipolar cells.

Unipolar Cells : All developing neuroblasts pass through a stage when they have only one process, the axon. In the adult mammals such true unipolar cells are very uncommon, but they are found in the *mesencephalic nucleus of the V cranial nerve.*

Bipolar Cells : These cells are typically spindle-shaped, having the axon at one pole an a dendron at the other. Many developing neuroblasts pass through this stage, and in the adult they are found in the *reina,* in the *spiral ganglion of the cochlea,* in the *vestibular ganglion,* in the *olfactory neuro-epithelium.*

Multipolar Cells : These are nerve cells with many processes. Found in the *central nervous system* and *autonomic ganglia,* multipolar cells are the most common type of nerve cells. They may be pear-shaped, stellate or pyramidal in form.

Nerver Fibres

Many exons acquire coverings or *sheaths* along most of their course. An axon and its sheaths are commonly called a nerve fibre. Never fibres tend to run in bundles. Such aggregations are called tracts in the central nervous system and nerves in the peripheral system. Nerve fibres are of two kinds :

(1) *myelinated* and

(2) *unmyelinated.*

Myelinated (Medullated, White) Nerve Fibres : The white matter of the central nervous system, cranial and spinal nerves consist largely of myelinated fibres. These fibres consists typically of a central core *(axon)* surrounded by a tubular *myelin* sheath. The unspecialized cytoplasm of an axon is called *axoplasm.* The surface membrane of the axoplasm is known as *axolemma* which is directly concerned with the transmission of the nerve impulse. The myelin sheath is composed of lipoprotein; in the fresh condition it appears white and highly refractile. Electron microscope studies of Betty Ben Geren-Uzman and Francis O. Shmitt have revealed that myelin is a lamellar structure. The lammellae appear to be arranged concentrically around the axon like a jelly roll.

In peripheral nerve fibres the myelin sheath is interrupted at regular intervals; these interruptions constitute the *nodes of Ranvier,* which appear as constrictions of the fibre when seen with the optical microscope. In the myelinated nerve fibre, resting and action potentials are limited to the nodes of Ranvier and this allows the type of nervous impulse transmission known as *saltatory conduction.* The myelin sheath is subdivided into *myelin segments* by oblique, circular *incisures* or clefts which arise as shearing defects. Myelin serves as an insulating jacket that improves the signalling efficiency of peripheral nerve fibres. The myelin sheath is lost near the peripheral termination of the fibre.

Myelinated nerve fibres of the cranio-spinal series are surrounded by another delicate sheath consisting of a series of separate, very much flattened nucleated *Schwann cells,* linked with each other. The part of a myelinated nerve fibre associated with one such cell, corresponds to an *internode i.e.,* the distance between two successive nodes of Ranvier. In other words, the node of Ranvier is the junction between two Schwann cells. A single oval nucleus is located about midway between two nodes causing a small depression in the myel and a little outwards bulging of the neurolemmal membrane. The function of Schwann cells is to produce the myelin by winding its cytoplasm tightly around the surface of the axon. In this way, a spiral envelop (myelin) of many turns develops around the axon.

External to the Schwann cells and closely adhering to them is a very delicate connective tissue membrane, the neurolemma (or neurilemma), derived from the endoneurium of the nerve.

Non-myelinated (Grey) Nerve Fibres : These fibres are very much smaller and do not possess a myelin sheath. The axon is surrounded by a syncytial; nucleated sheath of Schwann cells. The axons of autonomic ganglia are of this type.

Origin and Propagation of Nerve Impulse

Nerve Impulse is the sum total of *chemical* and *physical events* in the propagation of a wave of physiological activity along a nerve fibre.

OR

It is a wave of *electrical deploraization* and *repolarization* sweeping along a nerve fibre.

Propagation in a Non-Myelinated Fibre : When a nerve fibre is at rest, *i.e.*, when it is not carrying an impulse, the outside of the fibre (surrounded by Na^+) is Positively charged in relation to the inside. The inside of the fibre is negatively charged even though it contain K^+ because of the presence of Cl^- and a variety of LARGE organic anions (resulting from metabolism), such as acetate, pyruvate, lactate and amino acids - that are too large to diffuse easily through the axon membrane.

Extracellular Fluid (ECF)			*Intracellular Fluid*
CATIONS	Na^+	145	12
	K^+	155	
ANTON'S	Cl^-	120	3.8
HCO_s-	27	8	
	A-	7	155
POTENTIAL		0	—90 mV

1. *Origin of Resting Potential* : There is a differential distribution of ions between the axoplasm and the external medium. The extra cellular fluid (EFC) has a high concentration of Na^+ and Cl^- and the Intracellular fluid a high concentration of K^+ and of larger organic anions (A^-).

A potential difference arises when a membrane is selectively permeably to either anions or cations, and a difference in concentration of some non-diffusible ion exists between the inside and the outside of the membrane. Then a potential develops between the two sides, and the *solution of higher concentration becomes negative to the one of lower concentration.* This difference in electric potential on the two sides of the nerve fibre membrane is termed the membrane potential or resting potential which is about — .07 volt or — 70 mV. The nerve fibre membrane is said to be polarised in this state. The resting potential depends on a physico-chemical equilibrium condition known as the Donnan Equilibrium which predicts the movements of Na^+, K^+ and Cl^- ions into and out of living cells.

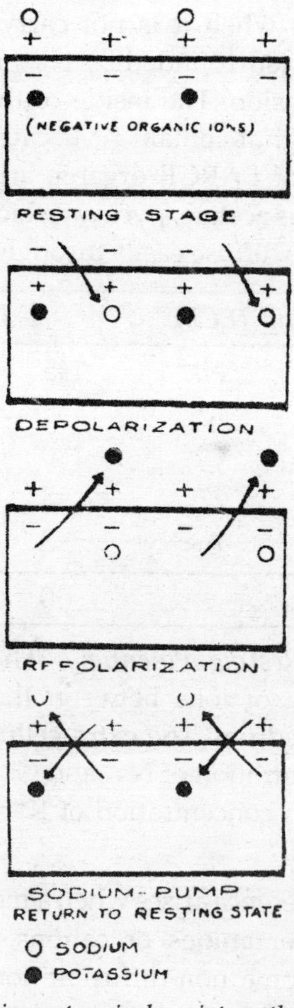

Movement of ions at a single point on the nerve fibre.

2. *Depolarization* : Stimulation of a nerve changes the permeability of the nerve fibre membrane. The stimulus may be of mechanical, electrical or chemical nature. As a result Na^+ quickly moves in. Since each Na^+ ion is positively charged, this movement results in a positive charge on the inside of the cell as compared with the outside. This reversal of electric charges is called depolarization.
3. *Repolarization* : K^+ ion with their positive charge, now begin to flow out faster than the Na^+ flows in. When these positive charges are moved to the outside of the cell, they initiate the process of replorization (return to the resting state and inside of the cell again becomes negative.
4. *Metabolic Pump : The active transport* of Na^+ out of the cell and the re-establishment of K^+concentration inside it restores the cell to a resting state. This mechanism is called Sodium Potassium Exchange Pump because it transports sodium out of the cell and at the same time also moves Potassium back into the cell, *against concentration, gradient by expenditure of energy derived* from cellular metabolism.
5. *The Action Potential* : As Na^+ rushes into the fibre, the outside of the fibre briefly becomes negatively charged at that point. Adjacent to this point, a part of the fibre remains positively charged. This difference in potential between the two areas is called the Action Potential. This is a temporary change between a stimulated and resting portions of a nerve fibre.

Since opposite charges attract one another, positive charges move towards the negative charges on both sides of the membrane. The result is a *partial neutralization* of the charge or a small *decrease* in the electric potential (a *depolarization)* of the membrane adjoining the site of the impulse. This depolarization causes the membrane to become more permeable. As a result Na^+ rushes into the cell ; the process ignites' and flares up to create the nerve impulse or action potential. Thus, action potential may be defined as a self propagating depolarization of the nerve membrane. Bit by bit, the impulse moves down the fibre. This is a regenerative process that leads to automatic self-reinforcement. The nerve impulse is thus a self-propagating wave of depolarization followed by repolarization moving down the nerve fibre.

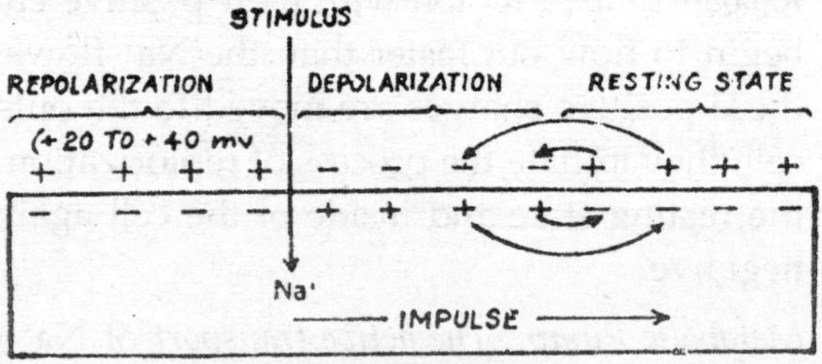

Local current low during the passage of an impulse along an un-myelinated nerve fibre.

The nerve impulse may be compared to the ignition of a string of fire-crackers. When a burning match stick is applied to the end of the string, its threshold is reached, and a spark is generated which begins to pass along the string. The energy required for this movement is derived not from the energy of the burning match stick, but from the chemicals contained in the fire-crackers. The spark does not weaken with the distance travelled. If the string branches, the spark passes along each branch with undiminished strength.

The nerve impulse travels much more slowly than an electric current and by a different means :

Electric Current and Nerve Impulse

Electric Current	*Nerve Impulse*
1. When an electric current travels along a conductor, the conductor serves only	1. During transmission of a nerve impulse ions are the only particles that

Contd...

Electric Current	*Nerve Impulse*
as a passive carrier of the electrons that move in the same direction as the current, because this movements of electrons is the current.	move. Moreover, they move at right angles to the movement of impulse, i.e., they move in the out of the fibre.
2. An electric impulse is conducted along the wire at the speed of the light.	2. In the neuron, an electrochemical reaction is simply moving down the length of the neuron.

Although a nerve impulse is not an electrical current flowing along a nerve fibre, the impulse is accompanied by measurable changes in electrical potential in the membrane.

Propagation in Myelinated Nerves : Myelin is a lipid-rich wrapping of high electric resistance arid functions as an effective insulator. But this insulation is broken as the nodes of Ranvier. Accordingly the current (Action Potential) flows from one node over the myelin to the next node. Because, the charges leap from node to node, this form of propagation is termed Saltatory Propagation in which the current spreads by *Jumping, skipping, dancing* or *leaping* from one node to another instead of travelling along the nerve membrane.

Saltatory propagation increases the speed of the impulse, because only the membrane at the nodes of Ranvier must undergo depolarization. As a result, impulse transmission is about 20 times faster in a myelinated than in an unmyelinated neuron.

Velocity of Propagation : It ranges from a few metres/sec for thin non-myelinated fibres to about 100 metres/sec in the thickest myelinated fibres. For example, in large mammalian nerve fibres. Highest speeds, equivalent to some 200 miles/ hour are found in the sensory and motor fibres concerned with body balance and fast reflex movements in non-myelinated axons, the greater the fibre diameter, the lower is the internal resistance and higher the conduction velocity. In myelinated fibres, the velocity is found to be directly related to diameter, because intemodal distance is proportion to the diameter.

Synapse and Synaptic Transmission

Structure : There is always more than one neurone involved in the transmission of a nerve impulse from its origin to its effector organ, whether it is a sensory or motor pathway.

The junction between two neurons is known as a synapse. Actually synapse is a gap, the synaple cleft, between the end of one nerve fibre and the beginning of the next. In most cases *post-synaptic neuron* forms terminal swellings or synaptic knobs on the *post synaptic neuron.* The size of the synaptic knob varies from 0.5 to 2 μ in diameter. The synapic knob contains numerous synaptic vesicles which are preformed "packets" of a chemical transmitter substance - the *acetylcholine.* This is formed in the cell body from where it is transported down the axon to the presynaptic terminals. The formation of acetylcholine is helped by a special enzyme, *choline acetylase.*

Action of Acetylcholine : The arrival of an impulse at the synaptic knob causes acetylcholine to be discharged into the synaptic cleft. Accumulation of acetylcholine in the synaptic gap reduces the polarization of dendrites, or cell body of the next, *follower,* neuron. If depolarization is sufficient, the threshold of the next neurone will be reached and a nerve impulse generated in it. In this way, acetylcholine serves as a *chemical transmitter* of the nerve impulse.

Removal of Acetylcholine : As nerve impulse after nerve impulse causes the secretion of acetylcholine into the synapse, it becomes highly concentrated and diffuses in all directions.

But as a matter of fact, it never gets very far or becomes very concentrated. An enzyme, *acetylcholinesterase* is present in the vicinity of syriapases which breaks down the acetylcholine thereby preventing continued depolarization of the next fibre or the inordinate spreading of the nerve impulse to other nerve fibres.

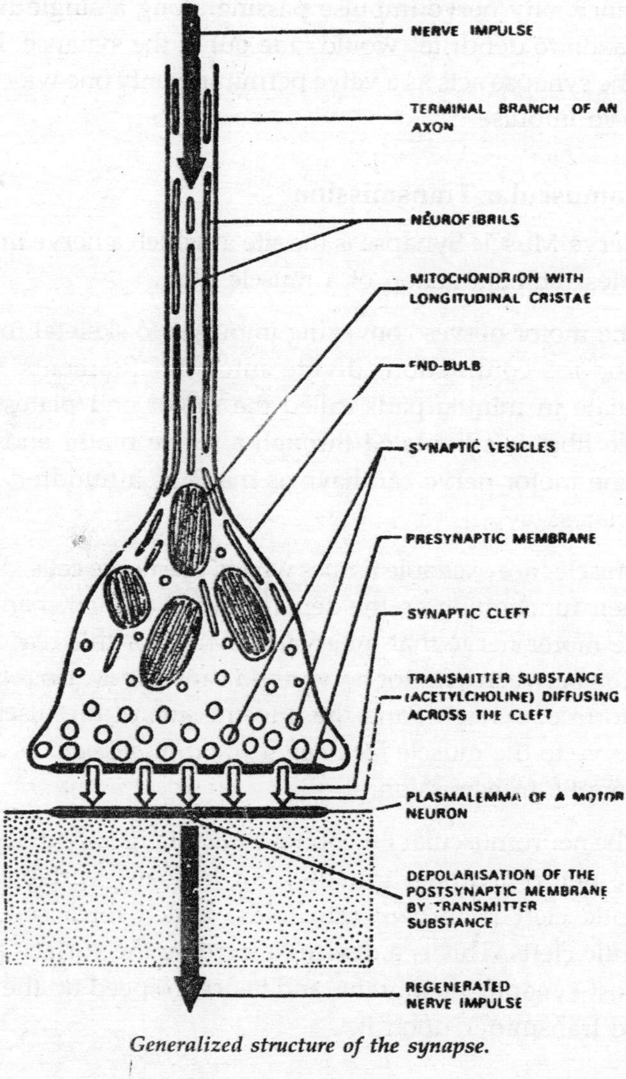

Generalized structure of the synapse.

Synapse Acts as a One Way Valve (One Way Conduction of Nerve Impulses) : If a nerve impulse was initiated somewhere in the middle of a nerve fibre, it would proceed to both ends. However, in the synapse, the propagation is strictly one-way, from the axon to the dendrite-never the reverse.

Since dendrites do not release a chemical transmitter substance, any nerve impulse passing along a single neuron from axon to dendrites would fade out at the synapse. In this way the synapse acts as a valve permitting only one way traffic of nerve impulse.

Neuromuscular Transmission

Nerve-Muscle Synapse is the site at which a nerve impulse activates the contraction of a muscle fibre.

The motor nerves conveying impulses to skeletal muscles to produce contraction, divide into fine filaments which terminate in minute pads called the motor end plates. Each muscle fibre is stimulated through a single motor end plate, and one motor nerve can have as many as a hundred motor end plates.

Muscles are excitable tissues which, like nerve cells, depend for their functioning on the depolarization of their membrane by the motor nerve that innervates them. For this reason, the junction between a motor nerve and its muscle is also regarded as a form of synapse, and the transmission of impulses from the nerve to the muscle *[neuromuscular transmission)* as a form of synaptic transmission.

The neuromuscular junction is similar in its essential details to the neuro-neuronal synapse, but in many vertebrates its synaptic cleft is thrown into folds, which form secondary synaptic clefts. This is a device to increase the surface area of the post-synaptic membrane, and hence to speed up the action of the transmitter upon it.

The motor end plate is populated with "synaptic vesicles" that may contain *acetylcholine,* which is released into the short synaptic gap by a nerve impulse and evokes electrical activity in the muscle.

Acetylcholine is not the only transmitter which may promote transmission across excitatory synapses. For example, *noradrenaline* acts as a transmitter at the neuromuscular junctions of mammalian smooth muscle, and at certain central synapses. *5—hydroxy tryptamine (serotonin)* and *gamma aminobutyric acid (GABA)* are other neurotransmitters of physiological importance.

Functional Properties of a Nerve Fibre Threshold : The stimulus applied to a nerve fibre must reach a certain minimum strength or threshold before it can initiate an action potential to travel along the nerve fibre.

The All or None Response : Once the threshold level has been attained or exceeded, the impulse is found to reach its maximum height independently of any further increase in the intensity of the applied stimulus.

Once started, the nerve impulse travels at a speed that is entirety independent of the intensity of the stimulus. A nerve impulse is, therefore, referred to as an all-or-non phenomenon: it requires a minimum intensity of stimulus of elicit it, but once initiated, it will proceed to its maximum size.

Refractory Period : The frequency with which impulses can pass along a nerve fibre is governed by the refractory period of the fibre. If an action potential is initiated by an applied electric shock, a second stimulus applied during the period when the action potential is reaching its peak, will not produce a second action potential, however intense the stimulus. This period of complete inexcitability is called the *absolute refractory period.*

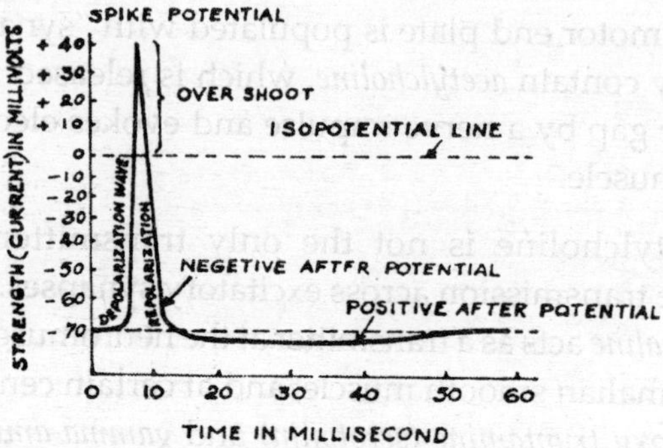

Diagrammatic representation of action potential.

The absolute refractory period determines the maximum frequency of conduction. In mammalian nerves the observed refractory period ranges from 1 to 6 milliseconds, which means that approximately 200 to 1000 impulses per second can be fired along a nerve.

Spike Potential : If the neuron is activated by an adequate stimulus, a marked alternation in the potential takes place. There is a very rapid change in potential from the resting level of — 70 mV to' zero (Isopotential Line) and then to + 20 or 40 mV. With almost equal suddenness, the potential then returns to the resting level.

The sharp rise and fall in the potential constitutes the spike potential. The reversal of polarity is called the overshoot.

Negative after Potential : There is a short section of the action potential just before it reaches the resting level where the rate of fall becomes abruptly slow. This slower fall is called the *negative after potential.*

Positive after Potential : During replorization, the potential returns to the resting level and in most cases, *beyond, i.e.,* to about — 75 mV. The segment of the curve that falls below the resting level is known as the *positive after-potential.*

Donnan Equilibrium

If a membrane separates two solutions of electroivtes in one of which there is a non-diffusible large on (P^-), a distribution of the other, diffusible, ions will result. This is because :

(1) The sum of cations on either side of the membrane must be equal to the sum of the anions on the same side. Thus, in the given example, 133.3 equals 33.3 + 100 ; and 66.6 equals 66.6 + 0.

(2) The products (x) of the concentrations of any pair of diffusible ions on one side of the membrane must be equal to the product of the concentrations of the same pair of ions on the other side. Thus, 133.3 x 33.3 equals 66.6 x 66.6.

The net result is that an equilibrium condition becomes established in which there remains some imbalance of ions on either side of the membrane, and the P^- side of the system is more negative than the other, *i.e.*, there is a potential difference across the membrane.

Electrical effect of the P^- ions will be not only to draw in positive ions, but to repel the Cl^- ions to an equal extent. At equilibrium, therefore, the ratios of Na^+ ions on the P^- side to those on the other side will be the reciprocal of the corresponding ratio for Cl^- ions.

Donnam Equilibrium

Initial State				*At Equilibrium*	
Side 1		Side 2		Side 1	Side 2
Na^+	P^-	Na^+	Cl^-	Na^+ Cl^- P^-	Na^+ Cl^- P^-
100	100	100	100	133.3 33.3 100	66.6 66.6 0

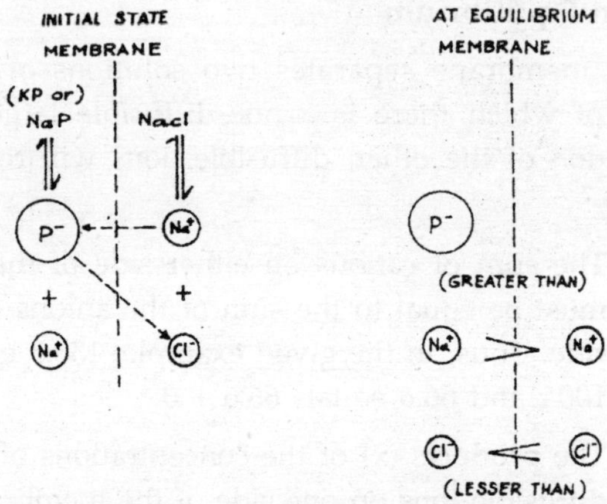

Diagram to illustrate the Donnan Equilibrium. In (A), the ionization of NaP and NaCl, placed initially on opposite sides of a porous membrane, are shown, the electrical forces exerted by the P-ions membrane, are shown. The electrical forces exerted by the P- ions are indicated by broken lines. The final equilibrium situation is shown in (B).

The Reflex Action and the Reflex Arc : Most of our body activities and reactions to environmental changes, both internal and external, are automatic, or in voluntary. Such responses which occur outside of our awareness, are immediate and rapid and are called reflexes. A reflex action can be defined as an automatic motor response to a sensory stimulus without the brain being immediately involved. Reflexes are usually classified in two broad subdivisions: (i) the *simple, of unconditioned reflex,* and (ii) the *conditioned reflex.*

Simple Reflex : A simple reflex is an inborn, unlearned response to a stimulus or change in environment. Most reflex actions are protective in function. There are a number of familiar examples.

1. The quick jerk of the leg if the patella tendon is sharply tapped ; this is termed the *knee Jerk.*
2. The quick closing of an eyelid when an object suddenly approaches the eye. This is known as the *corneal reflex.*

3. The rapid withdrawal of a hand when it is burned or pricked.
4. The quick recovery of the balance of the body to prevent falling, after a slip.
5. A sudden coughing attack if a crumb is inhaled.

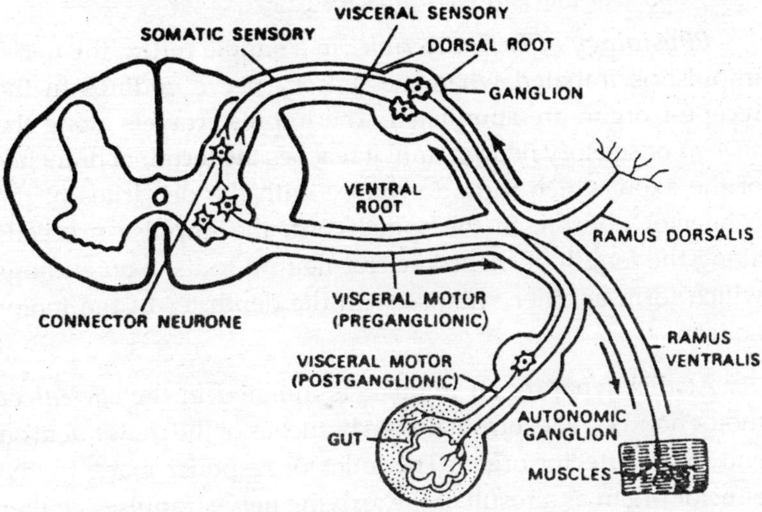

Schematic representation of a simple reflex arc.

Many other reflex actions occur within the body which never reach consciousness, for example, glandular secretions, movements of the stomach and small intestine, the changes in the calibre of blood vessels, breathing, rate of heart beat, and so on.

Reflex Arc : The anatomical basis of the simple reflex, called the reflex arc in an arrangement of atleast three neurons in a pathway that always passes through the central nervous system. The three components are as follows :

1. A sensory neuron which comprises the *sensory nerve endings* in a receptor organ, the *sensory nerve, posterior (dorsal) root ganglion cells and its nerve fibre* which passes to the dorsal horn of grey matter in the spinal cord.

2. A connector neuron which consist of a *nerve cell* its *dendrites* and *axon* in the spinal cord.

3. A motor neuron consisting of a *nerve cell* and its *dendrites* in the ventral horn of the grey matter. The *axon* of this nerve cell and the *motor end plates* terminate in the effector organ *e.g.,* muscles.

Physiology of Reflex Action : In a simple reflex, the nerve impulse is initiated when the sensory nerve endings in the receptor organ are stimulated. The impulse travels along the *afferent* or sensory neuron until it reaches the terminal branches of the axon which form a synapse with the dendrites of the connector neuron. In the spinal cord, the impulse proceeds along the length of this connector neuron to its axon endings which form another synapse with the dendrites of the motor neuron.

At this synapse, an impulse is initiated in the *efferent* or motor neuron. The terminal axon branches of the motor neuron end in the effector organ. The reflex or response made by the effector organ as a result of the arriving nerve impulse is either muscular contraction or glandular secretion, depending on the effector involved.

Coordinating Properties of the Reflex Arc : Reflex actions of higher animals almost always involve many reflex arcs, not just one. For example, if an individual receives a sudden, unexpected burn out his fingertip, he would not only jerk his hand as a result of a simple reflex, but would so respond in other ways. He would turn his head to the source of danger, cry out and very likely experience a series of sensations and reactions including pain, fear, and emotional tension.

The total reaction, therefore, would involve numerous muscles — including those of the arm, shoulder, neck, head trunk, tongue, farynx and respiratory tract. Besides, several mental processes participate in such an integrative response.

Thus, a simple stimulus *i.e.,* a mild, highly localized burn, evolves an extensive reflex behaviour involving the coordinated response of the entire nervous system.

Inhibitions of Reflex Actions : In the infant the act of micturition and defecation is carried out by reflex action and there is no voluntary control over these processes. As the child grows older, the sensory nerve tracts of the spinal cord become fully developed ; the impulses from the bladder and rectum reach the brain and can be controlled. Thus the act of micturition and defaecation becomes a conscious voluntary act in the adult.

Many reflex actions are inhibited by the body, thus preventing the overactivity of the muscle which otherwise would contract in response to every sensory stimulus, even although a response was unnecessary.

Certain reflex actions cannot be inhibited by the cerebrum, for example, the secretion of glands, vaso motor, respiratory, cardiac and alimentary reflexes.

If there is damage to the spinal cord in an adult due to injury or disease, the stimuli would not reach the sensory area of the brain. As a result, the individual will not be conscious of a full bladder or rectum. When the pressure reaches a certain level the bladder or rectum will be emptied by reflex action. This involuntary action is called incontinence of urine and faeces.

Reflex Centre : The term reflex centre refers to a group of neurons in the spinal cord or brain that controls the activity of a particular group of effectors. For example, the reflex centre for breathing located in the *medulla oblongata* of brain controls the impulses conducted to thoracic muscles responsible for breathing action. The cells of the reflex centre are in turn regulated by appropriate afferent impulses. Similarly, the reflex centre for heart-rate controls and adjusts the heart beat by motor impulses in response to sensory massages from aorta or the vena cava. Other reflex centres located in the medulla are those responsible for *vasodilation—vasoconstriction, vomitting, coughing, sneezing and swallowing.*

Conditioned Reflex : Conditioned reflexes, in contrast to simple reflexes, are not inborn but are acquired and important on past experience, training, and learning. The conditioned reflex was first clearly demonstrated in the early part of this century by the Russian physiologist Ivan Petrovitch Pavlov (1849— 1936). He showed that conditioned reflex can be developed experimentally to an extraordinary degree. Pavlov present a hungry dog with food, which elicited a reflex of salvation by the stimulation of the taste buds in the mouth. He then rang a bell each time the dog was given the food in order to associate the unconditioned response (salivation) with an additional stimulus (the bell). After a few trials, Pavlov was able to demonstrate that the dog salivated at the sound of the bell, without the sight or smell of the food. This is known as a conditioned reflex.

4

Respiration System

The term *respiration* includes both oxidative metabolism within the cell, and the means by which oxygen reaches the cell. Since all oxidative metabolism involves the production of carbon-dioxide, the transport and elimination of this gas is normally considered at the same time as the transport and utilization of oxygen. Respiration, therefore, includes oxidating of nutrient material in the cells to release energy, and breathing.

The respiration in animals is brought about in 2 phases :

External Respiration or Breathing or Ventillation : In this phase, blood takes up O_2 from the fresh air inhaled into lungs, and gives up CO_2 to it.

Internal or Tissue Respiration : In this second phase, blood distributes O_2 to the cells and receives CO_2 from through the tissue fluid. The internal respiration is chiefly concerned with the uptake of O_2 release of energy and the production of CO_2, by the cells. It is, therefore, also referred to as the *cellular respiration.*

Respiratory System

It consists of two divisions specialized for different purposes.

Respiratory Tract or Conducting Passage : Its function is piping air to and from the lungs. It comprises.

(i) Nasal cavity

(ii) Pharynx

(iii) Larynx

(iv) Trachea

(v) Two Primary Bronchi

(vi) Secondary, Tertiary Bronchi, and Terminal Bronchioles in lungs.

Respiratory Seat : The intimate interchanges between blood and air occur in this part. It consists of two lungs and their convertings, the pleura : containing within :

(i) Respiratory Bronchioles

(ii) Alveolar ducts

(iii) Atria

(iv) Alveolar sacs

(v) Alveoli

Nasal Cavity

The nasal cavity is the first of the respiratory organs and consists of large irregular cavity divided by a *septum.*

The *roof* is formed by the base of the skull.

The *floor* is formed by the roof of the mouth, which is composed of the maxillary bones, the palatine bones and the soft palate.

The *medial wall* is formed by the *septum* which is composed of perpendicular plate of the ethamoid bone superiorly, and the vomer-inferiorly. Anteriorly, the septum is formed by cartilage.

The *anterior wall* is formed by the nasal bones and cartilage covered with skin.

The *lateral walls* are formed by the superior maxillary bones and the inferior, middle and superior *Conchae or Turbinals* which are scroll-like (rabbit) or shelf-like projections (man), lined with respiratory epithelium. The conchae immensely increase the surface of the nasal chambers.

The *posterior wall* is formed by the posterior wall of the pharynx.

The nasal chamber are lined with a vascular, ciliated columnar epithelium, rich in mucous cells. This region of the respiratory tract acts as an air-conditioner because the incoming air is warmed, moistened, filtered and sterilized during its passage through the nose :

(1) *Warming :* Blood capillaries of the turbinals radiate heat in a similar manner as the hot-water pipes. The incoming air passing over them, therefore, gets warmed upto the body temperature.

(2) *Moistening :* The inspiratory air becomes moistened on coming in contact with the moist mucous secreted by glandular respiratory epithelium.

(3) *Cleaning :* Mucous is sticky in nature, therefore, particles of dust and other impurities in the air adhere to it. The cilia of the respiratory epithelium drive this mucous either towards the external nares or to the oropharynx where it is swallowed. Therefore, from a physiological standpoint, nasal breathing is superior to mouth-breathing.

(4) *Sterilization :* Mucous is antiseptic in action. To some extent, it kills the bacteria of the air going to the lungs.

The *anterior* or *external nares* form the anterior opening into the nasal cavity.

The *posterior internal nares* from the posterior opening from the nasal cavity into the pharynx.

The Pharynx : The *pharynx* is a cone-shaped tubular part which is common to the respiratory system and the alimentary canal.

For descriptive purpose the pharynx is divided into three parts : (a) the naso-pharynx, (b) the oro-pharynx, (c) the laryngo -pharynx.

(a) *Naso-pharynx :* The naso-pharynx is the part which lies behind the nose above the level of the soft plate. On its lateral walls are the two openings of the *auditory lubes* which lead from the naso-hparynx to the middle ear. In its posterior wall is the lymphoid tissue known as the naso-pharyngeal tonsil or, when enlarged, as *adenoids.* These are most prominent in the children upto approximately the age of 7 years and thereafter begin to degenerate.

(b) *Oro-pharynx :* The oro-pharynx is continuous in from with the buceal cavity and below with the laryngeal part of that pharynx. The *oral tonsils* which are collections of lymphoid tissue lie in its lateral walls.

(c) *Laryngo-pharynx :* This part of the pharynx extends from the oro-pharynx above the continues as oesophagus below.

The Larynx : The *larynx* or "voice box" is situated in the mid-line of the neck between pharynx above and the trachea below. It is placed in front of the oesophagus and corresponds with the levels of the 4th, 5th and 6th cervical vertebrae. After puberty there is a considerable enlargement in the size of the larynx in the human male as compared to the female. The prominent larynx in the male is often referred to as the "Adam's apple".

The larynx consist of a framework of the following hyaline cartilages :

(a) *The Thyroid Cartilage :* This is the largest cartilage and consists of two side wings known as the laminae which are fused together anteriorly to form the *laryngeal prominence* or Adam's apple.

(b) *The Cricoid Cartilage :* This is situated below the thyroid cartilage. It is shaped like a signet ring, completely encircling the larynx. The cricoid forms the lowest part of the larynx and connects with the trachea.

(c) *The Arytenoid Cartilages :* These are two pyramid shaped cartilages situated on top of the broad part of the cricoid cartilage. They give attachment to the posterior ends of the true vocal cords.

(d) *The Epiglottis :* This is a leaf-shaped plate of yellow elastic fibrocartilage situated in an upright position between the base of the tongue and the upper opening of the larynx. Its lower stalk-like part is attached by a ligament to the thyroid cartilage. The main function of the epiglottis is to prevent food from entering the larynx during deglutition or swallowing.

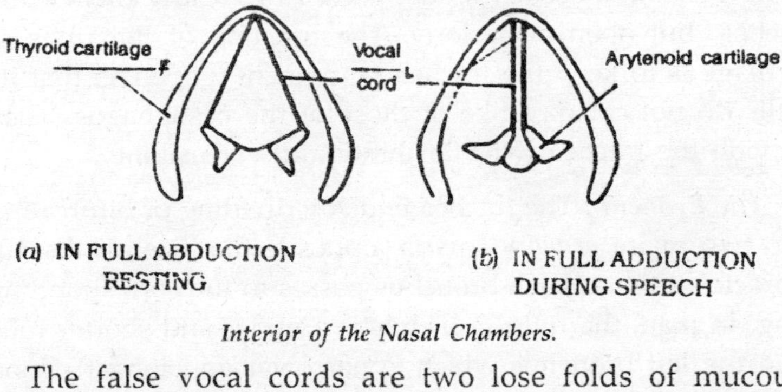

Interior of the Nasal Chambers.

The false vocal cords are two lose folds of mucous membrane situated above the true cords. They do not appear to play any special part in voice production.

Sound has the properties of pitch, loudness and quality.

(i) *The pitch* of the voice depends upon the *length* and *tightness* of the vocal cords. In the male, the cords are longer than in the female (2.3 cm as against 1.7 cm). Thus the male voice has a deeper pitch than the female.

(ii) *The loudness* of the voice depends upon the *force* with which the cords vibrate. The greater the force of expired air, the more vibration of the cords, thus the louder the sound.

The *quality and resonance* of the voice depend upon the shape of the mouth, the position of the tongue, the lips, the factal muscles and on the air sinuses in the bones of the face.

The Trachea : The *trachea* or wind pipe is a continuation of the larynx and extends to approximately are level of the 5th thoracic vertebra where it divides or bifurcates into the left and right primary bronchi. In man the trachea is a relatively thin walled rigid tube, about 40 inches long and 1 inch wide, it lies in front of the oesophagus.

The trachea is composed of from 16 to 20 incomplete, or C shaped rings *of hyaline cartilage,* which are joined together by fibrous and muscular tissue. These rings nearly encircle the trachea, but open posteriorly. The function of the rings of cartilage is to keep the trachea permanently open so that its walls do not collapse like of those of the oesophagus. They provide the trachea with flexibility and extensibility.

The Bronchi : The trachea ends by dividing or bifurcating into two main, *primary bronchi,* opposite the level of the 4th thoracic vertebra. Each bronchus passes to the corresponding lung. In man, the right bronchus is a wider and shorter tube than the left bronchus which is narrower and longer. Upon entering the lungs, bronchi divide into numerous smaller bronchi like the branches of a tree.'

The structure of the bronchi is similar to that of the trachea consisting of incomplete rings of hyaline cartilage. Both the trachea and bronchi are lined with pseudo-stratified, ciliated columnar epithelium containing *goblet cells* which secretes mucous. The mucous helps in holding the dust and bacteria, which are swept towards the pharynx by cilia.

As the bronchi become smaller by subdividing the cartilages become less well defined and more irregular in shape.

The Lungs : The lungs are a pair of conical-shaped organs occupying the greater part of the thoracic cavity. The lungs are separated from each other by *mediastinum* which contains the heart and great vessels, the oesophagus and, in its upper part, the trachea. Each lung is divided by deep fissures into lobes. In rabbit, the left lung is divided into 2 lobes; *left anterior* and *left posterior,* whereas the right lung has 4 lobes; *anterior azygos, right anterior, right posterior* and *posterior azygos.*

In man the left lung has 2 lobes: upper and *lower.* The right lung has 3 lobes : *upper, middle* and *lower.* They extend from the root of the neck above to the diaphragm below. Each lung is described as having a costal and a medial surface, an apex and a base. The *costal surface* is smooth and follows the shape of the chest wall. The *medial surface* of the lung is applied to the mediastinum. Its main feature is the presence of the *hilum* where the primary bronchus and pulmonary artery enter and the pulmonary veins leave the lung.

The *apex* rises into the root of the neck for about one inch above the clavicle. The *base* is concave and is related to the upper surface of the diaphragm.

The Pleura : Each lung is invested by a serous membrane—*the pleura* which consists of two layers : the visceral and parietal.

The Parietal Pleaura : This lines the interior of the chest wall and upper surface of the diaphragm.

The Visceral Pleura : This is firmly attached to the lung itself, completely covering its surfaces, except at the hilum where it is reflected over the structures entering the lung and becomes continuous with the parietal layer.

In normal health, the parietal and the visceral pleura are in *contact* separated only by a thin film of serous fluid. There is a *potential space* between the layers which is termed the *pleural cavity*. The *pleural fluid*, secreted by the epithelial cells of the pleura acts as a lubricant so that the two surfaces can slide smoothly over each other as the lungs expand the contract during the act of breathing.

The Interior of the Lungs : The lungs are composed of the bronchi, bronchioles, alveolar ducts, atria, alveolar sacs and alveoli; of which there are many thousands. The respiratory bronchioles, alveolar ducts, atria, alveolar sacs and alveoli constitute the *lobules* of the lungs. These lobules are separated from one another by the interlobular tissue which is composed of *areolar* and elastic tissue.

Inside the lung; *primary bronchus* divides repeatedly into a number of branches successively of thinner diameter which are called *secondary* and *tertiary bronchi*. These are lined by simple columnar epithelium with abundant cilia and goblet cells. The adventitious layer contains separate cartilaginous plates instead of C-shaped rings. Tertiary bronchi divide into still thinner tubes, the *terminal bronchioles*. In these tubes, an adventitia remains; but cartilage, glands and lymph nodules no longer occur. In the loss of a firm support, the mucosa is thrown into longitudinal folds. Terminal bronchioles are again subdivided in to *respiratory bronchioles*. About two arise from each terminal bronchiole; each subdivides once or twice. Arespiratory bronchiole is about 0.5 mm. or less in diameter. In this conduct, *alveoli* appear as little outpouchings that interrupt the main wall at intervals. They project beyond the ordinary wall as still thinner, box-like insets.

Each respiratory branchiole gives rise to several branching *alveolar ducts.* They are relatively long, thin walled fibro-elastic tubes, thickly beset with single alveoli. An alveolar duct terminates into a sort of antechamber, the *atrium.* Varying from 2 to 5 *alveolar sacs* open off from each atrium. Each alveolar sac is composed of many thin-walled, minor compartments, the *alveoli,* all of which open into the main lumen of the sac. An individual alveolus, or air cell, is like a hexagonal box with its top open. In this way, the interior of an alveolar sac presents a honey-combed appearance.

The alveoli have a very thin wall composed of squamous epithelium. It is lined by a layer of mucous and is closely surrounded by a network of capillaries. The blood in capillaries is separated from the air in the alveoli by just two layers of cells, *i. e.,* capillary endothelium and alveolar epithelium. This results in a quick and easy exchange of O_2 and CO_2 between the blood and alveolar air. Each alveolus is thus a miniature lung.

The human lungs have about 750,000,000 alveoli, which have a total surface area of about 100 sq. metres. This comes to about 50 times that of the skin surface.

Mechanism of Respiration

The mechanism of respiration is the process by which the lungs are expanded to take in air, then contracted to expel it. The cycle of respiration consists of 3 phases : (i) *inspiration,* (ii) *expiration* and (iii) *pause.*

This cycle of events occurs about 16 times per minute. The expansion of the chest during inspiration occurs as a result of muscular activity which is partly voluntary and partly involuntary. The main muscles of respiration in normal quite breathing are the *intercostal muscles* and the *diapharagm.*

The Intercostal Muscles : In man there are 11 pairs of intercostal muscles which occupy the spaces between the ribs.

They are arranged in 2 layers called : external and internal intercostal muscles.

(1) *The External Intercostal Muscles* : These extend in a *downwards* and *forwards* direction from the lower border of the rib above to the upper border of the rib below. Their action is to *raise the ribs upwards and outwards* thus increasing the size of the thoracic cage *laterally and dorsoventrally*. Because of the increase in the capacity of the thoracic cage, inspiration takes Place.

(2) *The internal Intercostal Muscle* : These extend in a *downwards and backwards* direction from the lower border of the rib above, to the upper border of the rib below, crossing external intercostal muscles at right angles.

These muscles are believed to have an action *antagonistic* to that of the external intercostal muscles during forced expiration, when they *draw the ribs downwards and inwards*. At the same time, the *abdominal recti muscles* contract forcibly pushing the stomach, liver etc. upwards against the diaphragm and reducing the volume of the thoracic cavity still further, from that direction.

The intercostal muscles are stimulated to contract by the *intercostal nerves*.

The Diaphragm : The diaphragm is the *main muscle of respiration*, and its action is only partly under control of will. It is a dome-shaped *musculo-fibrous structure* which separates the abdominal and thoraic cavities, thus forming a concave roof for abdominal cavity and a convex floor for the thoracic cavity.

The diapharagm is made up of *radiating muscle fibres* which are inserted at one end into a central *aponeurosis* (tendon) and originate from the circumference of the trunk. When the radiating muscle fibres contract, they pull on the central tendon and *flatten* the dome-shaped structure, thus increasing the *length* of the thoracic cage and, as a result, *inspiration* occurs.

When the muscle of the diaphragm is contracted, it increases the pressure within the abdominal and pelvic cavities, and so assists in *micturition* (the passing of urine) *defaecation* (the passing of faeces) and *parturition* (childbirth).

The diaphragm is supplied by the *phrenic nerves.*

Insipiration : It is described as an *active process* because it is the result of muscle contraction. During *inspiration* (breathing in), the *external intercostal muscles* and the diagphragm contract simultaneously and so the size of the thoracic cage is increased laterally, dorsoventrally and longitudinally at the same time.

As the lungs are held tightly against the thoracic wall. Increase in dimensions of the thoracic cavity results in expansion of the lungs, and pressure within the alveolar and in the air passages is reduced. This results in air being drawn into the lungs in an attempt to equalize the atmospheric and alveolar air pressures.

Expiration : This is a *passive process.* When the external intercostal muscles and diaphragm *relax,* the ribs fall back into place, the diaphragam ascends and again becomes dome-shaped. This is aided by the elastic recoil of lungs an thoracic wall.

With all this, the thoracic cavity decreases in dimensions, permitting the lungs to return to their original size. As they do so, air is forced out of them.

During *vigorous expiration* the antagonistic expiratory muscles—*the internal intercostal muscles,* come into play and decreases the size of the thoracic cavity still further.

After expiration, there is a short *pause.*

Special Respiratory Movements

(1) Cough is a forcible expiration usually proceeded by a prolonged inspiration. The sound of a cough is produced by forcing air though the narrow opening between the vocal cords.

(2) Hiccough is a noisy inspiration caused by muscular spasm of the diaphragm at irregular intervals. The noise is produced by the sudden sucking of air through the vocal cords.

(3) Sighs and yawns are types of prolonged inspiration.

Physiology of Respiration

Exchange of Gases : Between the process of inspiration and expiration the interchange of respiratory gases occurs. This interchange takes place between the blood in the capillary network which surrounds the alveoli, and the air in the alveoli of the lungs. In external as well as internal respiration, gases always tend to diffuse from a *high partial pressure* to a *lower pressure, i.e.,* down the concentration gradient.

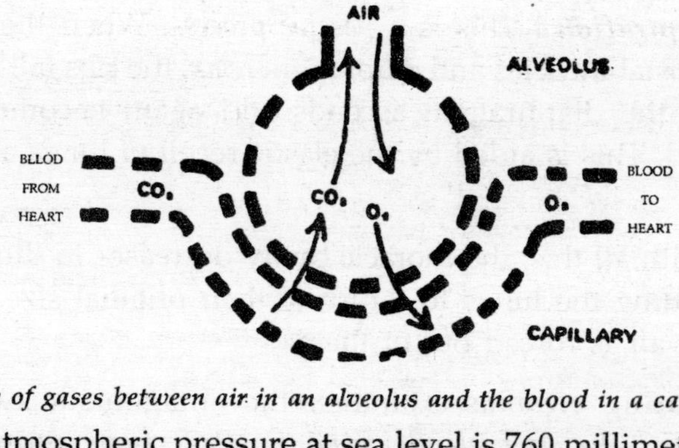

Interchange of gases between air in an alveolus and the blood in a cavillary.

The atmospheric pressure at sea level is 760 millimetres of mercury (mm Hg). This pressure is exerted by the mixture of following gases which make up the air :

Oxygen	20.84%	159.0 mm Hg
Nitrogen	78.62%	597.0
Carbon dioxide	0.04%	0.3
Water vapour	0.50%	3.7
	100%	760.0

Each one of these gases exerts a part of the total pressure depending upon its concentration in the mixture. The proportion of the total pressure provided by each gas is called its *partial pressure.* It is equal to the total pressure of the mixture divided by percentage of that gas in the mixture. For example: Partial pressure of oxygen = 20.84/100 of 760 = 159.0 mm Hg, Partial pressure of carbon dioxide = .04/100 of 760 = 0.3 mm Hg

Partial Pressures (mm Hg) Of Respiratory Gases.

Gas	*Inspired air*	*Alveolar air*	*Venous blood*	*Arterial blood*	*Expired air*
Oxygen	159	104	40	95	120
Carbon dioxide	0.3	40	46	40	27
Nitrogen	597	569	573	573	566

The gas pressure in the blood are usually expressed as gas "tensions", for example, the CO_2 "tension" (PCO_2) is the pressure of the dry gas (mm Hg) with which the dissolved carbonic acid in the blood is in equilibrium; similarly PO_2 (oxygen tension) is the pressure of the dry gas with which the dissolved oxygen in the blood is in equilibrium.

As gases diffuse down the concentration gradient, oxygen diffuses from the alveoli into the blood in the venous capillaries of lungs and carbon dioxide from the blood to the alveoli. This exchange of gases during external respiration is illustrated by the following :

I Oxygen tension in alveolar air =100 mm Hg

Oxygen tension in venous blood = 40 mm Hg

A pressure difference of 60 mm Hg serves to drive oxygen from the alveoli of the lung into the blood.

II CO_2 tension in alveolar air = 40 mm Hg

CO_2 tension in venous blood 46 mm Hg

A relatively small difference of 6 mm Hg is sufficient to drive CO_2 from the blood into the lung. This small difference in pressure is adequate because of the rapidity of the diffusion of CO_2 through the alveolar membrane.

By this mechanism carbon dioxide is excreted from the body and oxygen is absorbed and transported round the body in combination with haemoglobin in the erythrocytes. The blood therefore, now becomes arterial. It has an oxygen tension of about 100 mm Hg and a CO_2 tension of 40 mm Hg.

These gases are dissolved in the blood in simple physical solution. Exchange of gases between the blood and tissue cells during internal respiration takes place again in accordance with the usual physical laws of diffusion. Here, the concentration of O_2 is higher in the blood than in the tissue cells and concentration of CO_2 in greater in the tissue cells than in the blood. Oxygen, therefore, diffuses from the blood into the tissue cells, and CO_2 from tissue cells into blood, at the same time through the interstial (tissue) fluid. Oxygen, thus received by the tissue cells is utilized in oxidative metabolism producing energy and CO_2. This phenomenon maintains oxygen gradient between the blood and tissue cells and the CO_2 gradient between tissue cells and the blood.

The Transport of Oxygen by the Blood : Oxygen is transported in the blood by a respiratory pigment *haemoglobin,* which is a conjugated protein the *chmoprotein.* Haemoglobin consists of a large protein molecule, the *globin* consisting of 4 polypeptide chains (two α and two β chains), to each of which is attached a prosthetic *haem* group.

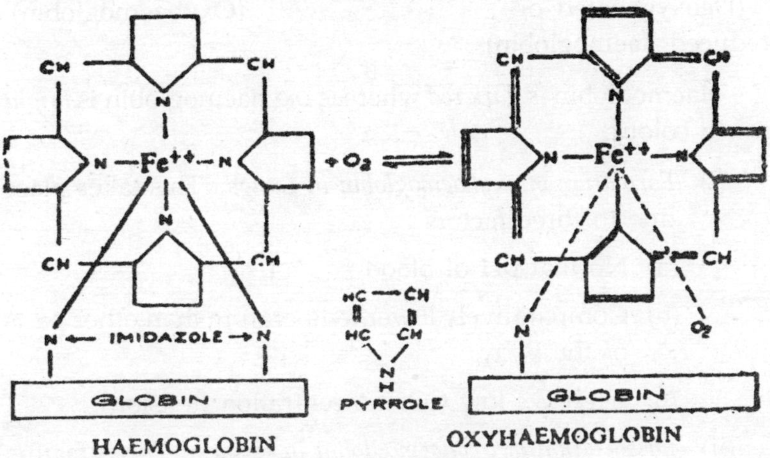

Structure and oxygenation of haemoglobin.

Haem is based on a structure known as a *porphyrin ring* which includes 4 *pyrrol groups* around a central ferrous iron Fe^{2+}. The iron is joined by four of its coordination bonds to N atoms of the porphyrin and by the two bands to *imidazole* N contained in histidine residues within the protein globin (coordination number of iron is 6).

The transport of oxygen by the blood from the lungs to the tissues is due mainly to the ability of haemoglobin to combine *reversibly* with oxygen. When haemoglobin is oxygenated, one of the imidazole N bonds is reversibly displaced by O_2. The reaction is thus a reversible "oxygenation" and not an oxidation. Similarly, reduction is not involved when O_2 is given up. Iron serves as a point of lose physical attachment for O_2 and remains in ferrous (Fe^{2+}) state throughout.

Since one molecule of haemoglobin has 4 haem groups, and each of them is capable of taking up one molecule of O_2; therefore, a mammalian haemoglobin should be depicted as Hb_4O_8.

$$\underset{\text{(Deoxygenated or reduced haemoglobin)}}{Hb} + 4O_2 \underset{\text{Tissues}}{\overset{\text{Lungs}}{\rightleftharpoons}} \underset{\text{(Oxyhaemoglobin)}}{-Hb_4O_8}$$

Haemoglobin is *dark red* whereas oxyhaemoglobin is *bright red* in colour.

(i) *Formation of oxyhaemoglobin in Lungs* : This takes place due to three factors :

(a) Normal pH of blood *i.e.*, 7.4.

(b) Comparatively lower temperature than other parts of the body.

(c) High O_2, low CO_2 concentration (tension).

(ii) *Dissociation of oxyhaemoglobin in tissues* : Three factors account for this process.

(a) Slightly acidic pH due to increase in H^+ concentration: CO_2 produced in the course of metabolism enters the blood where it is hydrated to from carbonic acid, which ionizes to form H^+ and HCO_3.

$$CO_2 + H_2O \rightleftharpoons H_2CO_3 \rightleftharpoons H^+ + HCO_3.$$

(b) Rise in temperature.

(c) High CO_2, low O_2 tension.

Oxygen-Haemoglobin Dissociation Curve : The oxygen-haemoglobin dissociation curve shown in fig. demonstrates a progressive increase in the percentage of haemoglobin bound with oxygen as the blood PO_2 increases, which is called the *per cent saturation of the haemoglobin.* As the blood leaving the lungs and entering the systemic arteries usually has a PO_2 of about 95 mm Hg, it can be observed from the dissociation curve that the *usual oxygen saturation of systemic arterial blood is about 97 per cent.* On the other hand, in normal venous blood returning from the peripheral tissues, the PO_2 is about 40 mm Hg, and *the saturation of the haemoglobin is about 75 percent.* Haemoglobin in the blood automatically delivers O_2 to the tissues at a pressure that is held tightly between about 15 and 45 mm Hg. The curve drawn with CO_2 at a tension of 40 mm Hg is considered as representative of the normal physiologic condition.

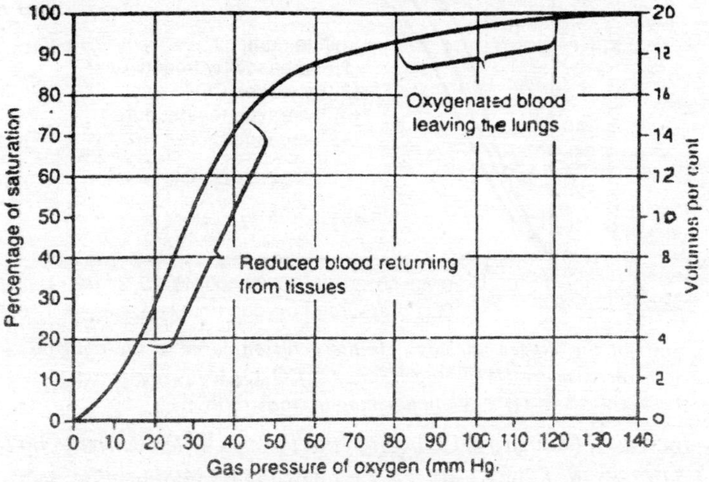

Oxygen-haemoglobin dissociation Curve.

Factors that Shift the Oxygen-Haemoglobin Dissociation Curve : The oxygen-haemoglobin curve of is for normal, average blood. However, a number of factors can displace the dissociation curve in one direction or the other in the manner shown in Fig. This figure shows that when the blood becomes slightly acidic, with the pH decreasing from the normal value of 7.4 to 7.2, the oxygen-haemoglobin dissociation curve shifts, on average about 15 percent to the right. Conversely, an increase in the pH from the normal 7.4 to 7.6 shifts the curve a similar amount to the left.

In addition to pH changes, several other factors are known to shift the curve. Three of these, all of which shift the curve to the *right, are* (1) increased carbon dioxide concentration (2) increased blood temperature, and (3) increased 2

3-diphosphoglhycerate (DPG), a metabolically important phosphate compound present in the blood but in different concentration under different metabolic conditions

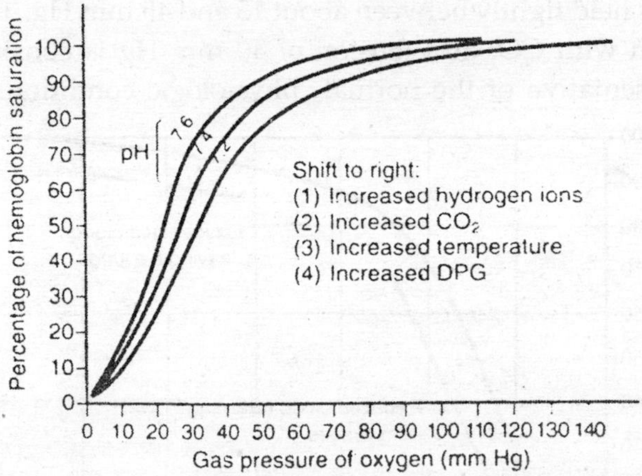

Shift of the oxygen-haemoglobin dissociation curve to the right by increases in (1) hydrogen ions (2) CO_2 (3) temperature, or (4) 2, 3- diphosphoglycerade (DPG).

Increased Delivery of Oxygen to the Tissues When Carbon Dioxide and Hydrogen Ions Shift the Oxygen-Haemoglobin Dissociation Curve—The Bohr Effect. A shift of the oxygen-heamoglobin dissociation curve in response to changes in the blood carbon dioxide and hydrogen ions has a significant effect in enhancing oxygenation of the blood in the lungs and then again in enhancing release of oxygen from the blood in the tissues. This is called the Bohr effect. Which can be explained as follows :

As the blood passes through the lungs, carbondioxide diffuses from the blood into the alveoli. This reduces the blood PCO_2, and decreases the hydrogen ion concentration because of the resulting decrease in blood carbonic acid. Both these effect shift the oxygen-haemoglobin dissociation curve to the left and upward. Therefore, the quantity of oxygen that binds with the heamoglobin at any given alveolar PO_2 now becomes considerably increased, thus allowing greater oxygen transport to the tissues.

Then, when the blood reaches the tissue capillaries, exactly the opposite effects occur. Carbon dioxide entering the blood from the tissues shifts the curve right-ward, which displaces oxygen from the heamoglobin and therefore delivers increased amounts of oxygen to the tissues at a higher PO_2 than would otherwise occur.

Effect of Increased Diphosphoglycerate : The normal DPG in the blood keeps the oxygen-haemoglobin dissociation curve shifted slightly to the right all the time. In hypoxic conditions that last longer than a few hours, the quantity of DPG in the blood increases considerably, thus shifting the oxygen-haemoglobin dissociation curve even farther to the right. This causes oxygen to be released to the tissues at as much as 10 mm Hg higher tissue oxygen pressure than would be the case without this increased DPG. Therefore, under some condition, the DPG mechanism can be important for adaptation to hypoxia, especially to hypoxia caused by poor tissue blood flow.

Respiratory Pigments : In animals, blood usually contains a *respiratory pigment* with a special affinity for oxygen, which increases the oxygen capacity of the blood many times. Four respiratory pigments are found in animals: *haemocyanin, haemerythrin, Chlorocuorin and haemoglobin.* The properties and occurrence of these pigments are summarised in Table.

The Transport of Co_2 in the Blood

CO_2 produced from metabolism in cells diffuses into the blood within the capillaries. It is transported in 2 forms :

(A) The Dissolved State

About 7% of CO_2 combines chemically with water of plasma forming carbonic acid

$$CO_2 + H_2O \rightleftharpoons H_2CO_3$$

Any increase in its concentration causes the dissociation of H_2CO_3 into a hydrogen ion and a bicarbonate ion

$$H_2CO_3 \rightleftharpoons H^+ + HCO_3$$

Transport Mechanism for oxygen within the Blood.

Pigment	*Metal component*	*Colour on deoxygenation*	*Colour on oxygenation*	*Animal Group*	*Occurrence in blood*
Haemocyanin	Copper	Colourless	Blue	Some snails, squids and Crustacea	Plasma
Haemoglobin	Iron	Colourless	Brown	Some polychaeate annelids sipunculids and branchiopods	Corpuscles
Chlorocruorin	Iron	Green	Reddish	Some polychaetes	Plasma
				Some molluscs	Plasma
Haemoglobin, is a conjugated protein. *cliromo-proteins*, which consists of a pigment *haem* (Iron porphyion). attached to the protein, *globin*. It has the ability to combine reversibly with oxygen Mol. Wl. 6. 450.	Iron (Ferrous state)	Purple-red	Orange-red	Some annelids	Plasma or Corpuscles
				Nemertine worms	Corpuscles
				Sea-cucumbes	Corpuscles
				Fishes	Corpuscles
				Amphibia	Corpuscles
				Reptiles	Corpuscles
				Birds	Corpuscles
				Mammals	Corpuscles

As Chemical Compounds

(1) *Carbamino Compounds :* About 23% of the total blood CO_2 is transported in combination with haemoglobin.

$$\underset{\text{(Free amino-group of haemoglobin)}}{CO_2 + Hgb.\ NH_2} \longrightarrow \underset{\text{(Carbamino haemoglobin)}}{Hgb.\ CO_2}$$

(2) *Bicarbonates :* About 70% of CO_2 produced as a result of oxidative metaolism in the body is carried as bicarbonates in the blood.

(i) *In the Erythrocytes :* CO_2 from the plasma enters in red blood cells and combines with water within the cells. The reaction is catalyzed by a zinc containing enzyme *carbonic anhydrase,* which is associated with the haemoglobin in red-cells (never in the plasma). *Carbonic anhydrase* catalyses the reaction between CO_2 and water, accelerating its reaction rate about 5000 fold. Instead of requiring many seconds or minutes to occur, as in the case of plasma, the reaction occurs so rapidly in the red blood cells that it reaches almost complete equilibrium within a very small fraction of a second. This results in the formation of carbonic acid which soon dissociates:

$$CO_2 + H_2O \xrightarrow{\text{Carbonic anhydrase}} H_2CO_3 \rightleftharpoons H^+ + HCO_3^-$$

A small amount of HCO_3^- is transported in the erythrocytes.

In the red-cells, haemoglobin remains combined with potassium and forms bicarbonate in the following way :

$$H^+ + HCO^-_3 + \underset{\text{(Potassium Haemoglobinate)}}{K.Hb} \longrightarrow KHCO_3^- + \underset{\text{(Haemoglobin acid)}}{H.Hb}$$

The hydrogen ions mostly combine with Hb for keeping the pH of blood (7.4) in steady state, because Hb is a powerful acid-base buffer.

(ii) *In the Plasma* : Most of the HCO_3^- diffuse into the plasma. Plasma transports CO_2 by three different mechanisms.

(a) *By phosphate Buffers* : Alkaline phosphates combine with carbonic acid in the plasma and form sodium bicarbonates.

$$\underset{\text{(Di-sodium hydrogen phosphate)}}{Na_2HPO_4} + H_2CO_3 \longrightarrow NaHCO_3 + \underset{\text{(Sodium dihydrogen phosphate)}}{NaH_2PO_4}$$

(b) By *plasma Proteins* : Proteins of all plasma mostly remain combined with Na and form bicarbonates in the following manner :

$$\underset{\text{(Sodium proteinate)}}{NaPr} + H_2CO_3 \longrightarrow \underset{\text{(Proteinic acid)}}{H.Pr} + NaHCO_3$$

(c) *By Chloride shift or Hamburger's Phenomenon* : CO_2 released from tissue metabolism enters the red cell (erythrocyte).

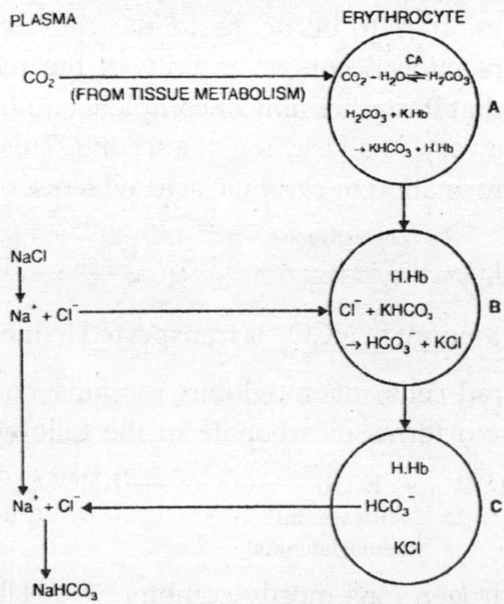

Under normal conditions the walls of the red cells are permeable to anions (Cl, HCO_3) but virtually impermeable to cations (Na^+, K^+) and to the large haemoglobin molecule. Under these circumstances chloride ions diffuse into the erythrocyte from the plasma and react with intracellular potassium bicarbonate. The bicarbonate ions which are thus released, now diffuse out of the erythrocyte into the plasma, whereas chloride ions are neutralized by the potassium in the erythrocyte. In the plasma, bicarbonate ions combine with sodium ions forming sodium bicarbonate ($NaHCO_3$). This phenomenon is known as chloride shift which permits the carriage of additional CO_2 as sodium bicarbonate by plasma.

All the reaction of chloride shift are reversible. At the lung, when the blood becomes arterial, chloride shifts back into the Plasma, thus, liberating intracellular potassium to buffer the newly formed oxyhaemoglobin and, in the plasma, neutralizing the sodium liberated by the removal of CO_2 during respiration.

***Anion Channel*:** The red-cell membrane contains an anion channel that plays a key role in the transport of CO_2 by blood and in the buffering of red cell pH. CO_2 entering red cells in tissue capillaries is converted into carbonic acid, H_2CO_3, by the action of carbonic anyhydrase. Carbonic acid then dissociates into H^+ and HCO_3^-. The released proton is taken up by deoxyhaemoglobin as part of the Bohr effect. Bicarbonate leaves the red cell through the anion channel in the exchange for Cl^-. The anion channel is very abundant, comprising about 1/3 of the total membrane protein. Each red cell contains some 10^6 copies of this integral membrane protein.

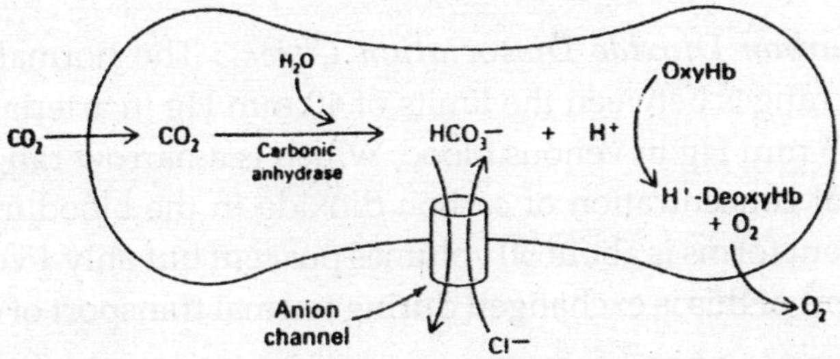

Anion channels in erythrocytes play an important role in the transport of CO² by enabling HCO³⁻ to be exchanged for Cl⁻

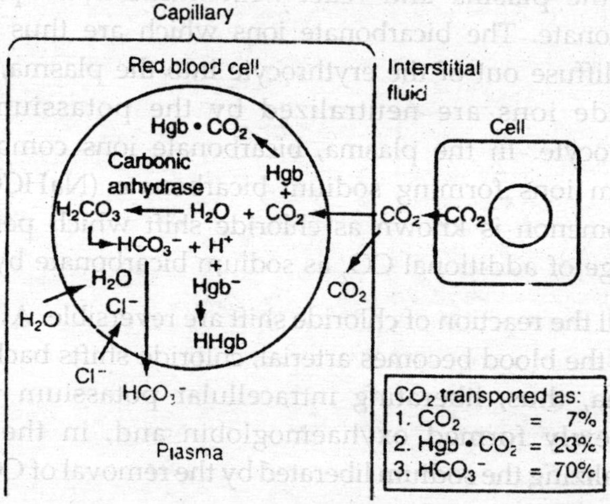

Release of Co_2 at the Respiratory Surface : Carbonic acid, bicarbonates of sodium and potassium and car-bamino-compounds are carried to the lungs where they break down under the influence of various factors and liberate free CO_2.

(i) $H_2CO_3 \xrightarrow[\text{anhydrase}]{\text{carbonic}} H_2O + CO_2\uparrow$

(ii) $2KHCO_3 \xrightarrow{\text{Acidity of Oxyhaemoglobin}} K_2CO_3 + H_2O + CO_2\uparrow$

(iii) $2NaHCO_3 \xrightarrow{\text{Acidity of Oxyhaemoglobin}} Na_2CO_3 + H_2O + CO_2\uparrow$

(iv) $Hb.NH.\ COOH \xrightarrow[\text{Concentration of } O_2]{\text{Reduction due to higher}} Hb.\ NH_2 + CO_2\uparrow$

Carbon Dioxide Dissociation Curve : The normal blood PCO_2 ranges between the limits of 40 mm Hg in arterial blood and 45 mm Hg in venous blood, which is a narrow range. The normal concentration of carbon dioxide in the blood in all its different forms is about 50 volumes per cent but only 4 volumes per cent of this is exchanged during normal transport of carbon dioxide from the tissue; to the lungs. That is, the concentration rises to about 52 volumes per cent as the blood passes through the tissues and falls to about 48 volumes per cent as it passes through the lungs.

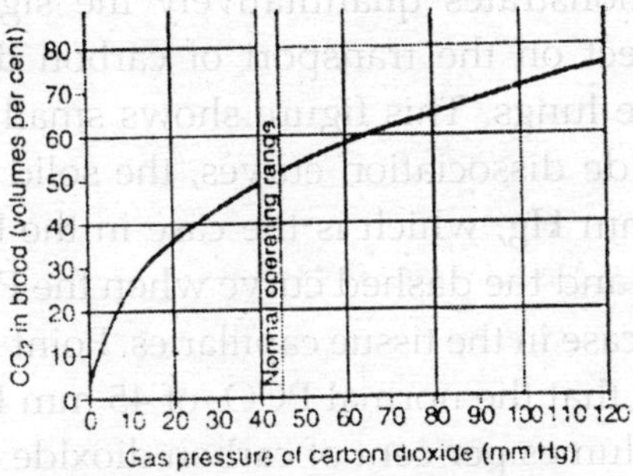

Carbon Dioxide Dissociation Curve

***The Haldane Effect*:** When oxygen binds with haemoglobin, carbon dioxide is released to increase CO_2 tranport. This effect called *Haldane* effect is quantitatively far more important in promoting carbon dioxide transport than is the Bohr effect in promoting oxygen transport.

The Haldane effect results from the simple fact that combination of oxygen with haemoglobin in the lungs causes the heamoglobin to become a stronger acid. This in turn displaces carbon dioxide from the blood and into the alveoli.

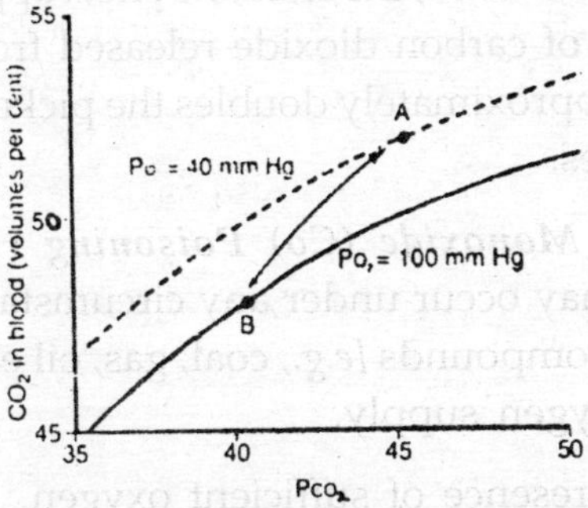

Portions of the carbon dioxide dissociation curves when the PO_2 is 100 mm Hg. The arrow represents the Haldane effect on the transport of carbon dioxide as discussed in the text.

In the tissue capillaries, the Haldane effect causes increased pickup of carbon dioxide because of oxygen removal from the hemoglobin and in the lungs, it causes increased release of carbon dioxide because of oxygen pickup by the haeamoglobin.

Fig. demonstrates quantitatively the significance of the Haldane effect on the transport of carbon dioxide from the tissues to the lungs. This figure shows small portions of two carbon dioxide dissociation curves, the solid curve when the PO_2 is 100 mm Hg, which is the case in the blood capillaries of the lungs, and the dashed curve when the PO_2 is 40 mm Hg, which is the case in the tissue capillaries. Point A on the dashed; curve shows that the normal PCO_2 of 45 mm Hg in the tissues causes 52 volumes per cent of carbon dioxide to combine with the blood. On entering the lungs, the PCO_2 falls to 40 mm Hg while the PO_2 rises to 100 mm Hg. If the carbon dioxide dissociation curve did not shift because of the Haldane effect, the carbon dioxide content of the blood would fall only to 50 volumes per cent, which would be a loss of only 2 volumes per cent of carbon dioxide. However, the increase in PO_2 in the lungs decreases the carbon dioxide dissociation curve from the dashed curve to the solid curve of the Fig. so that the carbon dioxide content falls to 48 volumes per cent (point B).

This represents an additional 2 volumes per cent loss of carbon dioxide. Thus, the Haldane effect approximately doubles the amount of carbon dioxide released from the blood in the lungs and approximately doubles the pickup of carbon dioxide in the tissues.

Carbon Monoxide (Co) Poisoning : Carbon monoxide poisoning may occur under any circumstances where carbon-containing compounds [*e.g.*, coal, gas, oil etc.) are burned with deficient oxygen supply.

In the presence of sufficient oxygen,

$$C + O_2 \longrightarrow CO_2$$

But if oxygen is deficient,

$$2C + O_2 \longrightarrow 2CO_2$$

The carbon monoxide poisoning cases occur when a person sleeps in a closed room with a burning coal-stove on a winter night, or when a person remains inside a closed motor garage for a long time with a running automobile engine. As there is insufficient oxygen in such a closed environment, CO is produced from the burning stove or from the running motor engine (exhaust gas) and proves fatal to the inhabitant.

In CO poisoning, haemoglobin combines with carbon monoxide even more readily than with oxygen (230 times as fast), to form a cherry-red compound *carboxyhaemoglobin*. This reduces the amount of haemoglobin available to carry oxygen. Moreover, CO is not easily released by the haemoglobin. When the CO in the inspired air is as low as 0.02%, headache and nausea will occur. If the CO concentration is only 1/230 that of oxygen in the air (approximately 0.1% CO), unconsciousness will occur in 1 hour and death in 4 hours. The skin of the victim of CO poisoning develops a cherry-red colour because this is the colour which haemoglobin attains when it combines with CO.

The treatment of CO poisoning is to bring about the break-down of HCO, replacing CO with oxygen. The application of O_2 and prompt removal of the victim from the fouled atmospherically save his life. To stimulate the respiratory centre of the victim to greater effort, 6%, CO_2 is added to the O_2.

Subdivisions of Air in the Lungs

Pulmonary Volumes/Aerodynamics : It is customary to divide the volume of air entering the lungs into various sub-division. These are as follows :

(i) *Tidal Volume* : The air normally inspired and expired at one breath is called the tidal volume. It amounts to about 500 ml.

(ii) *Complemental Air or Inspiratory Reserve Volume (IRV) :* After a normal inspiration, it is still possible to send more air into the lungs by forcible inhalation. The air inhaled in this manner is called the complemental air (3000 ml).

(iii) *Supplemental Air or Expiratory Reserve Volume (ERV) :* After a normal expiration, it is possible to expel more air by breathing out forcibly. The additional air pushed out like this is called the supplemental air (1100 ml).

(iv) *Vital Capacity (VC) :* The amount of air breathed in and out with the greatest possible effort is said to be the vital capacity of the lungs. It is equal to the sum of the tidal complemental and supplemental air (0.5 + 3.0 + 1.1 = 4.6 L).

(v) *Residual Air or Residual Volume (RV) :* Some air is left in the lungs after the most forceful expiration. This air is known as the residual air (1200 ml) and can be released only by opening the thoracic cavity.

(vi) *Dead Air Space :* The only place in the lungs where gaseous exchange takes place is in the alveoli. A part of the inspired air is left in the trachea and bronchial tree, the so-called dead air space, where no gaseous exchange occurs. This dead space air (about 150 c.c) is expelled at the next expiration. The expired air, therefore, contains fresh air from the dead space and foul air from the lungs.

	Volume (L)	
Tidal volume	0.5	quite breathing
Inspiratory reserve volume	3.0	forced inspiration
Expiratory reserve volume	1.1	forced expiration
Vital capacity	4.6	
Residual volume	1.2	always left in lung
Total lung capacity (TLC)	5.8	

(vii) *Total Lung Capacity (TLC)* : It can be seen from the above table that the total lung capacity (TLC) is the sum of the vital capacity (VC) and the residual volume (RV), *i.e.*,

$$TLC = VC + RV$$

(viii) *Pulmonary Ventillation* : The amount of air inspired per minute is called pulmonary ventillation. It is normally about 6L (500 mL/breath × 12 breaths/min).

(ix) *Maximal Voluntary Ventilation (MW)* : It is the *maximal breathing capacity* or the largest volume of gas that can be moved into and out of the lungs in 1 minute by voluntary effort. The normal MW is 125—170 L/min.

Respiratory Quotient : The amount of oxygen consumed and the amount of carbon dioxide given off during respiration, are often used as a measure of the extent of oxidation of fat, carbohydrate and protein by a particular animal. For example, in the oxidation of a hexose sugar the quantity of carbon dioxide formed is equal to the quantity of oxygen utilized :

$$C_6H_{12}O_6 + 6CO_2 \rightleftharpoons 6O_2 + 6H_2O$$

Here the ratio is $\xrightarrow{\frac{\text{volume of } CO_2 \text{ formed}}{\text{volume of } CO_2 \text{ utilized}}}$ equal to unity.

This ratio, which is known as the *respiratory quotient* (RQ) is different for each of the three major food stuffs. In the case of a fat such as triolein, the equation for its oxidation is as follows :

$$C_{57}H_{104}O_6 + 80O_2 \longrightarrow 57CO_2 + 52H_2O$$

In this case, the RQ is therefore equal to 57/80, or 0.71. Protein is not completely oxidised in the tissues, but its RQ can be determined by indirect means, and shown to be between 0.8 and 0.82.

Control of Respiration

The control of respiration is partly *nervous* and partly *chemical*. The two modes are closely linked to each other

Nervous Control : Located bilaterally in the medulla oblongata and pons varolii of the brain is a collection of highly specialized and widely dispersed groups of nerve cells collectively called the respiratory centre. From this centre, nerve impulses pass out in the *phernic nerves* to the *diaphragm* and *interocostal nerves* to the *intercostal muscles*. These impulses stimulate the muscles concerned to contract and increase the capacity of the throacic cavity with the result that inspiration occurs.

Chemical Control : The main chemical factors which influence the respiratory centre are *the amount of carbon dioxide (carbonic acid) and oxygen in solution in the blood*. In the carotid body, where the common carotid artery divides, and in the wall of the arch of the aorta there are cells which are sensitive to carbon dioxide excess and *oxygen lack. When these cells are stimulated, impulses pass in branches of the vagus and glossopharyngeal nerves to the respiratory centre* in the *medulla oblongata*.

The accumulation of carbon dioxide in solution in the blood is the most important stimulus to respiration. If the amount of carbon dioxide rises in the blood. For example during muscular exercise, the respiratory centre is stimulated. It therefore sends out impulses to the respiratory muscles to produce deeper and quicker breathing so that the carbon dioxide can be given out more rapidly by the lungs. In this way, the respriatory centre controls the rate and depth of breathing.

Physiology of Respiration : All living animals require some energy to perform their activities. The most common source of energy is oxidation of food by the available oxygen that is respired by the animal. *Thus, respiration can be defined as a*

catabolic process in which the respired oxygen is used in the oxidation of food resulting in the release of energy. Respiration, therefore, includes both breathing and oxidation. However, the process of taking in oxygen and giving off carbon dioxide is also, sometimes, called respiration. But this is only the breathing process and does not take into consideration the oxidation process, without which the main purpose of respiration cannot be achieved.

In respiration, the animals make use either of the oxygen of the atmospheric air (wherein it is about 21% of the air), or of the oxygen dissolved in the water.

There are two kinds of respiration.

Direct Respiration : Direct respiration is the exchange of oxygen and carbon dioxide between the cells of the organism and the surrounding medium. For example, small aquatic animals take up oxygen by diffusion from the small amount of this gas dissolved in water.

Indirect Respiration : Indirect respiration means the exchange of gases through some specialized structure (often called the respiratory organ) and involves two phases, external and internal respiration.

(a) *External Respiration* : External respiration includes the exchange of gases between the external environment and the blood of the respiratory organ.

(b) *Internal Respiration* : Internal respiration, or cellular respirations involves the exchange of gases between the bloodstream and the body cells. It is also known as *tissue respiration* and the oxygen consumption (microlitres O_2) per milligram of tissue per hour is denoted as Qo_2. Since the respiratory rate of each tissue depends on its demand for ATP, there are considerable variations in Qo_2 of various tissues of the body.

Internal respiration fulfills the ultimate object of the respiratory system in that it provides oxygen to cells and tissues of the body and rids them of carbon dioxide.

Respiration in Invertebrates

Invertebrates depend upon the following modes of respiration:

(a) *Direct Diffusion* : In protozoa, porifera, coelenterata and many worms oxygen is obtained by diffusion directly from the surrounding medium (water).

(b) *Integument:* In most annelids, gaseous exchange is made through a moist skin provided with blood vessels or other derivatives of the integument like parapodia which are highly vascular. A few crustaceans and molluscs also use their skin for respiration. In some molluscs *(e.g., Pila)* mantle is used and in *Palaemon,* there are three pairs of epipodites which function as respiratory organs. Some echinoderms make use of the finger-like dermal branchiae (papulae), as well as the tube feet.

(c) *Gills*: Gills are common in aquatic invertebrates. Some are simple in structure (as in marine annelids) and others are complicated (as in molluscs and arthropods). Numerous aquatic insects have gills through which oxygen passes directly to the tracheae.

(d) *Tracheae :* These are found in onychophora, insecta, myriapoda, and in some spiders and isopods. Tracheae usually open by spiracles that can be closed or opened according to the need.

(e) *Lungs* : The so-called lungs of invertebrates are a variety of structures. In land snails, mantle is modified as lung. Scorpions and spiders have booklungs opening on the ventral surface through small slit-like apertures called the stigmata.

(f) *Respiratory Trees* : Respiratory trees of holothurians are filled with sea water by cloacal pumping and contractions of the trees themselves, and gas exchanges with the coelomic fluid.

Respiration in Vertebrates

In vertebrates, the respiration is carried out by any of the following three devices.

(a) *Direct Surfaces such as the Skin and Pharyngeal Regions*: Many amphibians make use of their skin in respiration. Some salamanders depend on the skin and highly vascular pharyngeal region. Some fishes and eels also absorb oxygen through the skin.

(b) *Gills* : Many aquatic vertebrates have gills, either external (protochordates and some larval amphibians) or internal (cyclostomes and fishes).

(c) *Lungs* : Land vertebrates have the lungs for respiration. In amphibians, the lungs are simple ; but in higher vertebrates there is a gradual increase in the internal respiratory surface of the lungs, thus making them more efficient organs of respiration.

Respiration in Humans and other Mammals

Respiratory System : It consists of certain air channels and the lungs. The respiratory passage consists of nostrils (external nares), nasal chamber, posterior nares, pharynx, epiglottis, glottis, larynx or voice box, trachea or windpipe and two bronchi (Fig.). Within the lungs each bronchus divides and subdivides into smaller tubes (bronchioles) which lead to the *air sacs* and alveoli.

The human lungs are made up of some 7.5×10^8 alveoli, the total surface of which is about 100 square metres, or about fifty times that of the skin surface. Thus, each lung consists of its bronchial tree with its many air sacs and alveoli units together with other associated structures such as the blood vessels, nerves and pleura. All are supported and attached to one another by the connective tissue. Externally they are covered by the *visceral pleura* (a thin layer of smooth epithelium) which is also surrounded by another similar layer, the *parietal pleura* forming the inner lining of the wall of the chest. A small amount of pleural fluid for lubrication is found in the *pleural cavity.* The normal pressure of the fluid in the intrapleural space is -10 to -15 mm Hg. This unusually negative pressure in the intrathoracic space is caused by the continuous reabsorption of fluid into the pleural capillaries by a special lymph-pump system. The *diaphragm,* a dome-shaped partition forms the floor of the chest cavity and separates it from the abdomen.

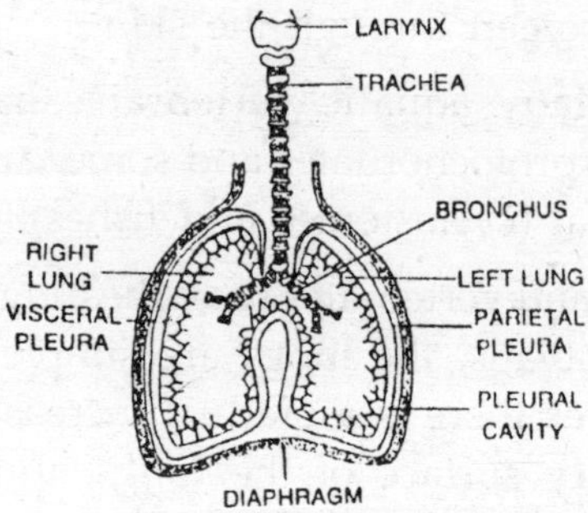

Lungs and diaphragm in man.

Mechanism of Breathing

Breathing is a mechanical process and is completed in two phases, inspiration and expiration. In *inspiration,* the ribs are elevated and the diaphragm contracted and flattened, the chest cavity is enlarged. This increase in the volume of the chest cavity and lungs causes the air pressure in the lungs to fall below the atmospheric pressure, and air passes through the air passageways to the lungs to equalize the pressure. In *expiration,* the ribs and diaphragm return to their original position, the chest cavity volume is decreased. The distended elastic lungs then contract and the air is forced out. Changes in the intrapleural pressure are responsible for air entering and leaving the lungs. In inspiration, expansion of the thorax, aided by descent of the diaphragm, decreases intrathoracic pressure from -4 to -10 mm Hg, and air pushes into the lungs. Thus, in inspiration the lungs are extending passively in response to the various mechanisms that result in an increase in thoracic volume. In expiration, the size of the thorax is decreased, the intrathoracic pressure is raised to-2 mm Hg, and air is forced out of the lungs (Fig.).

It should be noted that the diaphragm is the main muscle of inspiration. If the diaphragm descends 10 mm, it will increase the thoracic cavity volume by 250 ml. When it relaxes, passive expiration results. The contraction and relaxation of the diaphragm is controlled by the *phrenic nerves* arising in the neck from the third, fourth and fifth cervical nerves and passing down through the thorax to the diaphragm. Besides diaphragm, the *external intercostals* are the muscles mainly responsible for the elevation of the ribs in inspiration. They are inserted between two neighbouring ribs, sloping forward and downward and their relaxation brings about passive expiration. The *internal intercostals* form a deeper layer of muscle between the ribs with the fibres running in the opposite direction, from above downward and backward. On contraction, these muscles depress the ribs aiding in expiration during very deep breathing (active expiration).

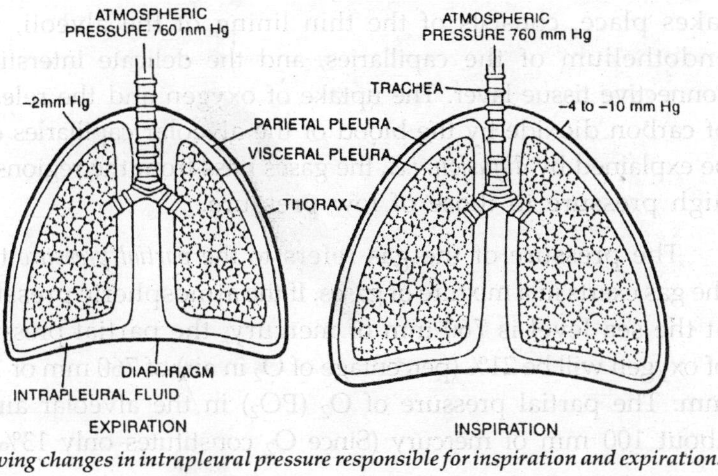

Showing changes in intrapleural pressure responsible for inspiration and expiration.

Generally, the following terms are used to describe respiration.

(a) *Eupnea* : normal respiration

(b) *Hypernea* : increase in respiratory rate and depth

(c) *Dyspnea* : irregularities of respiration

(d) *Apnea* : cessation of respiration.

The normal rate of respiration in the adult is 14 breaths/ minute , but in children it may be up to 30/minute. In exercise it is further increased. Each inspiration admits about 350 ml of new air to mix with the 2500 ml of old air present in the lungs. The quantity of new air that enters the lungs per minute is known as the *minute volume,* which in the average adult is about 4900 ml (350 × 14).

Physiology of Respiration : It may be discussed under the following heads :

(a) Gaseous exchange in the lungs; (b) Transport of gases by the blood; and (c) Cellular (internal) respiration and oxidation.

Gaseous Exchange in the Lungs (external respiration) : The respiratory membrane through which gaseous exchange takes place, consists of the thin lining of the alveoli, the endothelium of the capillaries, and the delicate interstitial connective tissue layer. The uptake of oxygen and the release of carbon dioxide by the blood of the alveolar capillaries can be explained by *diffusion, i.e.,* the gases pass from the regions of high pressure to those of low pressure.

The pressure of the gas refers to the *partial pressure* that the gas exerts in a mixture of gases. If the atmospheric pressure at the sea level is 760 mm of mercury, the partial pressure of oxygen will be 21% (percentage of O_2 in air) of 760 mm or 159 mm. The partial pressure of O_2 (PO_2) in the alveolar air is about 100 mm of mercury (Since O_2 constitutes only 13% of alveolar air) compared with that of 159 mm of mercury for the O_2 in the atmosphere. The partial pressure of O_2 in the blood of the alveolar capillaries is considerably lower, being only about 40 mm. Accordingly, the oxygen diffuses from the alveolar air into the blood. By the time the blood leaves the capillaries, its O_2 pressure has been raised to about 100 mm of Hg, the same as alveolar O_2.

In a similar manner, the CO_2 in the blood of the lung capillaries has a higher concentration (46 mm) than it has in the lung alveoli (40 mm). This accounts for the diffusion of CO_2 out of the blood into the alveoli. By the time the blood leaves the lungs, its CO_2 pressure has been lowered to approximately 40 mm. Thus, in the diffusion of both these gases there is a marked *diffusion gradient* that determines the direction of their flow.

The exchange of gases in the lung according to their partial pressures (concentration gradient) may be shown as follows:

Lung alveoli	100 mm Hg	40 mm Hg
	↓ O_2	↑ CO_2
Lung capillaries	40	46

The large volume of *inert gases* (mainly nitrogen), that is inspired normally passes into the blood with oxygen and carbon dioxide. As these gases are *chemically inactive,* they do not participate in the metabolic reactions of the body.

Transport of Gases by the Blood

Transport of Oxygen : The solubility of O_2 in water is rather low, but this shortcoming is overcome by the fact that the O_2 is bound to carrier substances in the blood. In mammalian blood, the O_2 carrier respiratory pigment is *haemoglobin.* The maximum amount of O_2 which the normal human blood can absorb is 20 ml per 100 ml of blood. When O_2 passes from the lung alveoli into the lung capillaries, it diffuses into the blood and unites with haemoglobin to form *oxyhaemoglobin.*

$$Hb_4 + 4O_2 \rightleftharpoons Hb_4O_8 \text{ or } Hb_4(O_2)_4 \text{ (Oxyhaemoglobin)}$$

As each haemoglobin molecule possesses 4 atoms of iron, it can transport 4 molecules of oxygen.

Under the normal conditions the arterial blood which has been exposed to the alveoli of the lungs is not quite completely oxygenated. With an O_2 tension of 100 mm of Hg, it is usually 98% saturated and therefore, contains 19.6 ml of O_2 (combined to haemoglobin) per 100 ml of blood. In addition to this there is about 0.2 to 0.3 ml of O_2 which is dissolved in the plasma.

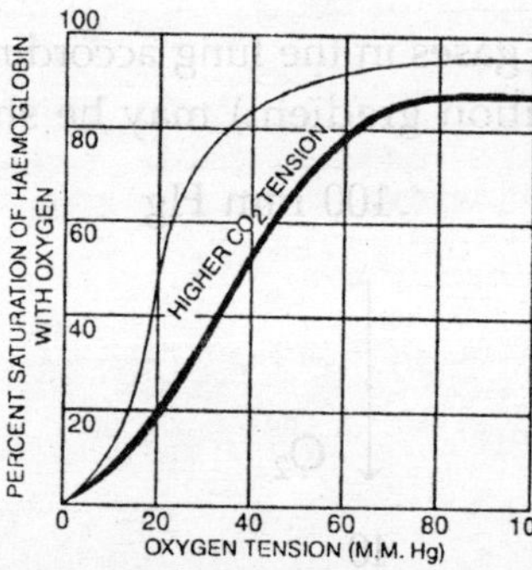

The oxygen-haemoglobin dissociation curve. Increased CO_2 concentration shifts the curve to the right.

The arterial blood and the alveoli have the same O_2 pressure (100 mm of Hg). But in the cells and tissues of the body the O_2 tension is considerably low (1 to 40 mm of Hg). The O_2 is accordingly liberated from the oxyhaemoglobin and diffuses out from the blood through the thin capillary walls into the cells. This is made possible by the important fact that the combination between O_2 and haemoglobin in the red blood cells to form oxyhaemoglobin is a *reversible* one.

The liberation of O_2 from the blood to the tissues is just as important as its rapid absorption by the blood during its passage through the lungs.

$$\underset{\text{Oxyhaemoglobin}}{Hb_4O_8} \longrightarrow \underset{\text{reduced haemoglobin}}{4Hb + 4O_2}$$

The reduced haemoglobin is further transported via blood to the lungs and the cycle is repeated, while the O_2 that has diffused into the cells is utilized in the oxidation of carbohydrates, resulting in the release of CO_2 and energy.

Oxygen-Haemoglobin Dissociation Curve

At high O_2 pressure, the haemoglobin combines with O_2 to form oxyhemoglobin. Each iron atom can bind one O_2 molecule, and when all sites are occupied, the haemoglobin cannot take on any more, since it is fully loaded or *saturated*. At low O_2 pressure, O_2 dissociates from its binding, and the haemoglobin will eventually give up all its O_2. At any given O_2 concentration, there is a definite proportion between the amount of haemoglobin and oxyhaemoglobin. In this way the *actual relationship between the partial pressure of* O_2 *and the degree of saturation of the haemoglobin with* O_2 *is shown by the remarkable oxygen haemoglobin dissociation curve* (Fig.). The curve shows that the haemoglobin is almost completely oxygenated (saturated) with O_2 at the O_2 partial pressure of about 100 mm Hg. At higher O_2 pressure, no more O_2 is taken up by the haemoglobin. At lower O_2 pressure, O_2 is given off and at 30 mm O_2 pressure, half the haemoglobin is present as oxyhaemoglobin. As the O_2 pressure decreases further, more O_2 is given off, and all is given up when the O_2 pressure reaches zero. Thus, the degree of haemoglobin saturation is lowered with the fall in the partial pressure of O_2. In the passage of blood through the tissues where the O_2 tension is low, rapid dissociation of oxyhaemoglobin occurs, yielding a comparatively large quantity of O_2 to the surrounding tissues and cells where it is most needed.

Lungs (P_{o2} = 100 mm Hg)

$$Hb + O_2 \longrightarrow HbO_2 \text{ (oxyhaemoglobin)}$$

Tissues (P_{o2} = 30 to 40 mm Hg)

$$HbO_2 \longrightarrow Hb + O_2$$

During exercise, there is a fall in tissue PO_2, an increase in Pco_2, and an increase in pH, local temperature and 2, 3-diphosphoglycerate concentration. All these factors promote the release of oxygen from oxyhaemoglobin (shifting the oxygen-haemoglobin dissociation curve to the right) and thus increasing the efficiency of oxygen delivery to the active tissues.

Factors Affecting the Oxygen-Haemoglobin Dissociation Curve

(i) *Effect of Partial Pressure of O_2* : The degree of saturation of haemoglobin completely depends on the partial pressure of O_2. An examination of the curve tells us that haemoglobin is almost completely oxygenated by an O_2 tension of 100 mm Hg, which is the same as that in the lungs. Therefore, an increase in the partial pressure of O_2 beyond 100 mm Hg would have nothing to do with the saturation of haemoglobin with O_2. However, the degree of haemoglobin saturation is lowered with the fall in the partial pressure of O_2. This explains the dissociation of oxyhaemoglobin at low partial pressure of O_2.

(ii) *Effect of Partial Pressure of CO_2* : The oxygen dissociation curve is also influenced by CO_2. An increase in the CO_2 shifts the curve to the right (Fig.), a displacement that is called the *Bohr effect* after the name of its discoverer. A higher CO_2 concentration, therefore, causes more O_2 to be given up at any given O_2 pressure, a result that is of significance in the tissues, for as the blood in the capillary gives off oxygen, CO_2 enters from the tissues and increases the amount of O_2 that is given off from the haemoglobin. Thus, *Bohr effect* facilitates the delivery of additional O_2 from the blood to the tissues. Conversely, the uptake of O_2 and the formation of oxyhaemoglobin in the blood coursing through the lungs greatly reduces the carbon dioxide-carrying capacity of the blood, thus facilitating the release of CO_2 to the alveoli.

(iii) *Effect of Temperature* : A rise in temperature also causes a decrease in the oxygen-carrying capacity of the blood. For example, during increased muscle activity, the rise in acidity due to CO_2 and lactic acid production and the localized increase in temperature results in a greater liberation of O_2 to the tissues. Thus, temperature affects the dissociation curve of O_2 by affecting the partial pressure of O_2.

(iv) *Effect of Haemoglobin* : It is apparent that the quantity of O_2 transported in the blood will depend on the haemoglobin content of the blood. Blood containing less haemoglobin (as in anemia) carries less O_2 to the cells and tissues.

(v) *Effect of pH* : The acidity of the blood (pH) also affects the degree of haemoglobin saturation with oxygen. An increase in the metabolic rate in the tissues increases the CO_2 and acid metabolites. The O_2 partial pressure is thus lowered, thereby more O_2 is liberated and made available to the tissues.

(vi) *Effect of 2, 3- diphosphoglycerate (DPG)* : A major fraction of the phosphate in erythrocytes is present as DPG which plays an important role in the release of oxygen to the tissues. An increase in DPG concentration is associated with decreased oxygen levels such as are found in anaemia or cardiac inefficiency. DPG reacts with haemoglobin to reduce the affinity of haemoglobin for oxygen and so makes more oxygen available to the anoxic tissues. In some genetic abnormalities, DPG is reduced and the red blood cells have a short survival time.

$$HbO_2 + DPG \rightleftharpoons Hb.DPG + O_2$$

Transport of Carbon Dioxide : Carbon dioxide is evolved in the body as a result of various metabolic activities of the cells and diffuses into blood. The total amount of CO_2 in the venous blood is about 60 ml per 100 ml blood, and the arterial blood contains about 50 ml total CO_2 per 100 ml. Thus, a relatively small amount of CO_2 is given off in the lungs. Carbon dioxide, that diffuses into the blood, is transported in the following three ways:

Transport of CO_2 in Physical Solution : As CO_2 enters the blood from the tissues, it combines with water of the plasma to form carbonic acid (H_2CO_3). Thus about 5% of CO_2 is carried in solution in the plasma as carbonic acid.

$$CO_2 + H_2O \rightleftharpoons H_2CO_3 \rightleftharpoons HCO_3^- + H^+$$

These ions then combine with the buffers of the blood.

Transport of CO_2 as Carbamino Compounds : About 10% of CO_2 is transported as carbamino compounds. In the red blood cells it combines directly with the amino groups ($—NH_2$) of the haemoglobin to form the so-called carbohaemoglobin.

$$CO_2 + Hb.NH_2 \rightleftharpoons Hb.NH.COOH$$

(carbohaemoglobin)

Transport of CO_2 as Bicarbonates : The rest, or about 85% of the total CO_2 is carried in the form of bicarbonates in both the plasma and red blood cells. As CO_2 enters the blood cells from the tissues, it combines with water to form carbonic acid (H_2CO_3), which dissociates to hydrogen ions (H^+) and bicarbonate ions (HCO_3^-). The latter diffuse into the plasma and with sodium or potassium ions in the plasma, form sodium or potassium bicarbonate.

$$Na^+ + HCO_3^- \longrightarrow NaHCO_3 \text{ (sodium bicarbonate)}$$

$$K^+ + HCO_3^- \longrightarrow KHCO_3 \text{ (potassium bicarbonate)}$$

It means at normal pH of mammalian blood, most of the total carbon dioxide will be present in the form of bicarbonate ion. According to Henderson-Hasselbalch equation, at pH 7.4, the ratio of carbonic acid to bicarbonate ion is 1 : 20 *i.e.*, for each one part of carbon dioxide present as acid, twenty times as much carbon dioxide is present as bicarbonate ion. The hydrogen ion formed as a result of the ionic dissociation of the carbonic acid is buffered by the various buffering substances in the blood, and the effect of carbon dioxide on the pH of the blood is therefore only moderate.

Thus, CO_2 relies on the blood salts for most of its transportation. The red blood cells contain an enzyme, *carbonic anhydrase,* which increases the speed of reaction between CO_2 and H_2O, resulting in the formation of H_2CO_3 which is rapidly converted to bicarbonates as indicated above. Thus, the enzyme is responsible in large part for the fact that most of the CO_2 in the blood is ultimately converted to the form of bicarbonate ions. But like all other enzymes, carbonic anhydrase also catalyzes the reversible reaction, that is the splitting of carbonic acid to water and carbon dioxide.

$$CO_2 + H_2O \underset{\text{(in lung capillaries)}}{\overset{\text{(in tissue capillaries)}}{\xrightleftharpoons{\text{carbonic anhydrase}}}} H_2CO_3$$

Thus, the enzyme also accounts for the rapid release of CO_2 from the blood during its passage through the lungs. Thus despite the low concentration gradient (tissue Pco_2 is 45 mm Hg, that of arteriolar capillaries is 40 mm Hg) carbon dioxide diffuses rapidly from the tissues into the blood. This is a good example of facilitated diffusion.

Carbon Dioxide Dissociation Curve

By plotting on the ordinate the total amount of CO_2 in the blood and on the abscissa the various CO_2 partial pressures, a CO_2 dissociation curve may be constructed (Fig.). However, the curves for oxygenated and deoxygenated blood differ somewhat. This is because oxyhaemoglobin is slightly more acidic than haemoglobin. Hence, oxygenated blood will bind slightly less CO_2. This phenomenon is closely related to the *Bohr effect* which shows the other side of the coin, *i.e.*, an increase in the CO_2 shifts the oxygen dissociation curve to the right giving more oxygen from the blood to the tissues. When CO_2 is added to the blood, it pushes the equilibrium between oxyhaemoglobin and haemoglobin in the direction of the weaker acid (haemoglobin). This, therefore, tends to release more oxygen which, in fact, is the Bohr effect.

In the body the actual dissociation curve for CO_2 will be the fully drawn line A-V, in which the point A stands for the normal arterial blood and the point V for the mixed venous blood which is never completely deoxygenated and this curve therefore is said to be the *functional curve* or the physiological dissociation curve.

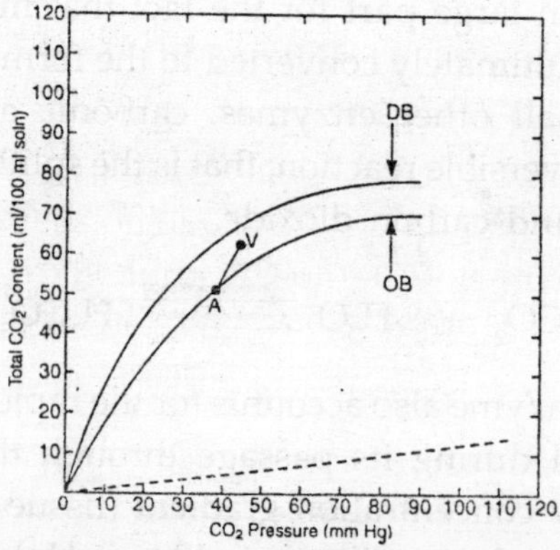

The carbon dioxide dissociation curve of mammalian blood (Winton and Bayliss, 1955). The dashed line indicates the amount of CO_2 held by physical solution in water. DB, curve for deoxygenated blood. OB, curve for oxygenated blood. A—V, physiological dissociation curve.

Isohydric Shift

When arterial blood arrives in the tissue capillaries, about 95% of the Hb is oxygenated. Because of the increased CO_2 tension and the decreased O_2 tension, the $HbCO_2$ dissociates; O_2 diffuses out into the interstitial fluid as CO_2 diffuses into the red blood cell. A small fraction of CO_2 is immediately bound as carbaminohaemoglobin. However, there still remains a large excess of CO_2 to be disposed of by other means. Under the influence of the enzyme *carbonic anhydrase,* this CO_2 is rapidly hydrated to H_2CO_3, which then dissociates. Then two opposing phenomena take place : (1) the H_2CO_3 would tend to lower the pH within the red blood cell, but (2) the transformation of HbO_2 to Hb involves an apparent change of pK from 6.2 to 6.6, which tends to raise the pH within the red blood cell. As a result, protons formed in the dissociation of H_2CO_3 are accepted by the groups in Hb participating in the Bohr effect described above. The net result of these two events is to maintain the pH essentially unchanged, and K^+ ions within the red blood cell, previously neutralized by HbO_2, are now neutralized by the newly formed bicarbonate (HCO_3^-) ions. Consequently, the major portion of the CO_2 that diffused into the red blood cell from the tissues leaves the capillary in venous blood as red cell HCO_3^-. This set of transformations is termed the *isohydric shift.*

Hamburgers Phenomenon or Chloride Shift : As shown above the dissociation of H_2CO_3 increases the number of bicarbonate ions in the red blood cells, and therefore the ions tend to diffuse away into the plasma. For each bicarbonate ion that comes out from the red cells, one negatively charged chloride ion present in the plasma moves into the red cells in order to maintain acid-base equilibrium for the blood and the electrical neutrality of the red blood corpuscles. Thus, *chloride shift involves the passage of chloride ions from the plasma into the red blood corpuscles to balance the bicarbonate ions that have passed from the red blood corpuscles into the plasma* (Fig.). The carbon dioxide capacity of the red blood corpuscles is further increased by the chloride shift, because the removal of bicarbonates, in this way, enhances their formation from the carbonic acid.

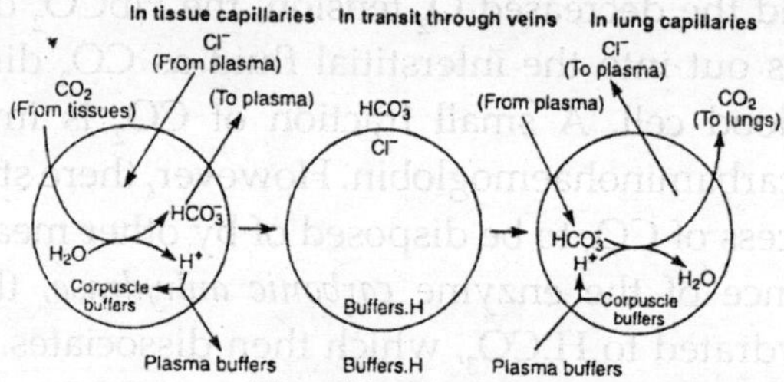

The Chloride shift.

In the red blood corpuscles most of the buffering is provided by the haemoglobin (Hb) itself. The latter is a negatively charged blood protein and combines with positively charged hydrogen ions (formed in the ionisation of (H_2CO_3), forming the haemoglobinic acid. This reaction is shown as follows:

$$BHb + H^+ \rightleftharpoons \underset{\text{(haemoglobinic acid)}}{HHb} + B^+$$

The resulting cation (B^+) from haemoglobin is balanced by the Cl^-.

Liberation of CO_2 in the Lungs

The CO_2 is carried by the blood-stream to the lungs in the form of carbonic acid, bicarbonates and carbamino compounds. Due to a higher tension of CO_2 in the lung capillaries (than the tension of CO_2 in the lung alveoli), CO_2 is given off from the blood and the reactions taking place are as follows :

(i) $H_2CO_3 \xrightarrow{\text{carbonic anhydrase}} H_2O + CO_2$

(ii) $NaHCO_3 \xrightarrow{\text{acidity of oxyhaemoglobin}} Na^+ + HCO_3^-$

or $2NaHCO_3 \xrightarrow{\text{acidity of oxyhaemoglobin}} Na_2CO_3 + H_2O + CO_2$

(iii) $HbNHCOOH \xrightarrow{\text{increasing oxyhaemoglobin}} HbNH_2 + CO_2$

The reciprocal effect of oxygenation on acid strength of the haemoglobin, the so-called *Haldane effect,* accounts for the CO_2 exchange. In the lungs, chloride ions move out of the red blood corpuscles and bicarbonate ions move back in. The enzyme carbonic anhydrase then promotes the rapid reformation of free CO_2, and this gas diffuses from the blood into the lung alveoli.

Cellular (internal) Respiration and Oxidation : The ultimate aim of all the respiratory activities is the cellular oxidation *i.e.,* the aerobic breakdown of the digested food materials and the release of energy thereby. The breakdown of carbohydrate may be represented by the following overall equation:

$$C_6H_{12}O_6 + 6O_2 \longrightarrow 6CO_2 + 6H_2O + \text{energy (686 Kcal/mole)}$$

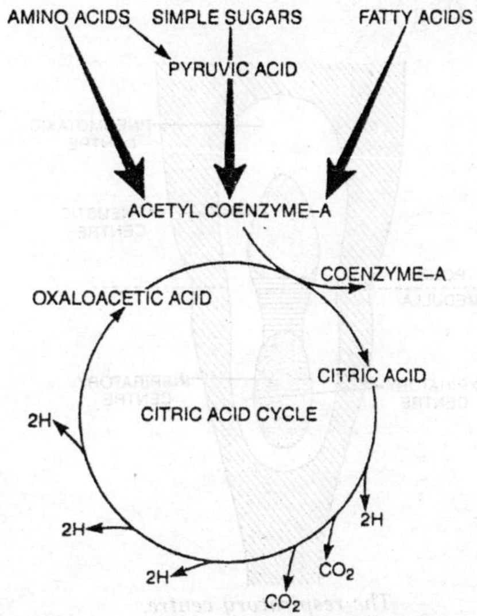

The interrelationship of metabolic pathways with Kreb's cycle.

The interrelationship of metabolic pathways of amino acids, simple sugars and fatty acids with the Kreb's cycle is summarized in Fig. Thus, within each cell glucose is first metabolized to pyruvic acid and then, if there is supply of O_2 to CO_2 and H_2O. It is when hydrogen atoms are combined with O_2 that most of the energy is released in the cell. By storing energy as ATP molecules, the cell can utilize this energy for its various activities as needed.

Nervous and Chemical Control of Respiration : Both the rate and depth of respiration can be regulated by chemical and neural factors. The process of breathing is essentially involuntary and automatic. However, it is influenced and controlled voluntarily within limits. Normally inspiration and expiration follow each other and both are basically muscular in action. The contraction and relaxation of the respiratory muscles (diaphragm and several chest muscles) depend on the arrival of nerve impulses originating from a group of neurons collectively forming the *respiratory centre* located in the medulla oblongata of the brain. The respiratory centre consists of the following parts (Fig.).

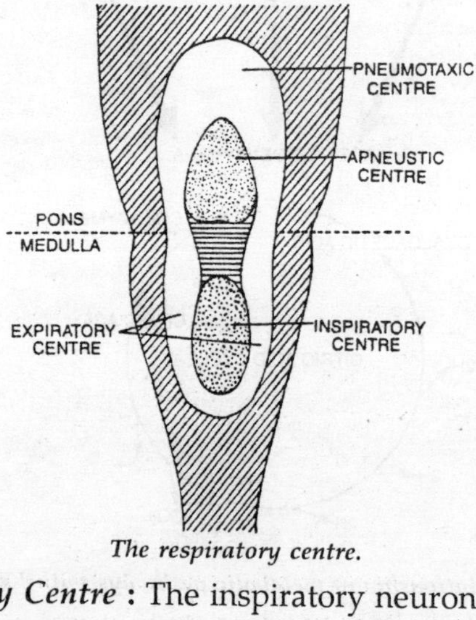

The respiratory centre.

Inspiratory Centre : The inspiratory neurons are grouped to form an inspiratory centre in the medulla oblongata. Motor neurons from this centre innervate the muscles of diaphragm and external intercostal muscles. The inspiratory centre probably originates the respiratory impulses which are modified by higher centres and also by impulses from spinal afferent nerves, specially those from the lungs.

Expiratory Centre : The expiratory neurons are grouped to form an expiratory centre in the medulla oblongata. Motor nerves from this centre innervate the abdominal muscles and internal intercostal muscles.

Pneumotaxic Centre : Located in the upper pons is a component of an oscillating circuit between the inspiratory and expiratory centre and itself. This centre does not originate impulses, but is stimulated by the activity of the inspiratory centre. When it is active, it stimulates the expiratory centre in the medulla oblongata and inhibits the other pontine centre (the apneustic centre). This facilitates expiration.

Apneustic Centre : The apneustic centre in the middle and lower pons stimulates the inspiratory centre in the medulla and causes inspiration. Prolonged electrical stimulation of the apneustic centre causes prolonged inspiration (without expiration). This condition is called *apneusis.* In normal respiration, the apneustic centre is rhythmically inhibited by the pneumotaxic centre, thus permitting normal expiration to occur.

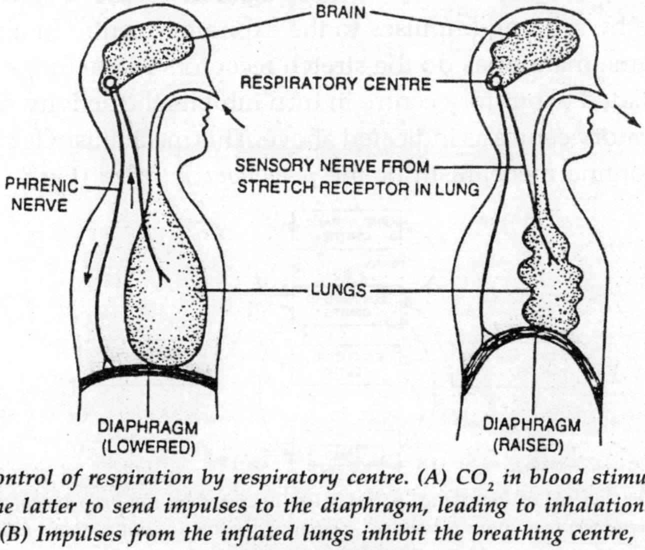

The control of respiration by respiratory centre. (A) CO_2 in blood stimulates the latter to send impulses to the diaphragm, leading to inhalation. (B) Impulses from the inflated lungs inhibit the breathing centre, leading to exhalation.

The reflex control of breathing depends mainly on the following reflexes (mechanisms):

Hering-Breuer Reflex : During inspiration, the inspiratory centre sends out nerve impulses which cause the contraction of the diaphragm and other respiratory muscles. As a result the thorax and lungs expand. Then some stretch-sensitive receptors (Hering-Breuer receptors) in the lung tissue are stimulated as soon as the expanding lungs attain a certain threshold tension. The nerve impulses thus initiated at the stretch-sensitive receptors in the lung are propagated by afferent neurons in the vagal nerves to the inspiratory centre which, as a result, is inhibited. The impulses may also go directly to the expiratory centre which responds to them by giving impulses that inhibit the inspiratory centre. The latter, therefore, stops discharging impulses to the concerned muscles which consequently relax causing expiration (Fig.). This mechanism is called the Hering-Breuer reflex, first described by them in 1868.

Pneumotaxic Reflex : The pneumotaxic or pontine centre is believed to function at the instigation of the inspiratory centre by sending impulses to the expiratory centre in almost the same manner as do the stretch receptors in the lungs. The stimulated expiratory centre in turn inhibits the activity of the inspiratory centre as indicated above. This mechanism is called the Pontine mechanism or the *pneumotaxic reflex* (Fig.).

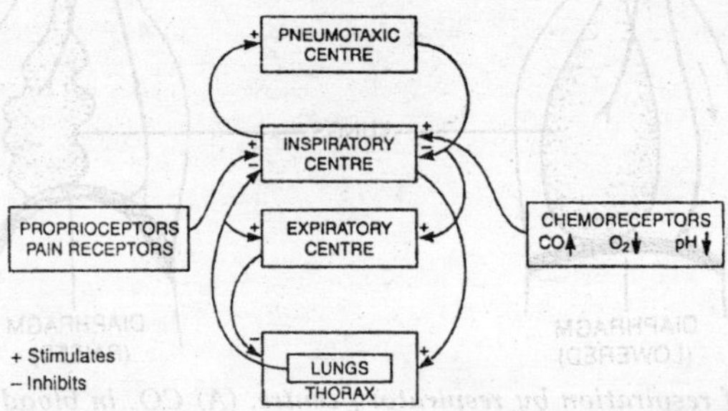

Reflex control of breathing. Reflexes are initiated by the proprioceptors, chemoreceptors, pain receptors, stretch of the lungs, and action of the inspiratory centre itself via the pneumotaxic centre or the expiratory centre.

Carotid and Aortic Body Reflexes (Chemoreceptor Respiratory Reflexes): *Heymans* (1936) demonstrated that *carotid bodies* (small vascular spherules located at the bifurcation of the external and internal carotids) and similar structures on the aorta, *aortic bodies,* were chemosensitive. Present evidence suggests that the chemosensitive cells of these bodies are active at the normal Po_2 of the arterial blood. However, a great increase in firing rate occurs when Po_2 is reduced to 50 mm Hg (normal blood Po_2 in the arteries is about 100 mm Hg). This fall in Po_2 increases the frequency and amplitude of nerve impulses from the carotid body nerves to the respiratory centres in the medulla. The resulting increase in alveolar respiration is a feedback mechanism that raises arterial Po_2. Thus, *peripheral chemoreceptors* (carotid bodies and aortic bodies) respond to changes in Po_2. The Pco_2 plays a role in the sensitivity of these cells to Po_2, because when Pco_2 is increased, the response to decreasing Po_2 is exaggerated. Thus, asphyxia, which reduces Po_2 and increases Pco_2 is a circumstance in which the receptors are greatly activated. Hypotension (low blood pressure) is another circumstance that may elicit chemoreceptor activity.

Central Chemoreceptors : The central chemoreceptor areas *(area lateralis)* of the medulla are found on either side near the entry of the 9th and 10th nerves. The cells of the area lateralis are activated by an increase in Pco2 of the blood and cerebrospinal fluid. Carbon dioxide diffuses freely between these fluids and causes a tremendous increase in respiratory rate and depth, thereby quickly removing the excess CO_2 from the body through the lungs.

As the blood-cerebrospinal barrier is practically impermeable to hydrogen ions, the response of central chemoreceptors to changes in pH is less direct. As CO_2 freely enters the cerebrospinal fluid, increased amounts of H_2CO_3 are formed. Consequently, as this acid dissociates, the pH of the cerebrospinal fluid is increased which serves as a stimulus for increased activity of the central chemoreceptors.

Respiration During Exercise

There is an increase in the rate and depth of breathing during physical exercise. In moderate exercise, ventilation increases in direct proportion to the amount of CO_2 formed. But during strenous exercise, lactic acid is also formed, and the resulting increase in arterial pH acts as an additional stimulus to pulmonary ventilation. These changes in Po_2, Pco_2 and pH *(humoral factors)*, however, can account for only a small amount of increased respiration during exercise.

By ingenious cross-circulation techniques exercise results in a marked increase in respiration even when the Po_2, Pco_2 and pH are artificially maintained at resting levels. This increase is caused by impulses reaching the respiratory centre from *proprioceptors* in the active skeletal muscles. Passive "pumping" of these limbs to excite these proprioceptors also increases pulmonary ventilation.

A third source of increased respiration during exercise is the *motor cortex,* which responds to the stress of exercise by exciting the respiratory centre by way of the reticular formation. This is similar to the role of the cortex in increasing cardiac output during excercise. As a result, the respiratory and circulatory adjustments to exercise are well integrated through these higher levels of the nervous system.

***Other Factors Affecting the Respiratory Centre* :** The respiratory centre is also influenced by some chemical and physical factors. The partial pressure of CO_2 in the blood is probably the most important, for the respiratory centre is highly sensitive to it. A little rise in the CO_2 tension stimulates the respiratory centre, resulting in an increased rate of breathing till the tension comes to normal. Increased acidity of the blood also has the same stimulating effect on the respiratory centre as does the CO_2. The increased concentrations of CO_2 and lactic acid during muscular exercise are responsible for the accompanying increase in the rate and depth of breathing. The breathing rate is also influenced by the body temperature and blood pressure. A rise in temperature stimulates breathing and vice versa. A decrease in blood pressure increases the breathing rate, whereas an increase causes a reverse effect.

In case of the voluntary control of breathing, impulses are transmitted from the higher centres of the brain (cerebral cortex) to the respiratory centre. Similarly, strong emotions, fright, excitement and other mental factors also have a striking influence on the breathing process.

Respiration at High Altitudes

At high altitudes the oxygen pressure may be quite low, 60 instead of 160 mm Hg. Consequently, there is a reduction in P_i, O_2 and symptoms of *hypoxia* (lack of O_2) appear. These symptoms include nausea, dizziness, loss of memory, cyanosis (blueness of lips and nails), and changes in heart rate and pulmonary ventilation. These symptoms are seen in varying degrees of severity in *mountain sickness,* depending on the altitude and the extent of acclimatization. These symptoms are exacerbated by excercise, which increases the tissue demand for oxygen. Therefore, it is usually necessary to breathe oxygen at 6500 m. In order to maintain the homeostatic mechanisms between respiration and blood, a mixture of O_2 (95%) and CO_2 (5%) is usually administered.

In respiration at high altitudes ventilation increases, which substantially reduces the difference between inspired Po2 and Po2 in the alveolar gas. Associated with this rise in ventilation there is also a reduction in Pa, CO2 which gives rise to *alkalosis* (fall in the alkali reserve of the blood) in new arrivals to a high atitude. Over a period by few days, the kidneys remove HCO_3- and blood pH is returned to normal. The diffusion capacity of the lung increases such that PA, O_2-Pa, O_2 is less in residents at high altitude than sea-level residents. This is at least partly related to an increase in the ratio of alveolor volume to total lung volume, and may also be the result of a reduction in the blood-air diffusion distance.

Adaptations also occur in the circulatory system. Both cardiac output and O_2 carrying capacity of the blood are higher in animals living at high altitude. The latter is related to a rise in haematocrit. The O_2-affinity of the blood of a number of mammals at high attitude, including human, decreases, and this is related to an increase in the level of 2, 3-diphosphoglycerate (DPG). This helps in the removal of O_2 from the blood at the tissues. The blood of vicuna and llama has a higher O_2 affinity than that of mammals living at sea level, so the response of the dissociation curve to hypoxia varies from animal to animal (*Lenfant*, 1973). At high altitudes, the number of red cells in the blood is increased, a condition termed *polycythemia* (see chapter on Physiology of Blood).

Respiratory Quotient

The ratio of the volume CO_2 given off to the volume of O_2 consumed during the same time is called the respiratory quotient (R.Q.). The R.Q. can be calculated for the oxidation of various food stuffs. In the reaction given below *i.e.*, for carbohydrate oxidation it is 1.00.

$$C_6H_{12}O_6 + 6O_2 \longrightarrow 6CO_2 + 6H_2O + \text{energy}$$

$$RQ = 6 \text{ volumes } CO_2/6 \text{ volumes } O_2 = 1.00$$

It varies with different foods, for fats the R.Q. is about 0.7; for proteins about 0.8, and for the average mixed diet about 0.85. It was formerly thought that the R.Q. of an organism or tissue gave direct information concerning the kind of food stuff being oxidised, or, in other words, of the kind of metabolism that is taking place in the body. But in practice the interpretation of *R.Q.* is not so simple, for cells can oxidise more than one of the three food stuffs at the same time, and certain other metabolic processes may use O_2 or produce CO_2.

Respiratory Pigments

These are the substances which combine reversibly with O_2, thus acting as a carrier or store of it. In this way, whenever a circulatory system transports O_2 from a respiratory surface to the body tissues, there is usually a transport pigment in the blood. One factor regarding the mode of functioning of these pigments is their O_2 affinity, which is indicated by the partial pressure of O_2 at which they are half-saturated with O_2. If the value (referred to as the p50) is high, the pigment has a low affinity, for it will give up its O_2 while this is still relatively plentiful in the surrounding medium. If the *p50* is low, the pigment has a high affinity, for it will only yield up its O_2 when the partial pressure of O_2 in the surrounding medium has become very low. All respiratory pigments contain a metal in an organic complex. A brief account of different respiratory pigments met within the animals is given below :

Haemoglobin : Haemoglobin is the best known O_2-carrying pigment. It occurs in the plasma of some animals and in the red blood cells of others. For example, certain annelids, molluscs and crustaceans carry haemoglobin in the plasma. A few annelids are known in which the pigment occurs both in plasma and blood cells. Other annelids, nemertine worms, sea cucumbers and vertebrates are among animals in which haemoglobin is found in the blood cells. With some very rare exceptions *e.g.*, the *leptocephalus* (larva of the eel), all classes of vertebrates possess haemoglobin in their red blood cells and their muscles contain *myoglobin* (muscle haemoglobin or *Mb).*

Haemoglobin molecules from wide variety of species differ qualitatively in such properties as solubility, crystal form, affinity for O_2, and absorption spactra. But all are composed of a colourless protein *globin* combined with *haeme.* Such differences in properties are due entirely to the different amino acid sequences and the conformations of the globin ; the haeme moiety is identical in all vertebrate and most invertebrate haemoglobins.

Haemoglobin is a conjugated protein *(chromoprotein)*. Each erythrocyte contains about 100 million haemoglobin molecules. Each unit of haemoglobin is actually a combination of an iron-containing *protoporphyrin (Heme)* and a specific protein globin. The unit molecular weight is approximately 17 to 18 × 10^3. The blood haemoglobin of higher vertebrates has four units *(i.e.,* molecular weight = 68 to 72×10^3).

The protein moiety or globin varies considerably in size, amino acid composition, solubility, and other physical properties from animal to animal. The iron content of mammalian haemoglobin is 0.336 per cent and the heme content is 4 per cent, the iron being normally in the ferrous state. In each unit of heme (Fig.), the iron atom is joined by four of its coordination bonds to the four nitrogen atoms of the protoporphyrin and one of the remaining six coordination bonds is joined to a molecule of globin. Thus, the chemical structure of haemoglobin consists of four molecules (or units) of heme combined with the globin molecule.

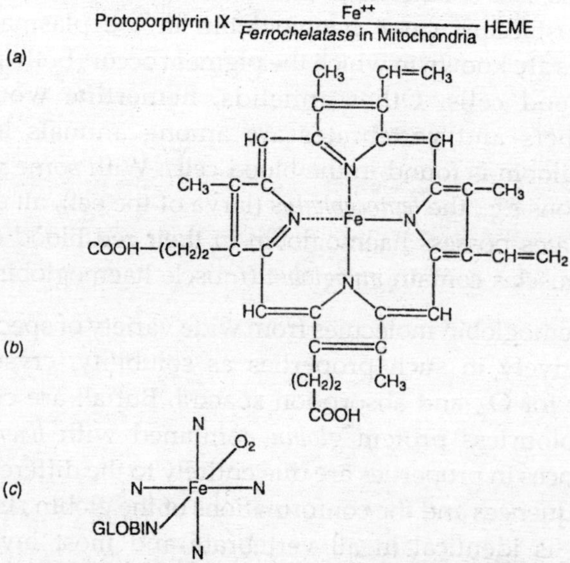

(a) Synthesis of Heme, (b) Structural formula of Heme,
(c) Linkage of ferrous iron, oxygen, and globin in a unit of haemoglobin.

Haemoglobin uncombined with O_2 and containing haeme with ferrous iron (Fe^{2+}) is called *ferrohaemoglobin, deoxyhaemoglobin,* or *reduced haemoglobin* and is abbreviated Hb. Human adult haemoglobin (HbA) or normal haemoglobin is composed of four polypeptide chains, two α-chains and two β-chains, and its subunit structure is designated $\alpha_2A\beta_2A$ (Fig.). Each unit can carry an O_2 molecule attached to the remaining coordination bond. Since each of the four hemes contains an iron that serves as the point of loose physical attachment for a single molecule of O_2, a mammalian haemoglobin can carry a maximum of four molecules of molecular oxygen. When haemoglobin (Hb) is fully oxygenated, it is called *oxyhaemoglobin* (HbO_2) which contains 4 molecules of O_2 per molecule of haemoglobin.

$$Hb_4 + 4O_2\ 5{=}\rightleftharpoons Hb4(O_2)_4 \text{ (oxyhaemoglobin)}$$

This reaction proceeds to the right when O_2 is present in excess, as in the lungs. Oxyhaemoglobin is carried in blood and thus reaches the body tissues. These are oxygen-poor, and the above reaction therefore shifts to the left. Free haemoglobin then reappears while the O_2 is taken up by the tissue cells.

Haemoglobin is adapted to its oxygen carrier function because: (1) it can pick up or release oxygen readily at partial pressure of oxygen normally found in the blood capillaries of the lungs or tissues, respectively ; and (2) its affinity for oxygen changes in response to variations in pH and CO_2 concentrations. The influence of pH and CO_2 on the affinity of haemoglobin for O_2 is called the Bohr effect.

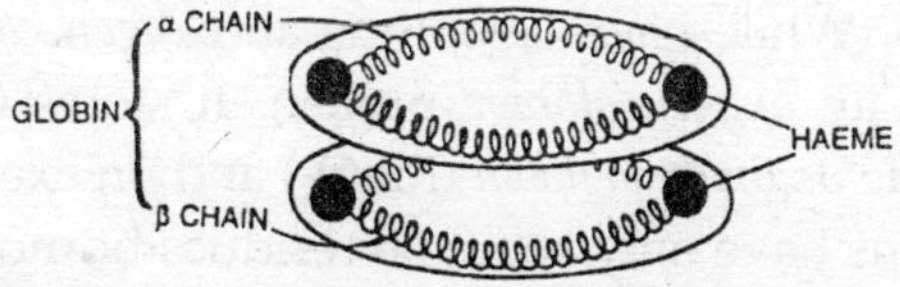

The structure of the haemoglobin molecule formed by four polypeptide chains and four molecules of haeme distributed to two indentical subunits.

Haemoglobin also functions as a carrier of small quantity of CO_2, conveying it as a waste product from the tissues to the lungs for excretion. However, the CO_2 is not transported by the haemoglobin's iron, but instead is carried in part in combination with the protein of the haemoglobin molecule.

Haemoglobin can also combine with 4 molecules of carbon monoxide (CO) to give carbon monoxide haemoglobin or *carboxyhaemoglobin* (HbCO), which is photosensitive and dissociates in light to release CO. The affinity of human Hb for CO is more than 200 times greater than its affinity for O_2. The brilliant cherry-red colour of HbCO is quite distinctive and is manifested in the skin and tissues of victims of CO poisoning.

Four molecules of NO also combine with one molecule of Hb to give the very stable HbNO. Many agents, *e.g.*, peroxides, ferricyanide, can oxidise the Fe^{2+} in Hb to Fe^{3+} to give *methaemoglobin* (met Hb), which binds neither O_2 nor CO. *Cyanomethaemoglobin* is formed on reaction of met Hb with CN^-

When haemoglobin content of the blood is reduced, the result is anemia. Haemoglobin is not only a carrier of O_2 and CO_2, but it is also one of the buffering agents in the blood. Some abnormal haemoglobins (haemoglobins, with altered solubility, altered O_2 affinity or unstable haemoglobins) also occur. For example, individuals homozygous for HbS, abnormal haemoglobin (sickle cell anaemia) have the severe disease *sickle-cell anaemia* and those heterozygous for HbS and HbA (normal haemoglobin) have *sickle-cell trait*, a condition usually without clinical symptoms.

Myoglobin : Myoglobin (Mb) of vertebrate muscle is very similar to Hb. While haemoglobin is an *oxygen carrier* protein, myoglobin is an *oxygen storage* protein. It stores O_2 in resting skeletal muscle as *oxymyoglobin* (MbO_2) and on exercise releases O_2. Myoglobins have only about 30 residues homologous with either the a or b chains of vertebrate haemoglobins, but their tertiary structures are strikingly like those of the Hb chains, with the same kinds of helical and non-helical secondary structures. The most important structural differences between Hb and Mb, however, are the distributions of R groups on the surfaces of the molecules. In haemoglobin, there are many surface R groups that provide complementary contacts between subunits and allow association of the chains to form the native quaternary structure, a_2Ab_2A. Since the R groups that provide these contacts in a and b chains are absent in Mb, in dilute solution Mb is a monomer and Mb cannot replace an a or b subunit in Hb.

In muscle, myoglobin binds O_2 reversibly. However, it exhibits no Bohr effect (which is exhibited by Hb) and its O_2 dissociation curve is hyperbolic (which is sigmoidal for human and frog haemoglobin). During muscle contraction, when the demand for O_2 is highest, O_2 dissociates from myoglobin and is available for oxidations. In humans, myoglobin is abundant only in cardiac muscle. In diving mammals, the myoglobin content of muscles is very high and facilitates submersion for long periods. The flight muscles of birds are also rich in myoglobin. At a venous Po_2 of 40 mm Hg, at which Hb is 66% saturated, myoglobin is still 94% saturated. At an O_2 tension of 10 mm Hg, Hb is 10% saturated whereas myoglobin is 80% saturated. Therefore, cytochrome oxidase can operate at Vmax when the medium provides O_2 at Po_2 of about 4 to 5 mm Hg. Thus, the affinities of these three proteins for O_2 are in the order cytochrome oxidase > myoglobin > Hb. As a result, myoglobin can accept O_2 from Hb and store it in the muscle cell for release to cytochrome oxidase when the O_2 supply is limited.

Erythrocruorin : Large haemoglobins, frequently called erythrocruorins, occur in the circulating plasma of many annelid worms and molluscs. The giant molecule of erythrocruorin from *Limnodrilus* (MW ≐ 3¥10^6), consists of 108 subunits (MW = 2.8×10^4), each bearing one haeme, arranged as a regular hexagonal cylinder having the dimensions 220 by 160 Å. This cylinder, in turn, consists of six lesser cylinders oriented in the same axis as the major cylinder axis. Thus, there are three layers of 36 protein subunits each in the overall packaged molecule. Since there are four different NH_2-terminal amino acids, it is assumed that there are four different subunit polypeptide chains in the molecule.

Chlorocruorin : Certain of the annelids, such as *Spirographis,* have green blood pigment, the chlorocruorin (MW = 3.4 ¥ 10^6). In this pigment, the porphyrin differs from protoporphyrin in that the 2-vinyl group is replaced by a formyl group. When such respiratory pigments are present in the coelomic fluid rather than in an organized circulation, they serve to store rather than transport O_2.

Haemerythrin : Haemerythrin is present in the blood of all sipunculid worms, a few polychaete forms, and the brachiopod *Lingula.* It is a rare, reddish-violet, iron containing respiratory pigment. The most carefully studied haemerythrin from *Sipunculus* has a molecular weight of 105,000 and contains 16 atoms of Fe^{2+}. Haemerythrin binds one O_2 per pair of iron atoms.

Some Characteristics of Respiratory Pigments

S.No.	*Pigment*	*Source*	*Location*	*Atoms of metal per mole*	*Molecular weight*
1.	Haemoglobin	Human being	Intracellular	4Fe	6.5×10^4
		Frog	Intracellular	4Fe	6.5×10^4
		Gastrophilus larvae	Tracheal cells	2Fe	3.4×10^4
2.	Myoglobin	Mammalian muscle	Intracellular	1 Fe	1.7×10^4
3.	Erythrocruorin	Annelids	Plasma	96 Fe	3×10^6
		Molluscs	Plasma	96 Fe*	1.5×10^6
4.	Chlorocruorin	Polychaete worms	Plasma	96 Fe	3×10^6
5.	Haemerythrin	Sipunculus	Intracellular	16 Fe	10.5×10^4
6.	Haemocyanin	Squid	Plasma	20 Cu*	3.75×10^6
		Snail	Plasma	200 Cu*	6.76×10^6

* approximately

Haemocyanin **:** The blood plasma of many molluscs and arthropods contains *haemocyanin,* blue pigment containing copper (instead of iron), but no haeme. The molecular weights of haemocyanins range from 0.5 ¥ 10^6 to 10^7. Squid haemocyanin is a decamer (MW = 3.75 ¥ 10^6, which on dilution or at high salt concentration dissociates into 5 dimers and then into 10 monomers. Each monomer contains 2 atoms of Cu^+ and, both in monomeric and polymeric form, binds one O_2 per two Cu^+. There is some evidence that O_2 binding and release involve oxidation and reduction of the copper.

Respiratory pigments change colour according to the degree of oxygenation (*e.g.*, haemoglobin is red when oxygenated, and purple when deoxygenated) and many have characteristic absorption spectra. These properties have facilitated their investigation. Some characteristics of respiratory pigments are shown in Table.

Anaerobic Respiration

When respiration goes on in the absence of O_2, it is called anaerobic. Fermentation in yeast and glycolysis in tissues and may microbes are typical examples of anaerobic respiration. The overall equation of fermentation can be written as follows :

$$C_6H_{12}O_6 \longrightarrow 2CH_3CH_2\,OH + 2CO_2 + \text{Energy (50 kcal/mole)}$$

The anaerobic breakdown of glucose in animal tissues differs from this in two respects : (i) it begins with glycogen stored in the tissues, for which reason it is called glycolysis, and (ii) it ends with lactic acid. The intermediate stages, however, are identical. The overall equation for glycolysis can be written as follows :

$$C_6H_{12}O_6 \longrightarrow 2CH_3\,CH(OH)\,COOH + \text{energy (36 kcal/mole)}$$

Such processes do not liberate as much energy as does aerobic respiration. Many organisms or parts of them respire anaerobically for sometime when their supply of O_2 is insufficient for aerobic respiration, in which case they may acquire an oxygen debt. A few bacteria are strictly anaerobic *i.e.*, they do not use free O_2 at all.

Oxygen Consumption and Metabolic Rate Determination

Basal Metabolic Rate : The term basal metabolic rate (BMR) refers not to the lowest possible energy metabolism of an organism but to a *minimum stable rate of metabolism* when the animal is in the resting condition. In birds and mammals, BMR is measured when they are in the resting and fasting state so that the digestive processes are not going on which would require the expenditure of energy. This rate can be determined in three ways :

(a) From the food consumption or by determination of the difference between the energy value of all the food taken in and the energy value of faeces and urine.

(b) From the energy released as heat, or from the total heat production of the organism.

(c) From the amount of oxygen used in oxidation process to obtain energy, provided the information is available about which substances have been oxidized and there is no anaerobic metabolism or external work performed.

But all the three methods are not equally satisfactory. However, the third method is technically easy and is widely used for this purpose. So when we say metabolic rate, we actually mean the rate of O_2 consumption. A simple apparatus to determine O_2 consumption is given in Fig.

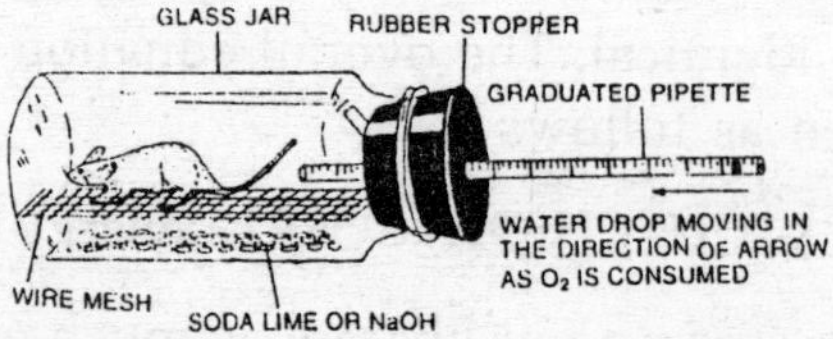

Simple apparatus to measure the rate of oxygen consumption.

As O_2 is used by the animal, the volume of the air in the jar decreases, for the CO_2 is absorbed by alkali as fast as it is formed. The movement of the drop in the graduated pipette shows the reduction in volume, and by timing its movement we get the number of ml O_2 used by the animal per minute. Further calculations can be made as shown below :

Weight of the apparatus without animal = *a* gm.

Weight of the apparatus including animal = *b* gm.

Weight of the animal *(b-a)* = *c* gm.

Time taken for the coloured drop of water to travel from 0.1 ml mark to 1.0 ml mark = *t* sec.

Thus, 0.9 ml of O_2 is consumed in *t* seconds by c gm of animal tissue (at room temperature A degree centigrade and B mm Hg pressure).

Therefore, O_2 consumption per hour by 1 gm of animal tissue (at A° centigrade and B mm Hg pressure).

$$= \frac{0.9 \times 60 \times 60}{t \times c} = \frac{3240}{tc} \text{ ml}$$

Standard Metabolic Rate

Poikilothermic animals have metabolic rates that are temperature dependent and therefore have no metabolic state corresponding with the avian or mammalian BMR. So for poikilothermic animals *the minimum fasting metabolism at a given temperature* is called the standard metabolic rate (SMR).

Metabolic Rate in Various Animals : For a comparison of the metabolic rate of two animals, their size difference is taken into consideration. The easiest way is to relate the O_2 consumption to body weight and calculate the amount of O_2 used per gram body weight in a given time, for example one hour. The metabolic rates (O_2 *consumption in* mm^3 */gram body weight /hour)* in various animals at rest is given below :

Octopus	80	Man	200
Squid	320	Rat	1500
Eel	128	Humming bird (at rest)	1000
Trout	226		
Frog	150	Humming bird (On flight)	13,000

The metabolic rate is calculated when the animal is in the resting condition, for any activity requires extra energy and the metabolic rate is naturally increased. A comparison of the metabolic rate of the Humming bird at rest and on flight makes this point clear. This is the reason that active animals always have high metabolic rates and therefore, some of the highest metabolic rates on record have been found in the flying insects and birds.

The metabolic rate is also related to body size as small animals have a high metabolic rate. For example, in a rat of 25 grams the oxygen consumption in mm^3/gram/hour is about 1580, in a rabbit of 2200 grams, 466, and in a man weighing 70,000 grams it further decreases to only 202. Thus, it is clear from these data that the metabolic rate increases with the decrease in body size (or weight). However, there should be certain limit of size within which this increase is possible.

The smallest existing mammals are some shrews, having a body weight of only 4 grams, but their metabolic rate is very high. To maintain this rate, a shrew has to eat constantly and is always on a search for food, consuming almost its own body weight of food per day. If a mammal were even smaller, it would have to eat at such a tremendous rate and spend so much energy in search of food that it simply could not exist. It can therefore, be concluded that there is a lower theoretical limit to the size of mammals.

However, since metabolic rate actually decreases with

the increasing body size, the limit to the body size of mammals is not set by the requirements for food. Land animals cannot increase to unlimited size because the mechanical strength of their muscles and bones could not support their enormous weight. The heavy bones and thick legs of elephant are evidence of the strength needed to support a huge body, and a further increase in weight is just not possible without the animal becoming overly cumbersome.

Metabolic Rate and Circadian Rhythms : One of the most remarkable attributes of organisms is the rhythmic or cyclic nature of most of their functions and activities. Many of these endogenous rhythms have a periodicity of 24 hrs. In an evolutionary sense this 24 hour periodicity is undoubtedly related to the period of the earth's rotation and hence the length of one light-dark cycle. In animals under natural conditions such a 24 hour periodicity is daily entrained or reinforced by the normal day-night cycle.

However, the absence of daily reinforcement does not lead to the loss or disappearance of biological rhythmicity. Majority of animals maintain a 24 hour periodicity even when kept in an unchanging environment in which they receive no clues about the natural day-night cycle. One of the most striking features of these rhythms is their insensitivity to environmental temperature. Even in poikilotherms wide variations in environmental temperature have only slight effect on the daily periodicity of behaviour or function.

In the absence of an external timegiver *(zeitgeber)* these endogenous rhythms rarely have a period that is exactly 24 hours. Instead, it varies between 22 and 26 hrs. Any given individual animal, however, keeps a constant periodicity to its daily rhythm that may be slightly shorter or slightly longer than the environmental cycle. *These daily endogenous cycles are called circadian rhythms (L. circum* = about, *dies* = *day) because they are about one day (though not exactly) in length.*

It is known that nearly any measurable physiological quantity, *e.g.*, body temperature, oxygen consumption, blood sugar, *etc.*, shows a striking daily cycle. Because all these and many other factors vary in a cyclic manner, time of day must be taken into account in virtually all physiological measurements. Oxygen consumption, activity, and body temperature can be monitored over long periods of time without disturbance, so the effects of circadian rhythmicity are well documented with regard to these functions.

Typically the body temperature of diurnal homeotherms is 1 to 3° higher during the day than at night. For nocturnal forms the cycle is reversed. The daily rhythms of the metabolic responses of vertebrates are complex than those of body temperature but they generally show the same fundamental pattern, with rates being high during the normal period of activity and low during the normal period of inactivity. The persistence of such endogenous rhythms allows the animal to be ready to function adequately before the environmental stimulus appears, *i.e.*, it is anticipatory and thus enhances the effectiveness of responses (*Aschoff,* 1960).

Biological Oxidations

Characteristics and Examples

(1) In contrast to explosive, inefficient, and uncontrolled oxidation in ordinary combustion of fuels, biological oxidation is stepwise, efficient and controlled, as in the latter case, energy is released gradually by breaking the chemical bonds in the food molecule one after another. A respiratory decomposition does not occur in a single large oxidation-reduction process, as just shown for methane. Instead, respiration takes place

through a series of consecutive redox reactions in which H is removed from fuel in several steps, each resulting in a successively higher oxidation level of carbon atoms. Such stepwise oxidation occurs generally in the respiration of all foods. The carbon groupings of food molecules are transformed successively from hydrocarbon-to-alcohol-to aldehyde-to acid-to anhydride. Such stepwise oxidations provide energy in smaller successive packs and each of these can become useful in the cellular metabolism.

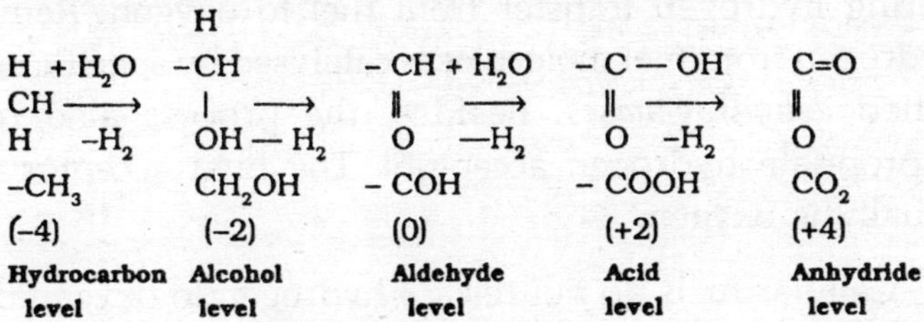

Foods and Oxidation Levels

Besides, the oxidation states of carbon here change successively from -4 to -2 to 0 to +2 to +4.

(2) Most of the energy of combustion dissipates as heat, v whereas a large fraction of the energy released in biological oxidation is captured in ATP. A very small amount of energy which is not captured, is liberated as heat. This too, need not be considered as "wasted", since in the warm-blooded animal it contributes to the maintenance of body temperature.

The Biological oxidation may take place in the following three ways:

(i) *By the addition of* O_2

$C + O_2 \rightarrow CO_2$

(ii) *By the removal of hydrogen*

$CH_3CHOH.COOH \rightarrow CH_3CO.COOH + 2H^+ + 2e^-$

Lactic acid Pyruvic add

(iii) *By an increase in positive valency*

$$2Fe^{2+}(Cl^-)_2 + Cl_2 \rightarrow 2Fe^{3+}(Cl^-)_3$$

Ferrous chloride — Ferric chloride

Biological oxidation is often synonymous with dehydrogenation i.e., removal of hydrogen atoms and electrons.

Hydrogen Transfer (Electron Transport System) : There are two kinds of ATP sources in respiration. One is nutrient oxidation by dehydrogenation, the other is creation of ATP during hydrogen transfer from fuel to oxygen. Removal of hydrogen from fuel molecules is catalysed by specific enzymes called *dehydrogenases.* Besides, the process also requires appropriate hydrogen acceptors. The final acceptor in cells usually is oxygen.

Cellular fuels do not release hydrogen to oxygen directly, but through a series of intermediate hydrogen carriers, the last of which finally releases hydrogen to oxygen. The advantage of such step by step transfers is that more usable energy may be obtained for metabolic use, than if hydrogen and oxygen were allowed to combine directly. The combination of hydrogen and oxygen is an energy-yielding process. Thus for every H_2 transferred by carriers from fuel to oxygen, a net total of 3 ATP moles is formed.

The sequence of enzymes and carriers responsible for transport of reducing equivalents (H or electrons) from substrates to molecular oxygen is known as the *respiratory chain* or *Electron transport System* (ETS). The respiratory chain is now known to be localised within mitochondria. The major components of the respiratory chain are arranged sequentially in order of increasing redox potential. Electrons flow through the chain in a stepwise manner from the more electronegative components to the more electropositive oxygen.

Every step in serial hydrogen transfer is itself a complete

redox process. The intermediate hydrogen carriers are alternately reduced and oxidised, *i. e.*, they operate in a cyclic manner, first accepting H (or electrons), then passing it on, and then again accepting new hydrogen etc. Hydrogen transfer by carriers yields energy at each transfer step, but at three of these steps, enough energy is liberated to generate ATP.

Electrons are transferred from NADH or $FADH_2$ to molecular oxygen (O) through a respiratory chain of 3 large protein complexes, linked by 2 mobile carriers (coenzyme Q and cytochrome C) on the inner mitochondrial membrane. In this sequence, NAD^+ (or $NADP^+$) and FAD are Vitamin B—derived coenzymes. Q is a coenzyme that consists of a substance *Ubiquinone,* which is a constituent of the mitochondrial lipids. CoQ has a structure that is very similar to Vitamin K and Vitamin E. It serves as a nighty mobile carrier of electrons between the flavoproteins and cytochromes of the electron transport chain. likewise, cytochrome C is a small soluble protein that functions as a mobile electron carrier.

The electron carriers between CoQ and molecular O_2 are *cyto-chromes.* A cytochrome is an electron-transporting protein that contains a haem prosthetic group. The iron atom in cytochromes oscillates between a reduced ferrous (Fe^{2+}) state and an oxidised ferric (Fe^{3+}) state during electron transport. The *haem group is a one-electron carrier,* in contrast with NADH, Flavo-proteins, and CoQ, which are *two-electron carriers.* Therefore, a molecule of reduced CoQ must transfer its two high potential electrons to two molecules of cytochrome b, the next member of the electron transport, chain.

There are 5 slightly different cytochromes which operate in succession in the respiratory chain. They are named in order, cytochromes *b,* c_1 *c,* a and a_3. The terminal cytochrome a_3 *cytochrome oxidase,* is responsible for the final combination of reducing équivalents with molecular oxygen. It releases electrons back to H^+, which has remained in the reaction medium ; whole H atoms then combine with oxygen, forming water.

The salient features of the respiratory chain are as follows:

(1) After dehydrogenation of food, either NAD or NADP may accept hydrogen; depending upon the nature of specific foods. In the mitochondria of a cell, where most respiratory oxidations occur, it is usually NAD which serves as hydrogen carrier after food substrate.

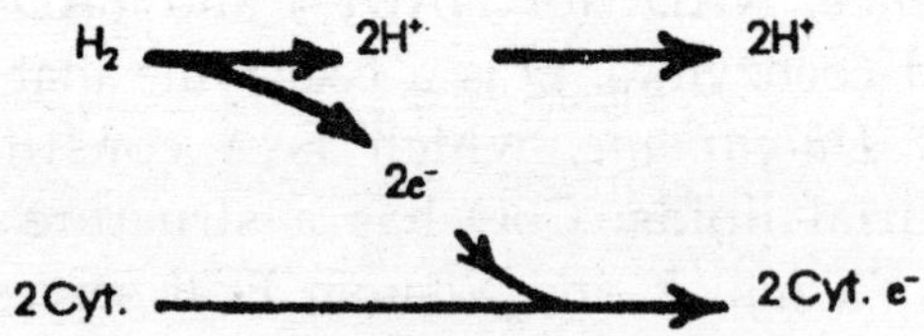

Transport of electrons by cytochromes.

NAD transports partly electrons, partly H atoms. Of each H_2 yielded by food, one whole H atom, and an electron from the other H atom join with electropositive NAD^+ (lacking one electron), resulting in neutral NAD. H. Thus NAD. H represents the reduced state and an H^+ ion remains in the medium. In hydrogen release, these reactions are reversed.

$$NAD^+ + H_2 \xrightarrow{\text{H ACCEPTING}} H^+ + NAD.H$$

$$NAD.H + H^+ \xrightarrow{\text{H DONATING}} H_2 + NAD^+$$

(2) FAD, coenzyme Q and molecular oxygen transport whole H atoms only.

(3) All Cytochromes transport electrons, not whole H atoms. In this case, both H atoms of a pair dissociate and each resulting electron is carried by one cytochrome molecule :

The H-carrying component of a cytochrome molecule is

the iron in the centre of its tetrapyrrole ring. This iron normally exists in the form of ferric ion (Fe^{3+}). When it accepts an electron, its electropositivity is reduced by one charge unit and therefore, it becomes a ferrous ion (Fe^{2+}).

$$Fe^{3+} + e^{-} \longrightarrow Fe^{2+}$$

(4) The energy generated by the transfer of electrons through the respiratory chain is utilized in the pumping of protons (H^+) from the matrix side to the cytosolic side of the inner mitochondrial membrane. The H^+ concentration becomes higher on the cytosolic side and an electric potential with that side positive (+) is generated.

Peter Mitchell in his chemiosmotic hypothesis (1961) proposed that a Proton Motive Force (PMF) is thus generated, consisting of (i) a *pH gradient* (cytosolic side acidic) and (ii) atrans-membrane electric potential (cytosolic side positive). Mitchell postulated that this PMF is used to drive ATP synthesis when protons flow back to the mitochondrial matrix through ATP synthase similarly as a flow of water from a higher to a lower level can be utilized to turn a water wheel.

Oxidative Phosphorylation

The NADH and $FADH_2$, formed in glycolysis, the citric acid cycle and fatty acid oxidation are energy rich molecules because they contain a pair of electrons which have a high transfer potential.

The concomitant transfer of electrons to molecular oxygen yields a large amount of energy which can be used to generate ATP. Once available, this energy can be utilized by the cell to drive all known endergonic process such as chemical syntheses, ion transport and mechanical exercise.

The generation of ATP molecules, as electrons are transferred from NADH or $FADH_2$ to O_2, by a series of electron earners is called *oxidative phosphorylation.* This is the major source of ATP in aerobic organisms. Some salient features of this process are :

(1) Oxidative phosphorylation is carried out by *respiratory assemblies* that are located in the *innermembrane of mitochondria. A* respiratory assembly contains numerous electron carriers such as the cytochromes. The reactions of the citric acid cycle and fatty acid oxidation, which supply most of the NADH and $FADH_2$, take place in the adjacent mitochondrial matrix.

(2) ATP is generated at three sites during oxidative phosphorylation as electrons flow through the Electron Transport System (ETS) from NADH to O_2. *Site 1* NADH dehydrogenase (complex I), *site 2* is Cytochrome reductase (complex III); and *site 3* is Cytochrome oxidase (complex IV).

(3) Oxidation and phosphorylation are *coupled processes.* The oxidation of NADH yields 3 ATP, whereas the oxidation of $FADH_2$ yields 2 ATP because the electrons from this carrier enter the chain at CoQ, after the first phosphorylation site.

The rate of respiration in mitochondria can be controlled by the concentration of ADP. This is because oxidation and phosphorylation are *tightly coupled* and ADP is an essential component of the phosphorylation process. Thus electrons do not normally flow through the electron transport chain to O_2 unless ADP is simultaneously phosphorylated to ATP. In addition to ADP, oxidative phosphorylation requires a supply of NADH, inorganic phosphate and O_2.

Aerobic and Anaerobic Hydrogen Transfer

(1) *Aerobic* : Organisms with ETS system that use atmospheric oxygen as a final hydrogen acceptor are

considered aerobic (with air). It is the standard, universal form among animals.

(2) *Anaerobic* : Anaerobic means "in the absence of molecular oxygen". Whenever hydrogen transport from NAD to oxygenis blocked (due to cyanide poisoning, Vitamin B—deficiencies or inadequate supplies of atmospheric oxygen), organisms begin to respire in a way that does not require oxygen. This type of respiration is called *anaerobic* (without air) or *fermentation.*

With the path from NAD to oxygen blocked, pyruvic acid, one of the compounds normally formed in the course of carbohydrate degradation, readily accepts hydrogen. The fate of pyruvic acid thus serving as a hydrogen carrier may be determined along two lines:

(a) *Formation of Lactic Acid* : When pyruvic acid reacts with hydrogen, lactic acid is formed in animals. The reaction is catalyzed by *Lactate dehydrogenase.*

$$\underset{\text{Pyruvic acid}}{\begin{array}{c}COOH\\ |\\ C=O\\ |\\ CH_3\end{array}} + NADH + H^+ \rightleftharpoons \underset{\text{Lactic acid}}{\begin{array}{c}COOH\\ |\\ HCOH\\ |\\ CH_3\end{array}} + NAD^+$$

Lactic acid then accumulates in the muscles causing fatigue and the animal must, therefore, stop intensive activity. During the later rest period, lactic acid is carried by blood from the muscles to the liver where it is converted to glucose. From the liver, glucose eventually returns via blood to muscles where it is transformed to glycogen. After the block to liver is no longer present or with extra oxygen now available, glycogen may be respired completely to CO_2.

(b) *Alcoholic Fermentation* : In plants and most micro-organism, pyruvic acid reacts with H_2 to form ethyl alcohol and CO_2. This process is called alcoholic fermentation. The great French biochemist Louis Pasteur called fermentation "life without air". Pasteur found that fermentation was an energy-producing mechanism used by organisms deprived of air, like single celled fungi, the yeasts. The reaction takes place in 2 steps :

(c) In contrast to the reversible formation of lactic acid in animals, pyruvic acid in yeast is irreversibly decarboxylated to acetaldehyde by *pyruvic decarboxylase,* which is not present in animal tissues.

$$\underset{\textbf{Pyruvic acid}}{CH_3COCOOH} \xrightarrow{Mg^{2+}} \underset{\textbf{Acetaldehyde}}{CH_3CHO} + CO_2$$

(ii) The acetaldehyde thus formed is reduced by NADH in a reaction catalyzed by *alcohol dehydrogenase.*

$$\underset{\text{Aceteldehyde}}{CH_3CHO} + NADH + H^+ \rightleftharpoons \underset{\text{Ethyl alcohol}}{CH_3CH_2OH} + NAD^+$$

Efficiency of Fermentation : Both lactic acid formation and alcoholic fermentation are inefficient processes. Not more than 5% of the energy stored in glucose is made available to the organism during anaerobic respiration, because fuel oxidation has stopped at the pyruvic acid stage.

Potential energy still contained in this acid remains unutilized and locked in the end products of lactic acid or ethyl alcohol. Nevertheless, under conditions of oxygen debt in animals, such as during intensive muscular activity, fermentation can proceed alongwith aerobic respiration to provide a little extra energy.

Energy Transfer

Cellular oxidations take place with very small rise in temperature, because the energy produced as a result, does not accumulate primarily as heat. Instead, a large traction of this energy is tapped in the form of high energy bonds within ATP.

There are two types of chemical bonds in organic molecules: low-energy and high-energy bonds.

(i) *Low-Energy Bonds* : Most bonds in fuel compounds are of the *low-energy* type like C—C, C—H or C—O. When one of these is broken, only a comparatively small amount of energy is released.

(ii) *High-Energy Bonds* : Chemical bonds between certain atomic groupings in cells, such as phosphate groups, require large amounts of energy to be formed. Since they contain *extra energy*, correspondingly large amount of energy is released when they are broken. Although a *low energy* phosphate bond —O—P may also be formed that joins O and P but a wavy symbol ~ in the O ~ P links signifies the presence of a *high-energy* phosphate bond.

Conversion of AMP to ADP and ADP to ATP in biological systems requires additional phosphate groups and large energy inputs. Respiration provides this energy through fuel oxidation in cells, by which high energy bonds are created in ADP and ATP. ATP passes on this energy to drive those processes that require energy. Thus, ATP may be designated as the energy rich end-product of respiration.

The creation of high-energy phosphate bonds —O - P takes place in 4 steps :

(a) Phosphorylation of a fuel molecule by the attachment of a phosphate group -O-P.

(b) Dehydrogenation of phosphorylated fuel substrate.

(c) Dehydrogenation yields energy but since an -O-P group is now part of the fuel molecule, the energy is not released as heat. On the other hand, it becomes redistributed within the fuel molecule, and low-energy phosphate bond -O-P is transformed into high-energy bond —O~P.

(d) This high-energy bond is then transferred to ADP, generating ATP.

The ultimate source of phosphate groups in oxidation phosphorylations is the phosphoric acid made available to the cells as a mineral nutrient, but the immediate phosphate source in cells usually is ATP itself. For example, in the phosphorylation of glucose in cells, ATP is the donor.

Significance of the ADP—ATP High-Energy Cycle : ATP is the source of high energy phosphate for nearly every energy requiring reaction in the cell. The hydrolysis of 1 mole of ATP to ADP yields 7000 cal of (potential) energy which is used in various activities of a cell. In degradation reactions, the terminal high-energy phosphate of ATP is split off with the formation of a low energy phosphate and the conversion of ATP into ADP.

5

The Blood

Blood may be defined as a highly complex, seemingly homogeneous fluid composed of a large variety of dissolved and suspended inorganic and organic substances as well as three kinds of cells or formed elements. The presence of these cells accounts for the designation of blood as a tissue (or a *liquid tissue)*.

In very small animals, oxygen penetrates the surface and diffuses to all parts of the organism. But in larger animals *(annelids, molluscs, arthropods, echinoderms* and *chordates)* a specialized transport system, the blood, becomes necessary. However, gas transport is not the only function of the blood. In higher animals, the blood has a number of functions stated below.

Functions of Blood

1. *Transport of Gases* : It transports oxygen from the lungs to the tissues and CO_2 from the tissues to the lungs.
2. *Transport of Nutrients* : It carries nutrients from the intestinal wall to various parts of the body, where they are used or stored.

3. *Transport of Waste Products* : Blood carries the excretory products from the cells or organs where they are formed to the excretory organs where they are eliminated.
4. *As a Communication System* : The blood also serves as a communication system within the body by transporting susbtances (vitamins, hormones, secretions, enzymes and other essential chemicals) that regulate the function of certain organs. For example, the hormone *secretin,* that is released from the wall of the small intestine into the blood, controls the secretion of pancreatic juice. It transports the secretions of endocrine glands to their places of activity.
5. *Defensive Action* : The blood acts as a great defensive mechanism against disease in two ways. *(a)* Its white blood cells engulf bacteria and foreign particles due to their phagocytic properties. *(b)* It develops *antibodies* which combat toxic agents.
6. *Coagulation* : The blood has the ability to coagulate and thus guards against *haemorrhage.*
7. *Functions of Plasma Proteins* : The plasma proteins of blood have various functions (see under the heading plasma proteins).
8. *Regulation of Body Temperature* : The water content of blood has three qualities which make it very suitable for this purpose, (a) *High-specific heat* : Due to this it can absorb a large amount of heat, thereby preventing sudden change of body temperature, (b) *High conductivity* : Thermal conductivity of water is higher than any other ordinary liquid. This helps in rapid and even distribution of heat, (c) *High latent heat of evaporation* : Latent heat of evaporation of water is very high and since water is constantly evaporated from skin and lungs, a large amount of heat is lost in this way.

9. *Regulation of Blood Pressure*: The blood has the ability to regulate its own pressure by changes in volume and viscosity (haematocrit value) of blood.
10. *Maintenance of Water Balance* : Among other factors blood also helps in maintaining an accurate balance between the water supply and loss.
11. *Maintenance of Acid-base Equilibrium* : By its efficient *buffering power (e.g.,* plasma proteins that are ionized, reduced, and oxidized to haemoglobin, etc.) and with the help of other organs it helps to maintain a constancy of acid-base content of all tissues including itself.
12. *Maintenance of Ion Balance* : The blood helps to maintain balance of ions between the cells and surrounding fluid.
13. *Functions of Mineral ions of Plasma* : The mineral ions of plasma function as nutrients and cellular control agents, and are essential regulators of the osmotic pressure of the blood (for details see under the heading Plasma).
14. *Maintenance of Own Integrity and Composition* : By regulation of its osmotic pressure, ion content, acid-base content, and by controlling the number of its cells, the blood has the unique ability to maintain its own integrity and composition. The uniformity in the pH of the blood of various vertebrates results from corresponding adjustments in the acid-base balance, achieved primarily through the amount of sodium.

Composition of Blood

The blood is a highly complex, liquid tissue composed of loose cells *or formed elements* and *plasma*. By volume, the blood is 55% plasma, and the remaining 45% formed elements or corpuscles. The formed elements have a specific gravity of 1.09 as compared with 1.03 for the plasma, thus accounting for their separation by gravitation or centrifugation. The blood *serum* is the fluid formed when the blood clots.

Blood Plasma : This fluid contains two groups of substances. One comprises the compounds that fluctuate in concentration in accordance with the changing rates of their supply and removal. In this category are a number of foods in transit, urea and other waste products in transit, hormones and vitamins in transit, and several other classes of materials present only occasionally. The second group includes the substances that are normally maintained at constant concentrations. In this category are water, most of the mineral ions, plasma proteins, and certain nutrients and waste products in transit.

(a) *Water* : The main constituent of blood plasma is water (about 90% by weight of the plasma) which has several functions . It maintains blood volume and blood pressure. The blood water functions as the *transport vehicle* for all other plasma components, and for respiratory gases, and blood cells.

(b) *Mineral Ions* : Dissolved inorganic materials carried in the plasma water comprise the *mineral ions*. These include mainly the positively charged Na^+, K^+, Ca^{++} and Mg^{++} ions and the negatively charged phosphate (H_2PO_4, HPO_4^{2-}-, PO_4^{3-}, chlorine (Cl^-), bicarbonate (HCO_3^-), and sulphate (SO_4^{2-}) ions. Over half of the total mineral content of plasma is sodium chloride. The ions are kept at constant concentration through mineral supply of food, turnover in tissue cells, excretion and other factors.

The mineral ions of the plasma function as nutrients and cellular control agents, and are essential regulators of the *osmotic pressure* of the blood. Another crucial regulating function of the plasma minerals is pH *control* and *pH* balance between blood and the tissue. Any small change of pH is readjusted to normal by the *buffering action* of the mineral ions. Some of the mineral ions exert other important effects. For example, calcium ions regulate the sensitivity of nerves and muscles. Calcium also influences the gel-sol states within cells, and it is a component of the cementing substances between cells. Since all cells are in contact with blood or lymph, the concentration of blood calcium evidently is of great significance.

(c) *Plasma Proteins* : Proteins make up about 7 to 8% of the plasma and are classified into four general subdivisions: *serum albumins* (4.2%), *serum globulins* (2.6%), *fibrinogen* (0.3%), and the remainder constituting 1% or less of the plasma. These compounds endow blood with some of its most important regulating functions, including the function of chemical body defence. They maintain the osmotic pressure of the blood, and since most of them are ionized, they also play a role in maintaining the pH of the blood, like mineral ions.

Some of the plasma proteins *(e.g., fibrinogen)* are involved in the coagulation of blood. Many of the proteins are active enzymatically. One of the catalytic agents is a clotting factor *prothrombin*. Others are enzymes such as are found in tissue cells generally and even in the intestine. For example, *trypsin* is usually present in the plasma. The catalytic functions of such enzymes in plasma are still largely obscure. Similarly obscure are the specific roles of plasma proteins called the *albumins*, so named because of their chemical resemblance to egg white. The activity of certain enzymes in plasma is a useful index of some abnormal conditions. For example, *serum amylase* is elevated in acute pancreatitis. Plasma levels of *glutamate—oxaloacetate aminotransferase*, *lactate dehydrogenase*, and other enzymes have proved to be of some value as diagnostic aids for assessing cardiac damage. Elevation of the plasma levels of these and certain other enzymes, *e.g., aldolase*, also occurs in liver disease.

Proteins called *globulins* are very important and include α ,

β, and γ globulins. Several proteins are designated as α_1 or α_2 globulins. Some are of unknown function, such as α_1- *acid glycoprotein,* which has 42% carbohydrate content, α-*foetoglobulin,* a glycoprotein is thought to serve functions in the foetus similar to those of albumin in the adult. *Retinol-binding protein,* which serves in retinal transport, forms an equilmolar complex with *transthyretin,* which also functions as a thyroxine-binding protein. *Ceruloplasmin* has eight sites that bind either Cu^+ or Cu^{2+}. *Haptoglobins* constitute about one-fourth of α_2-globulins and form complexes with haemoglobin. Such complexes are degraded by reticuloendothelial cells, and the iron is reutilized for haeme synthesis after degradation of globin and excretion of degraded haeme as bile pigment. Seven *protease inhibitors* have been found in human blood plasma, several of which are α-globulins.

The β-globulins of the plasma are a variety of proteins including the lipoproteins. Of them, *transferrin* constitutes about 3% of the total plasma proteins. It regulates the concentration of free iron in plasma, thereby preventing iron accumulation in tissues and urinary loss of iron. *Haemopexin,* another β-globulin, binds haeme and prevents its urinary excretion, thereby retaining haeme iron for further use. *C-reactive protein* promotes phagocytosis and β_2 *microglobulin* is the smaller subunit of the HL-A histocompatibility antigen complex, which regulates rejection of transplanted tissues.

The γ-globulin fraction contains the immunoglobulins, which are the *antibodies* and serve to destroy antigens. Included in the group are the so-called cryoglobulins, immunoglobulins that precipitate on cooling serum. Cryoglobulins are often found during inflammatory illnesses, *e.g.*, rheumatoid arthritis, as well as in multiple myeloma.

(d) *Glucose and Other Simple Sugars* : The concentration of glucose in the plasma normally averages about 0.1%.

It is absorbed by the capillaries in the intestine, stored principally as glycogen in the liver and muscles, and is carried to all parts of the body by the blood to be used as a principal source of energy in cellular metabolism.

(e) *Other Constituents of Plasma* : The remaining 2 to 3% of the plasma is made up of a variety of substances including amino acids, vitamins, lipids, hormones, urea, uric acid, and other metabolic waste products. In addition, plasma contains dissolved O_2 and CO_2 in transit to and from cells. All these substances fluctuate more or less widely in concentration in accordance with the changing rates of supply and removal.

Apart from the above constituents, blood plasma of certain animals also carries the blood pigments or respiratory pigments that transport the gases from the respiratory surface to the tissues and *vice versa*.

Blood Cells : The blood contains the following three kinds of formed elements or free cells which make up about 45% of the blood (Fig.).

(a) *Red Blood Cells* : The red blood cell or *erythrocyte* is by far the most prevalent of the formed elements. In man, each cubic millimeter of blood contains over 5 million red blood corpuscles. The enucleated mammalian erythrocytes lack DNA and the ability to synthesize proteins. Even with limited metabolic capability, these cells survive for weeks or months, depending on the species.

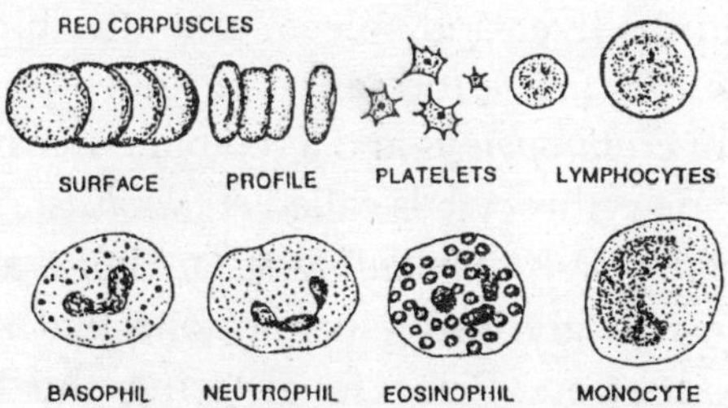

Formed elements of human blood.

Structure : In fishes, amphibians, reptiles, and birds the erythrocytes are oval, nucleated cells. Mammalian erythrocytes, on the other hand, are *non-nucleated* spherical discs, except in the camel group, whose red blood cells are oval.

Human erythrocytes average 7.5 μ in diameter, 2.4 μ in thickness at the edge, but only 1 μ thick in the centre. They are acidophilic and therefore, stained with eosin. Under electron microscope, mature erythrocytes lack an endoplasmic reticulum, Golgi apparatus, ribonucleoprotein and mitochondria. The human erythrocyte has a highly oriented structure of lipids, proteins and haemoglobin consisting of approximately 64% water, 28% haemoglobin (approximately 100 million haemoglobin molecules), 7% lipids or fatty materials, and the remaining 3% sugars, salts, enzymes and other proteins. Haemoglobin of the erythrocytes is responsible for the red colour of the blood.

Functions: Haemoglobin of the red blood corpuscles transports oxygen and enables our blood to carry approximately sixty times the amount of O_2 than an equal volume of water or plasma could transport in solution. Haemoglobin also functions as a carrier of a small quantity of CO_2, conveying it as a waste product from the tissues to the lungs for excretion.

Life History of the Red Blood Cell : The average life of a red blood cell is about 120 days. Under normal conditions they are continuously destroyed and new ones produced at the same rate of about 140 million per minute. In the embryo they are formed in the liver and spleen, and after birth, in the red bone marrow (Fig.). The formation of red corpuscles is called *haemopoiesis* or *erythropoiesis* and a red bone-marrow cell which gives rise to an erythrocyte is called *erythroblast*. The substance called the *erthropoiesis stimulating factor* (ESF), also known as *erythropoietin*, is a glycoprotein hormone that stimulates the red cell production. Prior to entering the bloodstream, erythrocytes appear as undifferentiated cells, each with a nucleus but without haemoglobin. As these cells mature, they undergo a series of changes which include disappearance of the nucleus (probably by extrusion or dissolution) and haemoglobin formation. Upon reaching maturity, they enter the bloodstream from the special connective tissue network of the red bone marrow in which they develop.

Control of Red Corpuscles Number : The major function of the red corpuscles is the transporting of respiratory gases. In the adult, red corpuscles are manufactured in the red marrow present at the ends of the long bones (*e.g.*, ribs, arms, legs). The liver and the spleen are the production sites in the embryo before the bones mature. After the skeleton is fully formed, the blood cell production is initiated in it, the spleen functioning principally a blood storing organ. It may contract like a sponge and squeeze reserve blood into circulation. The liver becomes the organ where the red corpuscles are destroyed.

Erythropoiesis is controlled by a feedback mechanism which depends upon the oxygen state of the blood. The reduced oxygen state of the blood affects the kidneys and the liver. As a result, the liver releases more globulin and the kidneys produce more *renal erythropoietic factor* (REF). The globulin and REF, interact in the blood to from ESF, which then stimulates greater red cell production with more haemoglobin, thus a greater quantity of oxygen can be carried which has a negative influence on the production of globulin and REF. At high altitudes, the low oxygen pressure stimulates the sequence just outlined, resulting in an eievated red cell count, a condition termed *polycythemia.* A persistently high oxygen concentration in the blood has the opposite effect on the liver and the bone marrow. Through this mechanism the number of red corpuscles is adjusted to oxygen requirements. Since such requirements and the composition of the atmosphere and hence the amount of oxygen normally remain fairly constant, the number of red corpuscles is also constant.

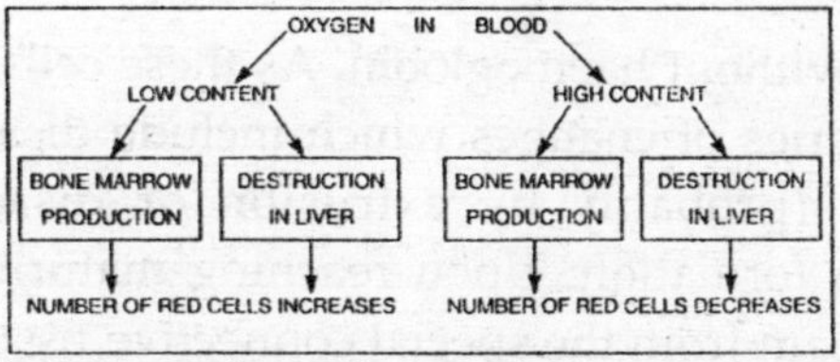

Control of red corpuscle number. Production in the bone marrow and destruction in the liver are geared to each other by the oxygen content of the blood.

When the number of red blood cells or their haemoglobin content or both are reduced significantly, the result is *anemia.* This may be brought about by excessive corpuscle destruction in an abnormal liver, or by extensive blood loss through wounding. In the latter case, production in bone marrow soon makes up the loss. Anemia may also arise as a result of inadequate production of corpuscles as, for example, when marrow cells are defective, or when the iron content of the diet is low.

However, the destruction of red corpuscles in the liver is random and nonselective *i.e.,* it affects young as well as old corpuscles. The iron of disintegrated haemoglobin becomes generally available in metabolism and may be used, for example, in the synthesis of cytochrome or of new haemoglobin. The tetrapyrrol fraction of destroyed haemoglobin is largely excreted in bile and urine.

White Blood Cells

White blood cells or *leucocytes* (leukocytes) are metabolically among the most active of the formed elements. About 7000 to 10,000 of these cells are found in each cubic millimetre of human blood. They are non-pigmented cells, *i.e.,* devoid of haemoglobin, contain a nucleus, and the majority are larger than erythrocytes, measuring from 8-15 μ in diameter depending on the type.

The term *leukopenia* describes a condition characterised by a low white cell count, which is usually associated with the viral diseases such as mumps, measles, chickenpox and poliomyelitis.

Structure and Types : Leucocytes are generally classified on the basis of the structural differences into two broad groups (Fig.): (i) leucocytes with a single nucleus (mononuclear) and a clear cytoplasm (agranulocytes), and (ii) polymorphonuclear leucocytes or granulocytes.

The first group consists of two types of cells : the *lymphocytes* and *monocytes,* both arising from the lymphatic tissue. White blood cells of the second group have their origin in the red bone marrow, contain cytoplasmic granules (hence the name granulocytes) and display a nucleus with two or more lobes. They are subdivided into three types (neutrophils, eosinophils, basophils, according to the staining reactions of their cytoplasmic granules.

(a) *Lymphocytes :* They range in diameter from 8 to 12 μ and constitute about 25% of the total number of leucocytes. They are small round cells with a more or less centrally placed nucleus. In a *small lymphocyte,* that is slightly larger than a red cell, nucleus is relatively large and occupies the major part of the cell. In a *large lymphocyte,* the cytoplasm is proportionally more and forms a wider zone around the nucleus.

(b) *Monocytes : A* monocyte is a large cell with a kidney or horse-shoe shaped nucleus. They range in size from 10-15 μ and are quite mobile. They represent 5 to 7% of the white blood cells.

(c) *Polymorphs or neutrophils :* A neutrophil has an irregular lobulate nucleus having 2 to 7 lobes. Their name is derived from the intense stain that the neutral dyes confer on their cytoplasmic granules (granules stain pale-violet with Leishmann's stain). Neutrophils are approximately 12-15 μ in diameter, circular in outline, and are the most active of all the white blood cells in terms of motility and ingestion of foreign particles. They constitute about 60 to 70% of the total number of leucocytes.

(d) *Eosinophils* : They are stained by the acid dye, eosin, and constitute 2 to 4% of the total white blood cells. They measure about 12-15 μ in diameter and contain a bilobed nucleus.

(e) *Basophils* : They are circular in outline and measure about 12-15 μ in diameter. Typically their nuclei contain 2 to 3 lobes, are centrally placed, and are stained relatively faintly. The cytoplasm is moderately abundant and cytoplasmic granules dominate the appearance of basophilic leucocytes. Unlike eosinophilic and neutrophilic granules, basophilic granules overlie and obscure the nucleus.

Leukocytes are also classified as the *lymphocytes* and the *phagocytes.* The latter can be divided into two categories, the *polymorphonuclear leucocytes* and the *monocyte-macrophages.*

Functions of Leucocytes : The primary function of the leucocytes is to protect the body against many types of bacterial infections. Polymorpho-nuclear leucocytes are the first cells to arrive at sites of inflammation and are an important component of the host defense machanism against bacterial infections. Macrophages function to scavange cellular debris generated by polymorphonuclear leucocytes.

They also destroy those antigens that the polymorphonuclear leucocytes are unable to digest. In addition, macrophages perform the critical immunologic role of processing antigens into a form that allows lymphocytes to recognize them. The ability of leucocytes to destroy microorganisms depends in part on their motility and their capacity to ingest foreign particles. In this respect *neutrophils* and *monocytes* are the most active. By forming pseudopods or streaming extensions of protoplasm, they engulf and destroy foreign particles, including bacteria. This ingestion process is called *phagocytosis*.

Lymphocytes too serve in body defence, but they also contribute to scar-tissue formation after internal or external injury and in this way they facilitate wound healing. When lymphocytes are stimulated by their specific antigens, they produce a variety of soluble products that function as mediators and modulators of inflammatory process. These diverse substances are called *lymphokines*. All of the lymphokines studied so far appear to be proteins. Numerous lymphocytes are normally present in the lymph nodes, where they act as lymph-purifying agents. For example, microscopic particles of dust, smoke, and other materials present in the atmosphere frequently get into the lungs and may become embedded in the lung tissue. Lymph then carries such particles to the lymph nodes where the lymphocytes engulf the particles and retain them permanently. However, lymphocytes are responsible at least in part for the manufacture of both the beta-and gamma globulins of the plasma, and therefore for the immunity substances *(antibodies)*.

Eosinophils and the *basophils* are not very mobile and show only a slight tendency to engulf foreign particles. Their role is not well established. However, they have some relationship to disease resistance since their concentrations in the bloodstream are specifically increased during certain infections. This phenomenon is very helpful in the diagnosis of certain diseases. For example, the eosinophil count is significantly increased during infection by hookworm, as well as in allergic states such as asthma.

The number of the white cells increases temporarily during infectious diseases, as would be expected in view of the phagocytic activity of the cells. An abnormally hight white cell count is called *leukocytosis.* An uncontrolled and abnormally rapid formation of white blood cells leads to the cancer of bone marrow or *leukemia.* This condition, which is eventually fatal, is accompanied by anemia because the bone marrow cells that normally give rise to red blood cells are displaced by leukemia cells.

Blood Platelets

Blood platelets or *thrombocytes* (Fig.) are small, oval fragments of specialized bone marrow cells, measuring from 2 to 4 μ in diameter and generally lacking nuclei. About 250,000 of these cells are found in each cubic millimeter of human blood. Blood platelets contain centrally placed blue-violet granular material, the *granulomere,* which may give a false impression of a nucleus. The periphery of platelets is clear of granules, stains light blue, and is called the *hyalomere.* Sometimes thread-like processes extend from blood platelets.

Function of Platelets : The metabolic activity of blood platelets has recently been found to be equal to, if not greater than, that of the white blood cells. Blood platelets play an essential role in the coagulation of blood. The platelets have the capacity to absorb many of the factors required for blood clotting and carry them to the site of blood coagulation. Platelets are necessary for the intrinsic clotting process. They release a phospholipid, the 'platelet factors 3,' which participates in the conversion of prothrombin to thrombin by blood clotting factors X (10th) and V (5th). Platelets may also function to plug small defects in a blood vessel. Platelet *serotonin,* a vasoconstricting agent, may induce constriction of small blood vessels at the site of injury and may function to reduce blood loss.

Platelets are required for *clot retraction.* It is likely that the ATP and actinomyosin-like compounds in platelets are required for this phenomenon. *Tocatins* (1936) has described clumps of platelets adhering to fibrin strands. As these clumps are formed, fibrin becomes twisted and bent and clot retraction takes place. Knots of platelets are found at the intersections of fibrin strands and may strengthen the clot. Intimate association of platelets and fibrin has been confirmed by electron microscopy.

Smith et al (1983) have summarized the roles of platelets in coagulation as follows:

(i) They form the platelet plug in injured vessels by adhering to exposed subendothelial cell structures.

(ii) They are the site of activation of some clotting factors.

(iii) They provide the surface upon which certain clotting factors bind.

(iv) They are a source of certain clotting factors, *e.g.*, factor XIII and phospholipids.

Main Constituents and Functions of Blood plasma.

S. No.	*Component*	*Function*
1.	Water	Maintains blood volume and pressure, forms lymph, water supply of cells, provides vehicle for other constituents.
2.	Mineral ions	Maintain osmotic balance, pH balance, buffer capacity, varied effects on tissue cells.
3.	Plasma proteins	All maintain osmotic and pH balance
	fibrinogen	Participates in blood clotting
	prothrombin	Participates in blood clotting
	albumins, enzymes	Functions obscure, basis of
	globulins	blood types, act as antibodies.
4.	Glucose and other metabolites	In transit to and from cells.
5.	O_2, Co_2, urea, various foods, hormones, vitamins, etc.	In transit to and from cells,

Cellular Constituents of Human Blood and their Functions

Component	*Origin*	*Size (diameter)*	*Appr. Number per cubic millimeter*	*Functions*
Erythrocytes	Spleen, liver (embryo) red, bone marrow (adults)	7 - 8 μ	5 million	Transport of O_2 and CO_2
Leucocytes	Red bone marrow	8 -15 μ	8000	Engulf bacteria, foreign bodies, aid in wound healing
Lymphocytes	Lymph nodes	8 -12 μ	20 - 25% of total white cells.	—do—
Thrombocytes (platelets)	Red bone marrow, lungs	2-4 μ	250,000 to 600,000	Release thromboplastin, aid in blood clotting

A deficiency of platelets is called *thrombocytopenia.* It can be caused by many toxic agents, including certain drugs, ionizing radiation, and some types of infections. In *thrombocytopenia purpura,* platelet levels in the blood are quite low, large extravasations of blood, or *purpura,* take place. Such blood disorders as leukemia may also be accompained by thrombocytopenia.

Physiology of Blood Coagulation : When blood is shed it loses its fluidity in a few minutes and sets into a semisolid jelly. This phenomenon is called *coagulation* or *clotting.* On keeping further, the clot retracts to a smaller volume and presses out a clear straw-coloured fluid, called the *serum.* Serum will not clot any more. The blood coagulation is a very complicated process and is under sensitive control. The coagulation is the property of plasma alone. The red blood corpuscles do not take part in it. They only become caught up in the meshes of the clot and are thereby removed. It is due to this fact that the clot has a red colour, and the serum is a clear non-cellular fluid. Blood platelets take part in coagulation of blood. They become sticky and rapidly adhere to the damaged blood vessels by binding to endothelial connective tissue.

The platelets also adhere to one another and finally form a plug, which is sufficient to stop bleeding if damage is small. The cessation of bleeding is called *hemostasis.* As the platelets aggregate, they release vasoactive amines, *e.g.*, serotonin and epinephrine, and prostaglandin metabolites, *e.g.*, thromboxane A_2, which stimulate vasoconstriction. Blood coagulation is then initiated around the platelets and the damaged tissues, leading to the formation of a *thrombus*, or blood clot, the major chemical defense against blood loss.

The following theories have been proposed to explain the mechanism of clotting of blood.

Best and Taylor's Theory : According to this theory, four substances are necessary for coagulation. These are : *prothrombin, calcium ions, thromboplastin* and *fibrinogen* (Fig.). Prothrombin is produced in the liver and is present in small quantities in the blood plasma also. Thromboplastin is a fatty compound, found in the tissues. Calcium occurs as an inorganic constituent in the plasma. The optimum concentration required is 5 mg per 100 ml and the normal blood contains about this level. Fibrinogen is a soluble protein, found dissolved in the plasma.

When blood is shed, thromboplastin (thrombokinase) is liberated from the damaged tissue and from the disintegrated blood platelets. It acts on prothrombin in the presence of calcium, converting it to active *thrombin.* Now thrombin reacts with fibrinogen, and converts it into fibrin, which is insoluble. The latter constitutes the blood clot or *thrombus* in which blood corpuscles are trapped.

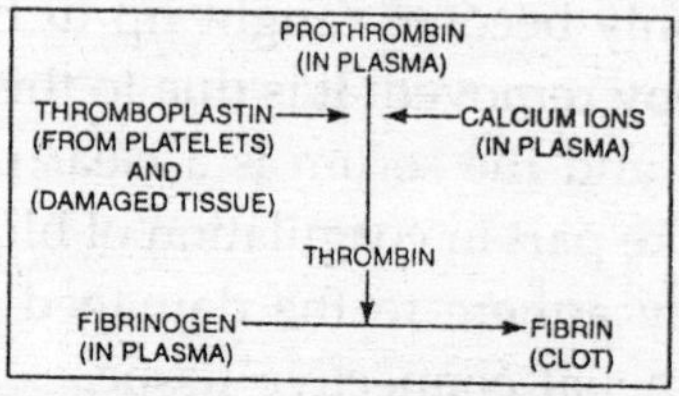

The main events of blood clotting (coagulation).

Blood within the blood vessels does not clot due to the lack of free thromboplastin, which is necessary to convert prothrombin into thrombin. If by chance thrombin is formed inside the blood vessels, it is immediately neutralized by another substance called *antithrombin*, which is normally present in the blood in very small quantities.

Howell's Theory **:** It states that clotting mechanism of mammalian blood takes place in two distinct stages:

(a) formation of thrombin from prothrombin.

(b) action of thrombin on fibrinogen.

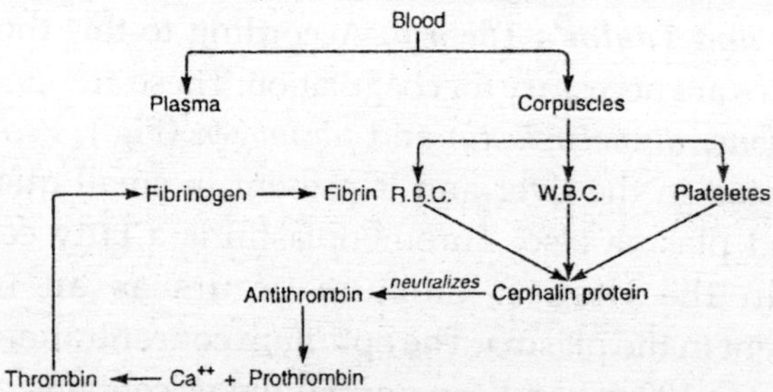

Mechanism of blood clotting (according to Howell).

However, it explains the formation of thrombin from prothrombin by the action of calcium alone. The blood does not clot in the blood vessels because of the presence of *antithrombin* which prevents the activation of prothrombin. On the injury of blood vessels, tissue cells produce a substance known as *cephalin protein* which neutralizes the antithrombin and thus allows calcium to react with prothrombin. As a result, thrombin is formed, which reacts with firbrinogen to form insoluble fibrin.

Fuld and Spiro's Theory : This theory differs from Howell's theory only in one important respect *i.e.,* it takes into account the presence of the enzyme *thrombokinase* which is produced especially by blood platelets, and reacts as follows :

(a) Thrombokinase + calcium ions + prothrombin → thrombin

(b) Thrombin + fibrinogen → Fibrin (clot)

Blood Coagulation Factors

In 1954, an International Commission on Blood Coagulation suggested a system of nomenclature of blood coagulation factors, which are designated by Roman numerals according to the recommendations of the commission. These factors are first dealt with in numerical order, which indicates the historical sequence of their discovery.

(i) *Factor I or Fibrinogen* : Plasma concentration of fibrinogen is 300 mg/dl and its molecular weight is 340,000. The molecule has two identical subunits, each consisting of three polypeptide chains, α , β , and γ . The chains are bound together by disulphide bonds at each end of the molecule (Fig.). It is globulin in nature and is distinguished from other plasma proteins by its property of clotting, during which *firbinogen* is converted into *fibrin.*

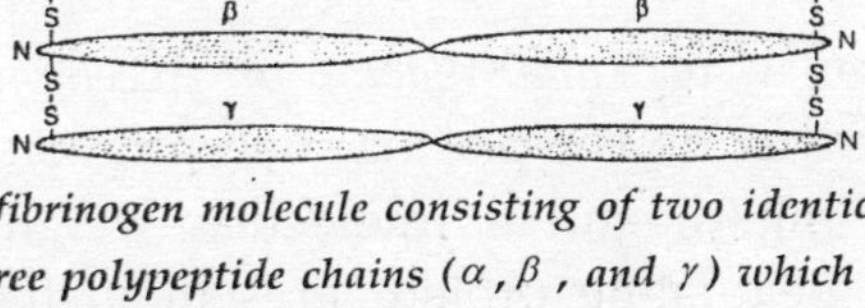

The structure of fibrinogen molecule consisting of two identical subunits, each having three polypeptide chains (α , β , and γ) which are bound together by disulphide bonds at each end of the molecule. The N-terminal amino acids are indicated by N.

(ii) *Factor II or Prothrombin* : Plasma concentration of prothrombin is 15 mg/dl and its molecular weight is

72500. It is protein in nature and is manufactured in the liver. However, vitamin K is essential for the formation of prothrombin. During clotting *prothrombin* is converted into *thrombin* by means of *prothrombin activator* which is also called thromboplastin.

(iii) *Factor III or Thromboplastin or Tissue Factor* : It is a protein having a molecular weight 43000. It is derived from two sources : intrinsic in the plasma, and *extrinsic* in the tissues. *Intrinsic thromboplastin* is formed in the plasma due to interaction between different plasma factors, *e.g.*, factor XII (Hageman factor), factor XI (PTA), factor IX (Christmas factor), factor VIII (antihaemophilic factor), calcium ions, factor V and factor X. *Extrinsic* or *tissue thromboplastin* is formed from different tissues, e.g., extracts of lungs, placenta, brain, etc., as a result of injury. So far it was known that prothrombin was converted into thrombin with the help of calcium ions and thromboplastin liberated from the damaged tissues. But recently it has been found that various plasma factors, *e.g.*, factor VII or *proconvertin,* are required for such conversion and the process is called *extrinsic thromboplastin formation.*

(iv) *Factor IV or Calcium* : The plasma concentration of Ca^{2+} ions is 0.5 mg/dl. Ionic calcium acts as a *cofactor* in coagulation. It is essential for the formation of both intrinsic and extrinsic thromboplastins and also in the conversion of prothrombin into thrombin.

(v) *Factor V or Labile Factor or Accelerator Globulin or Proaccelerin* : The plasma concentration of proaccelerin is 1 mg/dl and its molecular weight is 330,000. It is a heat-labile protein present in the plasma. This factor is necessary for the complete conversion of prothrombin into thrombin by the extrinsic or intrinsic thromboplastin.

(vi) *Factor VI or Accelerin or Va* : This factor is a hypothetical activation product of proaccelerin (factor V).

(vii) *Factor VII or Stable Factor or Proconvertin* : This factor is present in the plasma and is not used up during clotting. It is a protein associated with prothrombin. The plasma concentration of proconvertin is 0.05 mg/dl and its molecular weight is 45,500. It accelerates the formation of tissue thromboplastin, being activated by extract released from the damaged tissue. During blood clotting proconvertin is changed into convertin.

(viii) *Factor VIII or Antihaemophilic Factor (AHF) or Antihaemophilic Globulin (AHG) or Platelet Co-factor I* : The plasma concentration of antihaemophilic/von Willebrand factor is 1.6 mg/dl and its molecular weight is $1\text{-}2 \times 10^6$, having 15% carbohydrate content. This factor helps in the formation of intrinsic thromboplastin and intrinsic prothrombin coversion. It is present in the plasma and disappears when the blood clots. It is a protein closely associated to fibrinogen. This factor is antihaemophilic.

In *Haemophilia* (bleeder's disease) the defect is not in the platelets but it is due to the absence of this factor which helps in the breakdown of platelets and liberation of platelet cofactor I or thrombopastinogen. The deficiency of AHG resulting in classical haemophilia in the male is transmitted as a sex-linked recessive character. The body fails to synthesize this essential globulin due to the absence of the specific enzyme which is controlled by the mutant gene.

(ix) *Factor IX or Christmas Factor or Plasma Thromboplastin Component (PTC) or Platelet Co-factor II or Antihaemophilic Factor B* : The plasma concentration of Christmas factor is 0.3 mg/dl and its molecular weight is 57,000. It has

26% carbohydrate content. This factor is necessary for intrinsic thromboplastin formation. The absence of this factor leads to a disease known as *Haemophilia C.* This type of disease was first seen in a patient named *Christmas,* hence the name Christmas factor. This factor is not used up during clotting.

(x) *Factor X or Stuart-Prower Factor* : The international nomenclature has been given to this factor in 1959. The plasma concentration of this factor is 1.5 mg/dl and its molecular weight is 59,000. It has 15% carbohydrate content. It is a circulating plasma zymogen, which on activation by factor IXa serves to activate prothrombin to thrombin.

(xi) *Factor XI or Plasma Thromboplastin Antecedent (PTA) or Antihaemophilic Factor C* : The plasma concentration of factor XI is 0.6 mg/dl and its molecular weight is 160,000. It has 12% carbohydrate content. This is activated by the active Hageman factor (XIIa) and ultimately leads to the formation of thrombin. Deficiency of this factor leads to haemophiloid D and is transmitted as a sex-linked dominant in both the sexes.

(xii) *Factor XII or Hageman or Surface Factor:* The plasma concentration of factor XII is 2.9 mg/dl and its molecular weight is 76, 000. It has 15% carbohydrate content. This is a protein in nature. Inactive form is activated on surface contact. This is also known as the clot-promoting factor or glass factor.

(xiii) *Factor XIII or Fibrin-stabilising or Laki-Lorand Factor (LLF) or Fibrinase* : The active form along with calcium ions converts soft fibrin clot to a solid, fibrous state. The plasma concentration of factor XIII is 2.0 mg/dl and its molecular weight is 320,000. It has 2.5% carbohydrate content.

Other Factors : High-molecular-weight kininogen (HMW-K) or Fitzgerald factor :

This factor catalyses the initial reaction in which inactive factor XII is converted to active factor XII (XII a).

Pre-Kallikrein (Pre-K) or Fletcher Factor : Pre-Kallikrein or inactive plasma kallikrein is activated by pre-kallikrein activators, which are proteolytic fragments of the active form of clotting factor XII (Hageman factor).

Kallikrein : Plasma kallikrein also catalyses the activation of factor XII.

Platelet Phospholipid (PL) : Phospholipids from aggregated platelets (PL) and Ca^{2+} are necessary for full activation of factor X. This also catalyses the conversion of prothrombin to thrombin.

Protein C : Although not listed as clotting factors, protein C is often referred to as factor XIV. Its molecular weight is 62,000 . It is a ciruclating zymogen of protease which on activation by thrombin specifically inactivates factors V and VIII by proteolysis. Congenital absence of protein C leads to uncontrolled intravascular clotting and, in general, death in infancy.

Some features of human blood coagulation factors are summarized in Table :

The Human Blood Coagulation Factors and Some of their Properties.

Factor No.	*Name*	*Plasma concentration mg/dl*	*Molecular weight*	*Carbohydrate content %*
I	Fibrinogen	300	340,000	4
II	Prothrombin	15	72,500	8
III	Thromboplastin	...	43,400	...
IV	Calcium ions	0.5	....	...

Contd...

Factor No.	*Name*	*Plasma concentration mg/dl*	*Molecular weight*	*Carbohydrate content %*
V	Proaccelerin	1	330,000	?
VI	Accelerin	....	......	...
VII	Proconvertin	0.05	45,500	...
VIII	Antihaemophilic factor A	1.6	1-2 × 10^6	15
IX	Christmas factor	0.3	57,000	26
X	Stuart factor	1.5	59,000	15
XI	Plasma thromboplastin antecedent	0.6	160,000	12
XII	Hageman factor	2.9	76,000	15
XIII	Fibrin-stabilizing factor	2.0	320,000	2.5

Role of Thrombin : It is a homogeneous glycoprotein. It autocatalyses its own formation by converting plasma Ac-globulin into active serum Ac-globulin and by labilising platelets to accelerate thromboplastin generation and liberates vasoconstrictors. By acting as a proteinase, it splits only four peptide bonds in the conversion of fibrinogen to fibrin (Fig.) and these are all bonds between arginine and glycine residues.

Role of Phospholipid : Phospholipid *kephalin* (cephalin) helps in the formation of *prothrombinase.* In the intrinsic system it is in the platelet factor 3, a phospholipid which participates in the conversion of prothrombin to thrombin by factors X and V, and in the extrinsic in tissue thromboplastin.

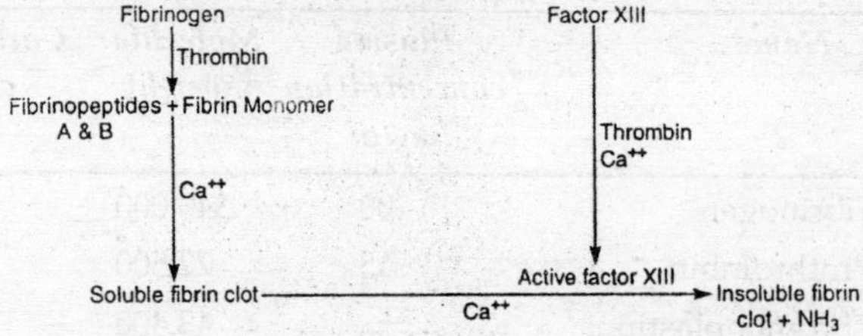

Showing the action of thrombin on fibrinogen.

Role of Vitamin K : Normal production in the liver of prothrombin as well as factors VII, IX and X requires adequate dietary intake of vitamin K *(koagulation* vitamin). Vitamin K is required to form the binding sites for Ca^{2+} in prothrombin as well as in factors VII, IX and X.

The Ca^{2+} binding sites are formed by clusters of y⁻ carboxyglutamate residues in the amino-terminal regions of each factor.

The antimetabolites of vitamin K, such as *dicumarol,* inhibit the synthesis of these factors. Vitamin K is an essential cofactor for a *carboxylase* located on the lumenal side of the rough endoplasmic reticulum that carboxylates glutamyl residues in newly synthesized vitamin K-dependent clotting factors.

Prothrombin Activator : This is formed in two ways : (i) as a result of tissue (extrinsic system) and (ii) by activation of an intrinsic system consisting entirely of the blood constituents. In both the cases the key reaction is the conversion of factor X to its active form (Xa). The latter then interacts with phospholipid and factor V to form prothrombin activator. In the extrinsic system the reactions are as given below :

$$\text{Factor X} \xrightarrow{\text{Tissue factor, Factor VII, } Ca^{2+}} X_a$$

$$X_a \xrightarrow{\text{Phospholipid, FactorV}} \text{Prothrombin activator}$$

(Prothrombinase)

However, in the intrinsic system the reactions leading to the formation of Xa are more complicated as shown in the following hypothesis.

Hypothesis of Davie and Ratnoff : *Davie* and *Ratnoff* (1965) have proposed the so-called 'waterfall' sequence hypothesis to explain the succession of events taking place in the coagulation of blood.

The sequence of events leading to coagulation depends upon whether *thromboplasin* is present or not. It is the active substance in aqueous extracts of all tissues. Thus, a wound that results in bleeding also disrupts tissue causing the liberation of thromboplastin. This substance then initiates the sequence shown in Fig. Because a substance is required that is not a normal constituent of blood, this sequence is referred to as the *extrinsic system of coagulation.* On the other hand when all the substances involved are normally found in the blood, this sequence is referred to as the *intrinsic system of coagulation.* In both the systems each coagulation factor exists in the plasma in an inactive (proenzyme) form and is activated sequentially until thrombin is formed, which then converts fibrinogen into fibrin (Fig.).

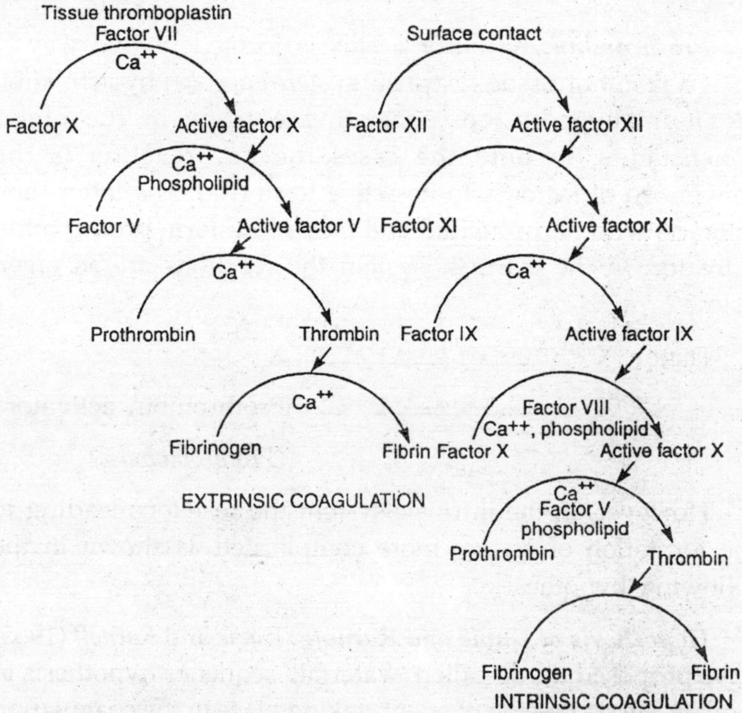

The "waterfall' schemes of extrinsic and intrinsic blood coagulation according to Davie and Ratnoff.

Enzyme Cascade Hypothesis

According to this hypothesis, suggested by *Macfarlane* (1967), the clotting mechanism consists of a large number of enzymatic reactions.

The term "cascade" expresses the relationship whereby an enzyme acts on a precursor, converting it to an enzyme that then acts on another precursor, converting it to an active enzyme and so on.

In the *intrinsic system*, the initial reaction is conversion of inactive factor XII to active factor XII (XIIa). This is catalysed by high-molecular-weight kininogen (HMW-K) and kallikrein (Ka). This activation usually occurs when blood is exposed to the collagen fibres underlying the endotheluim in the blood vessels.

Active factor XII (XIIa) then activities factor XI, and active factor XI (XIa) activates factor IX. Activated factor IX (IXa) forms a complex with active factor VIII, which is activated when it is separated from von Willebrand factor. Then, the complex of IXa and VIIIa in the presence of phospholipids from aggregated platelets (PL) and Ca^{2+} activate factor X.

After the formation of Xa both systems (intrinsic and extrinsic) pursue a common pathway till the formation of the insoluble fibrin clot. The *extrinsic system* is triggered by the release of tissue thromboplastin (TPL), a protein phospholipid mixture that activates factor VII. The TPL and factor VII activate factors IX and X.

Thereafter, in the presence of platelet phospholipid (PL), Ca^{2+} and factor V, activated factor X catalyses the conversion of prothrombin to thrombin. Finally, thrombin converts fibrinogen into fibrin and also catalyses the activation of factor XIII so that fibrin clot is stabilized (Fig.).

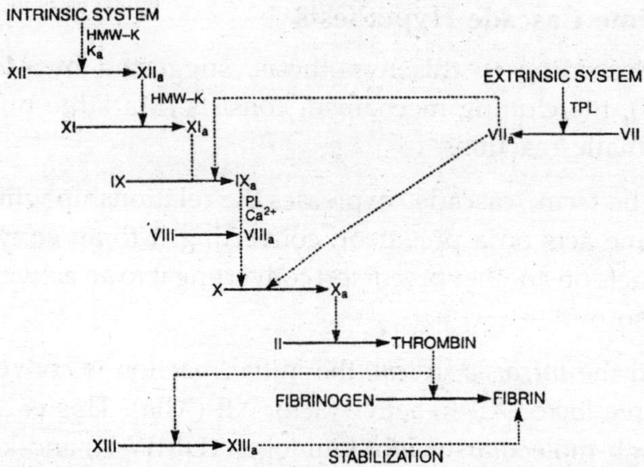

The scheme of reactions leading to blood coagulation by activation of intrinsic and extrinsic systems.

Another cascade, in which the enzyme *tissue plasminogen activator* converts the inactive precursor *plasminogen* to fibrin specific proteolytic enzyme *plasmin,* brings about the break down of the insoluble fibrin clot. This is called clot lysis pathway (Fig.).

One very important aspect of the enzyme cascade hypothesis is that it is an amplifying system so that minute amounts of earlier factors can lead to final rapid conversion of large amounts of fibrinogen to fibrin. For example, *Esnouf* and *Macfarlane* (1968) quote that the clotting of 1 ml of blood involves 3 *mg* of fibrinogen, 100 μg of factor II, 10 μg of factor X, 0.1 μg of factor VIII, and probably less of the earlier factors.

However, two comments on the enzyme cascade hypothesis must be made. Firstly, it is an over-simplification of an extremely complicated process. Because it has not yet been shown that all the reactions involved are enzyme-substrate in type. Secondly, there are some additional reactions. For example, thrombin exerts positive feedback effects on the factors concerned in the formation of prothrombin activator. It activates factor VIII, perhaps by releasing it from its attachment to fibrinogen. It also increases the activity of factor V and XIII, and it also promotes aggregation of platelets, thus increasing the amounts of available phospholipid. However, after activating these mechanisms, thrombin soon inactivates them. Thus, thrombin intensifies the reactions promoting the local clotting, but only for a short time, i.e., until a firm clot has been formed. The spread of clot formation from the site of injury does not normally occur, partly because of the reactions just mentioned which cut it short, and partly because blood contains inhibitors of probably all the factors known to be concerned with the mechanism of clotting.

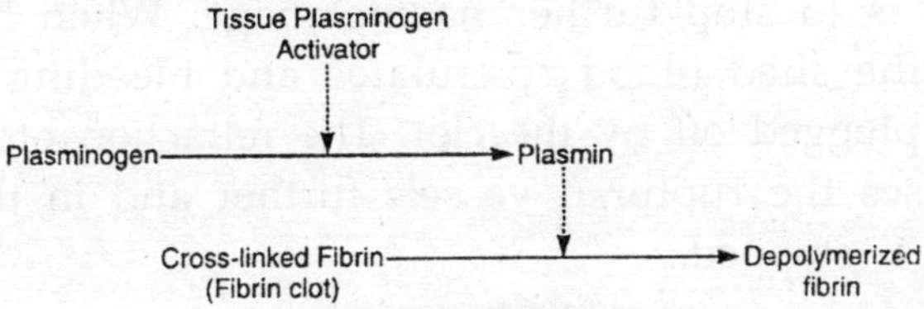

Clot lysis pathway.

According to Loewy et al (1991), clotting is an example of how post-translational modification—the covalent modification of proteins after the translation process—is used to regulate assembly processes.

Intravascular Clotting or Thrombosis : Thrombus is a clot formed in the living heart or inside the blood vessels from constituents of the blood. This is formed due to slowing of circulation and damage of the vascular endothelium. Masses of platelets are deposited on the damaged endothelium. Filaments of fibrin also form a network in this region. The platelets liberate thromboplastin. The fibrin, entangled in the lamellae of platelets, forms the thrombus or clot.

Intravascular thrombosis sometimes occurs in coronary and cerebral vessels which are called *coronary thrombosis* and *cerebral thrombosis* respectively.

Fibrinolysis : Clotted blood remains intact for several weeks if kept sterile. But if it is not kept sterile, the clot breaks up.

This fibrin breakdown in the clot is known *as fibrinolysis* and is brought about by a proteolytic enzyme in the plasma known as *plasmin* or *fibrinolysin*. The precursor of the enzyme, *plasminogen* (also called *profibrinolysin)* is activated to plasmin by activators present in the tissues, serum, urine, and some bacteria. Normally fibrinolysis is preveneted by the presence of *antiplasmin* in the blood. This remains attached to the plasma albumin.

Importance of Coagulation of Blood : The phenomenon of coagulation is of great physiological significance. Its purpose is to stop further haemorrhage. When bleeding occurs, the shed blood coagulates and bleeding vessels become plugged off by the clot. The retraction of the clot compresses the ruptured vessels further and in this way bleeding is stopped.

Blood and Lymph

Blood is a fluid connective tissue circulating in the body. It provides one of the methods of communication between the cells of different parts of the body. The blood carries oxygen and nutrients to the tissues and carbon dioxide and waste products from the tissues to the excretory organs. In addition it conveys antibodies to the site of injury or disease and hormones or chemical messengers from the endocrine glands to various target organs. The blood is the medium of transportation of all these substances.

The blood constitutes approximately 8% of the body weight or 5.6 litres in a 70 kg man.

Composition and Functions of Blood : The blood is composed of a faintly yellow transparent fluid known as the *plasma* and floating in this fluid are numerous *cells* or *corpuscles* of different kinds. The plasma constitutes approximately 55% and the corpuscles about 45% of the volume of blood.

I. The Cellular Contents of Blood

There are three varieties of cells or corpuscles present in blood.

A. *Erythrocytes or red blood corpuscles.*

B. *Leucocytes or white blood cells.*

C. *Blood platelets.*

Erythrocytes : As a matter of fact the word 'erythrocyte' is inappropriate for mammals because the adult corpuscle lacks a nucleus and, therefore, is not a complete cell. However, the absence of nucleus in mammals increases respiratory efficiency of the red blood corpuscle. All vertebrate, other than mammals have nucleated red cells.

Shape : These are circular, biconcave non-nucleated discs. The central part of the corpuscle is much thinner than the circumference, thus the term biconcave. In edge view the outline is like that of a dumb bell, this shape favours flexibility and the absorbing and releasing of gases quickly. RBC's of camel are oval in shape.

Size : In man, the living corpuscle is 8.5 μ in diameter. Rabbit has slightly smaller corpuscles than man. The extremes are 9.4 μ in elephant and 2.5 μ in musk deer. Amphibian RBC's are largest among the vertebrates.

Number : The normal erythrocyte count is usually slightly lower in women than in men.

Women : 4.5 to 5 million per cubic millimeter of blood.

Men : 5 to 5.5 million per cubic millimeter of blood.

The number varies with different pathological and physiological conditions. For example, the number is decreased in pernicious anaemia. On the other hand, continued living at high altitudes or exposure to carbon monoxide causes an increase in the number of circulating red blood corpuscles. An abnormal rise in RBC count is called *polycythemia.*

Colour : A single fresh corpuscle is pale-yellow, but when the corpuscles are seen in a mass, the haemoglobin gives the characteristic red colour.

Structure : The erythrocytes are surrounded by a thin peripheral membrane (0.02 μ) consisting of a lipid-protein complex. It serves as an elastic and semipermeable barrier. The nucleus and cytoplasmic organelles like endoplasmic reticulum, golgi apparatus, centrioles and mitochondria were lost in the later stages of development. The internal cytoplasm is a homogeneous colloidal sol.

Physical Properties : The red corpuscles are soft and flexible. They withstand much bending, squeezing and deformation as they are pushed through the narrow capillaries.

Rouleaux are columns of corpuscles, stacked one above the other like a pile of coins. The adherence is due to surface tension. These are formed spontaneously in a stagnant circulation or in a column of drawn blood.

Crenation is a shrivelling of the corpuscles due to concentration of the plasma by evaporation. It produces a prickly contour.

Haemolysis is the outward passage of haemoglobin from the corpuscle due to injury of plasma membrane caused by lipid solvents, bile salts and snake venom.

Agglutination is the clumping of corpuscles induced by various agents like salts solutions, glucose solution and antibodies.

Chemical Composition : The respiratory pigment of red corpuscles is *haemoglobin.* It is a complex protein having iron as one of the main constituents. The membrane of the erythrocyte is composed of 49% protein, 44% lipid, and 7% carbohydrate. 26.5 crore molecules of haemoglobin are packed in the intracellular frame-work of a red corpuscle.

Formation (Erythropoiesis) : The erythrocytes are formed in liver, spleen and lymphs in the *embryo* ; and in the red bone marrow (myeloid haemopoietic tissue) in the *adult*. The red bone marrow is and in the canacellous bone at the extremities of long bones and between layers of compact bone in flat and irregular bones such as the cranium, vertebrae, ribs, sternum, clavicles, scapulae and Pelvis. The process is known as *erythropoiesis* and is completed in 72 hours.

The development of a mature red corpuscles takes about a week, during which the endothelial cell enlarges, divides, forms haemoglobin and finally loses its nucleus. This process is called maturation and takes place along the following lines:

Stems cells or Myeloblasts ⟶ Proerythroblasts ⟶ Erythroblasts ⟶ Normoblasts ⟶ Reticulocytes ⟶ Erythrocytes.

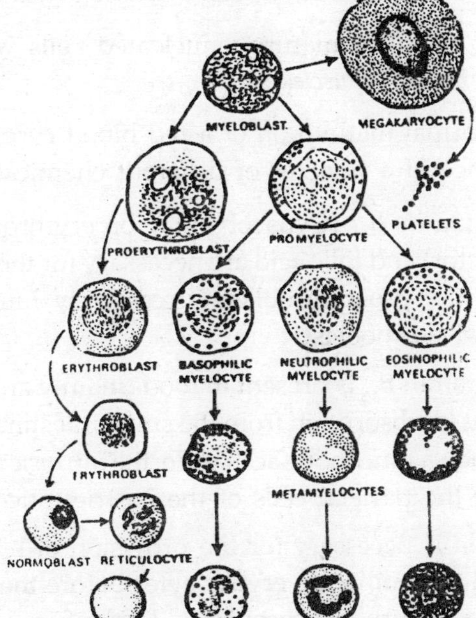

Erythropoiesis.

Myeloblast is an amoeboid cell with fairly abundant cytoplasm and a relatively primitive, undifferentiated nucleus.

Proerythroblast is larger than the myeloblast. Its nuclear network is slightly coarser and the cytoplasm is deeply basophilic due to the presence of ribonucleic acid (RNA).

Erythroblasts are half the size of the proerythroblasts due to mitosis in that stage. The nucleus is checkered with coarse chromation masses. The cytoplasm is losing its RNA and is acquiring haemoglobin.

Normoblasts are derived from erythroblasts by mitosis, hence they are smaller in size. The nucleus becomes progressively smaller and mitosis ceases after this stage. The cytoplasm stains strongly acidophilic due to haemoglobin.

Reticulocytes are young, immature erythrocytes with a delicate network of ribonucleic acid in the cytoplasm. The nuclei of normoblasts have been lost by extrusion in this stage.

Erythrocytes are mature, enucleated cells which are also called *red blood corpuscles.*

The normal maturation of a red blood corpuscle requires the presence of a number of different chemical substances :

(a) Vitamin B_{12} (Cynacobalmin or erythrocyte-maturing factor) and folic acid are necessary for the development of the proerythroblast successively into erythroblast and normoblast.

    Vitamin B_{12} is present in food (mainly animal proteins) but its absorption from the small intestine is dependent upon a mucopolysaccharide, the *intrinsic factor*, secreted by the parietal cells of the gastric mucosa.

(b) Iron is necessary for the provision of haemoglobin to fill the immature erythrocytes before they can become the mature red corpuscles. Lack of iron in the diet or loss of iron in bleeding causes iron-deficiency anaemia.

The normal haemoglobin levels in the blood are as follows :

Men 14 g per 100 ml blood

Women 12 g per 100 ml blood

(c) In addition to a normal diet, containing protein and ran small amounts of cobalt, copper, nicotinic acid, riboflavin and Vitamin C are also essential for erythropoiesis.

The development of red blood corpuscles is controlled by a feed-back mechanism. Deficiency of oxygen following haemorrhage or because an individual, lives at high altitudes where the oxygen pressure in the atmospheric air is reduced; and a hormone called *erythropoietin,* secreted by the kidneys are the two main factors that stimulate the bone marrow to increase its production of erythrocytes.

Metabolism : The mature mammalian erythrocyte lacks mitochondria ; hence the cytochrome system is absent and the tricarboxylic acid cycle is not evident. The energy of the mature erythrocyte is supplied primarily by anaerobic glycolysis and the phosphogluconate pathway.

Life Span and Disposal : The Life span of human erythrocytes is approximately 120 days (100 days in frog). When worn out, the cells break down in the spleen, which, however, saves and stores the iron for use in the preparation of fresh haemoglobin. The protein part of the erythrocyte is converted to a substance called *biliverdin* which is almost completely changed to *bilirubin,* a yellow pigment. While circulating in the blood, bilirubin is bound to the plasma protein albumin. It is carried by splenic and portal veins to the liver where it is again changed into a conjugated form arid excreted in the bile as bile pigment.

Functions : Haemoglobin in the erythrocytes carries oxygen from the lungs to all the cells of body for tissue respiration.

Leucocytes or White Blood Cells : The white blood cell or leucocyte contains a nucleus and is therefore, a true cell.

Structural Characteristics : Leucocytes are typical cells with amoeboid ability. There is a centriole near the nucleus but mitosis does not occur in the adult state. As in all true cells; mitochondria and Golgi apparatus is present.

The leucocytes differ from the erythrocytes in that they possess a nucleus, are fewer in number and larger in size. They vary in size from about 8 to 15 microns.

Number : The average normal limits are between 6,000 to 10,000 per cubic millimetre of blood. The count in children is higher than in adults. In pathogenic states, there are variations from the normal number.

An increase in the number of white blood cells is called leucocytosis : more than 12.000 per cu. mm. indicates some disease. This increase is out of proportion for one or more types.

(i) *Neutrophils* increase in pus-forming infections.

(ii) *Eosinophils* increase in allergies and parasitic infections.

(iii) *Basophils* increase in Chicken-pox.

(iv) *Lymphocytes* increase in whooping cough.

(v) *Monocytes* increase in tuberculosis.

A decrease below 6,000 is called leucopenia as in typhoid fever.

Classification : The leucocytes are divided into two main varieties :

(i) Granular or Polymorphonuclear leucocytes,

(ii) Agranular or Mononuclear leucocytes.

This classification is based on :

(i) the shape of the nucleus

(ii) the presence or absence of granules in the cytoplasm and

(iii) the type of granules, if present.

Granular Leucocytes

These cells develop in the red bone marrow from the same parent cells as the erythrocytes. *i.e., myeloblasts* in the red bone marrow. Before entering the blood stream, they develop through promyelocyte and myelocyte stages.

Myeloblast → Promyelocyte → [Basophilic myelocyte / Neutrophilic myelocyte / Eosinophilic myelocyte] Metamyelocytes

→ Juvenile → Basophil
→ Juvenile → Neutrophil
→ Juvenile → Eosinophil

As these cells develop, specific granules appear in larger numbers and they retain their nucleus.

The granulocytes constitute about 75% of total white Wood cells and three varieties are described :

(a) Neutrophils approximately 70%

(b) Eosinophils approximately 4%

(c) Basophils approximately 1%

Neutrophils (Phagocytic) : The cytoplasm is closely packed with fine, inconspicuous granules that absorb both acid and alkaline dyes thus producing a neutral purple colour. The nucleus has several (3 to 5) lobes connected by thin chromatin strands. Fewer lobes indicate less mature cells. About 3% of neutrophils from a female show the sex chromatin attached to one of the nuclear lobes by a thin stalk forming the so-called "drumstick chromosome".

Functions : Neutrophils protect the body against the

invasion of bacteria. They are attached in large numbers to any area of the body which has been invaded by micro-organism. Neutrophils leave the blood by squeezing themselves out through the walls of the capillaries in the infected area. This process is known as *diapedests.* Thereafter they ingest and kill the organisms by digesting them by means of various enzymes, a process known as *phagoctyosis.*

The *pus* which may exude from an infected area consists of destroyed tissue, live micro-organisms and dead neutrophils which have ingested more micro-organisms and dead neutrophils which have ingested more micro-organisms than they could digest.

The life span of neutrophils which remain in blood vessels is about 30 hours. After migrating from the blood capillaries into the inflamed tissue, they play a defensive role there and soon die.

Eosinophila (Non-phagocytic) : The diameter of the cell is nearly twice that of a red blood corpuscle. The cytoplasm is packed with coarse round granules which absorb an acid dye such as *eosin* which is red. The nucleus is usually bilobed, with a connecting isthumus.

Functions : Eosinophils engulf the particles which are formed when antigens and antibodies react with each other. In allergic conditions, such as asthma and parasitic infections of the digestive tract, they are increased in numbers.

Basophils (Non phagocytic) : The cytoplasmic granules absorb on alkaline or basic dye such as *methylene blue* and are stained dark blue to purple in colour. The granules are not as numerous as in eosinophils, but they are large, spherical and almost cover the nucleus which is generally S-shaped.

Function : Basophils are actively amoeboid and ingest small particles like carbon.

Agranular Leucocytes

They have a few non-specific or on granules in the cytoplasm and the nucleus is spherical to kidney shaped. They comprise about 25% of all leucocytes, and have two varieties: (a) Lymphocytes, large and small, (b) Monocytes.

(a) Lymphocytes (Non-phagocytic)

The large and small lymphocytes closely resemble each other and differ only in size. *Small lymphocytes* are most common in the blood. They are relatively larger than red blood corpuscles. A small lymphocyte has a relatively large nucleus, surrounded by a narrow rim of basiphilic cytoplasm. The nucleus, is nearly spherical, having a slight indentation. It is very dense and dark because of heavy chromatin masses.

Large lymphocytes are primitive, proliferative cells residing in the lymph nodes that give rise to the active small, lymphocytes. The nucleus in generally kidney-shaped.

Function : The lymphocytes are produced in the lymphatic issue of the body which is present in the spleen, lymph nodes thymus, tonsils and more scattered nodular masses.

(i) Lymphocytes are concerned with the process of immunity. They produce serum globulin β and γ - globulins), one of the plasma proteins, which gives rise to *antibodies* and *antitoxins.*

(ii) They contribute to scar-formation after injury and thus facilitate wound-healing.

(iii) Lymphocytes play an important role in the immunological reactions to tissue transplantation.

Monocytes (Phagocytic) : They are largest of the leucocytes and are relatively few in number. The nucleus varies from oval, round or indented to kidney shape and stains lightly than that of the lymphocytes. It tends to be eccentric in position. The abundant cytoplasm is lightly basiphilic.

***Function* :** The monocytes originate from a system of primitive cells, the *recticido endothelial system* which is found in organs such as the liver, spleen, lungs and lymphatic glands.

Their function closely resembles that of the neutrophils in that they are actively motile, phagocytic in action and will leave the blood capillaries to ingest micro-organisms and other foreign material that may be introduced into the tissue.

They are the prime scavangers of cells and tissue debris. They play a vital role in removing damaged tissue and thus preparing the way for regenerative processes of the body.

***Blood Platelets* :** These are protoplasmic disks that are found in mammalian blood (lower vertebrates have spindle-shaped cells named thrombocytes). Platelets arise as detached tips of protoplasmic processes extending from the cytoplasm of giant cells *megakaryocytes,* of red bone marrow. The shape is oval to round, often stellate. They are 2-3 μ in diameter. The protoplasm is granular and deeply basophilic in the centre but is pale and homogeneous on the periphery. There are approximately 300,000 platelets in a cubic millimetre of blood.

***Functions* :** Agglutinated platelets are associated with blood, clotting, both inside and outside of blood vessels.

Plasma

The plasma is composed of the following constituents :

1. *Water* 90 to 92%.
2. *Plasma Proteins* : Proteins in the plasma perform several important functions :
   (a) They exert an osmotic force of about 25 mm of mercury across the capillary wall which is called *oncotic pressure.*
   (b) They give viscosity to the blood, thus assisting to some extent in the maintenance of the *blood pressure.*

There are four principal plasma proteins :

(i) *Serum albumin* : It is formed in the liver.

(ii) *Serum globulin* : This is derived from the lymphocytes.

(iii) *Fibrinogen* : This protein is produced in the liver and is necessary for blood clotting. Plasma defibrinated by clotting or whipping is called *serum.*

(iv) *Prothrombin* : It serves as factor II in the blood clotting mechanism and is formed in the liver.

Mineral Salts (Inorganic Salts) : These include sodium chloride and sodium bicarbonate. ' Small amounts of potassium, magnesium, phosphorus, calcium, iron, copper and iodine are also found in the plasma.

Mineral salts are necessary for the *formation of cells, the contraction of muscles, the transmission of nerve impulse* and to maintain *the balance between acids and alkalis* in the body. In health, the blood is always *slightly alkaline* in reaction (pH 7.4).

Nutrient Materials : After the digestion of carbohydrates, proteins and fats, *amino acids, glucose, fatty acids, glycerol* and *vitamins* etc., are absorbed into the blood plasma from the alimentary tract. These are the nutrient materials which are required to maintain the body functions.

Organic Waste Products : *Urea, uric acid* and *creatinine* are waste products of protein metabolism which are formed in the liver. They are carried by the blood plasma to the kidneys for excretion.

Hormones: The "chemical messengers" secreted by the endocrine " glands are passed directly into the blood which transports them to the target organs.

Enzymes : Some enzymes are also present in the plasma.

Antibodies : These are protective substances consisting of complex proteins that are produced in the spleen and lymph glands by a special type of cells *plasma cells*. The antibodies are produced in response to the entrance of a foreign agent, the *antigen* into the body.

Gases : Small amounts of oxygen, carbon dioxide and nitrogen are dissolved in the plasma.

Summary of Blood Functions : The blood performs many important and vital functions in the animal body and is often described as "the river of life". Its various constituents carry out different functions as given under :

Erythrocytes

(1) *Transport of* O_2*:* The red corpuscles carry oxygen (in combination with haemoglobin) as oxyhaemoglobin from the lungs to the tissues for oxidation of food to release energy for various metabolic activities.

A small amount of CO_2 is also transported by erythrocytes from the tissues to the lungs as carbamino-haemoglobin for its disposal.

Leucocytes

(2) *Fighting Diseases :* The leucocytes act as soldiers and scavangers of the body.

  (i) The *neutrophils* are actively amoeboid and phagocytic. They appear in large numbers in and around the infected tissues and engulf live and dead bacteria, fungi and viruses.

  (ii) The *monocytes* are the prime scavangers of injured and dead cells. They devour these elements to clean the body.

  (iii) The *basophils* engulf particles of dust, smoke etc.

  (iv) The *lymphocytes* partly manufacture beta and

gamma globulins (antibodies) of the plasma. They also contribute to the scar formation after injury and thus facilitate in wound-healing.

(v) The *eosinophils* flock to the digestive and respiratory treats in allergies.

The dead leucocytes damaged by bacteria together with fragments of tissues form the pus.

Blood Platelets

(3) The blood-platelets are associated with blood-clotting.

Plasma

(4) *Transport of Nutrition* : It carries glucose and other organic metabolites from the alimentary canal and liver to various parts of the body for energy, growth and repair.

(5) *Disposal of CO_2* : Carbon dioxide is produced in the body as a metabolic waste. The blood plasma picks up CO_2 from all the tissues and brings it to the lungs for expiration.

(6) *Transport of Excretory Products* : Plasma carries nitrogenous waste products from various tissues to the liver where they are converted to urea and other harmless products.

The latter are transported from the liver to the kidney for removal as urine.

(7) *Distribution of Hormones and Vitamins* : It distributes hormones from the endocrine glands and vitamins to the specific place of their action in the body.

(8) *Regulation of Water-balance* : As the plasma supplies water to the tissues and receives excess water formed in metabolic processes. It regulates the water balance of the animals. Water provides vehicle for other

constituents whereas its excess it eliminated through kidney.

(9) *Osmotic Balance :* Plasma proteins and various mineral ions in the plasma maintain osmotic balance, pH balance, buffer-capacity and have varied effects on tissue cells.

(10) *Protection against Diseases :* Immunoglobulins of the plasma act as antibodies that neutralize the harmful effects of foreign agents in the body. They also form basis of blood types.

(11) *Regulation of Body Temperature :* Plasma regulates the body temperature by carrying heat from heat producing organs like muscles to those parts where heat is lost, *e.g.*, lungs (via expiration).

(12) *Blood-clotting :* Fibrinogen, prothrombin and other blood-clotting factors present in the plasma, participate in blood clotting, thus helping against haemorrhage.

(13) *Lymph Formation :* It participates in the formation of lymph and keeps all the tissues moist.

Blood Groups

The membranes of human red cells contain a variety of antigens called agglutinogens. Chemically, they are glycolipids and are inherited as Mendelian dominants. The most important and best known of these are the A and B of agglutinogens which constitute the ABO system of blood groups. According to this system, the individuals are divided into 4 major blood types, types A, B, AB, and O, on the basis of the agglutinogens present in their red cells. Besides, the other well-known blood group systems are Rh system, the McKell, lutheran, Duffy, Diego and Kidd systems.

Blood Groups Antibodies : Antibodies against agglutinogens are called agglutinins. Antibodies are produced mainly in the bone marrow and in the spleen, liver and lymph nodes. The blood group antibodies belong to the serum globulins and may be divided into two broad classes :

(i) *immune* and

(ii) *naturally occurring.*

Immune Antibodies : They may be produced by exposure to the red cells of another individual. This exposure may occur via (i) a transfusion or (ii) during pregnancy, when foetal red cells cross the placenta and enter the circulation of the mother.

(i) If the blood of a person containing Rh antigen is transfused into a person who lacks this antigen, the recipient is said to be immunized and the corresponding antibody (Rh-antibody) will appear in his serum.

(ii) Red cells from a foetus carrying an antigen. Inherited from its father but absent in the mother may also produce the equivalent immune antibodies by entering the maternal circulation through the placenta.

Immune blood group antibodies belong to the category of γ G globulins (IgG). Their molecular weight ranges from 150,000 to 160,000 and they have a unique ability to cross the placenta.

Naturally Occurring Antibodies : These antibodies are present in the serum of an individual who has never been immunized by the equivalent antigen. Such antibodies are first to be formed in a new born animal or human. Example are anti-A and anti-B agglutinins..

Naturally occurring antibodies are usually γ M globulins (IgM). They have a molecular weight of about 900,000 and do not cross the placenta.

The antigen-antibody relationship is one of great specificity like that between lock and key.

6

Blood Circulation

The blood vascular system includes the *heart* and *blood vessels.* In addition, there is a lymphatic system which returns lymph to the blood. Lymph is formed from blood plasma that has filtered from the blood vessels but is unable to go back directly into blood capillaries because of blood pressure. The lymph bathes the cells of the tissues directly and carries the digested foods and oxygen which the cells need. In return, the cells give to lymph their waste substances like urea and carbon dioxide.

In higher forms the circulatory system is of *closed type,* in which the blood is confined to the vessels throughout its course from the heart and back to the heart. In many invertebrates, with an *open type* of circulatory system, the blood vessels open into numerous irregular spaces or *lacunae* and blood comes directly in contact with the cells. From these intercellular spaces blood returns to the heart.

Circulation in Invertebrates

Among many lower invertebrates the role of a circulatory

system is performed by other structures. The development of a separate circulatory system is first seen in Nemerteans which have a few blood vessels in which the blood is propelled by the 'milking', action of the muscular walls of the vessels. The heart as a definite propulsive organ appears in the annelids, arthropods and molluscs. These are provided with well developed vascular systems, the arthropods and molluscs (except cephalopods) having the open type. In the earthworm in which a closed system is approached, the true heart is the dorsal blood vessel. The invertebrate heart is always located dorsal to the alimentary canal.

Circulation in Vertebrates

In vertebrates the circulatory system is of closed type and consists of a heart and blood vessels (arteries, capillaries and veins). The vertebrates have chambered hearts, having two, three or four chambers. The fish heart consists of two main chambers, auricle and ventricle. It is a *single-circuit heart*, the blood passes through it only once in each complete circuit. Amphibians and most reptiles have a 3-chambered heart. Birds and mammals have the 4-chambered heart and there are two circuits, one through the lungs *(pulmonary)* and the other through the body *(systemic)*. Starting with left ventricle, the blood passes through the aorta and its many divisions to all parts of the body and then into the capillaries within the tissues. Veins collect the blood from the capillaries and carry it back to the right auricle of the heart. This circuit from the left ventricle to the right auricle is the *systemic circulation*. From the right auricle the blood passes into the right ventricle, thence through the pulmonary artery to the lungs where the gaseous exchange is made. Then by a vein or veins the blood is carried to the left auricle and thence to the left ventricle. This short circuit from the right ventricle to the left auricle is the *plumonary circulation*. In this complete circulation the blood passes through the heart twice. The advantages of the 2-circuit over the 1-

circuit plan are many because there is no mixing of the oxygen-rich and oxygen-poor bloods. The blood in the aorta contains relatively more oxygen, a higher metabolic rate is thus possible and a resulting higher temperature can be maintained by the animal.

Physiological Anatomy of the Blood Vessels

(a) *Arteries* : All the blood vessels leading from the heart are called the arteries, whether they contain oxygenated blood as in the aorta or deoxygenated blood as in the pulmonary artery. The thick walls of the arteries consist of a heavy layer of elastic tissue. More small branches of an artery are called *arterioles,* having a diameter of about 0.1 mm or less.

(b) *Capillaries* : The arterioles lead into a branching series of microscopic vessels known as the capillaries that have a diameter of about 7-10 *microns.* The capillary walls are thin and allow an exchange of nutrients, gases and other substances between their blood and the cells through a process of diffusion or active transport.

(c) *Veins* : The capillaries in turn unite to form small veins that join to form larger and larger veins, which ultimately carry blood back to the right side of the heart. The veins do not have an inner layer of elastic membrane found in the arteries. Some of them possess internal valves along their length to prevent the backflow of blood.

Atherosclerosis and Arteriosclerosis

These are changes in the structure of the walls of the arteries, changes that ultimately impair their function. *Atherosclerosis* is considered to be the main cause of heart attack and stroke. It is caused by the deposition of lipid, usually cholesterol, in the smooth muscle cells of the artery wall. Cholesterol deposition is followed by some calcification and is topped by a fibrous cap, forming a plaque that protrudes into the lumen of the artery and cuts off blood to those parts of the body served by the diseased vessels. Although the precise role of cholesterol is uncertain, the formation of atherosclerotic plaques is accelerated by abnormally high blood cholesterol (130) and by other factors, such as cigarette smoking and high blood pressure, 180 systolic. The cumulative effects of these three major risk factors increase the likelihood of heart attack and stroke.

Arteriosclerosis is hardening of the arteries due to the deposition of calcium and fibrous tissue within their walls. This reduces the elasticity of the arteries and consequently causes increase in pressure within them. When blood reaches the brain under such high pressure it may rupture the small vessels, depriving the tissues of oxygen and resulting in a "stroke".

***Physiological Anatomy and Working of the Heart* :** The mammalian heart consists of four chambers, two auricles and two ventricles. Typically, one large blood vessel is connected to each of these chambers. The *aorta* leaves the *left ventricle* and the branches of this vessel supply all parts of the body with arterial or oxygen-rich blood. Venous blood collected from all body regions returns through the *vena cava* into the *right auricle.* The right auricle is connected with right ventricle through the *tricupsid valve,* an opening with three flaps. This valve allows blood to flow from the auricle to ventricle but not in the reverse direction. Venous blood collected in the right ventricle leaves this chamber via the *pulmonary artery* leading to the lungs where the blood is oxygenated and arterial blood returns through the *pulmonary vein* into the left auricle. A *bicupsid or mitral valve,* equipped with two flaps, separates the left auricle from the left ventricle. The bicupsid and tricupsid valves together may be referred to as the auriculo-ventricular valves or *A-V* valves. Smaller semilunar valves are situated where the aorta and the pulmonary artery leave the left and right ventricle respectively.

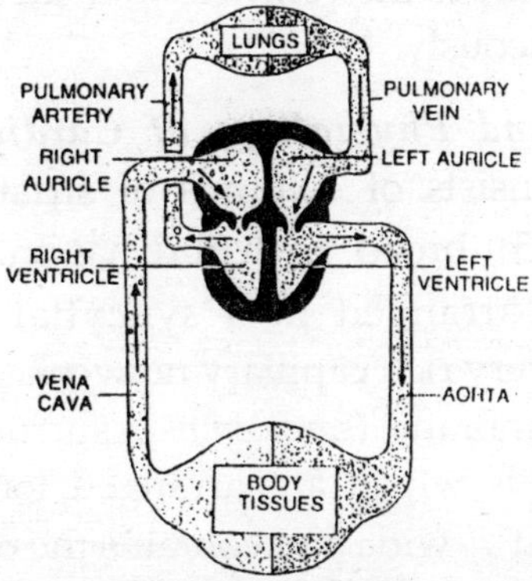

Course of blood circulation through the body.

The left chambers of the heart carrying the arterial blood are not connected directly with the right chambers, carrying the venous blood. In the mammalian heart the sinus has been incorporated into the right atrium. The wall of the heart in the adult is divided into three layers known as the endocardium, myocardium and epicardium. The *endocardium* consists of the endothelium which is continuous with that of the blood vessels and of subendothelial connective tissue. The *myocardium* is the muscle layer upon which depends the pumping action of the heart. It is thin in the auricles but very thick in the ventricles. In the left ventricle, which pumps blood into the farthest parts of the body, it is much thicker than in the right. The *epicardium* consists of the pericardial mesothelium together with underlying connective tissue. This layer is also called the *visceral pericardium* and with the *parietal pericardium* it bounds the pericardial cavity, forming a closed sac containing the pericardial fluid.

In the heart of humans and some other mammals there is a special conducting tissue concerned with the instigation and regulation of the heart beat. This specialized tissue is called *nodal tissue* and is necessary for the functional separation of the auricles from the ventricles. To produce a functional stimulation of the atrial and ventricular beats, this nodal tissue conducts impulses so rapidly to the ventricles that all their parts contract almost simultaneously.

Structure and Physiology of Cardiac Muscle : The myocardium consists of columns of striated muscle fibres, 100 μ long and 15 μ broad. Each fibre has only a single nucleus. The fibres are arranged in a syncytial fashion and are surrounded by very rich capillary network. Each fibre consists of an outer membrane *(sarcolemma)* surrounding numerous striated myofibrils which are arranged longitudinally. Each myofibril is about μ wide and contains the contractile proteins Stenosing and actin. It is interrupted by dark lines known as the Z lines at intervals of 1.2—2.5 μ. The distance between two Z lines is the *sarcomere,* the functional unit of the myocardium.

The ultrastructure of cardiac muscle is quite similar to that of skeletal muscle. The *sacromere* consists of alternate light and dark bands. The central dark band is 1.5 μ long and contains *myosin* filaments (l0m μ broad). The central band, known as the A band, is flanked by an I band on each side which appears light and consists of actin filaments (5m μ broad). The sacrolemma is invaginated by a system of transverse tubules which penetrate the sacromere at the Z line. The lumina of these tubules is continuous with the extracellular fluid. In addition, there are longitudinal tubules. At each Z line these longitudinal tubules approximate to but do not anastomose with the transversely disposed invaginations of the sarcolemma. The term 'triad' is used to describe the arrangement of the terminal dilations of two longitudinal tubules (known as cisternae) one oil each side of a transverse tubule.

Depolarization of the sacrolemma causes the development

of an electric potential which is conducted through the transverse tubule and then causes depolarization of the adjacent cisternae. This depolarization causes a release of Ca^{2+}, which diffuses into the myofibrils and activates myosin, which then splits ATP, yielding the energy necessary for the formation of actomyosin complex and for the contraction of the muscle. The Ca^{2+} ions so released are then actively transported back into the longitudinal tubules of the sarcoplasmic reticulum and relaxation occurs.

At the ends of each fibre, or cell, is a structure known as the *intercalated disc.* Electron microscopy indicates that these discs are actually cell junctions. Thus, they are cell membranes, often arranged in a zig-zag manner, with considerable amounts of amorphous dense material associated with them. The number of *intercalated discs* seems to increase with age. Human cardiac muscle shows a clear distinction between perinuclear and peripheral cytoplasm. Between the muscle fibres is connective tissue of reticular, collagenous, and elastic fibres.

The cardiac muscle differs quantitatively from skeletal muscle in having more mitochondria crowded between the myofibrils, and in being richer in the red-pigmented myoglobin than most skeletal muscle. Myoglobin is found in the sarcoplasm of cardiac muscle fibres. It is a conjugated protein containing an iron porphyrin prosthetic group closely related to haemoglobin. Myoglobin faciliates the transport of oxygen from cell membrane to the mitochondria where tissue respiration occurs. It should be noted that the heart muscle is not a structural syncytium but it behaves as if it were, the reason being that the electrical resistance offered by the intercalated disc is very low. Thus, when one cell is excited, the excitation process spreads easily across the intercalated disc to its neighbour. This allows the impulse to spread from one fibre to another throughout the whole atrium or ventricle causing them to respond as a single unit (all-or-none principle of heart beat).

Physiologically the hearts are of two types :

1. *Myogenic Heart* : It is a heart in which the *beat* originates *myogenically i.e.*, the heart beat is initiated and controlled by a special node of modified or unmodified heart muscles, the so-called sows *node* or *pacemaker*. Located in the wall of the right auricle, the pacemaker produces rhythmic signals leading to contraction of the heart. Because of this built-in mechanism, a myogenic heart exhibits an *intrinsic* beat. When isolated from the body and placed into an artificial nutrient medium, such a heart may for a time contract and relax in a slow rhythm. In the body, this inherent rhythm of the heart is regulated in both, amplitude (strength) and rate by control systems, which are nervous, physical, and chemical. Acetylcholine inhibits the heart beat of a myogenic heart. The hearts of *vertebrates* and *molluscs* are myogenic.

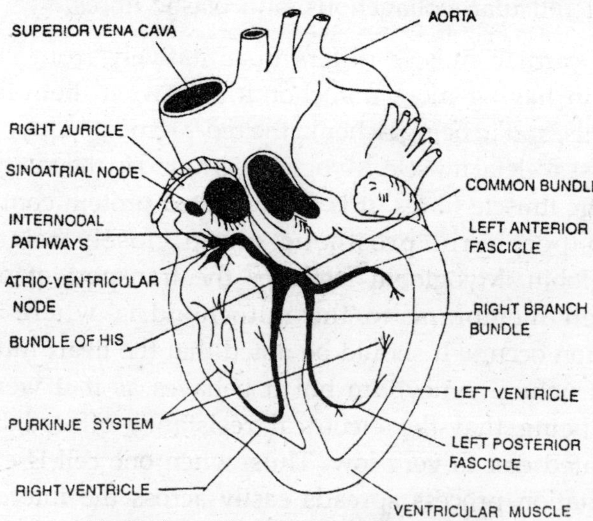

Mechanism of heart beat. Arrows indicate spread of the waves of excitation.

2. *Neurogenic Heart* : It is a heart in which the beat originates by nervous stimulation or *neurogenically i.e.*, the

contraction is initiated by the nerve ganglion situated in or close to the heart, the nerve cells acting as the pacemaker. Such a ganglion delivers rhythmic impulses to the heart muscle, and the removal of the ganglion stops the heart. Acetylcholine accelerates the heart beat rate of a neurogenic heart. The hearts of the most *arthropods* and some *annelids* are neurogenic.

Conduction System of the Heart

The cardiac conduction system consists of the following 3 elements :

1. *Sinoatrial Node* : The sinoaterial (SA) node is a specialized muscle fibre area in the posterior wall of the right atrium, below the opening of the superior vena cava. The nodal cells are smaller than the surrounding atrial cells (3 μm in diameter as compared to 5 to 11 μm for the a trial cells). The bundles of fibres from the SA node activate first the right and then the left atrium. The SA node is the normal cardiac *pacemaker* as its rate of discharge determines the rate at which the heart beats. Internodal atrial pathways exist between the SA node and the AV node.
2. *The Atrioventricular* (AV) *Node* : It is in the atrial septum, just above the tricupsid valve. The AV node is normally the only conducting pathway between the atria and ventricles.
3. *Bundle of His* : The AV node is continuous with the bundle of His, which gives off a left bundle branch at the top of the interventricular septum and continues as the right bundle branch. The left bundle branch divides into an anterior fascicle and posterior fascicle. The branches and fascicles pass down either side of the septum and come into contact with the *Purkinje system*, whose fibres spread to all parts of the ventricular

myocardium. Bundle of His is normally the only conducting tissue between the atria and the ventricles. This is the fastest conducting system in the heart. Conduction through the Purkinje fibres is 1.5 to 4.0m/sec as compared to 0.4m/sec through contractile myocardial fibres.

The conducting system is composed of modified cardiac muscle. The SA node and to a lesser extent the AV node also contain small round cells with few organelles, which are connected by gap functions. These are probably the actual pace maker cells, and therefore called P *cells.*

Heartbeat and its Physiology

The rhythmic contraction and relaxation of the cardiac muscles is known as the *heartbeat.* Each cycle of the *heartbeat* consists of the following two phases:

1. the contraction phase termed *systole* (atrial systole followed by ventricular systole).
2. the relaxation phase called *diastole.*

Thus, a complete heartbeat consists of one systole and one diastole, the whole beat lasting about 0.8 sec. in a normal human adult at rest. Normally the heart beats about 72 times per minute and pumps out about 60 to 110 ml of blood per heartbeat.

The origin and spread of cardiac excitation and the electrical activity of the heart may be considered under the following headings :

(i) *Origin of the Heartbeat :* The heartbeat originates in the sinoatrial node (S-A node) which, therefore, has been called the *pacemaker* of the heart. This node is situated at the junction of the superior vena cava and the right auricle.

(ii) *Spread of the Impulse* : From this node the impulse spreads over the auricles causing their contraction and then, it stimulates the A-V node. This in turn stimulates the rest of the conduction system causing the rest of the heart to contract. The auricular phase of the beat lasts about 0.11 sec. Then auricles relax for the remaining 0.69 sec. of the cycle and thereafter they become ready for the next beat.

With the contraction of the auricles and during their relaxations period, the ventricles are fully distended and contract in their turn. Their thick walls generate much more pressure than the walls of the auricles. A more rapid ventricular contraction rate is made possible by *Bundle of His* and purkinje fibre system which is specialized myocardium for conducting impulses through the heart. This produces simultaneous contraction of ventricles to insure efficient pumping of blood through the *aorta* and *pulmonary artery*, because the A-V valves are already closed and prevent backflow of the blood into the auricles. The ventricular systole (contraction) lasts about 0.33 sec and they relax for about 0.47 sec (ventricular diastole). As soon as the ventricular phase of the heartbeat is over, the auricular phase starts again.

(iii) *Electrical Events in the Origin of Heartbeat* : The answer of how actually the wave of excitation or the impulse originates in the pacemaker can be satisfactorily given by considering the electrical events responsible for it. Interestingly, the sodium, potassium and calcium ions play an important physiological role in generating an electrical impulse in the cells of the sino-auricular node. Therefore, the balance between these ions as well as their absolute concentration is important.

The pacemaker tissue is characterized by an unstable membrane potential that slowly decreases after each impulse until the firing level is reached and another impulse is generated.

Ordinarily (at rest) the inside of the pacemaker cell is electrically negative with respect to the outside which is positive. When this condition exists the cell membrane is said to be *polarized.* This is due to the predominance of Na^+ ions outside the cell which outnumber the K^+ ions inside the cell. At this time the potential difference is about -55 mV, and each cell thus becomes a small charged battery. However, the plasma membrane of pacemaker cells is partly permeable to Na^+ ions so that these ions continuously rush in the cells with a resultant decrease in the potential difference between inside and outside of the cells.

When the potential difference is reduced to about -40 to -45 mV, the Na^+ ions rush in quickly while some of the K^+ ions escape outside. This leads to a reversal of electrical charges so that the inside of the membrane of pacemaker cell now becomes positive with respect to the outside. It is generally believed that this initial depolarization represents the *pacemaker potential or prepotential.* Thus, the initial depolarization is due to Na^+ influx through rapidly opening Na^+ channels (the so-called Na^+ current, I_{Na})

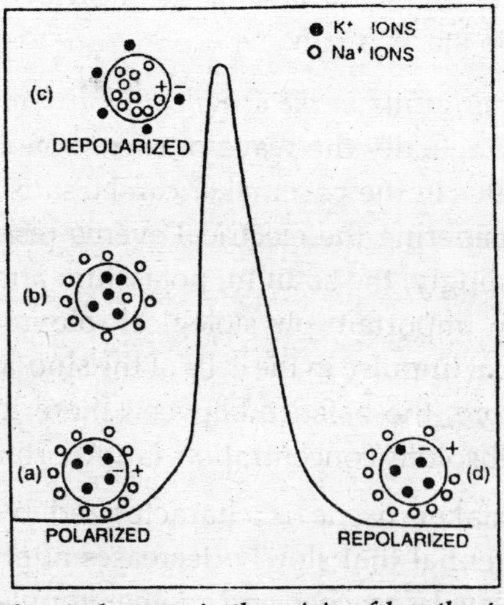

Electrical events in the origin of heartbeat.

When a certain level of depolarization is reached by the pacemaker potential, a spike is generated. In other words, depolarization causes the *action potential* and the electrical impulse generated spreads to adjacent cells by a kind of relay system so that the intensity of impulse does not decrease with the distance travelled. Depolarization initiated in the SA node spreads radially through the atria and then converges on the AV node. Atrial depolarization is complete in about 0.1 second. Because conduction in the AV node is slow, there is a delay of about 0.1 second *(AV nodal delay)* before excitation spreads to ventricles. From the top of the septum, the wave of depolarization spreads in the rapidly conducting purkinje fibres to all parts of the ventricles in 0.08-0.1 second.

The depolarization of the ventricular muscle begins at the left side of the interventricular septum and moves first to the right across the midportion of the septum. The wave of depolarization then spreads down the interventricular septum to the apex of the heart. It returns along the ventricular walls to the AV groove, proceeding from the endocardial to the epicardial surface. The parts of the heart to be depolarized in the last are the posterobasal portion of the left ventricle, the pulmonary conus and the uppermost portion of the septum.

At the peak of each impulse I_k begins and brings about *repolarization.* I_k then declines, and as K^+ efflux decreases, the membrane begins to depolarize, forming the first part of the *prepotential.* Ca^{2+} channels then open. These are of two subtypes in the heart, the *T channels* (T for transient) and the *L channels* (L for long-lasting). The calcium current due to opening of T channels (I_{ca}T) completes the prepotential, and the calcium current due to opening of L channels (I_{ca}L) produces the impulse. It should be noted that resting membrane potential of pacemaker tissue is somewhat lower than that of atrial and ventricular muscle.

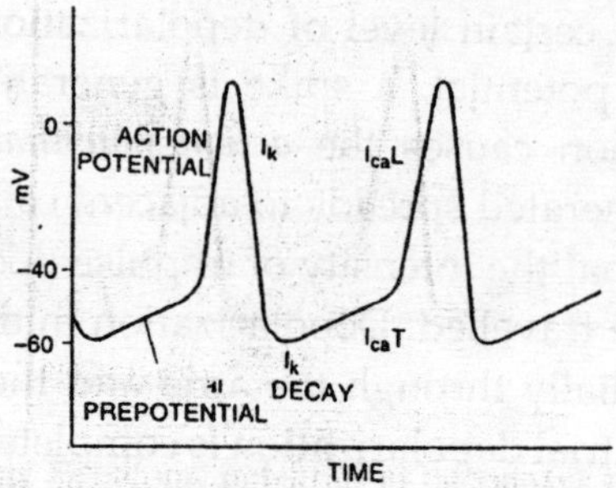

Diagram showing membrane potential of pacemaker tissue.

The depolarized cells soon repolarize by becoming negative inside and positive outside the plasma membrane. This is achieved by pumping out the Na^+ ions that had infiltrated earlier. Since the Na^+ concentration is already higher outside the cells, they have to be actively transported with the expenditure of the energy of ATP. The characteristic feature of the cardiac action potential is the long duration of the repolarization phase. The rate at which the electrical impulses are discharged depends on several factors : The vagus nerve tends to reduce it while the sympathetic nerves tend to increase it. The other factors which can bring a change in the stability of the membranes and thus alter the rate of cell discharge include the pH, the O_2 and CO_2 concentration of the blood, hormones and drugs, etc.

All-or-none Principle of the Heartbeat : In skeletal muscles the extent of contraction (within limits) is proportional to the strength of the stimulus. In the heart this is quite different. The heart, as a whole follows the so-called *all-or-none-principle, i.e., if the heart muscle beats at all, it does so to its maximum effect.* Thus, the strength of the contraction is independent of the strength of the stimulus and once the impulse has been initiated by the pacemaker cells the high conductivity of the cardiac muscle cells transmits it rapidly.

However, this principle holds true also for the *individual fibres* of the skeletal muscle, although the muscle as a whole is capable of graded contractions. The reason for this difference between the heart and the skeletal muscle lies in the relation of the muscle fibres to each other. In a skeletal muscle the individual fibres are isolated from each other by a *sarcolemma*, but in the heart the imperfectly cross-striated fibres are brought into intimate contact with each other by anastomosis. This permits the impulse to spread from one fibre to another throughout the whole atrium or ventricle, causing them to respond as a single unit. Although the contraction of the heart is always maximal, this does not mean that the extent of contractions are always the same. They are determined by the physiological state of the heart muscle.

Starling's Law of the Heart

The length-tension relationship in cardiac muscle is similar to that in skeletal muscle. As the muscle is stretched, the developed tension increases to a maximum and then declines as stretch becomes more extreme. Starling pointed this out in the form of a law and noted that the *"energy of contraction is proportional to the initial length of the cardiac muscle fibre"*. This is also called Frank-Starling law or mechanism.

When a large volume of venous blood arrives at the heart, the increased diastolic filling stretches the fibres of the right ventricle. Up to a certain point, this stretching increases the force of the contraction, enabling the ventricle to discharge a greater volume of blood. Thus, according to the Starling's law of the heart, within physiologic limits, the more the ventricles are filled with the incoming blood the greater is the force of their contraction at systole. Its physiological significance is that as blood is returned more rapidly to the heart during exercise, the increased volume stretches the walls of the ventricles and causes a more forceful propulsion of blood out of the heart. This is one of the main factors balancing the amount of blood entering and leaving the heart.

Starling's law of the heart also operates in the opposite condition to that just discussed. For example, in extensive haemorrhage the systolic and diastolic pressures are very low. The amount of blood delivered to the ventricle is greatly reduced, and, as a result the cardiac fibres are only slightly stretched. This results in a feeble heartbeat and in a much lower than normal systolic ventricular pressure. However, this force is sufficient to overcome the much reduced diastolic pressure in the aorta.

Electrocardiogram (ECG)

A heart beat can also be analysed in terms of electrocardiogram (ECG) obtained by electrocardiograph. Most electrocardiograph machines record the cardiac activity on a moving strip of graph paper. Typically the normal ECG consists of a series of slow waves, including upward (negative) deflections called P, R and downward positive deflections Q and S. Of them, (i) the P wave represents *atrial depolarization,* which precedes the contraction of the atria. The PQ interval represents the delay at the auriculo-ventricular junction, (ii) The QRS complex represents *ventricular depolarization* preceding ventriclular contraction. A normal QRS complex lasts about 0.08 sec. (iii) The T wave represents *ventricular repolarization.* It is much smaller than the QRS depolarization complex. There is no comparative atrial repolarization visible, for it is obscured by the large QRS depolarization of the ventricles, occuring at that time.

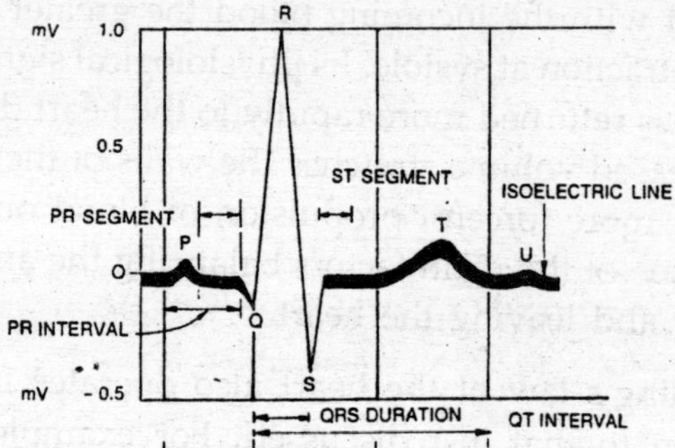

The waves of ECG.

An important parameter of the ECG is the *PR interval*, which indicates the length of time between the beginning of the atrial contraction and the beginning of the ventricular contraction, and thus the conduction time through the heart. A normal PR interval is 0.18 sec or less. In incomplete heart block, when conduction through the AV bundle is slowed, the PR interval may be almost doubled, and there may be 2 to3 P waves for each QRS-T complex. In complete heart block, the ventricles "escape" from SA and atrial control, and the P waves may be completely dissociated from the QRS-T complexes. Prolongation of the QRS complex, which normally lasts about 0.06 sec, indicates delayed conduction through the ventricles, often caused by ventricular hypertrophy. The QT interval coincides with the beginning and the end of ventricular contraction. It lasts about 0.40 sec. The electrocardiogram reflects the activity of various parts of the heart and thus aids in diagnosis of different heart ailments.

Heart Sounds

Two sounds are normally heard through a stethoscope.

The first sound, usually heard as the "lub" of the "lub dup, lub dup" of the heart, is very low pitched, and it is caused by the closure of the mitral and tricupsid valves before the beginning of the ventricular contraction. The first sound has a frequency of 30-80/sec and a duration of 0.05 sec. The *second sound*, "dup", is higher in pitch. It has a frequency of 150-200/sec and a shorter duration (0.025 sec). It is caused by the vibrations associated with closure of the aortic and pulmonary valves just after the end of ventricular systole.

The third and fourth sounds are less important and can be detected on the phonocardiogram. The *third sound* is due to vibrations of the cardiac walls produced by the rapid filling of the ventricles. The *fourth sound* occurs during atrial systole. It has low frequency and amplitude.

The characteristics of the sounds are clinically important. The cause of *cardiac murmurs* is disease or disorder of the heart valves.

Cardiac Arrhythmias and Heart Block : In the normal heart each beat originates in the SA node. This is called normal sinus rhythm (NSR). The heart beats about 70 times per minute at rest. The rate is slowed *(bradycardia)* during sleep and accelerated *(tachycardia)* by exercise, fever, emotion and many other factors.

Heart block is a condition in which the normal cardiac rhythm is disturbed *(cardiac arrhythmias)*, or there is defective conduction in some part of the heart. This might occur as *sinoatrial nodal block, atrioventricular nodal block, infranodal block* or *right* or *left bundle branch block.*

Sinus node dysfunction leads to marked bradycardia accompanied by dizziness and *syncope* (sick sinus syndrome). It is treated by implantation of pacemaker. When conduction between atria and ventricles is slowed but not completely interrupted, the condition is partial or *incomplete heart block.* In the form called, *first-degree heart block,* all the atrial impulses reach the ventricles but PR interval is very long. In the form called *second degree heart block,* not all atrial impulses are conducted to the ventricles. In another form of incomplete heart block, there are repeated sequences of beats in which the PR interval is progressively increased until a ventricular beat is dropped *(Wenckebach phenomenon).*

When conduction from the atria to the ventricles is completely interrupted, *complete* (third-degree) *heart block* results. The block may be due to disease in the AV node *(AVnodal block)* or in the corducting system below the node *(Infranodal block).* In case of AV nodal block the remaining nodal tissue becomes the pacemaker and the rate of the idioventricular rhythm (ventricles beat at a low rate) is about 45 beats per minute. In case of infranodal block due to disease in the bundle of His, the ventricular rate is 35 beats per minute or still lower.

If complete heart block develops suddenly, there may be a delay before the ventricles start beating at their own rate. Naturally, in this period the systemic blood pressure falls to a very low level and the blood supply to the brain is highly insufficient. It the ventricular standstill lasts for only a few seconds it may give rise merely to faintness and dizziness, but if it is more prolonged it may cause loss of consciousness, convulsions, and even death. An attack of complete heart block together with fainting constitutes the *Stork-Adams Syndrome.*

Cardiac Output

The cardiac output is the amount of blood pumped by the left ventricle into the aorta per minute. Because this is the volume of blood returning to the heart from the lungs, it is also the same volume that was pumped to the lungs by the right ventricle. As a result, in the normal heart, the right and left sides of the heart pump the same volume of blood. The normal cardiac output for the young adult male, reclining is about 5.6 litres/min, that for a female of the same age and size is 5 litres /min. Cardiac output is affected by changes in posture and by exercise.

The relationship between *cardiac output, heart rate* and *storke volume* is expressed as follows.

$$\text{Cardiac output} = \text{heart rate} \times \text{stroke volume}$$

If the heart rate is 72 beats/min and the volume of blood pumped out by each ventricle per beat (stroke volume) is 70 ml, then the cardiac output at rest is 5040 ml/min.

There is an integrated control of cardiac output. Regulation of cardiac output as a result of changes in the cardiac muscle fibre length is called *heterometric regulation.* Regulation due to changes in contractility independent of length is called *homometric regulation.*

Stroke volume depends largely on the volume of venous blood returning to the heart (Starling's law of the heart), the size of the heart, and its contractile strength. Trained athletes usually have hearts larger than normal individuals, because of hypertrophy of cardiac muscle fibres as a result of constant work against resistance. These muscle fibres have more contractile proteins and myoglobin than is found in untrained individuals, and consequently they contract with greater strength. This increase in the inotropic force of the heart increases the stroke volume and permits the athlete to achieve the same cardiac output with a slower heart rate. This is not the case in pathological enlargement of the heart, with which the athlete's heart should not be confused. An intersting problem is the machanism by which the athlete's heart rate is slowed by training. However, all individuals do not respond to the same extent, and it must be associated with a felicitous combination between the vagal inhibitory nerves and the inherent heart rate. The athlete's heart rate may be as low as 45 to 50 beats/min, but the increased stroke volume of 100 to 110 ml/ min results in normal cardiac output, the advantage of this is that the athlete has to increase heart rate proportionally less during strenous exercise than does the untrained individual. This permits better filling of the heart in diastole and contributes, together with the greater strength of the hypertophied heart, to an enhanced cardiac output..

Regulation of Cardiac Activity

(a) *Reflex Control (Nervous and Chemical Regulation)* : The pacemaker is the immediate regulator of the heart rate. But the action of the pacemaker is controlled by the *heart rate centre* located in the *medulla oblongata*. Two pairs of nerves lead from the heart rate centre to the heart. Impulses through one pair, formed by a sensory and a motor nerve, accelerate the heart beat, and impulses through the other pair also formed by a sensory and a motor nerve, slow the heart beat. The inhibitory nerve fibres (parasympathetic nerve fibres) pass from the heart rate centre into the large vagus nerves. Both the accelerator and inhibitor nerves terminate in the right auricle above the pacemaker. Sensory nerves lead to the heart rate centre from the walls of the vena cava and the aorta. When the pacemaker is stimulated, a wave of contraction spreads out from it through both auricles and ventricles. The vagus nerve fibres inhibit the heart rate by releasing *acetylcholine* (Ach). The accelerator nerves (sympathetic nerve fibres) increase the heart rate by releasing *sympathin.* The underlying mechanism is that the acetylcholine, which is released at the synapse (the small gap between the vagus nerve fibres and the pacemaker cell), increases the permeability of the cell membrane to the K^+ ions so that they pass into the cell increasing their negativity (hyperpolarization). Under such conditions, the time between two successive depolarizations would be longer and consequently the rate of heart beat slower.

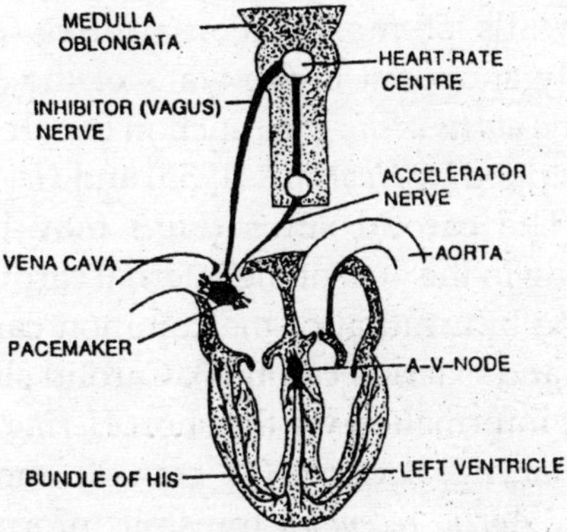

The motor innervation of the heart.

The firing rate of these two groups of motor fibres (sympathetic or cardioacceleratory, and parasympathetic or cardioinhibitory) is controlled by the antagonistic centres located in the medulla, the cardioaccelerator and cardioinhibitory centres.

The purpose of this control system is to increase heart rate whenever blood volume or blood pressure falls in the main arterial system and to decrease heart rate whenever blood volume or pressure increases above a normal level.

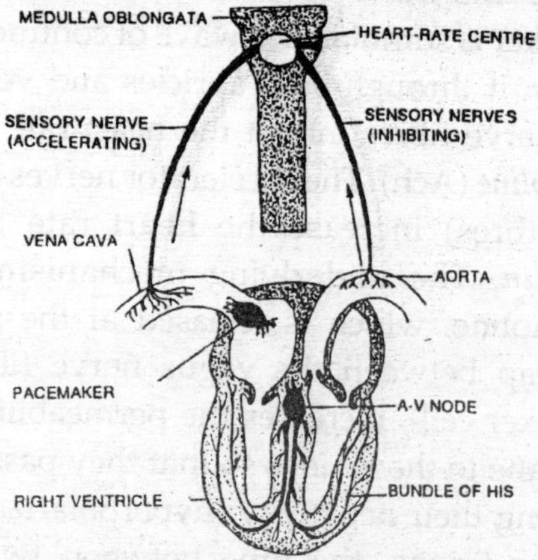

The sensory innervation of the heart.

(b) *Baroreceptor Reflexes (carotid sinus reflex and aortic reflex)*: Arterial baroreceptors are sensory nerve endings found in the walls of most major arteries, but they are especially abundant in the walls of the *carotid sinuses* and of the *aortic arch.* The function of carotid sinus was first described by *Heymans* (1955) and *Heymans and Neil* (1958). The carotid sinus (Fig.) may be seen as a thickening in the wall of the internal carotid artery (just above the bifurcation of the common carotid into the internal and external carotids). Carotid sinus receptors transmit information via the short *Hering's nerve* to the glossopharyngeal nerve (IX cranial) and then to the medulla. *Aortic receptors* transmit information to the medulla via the vagus nerve (X craninal). These baroreceptors are most sensitive within the normal physiological range of arterial pressure. Below about 60 mm Hg pressure the receptors do not fire at all and maximum rate of firing is attained at about 160 to 200 mm Hg. Thus their rate of firing is increased proportionally with increased stretch on the arterial walls (Fig.)

When the pressure in the aorta or carotid sinus goes above normal, there is an increased rate of firing of the sensory endings. This increased stream of impulses causes the cardioinhibitory centre to be activated. Nerve impulses are sent out along the parasympathetic vagus and the acetylcholine released by the nerve endings causes the heart rate to slow down. Simultaneously, inhibitory impulses are sent from the cardioinhibitory centre to the caridioaccelerator centre, and the stream of impulses from the sympathetic neurons to the heart is decreased . These activities thus slow down heart rate and cause a reduction in blood pressure. The reverse activities occur if the blood pressure drops below normal. As mentioned earlier baroreceptors respond most rapidly to changing blood pressures, the rate increases even during systole and decreases during diastole of the heart.

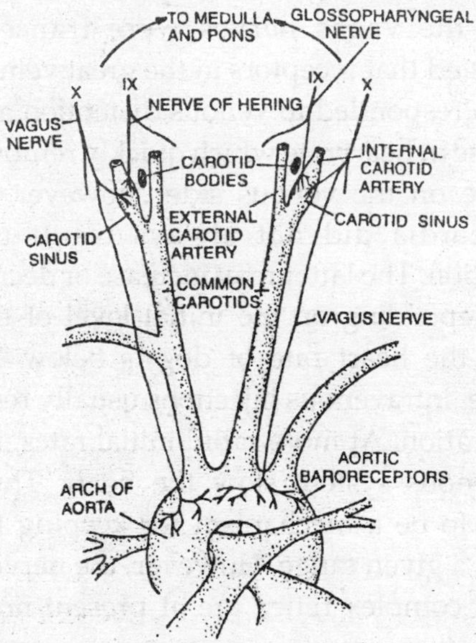

The arterial chemoreceptors; the aortic and carotid bodies.

The baroreceptors have sufficient sensitivity to regulate and maintain blood pressure constant during changes in body posture, as when a human sits down or stands up. It may also be noted that strong pressure exerted upon the carotid sinuses can result in unconsciousness due to the high pressures causing the heart to slow down or even stop. In the latter case, the heart usually begins to beat again after about 10 seconds due to the phenomenon of *vagus escape*—the vertebrate heart upon continued vagal stimulation is usually found to take up its own inherent rhythm due to the pacemaker and ignores the vagus excitation.

Chemoreceptor endings are found in the aortic arch and in the carotid arteries. These are primarily concerned with controlling oxygen and CO_2 levels in the blood by their actions on the respiratory system. But under severe stress conditions their stimulation will also change cardiac rates.

(c) *Bainbridge Reflex : Bainbridge* (1915) discovered that intravenous injections of blood or saline produced a tachycardia in dog. The tachycardia was not produced when the vagas nerves were transected. It was suggested that receptors in the great veins or in the left atrium responded to venous distention and caused an increase in heart rate which quickly removed the excess volume on the venous side. However, the expected tachycardia did not always result upon venous distention. The latter may increase or decrease the heart rate, depending on the initial level of the heart rate. When the heart rate of dog is below 140 beats per minute, intravenous injections usually result in cardiac acceleration. At more rapid initial rates, injections into the venous system slow the heart. Thus, the reflex seems to be a mechanism for keeping the heart rate within a given range. However, the nervous pathways of this complex reflex are at present not known.

(d) *Integration of Reflex Control*: There are connections between the cardioaccelerator and cardioinhibitory centres of the medulla and also connections between these centres and other regions of the brain, hypothalamus and cerebral cortex. The heart rate can be altered by activity in higher centres. It is well known that various emotional states can accelerate or slow down the heart rate, the reason being that states can accelerate or slow down the heart rate, the reason being that emotional centres in the cortex and centres in the hypothalamus provide inputs into the medullary centres. Thus, suprabulbar centres have been shown to be of utmost importance in the integration of reflex control of the cardiovascular system.

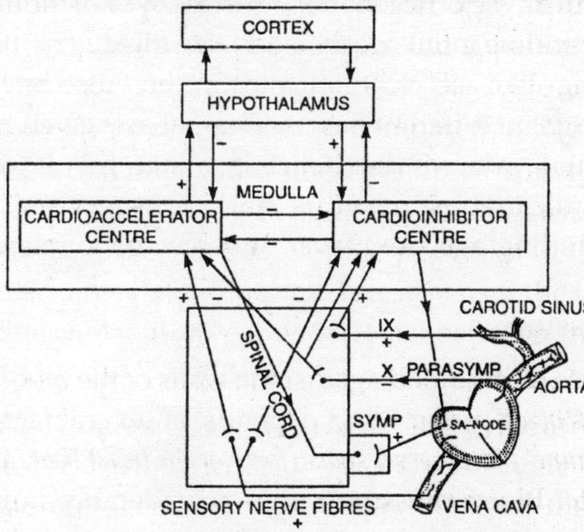

Integration of reflex control of cardiac activity.

(e) *Endocrine Control* : Endocrine glands play an important role in the control of heart rate and stroke volume. The release of adrenaline by the adrenal medulla is under the partial control of sympathetic nerves influenced by the cardioaccelerator centre in the medulla. In the flight or fight reflex of mammals, adrenaline is released into the bloodstream and among other actions increases the heart rate and also the contractile force of the cardiac muscle. The adrenal medulla and other endocrine glands also aid in maintaining the proper metabolic and ionic environment for correct activity of the heart.

(f) *Other Factors Regulating Cardiac Activity* : Besides the nervous control described above other factors also regulate the heartbeat. Most of them act directly on the heart muscle or on the pacemaker. Like any other metabolic process, the heartbeat is affected by temperature. High and low temperature accelerate and deaccelerate the heartbeat respectively. An acidic pH accelerates the heart, while the alkaline pH retards it. A high CO_2 concentration in the blood also increases the heart rate since it lowers the pH of the blood. The mineral ions in the blood and hormones like thyroxin, insulin, sex hormones, pituitary hormones, and adrenaline influence the heart rate directly or indirectly. Drugs like acetylcholine, nicotine etc., also have varied effects on different hearts. Thus, at any given moment, actual rate of the heart is a net result of many simultaneous nervous and non-nervous effects, including age, sex, fever and metabolic rates.

Blood Pressure

The force of the blood against the walls of the blood vessels is known as *blood pressure* and depends on several factors such as *blood volume, blood vessel space, force of the heart beat,* and *blood viscosity,* etc. Blood pressure varies considerably in different parts of the circulatory system. The maximum pressure is encountered in the *aorta* where it normally reaches to 140 mm of mercury. Then it gradually decreases in the arteries and arterioles, becoming lower in the capillaries and still lower in the veins. The arterial blood pressure undergoes a rhythmic change reaching a maximum during ventricular systole when blood is pumped into the arteries and decreasing during ventricular diastole. The former is called *systolic pressure* and the latter, the *diastolic pressure.* The normal systolic pressure in humans ranges from about 120 to 140 mm of mercury, whereas the diastolic pressure ranges from about 75 to 90 mm of mercury.

Factors Affecting Blood Pressure

Blood pressure is affected conciderably by numeous factors including age, sex, heredity, and other physical and emotional states. However, there is controversy about where to draw the line between normal and elevated blood pressure levels (hypertension), particularly in older persons. Because of nervousness, about 20% of hypertensive individuals have high blood pressure in the doctor's room than during their normal daytime. This is called white coat hypertension.

An important cause of the systolic pressure rise with age is decreased distensibility of the arteries. At the same level of cardiac output, the systolic pressure is higher in old persons than in young ones because there is less increase in the volume of the arteries during systole to accommodate the same amount of blood.

The *total blood volume,* affecting the blood pressure, is adjusted to some extent through the contractions of the spleen which bring stored blood into circulation. But adjustments of blood pressures by such means are comparatively slow. However, more effective adjustments are brought about by alterations in the force of the heart beat and the space within blood vessels. The force of the heart beat is determined by the *amount* of blood received and pumped out by the heart.

Regulation of Blood Pressure

Vasomotor Control

The space within the blood vessels is adjusted by contractions and relaxation of the muscles in their walls. These muscular activities are controlled by a *Vasomotor centre* in the brain, located in the medulla oblongata. The vasomotor centre consists of an *excitatory portion* that continuously fires sympathetic vasoconstrictor nerves to maintain vasomotor tone and an *inhibitory portion* that can inhibit sympathetic vasoconstriction, thereby allowing passive vasodilation. Nerves lead from the vasomotor centre to all blood vessels except the capillaries which do not have muscles. *Vasoconstriction* reduces the diameter of blood vessles, *vasodilation* increases it. Vasoconstriction and vasodilation together are referred to as *vasomotion.*

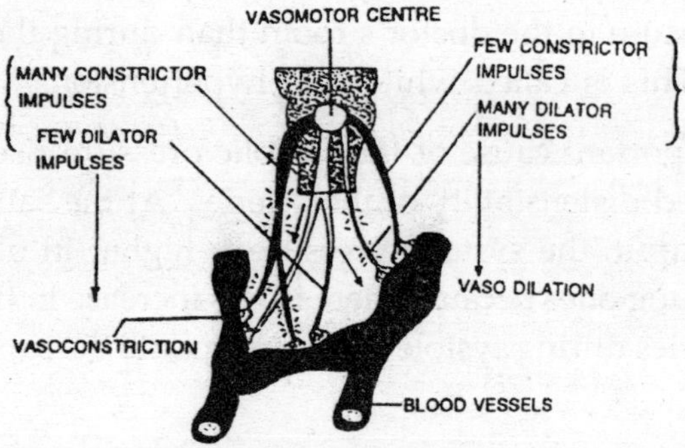

The pattern of vasomotion.

The pattern of vasomotion indicates that vasoconstriction occurs when a blood vessel receives many constrictor impulses and a few dilator impulses from the vasomotor centre. Vasodilation occurs when dilator impulses are many and constrictor impulses a few. Vasoconstriction and localized vasodilation follow through the numbered events shown,

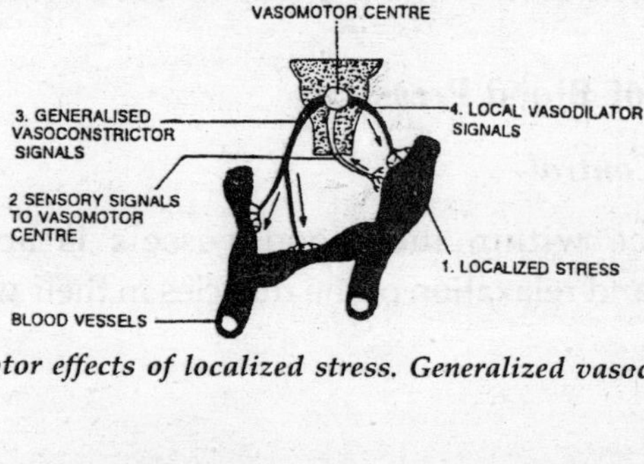

The vasomotor effects of localized stress. Generalized vasoconstriction.

raises the overall blood pressure, vasodilation lowers it. Pain, emotions and stresses generally all tend to increase the blood pressure through bodywide vasoconstriction. At the same time, in specific parts of the body which sustain a stress (like a wound etc.), the local blood pressure often falls through local vasodilation resulting in swelling . The adaptive advantage of this is clear. A rise in the overall blood pressure enables the animal to respond to the stress more effectively, while local vasodilation itself permits more blood to flow into that region, thus making available more nutrients and oxygen.

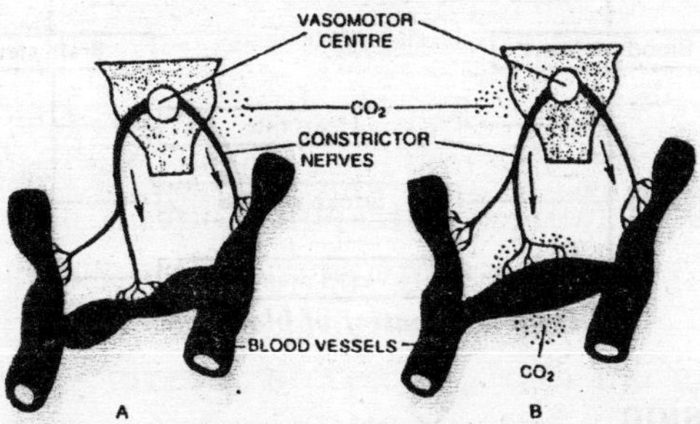

Vasomotor control by carbon dioxide. (A) generalized vasoconstriction as a result of stimulation of the vasomotor centre by CO_2 produced locally over-rides of the general vasoconstriction ordered by the vasomotor centre and brings about a local vasodilation.

Like the heart rate the blood pressure is normally prevented from varying excessively by stretch reflexes. Some sensory nerve fibres from the aorta lead into the vasomotor centre. When the aorta is greatly distended by blood spurting out from the heart *i.e.*, when blood pressure is high, a stretch reflex via the vasomotion centre brings about body-wide vasodilation. In this manner a high blood pressure is reduced to normal. Conversely, when the aorta distends very little or not at all, a stretch reflex leads to increased vasoconstriction and a rise of blood pressure.

The carbon dioxide of the blood also affects the vasomotor centre. If the blood reaching the vasomotor centre carries a high concentration of CO_2, then the centre transmits constrictor impulses throughout the body and the blood pressure will rise. Conversely, low CO_2 concentrations result in a fall of blood pressure.

The blood pressure is adjusted to by variations in the rate of tonic discharges of sympathetic nerves that constrict arterioles and veins and increase heart rate and stroke volume. This is called feedback control of blood pressure.

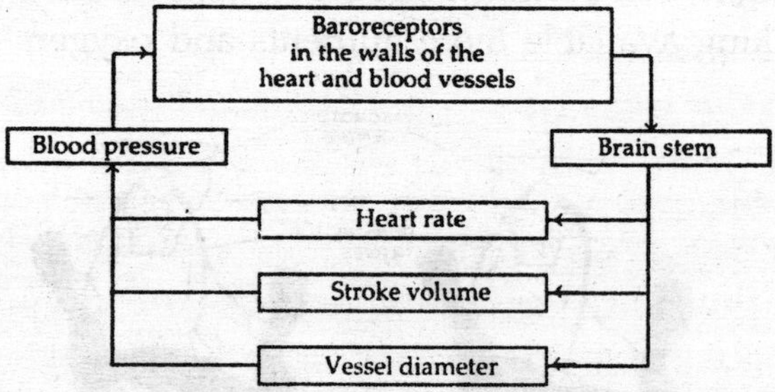

Feedback control of blood pressure.

Hypertension

The blood pressure in the aorta and in the brachial and other large arteries in an adult human rises to a peak value *(systolic pressure)* of about 120 mm Hg during each cardiac cycle and falls to a minimum value *(diastolic pressure)* of about 70 mm Hg. A sustained elevation of the systemic arterial pressure is called *hypertension.* In some cases *Pulmonary hypertension* can *also* occur, but the pressure in the pulmonary artery is relatively independent of that in the systemic arteries.

Some of the procedures that cause *experimental hypertension* in certain animals involve manipulation of the kidneys, nervous system, or adrenals. Some strains of rats develop hypertension quite spontaneously (SHR rats), or when given a high sodium diet (Dahl strain rat)

The hypertension that follows constriction of the renal arterial blood supply or compression of the kidney is called *renal or Goldblatt hypertension.* This may or may not be dependent on plasma renin activity. The decreased ability of the constricted kidney to excrete Na^+ is an additional factor contributing to renal hypertension.

Bilateral clamping of the carotid arteries proximal to carotid sinuses elevates the blood pressure and heart rate because the procedure lowers the pressure in the sinuses. If carotid sinus nerves are cut on each side, the same effect is obtained. The pressure rise following these two procedures is moderate as the aortic baroreceptors are still functioning normally. If baroreceptor afferents in the vagus nerves are also interrupted, the blood pressure rises to 300/200 mm Hg or higher and is unstable. Bilateral lesions of the nucleus of tractus solitarius (NTS), the site of termination of baroreceptor afferents in the medulla, cause severe hypertension that can be fatal. This form of experimental hypertension is called *neurogenic hypertension.*

In humans, sustained diastolic hypertension is a very common abnormality. It can be produced by many diseases including adrenocortical diseases, tumors of adrenal medulla or of catecholamine-secreting tissue elsewhere in the body, renal diseases, narrowing of the aorta, severe polycy-themia and by oral contraceptives ("pill hypertension"). Hypertension is also a prominent symptom of *toxemia of pregnancy,* which may be caused by a pressor polypeptide secreted by the placenta. Hypertension is also seen in Cushing's syndrome.

In most of the humans with high blood pressure, the cause of the hypertension is unknown and they are said to suffer from *essential hypertension.* In the early stage of hypertension, the blood pressure elevations are intermittent, but later the blood pressure elevation is continued. The baroreceptor mechanism is "reset" so that the blood pressure is maintained at the elevated level. In women, untreated essential hypertension is often benign, but it may also progress rapidly into malignant phase.

Chronic hypertension can enter an accelerated phase in which necrotic arteriolar lesions develop and there is a rapid downhill course with papilledema, cerebral symptoms and progressive renal failure. This syndrome is called *malignant hypertension* and is very fatal if not treated. However, its progression can be checked and it can be reversed by appropriate antihypertensive therapy.

Hypertension causes a number of serious disorders including hypertrophy of cardiac muscle or heart enlargement. There is an increased incidence of atherosclerosis in hypertension and myocardial infarcts are common even when the heart is not enlarged. Starling's law operates in hypertension, but there may be heart failure. Hypertensive individuals are also predisposed to thrombosis of cerebral vessels and cerebral hemorrhage. An additional disorder is renal failure. However, the incidence of heart failure, strokes, and renal failure can be considerably reduced by active treatment of hypertension, even when the hypertension is relatively mild.

At present, essential hypertension is a treatable but not a curable disease. Drugs that block α-adrenergic receptors, β-adrenergic receptors, and drugs that inhibit the activity of angiotension - converting enzyme, and calcium channel blockers that relax vascular smooth muscle can produce effective lowering of the blood pressure. In humans, several of the diseases known to cause hypertension are not only treatable but also curable. When fed on high sodium diet some humans show marked increase in blood pressure. As there is no easy test for salt-sensitivity (to distinguish salt-responsive from salt-resistant humans), the current emphasis on less sodium intake in all humans appears to be beneficial.

Hypotension

The term hypotension describes the condition or symptoms produced by sustained low blood pressure. The low pressure receptors are those in the great veins, right and left atria and pulmonary vessels.

Hypotension is also a common abnormality in humans. It can be produced by several disorders and diseases. Hypotension and hypovolemia produced by hemorrhage release large amounts of vasopressin resulting in water retention and reduced plasma osomolality. There is exponential increase in plasma vasopressin produced by decreases in blood pressure.

In cases of *Orthostatic (posturnal) hypotension,* which is common in patients using sympatholytic drugs, sudden standing causes a fall in blood pressure, dizziness, dimness of vision and even fainting which is called postural syncope. Postural hypotension may also occur in diseases such as diabetes and syphilis in which there is damage to the sympathetic nervous system. Another cause of postural hypotension is *primary autonomic insufficiency.* It may be in several forms including one in which there appears to be a cogenital deficiency of dopamine β-hydroxylase and consequently no production of epinephrine and norepinephrine.

Fainting or syncope, more commonly due to peripheral, vascular or cardiac abnormalities causing inadequate cerebral blood flow, produces hypotension together with bradycardia (slow heart rate). This condition is called *vasovagal syncope.* Patients with orthostatic (postural) hypotension also show *micturition syncope,* fainting during urination. It is due to combination of orthostasis and reflex bradycardia induced by voiding in these patients.

7

Muscular System

Muscular movement is a characteristic of the higher animals. Movement in lower animals is just a primitive version of the muscular activities found in higher animals. In addition to locomotion, muscles perform a number of functions associated with digestion, respiration, circulation, excretion and reproduction. Thus, muscular activity plays a key role in animal life.

Kinds of Muscles

The human body, like most higher animals, contains three distinct types of muscles : *smooth, skeletal* and *cardiac.*

(a) *Smooth Muscles* : These occur primarily in the wall of internal organs *(e.g.,* digestive tract, arteries, veins, etc.). They are spindle-shaped cells, each with a centrally located nucleus. Unlike skeletal and heart muscles, they are slow to contract, and do not have cross striations on their myofibrils. The contraction and relaxation of smooth muscles are directly controlled partly by the nervous system and partly by them. Unlike skeletal muscles, but like cardiac muscles, they are not under voluntary control.

(b) *Skeletal Muscles* : (also called Striated, Voluntary or Somatic muscles) contract most rapidly and vigorously. They are attached to the bones or tendons, so they are mainly responsible for the movements of the skeleton and therefore of the body. The tissue of skeletal muscle is formed by long cylindrical cells or fibres which are multinucleated. They have well developed cross striations. Each muscle fibre is composed of fibrils or myofibrils, having characteristically alternating light and dark regions. Many invertebrates, including even some coelenterates, also contain some striated muscles.

(c) *Cardiac or Heart Muscles*: These form a unique type of muscular tissue found only in the heart. They resemble with the skeletal muscles in being made up of striated multinucleated fibres. However, the nuclei of each fibre are not peripheral as in skeletal muscle, but are distributed within the cytoplasm. Cardiac muscle is functionally syncytial and contracts rhythmically in the absence of external innervation due to the presence in the myocardium of pacemaker cells that discharge spontaneously.

Fine Structure of the Skeletal Muscle Fibre : Each skeletal muscle fibre is an elongated multinucleate cell completely surrounded by a *sarcolemma,* the cell membrane. It consists of a semifluid cytoplasm called *sarcoplasm,* containing numerous longitudinal fibrils called myofibrils. The many nuclei scattered throughout the fibre are generally located immediately under the sarcolemma. The diameter of a muscle fibre ranges from 10 to lOO μ and they extend through the whole length of the muscle joining with the tendons at their ends. Sarcoplasm contains, in addition to several proteins and various enzymes, nuclei, numerous mitochondria, a small Golgi body, sarcoplasmic reticulum, lipid droplets, and glycogen. Because sarcosomes (mitochondria) are the sites of oxidativeone sarcomere within a myofibril energy metabolism, their presence in relatively large numbers in a muscle has usually been taken as an indication of a muscle that performs more work.

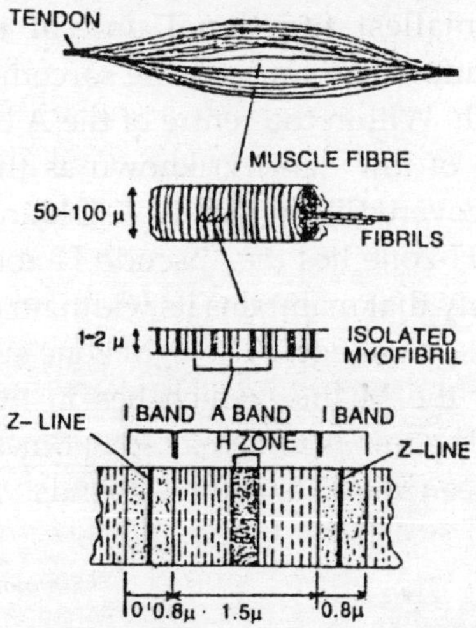

Schematic diagram of whole muscle fibre, fibril and banding pattern of one sarconere within a myofibril (Redawn from H. E. Huxley).

The *sarcaplasmic reticulum* is a system of tubules (sarcotubules) that does not open to the exterior. But the units of the other system, the *transverse tubules* open to the exterior of the fibre. These tubules are inwardly directed extensions of the sarcolemma and their cavities are thus in continuity with the fluid which surrounds the fibrils. Such continuity provides a simple means of conveying an electrical signal derived from the depolarization of the sarcolemma.

Each *myofibril* has a diameter from about 1 to 2 microns. The myofibrils are the actual *contractile elements* (units) of the muscle fibre, and it is the coincidence of light and dark bands of many myofibrils, which gives the muscular fibre its characteristically striated appearance. The striations or repeating pattern of light and dark bands are similar for each myofibril and are designated by letters. The dense (dark) bands are called *A* or *anisotropic bands,* while the lighter bands are called I *(isotropic)* bands. In the middle of the I band can be recognised the Z-band or Z-line (from the German, *Zwischenscheibe,* central membrane) as a narrow dark line.

The area between two adjacent Z lines is called a *sarcomere,* which is the smallest functional unit of the muscle. In uncontracted mammalian muscles the sarcomere is about 2.1 to 2.4 μ in length. Within the centre of the A band is a lighter region (a region of low density) known as the H *zone* (after *Hensen* who discovered it), which is visible in relaxed fibrils. In the centre of H-zone lies the "pseudo H-zone," a region of even lower density that maintains its width no matter, how the length of the muscle changes. The light zone surrounds a dark thin strip called the M line, which lies in the centre of the sarcomere. Finally a thin N *band* may sometimes be seen within the I band between the Z and the A bands.

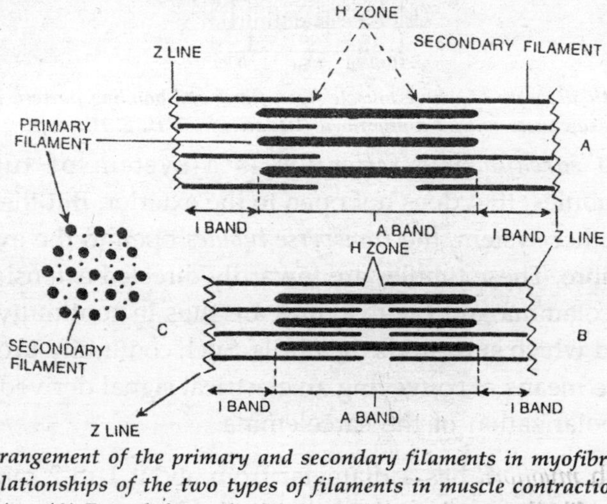

The arrangement of the primary and secondary filaments in myofibrils and the relationships of the two types of filaments in muscle contraction and relaxation. (A) Extended muscle, (B) Partially contracted muscle, (C) Cross-section through A band of myofibril (after H.E. Huxley).

Recent work of *H.E. Huxley* (1958,1965) has shown that each myofibril is composed of two types of filaments *(myofilaments)*, known as the *primary* or *thick* filaments and the *secondary* or *thin filaments*. In vertebrate muscles, the thick filaments are about 100 A thick and 1.5 μ long while the thin or secondary filaments are approximately 50 Å thick and 2.0 μ long. Thus, each myofibril is constructed of a very large number of each type of filaments, arranged so that each primary filament is surrounded by 6 secondary filaments, which are in turn shared with 6 surrounding primary filaments as shown in Fig. The M line, described above, is thought to be caused by a slight bulge in the centre of each thick filament.

Hoyle and his coworkers, after examining a variety of invertebrate and vertebrate striated muscles, have suggested that a *superthin filament* (called *T-filament)* runs the length of many sacromeres. These filaments, 25 Å in diameter, were observed by these workers in all bands of the myofibril and were seen to bridge the gap between the actin and myosin filaments produced when a muscle fibre was stretched. *McNeil* and *Hoyle* (1967) and *Walcott* and *Ridgway* (1967) have suggested that superthin filaments serve as elastic supportive elements of the myofibril.

The repeating band pattern in the myofibrils is due to the alternate overlapping of thick and thin filaments, the dark bands occurring where both kinds of filaments overlap, and the light bands where only one type of filaments overlap. The dense Z lines are continuous across the width of each muscle fibre and somehow serve to hold the myofibrils together. In this manner it accounts for the coincidence of A and I bands of the many myofibrils to give the whole muscle fibre its typically striated appearance. The two kinds of filaments are linked together by an intricate system of cross bridges which play an important role in muscle contraction (see ahead).

Chemical Composition of the Muscle Fibre : Each muscle consists of at least five kinds of materials. Of which about 75% is water, and nearly 20% protein. The remaining 5% includes glycogen, inorganic ions, and a variety of other organic substances like free amino acids and *phosphagens (phosphocreatine* in vertebrate muscles, and *phosphoarginine* of invertebrates). The phosphagens play an important role in muscle contraction. The muscle proteins are of 3 types.

1. *The Stroma Proteins* : They constitute about 1/5 of the total muscle proteins. They can be identified with collagen and elastin elements that bind muscle fibres together and transmit tension to tendons.
2. *The Ordinary Cellular Proteins* : They are not specifically characteristic to muscle cells and constitute another 1/5 of the total muscle proteins.

Most important of them are the enzymatic proteins which catalyse various chemical reactions of the muscle cells, and the proteins that supply energy to the contractile system.

3. *The Special Contractile Proteins* : The thick filaments are composed of the protein *myosin.* The thin filaments are made up of three different types of proteins—the globular proteins *actin* and *troponin,* and a long thin protein *tropomyosin.* One thin filament probably contains between 300 and 400 actin molecules, and about 50 tropomyosins and 50 troponins.

Proteins of the Myofibril

Protein	*Location*	*% of Myofibril Protein*	*MW*	*Chain Composition*
Myosin	Thick filament	43-45	521,000	2 heavy chains 4 light chains
Actin	Thin filament	21-23	41,900	1×41,900
Tropomyosin	In grooves of thin filament	5	70,000	2×35,000
Troponin	Thin filament at 400-Å intervals	5	70,500	1TN - T (1 × 30,500) 1TN-I (1× 22,000) 1 TN - C (1 × 18,000)
Actinin	Z-line	1	180,000	2×90,000

Contd...

Protein	*Location*	*% of Myofibril Protein*	*MW*	*Chain Composition*
Titin*	I-band	8-10	2500,000 to 3000,000	1×2500,000 to 1 × 3000,000
Nebulin*	I-band	3-5	700,000	700,000
M-line protein	M-line	2	165,000	1×165,000
C - protein	Thick filament	1-2	140,000	1 × 140,000

**Recently discovered by K. Wang, Myosin, Actin and Tropomyosin are mechano-chemical proteins ; Tropomyosin and Troponin are Regulatory proteins.*

Actin and myosin are absolutely essential for contraction. It is said that tropomyosin plays a vital role in sensitizing the contractile proteins (actin and myosin) to calcium ions. This sensitivity to calcium is essential for switching the muscle contraction on and off.

Among the three proteins of a myofibril, myosin is the most abundant (Table). Skeletal muscle contains twice as much myosin by weight as actin. By selective extraction procedures it has been possible to show that the thick filaments of the A-band are composed of myosin, while the thin filaments of the I-band are composed of actin and very likely tropomyosin. For example, after treatment of muscle with a suitable salt solution which is known to remove myosin, the A bands disappear. This is confirmed by the observation that similar treatment of myofibrils results in the removal of primary filaments. If actin of the myofibrils is extracted, most of the thin filaments and I bands are removed. The remaining fragments of I bands are possibly made up of tropomyosin.

(a) *Myosin* : Myosin molecules are tadpole-shaped with an overall length of 1520 ± 260 Å and a very high molecular weight about 521,000. Each molecule contains six polypeptide chains : *two heavy chains* (MW = 200,000) with identical sequences that account for most of the weight of the molecule, and *two pairs of light chains* that vary in size depending on the source (MW = 15,000 to 27,000). The NH_2 terminus of each heavy chain forms a *globular domain* to which the light chains are attached. This domain binds actin and contains the ATPase activity of myosin. The remaining part of each heavy chain is a fibrous, almost entirely α - helical structure that extends to the COOH terminus, coiled around the other heavy chain in the molecule.

Two of the myosin light chains can be dissociated from the globular domains at high pH, and are called *alkali light chains.* There are two different types of *alkali light chains :* α_{-1} (MW = 25,000) and A_2 (MW = 16,000). Both A chains have an identical COOH-terminal sequence of 141 residues preceded by an 8-residue segment that differes in five amino acid substitutions. The A_1 chain has an additional NH_2- terminal segment of 41 residues. Both A_1 and A_2 chains are required for myosin ATPase activity and binding to actin. The other pair of light chains (MW = 18,000), which can be dissociated from the globular domain by DNTB, 5,5'-dithiobis (2-nitrobenzoate) are designated DNTB *light chains* and have indentical sequences.

By brief digestion with trypsin, myosin molecules can be broken down into head and tail fragments. The head fragment is globular in shape measuring about 150-250 Å in length and 40 Å in diameter, with a molecular weight of about 350,000. This protein, also called *heavy meromyosin* (HMM), has two chemically different active sites. At one site binding to actin molecule and at the other hydrolysis of ATP occurs. The tail fragments of myosin molecules are called hydrolysis of ATP occurs. The tail fragments of myosin molecules are called *lightmeromyosin* (LMM). They are about 1000 Å long, 15-20 Å in diameter, and with a molecular weight of about 150,000. LMM is a rod-like structure probably consisting of two alpha helices coiled around each other like a two-stranded rope. In intact muscles, LMM and HMM perform entirely different functions.

Thick filaments are formed by packing of myosin molecules. About 400 molecules bind one another and their double heads project at interval distances of 14.3 nm along the fibre axis, staggered at 120° intervals so that every fourth pair of heads is aligned with each other at 43 nm. Myosin filaments are oriented in each direction from the M line lying in the middle of the H zone. A special protein *(M-protein)* at the M line marks the midpoint at which the myosin fibres meet tail to tail.

(b) *Actin* : Actin is a water-soluble, globular protein molecule consisting of 376 amino acid residues, including one residue of N^3-methylhistidine. As isolated, there is one molecule of ATP bound per actin. Actin also occurs in two forms. In the monomeric form the actin molecules are called globular actin molecules, or *G-actin.* They lack the ATPase activity, but each molecule possesses a molecule of ATP firmly bound to it. Each molecule of G-actin has a formula weight of about 45,000. In the polymeric form the actin molecules are fibre-like and are called fibrous actin, or F- *actin.* They occur when globular actin molecules undergo polymerisation in solutions of high ionic concentrations. The F-actin molecules are in the form of a two-stranded rope. The G-actin molecules can stick together only if the ATP bound to them becomes hydrolysed to ADP with the release of energy.

$$\text{G-actin-ATP} \longrightarrow \text{F-actin-ADP} + \quad \text{Pi}$$

In mammalian skeletal muscle tissue *myosin and F-actin* are arranged to form myofibrils. Along the main axis of the myosin unit are the projections know as the side chains. They are distributed around the myosin in a spiral manner and point toward actin units. In addition to its structural role, actin has the important property of activating the myosin ATPase, which catalyzes ATP hydrolysis. This function of actin is important in the hydrolysis of ATP by myosin that accompanies muscle contraction.

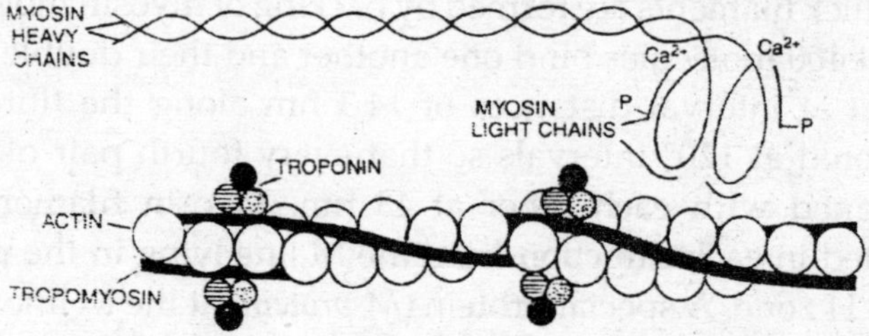

Diagrammatic representation of relative positions of Tropomyosin, troponin, and actin in thin filament of muscle. Each troponin comprises 3 different subunits. Tropomyosin lies in the groove of the helix, which is undisturbed. There is one tropomyosin and one troponin for each seven pairs of G-actin (after Adelstein and Eisenberg 1980).

(c) *Tropomyosin* : Tropomyosin is a rod-shaped molecule (MW = 66,000) about 41 nm long and consists of two nonidentical, α -helical polypeptide chains (α and β) twisted around each other in a coiled coil. The rod shaped molecules lie end to end in the grooves made by the two strands of actin filaments. This is present in lesser amount than actin and myosin. It lacks the ATPase activity. As said before, tropomyosin plays an important role in sensitizing the contractile proteins to calcium. Tropomyosin A (sometimes called Paramyosin) and Tropomyosin B are the two known forms of tropomyosin. Tropomyosin is found as a contaminant of highly purified actomyosin preparations and is associated with another protein, troponin that has been shown to combine with actin and exert through calcium an important effect on the reactions of the contractile system. This sensitivity is not shown by pure actin and myosin.

(d) Troponin: Troponin molecule (MW = 76,000) contains three different subunits, each named according to its function. T_n -T subunit binds tropomyosin, T_n-I inhibits the interaction between myosin and actin, and T_n-C subunit binds Ca^{2+} and regulates the interactions among T_n -T and T_n -I and their interactions with other components of the contractile system. Each subunit of troponin is noncovalently bound to at least two of the others, as shown here :

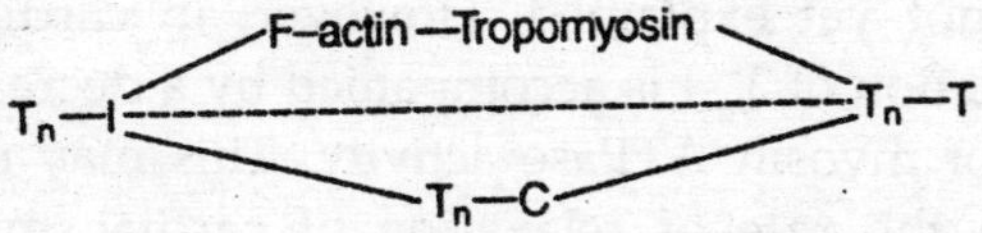

T_n-T (MW = 37,000) contains 259 amino acid residues in known sequence, half of which are either positively or negatively charged. The latter are clustered in the NH_2 terminus, the former in the COOH terminus. T_n-T binds strongly to tropomyosin as well as to T_n-C, and serves to anchor T_n-I and T_n-C to the actin-tropomyosin complex.

T_n-I (MW = 20,000) contains 179 amino acid residues, including a region rich in basic amino acids that contains the actin-binding site. Its NH_2- terminal contains the T_n-C-binding site. In resting muscle, T_n-I inhibits the interaction of myosin and actin and prevents actin activation of myosin ATPase activity.

T_n-C (MW = 18,000) contains 159 amino acid residues rich in glutamic and aspartic acids. Its sequence is highly homologous to that of calmodulin, and like which it contains four divalent ion-binding sites. Two of the divalent ion-binding sites of T_n-C are specific for Ca^{2+}, but Ca^{+} and Mg^{2+} compete for binding at the other two. Under physiological conditions,

T_n-C is thought to bind two Ca^{2+} and two Mg^{2+} ions. Indeed, T_n-C and calmodulin appear to participate equally well in glycogenolysis in skeletal muscle.

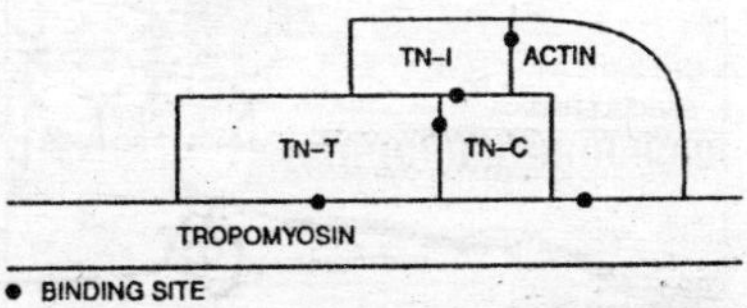

Diagrammatic relationship between actin, tropomyosin and three subunits of troponin molecule.

T_n-T, T_n-I, but not T_n-C, are covalently phosphorylated by cAMP-dependent protein kinase and ATP. The physiological significance of phosphorylation of T_n-I and T_n-T in skeletal muscle is not yet explained. However, in cardiac muscle, phosphorylation of T_n-I is accompanied by a decrease in actin activation of myosin ATPase activity. This may explain the increase in the rate of relaxation of cardiac muscle after administration of epinephrine.

Dystrophin-Glycoprotein Complex

A large protein called dystrophin (MW 427,000) forms a rod that connects actin to two other proteins in the sarcolemma and they in turn connect to the merosin subunit of the adhesion molecule laminin in the extracellular matrix (Fig.). These two proteins in the sarcolemma are α-DG (dystroglycan) and β-DG forming the dystroglycan complex. This complex is in turn associated with the sarcoglycan complex consisting of three glycoproteins, a, P, and y sarcoglycan.

It appears that the dystrophin-glycoprotein complex provides structural support and strength to muscle fibril, but it may have some additional function also. The complex is of special interest because congenital defects in it cause many of the different forms of muscular dystrophy (Worton, 1995). The dystrophin gene, which is one of the largest known genes, is also expressed in smooth and cardiac muscle.

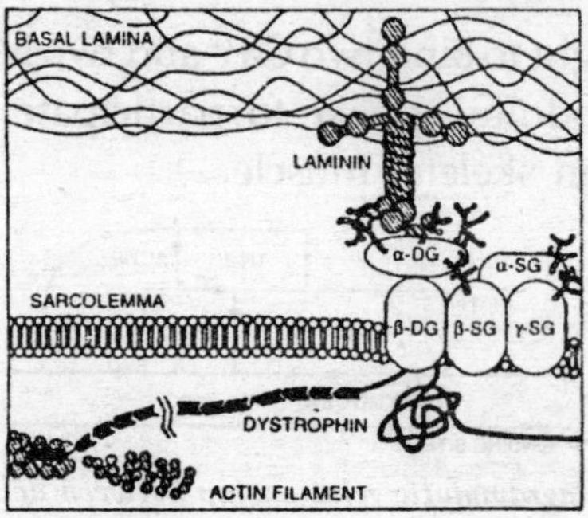

The dystrophin-glycoprotein complex.

Mutations in the gene coding for the various components of the dystrophin-glycoprotein complex cause muscular dystrophy, a syndrome characterised by progressive muscle weakness. Some forms of this disease cause great disability and are eventually fatal. In Duchenne's muscular dystrophy, which is x- linked, dystrophin is absent from the muscles. It occurs in about 1 of every 3000 male infants and is usually fatal by the age of 30. In Becker's muscular dystrophy, dystrophin is present but altered or reduced in amount. Autosomal recessive muscular dystrophy is due to a mutation in the gene for Y - sarcoglycan. In some cases mutations in one or another of the genes for all the three forms of sarcoglycans are associated with late-onset limb-girdle muscular dystrophy. However, the ways in which the various abnormalities of the dystrophin-glycoprotein complex cause dystrophy are at present unknown.

Cross-Bridges

The actin (thin) and myosin (thick) filaments are linked together by cross-bridges which play an important role in muscle activity. When a longitudinal section of muscle is examined under the electron microscope, it can be seen that the cross bridges project outward from both the ends of a myosin filament at fairly regular interval of 60 or 70Å. In the centre of the myosin filament there is a bridge-free region, about 0.15 to 0.2 μ long. This absence of cross-bridges is responsible for the pseudo H zone in the centre of A bands.

It has been recently shown by *H.E. Huxley* (1965) that the cross-bridges are nothing except the heads of the myosin molecules and all such heads (*zigzag* lines) or the cross-bridges have the enzymatic and actin-binding properties. During muscle activity each cross-bridge is attached to an active site on the actin filament at a definite rate and then detached from the same. Thus, cross-bridges play an important role in muscle contraction.

Physiology of Muscle Contraction : Muscles contract in reponse to a stimulus that is delivered to them through a motor nerve. So the first step in the excitation of a muscle is the arrival of a nerve impulse along the nerve fibre that is connected to the muscle at a junction called the *"end plate"*. In the resting state the outer membrane of the muscle fibre is electrically polarized, *i.e.*, its outer surface is positively charged with respect to the inner surface. The arrival of the nerve impulse brings about the release of *acetylcholine* at the neuromuscular junction. This substance depolarizes the muscle membrane by increasing its permeability to Na^+ ions and this depolarization in turn excites an *"action potential"* which propagates along the muscle fibre. The action potential sets off two independent electric currents in the muscle fibre, of which one is a weak longitudinal current and which, according to *Ove Knudsen* (1954) does not induce contraction. But the other current is a transverse wave that moves inward along the system of transverse tubules and is said to be responsible for contraction.

The following theories or models have been put forward to explain the mechanism of muscle contraction.

Sliding-Filament Theory

According to sliding-filament or Interdigitation theory, proposed independently by *A.F. Huxley* and *R. Niedergerke*, and by *J. Hanson* and *H.E. Huxley* (1958), the two kinds of filaments slide past each other so as to produce change in the length of the voluntary muscle. The theory proposes that shortening of the muscle fibre is caused by a *sliding movement of interdigitating filaments*, bringing about an increase in the overlap between them.

It has been found that during both contraction and stretching, the length of A-bands remains constant and the length of the I-bands changes in accordance with the length of the muscle. Since the length of the A-band is equal to the length of the thick filaments, so the length of these filaments, also remains constant. But the length of H-zone increases or decrease with the length of the I-band, so that the distance from the end of one H-zone through the Z-line to the beginning of the next H-zone remains almost the same. This distance is equal to the length of thin filaments, so they too do not change their length to a large amount. The only conclusion which has been drawn from these observations is that, when the muscle changes its length, the filaments themselves do not perceptibly change in length but the two sets of filaments slide past one another, *the thin filaments moving farther into the A-bands during contraaction and moving farther out of them during stretching.* So, in a powerful contraction the ends of thin filaments first meet each other and then they crumple up or overlap. Later on the same may happen with the thick filaments and the crumpling produces new band patterns in the muscle.

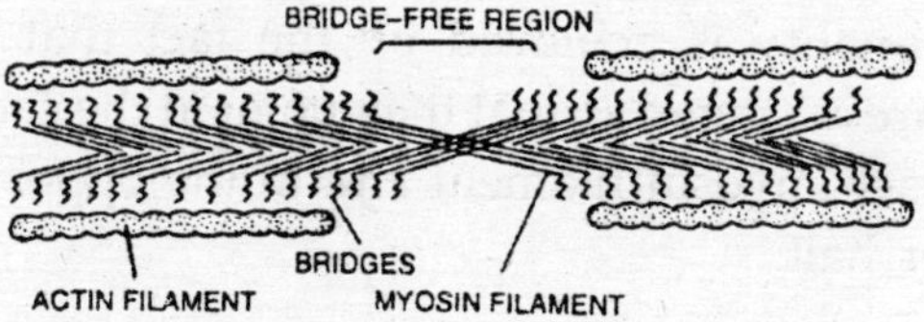

Schematic representation of the contact of actin and myosin in a voluntary muscle. The thin actin filaments above and below are so shaped that certain sites are closest to thick myosin filament in the middle. The heads of individual myosin molecules (zigzag lines) extend as cross bridges to the actin filament at these close sites (after H.E. Huxley, 1965).

The cross-bridges play an important role in muscle contraction and in fact the sliding movement is mediated by the bridges. In fact, mechanical linkage between the two types of filaments occurs via cross bridges. It has been concluded by *H.E. Huxley* (1965) that the heads of myosin molecules, which have the enzymatic and actin-binding properties, serve as the cross-bridges connecting the actin and myosin filaments in the muscle (Fig.). Besides, the myosin molecules possess an inherent directionality because they aggregate with their heads pointed in one direction along half of the filament and in the opposite direction along the other half of the myosin filament. This property of myosin molecules is very important because it explains a crucial feature of the sliding-filament theory at the molecular level. In a sliding-filament system in which a relative force is developed between actin and myosin molecules of the two types of filaments, it is essential that the appropriate directionality of sliding be built in some way. During contraction the thin filaments move toward each other in the centre of the A bands, so that it is required that all the elements of force generated by the cross-bridges in one half of the A band be oriented in the same direction, and that the direction of the force be reversed in the other half.

Thus, during interaction of actin and myosin the direction of the force developed depends on the orientation of actin and myosin molecules. On the basis of electron microscope observations it has been suggested by *H.E. Huxley* (1965) that this directionality is achieved by the fact that the myosin molecules are arranged so that they point in the same direction in half of each myosin filament and in the opposite direction in the other half.

During normal contractions the length of the sarcomere changes about 20 per cent and thin filaments in each half of the A band have to move about a quarter of a micron. Hence it seems physically impossible that the cross-bridges could remain attached to the same point on the actin filament throughout this process. It has been explained by *H.E.Huxley* (1965) that the bridges are attached to one site on the actin filament for part of the contraction, then detach and reattach themselves at a new site farther along. Moreover, at each site where cross-bridge and actin filament interact, one molecule of ATP is split to generate a sliding force between the two kinds of filaments.

The interaction of filaments during sliding has been

explained by giving the analogy of a ratchet, and the way in which the cross-bridges of myosin hook on to consecutive active sites on the actin filament further supports this analogy. As the skeletal muscle can contract by some 30 per cent, so each individual cross-bridge must detach itself from one site on the actin and reattach itself to another site further along, repeating the process 5 or 6 times, "with an action similar to a man pulling in a rope hand over hand" (*Wilkie*, 1968). However, another concept is that electric fields are established by the cross-bridges which attract the actin filaments thereby causing them to move.

Davies Theory

According to *R.E. Davies* (1963), the side chains on the myosin filament are important in this theory. Each of these side chains terminates in a protein-bound molecule of ATP. Near the base of the side chain is a fixed negative charge probably $-COO^-$ of the side group of glutamic or aspartic acid. Repulsion between this fixed charge and the net negative charge at the tip of ATP is thought to keep the side chain extended in the resting molecule. Otherwise the side chain polypeptide would tend to coil up. Should it do this, according to the theory, the phosphoric anhydride system would be drawn into a region of the myosin that actually provides a local catalyst for hydrolysing such anhydrides.

The side chain on myosin is called a cross-bridge because it çan be part of an ionic bond or bridge to a neighbouring actin molecule. This bridge is completed in the moment of activation when C^{2+} ions are released from a region of the cell called the sarcoplasmic reticulum. This ion with two plus charges can fit between the two sites with minus charges, one on myosin and one on actin. Once this happens, the minus charge at the ATP tip is neutralized. The force of repulsion between this tip and the fixed negative charge at the base of the side chain is removed.

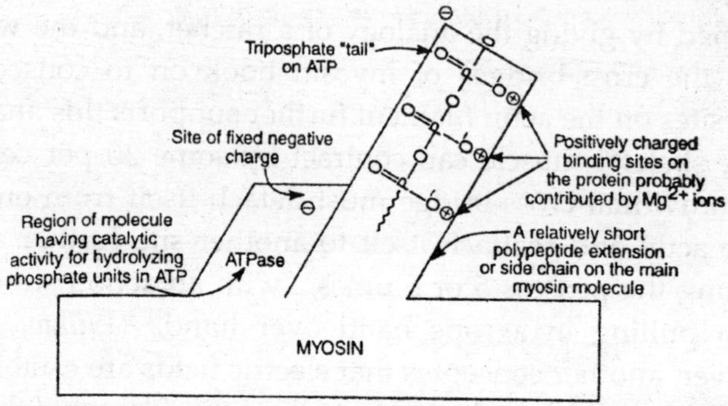

The cross-bridge in muscle myosin is depicted here in its resting or extended state. The net charge at the ATP tip is-1. Repulsion between it and the fixed charge keeps the side chain extended. The side chain is called H-meromyosin.

The side chain at once coils into a helix. It cannot do this, of course, without pulling on the actin molecule, for as long as the calcium ion is there the actin and myosin are for all practical purposes bound together indicates that this "pull" can only cause the actin and myosin filaments to slide by other, as *Huxley* said they must.

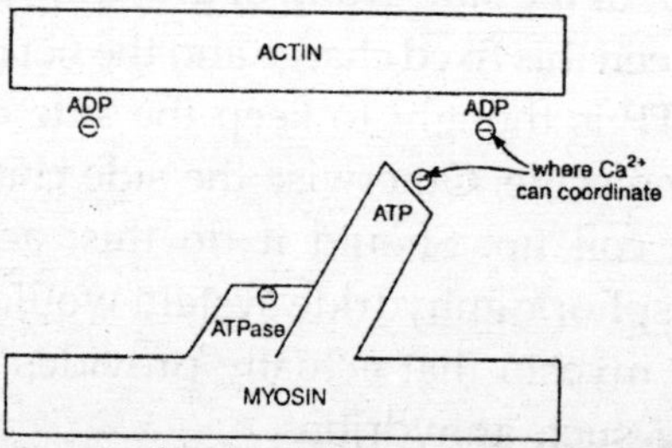

Showing the arrangement of actin and myosin in the resting state. When the muscle is activated, calcium ions are released from the sarcoplasmic reticulum, and they move to the negatively charged sites as shown.

When the side chain is in its coiled phase, the ATP tip is exposed to the action of the ATPase and a phosphate unit is clipped off by hydrolysis. Now comes a key step, putting a phosphate unit back on. The ATP is resynthesized at the ATP tip, now regenerated with its net negative charge, has this charge repelled by the fixed negative charge. This side chain uncoils and shoots back out ready for another cycle. Many rapid repetitions of these stages produce the overall effect of contraction. The presence of Ca^{2+} in the sarcoplasm activates a relaxing mechanism which pumps the calcium ions back into the sarcoplasmic reticulum. When the Ca^{2+} has been removed, the muscle is no longer able to contract.

The *Huxley* model of muscle contraction is thus given a molecular interpretation by the theory of *Davies.*

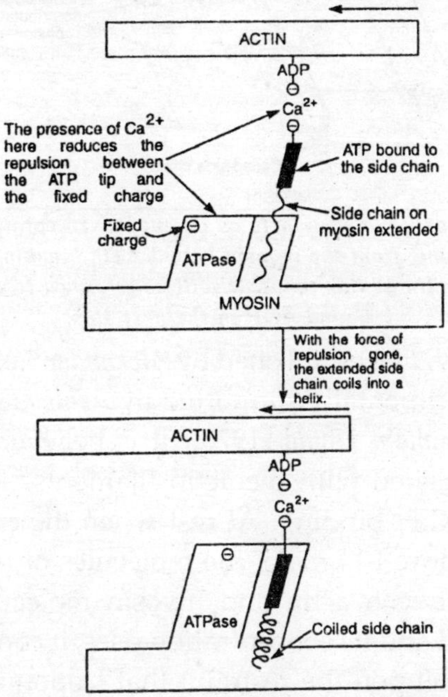

The calcium ion removes the force that keeps the side chain extended, and it coils back towards the main chain of myosin. As it does so, the actin filament is pulled in the direction of the arrows, causing it to slip along the myosin chain.

Calcium-Release Theory

The theory holds that contraction is initiated by Ca^{2+} ions released from a binding site in the muscle (sacs of the sarcoplasmic reticulum) by the action of electrical stimulus and relaxation results by subsequent uptake of these released Ca^{2+} ions by the sarcoplasmic reticulum. Thus, Ca^{2+} is considered to activate the contractile machinery.

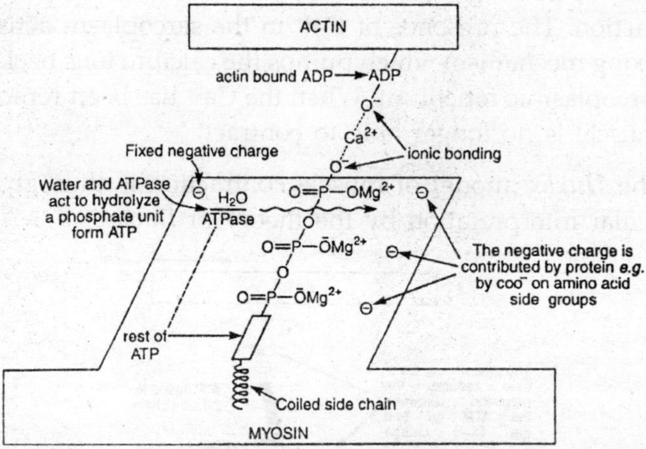

A site on the myosin molecule acts as an enzyme to catalyze the hydrolysis of a phosphate unit from the myosin-bound ATP, breaking the calcium-ion bridge to the actin at this point. It will be reformed to a different actin bound ADP in the next step.

This theory, first established by *Alexander Sandow* and others has been experimentally supported by *Ashley* (1967), *Winegrad* (1968), and *Graham Hoyle* (1970). It is believed that *troponin,* which is associated with the actin filaments, seems to play a major role in Ca^{2+} binding. At rest when the concentration of free Ca^{2+} is below 10^{-8} mole, the molecules of troponin inhibit interaction between actin and myosin molecules. But when Ca^{2+} is released and its concentration rises, it combines with the troponin. Thereupon the restraint that troponin places on the interaction of actin and myosin is released and the two types of filaments slide past one another causing the muscle to contract. Then the removal of Ca^{2+} during relaxation reverses the process.

However, there are two sources of Ca^{2+}. One is the *internal calcium* that supplements the other source, the *external calcium* entering the fibre during excitation. The possible sources of internal Ca^{2+} include the membranes of the fibre surface, the clefts, the invaginated tubules, the sarcoplasmic reticulum, and the mitochondria. The sarcoplasmic reticulum sacs adjoining the invaginated tubular system are considered as the major source of supplementary internal Ca^{2+}.

The release of the internal Ca^{2+} is caused by the initial stimulus which depolarizes the fibre membrane and the conduction of the negative depolarizing charge down the transeverse tubules. It has been shown by *Hygo-Gonzalez-Serratos* that the rate of inward conduction along the tubules of a frog fibre is about 8 cm/sec. The ions pass down the tubules and across the junction between the tubules and the sarcoplasmic rericulum.

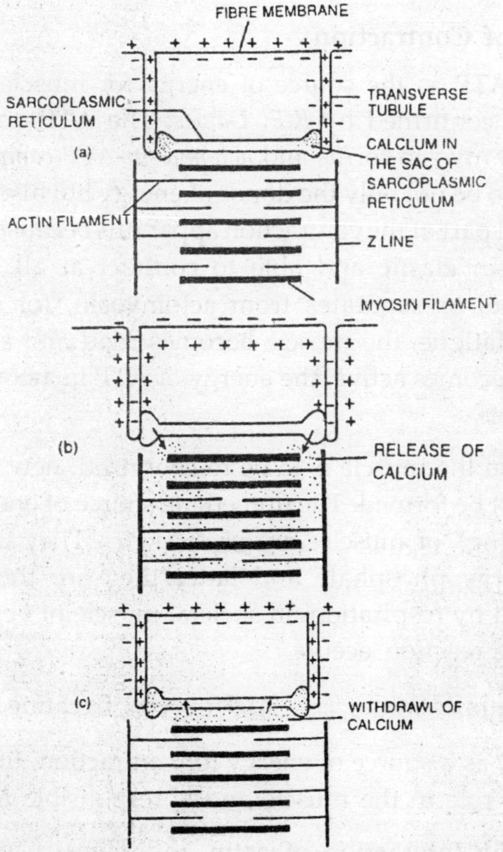

The role of calcium in muscle-activity (after Winegrad).

The process of relaxation depends on the return of the charge on surface membrane to its original condition. As the excitatory depolarization is followed by a rapid repolarization, the stimulus to the release of Ca^{2+} is quickly turned off and hence the release of Ca^{2+} is stopped. In other words, the relaxation is due to the diminishing amount of free Ca^{2+} as it is rebound to the reticulum. However, by the time the contraction reaches its peak, the free Ca^{2+} has already fallen to a low level. Thus muscle contraction is turned on by the release of Ca^{2+} and turned off by its withdrawl.

Energy of Contraction

The ATP as the source of energy for muscle contraction was first confirmed by *R.E. Davies*. The ATP, together with actin and myosin forms and *actomyosin-ATP complex*. Thus, it appears to be not only the donor of energy but also a necessary structural part of the contraction apparatus because ATP makes actomyosin elastic and able to contract at all. When ATP disappears or separates from actomyosin, for example, in extreme fatigue, the muscle becomes rigid and stiff. When a muscle becomes active, the energy of ATP in actomyosin-ATP is used up.

If then the muscle is to be re-energized, new actomyosin-ATP must be formed. The immediate source of energy for such "recharging" of muscles are *phosphagens*. They transfer their high-energy phosphate and latter they are themselves re-energized by respiration. In skeletal muscle of vertebrates the following reaction occurs :

$$\text{Phosphocreatine} + \text{ADP} \rightleftharpoons \text{Creatine} + \text{ATP}$$

Thus, as a source of energy for contraction, the ATP plays a central role in the muscle and is responsible for :

(i) the interaction of actin and myosin which causes shortening,

(ii) the synthesis of phosphocreatine from creatine and phosphate,

(iii) the synthesis of glycogen from lactic acid. This is necessary because the conversion of glycogen to lactic acid is another source of energy to synthesize ATP other than the phosphocreatine reaction.

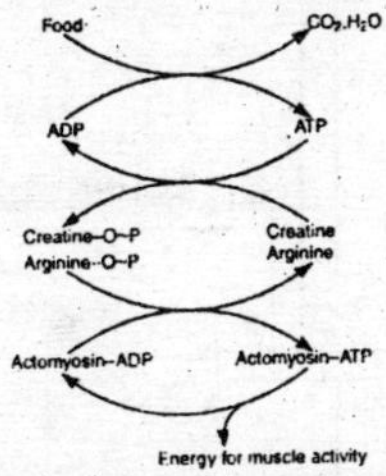

Showing that respiration supplies energy for muscle-activity via the phosphagens.

Chemical Reactions in Muscle Contraction : The various chemical reactions and the flow of energy in the muscle during contraction have been described by *Loewy* and *Siekevitz* (1966). Firstly myosin reacts with ATP to form a myosin-ATP complex. Then in the presence of K^+ and F-action, actomyosin-ATP complex is formed.

This complex in the presence of Ca^{++} and ATPase splits the ATP to inorganic phosphate and actomyosin-ADP complex, which then breaks down to free actin, free myosin and ADP. Thus, now it is known that the muscle contraction is turned on by the release of Ca^{++} and the hydrolysis of the ATP of actomyosin-ATP complex provides the necessary energy for the two kinds of filaments to slide past each other during contraction.

The relaxation is caused by binding Ca^{++} back into the sarcoplasmic reticular sacs, and thus inhibiting the breakdown of actomyosin-ATP complex. It has been also proposed that the Ca^{++}, as a trigger for muscle contraction, performs its work by combining with troponin which otherwise puts a restraint on the interaction of actin and myosin.

The muscle activity is continued by the continuous removal of ADP that accumulates in contraction, or in other words, by the conversion of ADP to ATP which is done in several ways as shown in Fig.

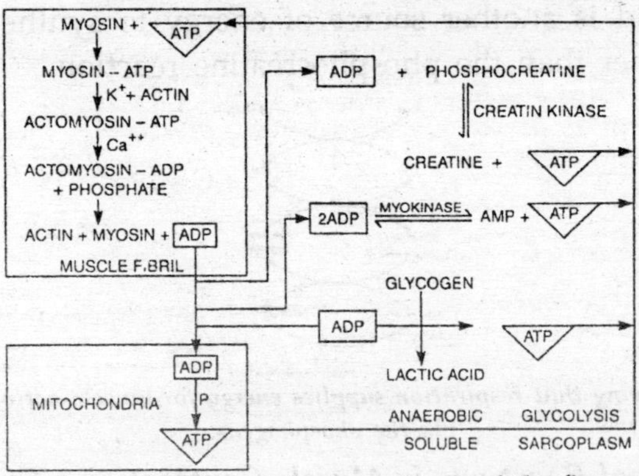

The chemical reactions in muscle contraction and the flow of energy.

Summary of the Present Concept : According to our present information the important events leading to muscle contraction are as follows.

(a) The arrival of an impulse

(b) Reduction of electric potential or depolarization of the fibre membrane

(c) Release of calcium (the activating substance)

(d) ATPase activation

(e) Hydrolysis of ATP at each site where cross-bridge and actin filament interact to generate a sliding force between the two kinds of filaments.

Similarly, the processes associated with relaxation are :

(a) Exhaustion of the nerve impulse

(b) Withdrawl of calcium

(c) Recreation of electric potential or repolarization of the fibre membrane.

(d) ATPase inhibition.

(e) Resynthesis of ATP.

Nerve impulse & Release of Ca^{++} Reduction of electric potential ATPase activation

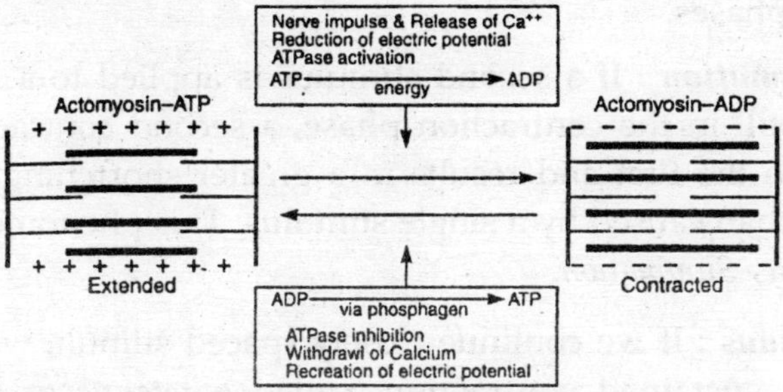

The summary of events leading to muscular contraction and relaxation.

The Single Muscle Twitch : A muscle twitch can be defined as a simple single muscular contraction caused by a single short stimulus. It can be recorded by *Kymograph* and consists of three successive phases. *(i) Latent period :* As its name implies. It is the lag period or time interval between the application of the stimulus and the actual beginning of visible muscle contraction. In the frog, the latent period lasts about 0.01 second and in the figure it is indicated by the distance from A to B.

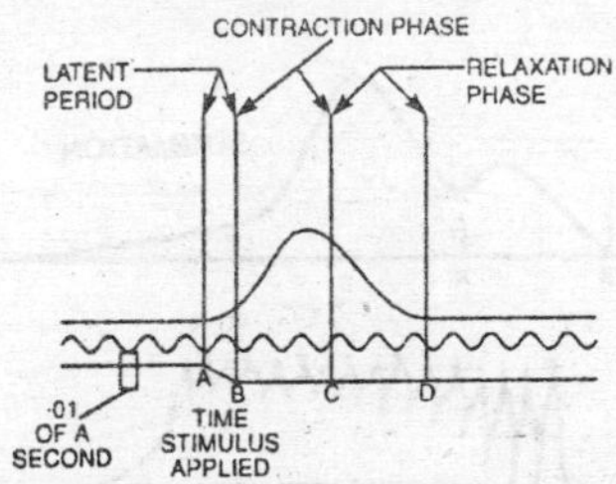

A single muscle twitch and its phases.

(ii) *Contraction Phase :* It follows the latent period, as indicated by the portion of the curve from B to C. It is the time interval in which the muscle reaches the peak of contraction and performs work. It has a duration of about 0.04 second.

(iii) *Relaxation Phase :* It is the final phase of the muscle twitch, or the time interval in which the muscle returns to its original length. This phase lasts about 0.05 second and is represented by the portion of the curve between C and D. Thus, it is the longest of the three phases.

Summation : If a second stimulus is applied to a muscle that is still in the contraction phase, a second contraction is added to the first and results in a greater shortening of the muscle than caused by a single stimulus. This phenomenon is known as *Summation.*

Tetanus : If we continue closely spaced stimuli, we get a smooth, sustained contraction, called a *tetanus* or tetanic contraction. A complete tetanus may be defined as a sustained contraction of a muscle due to the fusion of many twitches following each other in rapid succession.

In case the successive stimuli arrive at the muscle with long time intervals between them, the individual contractions can still be recognised since some degree of relaxation occurs between them to give a curve with a wavy plateau as shown in Fig. This condition is called *incomplete tetanus.*

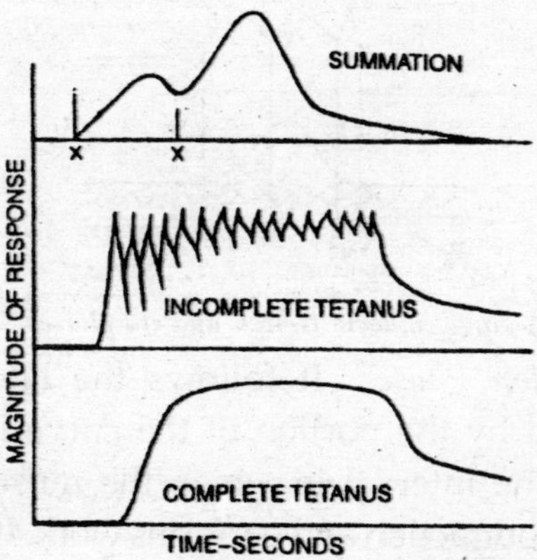

Diagram of summation, incomplete tetanus and tetanus.

Refractory Period

Skeletal muscle, like nerves, exhibits a refractory period. It is an extremely short time interval lasting about 0.002 to 0.005 second, during which a muscle will not respond to a second stimulus when given immediately after the first one. The second contraction is superimposed upon the first to give the summation effect.

It is thought to represent the time necessary for the occurrence of various physical and chemical changes which will allow subsequent contraction. Cardiac muscles have a longer refractory period than that of skeletal muscle and, therefore, lasts during the entire period of systole (the contraction of atria or ventricles of a heart) which means that cardiac muscle will not respond to a second stimulus while it is contracting.

Muscle Fatigue

If the energy is not continuously supplied to a contracting muscle, the actomyosin-ATP complex cannot be regenerated and the muscle becomes fatigued. Thus, fatigue is a condition or state whereby a musle experiences diminished ability to contract as a result of prolonged stimulation. However, it rarely occurs in intact muscle of the body, but can be easily demonstrated in excised muscles. Several factors are responsible for muscle fatigue, *e.g.*, lack of oxygen, lactic acid accumulation, and decreased level of phosphocreatine.

Rigor Mortis

When muscle fibres are completely depleted of ATP and phosphorylcreatine, they develop a state of rigidity called *rigor*. When this occurs after death, the condition is called *rigor mortis*. As we know soon after death the body passes into rigor mortis or the so-called stiffening of death. Normally, when muscle passes into rigor mortis its ATP and the glycogen content disappear and lactic acid accumulates in it. However, the mechanism by which muscle contracts after death (or in the absence of the action potential) is not known.

Types of Contraction

1. *Isotonic Contraction* : Isotonic means "same tension". It is that contraction which occurs against a constant load and there is a change in the length but the tension of the muscle remains practically the same.

The velocity of shortening as a function of load (force velocity relation) is given by the equation :

$$V = (P_o - P)\, b/(P + a)$$

where V is velocity of shortening, P_o is the maximum tension developed by the muscle, P is a given load, a is a constant with dimensions of a force, and b is a constant with dimensions of velocity. According to *Hill* (1953), the velocity of shortening depends on the difference between the force on the muscle and the maximal force it can develope. Thus, mechanical work is done in isotonic contraction.

The overall mechanical efficiency of skeletal muscle (workdone/total energy expenditure) ranges upto 50% while lifting a weight during isotonic contraction and is essentially 0% during isometric contraction.

2. *Isometric Contraction* : Isometric means "same measure" or length. When a muscle can contract without changing its length, it is called isometric contraction. In this case the tension increases sharply, but no mechanical work is done.

3. *Heat production in muscle* : In view of the laws of thermodynamics the energy supplied to a muscle must equal its energy output. The energy output appears in work done by the muscle, the energy-rich phosphate bonds formed for subsequent use and in heat. The

contraction of a muscle is accompanied by the liberation of heat in two phases :

(i) The *initial heat* is liberated during the contraction-relaxation cycle and is divided into several stages like *activation heat* (maintenance heat in Sustained contractions), *shortening heat* (liberated during the actual shortening of the muscle and is proportionate in amount to the distance the muscle shortens), and the *relaxation heat* (produced at the end of an isotonic contraction when a muscle has done work by lifting a load).

(ii) The *recovery heat* or *delayed heat* is liberated in the period immediately following the contraction cycle. This recovery heat is the heat liberated by the metabolic processes that restore the muscle to its precontraction state. The recovery heat is approximately equal to the initial heat, *i.e.*, the heat produced during recovery is equal to the heat produced during contraction.

The difference between the heat production of an isotonic and an isometric contraction is the shortening heat. The fact that shortening heat is independent of the load indicates that when a muscle performs more work, energy must be produced by the muscle. This is the *extra energy liberation,* and the phenomenon is often called the *Fenn effect.*

Physiology of Amoeboid and Ciliary Movements : Microbes and small multicellular organisms have other modes of locomotion, of which *amoeboid* and *ciliary* movements are the most important.

Amoeboid Movement

The characteristic movement of *Amoeba* by pseudopodia is called amoeboid movement. It is also found in the white blood corpuscles and in amoebocytes of sponges.

When a pseudopodium is formed, a blunt, finger-like projection called the *hyaline cap,* composed only of ectoplasm, first appears. A little later the *plasmasol* flows into this projection as it extends forward. At the start of the movement, *Amoeba* usually forms several small pseudopodia, of which one is gradually enlarged while the others disappear.

The protoplasm which enters the pseudopodium is naturally withdrawn from other parts, so that the animal not only changes its body shape but also its position. Thus, pseudopodia bring about a change in the shape and position of the *Amoeba.*

Theories of Amoeboid Movement

1. *Contraction Theory* : According to *Jennings* (1904), who worked on *Amoeba verrucosa,* a rolling movement of the plasmalemma accompanied by protoplasmic streaming, brings about locomotion. A similar contraction theory has been put by *Dellinger,* who worked on *Amoeba proteus,* suggesting a "walking" movement. In other species, *e.g., Difflugia,* there may be a walking movement, wherein the advancing tip is attached and the remainder is pulled forward.
2. *Change of Viscosity Theory* : This is also called *Gel-Sol-theory* of *Mast* and *Pantin.* According to this theory the two states of cyctoplasm are continuously converted into one another and bring about the formation of pseudopodia. Since the change of the gel form into the sol occurs continuously, a current of plasmasol is maintained toward the anterior direction.

   This flow itself seems to be brought about by the contraction of the plasmagel tube at the posterior end, thus forcing the plasmasol towards the anterior tip. During locomotion the elastic and contractile force of plasmagel is maximum at the sides, intermediate at the posterior end, and minimum at the anterior end. This results in an elongated shape of the animal and a forward extension of the anterior end to bring about locomotion.

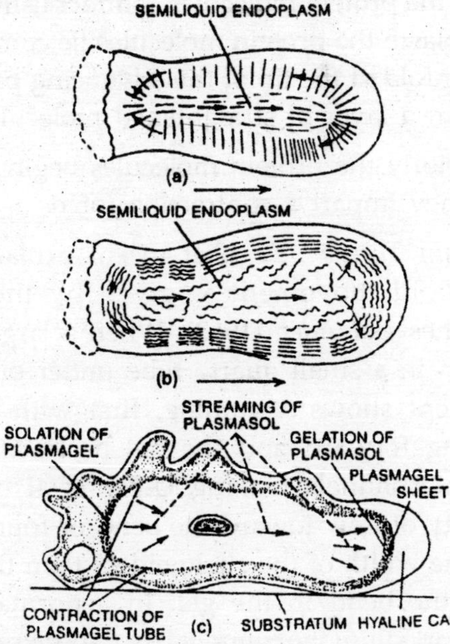

Theories of Amoeboid movement, (a) According to Allen's theory, the Amoeba is pulled forward by the contraction of cortical gel near the head, (b) According to Goldacre, contraction at the tail accounts for the movement, and (c) movement of Amoeba according to gel-sol theory of Mast and Pantin.

3. *Syneresis Thoery* : According to *Pantin*, water is removed by some sort of "syneresis" from protein at the posterior end and that the forward tide of syneretic fluid is important in locomotion. The tail of some amoeboid cells *(A. Umax)*, wrinkles during locomotion as if water was removed. Recent measurements of the refractive index of *A. proteus* and *A. dubia* show that in sustained locomotion the refractive index rises toward the tail, this region having 6 to 40% more (protein) than the advancing tip, and this gradient of refractive index disappears on stop of movement.

4. *Theory of Folding and Unfolding of Protein Molecules* : According to *Goldacre* (1951), the amoeboid movement is brought about by folding and unfolding of protein molecules and the energy for movement is supplied when the protein molecules contract using ATP. In the endoplasm the protein molecules lie compactly folded and unfold at the tip of the advancing pseudopodium to form a layer of straightened molecules.

   Posteriorly the protein molecules begin to fold again and they impart a contraction force.

5. *Fountain Theory* : A most recent explanation of the amoeboid movement is given by the "fountain" hypothesis of *Allen* (1962). When a large *Pelomyxa* is broken in a small quartz tube under oil the isolated fragment shows streaming, first with the axial sol flowing forward and the gel backward/then with several channels making U-bends at one end. The velocity of flow toward the bend or fountain is faster and the width of the channel less than the flow away from the bend in the gel. In a normally advacing pseudopodium the cylinder of sol shortens and thickens in the "fountain" zone.

It is postulated that gel strands, attached to the rim of the advancing tube, contract and pull the anterior sol forward. Water is removed to the hyaline cap by syneresis and passes back under the plasmalemma to be returned to the protein at the posterior zone of recruitment of sol.

Flow is chiefly at the boundary or shear zone between sol and gel. It is argued that contraction at a U-bend between sol and gel can equally well explain amoeboid movement and cyclosis.

Ciliary and Flagellar Movements : Cilia are the motor organelles differentiated at the cell surface. They are often classified into two types.

(i) *fagella,* which are relatively larger organelles occurring usually singly or in small numbers upon cells. Typical flagella characterize the class mastigophora (Gr. *mastix*-whip, *pherein* = to bear) of the phylum Protozoa.

(ii) *cilia* proper, which are relatively much smaller and occur characteristically in large numbers upon each cell. Cilia proper are seen in ciliated protozoans, coelenterates, and turbellarians.

Ciliary activity is restricted to an aqueous medium, and hence is found only on the surfaces which are submerged, or atleast covered by an aqueous film. In some cases cilia are concerned in one or more numerous functions, such as feeding (rotifers, protozoans, protochordates, etc.), circulation of body fluids (annelids),. cleansing (sea anemones), respiration (molluscs, protochordates, etc.) and the movements of materials within ducts and tubules (coelenterates, echinoderms, and molluscs.)

Functional Anatomy of the Cilia

Cilia of living cells represent a great uniformity in structure. Each cilium consists of a sheath containing an axial filament that usually extends distally beyond the limits of the sheath. The sheath may be circular or ovoid in cross-section and the axial filament pursues a straight or spiral course through the sheath. The proximal end of the axial filament is always provided with a basal granule, which is derived from the cell centrosome. Many cytologists have described a system of fibrils running between the basal granule and cell nucleus.

The ultrastructure of a cilium has been described with the aid of electron micrographs by a number of workers including *Roth* (1956) and *Porter* (1957). A cilium has nine parallel fibrils surrounding a pair of axial fibrils. The "9 + 2" arrangement of microtubules seen in cross section of a cilium (or flagellum) is one of the most conserved organelle structures in eukaryotic cells. The outer fibrils, with a diameter of 250 to 500 Å, are about twice as thick as the axial ones. All these observations, together with the fact that cilia show positive form and intrinsic birefringence, indicate that longitudinally oriented submicroscopic micelles compose the cilium. In this characteristic, therefore, cilia resemble with the contractile fibrils of muscle tissue.

Ciliary Control and Coordination

Cilia possess a well-developed *automaticity*. This automaticity of ciliary action resides entirely in the cilium and an immediately adjacent region of the cell. An organic connection with the basal granule is essential to the beat. On breaking the connection with the basal granule is essential to the beat. On breaking the connection between the cilia and granules, the ciliary activity is immediately stopped. It is not yet fully established whether the basal granules themselves, or some other functional elements in the same general area, are the essential elements for normal rhythmic activity.

Metachronism, of the ciliated surface is of great importance. In metachronal rhythm each preceding cilium beats a little ahead to its succeeding cilium. In *Opalina*, the direction of the metachronal wave changes freely with the direction of swimming, passing backward, to right or to left. Ciliary activity normally involves the use of oxygen, the rate of consumption varying with the ciliary rate.

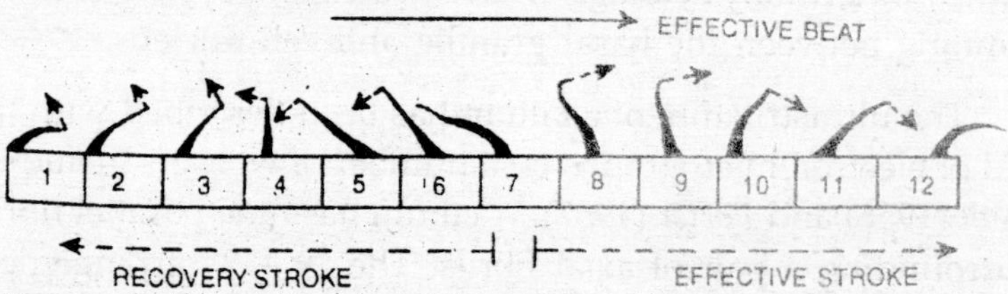

Diagrammatic representation of the metachronal rhythm. Cilia 1 to 12 are beating in sequence.

Types and Characteristics of the Ciliary Movements : According to *Gray* (1928), these are Pendular, Flexural, Undulatory and Funnelform or three-dimensional movements. Ciliary activity sometimes involves a combination of pendular and flexural activities (*e.g.*, the frontal cilia of the gill of *Mytilus*. Forward locomotion in such common flagellates as *Euglena*, in which the flagellum was formerly believed to pull the organism and hence function as a tractellum, appears to result from a bending of the whole flagellum, or at least of its active tip backward.

The undulatory activity typically passes from the base to tip around the flagellum, as well as along it. Such movement imparts rotational and gyrational components, along with a usual backward thrust. The resulting rotation and gyration of the body of the flagellate around an axis which constitutes the direction of locomotion is considered to provide an important force propelling the organism forward. The body proper acts under these circumstances as a screw propeller.

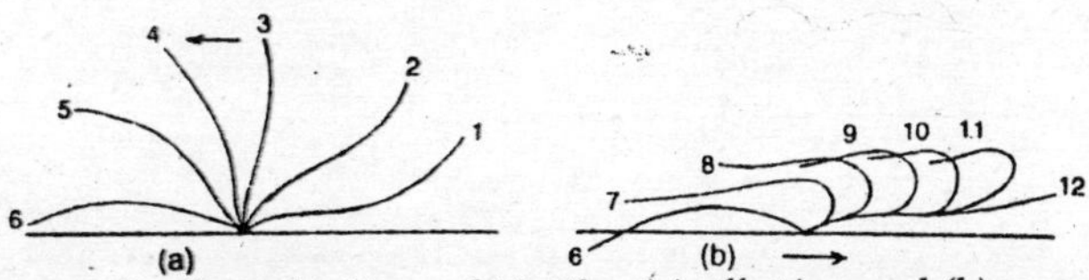

Movement of a frontal cilium of Mytilus (a) effective, and (b) recovery strok.

Factors Influencing the Ciliary Activity : Ciliary activity is influenced by a number of factors including temperature, hydrogen ion concentration, amount of oxygen, and other cations and anions in the medium.

8

Reproduction System

The reproduction may be defined as a self-directed process by which every kind of living organism produces new individuals of its own kind. There are two main types of reproduction : *asexual* and *sexual*. Sexual reproduction is of universal occurrence among all higher forms, including man. However, it has certain biological advantages and disadvantages over the asexual method. One of its advantages is that sexual reproduction permits the combining within a single individual of the best genes of its parents. The chief disadvantages of the sexual method are the hazards involved in the meeting of eggs and sperms and the possibilities of unfavourable growing conditions. But the fact that the sexual process is so extensively distributed in different organisms in such a vast variety of forms suggests that it serves in a role of broad biological significance in addition to its function in reproduction.

In both sexes, the gonads have a dual function : the production of *germ cells* (sperms by testes and ova by ovaries) and the secretion of sex *hormones* (androgens by testes and estrogens by ovaries.

Male Reproductive System

The male reproductive system in man consists of primary sex organs, the testes, and a set of accessory sexual structures (Fig.).

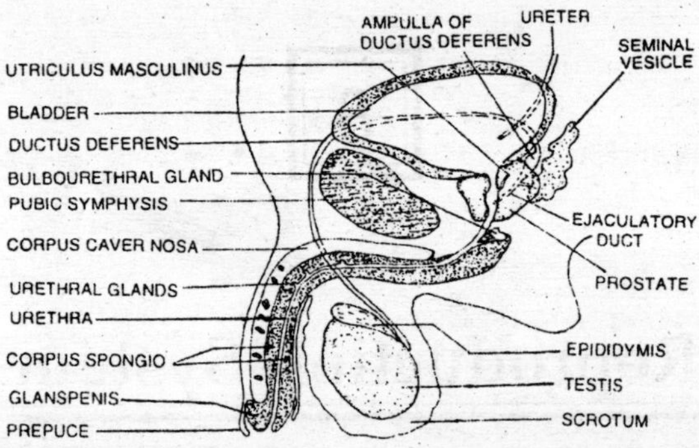

The male reproductive organs in median view (after Eberth).

Testis : Each of the two testes in the adult male is an ovoid body about the size of a walnut. Both are located outside the body cavity within a skin covered pouch called the *scrotum* or *scrotal sac.* The two principal functions of the testes are the formation of sperm cells by *spermatogenesis,* and the production of male sex hormone, *testosterone.*

The wall of each testis consists largely of connective tissue which is continuous with wall like partitions dividing the interior of the testis into approximately 250 small compartments called *lobules.* Each lobule contains from one to three tiny convoluted *seminiferous tubules* within which the spermatozoa are formed. *Sertoli cells* are believed to regulate the differentiation of spermatozoa and to nourish them. The special-cells situated among the seminiferous tubules are collectively called the *interstitial tissue,* which performs the important function of producing and secreting testosterone.

Hormonal Regulation of Testes

Both the structure and function of the testes are dependent upon hormonal stimulation by the pituitary *gonadotrophins,* which in turn are regulated by hypothalamic secretions. The hypothalamus of male secretes at least one, and perhaps two, gonadotropin-releasing hormones (GnRH): *luteinizing hormone-releasing hormone* (LHRH) and *follicle-stimulating hormone-releasing hormone* (FSHRH). These hypothalamic peptide hormones cause the release of LH and FSH from the anterior pituitary which perform the following functions.

(i) *Regulation of Testostreone secretion by LH*: The most important function of LH in the male is to stimulate the interstitial cells of Leydig to secrete *testosterone.* In the absence of LH (also known as ICSH or interstitial cell-stimulating hormone) the interstitial cells undergo severe atrophy and cease functioning. The testosterone (Fig.), an androgenic steroid hormone, is essential for puberal Fig. Testosterone, development and structural and functional maintenance of the male accessory reproductive organs.

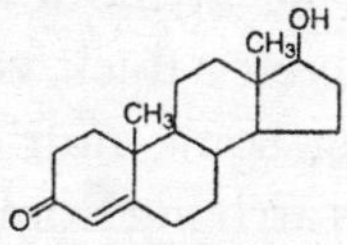

Testosterone.

(ii) *Regulation of Spermatogenesis by FSH:* The maintenance of the structure and the spermatogenic function of the seminiferous tubules is dependent upon the action of the *follicle-stimulating hormone* (FSH). It has been shown to bind specifically to the Sertoli cells, indicating the importance of FSH to spermatogenesis. However, FSH is known to stimulate spermatogenesis but cannot complete the maturation process into spermatozoa unless testosterone is present. Therefore, LH and FSH must be secreted simultaneously by the anterior pituitary for proper spermatogenesis to occur.

Accessory Structures of Male Reproductive System : The accessory ducts and glands of the male reproductive system include *vasa efferentia, epididymis, vas deferens, ejaculatory ducts, penis* and *glands.* The *prostate gland* secretes by way of numerous ducts leading into the urethra, an alkaline fluid which is believed to activate or increase the motility of the sperm and to neutralize the acid condition of the urethra and the female reproductive tract. In old man the prostate gland frequently becomes enlarged, thus constricting the urethra and making urination difficult.

Surgical removal of the gland is often used as a remedy for this condition. Slightly below the prostate gland is a pair of small glands (each about the size and shape of a small pea) collectively called the *Cowper's glands* or *Bulbourethral glands.* The mucous secretion of these glands drain into the urethra, serving as a lubricant. They also help to counteract acidity as does the prostatic secretion.

The paired *seminal vesicles,* are hollow glands about two inches in length which lie at the lower back surface of the urinary bladder. They secrete a thick, viscous, slightly yellowish seminal fluid containing some nutrients. The seminal fluid together with the sperms arriving from the testes and secretions by others glands are collectively called the *semen.* However, the secretory activity of seminal vesicles is dependent upon the action of testosterone.

The penis or the male copulatory organ consists of *erectile tissue* through which the urethra runs. During sexual excitement the sponge-like spaces of the erectile tissue become engorged with blood under pressure due to dilation of the tissue's arteries and constriction of its veins. The resulting turgidity causes the penis to become enlarged, hard and erect. In the sexual intercourse (also called mating, coitus, or copulation) stimulation of the sensory nerve endings in the penis as a result of friction against the lubricated walls of the vagina culminates in a short period of heightened paroxysmal sexual excitement called the *orgasm.*

The orgasm is accompanied almost simultaneously by *ejaculation*, the forceful coming out of semen from the penis. However, ejaculation is the result of several coordinated involuntary muscular contractions including those of the walls of the epididymis and vas deferens, propelling the sperms from the epididymis through the ducts into the urethra. The average volume of semen ejaculated during coitus is about 2.5 to 3.5 millilitre. There are normally about 100 million sperms per ml of semen. About fifty percent of men with counts of 20-40 million per ml and essentially all of those with counts under 20 million per ml are sterile. Sperms are usually released in the upper region of the vagina near the cervix, and are moved principally by muscular contractions of the female reproductive tract into the uterus and then into the oviducts.

Female Reproductive System

The reproductive system of human female consists of two female gonads (the ovaries) which produce the female gametes or egg cells, and associated ducts and structures. The latter include the oviducts (also called egg ducts or Fallopian tubes), uterus, vagina, and vulva. The female reproductive tract is concerned with: (1) the reception and transport of eggs and spermatozoa, making their union possible, (2) physiologic accommodation of the fertilized ovum and resulting foetus, and (3) the forcible delivery of the foetus at term.

The female sexual organs normally remain in an infantile state during approximately the first ten years of life. In the succeeding 2 to 4 years there occurs a gradual enlargement of the reproductive organs, accompanied by the growth of the breasts, changes in the contours of the body, and the appearance of axillary and pubic hair. This period of sexual development culminates in the appearance of the first menses *(menarche)* at an average of 14 years. Mature reproductive life, characterized by continually recurring menstrual cycles, lasts until about 45th to 50th years of life. The human female then enters a period known as the *menopause* during which the reproductive cycles become irregular, lengthen, and eventually cease. Thenceforth, in the *postmenopause,* the reproductive organs are functionless.

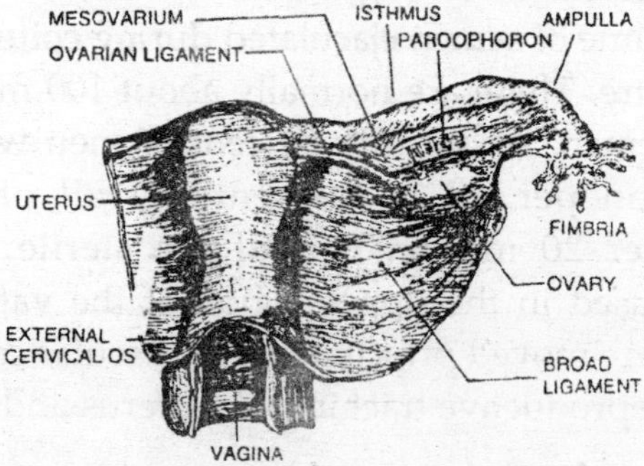

Reproductive organs of human female.

Ovary

Each of the two adult ovaries is about the size and shape of a large almond and is separately located in the pelvic region of the abdominal cavity. The surface of the ovary consists of a layer of columnar epithelial cells which at various intervals penetrate deeply as columns of cells into the underlying connective tissue. During embryonic development of female small groups of cells separate from these columns and become arranged into definite structures known as the *primordial* follicles, each consisting essentially of a large single cell surrounded usually by a single layer of cells.

At the time of birth, there are 2 million ova, but 50% of these are atretic. The atresia continues during development and the number of ova in both the ovaries at the time of puberty is less than 300,000. Only one of these ova per cycle (or about 500 in the course of a normal reproductive life) normally reaches maturity; the remaining degenerate.

Growth of the Follicles

The process that leads to discharge of an ovum from the ovary involves great increase in the diameter of the follicle. When this developmental process has been initiated the follicles are referred to as the *primary follicles.* The latter are 30 to 40 μ in diameter, whereas the diameter of the mature follicles may reach 10 to 12 mm. This increase in size is produced both by proliferation of the granulosa cells and by accumulation of the

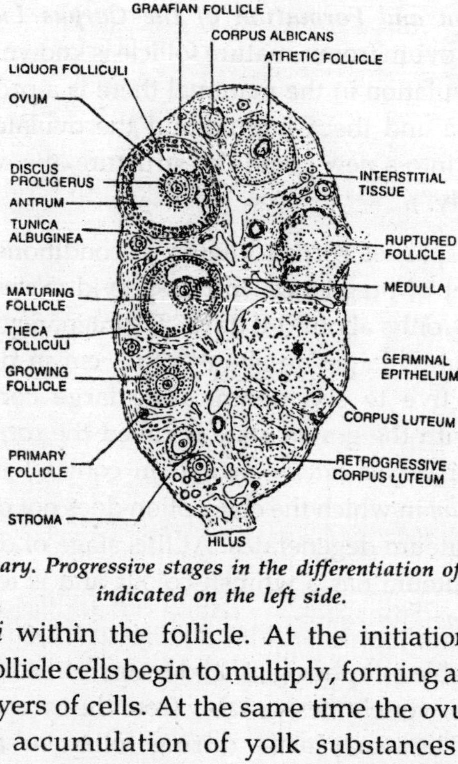

Mammalian ovary. Progressive stages in the differentiation of a follicle are indicated on the left side.

liquor folliculi within the follicle. At the initiation of follicle growth, the follicle cells begin to multiply, forming an increasing number of layers of cells. At the same time the ovum enlarges through the accumulation of yolk substances within its cytoplasm, and a new membrane, the *zona pellucida,* is deposited on the surface of the ovum.

As the follicle expands in diameter, it sinks deeper within the cortex and the neighbouring ovarian stroma are molded about it, forming the *theca folliculi.* The latter differentiates into two layers, the *theca internet* and the *theca externa.* When the follicle wall becomes six to nine cell layers thick, small lakes of follicular fluid form and almost immediately begin to coalesce, forming a crescent-shaped vesicle, the *antrum.* As the antrum enlarges through further accumulation of liquor, the ovum is displaced eccentrically, and the granulosa is disposed as a thin layer of cells, the *stratum granulosum,* lining the fluid-filled antrum. After the accumulation of liquor the follicle is termed as the *vesicular follicle* or the *Graafian follicle.*

Ovulation and Formation of the Corpus Luteum : The escape of the ovum from a mature follicle is known as *ovulation.* Following ovulation in the mammal there is a proliferation of the granulosa and theca interna and the ovulated follicle is transformed into a new glandular structure, the *corpus luteum* ("yellow body").

The life span of corpus luteum is conditioned upon the establishment of pregnancy. In non-gravid cycles the corpus luteum lasts only about 14 days. If pregnancy occurs, the corpus luteum enlarges to about 2 to 3 cm in diameter and persists for five to six months. The large corpus luteum associated with the gravid state is called the *corpus luteum of pregnancy* (or *corpus luteum verum)* in contrast to the *corpus luteum spurium* in which the conception does not occur. Finally the corpus luteum degenerates. At this stage of degeneration, the corpus luteum has a whitish colour and is termed as the *corpus albicans.*

Comparable morphological bodies are formed in the ovaries of many other vertebrates, ranging from the cyclostomes to the mammals. They may arise during follicular atresia (pre-ovulatory corpora lutea) or after ovulation (post-ovulatory corpora lutea). After many years of controversy *(Hoar* 1969), now it seems likely that these bodies produce hormones in some of the fishes and reptiles as well as in the birds and mammals.

Endocrine Functions of the Ovary

As in males, the gonadotropic hormones secreted by the adenohypophysis *i.e.*, FSH and LH, control the formation and growth of the follicles. The ovaries, in addition to producting germ cells, secrete the sex hormones *estradiol* and *progesterone.* Estradiol is produced primarily by the maturing follicles, and progesterone primarily by the corpus luteum. Estradiol induces growth and development of the female reproductive tract and the mammary glands. Progesterone or the so-called pregnancy hormone stimulates the uterine glands to secrete, renders the *endometrium* (the internal layer of the wall of the uterus) receptive to the ovum, inhibits the spontaneous contractions of the uterine musculature, and promotes breast development.

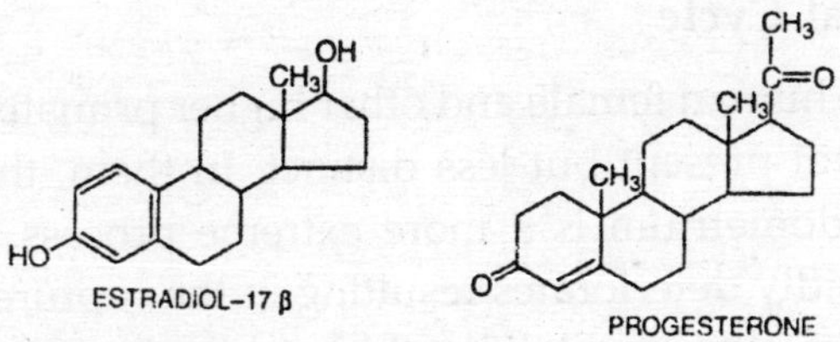

Structure of the important female sex harmonu

Estrous Cycle

The relatively brief periods within the breeding season when the female of the most mammalian species (with the exception of higher primates, including man) undergoes spontaneous ovulation and is most receptive of the male is known as the "heat" or *estrous.* The rhythmic or cyclic changes in the development, release and maturation of the egg cell, together with the accompanying structural and physiological changes in the reproductive system as well as behavioural changes, periodically prepare the female for mating, and are collectively termed the estrous cycle.

However, the endometrium undergoes a marked transformation during the estrous cycle. At the beginning of the estrous the relatively thin endometrium gradually thickens, its blood supply increases, and its uterine glands deepen as the follicular maturation proceeds. With the release of the egg cell and the subsequent appearance of the corpus luteum, the endometrium continues to thicken and build up even more rapidly. If fertilization occurs, the zygote implants itself in the endometrium. If fertilization does not occur the egg cell degenerates, the corpus luteum begins to disappear, and the endometrium regresses returning to its original thin state. The endometrium does not build up again until the beginning of the next estrous cycle. If pregnancy is absent the estrous cycle will be regularly repeated according to the mating pattern of the species.

Menstrual Cycle

In the human female and other higher primates the estrous cycle is still present but less distinct. In them, the regression of the endometrium is a more extreme process. The uterine lining rapidly deteriorates resulting in the rupture of its blood vessles, the flow of blood (which is predominantly arterial, only 25% being of venous origin) and other fluid, and the sloughing away of most of the endometrium, phenomena collectively called the *menstruation*. The cycle is thus uniquely marked by regularly occurring periods of menstruation. The period of normal menstruation (menses) which lasts about 4 or 5 days, occurs between the periods of ovulation and therefore, does not correspond to the estrous. The cycle is accordingly called *menstrual cycle* and in human female averages about 28 days from the start to the end, before the next cycle begins. Some women who are infertile have *anovulatory cycles, i.e.,* they do not ovulate but have menstrual periods at fairly regular intervals.

Changes in the endometrium of the uterus during the menstrual cycle are shown in Fig. The growth of the superficial and middle layers of endometerium occurs from the 5th to the

14th day of the cycle under the influence of *estrogen* produced by the growing ovarian follicle. At approximately this time ovulation occurs and the follicle is transformed into the progesterone-producing corpus luteum. In the second part of the cycle, following ovulation, when *progesterone* is the dominant sex hormone, a tremendous increase in growth and coiling of the superficial arteries occurs in preparation for the increased blood supply necessary for an implanting blastocyst. The stroma becomes loose and edematous, which facilitates implantation.

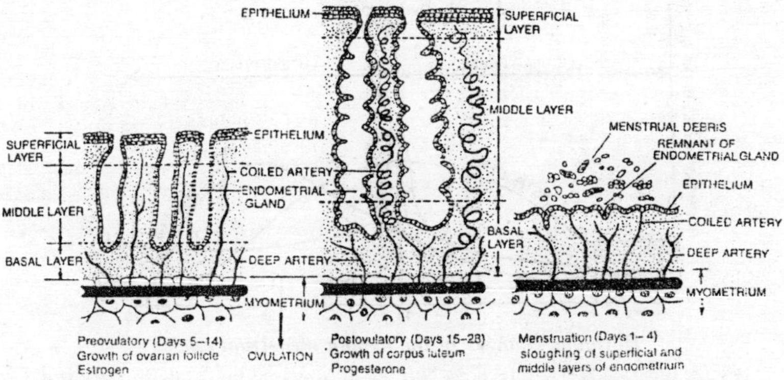

Changes in the endometrium of the uterus during the menstrual cycle.

The endometrial glands proliferate and become secretory, producing large quantities of glycogen and mucin. It has been suggested that these substances may be helpful in allowing the blastocyst to adhere to the endometrium and that the glycogen may be used as nourishment for a short while. Progesterone also slows the contractions of the myometrium, so that the blastocyst has a chance to adhere and implant itself in the richly vascular tissue that has been prepared so carefully for it. The implantation leads to a suppression of the menstruation.

Hormonal Regulation of Menstrual Cycle : The first two weeks of a menstrual cycle are dominated by FSH, estrogen and the follicle. This period is called the *follicular phase* and terminates with ovulation (Fig.). The second two weeks are dominated by LH, progesterone and the corpus luteum. This period is the *luteal phase* (secretory phase). If pregnancy does not begin, during the *luteal phase,* menstruation takes place in the course of the first few days of the next menstrual cycle. A new cycle starts when, in the absence of LH and progesterone, the pituitary resumes FSH production and a new follicle matures in the ovary. Thus FSH and estrogen are the components of one control cycle, LH and progesterone of another ; and these two occur alternately, the termination of one being the stimulus for the initiation of another.

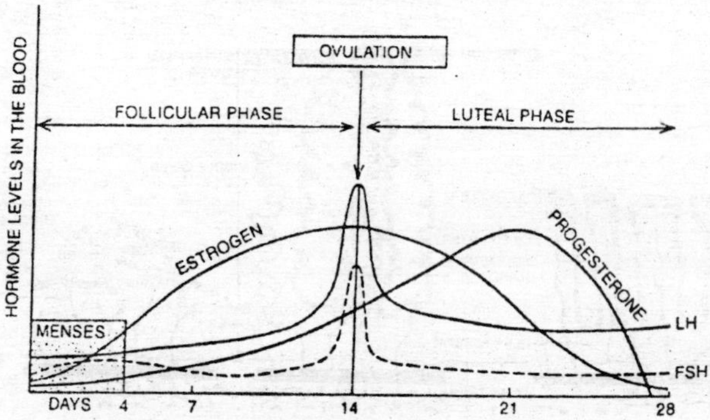

Hormone changes in the human menstrual cycle.

However, we may note that estrogen is not completely absent during the luteal phase. Some can be shown to be present in the blood even then, despite the absence of follicles at that time. Similarly, some progesterone is present during the follicular phase of the menstrual cycle (Fig.). Exactly where these hormones come from is still undetermined. Immature follicles which will attain maturity only during future cycles might secrete some estrogen during a given luteal phase. Analogously, corpora lutea from the past cycles and in the process of degeneration might secrete some progesterone during a given follicular phase. Or other ovarian tissues might secrete the hormones. Yet, although both female sex hormones are present at every stage of the menstrual cycle, their quantities do fluctuate sharply. Estrogen reaches a definite peak late during the follicular phase and at the time of ovulation, whereas progesterone reaches the peak concentrations during the luteal phase of the cycle.

This fluctuation in the concentrations of estrogen and progesterone has far-reaching consequences. In addition to its effect on the uterus, progesterone promotes the development of the duct system in the mammary glands. It induces differentiation of estrogen-prepared ductal tissue and supports the secretory function of the breast during lactation. Body temperature increases somewhat during the follicular phase, then falls during the luteal phase. Sex drive is likely to be more pronounced during the follicular phase, since estrogen maintains it. Since estrogens, like androgens in males, affect mental processes, it is possible that the monthly fluctuation in these hormones contributes to the emotional fluctuations rather characteristic of the females.

Fertilization and Implantation

In humans, the time of greatest fertility coincides roughly with the time of ovulation. The fertilization of the ovum by the sperm usually occurs in the midportion of the uterine tube. Sperm binds to a sperm receptor called ZP_3 in the zona pellucide, a membrane that surrounds the ovum and this is followed by the acrosomal reaction, the breakdown of the acrosome (the lysosome like organelle on the head of the sperm). During the process some enzymes are released, including the trysin-like protease *acrosin.* Acrosin facilitates the penetration of the sperm through the zona pellucida. The fusion of sperm with the membrane is mediated by *fertilin,* a protein on the surface of the sperm head that resembles the viral fusion proteins which permit viruses to attack cells. The fusion provides the signal that initiates development.

The developing embryo, now called a *blastocyst,* moves down the tube into the uterus. This journey takes 3 to 4 days during which the blastocyst reaches the 8 or 16 cell stage. After coming in contact with the endometruim, the blastocyst is surrounded by an outer layer of *syncytiotrophoblast*, a multinucleate mass with no cell boundaries, and an inner layer of *cytotrophoblast* made up of individual cells. The syncytiotrophoblast erodes the endometrium and the blastocyst burrows into it. This is called implantation. The implantation site is usually on the dorsal wall of the uterus. A placenta then develops and the trophoblast remains associated with it.

Gestation

The pregnant female contributes vitally to the offspring development. The most extensive maternal contribution is made while the offspring is retained within the environment of the uterus *i.e.*, during the period of *gestation*. A final contribution is made when the course of pregnancy terminates, during the process of birth.

(a) *Luteal Phase of Gestation* : When the zygote reaches the uterus a few days after fertilization, it is actually already an embryo composed of several hundred cells. This embryo is deposited on the endometrium which has been readied for its reception by the progesterone. The embryo next becomes *implanted* in the wall of the uterus *i.e.*, it is gradually surrounded by the uterus tissue. Such implantation leads to a suppression of the menstruation. Through still poorly identified nervous pathways, the presence of an embryo in the uterus wall is signalled *via* the brain to the pituitary which continues to produce LH as a result. The corpus luteum consequently continues to secrete progesterone. Thus the luteal phase of the menstrual cycle in which fertilization has occurred now does not terminate as usual but continues directly as a luteal phase of gestation. Moreover, further menstrual cycles are suppressed throughout the pregnancy.

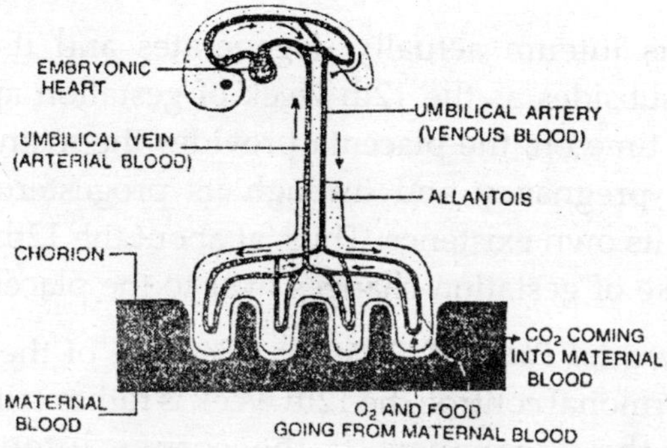

The embryonic blood circulation and the placenta.

In the human female the corpus luteum persists for about the first 12 weeks of pregnancy. During this luteal phase, progesterone output increases and the uterine lining becomes even more glandular and vasculansed. The early embryo, embedded and held fast within the wall of the uterus soon comes to consist of two groups of cells. The central group of cells forms the embryo proper. Surrounding this central group are the cells which form the so-called *extraembryonic membranes* (chorion, amnion, allantois and yolk sac). In the human female the chorion, as in the reptiles and birds, forms as an outer enclosure around the other membranes and the embryo being in direct contact with the tissue of the uterus. In one region the chorion develops numerous finger-like outgrowths which branch extensively and erode paths through the spongy uterine wall. In this manner the tissues of the chorion and the uterus become firmly attached to each other. These interfingering and interlacing tissues are known as the *placenta*. When fully developed, the placenta functions both as a mechanical and a metabolic connection between the embryo and the female body. However, the embryonic and maternal bloods do not mix, being separated by the chorionic and allantoic membranes.

The placenta also functions as an endocrine organ. As it grows and develops, it manufactures slowly increasing amounts of estrogen and progesterone. In fact the progesterone output eventually becomes far greater than that of the corpus luteum.

The corpus luteum actually degenerates and its hormone secretion subsides as the 12th week of gestation approaches. From that time on, the placenta provides the main hormonal control of pregnancy and through its progesterone output maintains its own existence. Thus, at about the 12th week, the luteal phase of gestation changes over to the placental phase.

(b) *Placental Phase of Gestation* : Because of the change in hormonal control, the 12th week is rather critical period of the pregnancy. If the corpus luteum should degenerate a little too soon and if the placenta should reach full development a little late, then the amount of the available progesterone during this gap is likely to be inadequate. In the absence of progesterone, the placental tissue could not be maintained. Just as at the end of a menstrual cycle, the uterine lining would disintegrate and the embryo, no longer anchored securely, would be aborted. Miscarriages occur frequently near the end of the 3rd month of the pregnancy. Such mishaps due to hormones deficiency can be prevented by injecting some progesterone into the pregnant female. The placental phase continues up to about 9 months during which all the bodily systems are formed.

Parturition : In correspondence with the growth and development of the foetus the amnion gradually enlarges and the uterus stretches. Under the stimulus of the increasing quantities of sex hormones from the placenta, the mammary glands enlarge markedly during the last months of pregnancy. The process of birth normally starts when the chorion and the amnion rupture, escaping the amniotic fluid to the outside. Once labour is started, the uterine contractions dilate the cervix, and this dilation in turn sets up signals in the afferent nerves that increase *oxytocin* secretion. Oxytocin increases uterine contractions in two ways : *(a)* it acts directly on uterine smooth muscle cells to make them contract, and *(b)* it stimulates the formation of prostaglandins in the decidua, which further enhance the oxytocin-induced contractions. During labour, spinal reflexes and voluntary contractions of the abodominal muscles ("bearing down") also aid in delivery.

Even though the source of progesterone is removed with the expulsion of the placenta, normal menstrual cycles in human female are generally not resumed as long as the nursing continues. The manufacture of lactogenic hormones by the adenohypophysis apparently inhibits FSH production. But once the child is weaned, FSH is formed again in quantity and a new ovarian follicle begins to mature.

Endocrine Regulation of Reproduction : Hormones control the development of sexual structures and the time of reproductive events. In the vertebrates the adenohypophysis, through its gonadotropic hormones, links the neurosecretory centres of the brain to the endocrine tissues of the gonads. Thus, neurosecretory centres, gonadotropins and gonadal hormones form a delicately balanced system for the timing of reproduction, the development of the gonads, and the differentiation of the secondary sexual characters. These various hormones together with the nervous system regulate the behaviour associated with the reproduction.

As mentioned earlier in this chapter the endocrine tissues (androgen-producing tissue, and estrogen and progesterone-producing tissue) are found in the gonads of all vertebrates. From the most primitive to the most advanced forms they produce a series of steroid hormones whose synthesis is pituitary-regulated. This concert of hormones from the pituitary and the gonads performs together to regulate the reproductive process, and, while the pituitary gonadotropins stimulate the gonadal tissues, the gonadal steroids through a feedback mechanism serve to control the activity of the gonadotropic cells of the pituitary.

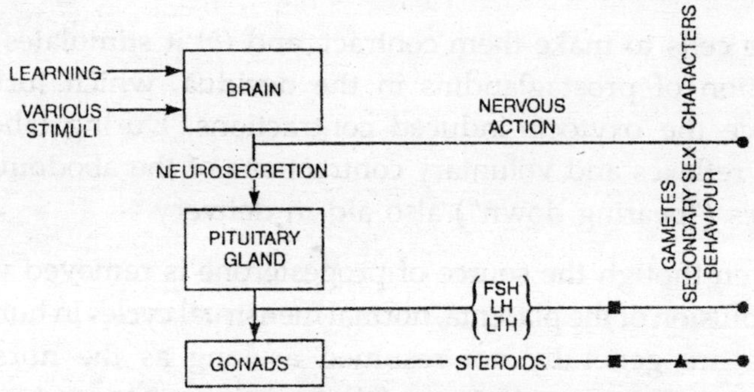

The control of reproduction in the vertebrates.

In brief, the endocrine regulation of the vertebrate reproduction is brought about by the following hormones (see next chapter for details).

(a) Gonadotropins

1. Follicle-stimulating hormone (FSH)
2. Luteinizing hormone (LH)
3. Lactogenic hormone (prolactin)
4. Prolactin-like substances responsible for luteotropic action in rodents, stimulation of pigeon's crop glands for the production of pigeon's "milk", etc.

(b) Gonadal steroids

1. Androgens
2. Estrogens
3. Progesterone

(c) Placental hormones

1. Placental gonadotropins (Human chorionic gonadotropin HCG, Pregnant mare serum gonadotropin PMSG)
2. Chorionic thyrotropin
3. Chorionic growth hormones

(d) Relaxin (isolated from the blood and reproductive tissues)

Thus, it may be concluded that the gonads are concerned with the later events of reproduction (such as the differentiation of the secondary sex characters and sexual behaviour) while the pituitary is more directly involved in the earlier phases (such as the gametogenesis and pre-breeding migrations). However, strictly independent action of the pituitary and gonadal hormones is unlikely in the intact animal.

Secretion of the gonadal hormone is stimulated by the pituitary, while secretion of the pituitary gonadotropins is inhibited by the gonadal hormones. Within this broad framework of reproductive controls, interrelationships and their actions on the specific target organs are almost as våriable as the animals themselves.

Contraception : Once conception has occurred, abortion can be produced by progesterone antagonists such as the synthetic steroid *mifepristone* (RU-486). In some countries mifepristone combined with a prostaglandin is used to produce selective abortions.

These days several methods are used to prevent conception. *Vasectomy* (bilatral ligation of the vas deferens) is considered to be a relatively safe and convenient procedure of male contraception. *Tubal ligation* is a procedure for female contraception. Intra uterine implantation of metal or plastic pieces (Copper 7, Loop D), which are called *intrauterine devices* (IUDs), also act as effective contraceptive method. Although the mechanism of action of IUDs is still debated, there is evidence that at least those containing copper exert a spermicidal action. However, the usefulness of IUDs is limited by their tendency to cause intrauterine infections.

Oral contraceptives are made of small as well as large doses of estrogen and progestin (> 50 μg estrogen and progestin, and < 50 μg estrogen and progestin). Progestin alone can also be used for contraception, although it is more effective when combined with estrogens. An orally active estrogen such as *ethinyl estradial* is often combined with synthetic progestin such as *norethindrone*.

The pills are administered for 21 days, then withdrawn for 5 to 6 days to permit menstrual flow and started again. Women undergoing long term treatment with relatively large doses of estrogen do not ovulate, probably because they have low FSH levels and multiple irregular bursts of LH secretion rather than a single midcycle peak. Women treated with similar doses of estogen plus a progestin also do not ovulate because the secretion of both gonadotropins is suppresed. Besides, progestin also makes the cervical mucous thick and unfavourable to sperm migration, and it may also interfere with implantation.

Among other methods of contraception are use of diaphragm, condom and spermicides, and withdrawl and rhythm.

Women are sometimes given large doses of estrogens for 4 to 6 days to prevent conception after coitus during the fertile period (postcoital or "morning after" contraception). For emergency contraception, emergency contraceptive pills are used. These are basically regular oral contraceptive pills. The dosage is simple, 2 pills as soon as possible and 2 pills 12 hours later. This therapy, known as the Yuzpe regimen, is named after the Canadian doctor who discovered it, and is widely used in developed countries.

Bibliography

Alcock, John : *Animal Behavior : An Evolutionary Approach*, Sinauer, 1998.

Altmann, S. A. and Altmann, J. : *Baboon Ecology : African Field Research*, Oxford, New York, 2000.

Asdell, S. A. : *Patterns of Mammalian Reproduction*, Comstock, New York, 1946.

Aslam, M. S. : *Genetics-Behavioural and Molecular*, Campus Books, New Delhi, 2003.

Aubrey, Manning : *An Introduction to Animal Behaviour*, ELBS Publication, New York, 2001.

Bahr, G. F. : *Chromosomes and Chromatin Structure*, Academic Press, New York, 2001.

Barnett, S. A. : *Study in Behaviour*, Science Paper Backs, London, 2001.

Bendall, J. R. : *Muscles, Molecules and Movement*, Heinemann Educational Books, London 1969.

Benjamin, Gutfreud : *Enzymes : Physical Principals*, John Wiley, London, 1972.

Bernhard, : *The Structure and Function of Enzymes*, Routledge, New York, 1968.

Bhaskar, Harsh Vardhan : *Animal Ecology*, Campus Books, New Delhi, 2004.

Bogert, L. J. and Calloway D.H. : *Nutrition and Physical Fitness*, Saunders, London, 1973.

Bolhuis, J. J. and Hogan, J.A. : *The Development of Animal Behaviour : A Reader*, Blackwell, 1999.

Boulenger, E. G. : *An Introduction to Animal Behaviour*, Discovery Publishing, New Delhi, 2003.

———— : *Animal Mysteries*, Discovery Publishing, New Delhi, 1993.

Boyd, J. D. and Hamilton, W.J. : *The Human Placenta*, Cambridge, London, 1978.

Buffon's, B. : *Natural History of Birds, Fishes, Insects and Reptiles*, Discovery Publishing, New Delhi, 2003.

Carpenter, K. : *Dinosaur Eggs and Babies*, Cambridge, New York, 1996.

Castillo, J. : *Progress in Biophysics and Biophysical Chemistry*, Prentice Hall, London, 1956.

Chauvin, Remy : *Ethology : The Biological Study of Animal Behaviour*, International Universities Press, New York, 1999.

Damron W. Stephen : *Introduction to Animal Science : Global, Biological, Social and Industry Perspectives*, Prentice Hall, London, 2000.

David, M. : *The Marvels of Animal Behaviour*, National Geographic Society, Washington, 2000.

David, McFarland : *Animal Behaviour : Psychobiology, Ethology and Evolution*, ELBS Publication, New York, 2000.

Dickerson and Geis : *The Structure and Action of Proteins*, Academic Press, New York, 1973.

Ernster and Drahota : *Mitochondrial Structure and Function*, Academic Press, New York, 1969.

Filley, G. F. : *Acid-Base and Gas Regulation*, Lea & Febiger, New York, 1971.

Finn, Frank : *Classification of Birds*, Discovery Publishing, New Delhi, 1994.

Florey, E. : *An Introduction to General and Comparative Animal,* Oxford University Press, New York, 1989.

Freeman, W. H. : *Animal Societies and Evolution,* A Scientific American Publication, New York, 2001.

Gene and Reynolds : *Water-soluble Vitamins,* John Wiley, London, 1963.

Gorbman, A. and Bern, H.A. : *A Text Book of Comparative Endocrinology,* John Wiley and Sons, New York, 1962.

Greenberg, Rodwell : *Amino Acid Metabolism,* Cambridge University Press, Cambridge, 1964.

Gyorgy, Pearson : *The Vitamins,* Academic Press, New York, 1967.

Houssay, Frederic : *Industries of Animals,* Discovery Publishing, New Delhi, 1999.

Huxley, H. E. : *The Mechanism of Muscular Conduction,* Spring Harbor, Harbor, 1972.

James, F. : *Animal Social Behaviour,* Wittenberger Duxbury Press, Boston, 1999.

Jencks, : *Catalysis in Chemistry and Enzymology,* McGraw Hill, New York, 1961.

Jennifer, Owen : *Feeding Strategy,* Oxford, New York, 2002.

John, Alcock : *Animal Behaviour: An Evolutionary Approach,* Sinaver Associates, New York, 2002.

Johnson, Christopher B., Franz, Chlodwig : *Breeding Research on Aromatic and Medical Plants,* The Howorth Herbal, London, 2002.

Johnston, R. G. : *Introduction to Sheep Farming,* Granada, 1983.

Jorge, Peter and Ewert Verlag : *Neuroethology : An Introduction to the Neuropathy Sociological and Fundamentals of Behaviour,* Oxford, New York, 2001.

Kamra, D. N. and Pathak, N. N. : *Nutritional Microbiology of Farm Animals,* Vikash Publishing House, New Delhi, 1995.

Kay, Ian : *Introduction to Animal Physiology,* BIOS, London, 1998.

Khanna, D. R. : *Animal Distribution,* Discovery Publishing, New Delhi, 2004.

Konrad Lorenz : *King Solomon's Ring,* MacMillan, London, 2001.

Konrad, Z. and Lorenz, Verlag : *Messengers to the Brain : Our Fantastic, Five Senses,* National Geographic Society Publication, Washington, 2001.

Konrad, Z. and Lorenz, Verlag : *The Foundations of Ethology,* New York, 1999.

Krishnan, M. : *The Hand Book of India's Wildlife,* Oxford, New York, 2002.

Kumar, Ashok : *Animal Physiology,* Discovery Publishing, New Delhi, 2003.

———— : *Aviculture,* Discovery Publishing, New Delhi, 2003.

———— : *Movement and Locomotion in Animals,* Discovery Publishing, New Delhi, 2003.

Kumar, Bipin : *Histology,* Campus Books, New Delhi, 2001.

Kung, S. D. and Yang, S.F. : *Discoveries in Plant Biology,* MacMillan, London, 2000.

Linville and Kelly : *A Text Book in General Zoology,* Discovery Publishing, New Delhi, 1993.

Lister, Harry : *An Introduction to Zoology,* Discovery Publishing, New Delhi, 1999.

Lubibiecki, A. S. : *Large-Scale Mammalian Cell Culture Technology,* Dekker, 1990.

Maheswari, S. K. : *A Textbook of Life Science,* Dominant Publishers, New Delhi, 2004.

Mary, E. Torrence and Issacson : *Microbial Food Safety in Animal Agriculture,* Blackwell, London, 2003.

McCance and Widdowson : *The Composition of Food,* Her Majesty's Office, London, 1960.

Merler, E. : *Immunoglobulins : Biological Aspects and Clinical Uses,* Natural Academy of Sciences, Washington, 1970.

Michael, A. M. : *Irrigation : Theory and Practice,* Vikash Publishing House, New Delhi, 1997.

Morgan, S. J. : *Animal Cell Culture : Introduction to Biotechniques,* BIOS, London, 1993.

Nama, Shinde and Jadhav : *The Nematodes of Indian Mammals,* Dominant Publishers, New Delhi, 2004.

Openheimer, J. H. and Others : *Nuclear Receptors and the Initiation of Thyroid Jormone Action,* Little Brown and Company, Boston, 1976.

Pathak, Nityanand : *Textbook of Feed Processing Technology,* Vikash Publishing House, New Delhi, 1997.

Price, M. R. S. : *Animal Re-Introduction: The Arabian Oryx in Oman,* Cambridge, 1989.

Prosser, Brown, : *Comparative Animal Physiology,* Saunder, Philadelphia, 1961.

Quirk, R. P. : *Chemistry of Muscular Contraction,* Academic Press, New York, 1970.

Ranjhan, S. K. : *Animal Nutrition and Feeding Practices,* Vikash Publishing House, New Delhi, 1995.

——— : *Animal Nutrition in the Tropics,* Vikash Publishing House, New Delhi, 1998.

——— : *Textbook on Buffalo Production,* Vikash Publishing House, New Delhi, 1996.

——— : *Textbook on Quality Meat Production,* Vikash Publishing House, New Delhi, 1999.

Rao, D. Bhaskara : *Animal Kingdom,* Discovery Publishing, New Delhi, 1995.

Robert, Burton : *Animal Behaviour,* Time Life International, Nederland, 1999.

Roger, Tory : *The Birds : Life Nature Library,* Peterson, London, 1999.

Ronald, H. Bailey : *The Role of the Brain*, Time-Life Books, New York, 2001.

———— : *The Secret World of Animals*, A National Geographic Society Publication, Washington, 2002.

Saharia, V. B. : *Wildlife in India*, Natraj Publishers, Dehradun, 1999.

Samuel, K. Wasser : *Social Behaviour of Female Vertebrates*,Academic Press, New York, 2002.

Samuel, Israel and Toby, Sinclair : *Indian Wildlife*, Hans Johannes Hoefer, Singapore, 2001.

Sandhu, G. S. : *A Textbook of Fish and Fisheries*, Dominant Publishers, New Delhi, 2005.

———— : *Analytic Genetics of Animal Development*, Campus Books, New Delhi, 2004.

———— : *An Introduction to Birds*, Campus Books, New Delhi, 2003.

Sarkar, Amira : *Development of Animal Behaviour*, Discovery Publishing, New Delhi, 2003.

———— : *Fundamentals of Animals Behaviours*, Discovery Publishing, New Delhi, 2003.

———— : *Social Behaviour in Animals*, Discovery Publishing, New Delhi, 2003.

Schmidt, Nelson K. : *Animal Behaviour*, Prentice Hall, New Jersey, 1970.

Sela, M. : *New Perspectives in Biology*, American Elsevier, New York, 1963.

Shreeve, W. W. : *Physiological Chemistry of Carbohydrates in Mammals*, Saunders, New York, 1974.

Singh, Gurdarshan, Bhaskar, H. : *An Introduction to Fishes*, Campus Books, New Delhi, 2003.

Spater, P. J. B. : *An Introduction to Ethology*, Cambridge University Press, New York, 1999.

Tikader, B. K. : *Threatened Animals of India*, Zoological Survey of India Publication, Calcutta, 1999.

Tortora and Anagnostrakes : *Principles of Anatony and Physiology*, Harper and Row Publishers, New York, 2002.

Tullts, L. P. : *Blood Cells and Plasma Proteins : their State in Nature*, Academic Press, New York, 1953.

Underwood, E. J. : *Trace Elements in Human and Animal Nutrition*, Academic Press, New York, 1971.

Van, Grvisen : *Wildlife of India*, Guide Book Company Ltd., New York, 2002.

Wadsworth, Cecie Starr : *Basic Concepts in Biology*, Wadsworth, New York, 1997.

Whelan : *Control of Glycogen Metabolism*, Academic Press, New York, 1968.

Whitfield, Dr. P. : *Longman World Guide to Mammals : A Visual Who's Who of the Mammal Kingdom*, Longman, London, 1985.

Williams, C. A. and Chase, M. W. : *Methods of Animal Behaviour*, Academic Press, New York, 1968.

Williams, S. R. : *Animal Behaviour*, MacMillan, London, 1973.

Yadav, Manju : *Applied Entomology*, Discovery Publishing, New Delhi, 2003.

Yadav, P. R. : *Environmental Biodiversity*, Discovery Publishing, New Delhi, 2003.

Yapp, W. B. : *An Introduction to Animal Behaviour*, Clarendon Press, New York, 1960.

Zhukov, M. F. : *Organic Chemistry of Biological Compounds*, Englewood Cliffs, Prentice Hall, 1971.

Zumdahl, S. S. : *An Introduction to General and Comparative Animal Behaviour*, Oxford University Press, New York, 1981.

ENCYCLOPAEDIA OF
ANIMAL BEHAVIOUR

ENCYCLOPAEDIA OF ANIMAL BEHAVIOUR

(Vol. 4)

Dr. A. Tyagi

ANMOL PUBLICATIONS PVT. LTD.

NEW DELHI - 110 002 (INDIA)

ANMOL PUBLICATIONS PVT. LTD.
H.O.: 4374/4B, Ansari Road, Darya Ganj,
New Delhi-110 002 (India)
Ph.: 23261597, 23278000

B.O.: No. 1015, Ist Main Road, BSK IIIrd Stage
IIIrd Phase, IIIrd Block,
Bangalore - 560 085 (India)
Visit us at: www.anmolpublications.com

Encyclopaedia of Animal Behaviour

First Published, 2006

ISBN 81-261-2675-2 (Set)

PRINTED IN INDIA

Published by J.L. Kumar for Anmol Publications Pvt. Ltd., New Delhi - 110 002 and Printed at ARORA Offset Press, Delhi.

Contents

Preface

Behaviour could be most simply defined as movement or a way one conducts oneself. Animal behaviour is, obviously, the way in which animals-non human-lead their life, such as, feeding, migration to different parts in different seasons, sexual life, gregariousness etc.

Different factors need to be studied to have right knowledge about animal behaviour. One of them is the concept of "Natural Selection" propounded by Darwin. He observes that species compete with each other for survival.

Apart from animals, there also occurs a peaceful contest among birds. For example, the severest rivalry between the males of many species to attract their female counterpart is one of the patterns of animal behaviour. It also includes a genetic component similar to the morphological and psychological traits.

The present multi-volume encyclopaedia is a painstaking research work on animal behaviour. It would certainly prove to be an asset for psychologists and those who seek thorough knowledge on the subject.

Editor

1

The Fundamentals

Physiology (Gr. *Physiologikos* = discourse on natural knowledge) is the study of normal function in living organisms. The term was introduced by the French physician *Jean Femel* in 1552. As the living organisms are composed of chemical substances, physiology involves the application of the laws of chemistry and physics.

Physiology studies not only the functional activities themselves, and their inter-relationship, but it seeks to explore their mechanisms. It defines the characteristics of bodily activities and explores the factors which influence them. It studies the conditions which favour these activities, the mechanisms involved in them and the way in which various processes modify each other.

The field of physiology is often subdivided into 2 divisions : *General Physiology*—the study of fundamental processes in terms of chemistry and physics, and Comparative Physiology—the attempt to discern functional similarities and differences between gross systems of whole organisms.

Functional Plan of the Body

The energy used by the effector muscles and by the mechanisms responsible for maintaining the fabric of the body is derived from food stuffs. A food stuff is a chemical substance which can be taken in or absorbed by a cell and serves it as a source of chemical energy. Many food-stuffs are available to man from both plant and animal sources. They are taken into a long coiled tube, the *gastrointestinal tract,* in appropriate amounts which are normally automatically regulated to the needs of the individual.

Once the food-stuffs have entered the digestive tract they are broken down into simpler constituents by the action of enzymes which are formed by glands opening along its length. These are salivary glands, liver pancreas and other intestinals glands. The products of digestion are in a form suitable to pass through the wall of the digestive tract into the blood stream; the unwanted residues are rejected. The nutrients are then carried by the blood to all parts of the body where they are either used for energy, converted into building materials or stored as food reserves for future use.

The liberation of energy from food involves a chemical process *oxidation,* which is analogous to the burning of wood or coal in air with the production of heat, ash and vapour. In the body, oxidation depends upon the activity of enzymes, but it requires oxygen and gives off carbon dioxide and waste products, which must be supplied and removed adequately or the organism rapidly dies. The exchange of gases is the function of the lungs which are spongy organs where gases can pass rapidly to and from air and blood and through which all the blood passes at frequent intervals. The carbon dioxide is breathed way but the oxygen is carried by the haemoglobin of the blood cells to the tissues, where reverse process of discharge of oxygen and absorption and chemical processes involved are so dovetailed and adjusted that these complex exchanges continue with the utmost precision and efficiency for very long periods. They are self-compensating and adjusting, like all other physiological activities.

The elaborate metabolic processes supporting life lead to the production of unwanted *waste products* which have to be excreted if the organism is not to be poisoned by its own products. These are carried by the blood to the kidneys, which are filters with the most accurate powers of removing unwanted material and returning chemicals of value to the body. Small amounts of metabolic wastes are excreted In the breath, the sweat and the faces. The skin also helps in the regulation of water balance in the body.

The ceaseless traffic of gases, food, waste products and chemical messengers, hormones, in the blood stream is made possible by the heart and *circulation.* The heart consists of two pumps made of muscle, each joined to the other through an immensely complex closed system of branching tubes. One sets of tubes runs to and from all the body organs, muscle and brain, and is called the *systemic circulation.* The other set, the *pulmonary circulation* conveys blood to and from the lungs only. The heart pumps steadily away throughout life and the blood flows continuously through these closed circuits. The tissue cells are never touched by the blood itself; all the interchange of gases and chemicals occur by solution and diffusion in the tissue fluid, which permeates the whole body and infiltrates between the cells.

The *lymphoid tissue* consists of lymphoid cells, lymphatic channels and the lymph nodes. It is distributed throughout the body *e.g.,* lymph nodes, pharyngeal tonsils, spleen, Peyer's patches, vermiform appendix etc. The cells of the *reticulo-endothelial system* carry out as diverse functions as phagocytosis, formation and destruction of red blood cells etc.

The effector organs, especially the muscles, and the nervous integrate and co-ordinate all other activities of the body. The *muscles* carry out the conscious needs and intentions of the individual but this is only a mall part of their function. They help in movement of the limbs, locomotion and liberation of heat and energy. To a great extent the contraction of muscles is unconscious and automatic, forming what are called reflexes. A *reflex* is an involuntary response of a muscle or gland to impulses conveyed from another part of the body by nervous pathways. Reflex responses may attain elaborate patterns of interlocking relationship such as the mating and breeding behaviour of birds, but they are also responsible for a large series of mundane activities like standing, walking, running, breathing, micturition and defecation.

The *brain* is the site of consciousness, the seat of memory and the organ to which, above all, man owes his supremacy over all other living forms. The faculty of learning is found in animals but only man is known to possess the powers of creative thought and abstract thinking which have led to the development of civilization. The *autonomic nervous system* consists of *sympathetic* and parasympathetic nerves. Hypothalamus is the higher centre for the autonomic nervous system.

Hormones are the chemical secretions of ductless glands like pituitary, thyroid, parathyroids, adrenals, islets of langerhans in the pancrease and the gonads. Hormones circulate in blood stream and help in various metabolic activities.

The *sense organs* mainly consist of taste, smell, vision and hearing.

Finally, comes the capacity for *reproduction*, without which all else would be futile. The outlines of the mechanism of fertilization and the elaborate hormonal controls of the profound bodily changes occurring in the mother are fairly clear. The mechanism of heredity and development, the way in which the unparalleled complexity of the adult body derives from the minute speck of nucleoprotein in the fertilized ovum, is still only dimly comprehended and in comparison, the intricate interplay of function in the adult body is a simple problem.

The student of physiology is at first rather like a bee outside a hot-house. He sees all sorts of mysterious and exciting flowers inside but every entry is stopped by the glass of unforeseen problems. Each new approach seems blocked by its own difficulty, every description seems to depend upon a new obscurity. A door must be opened somewhere, a start must be made.

2

Regulation of Temperature

The word 'temperature' may be defined as the degree of hotness or coolness of any thing, or, specifically as the degree of heat of a living body, usually as measured by a thermometer.

However, active life is limited to a narrow temperature range from about -2°C to about + 50°C. Beyond this range many animals can survive in an inactive or torpid state. A large number of animals, including all aquatic invertebrates, have the same temperature as their surroundings. But birds and mammals generally maintain a nearly constant temperature in spite of wide variations in the surrounding temperature. Thus, animals are of the following types :

Cold-blooded (Poikilothermic) Animals : A 'cold-blooded' animal is not necessarily cold, a desert lizard or an insect sitting in the sun may have a higher body temperature than a mammal; but it has nearly the same temperature as its surroundings (Fig.). For example, a fish has the temperature of the water it swims in, and an earthworm has the temperature of the soil it lives in. However, the corresponding term 'poikilothermic' (Greek *poikilos* = changeable) is really more descriptive, for it signifies changeable temperature. In other words, it refers to the fact that the temperature of the cold-blooded animal fluctuates with that of their surroundings.

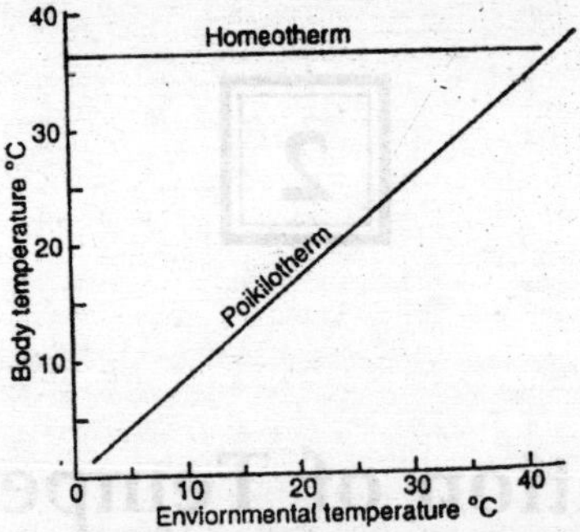

Warm-blooded (Homeothermic) Animals : A 'warm-blooded' animal is not necessarily always warm. A few mammals and some birds undergo hibernation during which their temperature may decrease to near the freezing point of water, with no harm to the animal and in that state it seems odd to call them warm-blooded. The corresponding term 'homeothermic' (Greek *homos* = like, and *homoios* = similar) or homothermic or homoiothermic is a better expression than 'warm-blooded' for it signifies the constant temperature. In other words, mammals and birds are called homeothermic because they maintain a quite constant body temperature in spite of wide variation in the surrounding temperature (Fig.). Thus, a polar bear, a tropical monkey and a walrus have almost the same body temperature of about 38°C.

Heterothermic Animals : The term 'heterothermic' (Greek *heteros* = different) is used to refer to the animals that at times have a high and well-regulated body temperature, but at other times are more like the cold-blooded animals.

Ectothermic and Endothermic Animals : By exposing itself to the sun, a lizard can maintain its body temperature far above that of the surrounding air. Hence, to distinguish the lizard from the birds and mammals, which maintain their high body temperature by metabolic heat production, we use the term ectothermic and endothermic. *Ectothermic* animals are those which depend on external heat sources, primarily the solar radiation, to maintain their body temperature. *Endothermic* animals are those which are able to maintain a high body temperature by internal heat production.

However, these terms and definitions also have their limitations. For example, a tuna fish may keep the temperature of its muscles 10 to 15°C above that of the water. The heat is derived from muscular metabolism, but we do not call the tuna an endothermic animal in the same sense as the birds and mammals.

Temperature Regulation in Poikilothermic Animals : Although the temperature of a poikilothermic animal depends on that of the surroundings, this does not mean that the animal has no control whatever over its body temperature. Some poikilothermic animals can and indeed do stay warmer than the air or water in which they live.

The heat balance equation for a poikilothermic animal may be written as follows:

$H_s = H_{tot} \pm H_r$ — He—H_e where

H_s = storage of heat in the body

H_{toi} = total heat production

H_r = heat exchange by radiation

H_c = heat exchange by conduction and convection

H_e = heat exchange by evaporation

If the body temperature is higher than the surrounding medium, the terms for conduction (H_e) and evaporation (H_e) will signify heat losses and be negative. Hence, if the animal has to increase the body temperature, it has to do something to maximize the heat storage (H_s). The mechanisms vary in different groups of poikilotherms and are as follows:

Aquatic Poikilothermic Animals

For an aquatic animal the heat balance equation may be written as follows:

$$H_s = H_{tot} — H_e$$

because in water there is no heat exchange by evaporation and radiation. In order to increase heat storage, either the total heat production is to be increased, or the heat loss by conduction is to be reduced. Because of high thermal conductivity and high heat capacity of water, small animals lose heat rapidly and are unable to attain a body temperature very different from the medium.

However, some animals are able to attain a high body temperature with the help of a *'heat exchanger'* placed between the gills and tissues. For examples, the tuna and some sharks maintain a high temperature in their swimming muscles, independently of the water in which they swim. In the ordinary fish the swimming muscles are supplied with blood from the dorsal aorta. But in the tuna the muscles used for steady, fast swimming (the so-called dark red muscles) receive their supply from the blood vessels running along the side of the fish just under the skin. From these major vessels arise a large number of parallel fine blood vessels which form a slab in which arteries are densely interspersed between veins running in the opposite direction. This puts the cold end of the heat exchanger at the surface of the fish and the warm end deep in the muscles.

The arterial blood from the gills is at water temperature, and as this blood runs in the fine arteries between the veins, it picks up heat from the venous blood which comes from the muscles. The exchange of heat is facilitated by the small diameter (about 0.1 mm) of the vessels. When the venous blood has reached the larger veins under the skin, it has lost its heat, which is returned to the muscle via the arterial blood. Consequently, the tuna can maintain muscle temperatures as much as 14°C higher than of the water in which it swims.

Thus, with the help of *heat exchangers* the swimming muscles are kept warm. The advantage is that the high temperature increases the power output of the muscles. At higher temperature the muscle contracts faster, and since the number of contractions per unit time increases, the work per unit time (power output) increases accordingly. A high power output gives the tuna a high swimming speed relatively independent of the water temperature, and this in turn makes it possible to pursue successfully the prey which otherwise swims too fast to be caught, such as the squid or the mackerel. In the tuna, separate heat exchangers permit the digestive organs and the liver to be kept at a higher temperature. This increases the rate of digestion and in turn the rate of fuel supply to the muscles for their high power output.

Fish living in cold arctic waters appear to have an *antifreeze substance* in their tissues that prevents freezing when temperatures fall below those at which their body fluids would normally freeze.

Terrestrial Poikilothermic Animals and Insects : In order to raise the body temperature in terrestrial animals the evaporation and heat loss by conduction are to be reduced, and the heat gain by radiation and metabloic heat production are to be increased. The only commonly available radiation source is the sun, and in the absence of solar radiation, the metabolic heat production is the only way to increase the heat gain. Solar radiation is used especially by the insects and reptiles. To increase the amount of radiant energy absorbed, they depend both on their colour and on their orientation relative to the sun. Many reptiles can change their colour by dispersion or contraction of the black pigments in their skin. Since about half of the solar radiation energy is in the visible light, a dark skin increases the amount of solar energy that is absorbed rather than reflected.

The other way of increasing the heat gain from solar radiation is to increase the exposed area. This is done by orienting the body at right angles to the direction of the sun rays, and by flattening the body and spreading the legs. In this way lizards can attain the temperature much higher than the surrounding air. When a suitable temperature has been attained, further heating is avoided by lightening the skin colour and changing the posture to avoid direct sun rays.

Lizards and other reptiles burrow at night when the desert or mountain temperatures are at their lowest. Some heat from the day remains underground and the reptile's body is prevented from freezing. At sun rise the lizard, which may have reached a body temperature as low as 5°C during the night slowly emerges from its burrow to orient itself to the sun light so that the maximum heat is obtained. By proper orientation lizards can attain body temperatures that are as much as 30°C above that of the air.

In terrestrial poikilotherms, internal heat production can also be used to attain a high body temperature. It has been reported that some female snakes do this while incubating their eggs and it is done by increasing the rate of oxygen consumption. The increase in oxygen consumption seemed to be caused by strong muscular contractions reminiscent of shivering in mammals.

Insects have a special thermal problem because of their high rate of metabolism during the flight. An insect can increase the temperature of the flight muscles by a process similar to shivering in man. Such heating of the muscles before take-off occurs mainly in large insects such as the locusts, large moths, butterflies, and bumblebees, and also in the wasps and bees which can fly rapidly. *Heinrich* and *Bartholomew* (1971) have made an analysis of the pre-flight warm-up in the sphinx moth, *Manduca sexta*. During the warm-up the animal seems to shiver as the wings vibrate lightly. The heat is generated by the contraction of muscles, and the reason that the wings do not flap is that the muscles for the upstroke and downstroke contract simultaneously, instead of alternately as in the flight.

Honeybees regulate not only their own individual temperature, but that of the entire hive as well. In the summer, the temperature of a beehive is maintained at about 35°C and fluctuates very little. This provides the colony with an optimum temperature for the development of the brood.

If the bee-hive temperature rises above the regulated level, the bees spread droplets of water over the comb, and evaporation is aided by air currents created by a large number of bees fanning their wings. Below 30°C honeybees *Apis mellifera* do not usually consume any water, but at higher temperatures their water intake increases considerably as the water is used for cooling. (*Tree* and *Spencer-Booth*, 1950).

The *clustering* of bees is a heat conservation mechanism, and bees reflexively cluster upon stimulations involving the sight, sound, odours of abdominal scent glands and the temperature. The minimum number of bees that will cluster depends upon the temperature. In fact, there is a tendency for the cluster temperature to be higher at the very lowest ambient temperatures. For each 5°C fall in the outside air temperature, the cluster temperature increases by about 1°C.

At lower temperatures a smaller number of bees can be stimulated into clustering than at higher temperatures, where perhaps 50 bees are a minimum number that can be induced to cluster. This thermal regulation of the hive, which is needed for proper reproduction and for the development of young bees, is based on both physiological heat production and behavioural mechanisms associated with clustering and wing fanning. Bees may also cool a hive during warm weather by standing at the hive entrance and fanning their wings to produce a current of air through the hive. Ants and other social insects exhibit similar behaviour in maintaining colonies at proper temperatures.

Temperature Regulation in Homeothermic Animals : For a body to maintain a constant temperature, the heat loss must be equal to the heat gain. The "Du Bois temperature balance" (Fig.) shows graphically how the physiological and metabolic reactions which produce heat must be matched against those which radiate or conduct it away in order to provide a constant body temperature.

Heat Loss : It occurs by the following means.

(a) *Conduction* : Conduction of heat consists of direct transfer of the kinetic energy of molecular motion, and it always occurs from a region of higher temperature to one of the lower temperature. The loss of heat by *conduction to air* represents a sizable proportion of the body's heat loss (about 15 per cent).

(b) *Convection* : The transfer of heat in fluids is almost invariably accelerated by the process of convection, which refers to mass movement of the fluid. The free or *natural convection* is caused by temperature difference. Convection caused by the external forces, such as the wind, water currents, etc., is referred to as the *forced convection.*

(c) *Radiation* : Heat transfer by Radiation takes place in the absence of direct contact between the objects. All objects at a temperature above absolute zero emit electromagnetic radiation. The intensity and the wavelength of this radiation depend upon the temperature of the radiating surface and its emissivity. The human body radiates heat in all directions.

(d) *Evaporation* : The amount of heat required to achieve the phase change from liquid water to vapour is known as the *heat of vaporization (Hv)*. The heat of vaporization changes slightly with the temperature at which the evaporation takes place. Thus, at 0°C the Hv is 595 cal/gram water, at 22°C it is 584 cal/gram, and at 100°C it is 539 cal/gram. In physiology, it is customary to use the figure 580 cal/gram, which is an approximation of the value for vaporization of water at the skin temperature of a sweating man, about 35°C. Heat loss due to evaporation takes place as follows.

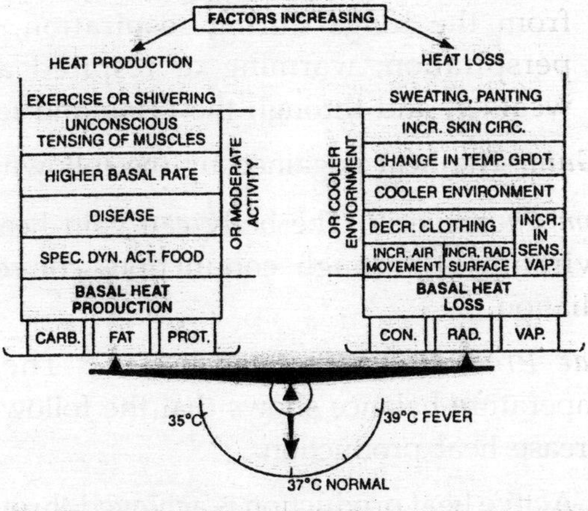

The Du Bois temperature balance (balance between the factors increasing heat production and heat loss).

(i) *Active Heat Loss* : When a man is exposed to hot surroundings he cools himself by evaporation of sweat from the general body surface *(sweating)*. Stimulation of the anterior hypothalamus (preoptic area) by excess heat causes sweating. The impulses from this area are transmitted in the autonomic pathways to the cord and then through the sympathetic outflow to the skin. The sweat glands are innervated by sympathetic *cholinargic* nerve fibres.

However, a dog evaporates most of the water from the respiratory tract by panting, a very rapid, shallow breathing which increases evaporation on by the thermoregulatory centres of the brain (hypothalamus). The actual panting process is controlled by a panting centre that is closely related to the pneumotaxic respiratory centre in the pons.

Some animals use a third method for increasing evaporation. They spread saliva over their fur and lick their limbs, thus achieving cooling by evaporation.

(ii) *Basal Heat Loss* : This includes evaporation of water from the lungs during respiration, insensible perspiration, warming of respired air in cold weather, and through the urine and faeces.

Heat Gain: The heat is gained by the following means.

(i) *From Environment:* The body can gain heat from the environment through conduction, convection and radiation.

(ii) *Heat Production (thermogenesis)* : The Du Bois temperature balance shows that the following factors increase heat production.

(a) Active heat production is achieved through *exercise* and *shivering*.

(b) Basal heat production depends on (i) unconscious tensing of muscles, (ii) disease, and (iii) specific dynamic action of food. The extent to which a particular food item influences the rate of metabolism is referred to as its specific dynamic action. Proteins have a higher specific dynamic action. As they raise the metabolic rate by about 30%, they help in warming up the body in cold weather.

In homeothermic animals the thermogenesis depends on the following:

1. *Shivering Thermogenesis :* A man, when appropriately acclimated, may show an eighteen-fold increase in heat production within 12 minutes if suddenly immersed in a bath at 4°C. This record heat production is associated with the shivering or the *involuntary muscle contraction.*

Shivering can be evoked either by a slight decrease in the core temperature or by a decrease in the skin temperature. Thus, its control can be either central or peripheral or both. The conspicuousness and importance of shivering varies from animal to animal. Among mammals, the most obvious shiverers are probably the fruit bats or flying foxes of the Old World. Some flying foxes of the genus *Pteropus* shiver strongly and visibly at all temperatures below their thermal neutral zone.

2. *Nonshivering Thermogenesis :* The heat produced from extramuscular sources is termed nonshivering thermogenesis. During cold stress, acclimated rats produce as much as 65% of their heat in fat and visceral organs.

Nonshivering thermogenesis plays an important role in the acclimation of mammals to the low temperatures and in arousal from the hibernation and daily torpor. It is clear that the capacity for nonshivering thermogenesis acquired during acclimation to cold allows rats to be active at much lower temperatures than would be the case if they had to depend on shivering alone, for exercise elimnates shivering but does not affect the nonshivering thermogenesis. Exercise and nonshivering thermogenesis are additive, exercise and shivering are not. Nonshivering thermogenesis also significantly increases metabolic scope for activity. In cold-acclimated rats with a high level of nonshivering thermogenesis, the metabolic scope below

30°C is almost the twice that of the warm-acclimated rats at the same temperature.

The increase in nonshivering thermogenesis associated with the cold adaptation in animals involves shifts in the pathways and relationships of the intermediary metabolism. Although the situation is not completely understood, it seems probable that the cold-acclimated rats have increased capacity for the formation and utilization of glucose.

It also seems clear that some nonshivering calorigenesis is evoked in part through the hormonal stimulation of lipid metabolism. For example, the colorigenic response to injection of norepinephrine in cold-acclimated rats involves a marked increase in oxygen consumption and a rise in body temperature. Both these responses can be related to in increased capacity of the liver to form and mobilize nonesterified fatty acids. The thyroid hormone plays an important role in the long term development of mammalian nonshivering thermogenesis. One of the important sites of shivering thermogenesis is the brown fat (see below).

3. *Special Thermogenic Tissue* : Many mammals have a special thermogenic tissue known as the *brown fat* or the brown adipose tissue. Brown fat was first described in hibernating marmots, *Muris alpinus.* It was called the hibernating gland and assumed to play an active role in hibernation. The tissue is localized in the neck and thoracic regions in relation to the major blood vessels so that its heat is quickly transported to those organs (heart and brain) whose continuous high temperatures are vital for a homeotherm. Histologically, the brown fat is an extremely vascular tissue with relatively high concentrations of myoglobin, cytochromes, and flavin compounds. These give it a pale buff to dark reddish brown colour which contrasts with the ordinary white adipose tissue. It also differs cytologically from the

white fat in that the cells are smaller, polygonal, with central nuclei and numerous fat droplets which are dispersed rather than concentrated in a single vacuole; mitochondria are small, numerous and closely associated with the fat droplets. Brown fat also differs biochemically from the white fat, with higher concentrations of water, protein, phospholipid, cholesterol, and certain mitochondrial enzymes. Details of mitochondrial respiration in the brown fat are speculative but heat, rather than ATP, appears to be the main end product. This tissue can be thought of as a rapidly firing furnace that warms up some mammals quickly in emergency, without muscular activity or shivering.

The brown fat is most conspicuously developed in many newborn mammals (including the human), in hibernating animals at the time of arousal, and in cold-acclimated animals. Like a smoothly operating automatic furnace, it responds to cold stress with a large burst of heat. The physiological mediator for the stimulation of brown fat appears to be the noradrenaline. Sympathetic stimulation activates the adenyl cyclase system with rapid lipase hydrolysis and oxidation of fats.

Heat Balance: Body temperature is controlled by balancing heat production against heat loss. After considering the means of heat loss and heat gain, the heat balance equation for a homeothermic animal may be written as follows :

$$H_{tot} = \pm \text{Hc} \pm Hr \pm He \pm \text{Hs}$$

where H_{tot} = metabolic heat production (always positive)

H_c = conductive and convective heat exchange

H_r = net radiation heat exchange

H_e = heat loss in evaporation

H_s = storage of heat in the body

The three components of heat exchange, conduction, radiation and evaporation depend on the external factors ; of them the most important single factor is the temperature. It is obvious that the heat losses increase when the external temperature falls. If, on the other hand, external temperature rises, the heat losses decrease, and if external temperature exceeds the body surface temperature, heat exchange by both conduction and radiation may be from the environment to the organism. The total heat gain is then the sum of heat gain from metabolism plus heat gain from the environment. This situation still permits the maintenance of a constant body temperature, provided that evaporation is increased sufficiently to dissipate the entire heat load.

Physiological Response to Cold

Temperature regulation in cold is for keeping the body warm. To maintain constant body temperature we must satisfy the steady state condition in which the rate of metabolic heat production (H) is equal to the rate of heat loss (Q). In cold the heat loss takes place only through conduction (including convection) and through radiation, and that heat loss through evaporation can be disregarded. The following equation relates heat loss to ambient temperature :

$$H = Q = C(T_b - T_a)$$

It says that the rate of metabolic heat production (H) equals the rate of heat loss (Q), which in turn is proportional to the temperature difference seen the environment (T_b-T_a) The term C is a conductance term. Among the terms in the above equation there is very little an animal can do about the ambient temperature (T_a). Hence, to adjust to an unfavourable T_a an animal has only the three remaining terms, i.e., the heat production (H), the conductance term (C), and the body temperature (T_b) that could possibly be adjusted. Since we are concerned with the maintenance of the body temperature (T_b), we are left with adjustments to be made either in the heat production (H), or in the conductance term.

Increase in the heat production is achieved by (i) muscular activity and exercise, (ii) shivering, and (iii) nonshivering thermogenesis. Conservation of heat is brought about *by piloerection peripheral vasoconstriction* and *countercurrent* warming of blood. Conductance, in the sense it has been used above, is a measure of heat flow from the animal to the surroundings. It includes the flow of heat from deeper of the body to the skin surface and from the skin surface through the fur to the environment. When conductance is low, it means that the insulation value is high. The insulation values of the fur from various animals differ considerably.

The insulation value increases with the thickness of the fur and is maximum in the animals which have a thick fur, such as the white fox. Among the smaller animals there is a clear correlation between the thickness (and insulation) and the size of the animal. They must have relatively short and light fur or the animal could not move about. This is particularly true for small rodents and shrews, the smallest mammals. Because of their relatively poor insulation, they either take advantage of microclimates *(e.g.,* by living in burrows), or escape from the problem of keeping warm by hibernating. It is also well known that the thickness of the fur changes with the seasons, and that the winter fur is heavier and presumably affords better insulation than the thinner summer fur.

In arctic land mammals the temperature gradients are different, the surface temperature of the body skin under the fur is regularly within a few degrees of the core temperature *(Irving* and *Krog,* 1955) and most of the insulation, therefore, resides outside the skin surface. The polar bear, which has a fur that loses most of its insulation value in water, also has a substantial layer of blubber without which the semi-aquatic way of life would be impossible. The main difference in the insulation between aquatic and terrestrial mammals is that the insulator of the aquatic mammal (the blubber) is located internally to the surface of heat dissipation. Therefore the blood can bypass the insulator, and heat dissipation during heavy exertion or in warm water can be independent of the insulator. In contrast, terrestrial animals have the insulator located externally to the skin surface. Hence they cannot modulate the heat loss from the general skin surface to any great extent.

Conservation of Heat

It is achieved by the following methods :

Piloerection : It means the erection of hair. In hot weather the hair of homeothermic animals lie flat; but in cold, hair are erected and the animal appears more or less coated with fur. The latter acts as an insulator and thus heat loss is decreased. The erection of hair is achieved by small muscles at the root of the hair and is controlled by the central nervous system in response to the information received from the thermal receptors of the skin. Thus, piloerection is an important physiological response to cold in all homeotherms, including man.

Peripheral Vasoconstriction : Conservation of heat is also achieved by peripheral vasoconstriction, *i.e.*, the constriction of skin blood vessels. It reduces the amount of blood flowing through the skin and, therefore, the heat loss from the skin surface is reduced. Vasoconstriction usually takes place in the covered parts of the body.

Countercurrent Heat Exchanger : The countercurrent heat exchanger is based on a particular arrangement of the circulatory vessels in which the blood flows in the opposite direction. In the peripheral regions of the body, for example, the arterial blood at the higher core temperature flows to a region of lower temperature.

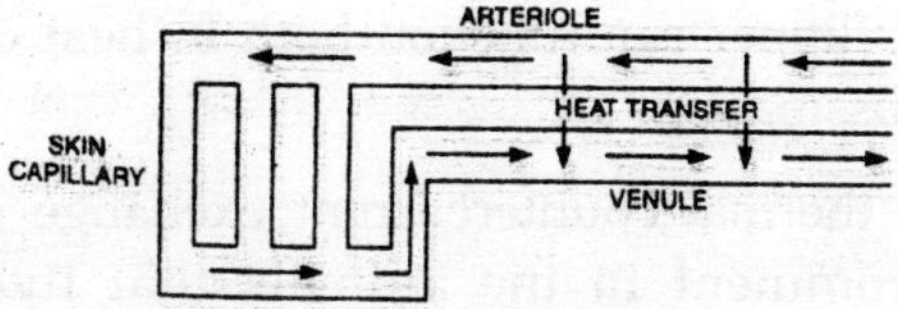

Diagrammatic representation of the thermal countercurrent exchange system.

The venous blood that has passed through the peripheral region is at a lower temperature than the arterial blood because of heat loss to the environment. Because of the contact of the vein with the artery, the heat of the arterial blood is passed to the returning venous blood (Fig.). Through such conduction of heat from the artery to the vein, the blood that reaches the periphery does not lose as much heat to the environment as would arterial blood that had not transferred some heat to the venous blood. Such a system provides for an exchange of heat to core-directed blood but still permits the blood to carry essential nutrients and other materials to peripheral cells. The system also ensures that the venous blood will not have a temperature low enough to act as a stimulus for the thermogenic machinery.

The countercurrent heat exchangers are present in many regions of the mammalian body. They are found, for example, in the circulation of the digestive tract and are needed there so that the cold liquids or foods enterning the digestive system do not remove too much heat from the lining of the tract. The countercurrent exchange systems also operate in the brown adipose tissue. In the flippers and flukes of whales and seals, major arteries are completely surrounded by veins, allowing for a maximal heat exchange between the arterial and venous bloods. This exchange prevents too much body heat being lost from the broad surfaces of their flippers and flukes to the cold marine waters in which these animals live. When it is necessary for heat to be lost, blood is shunted to other venous pathways that are not in contact with arteries. In this way the circulatory system of the flipper can function both in heat conservation and in heat dissipation.

Although thermal countercurrent exchange systems are especially prominent in the animals that live in colder environments, such arrangements are found in the animals of both the temperate and tropical environments as well. For example, the sea cows, which live in tropical and subtropical water, have similar heat exchangers in their appendages. Heat exchangers in the limbs are not restricted to aquatic animals. Even in the limbs of man some heat exchange takes place between the main arteries and the adjacent larger veins located deep within the tissues. In birds, heat exchange in the legs is very important, especially for the birds that stand or swim in water. Unless the blood flowing to these thin-skinned peripheral surfaces went through a heat exchanger, the heat loss would be very great indeed. A gull placed with its feet in ice-water for two hours lost only 1.5% of its metabolic heat produced from the feet *(Scholander*, 1955).

Thus, due to the heat exchanger the heat loss is quite insignificant. Some tropical animals like sloth also have heat exchangers in their arms.

Physiological Response to Heat

It has been described that the birds and mammals increase their heat production in order to stay warm at the ambient temperature below a certain point, the lower critical temperature. Above the critical temperature the metabolic heat production remains constant through a range usually known as the thermoneutral range. So, for the body temperature to remain constant, metabolic heat must be lost at the rate it is produced. In the heat balance equation:

$$H = C(T_b - T_a)$$

if H and T_b, remain constant and T_a changes, the conductance term (C) must also change. The conductance term refers to the total heat flow from the organism, and it can be increased in several ways.

In heat regulation in cold the role of evaporation was, however, disregarded because it made up a fairly small fraction of the total heat exchange. But in hot climate (at higher temperature) evaporation is the key factor in the heat balance. When some mammals and birds are kept in air that equals their body temperature, those that are able to keep their body temperature from rising evaporate an amount of water which is equivalent to their metabolic heat production. When the ambient temperature exceeds the body temperature, as in the desert, the evaporation is used to dissipate the sum of metabolic heat production and the heat gain from the environment.

In humans, *sweating* is the means for evaporation and they sweat to increase cooling by evaporation. However, several furred animals, including cattle, large antelopes and camel also depend mainly on sweating for evaporation. As dogs have a few sweat glands, they cool primarily by *panting*, a very rapid shallow breathing which increases evaporation from the upper respiratory tract. A number of smaller ungulates, goats, sheep, small gazelles and many carnivores also depend on panting for evaporation. In some animals, for example, marsupials, a third method is used to increase evaporation. They spread saliva over their fur and lick their limbs, thus achieving cooling by evaporation. Birds, in the absence of sweat glands, increase evaporation either by panting or by a rapid oscillation of the thin floor of the mouth and upper part of the throat, a mechanism known as the *gular flutter*. They may use one or both the mechanisms at a time (*Lasiewski* and *Crawford*, 1968).

Some animals, such as the African gazelles, and domestic ungulates, etc., have the mechanism for selective cooling of blood to the brain. Some reptiles, when exposed to heat, increase their respiration frequency and breathe with an open mouth, a situation reminiscent of panting.

Besides, some animals are able to maintain different parts of the body at different, well-regulated temperatures. For example, in the desert lizard, *Sauromalus obesus*, the carotid arteries run very close to the surface of the pharynx. As the arteries pass right under the moist surfaces where evaporation takes place, the blood is cooled before it enters the brain. Similarly as mentioned above, in the gazelles and ungulates the brain is supplied with cool blood.

In these animals most of the blood to the brain flows in the external carotid artery, which at the base of the skull divides into hundreds of small arteries, which then rejoin before passing into the brain. These small arteries lie in a large sinus of venous blood which comes from the walls of the nasal passages where it has been cooled. The blood that flows through these small arteries is, therefore, cooled before it enters the skull, and as a result the brain temperature may be 2 or 3 degrees lower than the blood in the carotid artery and the body core.

Body Size and Heat Regulation

In hot conditions the total heat gain consists of : heat gain from the environment, and metabolic heat gain. The heat gain from the environment by conduction and convection, and radiation from the sun and from the ground, are surface processes, and the total environmental heat load, therefore, is directly related to the surface area. Therefore, the total heat load or the sum of environmental and metabolic heat gain, is roughly proportional to the body surface area.

According to *Schmidt-Nielsen* (1964), for a mammal to maintain a constant body temperature under desert conditions, water must be evaporated in proportion to the heat load. Because of the larger relative surface area of small animals, the heat load, and therefore the estimated evaporation in relation to the body size, increases rapidly in the small animal. For example, if a man in the desert evaporates at a rate of one litre per hour (which equals about 0.60 litre per metre^2 body surface per hour), small rodents weighing between 10 to 100 g will have to evaporate at a rate of 15 to 30 per cent of their body weight per hour (Fig.). Since a water loss of 10 to 20 per cent is fatal for mammals, such water loss is impossible.

Hence, desert rodents, because of their small size alone, must escape from heat, which they do by going inside their underground burrows during the day. However, the small squirrels of the North American deserts go on moving in and out of their cool burrows in the daytime. By doing so they can rapidly pick up heat and unload it again and again without using water for evaporation. This permits them to be active during the day.

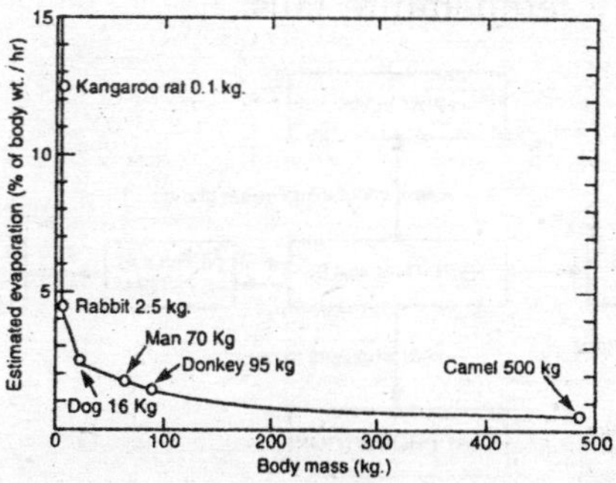

Estimated evaporation in relation to the body size. The curve is calculated on the assumption that heat load is proportional to body surface.

A large animal, such as a camel, evaporates less water and derives a considerable advantage simply from being big. As the camel uses several methods to reduce heat gain, a lower-than-predicted (from body size alone) water loss is due to a number of reasons : (a) heat is stored because of an increase in body temperature (in a camel deprived of water the morning temperature may be as low as 34°C and the late afternoon temperature may be about 41 °C, this 7°C increase in body temperature corresponds about 29000 kcal of heat which equals a saving of 5 litres of water), (b) the increase in body temperature decreases the heat flow from the environment, (c) the fur reduces the heat gain from the environment, and (d) it can tolerate a greater degree of water depletion of the body, and when water is available, it may drink more than one third of its body weight.

Role of Hypothalamus in Temperature Control in Humans: The regulation of body temperature is very precise and is controlled by the *temperature-regulating centres* located in the *hypothalamus*. It has been found that the anterior hypothalamus contains heat-sensitive neurons and the posterior hypothalamus contains cold-sensitive neurons. The heat-sensitive neurons increase their firing rate as the temperature rises. The cold-sensitive neurons, on the otherhand, increase their firing rate when the body temperature falls.

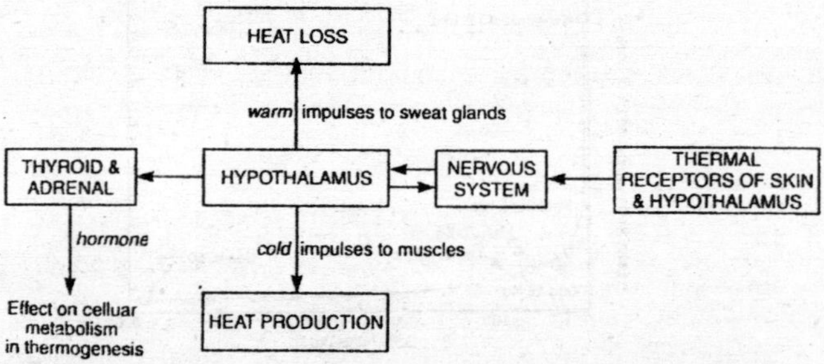

Diagrammatic representation of the role of hypothalamus in temperature regulation. The hypothalamus receives information about the temperature of body surface via afferents from thermal receptors in the skin, and it also detects changes in the temperature of the blood by means of receptors in the hypothalamus itself.

When the body needs more heat, the hypothalamus sends nerve impulses to the muscles, causing contractions and shivering. On the other hand, if the body needs cooling, impulses from the hypothalamus to the sweat glands induce sweat production. The temperature-regulating centre, therefore, acts like a thermostat. Thus, hypothalamus has two roles in thermoregulation. It integrates all of the sensory information available to the nervous system about internal and external temperature conditions and also acts by itself as a sensory mechanism for determining the core temperature. The hypothalamus also acts in integrating homeostatic functioning. Temperature regulation, for example, cannot cause such losses of water by sweating or panting that would disturb the osmoregulation.

The hypothalamus exerts its effects in temperature regulation not only by causing the output of impulses along various motor neurons, as mentioned above, but also affects the thyroid and adrenal glands whose hormonal secretions can influence the cellular metabolism in thermogenesis. The increased thyroxine output by the thyroid gland increases the rate of cellular metabolism throughout the body, which is yet another machanism of *chemical thermogenesis*. Thus, the integrating role of hypothalamus in temperature regulation is further complicated by its neuroendocrine activities.

Any disturbance in the temperature-regulating centre results in fever *(pyrexia)*. It is a condition in which a higher than normal temperature is maintained by the hypothalamus. It is *pyrogens* (fever-producing substances) released grain toxic bacteria or pyrogens released from degenerating tissues of the body that cause fever during disease conditions.

The onset of fever is marked by all the reactions normally seen when a man is exposed to cold. There is shivering, piloerection, cutaneous vasoconstriction, and an increase in thermogenesis. The rate of heat loss is decreased, the rate of heat production rises and thus the core temperature is increased. On the other hand, termination of fever is marked by those reactions normally seen upon exposure to high temperature as the body cools off. Drugs such as aspirin that reduce the level of fever are called antipyretics. They do not reduce normal temperature, but they probably act by antagonizing the actions of *pyrogens* on the hypothalamus. They reset the hypothalamic thermostat and the temperature returns towards normal as a result of increased heat loss brought about by peripheral vasodilation and sweating.

Behavioural and Adaptive Temperature Controlling Mechanism : Besides the physical and chemical mechanisms behavioural mechanisms are used by a variety of organisms, including poikilotherms, to at least periodically maintain a body temperature higher than that of the environment.

In humans, *behavioural control of body temperature* is yet another temperature-controlling mechanism that is even more important. Whenever the internal body temperature becomes too high, signals from the hypothalamus give the person a psychic sensation of being overheated. Conversely, whenever the body becomes too cold, signals from the skin and other receptors elicit the feeling of cold discomfort. Therefore, the person makes appropriate environmental adjustments to regain comfort, such as moving into a heated room in freezing weather. This is the only really effective mechanism for body heat control in extremely cold environs.

Some behavioural adaptations are especially interesting because they extend the resources of the organism beyond the range of its normal physiological reactions. They are as follows:

Hibernation : It is difficult to give a completely satisfactory definition of the term hibernation. The word hibernation (from Latin *hiberna*- winter) is sometimes used to designate an over-wintering, inactive stage of poikilothermic animals such as the insects and snails, etc. The hibernation of homeothermic animals may be defined as a state or a torpid condition in which the body temperature is greatly decreased and the metabolism, respiration, heart rate, etc., are considerably reduced. Thus, the state of hibernation allows the animal to remain dormant during the winter when the food is less readily available.

Mammals that hibernate belong to three orders — insectivora (e.g., the European hedgehog *Erinaceous europaeus),* rodentia (e.g., the golden hamster *Mesocricetus auratus,* the European dormouse *Glis glis,* and the little pocket mouse *Perognathus longimembris,* the American woodchuck *Marmota monax,* the European marmot *Marmota marmota,* and a variety of ground squirrels including *Citellus beecheye. C. franklini, C. lateralis,* etc.), and chiroptera *(e.g.,* the little brown bat *Myotis lucifugus).* Hibernators are also found among the Australian marsupials, such as the pigmy opossum. Hibernating birds include hummingbirds, the insect eating swifts, and some mouse birds *(e.g.,* the African genus *Colis).*

Hibernation is a phenomenon that is under accurate physiological·control. The hibernator is in a state comparable to that of a poikilotherm and its body temperature is close to that of the surroundings and rises and falls with it. When the temperature increases appreciably, the animal warms up again and returns to the homeothermic state. During hibernation the oxygen consumption may decrease to 1/30 to 1/100 of the normal, the heart rate goes down to a few beats-per minute, and the respiratory movements become very slow. It may appear that the major change in a hibernator is the failure of the temperature-regulating centre. But this is not the case because it has been found in some hibernators that if the temperature drops to below freezing, heat production in the animal increases to prevent its tissues from freezing. Thus, the hibernator does not behave as a poikilotherm and hibernation is a well regulated physiological state that permits survival during the colder seasons when the food is in short supply.

However, bears are not considered true hibernators, for although they sleep during the major part of the winter, their body temperature does not decrease more than a few degrees and they show only a moderate drop in metabolic rate and other physiological functions. Bats also differ from other hibernators in that they do not restrict hibernation to the winter.

In summer, when they rest during the day, their body temperature decreases to the ambient temperature. Thus, they behave as hibernators every day, and the long winter-sleep can be considered as an extension of their daily sleep.

The process of hibernation and the factors that induce or inhibit it vary considerably from one species to another. In nature the beginning of the hibernation cycle is usually associated with the time of the year, but it is not necessarily induced by low temperature or lack of food. The annual cycle of hibernation is influenced by the duration of the daily light cycle and is also associated with endocrine cycle. For example, it is often impossible to induce even a good hibernator to begin a cycle of hibernation during early summer and especially during the reproductive period. In arousal the hibernator rapidly activates both heat-conserving and heat-producing mechanisms. The rewarming during arousal is metabolically the most expensive part of the hibernation cycle. However, not only the muscles but also the brown fat is important in the re-warming process. As the brown fat can consume oxygen at a high rate, it is particularly prominent in hibernators and it is considered that mere is a direct connection between hibernation and the amount of brown fat. However, it is most surprising that the brown adipose tissue, which is so important for mammalian hibernators, appears to be absent from a number of birds that regularly hibernate.

Aestivation : Some animals enter the state of torpor that are not necessarily induced by low temperatures and in which most physiological activity appears normal although body temperature may be several degrees below the normal. This is the phenomenon of aestivation. Thus, aestivation (Latin *aestas* = summer) refers to inactivity during the summer and is shown by some mammals and other groups of animals.

It may be applied to snails that become dormant and inactive in response to drought, or it may refer to ground squirrels that during the hottest months disappear into their burrows and remain inactive. However, the Columbia ground squirrel begins to aestivate in the hot month of August and remains inactive throughout the autumn and winter and does not appear again until the following May. Does it aestivate or hibernate, and when does one state change into the other is not fully known.

According to *Schmidt-Nielsen* (1975), no clear physiological distinction can be made between the two states and torpor and hibernation may be treated as one coherent physiological phenomenon.

Regulation of Body Temperature

Human beings are warm-blooded animals and the body temperature is maintained at an average of 36.8°C (98.4°F). In health, variations are usually limited to between 0.5 and 0.75°C, although it may be found that the temperature in the evening is a little higher than in the morning. This is the optimum temperature for the many complex chemical processes to occur. If the temperature is raised the metabolic rate is increased and if it is lowered the rate of metabolism is reduced.

To ensure this constant temperature a fine balance is maintained between heat produced in the body and heat lost to the environment.

Heat Production

A variety of basic chemical reactions contribute to body heat production at all times. Ingestion of food increases heat production because of the specific dynamic action of the food, but the major source of heat is the contraction of skeletal muscle (Table). Heat production can be varied by endocrine mechanism in the absence of food intake or muscular exertion. Epinephrine and norepinephrine produce a rapid but short-lived increase in heat production; thyroid hormones produce a slowly developing but prolonged increase.

Furthermore, sympathetic discharge is decreased during fasting and increased by feeding.

Body Heat Production and Heat Loss

Body heat is produced by :	
Basic metabolic processes	
Food intake (specific dynamic action)	
Muscular activity	
Body heat is lost by :	Percentage of Heat Lost at 21°C
Radiation and conduction	70
Vaporization of sweat	27
Respiration	2
Urination and defection	1

A source of considerable heat, particularly in infants, is a special type of fat, brown fat, that is located between and around the scapulas and in other parts of the body. This fat has a high rate of metabolism, and its function has been likened to that of an electric blanket.

Heat Loss

Cutaneous Vasodilation : The amount of heat lost from the skin depends to a great extent on the amount of blood in the vessels in the dermis. As the amount of heat produced in the body increases, the artertoles become dilated and more blood pours into the capillary network in the skin. In addition to increasing the amount of sweat produced the *temperature of the skin is raised*. When this happens there is an increase in the amount of heat lost by :

Radiation
Conduction
Convection

In *radiation* the exposed parts of the body radiate heat away from the body.

In *conduction* the clothes in contact with the skin conduct heat away from the body.

In *convection* the air passing over the exposed parts of the body is heated and rises, cool air replaces it and convection currents are set up. Heat is also lost from the clothes by convection.

If the external environmental temperature is low or if heat production is decreased, the blood vessels, under the influence of the sympathetic nerves, constrict, thus decreasing the blood supply to the skin and so preventing heat loss.

Birds have a layer of feathers next to the skin, and most mammals have a significant layer of hair or fur. Heat is conducted from the skin to the air trapped in this layer and from the trapped air to the exterior. When the thickness of the trapped layer is increased by fluffing the feathers or erection of the hairs (horripilation), heat transfer across the layer is reduced and heat losses (or. in a hot environment, heat gains) are decreased. "Goose pimples" are the result of horripilation in humans; they are the visible manifestation of cold-induced contraction of the piloerector muscles attached to the rather Meager hair supply. Humans usually supplement this layer of hair with a layer of clothes. Heat is conducted from the skin to the layer of air trapped by the clothes, from the inside of the clothes to the outside, and from the outside of the clothes to the exterior.

Dark clothes absorb radiated heat, and light-coloured clothes reflect it back to the exterior.

Sweating : The other major process transferring heat from the body in humans and those animals that sweat is vaporization of water on the skin and mucous membranes of the mouth and respiratory passages. Vaporization of 1 g of water removes about 0.6 kcal of heat. A certain amount of water is vaporized at all times. This insensible water loss amounts to 50 mL/h in humans. When sweat secretion is increased, the degree to which the sweat vaporized depends upon the humidity of the environment. It is common knowledge that one feels hotter on a humid day. This is due in part to the decreased vaporization of sweat, but even under conditions in which vaporization of sweat is complete, individual in a humid environment feels warmer than an individual in dry environment.

The reason for this difference is not known, but it seems related to the fact that in the humid environment sweat spreads over a greater area of skin before it evaporates. During muscular exertion in a hot environment, sweat secretior reaches values as high as 1600 mL/h, and in a dry atmosphere, most of this sweat is vaporized. Heat loss by vaporization of water therefore varies from 30 to over 900 kcal.

Panting : Some mammals lose heat by panting. This rapid, shallow breathing greatly increases the amount of water vaporization in the mouth and respiratory passages and therefore the amount of heat lost. Because the breathing is shallow, it produces relatively little change in the composition of alveolar air.

Temperature-regulating Mechanisms

The reflex and semireflex thermoregulatory responses are listed in Table. They include autonomic, somatic, endocrine, and behavioral changes. One group of responses increases heat loss and decreases heat production ; the other decreases heat loss and increases heat production. In general, exposure to heat stimulates the former group of responses and inhibits the latter, whereas exposure to cold does the opposite.

Curling up "in a ball" is a common reaction to cold in animals and has a counterpart in the position some people assume on climbing into a cold bed. Curling up decreases the body surface exposed to the environment. Shivering is an involuntary response of the skeletal muscles, but cold also causes a semiconscious general increase in motor activity. Examples include foot stamping and dancing up and down on a cold day. Increased catecholamine secretion is an important endocrine response to cold; adrenal medullectomized rats die faster than normal controls when exposed to cold.

TSH secretion is increased by cold and decreased by heat in laboratory animals, but the change in TSH secretion produced by cold in adult humans is small and of questionable significance. It is common knowledge that activity is decreased in hot weather—the "it's too hot to move" reaction.

Temperature-Regulating Mechanisms

Machanisms activated by cold :	
Shivering Hunger Increased voluntary activity Increased secretion of norepinephrine and epinephrine	Increase heat production
Cutaneous vasoconstriction Curling up Horrpilation	Decrease heat loss
Machanisms activated by heat :	
Cutaneous vasodilation Sweating Increased respiration	Increase heat loss
Anorexia Apathy and inertia	Decrease heat production

Thermoregulatory adjustments involve local responses as well as more general refles responses. When cutaneous blood vessels are cooled they become more sensitive to catercholamines and the arterioles and venules constrict. This local effect of cold directs blood away from the skin.

Another heat-conserving machanism, that is important in animals livingin cold water is heat transfer from arterial to vanous blood in the coming from the extremities (countercurrent exchange). This keeps the tips of the extremities cold but conserves body heat.

Nervous Control : The centre controlling temperature is situated in the *cerebrum* and involves a group of nerve cells in the *hypothalamus* called the *heat regulating centre.* There is also a group of nerve cells in the *medulla oblongata* known as the *vasomotor centre* which controls the calibre of the blood vessels, especially the small arteries and the arterioles, and they control and amount of blood which circulates in the capillaries in the dermis. Stimulation of the anterior hypothalamus causes cutaneous vasodilation and sweating. Posterior hypothalamic stimulation causes shivering.

The heat regulating centre and vasomotor centre are thought to be extremely sensitive to the temperature of the blood and any significant *change* stimulates them to activity. From these centres sympathetic nerves convey impulses to the *sweat glands, arterioles* and the *arrector muscles* of the hairs in the skin.

3

Transfer of Energy

In many living organisms, including man, the *oxidative breakdown* of carbohydrates, fats and proteins is the principal source of energy for many activities of a cell. Oxidation originally means reaction with oxygen, but today, in a chemical context, it means *the loss of electrons by a molecule* or *atom*. Reduction, on the other hand, means the gain of electrons by the molecules or atoms. Oxidation of one substance must necessarily accompany the reduction of another. Hence, a single term *"redox reaction"* is used for any oxidation-reduction reaction. In any redox reaction, the number of electrons lost by the reducing agent must be equal to the number of electrons gained by the oxidizing agent.

Biological Oxidation

Biological oxidation means the oxidation of foods and substrates, that is carried out inside the living cells. In biological oxidation, the food molecules are broken down into smaller fragments and energy is released step by step. Although ordinary combustion or burning of fuels is similar in some way to the biological oxidation, yet there are certain fundamental differences as mentioned below:

1. In combustion, many chemical bonds are broken simultaneously releasing the energy with a great deal of heat. In biological oxidation, on the other hand, energy is released gradually by breaking bonds one after another and temperature is kept low.
2. Most of the energy of combustion is dissipated as heat, whereas the energy released in biological oxidation creates new chemical bonds and only a little energy is lost as heat. By placing the energy in small packages (chemical bonds) cellular metabolism can transfer energy to wherever a cell needs it.
3. Another important feature of biological oxidation is that it occurs in a series of many successive and coordinated enzymatic reactions and not in a single chemical step. Thus each of the many steps constituting biological oxidation is catalyzed by a specific enzyme.
4. Oxygen is not directly necessary in biological oxidation. However, in the two processes of combustion and cellular oxidation, the potential energy within a given amount of an organic compound will always yield the same amount of energy , whether by one process or by the other. This is an accordance with the first law of thermodynamics.

The biological oxidation may take place in the following three ways :

(i) *By addition of oxygen*

$$C + O_2 \longrightarrow CO_2$$

(ii) *By removal of hydrogen*

$$CH_3CHOH.COOH \longrightarrow CH_3.CO.COOH + 2H$$

Lactic acid Pyruvic acid

(iii) *By an increase in positive valency*

$$2Fe^{++}\,(Cl^-)_2 + Cl_2 \longrightarrow 2Fe^{+++}\,(Cl^-)_3$$

In this redox reaction, the ferrous ion is oxidized to the ferric state by losing an electron to chlorine.

In a simple form of biological oxidation, an enzyme *dehydrogenase* catalyzes the transfer of two hydrogen atoms (or electrons) from a substrate to molecular oxygen. The hydrogen atoms unite with oxygen, which is reduced to form hydrogen peroxide. Since H_2O_2 is toxic, it is immediately converted into H_2O and O_2 by the enzyme *catalase*, which is present in all cells.

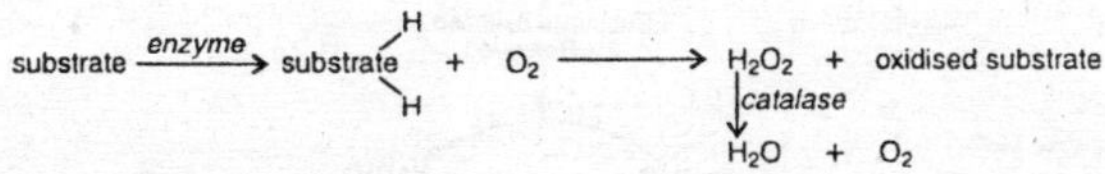

In the above reaction, the removal of activated hydrogen from the substrate is a simple process. But in most biological reactions, the process involves several intermediate steps.

All biological oxidations are actually *dehydrogenations i.e.,* removal of hydrogen atoms and electrons. However, it is not easily possible to remove electrons because they cannot exist freely and as soon as they are removed they must be transferred to another molecule which becomes reduced.

Thus, all biological oxidations are oxidation-reduction reactions that are catalyzed by the enzymes called *dehydrogenases.* In these reactions one molecule is every time oxidized and the other is reduced. In all living cells there are a number of organic molecules functioning as *hydrogen* or *electron acceptors.* They are called *coenzymes.* The most important coenzymes are nicotinamide adenine dinucleotide (NAD), nicotinamide adenine dinucleotide phosphate (NADP), flavin mononucleotide (FMN) and flavin adenine dinucleotide (FAD). These coenzymes attach themselves to *dehydrogenases* only at the time of reaction and are present in the medium rest of the time.

Every coenzyme may function again and again as it is not used up in the course of functioning. This is because the hydrogen atoms or electrons transferred to them do not remain permanently attached but are transferred in the next reaction usually to another coenzyme, and finally to the oxygen itself. The following reaction shows how NAD participates in the biological oxidation of lactic acid to pyruvic acid.

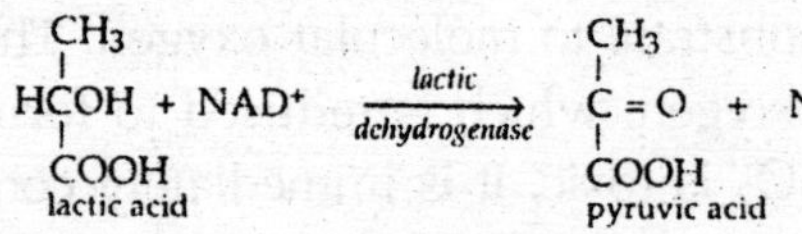

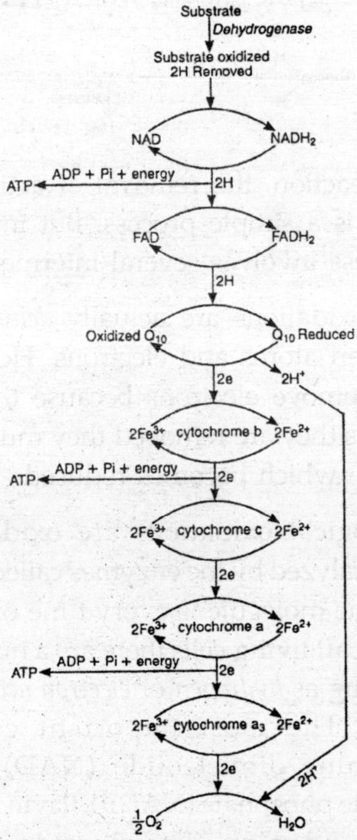

The electron transport and formation of ATP by oxidative phosphorylation.

Steps of Biological Oxidation

1. The first step in biological oxidation, consists of the activation of two hydrogen atoms and their electrons from a substrate by the enzyme dehydrogenase.
2. The hydrogen atoms and their electrons are then transferred to the first hydrogen acceptor, the NAD, which is then reduced.
3. From the reduced NAD or $NADH_2$ the hydrogen atoms and their electrons are passed on to the FAD, which is in turn reduced to $FADH_2$.
4. In the next step $FADH_2$ is oxidized when it delivers the hydrogen atoms and their electrons to the coenzyme Q_{10}.
5. From reduced coenzyme Q_{10} hydrogen atoms are released in the cytoplasm and electrons are now passed on to a series of cytochromes (b, c, a and a_3). In this process, each cytochrome is alternately reduced and oxidized.
6. In the last step, the electrons are finally transferred from cytochrome a_3 (also called cytochrome oxidase) to the final acceptor or the molecular O_2. Oxygen then unites with hydrogen atoms (protons) set free earlier in the cytoplasm to form water.

$$\text{2cytochrome } a_3\text{–}(Fe^{++}) + 2H^+ + 1/2\ O_2 \longrightarrow$$
$$\text{2cytochrome } a_3\text{–}(Fe^{+++}) + H_2O$$

The above electron transport chain (Fig.) is present in the mitochondria in which the biological oxidations are carried out. However, it may be mentioned that most of the hydrogen atoms and their electrons that are passed on to the electron transport chain, are removed by different dehydrogenases from different substrates of Kerbs cycle and only 2 pairs of hydrogen atoms are released in glycolysis.

Oxidative Phosphorylation and Formation of ATP : In biological oxidation, the energy released in the oxidation of a substrate is used to create new ATP molecules. The formation of ATP is possible by *oxidative phosphorylation,* a process in which oxidation is coupled with phosphorylation (Fig.) and which occurs in the mitochondria of living cells.

The general plan of *oxidative phosphorylation* has been explained by *chemiosmotic* or *proton-motive* mechanism (hypothesis) proposed by P. *Mitchell* in 1961. The hypothesis proposes that the primary product of electron transport is the formation of an *electrochemical proton gradient* across the inner mitochondrial membrane (Fig.).

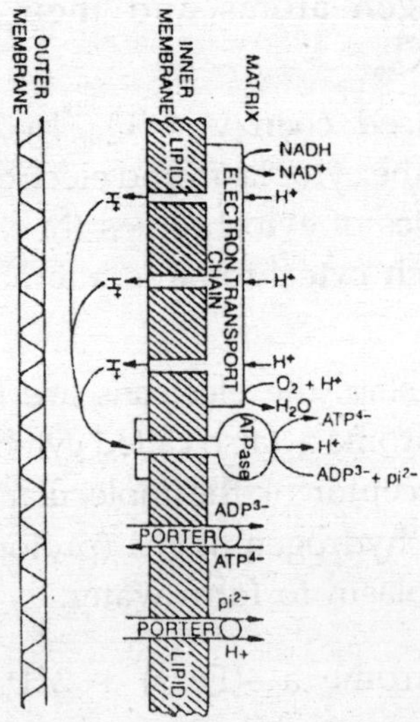

Oxidative phosphorylation by proton-motive mechanism.

ATP synthesis results from the spontaneous flow of protons down this gradient; the path leads through the membrane via a complex of proteins known as the *coupling factor.* The gradient is established by an obligatory movement of protons during respiration and is preserved by the presence of a *membrane with generally low permeability to protons,* except at special locations where the energy of the proton gradient is harnessed to do cellular work. The critical proton barrier is the inner mitochondrial membrane. Exactly how the electro-chemical gradient of protons, *the proton-motive force* (pmf), as *Mitchell* first called it, is used to phosphorylate ADP remains a central mystery of energy metabolism. The following reaction is derived by the proton current.

$$ADP^{3-} + Pi^{-} + H^{+} \longrightarrow ATP^{4-} + H_2O$$

The source of protons delivered to the outside of the membrane is either protons generated as certain components (*e.g.*, NADH, $FMNH_2$) of the electron transport chain are oxidized, or through a vectorially arranged process in which various electron carriers, in their reduced states, would dissociate one or more protons, releasing them to the outside of the membrane and recapture protons from the matrix when reoxidized. There is secretion of $4H^+$ per electron pair at each phosphorylating site. This is sufficient to synthesize ATP and to titrate Pi as it is transported into the matrix. When the protons have passed through the ATPase and return to the matrix, they have completed a proton circuit. This flow of protons from high protic potential to low protic potential is analogous to an electric circuit (*Smith* et al, 1983).

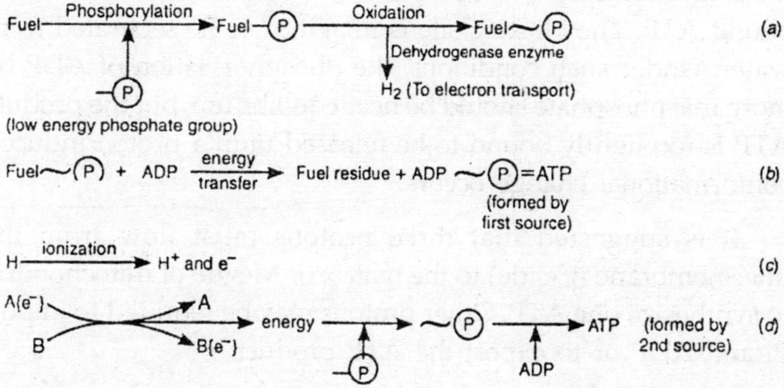

Fig. Showing phosphorylation, oxidation (dehydrogenation) and two sources of ATP formation (one being the oxidation of phosphorylated food by dehydrogenation and the other being the dehydrogenation energy or the creation of high-energy phosphate bond during electron transfer).

Another source of ATP is dehydrogenation energy (Fig.), the energy liberated in the transfer of electrons. This energy is also used in the generation of more ATP molecules by oxidative phosphorylation. It has been stated that NAD works as the first hydrogen acceptor. For every molecule of $NADH_2$ that is oxidized via the electron transport system, 3 molecules of ATP are formed.

$$NADH_2 + 1/2O_2 + 3ADP + 3Pi \longrightarrow NAD + 3ATP + H_2O$$

Of the 3 ATP molecules, one is formed during the passage of hydrogen atoms from $NADH_2$ to FAD, and the rest 2 are formed during the transfer of electrons from coenzyme Q_{10} to molecular O_2.

Thus, the most significant product of biological oxidation is ATP and its formation depends upon the coupling of oxidation with phosphorylation. The possible mechanism of oxidative phosphorylation is the *conformational coupling hypothesis* put forward by P. *Boyer* (1977). The hypothesis proposes that protons passing into the F_1 knobs of mitochondria would bend and alter the conformation of proteins there, allowing the release of a tight bound ATP. The binding site is imagined to be separated from water. Under such conditions, the phosphorylation of ADP by inorganic phosphate should be near equilibrium, but the product ATP is too tightly bound to be released until a proton induced conformational change occurs.

It is suggested that three protons must flow from the intermembrane (C-side) to the matrix or M-side of mitochondria to synthesize one ATP. Other protons may be required to import Pi and ADP or to export the ATP product.

Transfer of Energy in Cellular Metabolism : ATP plays the most important role in the transfer of energy. Besides ATP a number of compounds are used in energy transfer in cells.

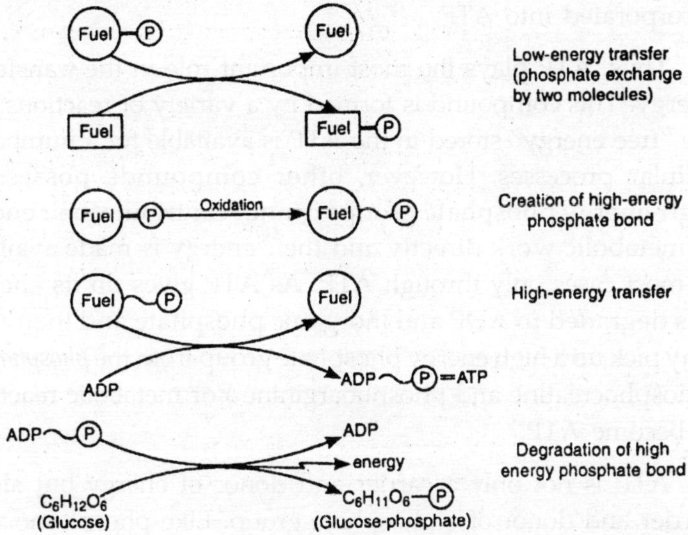

The four ways in which the energy of the phosphate bond is transferred.

Low-energy Transfer : It means exchange of phosphate by two molecules. The most important high-energy bond is the *phosphate bond i.e.,* the bond which joins a molecule to a phosphate group (phosphorylation). Once an organic molecule is so phosphorylated, its phosphate group may then be transferred in turn to various other organic compounds. This is called *low energy transfer, i.e.,* the low energy value of the phosphate bond does not change as in phosphate exchange by two molecules (Fig.).

High-energy Transfer : In many cases the transfer of energy occurs between organic molecules possessing the high-energy phosphate bond and the ADP/ATP system which harvests respiratory energy. This is called *high-energy transfer.* For example, when a fuel molecule is phosphorylated and oxidized, it may possess a high-energy phosphate bond. If the latter is then transferred to ADP, the ADP becomes ATP by accepting a high energy phosphate bond from the fuel (Fig.). Through this type of transfer, some of the energy of a fuel molecule becomes incorporated into ATP.

Thus, ATP plays the most important role in the transfer of energy. This compound is formed by a variety of reactions and the "free energy" stored in the ATP is available for a number of cellular processes. However, other compounds possessing high-energy phosphate groups cannot contribute their energy to metabolic work directly and their energy is made available in most cases only through ATP. As ATP gives up its energy, it is degraded to ADP and inorganic phosphate and then ADP may pick up a high energy phosphate group from the *phosphagens* (phosphocreatine and phosphoarginine) or metabolic reactions to become ATP.

ATP is not only a carrier and donor of energy but also a carrier and donor of a phosphate group. Like phosphoric acid, ATP is often used in phosphorylation in the cells. In such cases, it may react with a fuel molecule, and donate its third phosphate to the molecule, *e.g.*, the phosphorylation of glucose (Fig.). To complete such reactions, only a little energy of ATP is used and the rest is dissipated as heat.

Besides ATP, several other compounds (*e.g.*, coenzyme A, $NADH_2$, and phosphagens) also play a role in energy transfers in cellular metabolism.

4

The Solution

A mixture composed of particles that are dispersed in another medium may be one of the three categories : *solution, suspension* or *colloid.* In a solution, the individual molecules or ions of a substance, the *solute,* are small enough to dissolve in a liquid medium, the solvent. These molecules or ions are so small (less than 1×10^{-7}) that they never settle out under the influence of gravity. The mixture is homogeneous, *i. e.,* perfectly uniform in appearance. Molecules or ions of the solute in a solution cannot be filtered out. If a solution is heated vigorously the solvent evaporates and the solute particles separate out as crystals. Therefore, solution is also sometimes referred to as a *crystalloid* (Thomas Graham, 1861).

Sugar dissolved in water provides the best example of a true solution. When a solid is dissolved in water upto the limiting concentration, it is called a *saturated solution.*

Suspension

A mixture like that of sand and water is called a suspension. The sand particles (crystals of SiO_2) are so large (more than

2×10^{-5}) that after a period of time, they settle by gravity to the bottom of the container. It is easy to separate sand and water by filtration.

Colloids : The term *colloid* is derived from Greek *koli*, glue and *eidos*, appearance. Colloids are suspensions of particles that are larger than those in true solution but still too small to settle out by gravity. They are of intermediate size, from about 1/1,000,000 to 1/10,000 mm (1×10^{-7} to 2×10^{-5}) in diameter. They cannot be filtered out by ordinary filters. The white of egg is a familiar colloid.

Types of Colloids : There are eight general types of colloidal systems :

(A) A colloidal *gas* dispersed in

(i) a solid, *e.g.*, cheese : colloidal air in fat-protein.

(ii) a liquid, *e.g.*, Campa Cola, Thums Up etc.

(B) A colloidal *liquid* dispersed in

(iii) a liquid *(emulsion)*, *e.g.*, milk : colloidal fat in water.

(iv) a solid *(gel) e.g.*, gelatin : colloidal water in protein.

(v) a gas *(aerosol)*, *e.g.*, fog : colloidal water in air.

(C) A colloidal *solid* dispersed in

(vi) a liquid *(sol)*, *e.g.*, milk of magnesia : colloidal $Mg(OH)_2$ in water, and starch dispersed in water.

(vii) a solid, *e.g.*, colloidal gold in glass (rubi glass).

(viii) a gas *(aerosol)*, *e.g.*, cigarette smoke : colloidal ash in air.

Nature of Colloids : A colloidal system consists of two heterogeneous components: a *dispersed phase* and a *dispersion phase* or medium.

(i) *Dispersed or discontinuous phase* comprises macromolecular solids like proteins and nucleic acids; and liquids such as oily fats. The dispersed particles of the discontinuous phase are called *micelles*.

(ii) *Dispersion or continuous phase* is the medium in which insoluble materials of colloidal size are dispersed. Dispersion medium can be solid, liquid or gas. In protoplasmic colloidal systems, this medium is water. The particles of the dispersion phase are called the *intermicellar particles.*

Properties of Colloids :

(1) *Electrical Properties :* The particles of a colloidal solution carry similar electric charges which may be positive or negative (colloidal ions). These charges make the particles repel one another and this keeps them suspended. Every ion in the dispersion medium has an opposite and equal charge so that the system as a whole is electrically neutral.

(2) *Brownian Movement :* The dispersed particles of a colloidal system are constantly moving in random zig-zag paths. This haphazard motion of the dispersed particles in the dispersion medium is called the Brownian movement (after Robert Brown, 1828). The smaller the particle size the more violent the Brownian movement. This type of movement is due to the bombardment of the colloidal particles by molecules of dispersion medium.

(3) *Tyndall Phenomenon :* Light shining through a colloidal mixture gives it a cloudy or opalescent appearance as opposed to a true solution which is optically clear. This is called the Tyndall phenomenon. It results from the light scattering properties of colloidal dimensions. When a beam of light is passed through a colloidal solution, the path of the beam of light can be observed from the side because the dispersed particles are big enough to scatter the light. This effect is used to distinguish colloids from true solutions.

(4) *Sol and Gel :* A colloid may remain in two states, as a

liquid or as a solid. The colloidal solution or *sol* is a system of solid particles of colloidal dimensions dispersed in a solvent medium. The *sols* are divided into two classes :

(i) *Lyophilic (Liquid Loving)* : These are the reversible colloids which after coagulation can be again converted into colloidal state by adding the solvent. Examples are glue, starch and gelatine. Formerly, the sols in which water is the dispersed medium were called hydrophilic (water loving), but now-a-days the more general term "lyophilic" is used because there are several colloidal solutions in which dispersion media are other than water.

(ii) *Lyophobic (Liquid Fearing)* : There are the irreversible colloids which can readily be precipitated from the solution. They do not regain their colloidal state even if the solvent is added. For example, the colloidal solution of gold, once coagulated, cannot be reconverted into colloidal state by dipping it into the solvent.

If liquid particles are dispersed in a solid medium the colloidal system is known as a *gel* which is like a quasi-solid, flexible gelly, as in gelatin and fruit jellies. The semisolid structure of a gel is due to the solid containing a liquid arranged in an intricate meshwork extending throughout the system. Many long-chain proteins and high molecular nucleic acids are commonly found as gels.

The cytoplasm of a living cell is normally in the gel state but it undergoes continuous, dynamic, localized and reversible solgel transformations, also called *phase reversals.* In living materials, a region remains more or less permanently in a gel state, like skin, or in a sol state, like blood plasma. However, the latter becomes a gel on clotting. Gels maintain definite forms and shapes despite high water contents.

Increased temperature may convert a gel into a sol. This change is called *solution*. Conversely, lower temperatures promote conversion of sols to gels, a process known as *gelation*. Many other physical and chemical influences, such as low or high pH or pressure, affect solgel transformations.

(5) *Dialysis* : If a mixture of crystalloids and a colloid is placed on one side of parchment membrane facing water on the other side, the colloid particles would be retained on the same side, whereas crystalloids and a great deal of dispersion medium would pass out through the parchment membrane to the other side. This phenomenon is called "dialysis" which is employed in "artificial kidney" to remove urea and other unwanted small metabolic wastes from plasma proteins.

(6) *Ageing* : Colloids undergo ageing. The particles in a fresh young colloidal system are surrounded by layers of *bound water*. Colloid particles *imbibe* a good amount of water. The molecules of water are *absorbed* to the surface of the colloidal particles by electric attraction. Adsorption is a peculiar form of combination in which substances adhere together by surface contact. It is not a true chemical reaction. Water in the gel, therefore, does not "run out" through its lattice work. As time passes, binding capacity of the particles diminishes, and some of the water runs out. The colloid contracts and undergoes gelation gradually. Examples are exudation of water from long-standing milk curd and custard.

(7) *Flocculation or Coagulation* : In colloidal solution all particles have similar charge and they remain separate due to forces of repulsion. When the charge on the particles is neutralized by the addition of ions or micelles of opposite charge, the particles undergo collision because of Brownian movement and *agglomeration* of

the individual particles takes place. It results in the formation of larger particles which settle down out of the system. This phenomenon, is called flocculation or precipitation.

(8) *Formation of Boundary Membranes* : An interfacial boundary membrane is formed between a colloidal system and a different medium such as air, water or solid surfaces, *e.g.*, the 'skins' on boiled milk, pudding, custard etc.

Biological Significance of Colloids : Colloids are of immense biological value. Some examples are as follows :

(1) *Cell protoplasm* in every cell exists in a colloidal state.

(2) *Blood plasma, Lymph* and *milk* are all colloidal emulsions.

(3) *Blood clotting* is a colloidal phenomenon in which plasma (sol) is converted into a cloth (gel).

(4) *Imbibition of Water* : Since colloid particles imbibe water, a good amount of water remains stored in the body.

(5) *Interfacial Reaction* : Colloids are dispersed in the form of minute particles, hence they provide a very large surface area for various reactions to occur, such as enzyme action, adsorption, surface tension etc.

Hydrogen Ion Concentration (ph) (Potentia Hydrogen) : The number of hydrogen ions present in a solution is a measure of the acidity of the solution.' The maintenance of the normal hydrogen ion concentration within the body is an important factor in the environment of cells.

A standard scale for the measurement of the hydrogen ion concentration in solution has been developed. All acids do not ionise completely when dissolved in water, that is, all the molecules of acid in solution do not ionise and exist in the solution as electrically charged particles. The hydrogen ion concentration is a measure, therefore, of the amount of *dissociated acid* rather than of the amount of acid present. Strong acids dissociate more freely than weak acids, while carbonic acid dissociates much less freely into H^+ and Cl^-, while carbonic acid dissociates much less freely into H^+ and CO_3^-. The number of *free hydrogen ions* in a solution is a *measure of its acidity* rather than an indication of the type of molecule from which the hydrogen ions originated.

The alkalinity of a solution is dependent upon the number of hydroxyl ions (OH^-) present. Water is a neutral solution because every molecule contains one hydrogen ion and one hydroxyl radical. For every molecule of water (H.OH) which dissociates, one hydrogen ion (H^+) and one hydroxyl ion (OH^-) is formed, each one neutralizing the other.

The scale for measurement of pH was developed taking water as the standard. It was found by experiment that 1 mole in 550,000,000 moles of water ionises into a hydrogen ion and a hydroxyl ion. This is the same proportion as 1 gram hydrogen ion in 10,000,000 litres of water. Therefore, 1 litre of water contains $\frac{1}{10,000,000}$ of a gram of hydrogen ion. In order to make these figures more manageable this can be written :

1 litre of water contains $\frac{1}{10^7}$ g hydrogen ion.

$$(10 = 10^1 \; ; \; 100 = 10^2 \; ; \; 10,000 = 10^4).$$

The fraction $\frac{1}{10^7}$ may be written 10^{-7} the negative power indicates that this is a fraction, $\frac{1}{10,000,000}$ may be written $\frac{1}{10^7}$ or 10^{-7}.

Later, for every day use, only the power' figure was used and the symbol pH placed before it.

Thus a neutral solution such as water, where the number of hydrogen ions is balanced by the same number of hydroxyl ions, the pH = 7. The range of this scale is from 0 to 14. If the pH is 0 it would mean that 1 litre of water contained 1/1 = 1 gram of hydrogen ion ; or at the other end of the scale if there were no hydrogen ion present it would be written $\frac{1}{10^{14}}$ or pH 14. It will be noted that a change of pH of 1 at any level in the scale means an increase or decrease by a factor of 10 hydrogen ion concentration because the pH scale is a logarithmic one.

The pH scale thus runs from 1 to 14, neutrality being at pH 7.0. Numbers lower than 7.0 denote an excess of hydrogen ions and an *acid* solution ; numbers *higher* than 7.0 denote an excess of hydroxyl ions and an *alkaline* solution.

pH Values of the Body Fluids : All body fluids have pH values which must be maintained within relatively narrow limits and within which the cells function normally. These pH values are not the same in every part of the body, for example, the pH values of the following secretions are :

Blood	7.4 to 7.5
Bile	7.8 to 8.6
Pancreatic juice	8.0
Saliva	6.4 to 6.8
Urine	5.5 to 7.0
Gastric juice	1.5 to 1.8

It can be seen that there is a wide variation of pH values within the body. The individual pH in an organ is produced by it's secretion of acids or alkalies which will establish the optimum level. The highly acid pH of the gastric juice is maintained by the secretion into the stomach of hydrochloric acid produced by chemical changes which take place in parietal cells in the wall of the stomach. The low pH value in the stomach provides the environment best suited to the functioning of the enzyme pepsin which is present in gastric juice. Saliva has a pH of between 6.4 and 6.8, which is the optimum value for the action of ptyalin. Ptyalin action is inhibited when food containing it reaches the stomach and is mixed with highly acid gastric juice.

It will be noted that the blood has a pH value of between 7.4 and 7.5. Thus, therefore, it is the general pH level in the body which may be altered in an individual organ such as the stomach, to meet the needs of the special functions of that organ. The range of pH of the blood which is compatible with life is 7.0 to 7.8. The metabolic activity of the body cells produces certain acids and alkalies which tend to alter the pH of the tissue fluid and the blood. To maintain the pH within the normal range there are substances present in the blood which act as *buffers.*

Buffers

Buffers are chemical substances which are able to 'bind' free hydrogen ion or hydroxyl ion without change in pH. Three major buffer systems are found in animals, and these involve *phosphate, bicarbonate,* or *amino acids.* A certain amount of all three is normally present in the body and cellular fluids, and so the maintenance of a constant pH depends on a complex system.

(i) *Phosphate System* : A solution of sodium dehydrogen phosphate ionizes in solution in the following manner :

$$NaH_2PO_4 \rightleftharpoons Na^+ + H_2PO_4^+ \rightleftharpoons H^+ + HPO_4^{2-} \rightleftharpoons H^+ + PO_4^{3-}$$

The final ionization in this series is only half complete at a pH of nearly 13.0, and may be ignored from a physiological viewpoints. The predominant phosphate ions in solution at a physiological pH will therefore be $H_2PO_4^-$ and HPO_4^{2-}. Phosphate buffers are, in fact, usually made up as a mixture of the two salts, sodium dihydrogen phosphate NaH_2PO_4, and disodium hydrogen phosphate Na_2HPO_4. It is evident from the scheme of ionization given above that the second of these two salts act as a weak acid, since it produces some hydrogen ions in solution, whereas the first salt represents the salt of phosphoric acid H_3PO_4 with the strong base NaOH. The mixture of these two salts will behave like the generalized buffer. Thus, if hydroxyl ions are added, water is formed :

$$NaH_2PO_4 + NaOH \rightarrow Na_2HPO_4 + H_2O$$

The addition of acid results in the formation of the basic salt :

$$Na_2HPO_4 + HCl \rightarrow NaCl + NaH_2PO_4$$

(ii) *Bicarbonate System* : The bicarbonate ion HCO_3^- can take up hydrogen ions from water to form carbonic acid :

$$HCO_3^- + H^+ \rightleftharpoons H_2CO_3$$

Although little carbonic acid is formed under normal conditions, the addition of alkali to this solution will cause it to react with the carbonic acid, remove it through the formation of water, and so facilitate the production of more carbonic acid:

$$H_2CO_3 + OH^- \rightarrow H_2O + HCO_3^-$$

The net effect is that the added hydroxyl ions have been removed from solution, with the formation of more bicarbonate ion. If acid is added, it will react with the bicarbonate ion to produce carbon dioxide and water :

$$H^+ + HCO_3^- \rightarrow H_2O + CO_2 \uparrow$$

Some of the carbon dioxide will go into solution and form carbonic acid, but since this is a weak acid, it's effect is negligible. Thus bicarbonate can act as a buffer because it forms a weak acid in solution, and it's action conforms to the generalized scheme previously described.

Sodium bicarbonate ($NaHCO_3$) is normally present in blood

and acts as buffer to acid radicals like lactate, phosphate and sulphate which are stronger than carbonate.

(iii) *Amino Acids System* : Amino acids contain in their molecule both a basic -NH_2 group and an acidic -COOH group. They can be visualized as existing in the form of a neutral *zwitterion* in which a hydrogen atom can pass between the -NH_2 and -COOH groups, for example in the case of *glycine*, the molecule can be depicted as

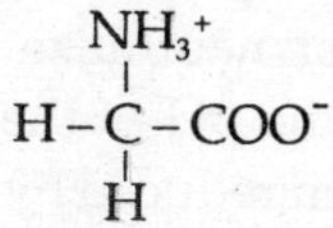

By the addition or subtraction of a hydrogen ion to or from the zwitterion, either the cation form or the anion form will be produced :

$$\underset{\text{cation form}}{H_3N^+ - CH_2 - COOH} \underset{}{\overset{+H^+}{\rightleftharpoons}} \underset{\text{zwitterion}}{H_3N^+ - CH_2 - COO^-} \underset{-H^+}{\rightleftharpoons} \underset{\text{anion form}}{H_2N - CH_2 - COO^-}$$

Thus, when OH′ ions are added to the solution of amino acid, they take H^+ ions from it to form water, and the anion is produced. If H^+ ions are added, they are taken up by the zwitterion to produce the cation form. In practice, if NaOH was added, the salt $H_2N—CH_2—COONa$ would be formed, and the addition of HC1 would result in the formation of the hydrochloride $ClH—H_3N—CH_2—COOH$, but these substances would ionize in solution to some extent, to form their corresponding ions. *Haemoglobin* and *plasma proteins* act as buffers in a similar way.

Amino acids differ in the degree to which they will produce the cation or anion form. In other words, a solution of an amino acid is not neutral, but is either predominantly basic or acidic, depending on which form is present in greater quantity. For this reason, different amino acids may be used as buffers for different pH values, and a mixture of them possesses a wide buffer range.

The acids which are constantly added to the tissue fluids by the cells are first buffered and then disposed of in various ways. Lactic acid resulting from muscular contraction is partly converted back into glycogen by the liver. Phosphates ; sulphates and chlorides are excreted by the kidneys. Carbon dioxide, which is continually produced by all cells, is removed in the lungs which thus exert an accurate and rapid control over the acid-base balance.

The ability of the complex buffer systems of the blood to 'bind' hydrogen ions or neutralize acids is called the *alkali, reserve* of the blood. If the pH of the blood falls below 7.4 the reserve of alkali has been reduced by an increase in production of hydrogen ions by the cells, and the condition of *acidosis* exists. When the reverse situation pertains and the pH is raised above 7.5 the increased alkali produced uses up the *acid reserve* and a state of *alkalosis* exists.

Ionic Dissociation

Almost all chemical reactions of biological importance occur in aqueous solution. Compounds which give rise to ions in solution are called *electrolytes*. For example, when solid NaCl (an ionic compound) dissolves in water, Na^+ and Cl- ions leave the crystal and are available among the water molecules. This separation into free ions is termed *ionic dissociation*. For example,

$$\underset{\text{sodium chloride}}{NaCl} \xrightarrow{H_2O} \underset{\text{sodium ion}}{Na^+} + \underset{\text{chloride ion}}{Cl^-}$$

$$\underset{\text{acetic acid}}{CH_3COOH} \xrightarrow{H_2O} \underset{\text{acetate ion}}{CH_3COO^-} + \underset{\text{hydrogen ion}}{H^+}$$

$$\underset{\text{ammonium hydroxide}}{NH_4OH} \xrightarrow{H_2O} \underset{\text{ammonium ion}}{NH_4^+} + \underset{\text{hydroxyl ion}}{OH^-}$$

Some *strong electrolytes* always exist in ionic form even in the solid state, *e.g.*, sodium chloride (Na^+Cl-), sodium bicarbonate (Na^+HCO_3-). Other electrolytes dissociate completely into ions when in solution, *e.g.*, hydrogen chloride in water (H^+ + Cl-). Besides, there are some *weak electrolytes,* which are only partially dissociated into ions, and their solutions contain undissociated molecules as well as the corresponding ions.

When ionic dissociation produces equal amounts of positive and negative electric charges, solutions containing dissociated ionic compounds remain electrically neutral. If placed in an electrical field, positively charged ions move towards the negative pole (Cathode). They are therefore called *cations.* On the other hand, negatively charged ions move towards the positive pole (anode); they are therefore called *anions.* Some physiologically important cations and anions are as follows:

Cations : hydrogen ion (H^+), ammonium ion (NH^+_4), metallic ions like Na^+, K^+, Ca^{2+}, Mg^{2+}, etc.

Anions : Chloride ion (Cl-), hydroxyl ion (OH-), bicarbonate ion (HCO_3-), sulphate ion (SO_4^{2}-), dihydrogen phosphate ion (H_2PO_4-), hydrogen phosphate ion (HPO_4^{2}-), etc.

It should be noted that many of the organic solutes in the body fluids like glucose and urea, are *non-electrolytes* and exist in solution only as unionized molecules. Water is also an excellent solvent for many non-electrolytes.

Dissociation Constant

The acetic acid dissociates into ions according to this equation:

$$CH_3COOH \rightleftharpoons CH_3COO- + H^+$$

The equilibrium constant for this reaction is :

$$K = \frac{[CH_3COO^-][H^+]}{[CH_3COOH]} = 1.78 \times 10^{-5} = 10^{-4.75}$$

In this case *K* describes the electrolytic dissociation of acetic acid and is called the *dissociation constant*. The higher the dissociation constant, the stronger the ionization of the compound in question. Weak acids are those with small dissociation constants. As the exponential form is inconvenient, it is useful to express the dissociation of the weak acid in terms of the negative logarithm of the equilibrium constant.

$$pK \text{ (acetic acid)} = -\log K = \log \frac{1}{K} = 4.75$$

All weak electrolytes have one or more pK values, which are written as pK_1, pK_2 etc. as shown below :

pK values of some electolytes

Acid	*Conjugate base*	*pK*
CH_3COOH	CH_3COO^-	4.75
H_3PO_4	$H_2PO_4^-$	2.0 (pK_1)
$H_2PO_4^-$	HPO_4^{2-}	6.7 (pK_2)
HPO_4^{2-}	PO_4^{3-}	11.7 (pK_3)
NH_4^+	NH_3	9.3
H_2CO_3	HCO_3^-	6.4 (pK_1)
HCO_3^-	CO_3^{2-}	10.2 (pK_2)

It should be noted that there is a special relationship between the *degree of dissociation* α (the value showing what that of the dissolved moles of the electrolyte has fallen into ions) and the dissociation constant. This relationship is described by the dilution law formulated by Ostwald :

$$\frac{\alpha^2}{1-\alpha} c = K$$

Where α is the degree of dilution; c is the molar concentration of the electrolyte dissociating into two ions;

and K is the dissociation constant. Once two quantities in this equation are known, the third can be determined for the given conditions.

Such calculations give constant values of K only for weak electrolytes. For strong electrolytes K is not constant and increases with the concentration of the solution.

Dissociation degree and dissociation constant of electrolytes at 18°C.

Electrolyte	*Concentration*	α, %	*K*
CH_3COOH	0.1N	1.32	1.76×10^{-5}
CH_3COOH	0.01N	4.10	1.75×10^{-5}
KC1	0.1N	86.20	0.536
KC1	0.01N	94.20	0.152

Since the dissociation constant of weak electrolytes is a constant value, the concentration of various ions in a solution can be changed.

Acids

An acid is defined as any compound that can liberate a hydrogen ion.

$$\underset{\text{acetic acid}}{CH_3COOH} \rightarrow \underset{\text{acetate ion}}{CH_3COO^-} + \underset{\text{hydrogen ion}}{H^+}$$

This equation shows that *acetic acid* dissociates in such a way that hydrogen ions are formed. This is what actually makes acetic acid an acid. Some acids in aqueous solutions are fully ionized at all times. They are called *strong acids, e.g.,* hydrochloric acid.

$$HCl \rightleftharpoons H^+ + Cl^-$$

On the other hand, *weak acids* are only partially ionized, only a small proportion of their replaceable hydrogen being in the ionic form. For example,

$$\underset{\text{carbonic acid}}{H_2CO_3} \rightleftharpoons H^+ + HCO_3^-$$

$$\underset{\text{acetic acid}}{CH_3COOH} \rightleftharpoons H^+ + CH_3COO^-$$

In all the above cases H^+ ions are produced. But water breaks the bonds of different ion-producing compounds to different degrees. In the case of HC1, virtually all bonds of H are broken ; in the case of H_2CO_3 and CH_3COOH, only some are broken. Consequently, a solution of acetic acid will contain partly the whole compound CH_3COOH and partly the free acetate and hydrogen ions.

Bases

A base is defined as any compound which can accept hydrogen ions or which is characterized by the ability to combine with the H^+ ions furnished by an acid. Some *strong bases* (alkalies), like sodium hydroxide, are completely ionized in solution at all times.

$$NaOH \longrightarrow Na^+ + OH^-$$

Thus, any compound which dissociates to yield *hydroxyl ions* (OH^-) is a base or an *alkali.* On the other hand, *weak bases* owe their basicity to their power to capture H^+ as shown below :

$$NH_3 + H^+ \rightleftharpoons NH_4^+$$

$$HPO_4^{2-} + H^+ \rightleftharpoons H_2PO_4^-$$

A solution of ammonia (ammonium hydroxide) is alkaline.

$$NH_3 + H^+ + OH^- \rightleftharpoons NH_4^+ + OH^-$$

It should be noted that acids and bases can be either neutral molecules (*e.g.,* HC1 and NaOH) or charged ions (for example, NH_4^+ - and HPO_4^{2-}-). Most of the acids and bases of biological interest are generally weak.

Salt

A salt is defined as a compound that results from the chemical interaction of an acid and a base. Sodium chloride is a salt because it is formed by the interaction of hydrochloric acid and sodium hydroxide:

$$\underset{\text{acid}}{HC1} + \underset{\text{base}}{NaOH} \longrightarrow \underset{\text{salt}}{NaCl} + H_2O$$

Salts formed from a strong acid and a strong base are fully ionized even in the solid state, for example sodium chloride ($Na^+ Cl^-$). Salts from either a strong acid with a weak base or a weak acid with a strong base are also fully ionized when dissociated in water, but this does not mean that their solutions will be neutral. In the case of the salt of a weak acid with a strong base, the anion will react partially with H^+ from the water to give the undissociated weak acid, leaving an excess of OH^-. The solution will be alkaline with a pH above 7. For example, sodium bicarbonate in water.

$$Na^+ + HCO_3- + H^+ + OH- + H_2O \rightleftharpoons Na^+ + H_2CO_3 + OH^- + H_2O$$

Similarly the salt of a weak base with a strong acid gives an acid solution because of the reversible loss of H^+ from the cation. For example, ammonium chloride in water :

$$Cl^- + NH_4^+ + OH^- + H^+ + H_2O \rightleftharpoons Cl^- + NH_3 + H_2O + H^+ + H_2O$$

All salts of both strong and weak acids dissociate completely in water to free cations and anions.

pH (Hydrogen ion concentration) : pH is defined as the negative logarithm of the hydrogen ion concentration in an aqueous solution (at a given temperature).

$$pH = -\log [H^+] \text{ moles/l}$$

$$= \log \frac{1}{[H^+]}$$

where the logarithms are taken to the base ten.

The following points describe the important characteristics of the pH scale :

1. The pH scale is inversely related to the hydrogen ion concentration [H^+] ; that is, the lower the pH the higher the hydrogen ion concentration.

2. The pH scale is *exponential* rather than linear. A change of lpH unit corresponds to a 10-fold change of H^+ ion concentration of the solution. Thus, a change of pH from 7 to 6 represents a 10-fold increase in concentration of H^+ ions. Similarly a change from 7 to 8 represents a 10-fold increase in concentration of OH^- ions. In other words, pH 6 is 10 times more acidic than pH 7 and pH 5 is 100 times more acidic than pH 7 and so on, or pH 8 is 10 times more alkaline than pH 7 and pH 9 is 100 times more basic than pH 7.

3. For solutions of strong acids which dissociate completely (such as HC1), the pH is approximately equal to the logarithm of the reciprocal of the acid concentration. Thus, the pH of a 0.001 M HC1 solution is close to 3.

It follows that if the pH of a solution is 4, its hydrogen ion (H^+) concentration is 10^{-4} and its OH^- ion concentration is 10^{-10} at a given temperature. The pH scale is normally taken as extending from 0 to 14. A neutral solution has a pH value of 7, an acid solution has a pH value less than 7, and a basic solution has a pH value greater than 7 as indicated below :

	acidic →	*neutral* ↓	*alkaline* →
pH	1 2 3 4 5 6	7	8 9 10 11 12 13 14
	Very acid	↑ *Neutral*	Very basic
	(most H^+ released)		(most OH^- released)

The reaction of neutral water and of any neutral solution is expressed as pH7

$H^+ \times OH^- = K_w = 10^{-7} \times 10^{-7} = 10^{-14}$

$= K_w = 10^{-14}$

(K_w means ionic product)

The concentration of H^+ and OH^- in neutral water is always 10^{-14}. Measurement shows that in a litre of pure water 0.0000001 g of H^+ ions is present. The pH of pure water therefore is:

pH (water) = log 1/0.0000001) = 7

Relationships of hydrogen ion concentrations are expressed in various ways. Living matter, containing a mixture of variously dissociated acids, bases, and salts, usually has a pH very near to neutrality. For example, the pH of human blood is generally 7.4. All enzymatic reactions are carried at an optimum pH.

Buffers

A buffer solution is one which can accommodate the addition to it of moderate amounts of acid or base without marked changes in its pH value. The capacity of a solution to resist alteration in its pH value is known as *buffer capacity.*

The general form of the buffering reaction is :

$$\text{Buffer} + H^+ \rightleftharpoons \text{H Buffer}$$

In this example, a free H^+ combines with the buffer to form a weak acid (H Buffer) that can either remain as an undissociated molecule or dissociate back to Buffer and H^+. When the hydrogen ion concentration increases, the reaction is forced to the right and more hydrogen ions bind to the buffer, as long as available buffer is present. Conversely, when the hydrogen ion concentration decreases the reaction shifts toward the left and hydrogen ions are released from the buffer. Thus, changes in hydrogen ion concentration are minimized.

Relationships of H^+ ion Concentrations Expressed in Various Ways.

Reaction	*Fraction of normality*	*H^+ions/l (grams)*	*Log of (H^+ions)*	*Expressed as pH*
Acid	N/1	1.0	-0	0.0
Acid	N/10	0.1	-1	1.0
Acid	N/100	0.01	-2	2.0
Acid	N/1000	0.001	-3	3.0
Acid	N/10,000	0.0001	-4	4.0
Acid	N/100,000	0.00001	-5	5.0
Acid	N/1000,000	0.000001	-6	6.0
Neutral	Pure Water	0.0000001	-7	7.0
Alkaline	N/1000,000	0.0000000C1	-8	8.0
Alkaline	N/100,000	0.000000001	-9	9.0
Alkaline	N/10,000	0.0000000001	-10	10.0
Alkaline	N/1000	0.00000000001	-11	11.0
Alkaline	N/100	0.000000000001	-12	12.0
Alkaline	N/10	0.0000000000001	-13	13.0
Alkaline	N/1	0.00000000000001	-14	14.0

Buffer System

A buffer system has the property of resisting pH changes despite additions of acid or base. An adequate buffer system can be obtained from a weak acid mixed with the s. lt from that acid and a strong base. For example,

$$\underset{\text{weak acid}}{HA} + \underset{\text{salt}}{B^+ A^-}$$

where B^+ = the cation of a strong base *e.g.*, Na^+

A^- = anion from weak acid

Relationships of H^+ and OH^- ions and pH.

H^+ ions	*OH^- ions*	*pH*	*Reaction*
10^{-0}	10^{-14}	0	Acid
10^{-1}	10^{-13}	1	"
10^{-2}	10^{-12}	2	"
10^{-3}	10^{-11}	3	"
10^{-4}	10^{-10}	4	"
10^{-5}	10^{-9}	5	"
10^{-6}	10^{-8}	6	"
10^{-7}	10^{-7}	7	Neutral
10^{-8}	10^{-6}	8	Alkaline
10^{-9}	10^{-5}	9	"
10^{-10}	10^{-4}	10	"
10^{-11}	10^{-3}	11	"
10^{-12}	10^{-2}	12	"
10^{-13}	10^{-1}	13	"
10^{-14}	10^{-0}	14	"

This corresponds to a partially neutralized solution of weak acid (HA$\rightarrow$ H^+ + A^-) with some reasonable proportion of the acid converted to the anion. Its pH will be higher than the pH of the weak acid alone. The pH of a buffer solution is determined not on the absolute concentrations of buffer constituents but on their ratio, *i.e.*, on the ratio of the amount of salt to weak acid as stated in the *Henderson-Hasselbalch equation* :

$$pH = pK_a + \log\frac{[salt]}{[acid]}$$

where K_a is the dissociation constant of the acid and pK_a is the pH at which the acid is half neutralized and equal amounts of acid and ion (salt) are present. This equation is applicable to all the buffer systems. The buffer capacity, measured as the ability of the solution to minimize changes in pH due to addition of base, is strongest near the midpoint of a titration, when $[A^-] = [HA]$ and $pH = pK_a$. The most important examples of biological buffer systems are as follows:

Bicarbonate Buffer : The major buffer of extracellular fluid is the HCO_3^-/H_2CO_3 system. This results from a number of factors. (0 There is considerably more HCO_3^- present in extracellular fluid than any other buffer component, (ii) There is an unlimited supply of CO_2, (iii) Physiological mechanisms maintain normal extracellular pH by controlling either the HCO_3^-or the CO_2 concentration of extracellular fluid, and (iv) this buffer system operates in conjunction with Hb.

The Henderson-Hasselbalch equation for this buffer system may be written as :

$$pH = pK + \log \frac{[HCO^-_3]}{[H_2CO_3]}$$

In this case pK H_2CO_3 = 6.1, the ratio $\frac{[HCO^-_3]}{[H_2CO_3]} = \frac{20}{1}$ for

$$pH = pK + \log \frac{[HCO^-_3]}{[H_2CO_3]}$$

$$7.4 = 6.1 + 1.3$$

$$1.3 = \log 20$$

Moreover in true plasma total CO_2 content is equal to $[H_2CO_3] + [HCO_3^-]$. Hence, if the total CO_2 content of true plasma is known and the pCO_2 of alveolar air, *i.e.*, arterial pCO_2 is determined, the pH of the true plasma can be calculated. This method of determining blood pH (true plasma pH) was used before the advent of commercial pH meters. The Henderson-Hasselbalch equation is important, for it enables us to appreciate that the blood pH is dependent on the *ratio* of the concentration of free acid to the concentration of the buffer anion (bicarbonate). Therefore, acidaemia or alkalaemia could be caused by either a change in the nonvolatile component (HCO_3^- concentration) or in the volatile component (H_2CO_3 concentration and hence pCO_2).

Since the $[H_2CO_3]$ is fixed only by the alveolar CO_2 tension, if the gas tension is equivalent to that normally present in the blood, the HCO_3–H_2CO_3 system is more useful at pH 7.4, with a ratio of 20, than it would be at its pK 6.1 where the HCO_3^- would be exhausted by addition of 1.25 meq per litre of acid, and addition of this amount of alkali would result in a rise of 0.3 pH unit.

The buffer efficiency of HCO_3^-/H_2CO_3 is further enhanced by the presence of erythrocytes. As CO_2 diffuses into cells, the H_2CO_3 reacts with Hb forming HCO_3-, which then enters the plasma in exchange for Cl^-. This is not contingent upon deoxygenation of Hb, but is achieved more readily and with even less pH change when deoxygenation occurs simultaneously. Conversely, lowering the CO_2 tension results in a reversal of this process, with consequent diminution of plasma $[HCO_3$-]. Although the situation does not arise under physiological conditions, only in the presence of red cells does the total CO_2 content of plasma fall to zero at a pCO_2 of zero mm. This is possible because there is sufficient Hb in whole blood to permit the following series of reactions.

$$H^+ + HbO_2 \longrightarrow O_2 + HHb^+$$

$$HHb^+ + HCO_3^- \rightleftharpoons Hb + H_2CO_3 \rightleftharpoons H_2O + CO_2$$

The normal value of 20 : 1 of the ratio of HCO_3-/H_2CO_3 is also maintained through respiratory regulation of the pH of extracellular fluid. In addition to the automatic self adjustments made possible by intracorpuscular Hb (as described above), the body possesses two further safeguards (the respiratory apparatus and the kidneys) which by their control of plasma [H_2CO_3] and [HCO_3^-] respectively serve in auxiliary fashion to maintain constant pH of blood plasma. Unlike the [HCO_3^-] the [H_2CO_3] is determined solely by the partial pressure of CO_2 in the alveolar air in equilibrium with extracellular fluid.

This, in turn, is dependent upon the rate at which CO_2 leaving pulmonary blood is diluted with atmospheric air and hence upon the rate and depth of respiration. The latter are regulated by the nervous system at the respiratory centre which is sensitive to pH and pCO_2 of the extracellular fluid. When the pH falls below the normal because of diminished [HCO_3^-], respiration is stimulated, lowering alveolar pCO_2 and hence extracellular [H_2CO_3]. This tends to return the HCO_3–H_2CO_3 ratio to its normal value of 20 : 1 and thus to restore pH toward 7.4. With a high plasma pH, the respiratory rate falls, alveolar pCO_2 and hence plasma [H_2CO_3] rise, and the pH moves toward 7.4.

But perfect compensation is not attained since the increased plasma [H_2CO_3] opposes the effect of elevated pH on the respiratory centre. If the respiratory rate falls sufficiently, the diminished pO_2, sensed by the carotid body chemoreceptors, could be a stimulus for increased respiratory activity.

As mentioned earlier the pH is dependent not on absolute concentrations but solely on the HCO_3^-/H_2CO_3 ratio. The buffer systems of plasma can withstand the addition of 16 meq of acid or 29 meq of alkali per litre and still maintain the pH within the range compatible with life. With pulmonary compensation the normal pH range can be maintained despite addition of 23 meq of acid or 80 meq of alkali per litre of plasma.

Phosphate Buffer : Although the contribution of phosphate buffer ($HPO_4^{2-}/H_2PO_4^-$) to the buffering power of the plasma is negligible because their plasma concentration is very low

(3 mg/100ml), this buffer system is present (as is HCO_3^-/H_2CO_3) in the tubular fluid of the kidney. The Henderson-Hasselbalch equation for this buffer system may be written as :

$$pH = pK + \log \frac{[salt]}{[acid]}$$

$$pH = pK_{phos} + \log \frac{[HPO_4^{2-}]}{[H_2PO_4^-]}$$

$$7.4 = 6.8 + 0.6$$

$$0.6 = \log 4$$

Thus, phosphate equilibrium in plasma is such that 4 parts of the so-called *basic phosphate* (HPO_4^{2-}) exist to 1 part of *acidic phosphate* ($H_2PO_4^-$) at a pH of 7.4, for pK_{phos} is 6.8.

At a urinary pH of 5.8, ten parts of phosphate are in the 'acidic' form and one part in the 'basic' form ($5.8 = 6.8 + 1^-$), for each phosphate ion excreted as $H_2PO_4^-$ one Na^+ ion is saved and one H^+ ion is excreted. The phosphates are by far the most important buffers in the urine, although in acid urine the formation of NH_4 from NH_3 and H^+ helps to keep the $[H^+]$ lower than it would otherwise be.

Thus, phosphate buffer system contributes little to the total buffering capacity of blood because the blood phosphate concentration is very low compared with the amount of protein present. But the phosphate buffer pair is a major outlet for H^+ via the urine which has a relatively high phosphate content. Hence the pH of the urine is provided by the excretion of H^+ into the tubular fluid.

Oxyhaemoglobin Buffer : The buffer pair formed by HbO_2^- /$H.HbO_2$ *i.e.*, oxyhaemoglobin anion/oxyhaemoglobin (acid) is also important. It is well known that uptake of CO_2 from the tissue depends upon the following reactions :

$$CO_2 + H_2O \rightleftharpoons H_2CO_3 \rightleftharpoons H^+ + HCO_3^-$$

Thus, the uptake of CO_2 from the tissues is dependent upon the supply of hydrogen acceptors so that the reaction can proceed towards the right. The most important hydrogen acceptor in blood is the haemoglobin ion. Most of the power of haemoglobin in mopping up hydrogen ions in the range pH 7.0 to 7.70 comes from the dissociation of its imidazole groups [haemoglobin contains 38 molecules of histidine (β-imidazole α-amino-propionic acid) per molecule]. The dissociation tendency of the imidazole group like that of any weak acid is largely dependent upon the pH of the solution. As the pH rises, the dissociation is reduced and as a result the H^+ ions added are 'mopped up'.

Oxyhaemoglobin dissociates more completely than does reduced haemoglobin. As a result, reduced haemoglobin produces less H^+ at a given pH than does oxyhaemoglobin. Thus, when oxyhaemoglobin solution is reduced the pH of the solution falls, *i.e., reduced haemoglobin* is a better H^+ acceptor than oxyhaemoglobin. This is of great importance in blood physiology because the entrance of CO_2 from the tissues to the capillary blood is accompanied by the simultaneous reduction of oxyhaemoglobin. As the uptake of CO_2 depends upon H^+ acceptors, this increase in H^+ acceptance by the reduced form facilitates the uptake and buffering of CO_2. For each mMol of oxyhaemoglobin reduced, about 0.7 mMol of H^+ can be taken up ; as a result 0.7 mMol CO_2 can enter the blood without any change in pH.

As arterial blood enters the tissues, CO_2 diffuses into the erythrocytes, thus potentially lowering the pH and the affinity of Hb for O_2 as shown in the following reactions.

$$CO_2 + H_2O \rightleftharpoons H_2CO_3 \rightleftharpoons H^+ + HCO_3^-$$

$$H^+ + HbO_2 \rightarrow O_2 + HHb^+$$

On the other hand, in the lungs the loss of CO_2, which would potentially raise the pH, increases the affinity of the

Hb for O_2, thus permitting the saturation of Hb with O_2 at a lower pO_2.

$$HHb^+ + O_2 \rightleftharpoons HbO_2 + H^+$$

It should be noted that oxygenation of Hb results in a shift in the apparent pK_a of some acidic groups on the peptide chains from 7.71 to 6.17; thus HbO_2 acts as a stronger acid than Hb. Approximately 0.7 mol of H^+ is released as 1 equivalent of O_2 is bound.

Importance of Buffer Systems

Buffer mixtures are very important in living organisms and the mineral world. An example of a natural buffer is the blood of mammals. It always contains free carbonic acid and sodium carbonate. Therefore the pH of blood is always maintained at 7.4.

The buffering action of soils is very important in agriculture, because plants absorb artificial fertilizers from the soil to change the pH in solutions that they extract from the soil in an unfavourable direction. An imbalance in the buffering action of soil is detrimental to useful micro-organisms living in it.

Buffer solutions are very important in the treatment of domestic sewage, because the micro-organisms which mineralize their organic matter thrive better in a neutral medium. A shift towards acidity or alkalinity inhibits the vital processes in the microbes, thus adversely affecting the working of sewage treatment plants.

Buffers play an important role in the chemical treatment of water to separate it from suspended matter of coagulation. The higher the buffer capacity of the treated water, the more efficient its purification with a hydrolyzing coagulant. The buffer capacity of natural water accounts for its neutralizing power.

Buffers are widely used in volumetric analysis. For example, an ammonium buffer ($HH_4OH + HH_4CI$) is used to determine Ca and Mg ions in water (trilonometric method). Buffer mixtures are used to determine the pH of solutions colorimetrically and potentiometrically.

In cellular physiology, buffers are an important way for the cell to maintain constant or smoothly changing conditions. Since cellular metabolism is constantly producing and consuming protons, the pH of the cell in the absence of buffers would change rapidly between low and high levels.

5

The Hormones

Hormones are chemical messengers that coordinate the activities of different cells in multi-cellular organisms. The term *hormone* (from Greek, meaning to spur on) was first used in 1904 by William Bayliss and Ernest Starling to describe the action of secretin a molecule secreted by the duodenum that stimulates the flow of pancreatic juice. Some useful concepts emerged from their work were as follows :

1. Hormones are molecules synthesized by specific tissue, the *glands*.
2. They are secreted directly into the blood which carries them to their sites of action.
3. They specifically alter the activities of responsive tissues which are called target organs or *target cells*.

Categories of Hormones

Hormones are chemically diverse. We can classify them into 4 categories on the basis of their chemical structure :

1. *Small Molecules Derived from Amino Acids* : epinephrine, thyroxine.

2. *Polypeptides* or *Proteins:* oxytocin, insulin, thyroid-stimulating hormone and many growth factors.
3. *Steroid hormones :* made from cholesterol, *e.g.*, cortisol.
4. *Eicosanoids :* formed from arachidonate, a C_{20}-poliunsaturated fatty acid, *e.g.*, prostaglandin PGA,.

Role of Cyclic Amp as a Second Messenger in Hormone Action :

Discovery of Cyclic AMP : Earl Sutheland, Leon Heppel, and David Lipkin working independently in 1950s discovered a new nucleotide adenosine 3', 5'-monophosphate, now commonly referred to as *cyclic AMP* or cAMP.

Synthesis of Cyclic AMP : Cyclic AMP is formed from ATP by the action of *adenylute cyclase,* an integral membrane protein.

$$\text{ATH} \xrightarrow{Mg^{2+}} \text{cyclic AMP + PPi + } H^+$$

This reaction is slightly endergonic having a AG°' of about 1.6 kcal/mol. It is driven in the direction of the synthesis of cyclic AMP by the subsequent hydrolysis of *pyrophosphate.*

Degradation of Cyclic AMP :

Cyclic AMP is degraded by a specific *phosphodiesterase,* which hydrolyze it to AMP.

$$\text{Cyclic AMP + } H_2O \xrightarrow{Mg^{2+}} \text{AMP + } H^+$$

This reaction is highly exergonic having a Δ G°' of about—12 kcal/mol. yet cyclic AMP is a very stable compound in the absence of the phosphodiesterase.

Cyclic AMP is a Second Messenger in the Action of Many Hormones : Sutherland's work led to the concept that cyclic AMP is a *second messenger* in the action of some hormones. The first messenger is the hormone itself. The essential features of this concept are :

1. The plasma membrane contains receptors for hormones.
2. The combination of a hormone with its specific receptor leads to the activation of a *denylate cyclase,* which is also bound to the plasma membrane.
3. The increased activity of adenylate cyclase increases the level of cyclic AMP in the cytosol.
4. Cyclic AMP then acts inside the cell to alter the rate of one or more processes.
5. *The most important feature of the second messenger model is that the hormone need not enter the cell.* Its impact is made at the cell membrane. The biological effects of the hormone are mediated inside *the* cell by cyclic AMP rather than by the hormone itself.

The second messenger model makes several predictions that can be experimentally tested :

1. Adenylate cyclase in a target cell is stimulated by hormones affecting that cell. Cells that do not show a characteristic biological response to a particular hormone do not increase their cyclase level in its presence.
2. The change in concentration of cyclic AMP in a target cell occurs at the same time as the response to hormonal stimulation. Variations in hormone levels match by variations in the concentration of cyclic AMP.
3. Inhibitors of the phosphodiesterase such as theophylline and caffeine act synergistically with hormones that use cyclic AMP as a second messenger.
4. The biological effect of a hormone are mimicked by the addition of cyclic AMP or a related compound to the target cells.

Experiments based on these criteria have proved that *cyclic AMP is a second messenger for many hormones in addition to epinephrine and glucagon.* These are as follows :

Hormones using cyclic AMP as a second messenger

1. Calcitonin
2. Chorionic gonadotropin
3. Corticotropin
4. Epinephrine
5. Follicle-stimulating hormone
6. Glucagon
7. Luteinizing hormone
8. Lipotropin
9. Melanocyte-stimulating hormone
10. Norepinephrine
11. Parathyroid hormone
12. Thyroid stimulating hormone
13. Vasopressin.

Mechanism of second messenger concept of cyclic AMP

1. *A guanyl-nuceotide-binding protein is an intemediary in the activation of adenylate cyclase.* This signal coupling protein was named G *protein* (G).
2. The hormone-receptor complex does not directly stimulates adenylate cyclase. Rather, the activated receptor stimulates the G protein, which carries the excitation signal to adenylate cyclase. This signal-transduction process takes place in the plasma membrane.
3. Alfred Gilmans has shown that the G protein is a peripheral membrane protein consisting of α β and γ subunits. It interconverts between a GDP form and a GTP form. The GTP form activates adenylate cyclase, whereas the GDP form does not in the absence of hormone, nearly all of the G protein is in the inactive GDP form.

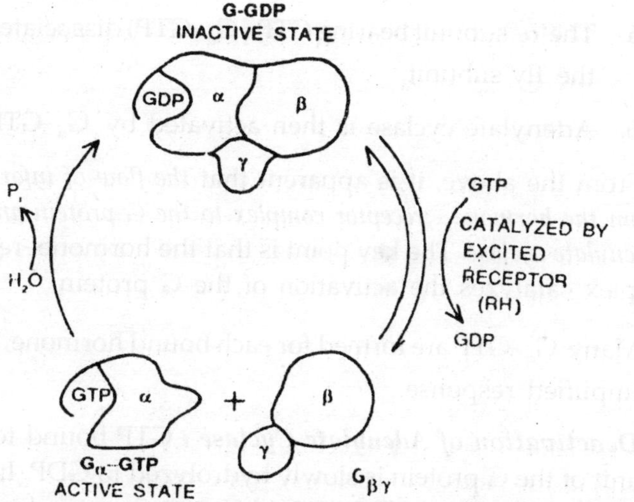

G protein inter-converts between an inactive GDP form and an active GTP form. The exchange of GTP for bound GDP is catalyzed mediated by a hormone-receptor complex.

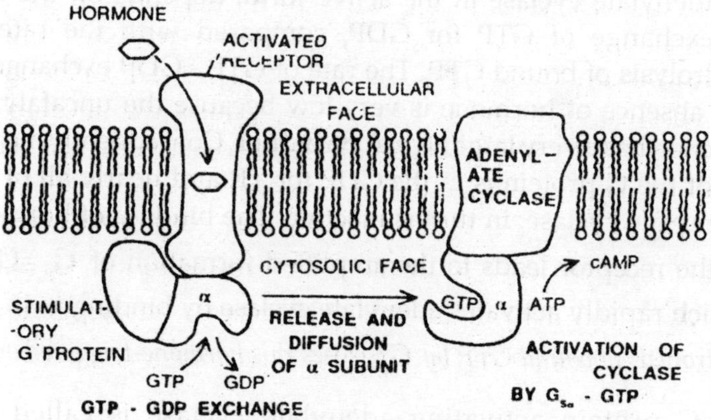

The activation of adenylaic eyclase mediated by a hormoncreceptor complex.

4. *The binding of hormone to the receptor triggers the exchange of GTP for bound GDP.* In this process, the hormone-receptor complex (i) binds to the G protein, (ii) induces the release of bound GDP, and (iii) allows GTP to enter.

5. The α subunit bearing GTP (G_α-GTP) dissociates from the fly-subunit.

6. Adenylate cyclase is then activated by G_α-GTP.

From the above, it is apparent that *the flow of information is from the hormone - receptor complex to the G protein and then to adenylate cyclase.* The key point is that the hormone -receptor complex catalyzes the activation of the G protein.

Many G_α-GTP are formed for each bound hormone, giving an amplified response.

Deactivation of Adenylate Cyclase : GTP bound to the a subunit of the G protein is slowly hydrolyzed to GDP. In other words, the *G protein is GTPdse.* In this way *G protein has a built in device for switching off.*

The proportion of G protein in the GTP state, and hence of adenylate cyclase in the active form, depends on the rate of exchange of GTP for GDP, compared with the rate of hydrolysis of bound GTP. The rate of GTP - GDP exchange in the absence of hormone is very low because the uncatalyzed reaction has a very large activation barrier. Consequently, nearly all of the G protein is in the GDP for TR and nearly all of the adenylate cyclase, in turn is inactive. The binding of hormone to the receptor leads to the amplified formation of G_α=GTP, which rapidly activates adenylate cyclase by binding to it. *The hydrolysis of bound GTP by G_α closes this hormone-triggered cycle.*

G protein activating adenylate cyclase is called the *stimulatory G protein* (Gs).

7. The activated adenylate cyclase in the plasma membrane catalyzes the formation of cyclic AMP from ATP.

$$\text{ATP} \xrightarrow[\text{Mg}^2]{\text{Adenylate cyclase}} \text{Cyclic AMP} + \text{PPi} + \text{H}^+$$

8. The increased intercellular level of cyclic AMP activates a *protein kinase.* This kinase is inactive in the absence of cyclic AMP. The binding of cyclic AMP allosterically stimulates the protein kinase.

9. In skeletal muscle, this protein kinase phosphorylate both *glycogen synthase* (rendering it inactive) and *phosphorylase kinase* (rendering it active). In this way, cyclic AMP stimulates glycogen break down and stops glycogen synthesis. A similar mechanism operates in the liver.

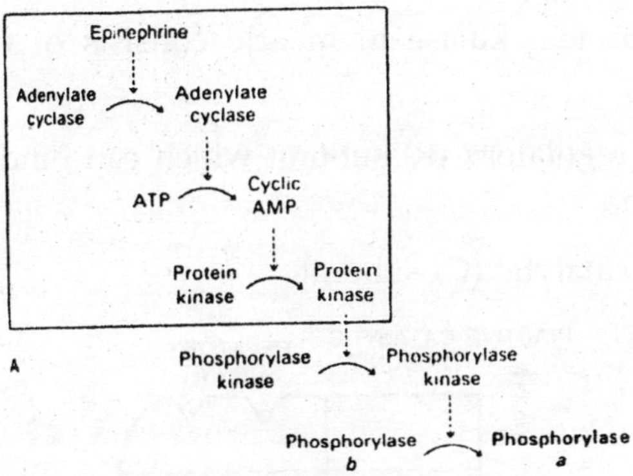

A. Glycogen degradation

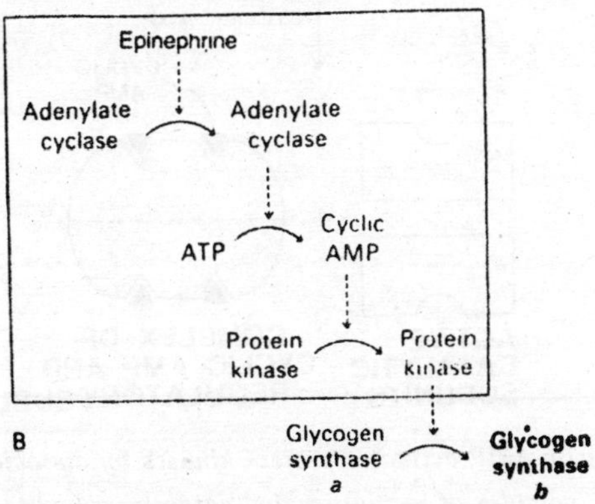

10. All *known effects of cyclic AMP in eukaryotic cells results from the activation of protein kinases.* These protein kinases, which are activated by cyclic AMP concentrations of the order of 10^{-8}m. modulate the activities of target proteins by phosphorylating seine residues in Arg-Arg-X-Ser-X sequences (X denotes any residue). Phosphorylation of target proteins generally changes their conformation and, in enzymes, increases or decreases their activity, thus producing changes in metabolism.

Mechanism of activation of the cyclic-AMP controlled kinase.

The protein kinase in muscle consists of two kinds of subunits :

(i) a regulatory (R) subunit which can bind cyclic AMP and.

(ii) a catalytic (C) subunit,

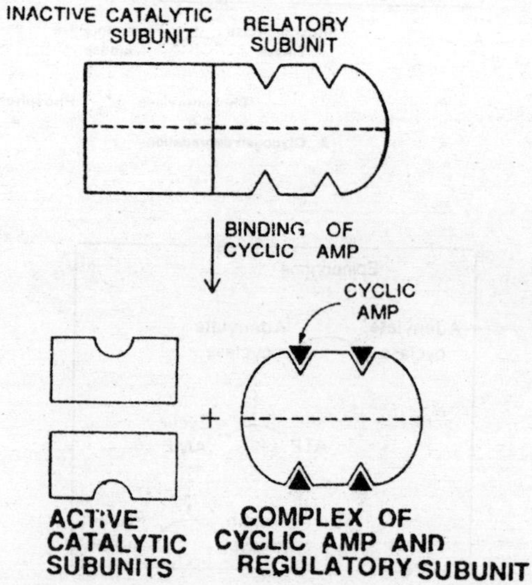

Cyclic AMP activates protein kinases by dissociating the complex of regulatory and catalytic subunits.

In the absence of cyclic AMP, the regulatory and catalytic subunits form an R^ complex that is enigmatically inactive. The binding of two molecules of cyclic AMP to each of the regulatory subunits leads to the dissociation of the R^ complex into an R, subunit and two C subunits. These free catalytic subunits are then enzymatically active. It is inferred, therefore, that *the binding of cyclic AMP to the regulatory subunit relieves its inhibition of the catalytic subunit. Cyclic AMP acts as an allosteric effector.*

The adenylate cyclase cascade amplifies the hormonal signal

The concentration of many hormones in the blood is of the order of 10^{-10}m. The adenylate cyclase cascade greatly amplifies such weak hormonal signals. Three steps in the cascade provide a high degree of amplification :

1. Each hormone-receptor complex catalyzes the formation of many G_{α}-GTP.
2. Many molecules of cyclic AMP are formed by an activated adenylate cyclase
3. Each cAMP-activated protein kinase can alter the activity of many molecules of each target protein

Cyclic AMP affects a very wide range of cellular processes.

For example :

1. It enhances the degradation of storage fuels.
2. It increases the secretion of acid by the gastric mucose.
3. It leads to the dispersion of melanin pigment granules.
4. It diminishes the aggregation of blood platelets.

Hormone-like Substances

Besides hormones, a number of other hormone-like substances are also produced. These are *(A)* prostaglandins, (B) pheromones, (C) neurohormones, (D) neurotransmitters, and *(E)* neuropeptides. However, they differ from hormones in several ways and their effects are wide ranging.

Prostaglandins: Prostaglandins (PG) are a group of biologically active lipids which appear to play important roles in many reproductive processes. These substances were first found in human seminal fluid and the name 'prostaglandins' was given to them by *von-Euler* (1933-34). Actually the term was a misnomer because the main source of prostaglandins in semen is the seminal vesicles and not the prostate gland. Since then prostaglandins have also been found in the semen of other vertebrates like sheep, monkey and goat. In human beings they occur in other body fluids also including the menstrual fluid of women, and in tissues of pancreas, stomach, lung, kidney, brain, spinal cord, iris, and placenta, etc.

Prostaglandins differ from all other hormones in being fatty acids (that is, incorporating the COOH group) and all those so far examined contain 20 carbon atoms. The primary prostaglandins (PGE and PGF) are synthesized in the body from some polyunsaturated fatty acids by the formation of a five-member ring and the incorporation of 3 oxygen atoms at certain positions. Thus, all prostaglandins have the same basic "skeleton", *prostanoic acid* and are derived from essential fatty acids by cyclization and oxidation (Fig.). There are two main sets of prostaglandins, the E

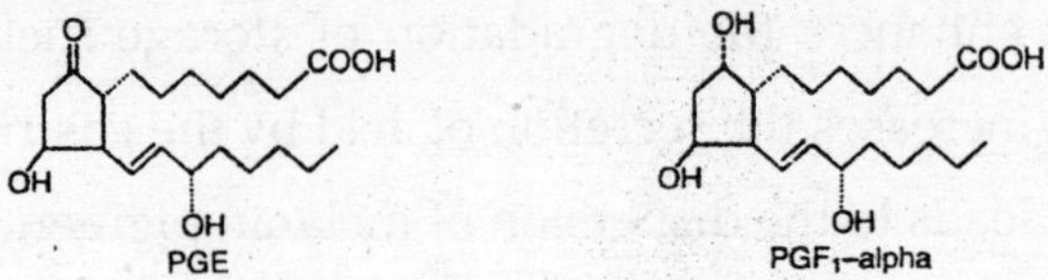

The structure of PGE$_1$ and PGF$_1$ prostaglandins.

and F series, each having three members E_1, E_2 and E_3 and $F_1\alpha$, $F_2\alpha$, and $F_3\alpha$. The two series differ from each other in that the PGE forms contain a keto oxygen at C-9 and a hydroxyl group at C-ll, whereas the PGF structures have a hydroxyl group at both positions in the cyclopentane ring. Prostaglandins of the 1, 2 and 3 series possess 1, 2, and 3 double bonds respectively. The primary structures can thereafter be converted into other members (PGA, PGB) of the prostaglandin family.

Biological Role of Prostaglandins : Prostaglandins interact with specific receptors that are localized in the plasma membranes of cells. The receptors for the PGE forms are different from those for the PGF. Prostaglandins are among the most potent of all the known biological substances producing marked effects in extremely small doses. In general, their effects are based on the regulation of the activity of smooth muscles, of secretion (including some endocrine gland secretions), or of blood flow. It seems possible that prostaglandins show their effects by affecting adenylcyclase and cyclic AMP. Through these actions prostaglandins are capable of affecting many aspects of human physiology, and this accounts for their importance in pharmacology.

According to *Pike* (1971), prostaglandins may be formed by the lungs during anaphylaxis, by the kidney when its blood supply is restricted, by the skin during human allergic contact eczema and certain experimental inflammatory conditions. This suggests that they may play an important role not only in normal physiological activities but also in certain pathological conditions. The wide range of their physiological effects also includes blood pressure regulation, lipid and carbohydrate metabolism, and tissue inflammation in response to disease or wounds.

Effects on Female Reproductive System : The effects of prostaglandins on female reproductive system are very striking. *Bygdeman* (1971) has shown that an intravenous injection of a very low dose of either PGE_2 or PGF_2 alpha stimulates the contration of the uterus and can be used to initiate labour in women or to produce abortion. Drs. *Sune Bergstom* and *Bengt Samuelson* have won the 1982 Nobel prize with *Dr. John Vane* for the development of a home abortion treatment based on prostaglandin at Stockholm's Karolinska Institute. The technique is an abortion suppository which involves the prostaglandin's chemical action causing cramping of the uterine muscles and the abortion of the foetus. However, the technique has been criticised by Dr. *P.H. Petersen.* Now there seems to be a strong possibility that prostaglandins or their synthetic analogues may be used as important agents to control population.

Effect on Blood Flow Control and Immune Responses: Prostaglandins are released into the local tissue fluids and the circulating blood under both physiological and pathological conditions. Although some of them cause vasoconstriction, most of men seem to be mainly vasodilator agents. Administration of PGE_1 or PGE_2 intravenously augments blood flow by inducing a generalized vasodilation, with decreased peripheral resistance. The inhibitory actions of prostaglandins or immune responses are due to inhibition by PGE_1 of antigen activation of thymus-dependent lymphocytes, a process of importance in immunologic responses.

Effect on Intestinal Secretions: The prostaglandins inhibit gastric secretion but stimulate pancreatic and mucous secretion in the intestine and markedly increase intestinal motility. *Andre Robert's* experiments on rats and dogs have shown that the stomach normally produces E prostaglandins which serve to regulate gastric secretion under normal circumstances and thus protect the stomach against ulceration. If his studies are confirmed in man, the administration of E prostaglandins may be helpful in treating ulcer patients who lack this normal protection. Prostaglandins also appear to enhance ion flux across epithelial membranes.

Effect on Nerve Fibres : Recently it has been shown that production and release of prostaglandins can be evoked by stimulation of nerves and there exists an interrelation between the prostaglandins and the substances involved in the transmission of nerve impulses. Normally, prostaglandins play a negative feedback role on the release of noradrenaline by sympathetic nerve fibres. They also act as modulators of the actions of chemical transmitters released from nerve endings and of hormones from endocrine glands.

Effect on inflammation : Among the many tissue products responsible for tissue changes during inflammation, prostaglandins are considered important. Prostaglandins produced at a local site as a result of injury, burns, etc., act as mediators of the inflammatory response.

Effect on Endocrine Glands : It has been also suggested that prostaglandins act as controlling agents which modify the action of hormones on target cells. Several endocrine glands such as the thyroid, adrenal cortex, ovary, and parathyroids respond to prostaglandin administration with increased secretion of their hormones. In addition, through their effects on adenylate cyclase the prostaglandins may modulate hormonal actions. The effects of many hormones that increase the activity of the adenyl cyclase system are antagonized or enhanced by PGE_1 or PGE_2. For example, LH, by mobilizing triglycerides and phospholipids in the ovary, might increase the synthesis of prostaglandins through a greater concentration of their precursor unsaturated fatty acids. The prostaglandins so synthesized could modify the metabolic consequences of LH on the ovarian cell.

Effect on Reproduction : The secretory epithelium of seminal vesicles secretes a mucoid material containing fructose, citric acid and other nutrient substances as well as large quantities of prostaglandins and fibrinogen. The prostaglandins are believed to aid fertilization in two ways. First, by reacting with the cervical mucus to make it more receptive to sperm movement. Second, by possibly causing reverse peristaltic contractions in the uterus and fallopian tubes to move the sperm toward the ovaries.

Pheromones: The term *'pheromone'* was originally applied to the sex *attractants* of insects but now it includes all kinds of chemical substances released into the environment to aid in the integration of behaviour among members of the same animal species. According to *Karlson* (1960), chemical signals exchanged between individuals of the same species are pheromones, which cause a specific reaction either in the form of a behavioural response or in a particular developmental process.

The pheromones differ from hormones in several respects :

(a) they are produced by the exocrine glands,

(b) they are transmitted via the external environment,

(c) they are typically more species-specific than the hormones, and

(d) they evoke specific behavioural, developmental, or reproductive responses in the bodies of other individuals of the same animal species, whereas the hormones regulate organism's own physiology.

Thus, pheromones are not hormones but the activity of the glands that produce them often depends on hormonal stimulation. The pheromones act, in general, on the hypothalamus via olfactory pathways and thereby alter the behaviour or the physiology of the recipients. The source of such substances, also called ectohormones, are a variety of glands found in the skin or other places in insects and mammals.

On the basis of their mode of action pheromones are grouped into two categories. These are :

Signaling or Releaser Pheromones : The pheromones that produce a simple "releaser" effect *i.e.*, a single, rapid, specific response mediated directly by the central nervous system, are called *releaser pheromones*. To this category belong the sex attractants (sex pheromones), and the trail and alarm substances.

The ant *Iridomyrmex pruinosus* uses 2-heptanone as an alarm pheromone. Among insects the sex attractants have been identified in silk moth, gypsy moth, cockroach, honey bee and mosquitoes. Female silkworm moths *(Bombyx mori)* secrete a sexual attractant, *bombykol :*

$$CH_3—(CH_2)_3—CH = CH —CH = CH—(CH_2)_8—CH_3OH$$

Only the antennal chemoreceptors of the male are excited by this substance, those of the female are insensitive. When sensed by the male, a behaviour pattern is initiated such that the male flies to the female.

Many studies have been made on the sexual attractants of the gypsy moth *(Porthetria dispar)* whose females secrete the pheromone *glyplure*:

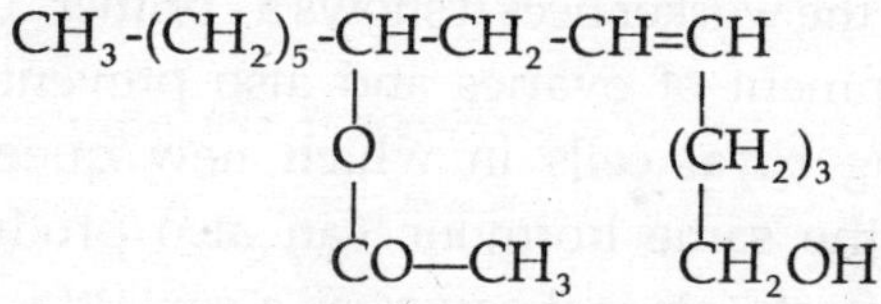

Depending on the wind speeds, gypsy moth females can attract males from up to 4500 m away.

With respect to communication, pheromones cause a variety of actions on the part of individuals. They may cause the dispersal or aggregation of a population. When populations of flour beetle *Tribolium confusum* become dense, *quinones* released from the abdominal glands cause the individuals to disperse. Insect aggregation, through the action of pheromones, is examplified by the behaviour of honey bees and other social insects *(Renner,* 1960; *von Frisch,* 1967).

When worker honey bees find a good source of food, they release pheromones from an abdominal scent gland to aid other workers in finding the food. Similar activity is found when a new hive is started. One component of the pheromone secretion is *geraniol,* but at least four other compounds are present *(Bock* and *Schearer,* 1962).

Primer Pheromones : The pheromones which trigger a chain of physiological events, or those that activate a long series of neuroendocrine events that develop slowly and require prolonged stimulation, are called *primer pheromones*. The primer pheromones often act by physiological inhibition but they can produce the opposite effect. For example, the adult males of the migratory locust secrete a volatile substance from their skin that accelerates the growth of young locusts and thus plays an important role in the formation of migratory locust swarms.

An important example of primer pheromones is the secretion produced by the mandibular glands of the honey bee queen. This substance has been recently identified as *9-ketodecanoic acid* and synthesized (Fig.). When this pheromone ingested by the worker bees it shows a "primer" effect, inhibiting the development of ovaries and also preventing them from constructing royal cells in which new queens are reared. However, the same hormone can also produce releaser or signaling effects, since it serves as a sex attractant during the queen's nuptial flight.

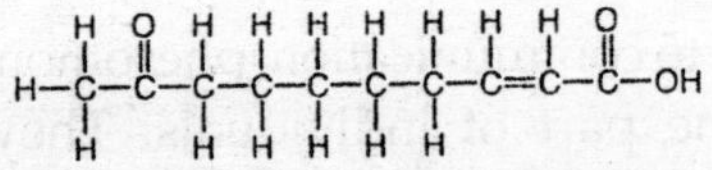

Honey bee queen substance (9-ketodecanoic acid).

Primer pheromones and sex attractants have been identified in a wide variety of terrestrial animals, including mammals. The sex attractants of civet *(Viverra zibetha)* and muskdeer *(Moschus moschiferus)* are called *civetone* and *muskone* respectively. These high molecular weight compounds contain about 16-17 carbon atoms and are highly volatile and diffusible. The civetone is secreted from the para-anal glandular pouch, and is used for defence and also a territory marker (*Wilson* and *Bossert*, 1963).

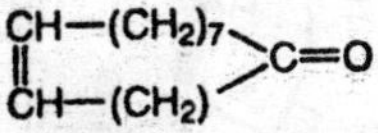

Civetone

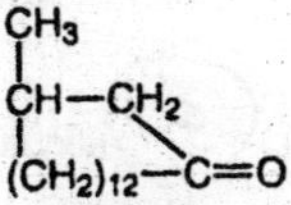

Muskone

The muskone functions as a sexual attractant. Another example of a mammalian pheromone is the one produced by male mice. *Helen Bruce* has described that the odour of a strange male of a different strain blocks the pregnancy of a newly impregnated female mouse *(Bruce effect).*

Pheromones in Humans : Now it is known that humans also produce sex pheromones. *Le Magnen* has reported that the odour of *Exaltolide* (a synthetic analogue of mammalian sex athractants, 14-hydroxy tetradecanoic acid) is perceived only by the sexually mature females.

The main source of human sex pheromones are some odoriferous glands in the skin, including the mammary glands, axillary glands, circumoral glands, circumanal glands, eyelid glands, glands of the exteral auditory canal and the glands of public region, etc. Saliva, urine, tears and vaginal secretions are other probable sources of human pheromones. It appears that pheromones also operate in the regulation of sexual behaviour in primates.

Neurohormones: The chemical messengers released by neurosecretory cells are called neurohormones. The neurohormones of the vertebrate neurohypophysis are known as *oxytocin* and *vasopressin. Harris* (1966) has suggested that the neurohormones are released by the neurosecretory cells in the hypothalamus and conveyed to the anterior pituitary by portal vessels where they regulate the output of hormones. Seven different neurohormonal substances that are peptides have been defined. They promote the release of thyrotrophin, ACTH, growth hormone, FSH and LH respectively, and inhibit the release of prolactin and melanocyte-stimulating hormone from the adenohypophysis.

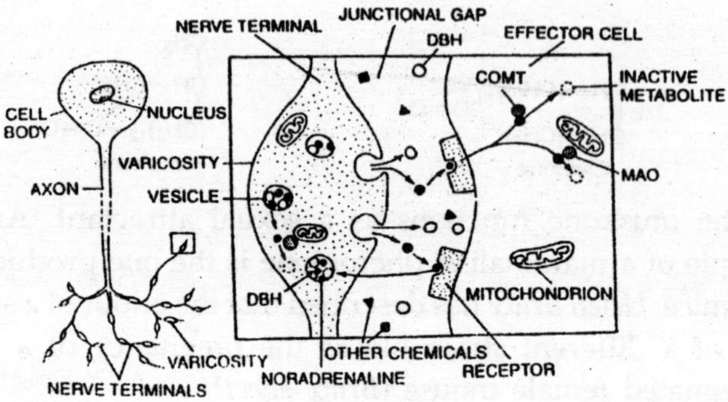

The release and inactivation of the neurotransmitter, noradrenaline.

Thus, neurohormones may perform regulatory functions directly by employing nonendocrine targets such as the contractile elements of the mammary glands, uterine muscle or kidney tubules, or indirectly by activating the endocrine glands, which in turn bring about adjustments by their own hormones.

Neurotransmitters: A neurotransmitter (neurohumor) is a chemical substance that is discharged from a nerve fibre ending, and then it reaches and is recognized by a receptor on the surface of a postsynaptic nerve cell (or effector cell) and either stimulates or inhibits the second cell. In many instances the neurotransmitter is a *monoamine* (a substance the nerve cell synthesizes by making minor changes in an amino acid). Sometimes the neurotransmitter is the amino acid itself.

In any case its release and reception takes milliseconds, whereas the release of the hormone, its circulation in the blood and its influence on the target cells takes more time (minutes or hours). Thus, neurohumors are chemical messengers, but they differ from ordinary hormones because they do not depend on blood for their circulation. They also differ from neurohormones with respect to their source ; they are the product of the conventional nerve cells instead of coming from the neurosecretory cells.

However, a few differences mentioned earlier between the neurotransmitter and the hormone do not serve the whole purpose. For example *norepinephrine,* a monoamine neurotransmitter, turns out to be also a hormone (it is released by the gland cells in the medulla of the adrenal gland). Conversely, *vasopressin,* a peptide hormone, turns out to be also a neurotransmitter (nerve cells in the hypothalamus rely on vasopressin to signal other nerve cells in the brain).

According to *Julius Axelrod* (1974), a neurotransmitter should fulfil the following criteria.

(a) Nerve cells should have the enzymes required for the synthesis of the neurotransmitter.

(b) On stimulation nerve cells should liberate the neurotransmitter which should then react with a specific receptor on the postjunctional cell and produce a response.

(c) There should be a mechanism to terminate the action of the neurotransmitter rapidly.

On the basis of these criteria *acetylcholine* and *noradrenaline* (norepinephrine) have been established as the neurotransmitters and the nerve fibres that produce them are called "cholinergic" and "noradrenergic" respectively. Of the neurotransmitters secreted by enteric neurons, acetylcholine most often excites gastrointestinal activity; norepinephrine, on the other hand, almost always inhibits gastrointestinal activity. However, some other chemicals fulfil the above criteria only partly. They are dopamine, adrenaline, serotonin, etc.

Neurohumors are involved in a variety of physiological processes, some of which are given below :

(i) They make the heart beat faster (sympathin or norepinephrine) or slower (acetylcholine).

(ii) They are responsible for synaptic transmission and neuromuscular transmission.

(iii) They cause the glands to synthesize hormone-producing enzymes or to secrete the hormones.

(iv) Through these chemicals the brain regulates movement and changes in the mood and behaviour.

(v) Perhaps they also play a role in the diseases such as depression, schizophrenia, motor disorders, and hypertension, etc.

Neuropeptides: Recently *Bloom* (1981) has described that the *neuropeptides* are short chains of amino acids that are active in the nervous system. In some cases they transmit signals between nerve cells and also serve the body as a hormone. They differ from small-molecule, rapidly acting transmitters (acetylcholine, etc) in their mode of action. They usually cause more prolonged actions; for example hypothalamic-releasing hormones (LH-releasing hormone, thyrotropin-releasing hormone, somatostatin) and pituitary peptides (ACTH, oxytocin, vasopressin etc.). Some neuropeptides that act on gut and brain are *Leucine enkephalin, neurotensin,* etc.

Bloom (1981) has described that the nerve cells containing *endorphin* are found in the rat in the arcuate nucleus of the hypothalamus. The axons of the cells project along pathways in the brain similar to those of the cell containing oxytocin. However, the nerve cells containing enkephalin are widespread in the brain of the rat.

6

The Pheromones

Pheromones are chemical substances secreted by an animal externally with specific effects on the behaviour or physiology of another individual of the same species. The term pheromone was coined by *Karlson* and *Luscher* in 1959.

The immediate and invariable response produced by pheromones (from the Greek *pherein,* to carry; *hormone,* to excite) is comparable to the action produced by hormones. But unlike hormones, which work *within* an individual, these chemicals also known as ectohormones, work *between* the individuals.

When these chemicals are used to communicate with other species, and the receiver benefits, they are called *Kairomones.* If it is the sender who benefits, the signal is an *Allomone.*

These substances may be secreted rather generally by the skin or by specialized glands and, similarly, their detection by the recipient individual may be simply by ingestion or by specialized chemi-receptors such as those on the antennae of the silkworm moth.

Characteristics of Pheromones :

(1) Pheromones secreted by most of the animals are proteins of low molecular weight or fatty acid derivatives.

(2) Pheromones are generally volatile substances and spread in the nature by vaporization and diffusion.

(3) The maximum effect of pheromones depends on their concentration and vicinity to the animals effected.

(4) Pheromones having a long carbon chain serve as sex attractants whereas small carbon chained pheromones alert the animals against dangers.

(5) Sex attracting pheromones are generally released during the night and they can travel along the wind upto several kms.

Pheromones in Insects

E.F. Knipling and M. Viroza were the pioneers who identified and extracted pheromones. They also synthesized artificial pheromones and used them in insect control.

Classification of Pheromones : Pheromones are classified (i) according to the response of the recipient and (i) on the basis of their biological function.

According to Response of the Recipient

(i) *Releaser Pheromones :* These are typically odorous and act directly on the central nervous system of the recipient. Releaser pheromones produce an immediate change in the behaviour of the recipient, *e.g.*, the sexattractants of moths, and the trail pheromones and alarm substances released by ants.

(ii) *Primer Pheromones :* These act slowly and bring about a chain of physiological events in the recipient without any immediate change in behaviour. In a true sense, the body is "primed" for new biological activity, and control the number of reproductive members in social insects like termites, ants and bees.

On the basis of Biological Function

(i) *Trail Pheromones* : The social insects like ants and bees secrete a trail pheromone. The workers of ants such as *Solenopsis,* on their way back after finding a source of food, release a highly specific, volatile "trail pheromone" marking the trail in the form of a series of scent spots so that fellow ants find their way to the source of food. The pheromone is produced in Dufour's gland (poison gland), from Pavan's gland or from the hind gut. Accordingly, the secretion is released through the sting, the edge of the 6th abdominal sternite or the anus, as the case may be.

Worker bees, on finding a source of food, secrete a C_{10} alcohol called *geraniol* which leads other worker bees to the food.

(ii) *Alerting or Alarm Pheromones* : Many species of ants secrete "alarm pheromones" if the colony is disturbed. The workers are stimulated by a pheromone from the mandibular or anal glands which is released by any individual of the colony being disturbed. The pheromone transmits the alarm to the other ants in the area. As a result, they react by rushing about in a very agitated manner or assume an aggressive posture.

(iii) *Sex Pheromones* : Olfactory sex pheromones, also called "sex attractants" have now been established in more than thirty species of moths, butterflies, flies, beetles and cockroaches. In general males are attracted to females, but in the case of boll weevil, *Anthonomus grandis,* the males produce the attractant. In some other cases, aggregations occur in which both sexes are brought mutually together by these substances. For example, aphrodisiacs of many *Lepidoptera* and *Hymenoptera* are usually produced by we male after it has responded to the attractant of the female. Pheromones of the male then stimulate the copulatory behaviour of the female which is essential for successful mating.

The recipients of the sex pheromones respond in one of the two ways

1. They may be stimulated to approach the releasing insect from some distance.

Or

2. They may be stimulated to perform some close range sex foreplay like courtship.

A few examples of sex pheromones are as follows :

1. *Bombykol* : Female silkworm moth *Bombyx mori* produces a pheromone called *bombykol* which attracts males from astonishing distances. It is detected by the male with the help of very fine alfactory sensory cells present on its antennae. As a result, the male starts flying upwind towards the female that is producing the pheromone. When male locates her, they copulate. *Bombykol* active at 10^{12} μ g, is an unsaturated C_{16} fatty alcohol with two double bonds :

$$CH_3 - (CH_2)_2 - CH = CH - CH = CH - (CH_2)_8 - CH_2OH$$

Bombykol

2. *Disparlure* : The female gypsy moth, *Porthetria dispar* secretes a sex pheromone called *disparlure*. Like bombykol, it, too is active to 10^{-12} μ g. Depending on wind speed, gypsy moth female can attract male from upto 4500 metres away.

3. *Antiqueen Pheromone* : In social insects such as honey bees, the queen secretes an antiqueen pheromone which acts on the corpora allata of the developing females and prevent their development into queens.

When the queen dies, the antiqueen pheromone is no longer present, and consequently one or more of the nymphs develop into queens. The same pheromone, however, serves, as a sex attractant to the drones during nuptial flight.

$$CH_3.\ CO(CH_2)_5\ CH = CHCOOH$$

Honey bee antiqueen substance

4. *Cabbage looper sex pheromone.*

$$CH_3(CH_2)_3\ CH = CH\ (CH_2)_6\ O — COCH_3$$

Sex pheromone of cabbage looper, *Trichoplusiani*

5. *Cotton pink boll worm sex pheromone.*

$$(CH_3CH_2CH_2)_2\ C = CH — (CH_2)_2 — CH = CH—(CH_2)_4\ OCOCH_3.$$

Sex pheromone of cotton pink boll worm,

Pectinophora gossypiella

Sex pheromones have been reported in still some other insects like peach stem borer, *sanninoida exitiosa* ; tobacco horn worm, *Protoparce sexia*; sugar cane stem borer, *Diatraea saccharialis*; cucumber beetle, *Biabrotica balteata* and coddling moth, *Carpocapsa domorella.*

(iv) Aggregating pheromones

(v) Area making (terriotoriality) pheromones. Some animals use pheromones to mark their territorial and home ranges.

(vi) Recruiting pheromones.

Pheromones in Mammals

Pheromones are known to play a role in the reproductive physiology of mammals. The earliest chemical studies on mammalian pheromones were carried out on the musk deer. *Muscopyridine* has been found to be the active principle of the odour emitted by the musk deer.

Pheromones in Mice : It has been found experimentally that when female mice are placed in groups of four or more per cage, they show a higher incidence of anoestrus and psuedo pregnancy than when caged individually. If one male mouse is kept in the cage, the odour emitted by the male reduces the frequency of reproductive abnormalities. Further, the smell of the urine of a male, obviously a pheromone, terminates the pregnancy in a newly impregnated female and it becomes sexually receptive again.

Pheromones in Monkeys : In monkeys, the sex drive of the male is greater when he is exposed to a female at the time of ovulation than when he is exposed to female at another time of her cycle. The "pheromone message' sent by the female to the male in this situation is olfactory. This pheromone is vaginal in origin.

Pheromones in Humans : Only little information is available with reference to human pheromones. Certain fatty acids exhibiting pheromonal properties are found in increased amounts in human vaginal secretions at about midcycle. Pleasant, musk-like odour of a human male pheromone, *14-hydroxytetradecanoic acid*, is perceived only by sexually mature women, particularly at the time of ovulation. Males and young girls are relatively insensitive to this substance. Similarly, *copulins* have been identified in human female vaginal secretions that have an arousing effect on the males.

7

The Protoplasm

The term *'protoplasm'* was introduced by *Purkinje* in 1839. *Huxley* (1868) has defined protoplasm as 'the physical basis of life'.

Protoplasm (Gr. *protos* = first, *plasma* = form) is the general term for the living substance that forms the cells. It may be defined as the complex colloidal physicochemical system that constitutes the living matter. However, not everything found in the protoplasm is alive, for it contains many non-living materials. Hence, the mystery of the protoplasm as the living matter of all animal and plant cells does not lie in its chemical elements but in the way they are linked together into compounds and are organised into complex molecules (macromolecules).

Physical Nature and Properties

The physical nature of protoplasm has been a subject of study for many scientific workers of the nineteenth century. The old theories proposed to describe the physical organisation (Fig.) of protoplasm are as follows:

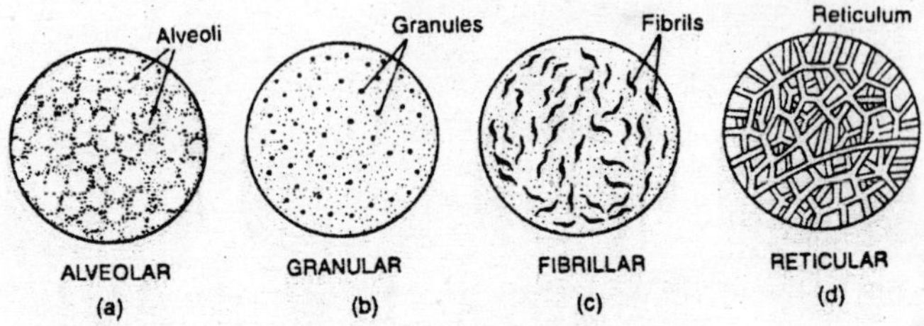

Physical nature of the protoplasm.

Alveolar Theory : According to this theory, proposed by *Butschin* in 1892, protoplasm consists of alveoli or bubbles of a fluid of lesser density distributed in a fluid of greater density. Therefore, protoplasm looks like a foam or an emulsion composed of two liquids.

Granular Theory : This theory was proposed by *Altmann* in 1893. According to him, the protoplasm is made up of tiny granules dispersed uniformly in a homogeneous fluid medium.

Fibrillar Theory : According to this theory, suggested by *Flemming* in 1897, the protoplasm consists of numerous minute fibrils or thread-like structures dispersed in a fluid medium.

Reticular Theory : According to *Klein, Leydig,* and *Heitzmann,* the protoplasm consists of a reticulum or network of fibres, the meshes in the net containing the fluid.

Colloidal Theory

According to the modern theory, the protoplasm is a complex *"colloidal system"*. It is seemingly homogeneous but it consists of a variety of inorganic and organic materials some of which are dissolved but others are suspended in its liquid medium (matrix). Colloidal particles are suspended due to electric charges and *Brownian movement.*

However, protoplasmic colloids have the capacity to assume different consistencies, or change `phase' (state). When solid particles are uniformly distributed, protoplasm has a fluid-like consistency and is said to be in a *sol state* (Fig.). But when colloidal particles adhere together to form a continuous network with liquid particles enclosed in its meshes, protoplasm becomes semisolid or jelly-like and is said to be in a *gel state.* However, under appropriate conditions the sol and gel states are reversible.

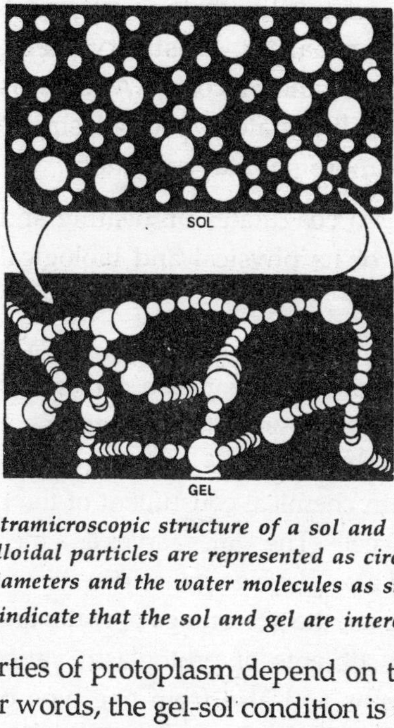

The ultramicroscopic structure of a sol and a gel. The colloidal particles are represented as circles of different diameters and the water molecules as small dots. The arrows indicate that the sol and gel are interconvertible.

Many properties of protoplasm depend on the fact that it is a colloid. In other words, the gel-sol condition is the basis for the physical properties and the behaviour of the protoplasm. Some important physical properties are as follows:

(i) Protoplasm is a semitransparent, viscous, and elastic fluid material.

(ii) Under normal conditions the protoplasm of a living cell is in a state of constant motion (brownian movement, amoeboid movement and cyclosis or cytoplasmic streaming).

(iii) Protoplasm responds to the action of external stimuli such as needle or pin-prick, electric shock, sudden variation of temperature or light, or application of particular chemicals, etc. On stimulation protoplasm contracts but expands again when the stimulating agent is removed. The "contractility" is the most important of all the physical properties of protoplasm.

(iv) Besides irritability, other physical properties of protoplasm are : elasticity, tensible strength, cohesiveness and viscosity. A high viscosity indicates a more gel-like state of protoplasm, and low viscosity shows a more sol-like condition.

(v) Protoplasm coagulates on heating and when killed, it loses all of its physical and biological properties.

Chemical Nature

Normally the active protoplasm consists of more than 75% of water and less than 25% of other substances. The latter include proteins, carbohydrates, lipids and inorganic materials. Thus, water is the main chemical constituent of the protoplasm of all animal and plant cells. The water serves as a medium for all other constituents and also plays a vital role in the metabolic activities of the cell. However, the percentage of water and other inorganic and organic constituents of protoplasm varies in the cells of different organisms and also from one type of cell to another within a single organism.

Inorganic Constituents

There is a large variety of inorganic materials in the protoplasm. They occur freely in solution as ions or combined with other compounds. The most abundant positively charged inorganic ions are H^+, Ca^{2+}, Na^+, K^+ and Mg^{2+} ions. The negatively charged ions that are found in the protoplasm in abundance are OH^-, CO_3^{2-} HCO_3^-, PO_4^{3-}, SO_4^{2-} and Cl^- ions. The protoplasm also contains a large number of mineral or *inorganic salts* such as sulphates, phosphates, chlorides and carbonates of sodium, potassium, magnesium, calcium and iron, etc.

Some gases, O_2 and CO_2, are also dissolved in the water of the protoplasm. For their transport these gases depend on some carriers, water, or salts present in the protoplasm.

The ions and salts of the protoplasm are very essential for life. The distribution of the ions of various salts inside and outside the cell is important in maintaining the osmotic balance in the cell. Certain ions are of particular importance in the organization of protoplasm and its metabolic activities. One of the most significant of them is the phosphate group (PO_4^{3-}), which is associated with phosphoprotein, phospholipid and nucleotides. The main source of energy for the living cells is ATP, which has two high-energy phosphate bonds. Another inorganic constituent, sulphur, is important in the organization of several organic compounds, such as the amino acids, and therefore of several enzymes. Sulphur also provides linkages in the spindle fibre during cell division. The inorganic ions like iron, magnesium, and manganese may serve as co-factors for the activity of specific enzymes. The calcium plays important roles in muscle contraction and blood coagulation.

Organic Constituents

The organic constituents of the protoplasm are as follows :

1. Proteins.
2. Carbohydrates.
3. Lipids
4. Nucleotides.
5. Hormones and Vitamins.

Proteins

The term *'protein'* was derived by *Mulder* in 1840 from a Greek word *proteios* which means "of the first rank" and was defined as the most important of all the known substances in the living cell through which the chief phenomena of life are produced. The most remarkable fact about proteins is that they are not only the main building material of the cell, but they also regulate almost all its activities.

Proteins are high-molecular-weight organic compounds made up of smaller units, which the famous chemist *E. Fischer* (1852-1919) named "amino acids". *Fispher* was awarded Nobel Prize in 1902 for his biochemical work on proteins. The polymers containing a small number of amino acids (2, 3, 4 to about 100) are commonly called *peptides,* whereas those made up of a large number of units (up to 10000, or more) are called proteins.

Structure and Classification of Amino Acids : All amino acids conform to the general formula shown below.

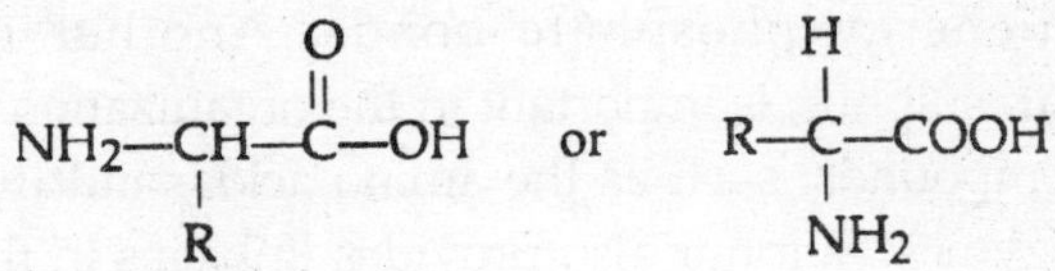

The alpha carbon atom of each amino acid is covalently bonded to four groups :

(1) a hydrogen atom,

(2) an amino group,

(3) an acid group, and

(4) a side chain called an R group.

It is the specific chemical nature of the R group that distinguishes one amino acid from another. The four groups that are attached to the alpha carbon atom of an amino acid are all different except in the case of *glycine,* where R is a hydrogen atom. Therefore, all amino acids except glycine are optically active. The four groups may be considered to lie at the four corners of a regular tetrahedron with the alpha carbon atom at the centre. Accordingly, these groups may be arranged in either of two ways, giving the L and D isomers. All proteins synthesized on ribosomes are formed exclusively from L-amino acids. D—amino acids occur in some antibiotics.

The common amino acids are classified as follows :

(i) *Neutral amino acids* : Glycine (Gly), Alanine (Ala), Valine (Val), Leucine (Leu), Isoleucine (Ile), Serine (Ser), Threonine (Thr).

(ii) *Acidic amino acids* : Aspartic acid (Asp), Glutamic acid (Glu).

(iii) *Amidic amino acids* : Asparagine (Asn), Glutamine (Glu)

(iv) *Basic amino acids* : Histidine (His), Arginine (Arg), Lysine (Lys)

(v) *Aromatic amino acids* : Phenylalanine (Phe), Tyrosine (Tyr), Tryptophan (Trp).

(vi) *Sulphur-containing amino acids* : Cysteine (Cys), Methionine (Met).

(vii) *Secondary amino acids* : Proline (Pro), Hydroxyproline (Hyp).

In addition to the common amino acids there are some rare amino acids. For example, Beta-alanine is found in the Vitamin pantothenic acid; gamma-aminobutyric acid is found in brain tissue and citrulline, ornithine, and homoserine occur as intermediates in metabolism. But these are rarely, if ever, incorporated into proteins.

Some representative amino acids are given below :

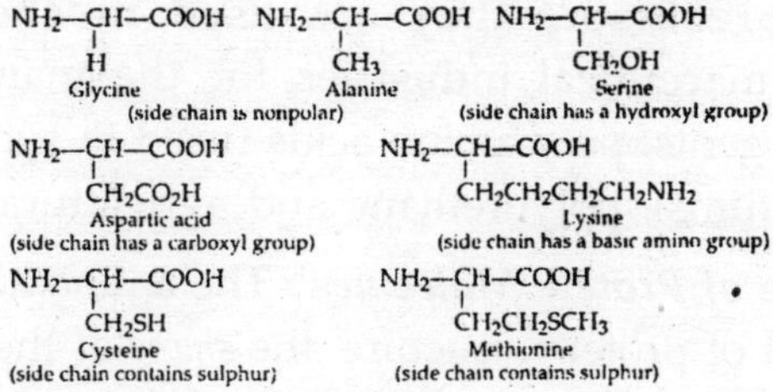

Properties and Uses of Amino Acids : The amino acids and proteins are typical amphoteric compounds because they can react with both acids and bases. The *dipolar ionic structure (Zwitterion)* of an amino acid may be written as :

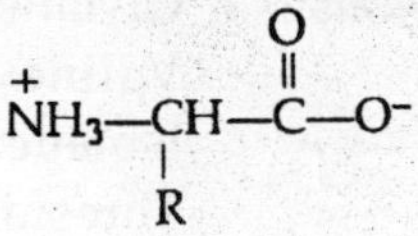

Therefore, amino acids have the ability to neutralize both stronger acids and stronger bases and thus serve as buffers, keeping the pH of their solutions constant.

All the natural amino acids (except glycine) and proteins are optically active and have the L configuration. Our enzymes which themselves are made from the L-forms of amino acids, work with L-forms only.

Amino acids have a wide variety of applications some of which are as follows :

(1) They are used as nutritional supplements in foods, *e.g.*, lysine, tryptophan and threonine are used to enrich vegetable protein; methionine is used to enrich soyabean meal.

(2) In food processing they are used as flavour enhancers and additives, *e.g.*, glycine is used as a sweetener, bacteriostatic agent and antioxidant.

(3) Amino acids are used therapeutically in infusions and some analogues are used as psychotherapeutic agents.

(4) As precursors they are used in chemical and pharmaceutical industries for the manufacture of detergents, poly amino acids (used in synthetic fibres and films), polyurethane and agricultural chemicals.

Structure of Protein Molecules : The amino acid sequence is one aspect of protein structure, the *shape* of the molecule is another. In a peptide chain two adjacent amino acids are linked to each other by a *peptide bond* (amide linkage).

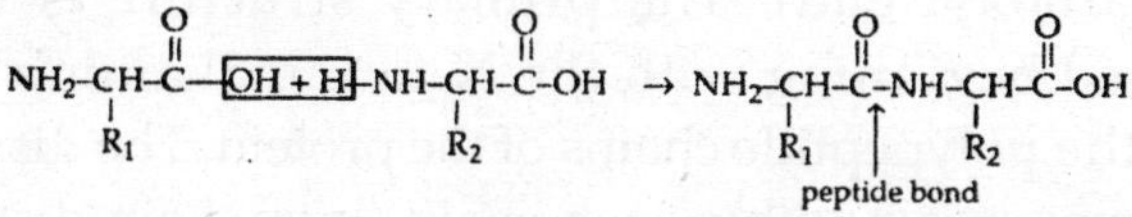

These bonds are formed, in effect, by the elimination of one molecule of water from the two amino acids sharing the bond, such that the alpha-carboxyl carbon atom of one amino acid is linked to the alpha-amino nitrogen atom of the next amino acid. This type of reaction is called *condensation reaction* or, in this particular case, a *dehydration synthesis*. When only two amino acids are linked together, it is called a dipeptide, when three, a *tripeptide* and when a large number of amino acids are linked together, it is called a *polypeptide*.

Therefore, the word "polypeptide" is sometimes used as a synonym of the word 'protein'. The sequence of amino acids in polypeptides is generally written as the α -amino or N-terminal group on the left and the α -carboxyl or C-terminal group on the right and a three letter abbreviation for each amino acid is used.

Four levels of protein structure may be described, which are as follows:

Primary Structure : The sequence of amino acids in a protein makes up the primary structure. The binding of two polypeptide chains by the disulphide bridge (—S—S—) of a cysteine molecule is part of the primary structure. All of the alpha-amino and alpha-carboxyl groups of the amino acids in a polypeptide chain participate in the formation of peptide bonds except two, which are located at either end of the polypeptide chain.

The end of the polypeptide that bears the amino acid with the free alpha-amino group is called the *N-terminus* (or α -amino end), while the end that has the amino acid with the free alpha-carboxyl group is called the *C-terminus* (or α -carboxyl end). The primary structure is usually delineated beginning with the N-terminal amino acid of each of the polypeptide chains of the protein. The disulphide bridges may occur within each chain (intrachain disulphide bridges) and also covalently link the chains together (interchain disulphide bridges).

An understanding of the primary structure of a protein is important for two reasons. First, the primary structure of a protein is determined uniquely by the nucleotide sequence of the corresponding DNA gene, through the process of transcription and translation. Therefore, it is necessary to know the primary structure of a protein in order to understand the overall process whereby the hereditary material controls the synthesis of proteins. Second, it is the primary structure of a protein that determines the final three-dimensional structure of the molecule and hence the physiological role it plays.

According to *Loewy et al* (1991), the primary structure is the fundamental link between the genetic material on the one hand and the physiological function on the other. Another interesting biological result of the study of primary structure is the discovery that proteins with similar functions have similar primary structure although they may occur in quite unrelated organisms. For example, cytochrome *c*, which occurs in almost all cells, exhibits only minor modifications in its primary structure from the yeast cell to mammals.

Secondary Structure: *Pauling* and *Corey* (1951) provided the first clue to understanding the secondary structure of a protein. All protein molecules are not simple long polymeric chains. Instead the chains are coiled into a spiral called an α -helix. As indicated below, the opportunity for carbonyl oxygen to form *hydrogen bond* to amide hydrogen provides the forces that cause the coiling or the secondary structure of the protein.

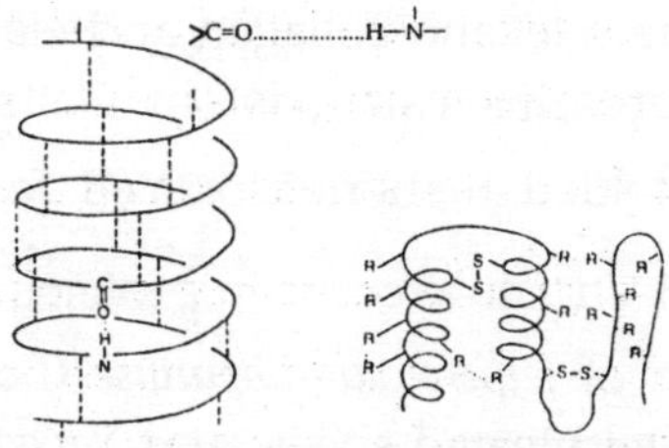

Schematic representation of the structural levels in proteins. 'R' denotes amino acid chains, dots (...) indicates non-covalent interactions.

Thus, a protein's secondary structure identifies : (1) the position and extent of those regions of the polypeptide chain (or chains) that are twisted to form helices, (2) the nature or type of helix that is present, and (3) the position, extent and nature of any non-helical regions.

The non-helical region at the N-terminus (if N-terminus is not part of a helix) is denoted by NA and its amino acids numbered consecutively (NA_1, NA_2, NA_3, etc.). If there is a non-helical part at the C-terminus, it is identified on the basis of the last helix. For example, in a polypeptide chain containing three helices (A to C), a non-helical segment at the C-terminus would be identified as CC and its amino acids numbered as CC_1, CC_2, CC_3, etc. Beginning at the N-terminus , the helical regions are denoted by the letters, A, B, C, D, etc., and the amino acids within each helix assigned numbers *(e.g.,* A_1, A_2, A_3, etc.).

The interhelical regions of each chain are denoted by the letters of the adjoining helices *(i.e.,* non-helical regions AB, BC, CD, etc.) and the amino acids within these regicns are also assigned numbers *(i.e.,* AB_1, AB_2, AB_3, etc., or BC_1, BC_2, BC_3, etc.). Using this type of nomenclature the specific position of any amino acid can be identified . The lengths of α -helices vary in proteins, but on the average there are about 10 residues per helical segment, the exact number depending on the amino acid sequence.

The secondary protein structure is a characteristic of fibrous and contractile proteins such as keratin of hair, fibrin of blood clots, myosin of muscle, and collagen and elastin of connective tissue. Fibrous proteins can exist in a stretched-out form (α-form), or in a folded or shorter form (β-form). Two types of pleated sheet or β-structure occur in proteins. For example, the polypeptide chain of *superoxide dismutase* (bovine erythrocyte) contains antiparallel pleated sheets and *flavodoxin* is formed of parallel pleated sheet strands.

The β-pleated sheet structure of fibres like silk is stabilized by three kinds of bonds. Covalent bonds account for the fibre's great tensile strength along the fibre axis. Perpendicular to the fibre axis, there are hydrogen bonds and in the third dimension, Van der Waals interactions. The latter two interactions are weaker than covalent bonds and therefore account for the great flexibility of these fibres.

In some proteins other secondary structural features exist, which differ from the α-helix and the β-sheet in not having a regular hydrogen bonded structure. The two such features involving continuous chains of amino acids are the *reverse turn* and the *omega loop*. The reverse turn structure allows a polypeptide chain to change direction over a short distance. It can occur in the interior of proteins as well as at the surface. This structure involves only 3-5 amino acid residues and the backbone groups pack together tightly, having the side chains projecting outward. The *omega loop* involves 6 to 16 amino acid residues and the ends of the loop are relatively close together. Omega loops generally contain hydrophilic residues and are found almost exclusively at the surface of proteins.

The X-ray diffraction analysis indicates that of the various kinds of secondary structures, *helices* account for 26% of the amino acids residues, *sheets* for 19%, *turns* for 15%, and *omega loops* for 21%. A recent study indicates that formation of these structures depends on the interaction between certain neighbouring amino acid residues.

Tertiary Structure : The tertiary structure of a protein results when a helix is folded or assumes some other configuration. In globular proteins (*e.g.*, haemoglobin and myoglobin) it is the non-helical regions that permit the folding which is achieved by a variety of interactions between one part of the polypeptide chain and another and between the polypeptide and neighbouring water molecules. Such interactions include (1) ionic bonds, (2) hydrogen bonds, (3) weaker hydrophobic bonds, and (4) covalent disulphide bonds. Such interactions contribute to the stability of a specific tertiary structure. (Fig.)

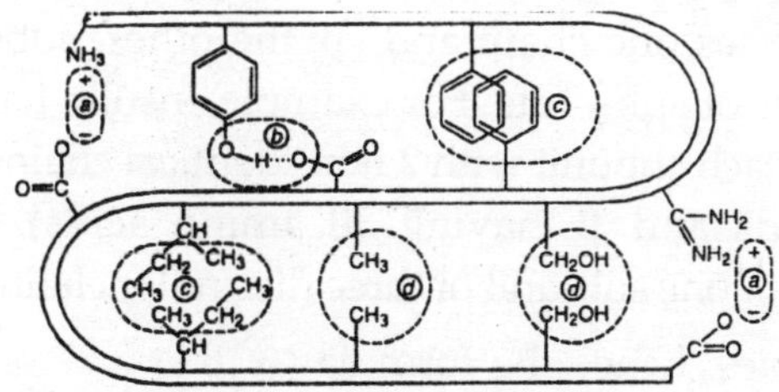

Types of non-covalent bonds that stabilize the protein structure, (a) ionic or electrostatic bonds, (b) hydrogen bonds, (c) weaker bonds, and (d) Van der Waals forces.

Many proteins such as immunoglobulins are divided into domains, *i.e.*, a single polypeptide chain is folded in such a way that two or more globular domains with similar tertiary structures are formed. In a thermodynamic sense, the tertiary form is a very stable form of the protein molecule.

Quaternary Structure: The arrangement, with respect to one another, of polypeptide chains and sub-units in a protein is called its *quaternary structure.* The chain composition of a protein can be determined by treating a protein with an unfolding reagent (urea), or a detergent (SDS), or both, in the presence of a disulphide reducing agent (mercaptoethanol), and subjecting the resulting mixture to acryl amide gel electrophoresis. Following Table indicates the quaternary structure of a few proteins.

Quaternary structure of some proteins

Proteins	Formula	Molecular Weight
Insulin	$(\alpha \beta)_2$	11,460
Haemoglobin	$\alpha_2\beta_2$	64,500
Ig G Immunoglobulin	$(\alpha_2\beta_2)$	150,000
IgM Immunoglobulin	$(\alpha_{10}\beta_{10}\gamma)$	900,000
Fibrinogen	$(\alpha_2\beta_2\gamma_2)$	340,000
Myosin	$(\alpha_2\beta\,\beta'\gamma_2)$	340,000

α denotes one chain and (β the other; subscript denotes the number of sub-units. For example, insulin has two identical sub-units, each subunit with 2 non-identical chains (α having 21 amino acids and β having 30 amino acids). Fibrinogen is composed of one sub-unit of three dissimilar chains, each present in two copies, hence the formula $(\alpha_2\beta_2\gamma_2)$.

In proteins that are made up of two or more polypeptide chains, the quaternary structure refers to the specific orientation of these chains with respect to one another and the nature of the interactions that stabilize this orientation. In some proteins the sub-units are polypeptide chains whose primary structures are identical to each other *(e.g.,* in L-arabinose isomerase all the 6 sub-units are identical), while in others the sub-units are different *(e.g.,* immunoglobulin G has 4 sub-units, 2 light chains and 2 heavy chains).

It may be noted that proteins can contain either a small number of large sub-units *(e.g., thyroglobulin* has only 2 identical sub-units each having a molecular weight 335,000), a large number of small sub-units *(e.g., apoferritin* has 24 identical sub-units each having a molecular weight 19,000), or any intermediate combination. Human *haemoglobin* A has 4 sub-units, 2 Alpha chains, each having a molecular weight of 15700, and 2 beta chains each having a molecular weight of 16500.

Electron microscopy provides a useful means for examining the quaternary structure of proteins having many sub-units.

Properties of Proteins

1. Proteins occur in a wide variety of molecular sizes from small molecules such as the *adrenocorticotrophic hormone* (ACTH) which consists of only 39 amino acids and has a molecular weight of 4500, to extremely large molecules such as the *haemocyanin,* invertebrate blood pigment which consists of 8200 amino acids and has a molecular weight of 910,000.
2. Molecules of proteins, because they usually have several amino and carboxyl groups distributed among side chains, have ionization properties like those of simple amino acids. They normally possess several separate negatively charged and positively charged sites and the net charge may be negative, positive, or zero, depending upon the specific protein, the pH of the solution, and the presence or absence of some metallic ions that may affect the net electric charge. Protein molecules with no net electric charge are said to be in an isoelectric state. Therefore, each soluble protein has its own unique isoelectric point, the pH at which its net charge is zero, and exhibiting no migration in electric field.
3. A number of agents *(e.g.,* heat) and chemicals *(e.g.,* ethyl alcohol, etc.) destroy the nature and the activity of the protein. This is called *protein-denaturation.* After denaturation the protein usually coagulates.
4. The digestion of the protein by a proteolytic enzyme occurs by the hydrolysis of peptide bonds and the protein is digested according to the following sequence:

Protein → proteoses → peptones → polypeptides → amino acids

In the complete hydrolysis of a polypeptide only carbonyl-to-nitrogen bonds are ruptured (see arrows in the following reaction). A hydroxyl group, from water becomes attached to each carbonyl carbon, and a hydrogen atom (also from water) becomes attached to each nitrogen.

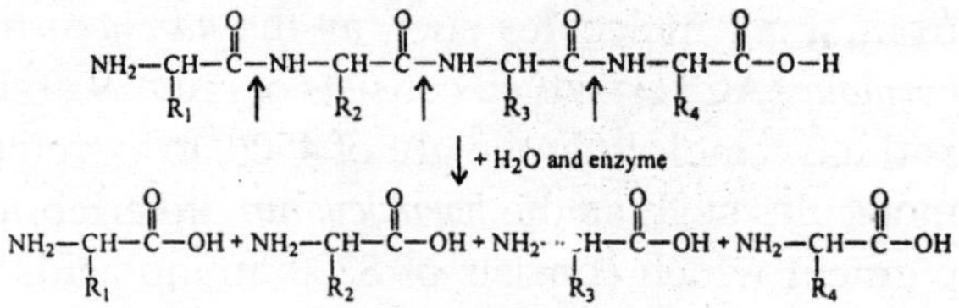

Classification of Proteins

Proteins are classified in various ways. On the basis of their gross structure (molecular organization) and solubility, they are classified into two groups : fibrous proteins and globular proteins. *Fibrous proteins* are insoluble in aqueous media and are further divided into :

1. *Collagens* : or the proteins of the connective tissue. The heat converts them into more soluble *gelatins.*
2. *Elastins* : or proteins of the elastic tissues (tendons, etc.).
3. *Keratins* : or the cysteine-rich protein making wool, hair, hoofs and nails, etc.
4. *Myosins* : or the muscle proteins directly involved in muscle activity.
5. *Fibrin* : or the protein that forms the clot.

The *globular proteins* are soluble in aqueous media and are easily denatured. They are further classified as :

1. *Albumins* : They are soluble in water and are easily coagulated by heat. Egg albumin is the familiar example.
2. *Globulins* : They are soluble in solutions of electrolytes and also coagulated by heat. The g-globulins of our blood function as antibodies.

On the basis of their chemical nature, proteins are generally classified as : simple proteins and conjugated proteins.

Simple Proteins : The proteins that yield only amino acids on hydrolysis are called simple, proteins, *e.g.*, albumins, globulins, etc.

Conjugated Proteins : Simple proteins bound to a non-protein group or a *prosthetic group* (Gk. *prosthesis*, an addition) are called conjugated proteins. These proteins are further classified on the basis of the nature of the prosthetic group.

(a) *Nucleoproteins* : In which the prosthetic group is the nucleic acid, *e.g.*, histones and non-histones. Histones contain about 25% arginine and lysine and are quite similar in all animal and plant cells. Their highly basic nature accounts for the close associations they form with the nucleic acids (deoxyribonucleoprotein in eucaryotic cells and ribonucleoprotein in eukaryotes and prokaryotes). The non-histones are more heterogeneous in amino acid composition and have acidic properties.

(b) *Glycoproteins* : In which the prosthetic group is a carbohydrate, *e.g.*, mucin in saliva and many of the blood plasma proteins and a large number of hormones and enzymes. From a large number of carbohydrates, only about 9 (nine) monosaccharides occur as regular constituents of glycoproteins and the amount of carbohydrate present in them varies from about 1% (*e.g.*, in ovalbumin) to 85% (*e.g.*, in blood-group glycoproteins). In ovalbumin (egg white) having a molecular weight of 45,000, there is only one monosaccharide per molecule, while in mucin having a molecular weight of about 1 million, about 800 monosaccharides are present. The carbohydrate moieties of glycoproteins are usually bound to the protein through covalent bonds with either serine, asparagine, threonine, hydroxylysine or hydroxyproline. The carbohydrate attached to a site of the protein may be a single monosaccharide unit or a linear or branched chain of several monosaccharides called an oligosaccharide.

(c) *Phosphoproteins*: In which the prosthetic group is some phosphorus containing substance other than phosphoglycerides or nucleic acids *e.g.*, *Casein* in milk.

(d) *Chromoproteins* : In which the prosthetic group is a pigment, *e.g.*, haemoglobin, myoglobin, cytochromes, and haemerythrins. The prosthetic groups of the chromoproteins such as the haemoglobin and cytochromes, are bound to the polypeptide portion of the molecule through a combination of covalent and non-covalent bonds.

(e) *Lipoproteins* : In which the prosthetic group is a lipid. This class includes some of the blood plasma proteins and also a large number of membrane proteins. The lipid content of lipoproteins is often very high, accounting from 40% to 90% of the total molecular weight. The amount of lipid present affects the density of the molecule and this property is often used as the basis for their classification. Lipoproteins may be less dense than water, their density being less than 0.9, whereas uncomplexed proteins have a density of about 1.3. The interactions between the protein and lipid portions of a lipoprotein usually involve similar functional groups. For example, the hydrophobic portions of fatty acids, sterols and glycerides form van der Waals interactions with the hydrophobic side chains of the non-polar amino acids.

Functions of Proteins

Proteins are indispensable for life and perform a number of functions. On the basis of their functions, proteins can be put into the following categories.

(i) *Structural Proteins* : These are fibrous or thread-like proteins such as *collagen* and *fibrin* and are used as construction materials. Collagen is a secretory product of many types of cells, such as fibroblasts, bone, tendon, cartilage, and the various epithelia of basement membranes. It is the most abundant protein in mammals.

(ii) *Contractile Proteins* : These proteins (Actin and Myson) have been isolated from various cell types. They are directly involved in the contraction of muscles through which the mechanical work is performed.

(iii) *Blood Proteins* : The blood contains a number of proteins with different functions. One group of blood proteins (serum albumins) control the osmotic pressure and the pH of the blood. The second important protein is *fibrinogen*. It is a soluble precursor of fibrin which plays a vital role in blood coagulation. The third protein of the blood is the *haemoglobin*, which transports oxygen from the lungs to all tissues of the body.

(iv) *Antibodies* : In addition to the above three types of proteins, the g - globulins of our blood function as the antibodies. The body makes these proteins to destroy foreign proteins (antigens).

(v) *Enzymes* : These proteins work as the biological catalysts. Most enzymes are characterized by a high degree of specificity, *i.e.*, they will catalyze one particular reaction but not another and require an optimum pH and temperature to work. Enzyme molecules themselves are not consumed in the chemical reactions that they catalyze.

(vi) *Hormones* : Many hormones are proteins and play important role in metabolic reactions, *e.g.*, insulin, oxytocin, etc.

(vii) *Toxins* : These proteins are produced in the living organism by bacteria and act as poisons to that organism.

(viii) *Nutrient Proteins* : Proteins of our food serve as a source of the amino acids. These essential amino acids are readily synthesized in plants and in some animals and are ingested as proteins by man.

Some of the 20 naturally occurring amino acids found in proteins are essential and must be taken in the diet. However, detailed amino acid requirements have been determined in a few animals but they have been found to be similar in a wide range of forms including mammals (man and rat), birds (chicken), insects *(Tribolium confusum)* and the ciliate protozoan *Tetrahymena*.

Carbohydrates: Carbohydrates are a group of universally occurring compounds having the general formula $(CH_2O)n$. They are the compounds of carbon, hydrogen and oxygen derived more or less directly from CO_2 and water in photosynthesis.

$$CO_2 + H_2O + \text{solar energy} \xrightarrow[\textit{plant enzymes}]{\textit{chlorophyll}} \underset{\text{Basic unit in carbohydrates}}{(CH_2O)} + O_2$$

Structure and Classification : Structurally, the sugars are polyhydroxy aldehydes or polyhydroxy ketones. Carbohydrates are generally classified as monosaccharides, oligosaccharides (composed of 2 to 9 monosaccharide residues) and polysaccharides (consisitng of more than 9 monosaccharide residues)

Monosaccharides : Carbohydrates that cannot be hydrolyzed to simpler molecules are called simple sugars or *monosaccharides*. A monosaccharide may have, 3, 4, 5, 6 or more carbon atoms and accordingly it is called a triose, tetrose, pentose, and hexose respectively.

If the monosaccharide has an *aldehyde* group, it is called an aldose, *e.g.*, glyceraldehyde (a triose), erythrose (a tetrose), ribose (a pentose), glucose (a hexose), mannose (a hexose) and galactose (a hexose). If a *keto* group is present in a monosaccharide, it is called a ketose, *e.g.*, dihydroxy acetone (a triose), erythrulose (a tetrose), ribulose (a pentose), xylulose (a pentose), and fructose (a hexose). Each of the carbon atoms of the chain is assigned a number beginning with the end that is closest to the carbon bearing the double-bonded oxygen. In the aldoses, the number 1 carbon (the ultimate carbon) forms the double bond with oxygen, whereas in the ketoses, it is usually the penultimate carbon. All the aldoses may be considered to be derivatives of glyceraldehyde, while the corresponding ketoses are derivatives of dihydroxy acetone.

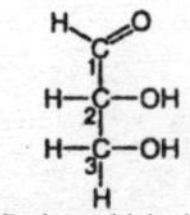
D-glyceraldehyde

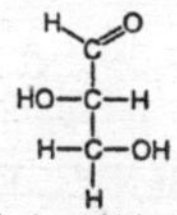
L-glyceraldehyde

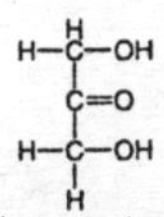
Dihyroxy acetone (a triose)

Some of the monosaccharides can occur in either of two forms : *open chain* and *cyclic* or *ringed*. Cyclic forms that contain a six-member ring are called *pyranoses* and those that contain a five-member ring are called *furanoses*. The two ring forms of glucose are called the α and β anomers. In aqueous solutions of glucose, the β anomer accounts for about 63% of the molecules present, the α form 36%, and the open chain form about 1%.

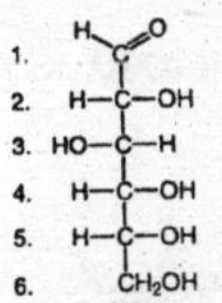
D-Glucose (Open chain structure)

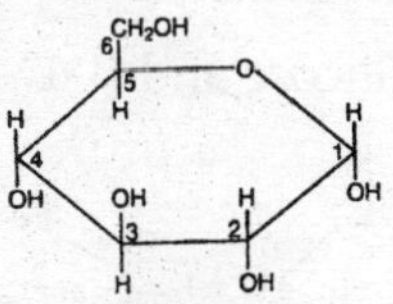
α-D-Glucose, (Haworth projection)

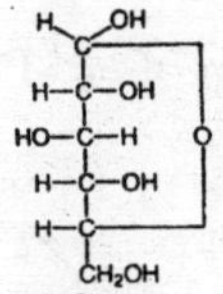
α-D-Glucose, (ring form)

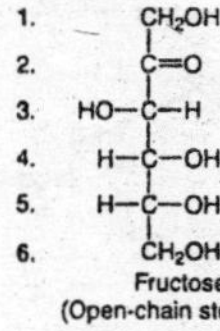
Fructose (Open-chain structure)

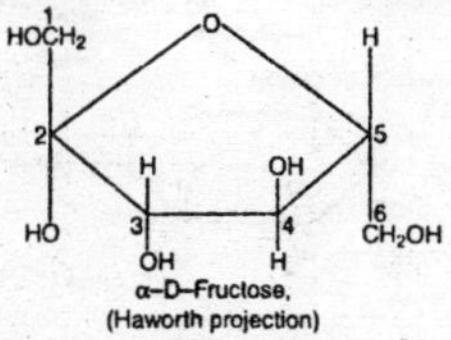
α-D-Fructose, (Haworth projection)

Disaccharides : Carbohydrates that can be hydrolyzed into two or more monosaccharides are called an *oligosaccharide.* The most important of these are *disaccharides* like sucrose, maltose and lactose.

The common disaccharides have the general formula, $C_{12}H_{22}O_{11}$ and yield hexoses on hydrolysis:

$$C_{12}H_{22}O_{11} + H_2O \longrightarrow 2C_6H_{12}O_6$$

The bonds that unite the adjacent monosaccharides are called *glycosidic bonds* and are formed by the condensation of a hydroxyl group of carbon atom number 1 of one monosaccharide with the hydroxyl group of either the number 2, 4, or 6 carbon atom of another. The formation of common disaccharide *maltose* from two molecules of glucose is shown below :

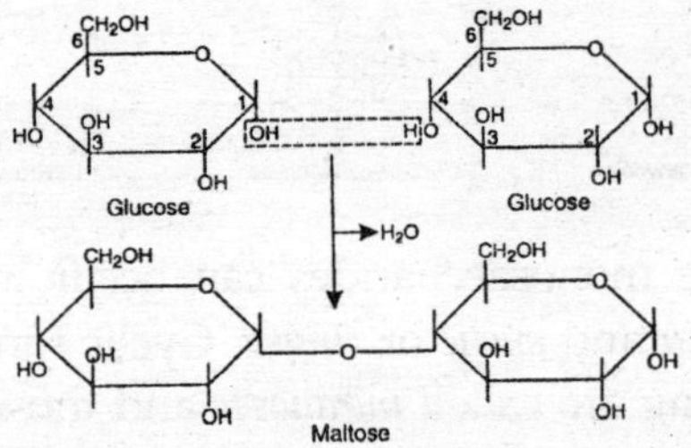

In *maltose,* the oxygen bridge is formed between the number 1 carbon atom of one α -D-glucose unit and the number 4 carbon atom of the other. The bond is referred to as α 1-4 glycosidic bond. Another important disaccharide *sucrose* is formed by the condensation of α -D-glucose and β -D-fructose as given below:

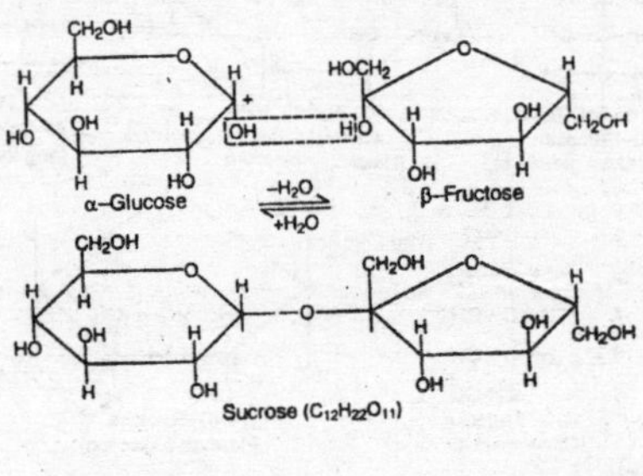

The products of hydrolysis of the more common disaccharides are as follows :

Maltose	+	H_2O	=	glucose + glucose
Lactose	+	H_2O	=	glucose + galactose
Sucrose	+	H_2O	=	glucose + fructose
Cellobiose	+	H_2O	=	glucose + glucose

Polysaccharides : Carbohydrates that yield a large number of monosaccharides on hydrolysis are called polysaccharides. They can be divided into two main functional groups :

(a) *Structural Polysaccharides* : They serve as extracellular or intercellular supporting elements such as *cellulose* (found in plant cell walls), *mannan* (found in yeast cell walls), *chitin* (in exoskeleton of arthropods), *hyaluronic acid* (in connective tissue) and the *peptidoglycans* of bacteria.

(b) *Nutrient Polysaccharides* : They serve as reserves of monosaccharides and are in continuous metabolic turn over. This group includes *starch* (plant cells and bacteria), *glycogen* (animal cells) and *paramylum* (in certain protozoa).

Polysaccharides can also be divided into two broad classes : the *homopolysaccharides* and the *heteropolysaccharides.* In the former, all of the constituent sugars are the same. This class includes cellulose, starch, glycogen and paramylum. In the heteropolysaccharides, the constituent sugars may take different forms as in hyaluronic acid. Some important polysaccharides are described as follows :

Starch : The starch granule is made up of a mixture of two different polysaccharides—*amylose* and *amylopectin,* and the relative amounts of these two polysaccharides vary according to the source of the starch. Natural starches are about 80 to 90% amylopectin and 10 to 20% amylose. The amylose component consists of an unbranched $\alpha 1 \rightarrow 4$ polymer of glucose as shown below :

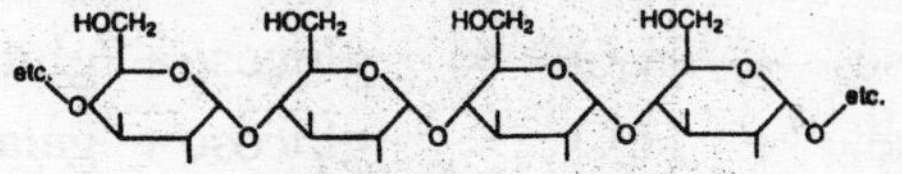

Amylose
(a polymer of α-glucose units)

The amylopectin is a branched polymer of glucose containing $\alpha 1 \rightarrow 4$ and $\alpha 1 \rightarrow 6$ linked glucosyl units.

Glycogen : Glycogen is often referred to as "animal starch". The glycogen found in the liver and muscle tissue is also a polymer of the α-form of glucose. Liver glycogen has about 30,000 glucose residues and molecular weight of 5×10^6. In the glycogen molecule, glucosyl units are linked by $\alpha 1 \rightarrow 4$ glycosidic bonds to form long chains, and these chains are interconnected by $\alpha 1 \rightarrow 6$ glycosidic bonds. It differs from starch by the apparent absence of any molecules of the unbranched, amylose type. Instead, it is branched very much like the amylopectin. It should be noted that in a glycogen molecule there is only one glucose unit whose number 1 carbon atom bears a hydroxyl group. This single free 1—OH group is called the "reducing end" of the molecule and is denoted by the letter R (Fig.). In contrast, numerous "non reducing ends" are present (*i.e.*, free 4—OH and 6—OH groups) at the terminals of the outermost chains.

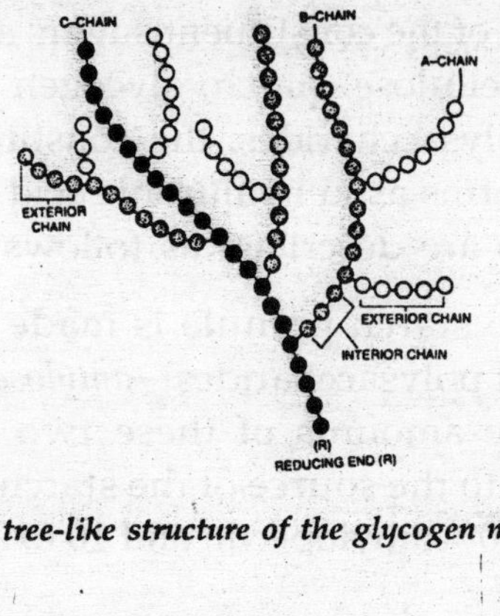

Bush or tree-like structure of the glycogen molecule.

In man and other vertebrates, glycogen is stored primarily in the muscles and is the principal form of stored carbohydrate.

Cellulose : Cellulose is the major component of plant cell walls, where it plays *a* structure role. Cellulose is an unbranched polymer of glucose in which the adjacent monosaccharides are joined by $\beta\ 1 \rightarrow 4$ glycosidic bonds as shown below. Chain lengths may vary from several hundred to several thousand glucosyl units, with a molecular weight of 2.3×10^6. The molecule is insoluble because the hydroxyl groups interact with one another to cross-link to molecule.

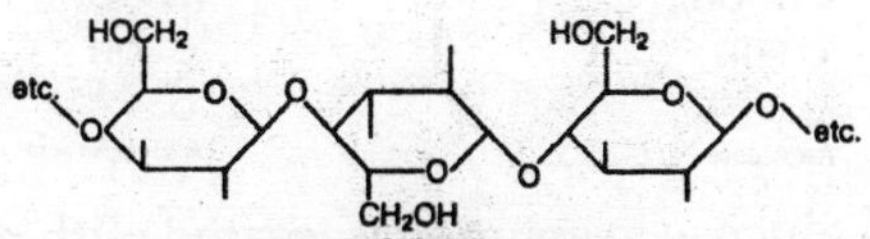

Cellulose
(a linear polymer of β–glucose)

Peptidoglycan : In the bacterial cell wall, the polysaccharide chains are cross-linked by short peptide chains, which are polymers of amino acids. Because it contains both polysaccharide and peptide parts, the structure is known as a *peptidoglycan.* However, there is a great deal of variation in the details of these structures among different types of bacteria.

Properties of Carbohydrates

1. Monosaccharides and disaccharides are soluble in water, but polysaccharides are insoluble. When glucose is dissolved in water, it shows a specific rotation (mutarotation).
2. All aldoses and ketoses give positive tests with alkaline reagents (Tollen's reagent, Benedict's or Fehling's solutions), hence they are often called *reducing sugars.*
3. In the presence of a mild oxidising agent, a monosaccharide is readily oxidized to an acid. The

resulting monocarboxylic acid bears the family name *glyconic* acid (*e.g.*, gluconic acid from glucose).

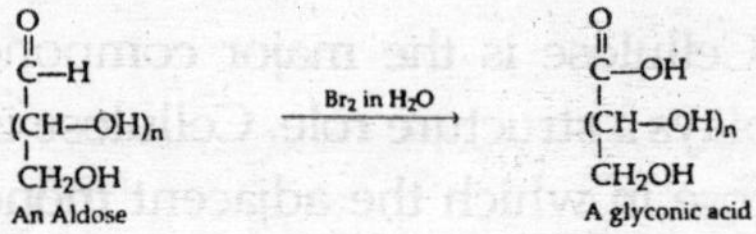

4. A stronger oxidizing agent converts aldoses into dicarboxylic acids having the general name *glycaric acids* (*e.g.*, glucaric acid from glucose, etc.)

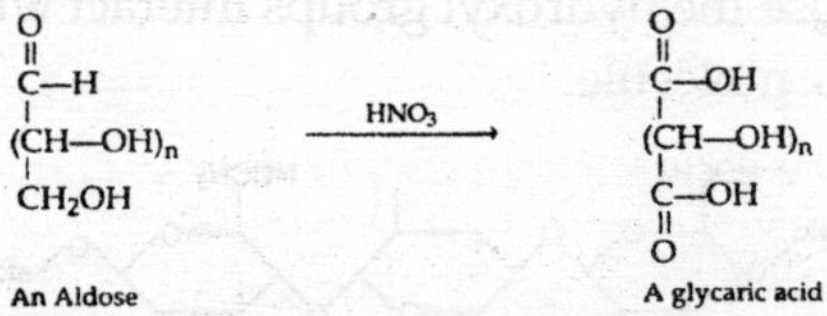

5. When glucose in solution is treated with methyl alcohol and HCl, *glucosides* are formed.

$$C_6H_{12}O_6 + CH_3OH \xrightarrow{HCl} \underset{\text{methyl glucoside}}{C_6H_{11}O_5{-}OCH_3} + H_2O$$

6. Monosaccharides react with phenylhydrazine to form phenylhydrazones. The phenylhydrazone first formed further reacts with 2 more molecules of the reagent to produce sparingly soluble yellow *osazones* which have characteristic crystalline shapes and definite melting points and serve for the identification of sugars.

   The reaction leading to the formation of *glucosazone* is indicated by the following equations:

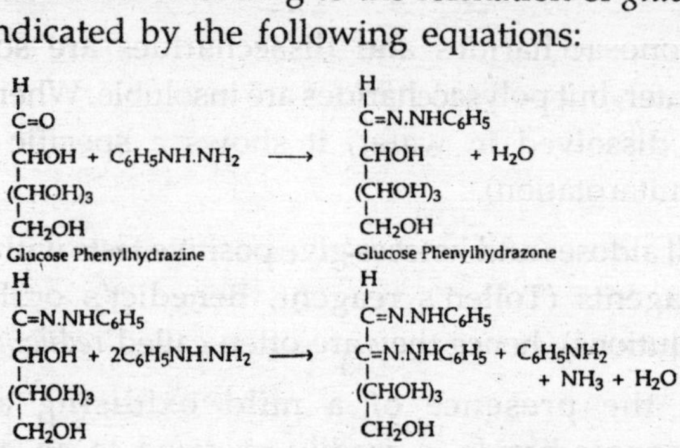

7. Cane sugar does not reduce Fehling's solution, but does so after heating with dilute acids (*i.e.*, after inversion).

8. Sucrose is fermented to alcohol by many enzymes.

$$\text{Sucrose} \xrightarrow{invertase} \text{Invert-sugar} \xrightarrow{zymase} \text{Alcohol}$$

9. Starch does not reduce Fehling's solution or react with phenylhydrazine. However, amylose gives a blue colouration with dilute iodine solution (in KI), which disappears on heating but reappears on cooling. In this reaction an unstable compound is possibly formed between starch and iodine or it is an adsorption phenomenon.

10. During digestion, the starch is first hydorlyzed by the enzyme amylase (ptyalin) to maltose. The digestion of maltose and other disaccharides (lactose and sucrose) is completed by maltase, lactase and sucrase respectively.

Function or Uses of Carbohydrates : Carbohydrates perform both functional and structural roles in the cell. They are very important biologically as a source of energy for the cell. The main source of energy is glucose. It is the chief form in which the carbohydrates are transported from cell to cell by the blood in the animals and by the sap in the plants. Carbohydrates are also used as structural building blocks. For example, *cellulose* is the major component of the plant cell walls. *Chitin* is a structural substance of the cell wall of fungi and the exoskeleton of arthropods.

Some carbohydrates may serve as the prosthetic group of certain proteins. Glycogen is the chief form in which the carbohydrates are stored in the animal cells and starch is the major storage material in the plants.

Lipids: Fats and their derivatives are known collectively as lipids. Their molecules consist of esters of the trihydroxy alcohol, glycerol, with various long-chain carboxylic acids (called fatty acids). A simple typical fat contains three ester groups (COO) per molecule, and it is called a *triglyceride* (neutral fat).

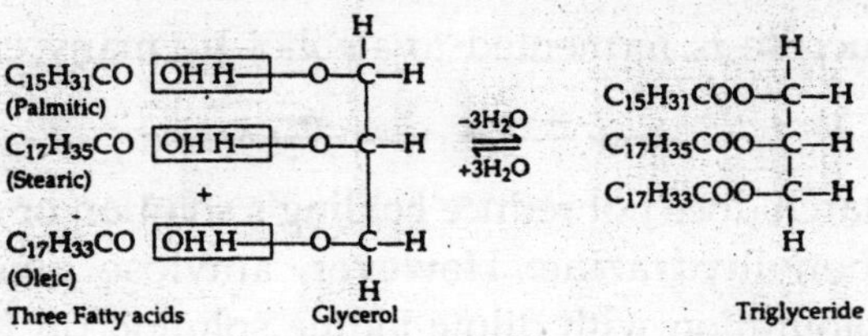

Fatty acids are compounds consisting of a long hydrocarbon chain with a carboxylate group at one end. The general formula is $CH_3(CH_2)_nCOOH$.

They are obtained from the lipids of most plants and animals and have the following characteristics :

(i) They are usually monocarboxylic acids, R—CO_2 H.

(ii) They possess an even number of carbon atoms.

(iii) They R group is usually an unbranched chain.

(iv) The R group may be saturated, or it may have 1, 2 or 3 double bonds.

Fatty acids are divided into 2 classes depending upon whether or not the carbon chain carries the maximum possible number of attached hydrogens. If all carbon atoms are fully saturated with hydrogen, the fatty acid is called *saturated* and has the structure like.

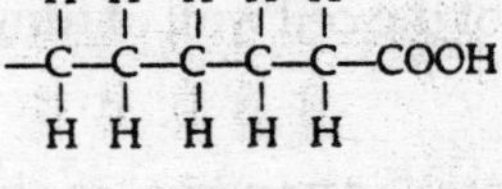

The most abundant saturated fatty acids are palmitic acid, $CH_3(CH2)_{14}COOH$, and stearic acid, $CH_3(CH_2)_{16}COOH$. These and some other acids along with glycerol form the bulk of the body fat in most organisms. When all the carbon atoms are not fully saturated with hydrogen, they are joined by double bonds and the fatty acid is called *unsaturated*.

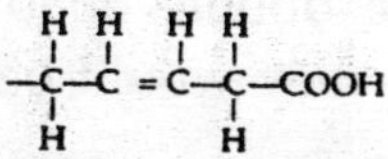

Typical unsaturated fatty acids are oleic acid, $CH_3(CH_2)_7$ $CH=CH(CH_2)_7$ COOH or $C_{17}H_{33}COOH$ and linoleic acid $CH_3(CH_2)_4CH=CHCH_2CH=CH(CH_2)_7COOH$. The melting point of a fatty acid depends on its nature or kind, the greater the degree of unsaturation, the lower the melting point. For example, the melting point of the *stearic acid* (saturated) is 70°C and that of *oleic acid* (one double bond) and *linoleic acid* (2 double bonds) is 4°C and —5°C respectively. For this reason most animal fats are solids and most vegetable oils are liquids at room temperature. The oils are more unsaturated than the fats because they are composed of the glycerides of unsaturated fatty acids. The hydrogenation or the double bonds of the fatty acid present in oils (converting them into saturated acids) raises the melting point of the glycerides. The "hardening" of vegetable oils such as the cotton seed or groundnut oil is an important commercial process by which the cooking fats like "Dalda", *etc.*, are produced.

$$\text{Oils} + H_2 \xrightarrow[\text{catalyst}]{\text{Ni or Pt}} \text{Saturated fats}$$

Properties of Triglycerides

1. They are generally insoluble or sparingly soluble in water but freely soluble in organic solvents like ethyl alcohol and ether, etc.

2. Upon hydrolysis, a simple lipid yields glycerol and fatty acids.

$$\text{Triglyceride} + 3H_2O \xrightarrow{\text{Lipase}} \text{Glycerol} + 3 \text{ Fatty acids}$$

3. When the ester links in triglycerides are saponified by NaOH or KOH, glycerol plus the salts of the fatty acids are produced. These salts are soaps.

$$\text{Triglyceride} + 3NaOH \xrightarrow{\text{Heat}} \text{Glycerol} + \text{A mixture of three salts}$$

4. If some of the double bonds in vegetable oils are hydrogenated, they would become like animal fats

$$\text{Vegetables oil} \xrightarrow{\textit{Hydrogenation}} \text{Saturated fat}$$

5. When the oils and fats are left exposed to warm, moist air for sometime, they become *rancid,* giving disagreeable flavours and odours. The *rancidity* is due to two kinds of reactions : hydrolysis of ester links and oxidation of double bonds.

Classification

Lipids are usually classified into :

Simple Lipids : They contain only carbon, hydrogen and oxygen. Triglycerides (fats and oils) and waxes are simple lipids. The hydrolysis of triglycerides yields fatty acids and glycerol, while the hydrolysis of wax *(e.g.,* Bees wax) yields fatty acids and long-chain alcohols.

Compound Lipids : Compound lipids contain other elements such as sulphur, phosphorus, or nitrogen, in addition to the carbon, hydrogen, and oxygen of the simple lipids. Some compound lipids are described as follows:

Phospholipids : These are compound lipids, containing two fatty acids (diglycerides) attached to the glycerol molecule. Instead of the third fatty acid, they have a *phosphate* molecule, usually with an additional water-soluble molecule attached to it. Hence they are also called *gly-cerophosphatides.* This chemical constitution makes the phospholipid molecule *amphipathic,* with a nonpolar, water-insoluble part (the fatty acids) and a highly polar, water-soluble part (the rest of the molecule) as shown below :

glycerol + two fatty acids + phosphate + water soluble molecule

Thus phospholipids are all phosphate esters of glycerol, with the following general formula:

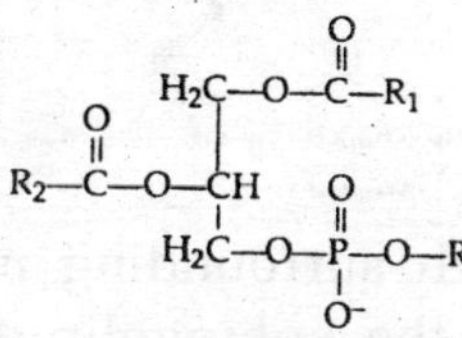

Phospholipid

R_1 and R_2 are fatty acids; R_3 represents a variety of residues attached to phosphoric acid.

In cell membranes there are five main phospholipids which differ according to the type of water-soluble molecule attached to the phosphate. The most abundant form is *lecithin* in which this additional molecule is *choline,* a vitamin B. Therefore, lecithin is called *choline phosphoglyceride.* The phospholipids are bound into the structure of the cell membrane.

When the phosphate group has no attached molecules, the phospholipid is slightly less polar. This group of phospholipids includes the phosphoglycerides derived from *phosphatidic acid.*

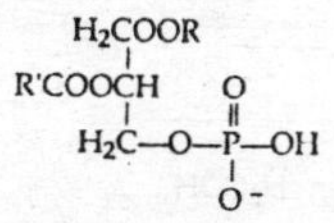

L-Phosphatidic acid

It has been suggested that it functions as a movable molecule within the cell membrane, acting as a carrier for ions.

Sphingolipids : These are derivatives of *sphingosine,* an amino alcohol possessing a long unsaturated hydrocarbon chain.

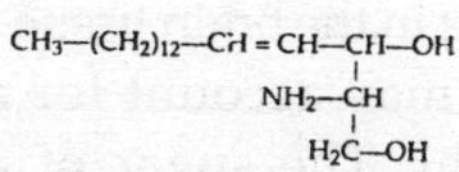

Sphingosine

Thus, the structure common to all sphingolipids is a sphingosine residue on which two other residues (R_1 and R_2) are substitued as shown below (R_1 may be a fatty acid while R_2 can vary widely).

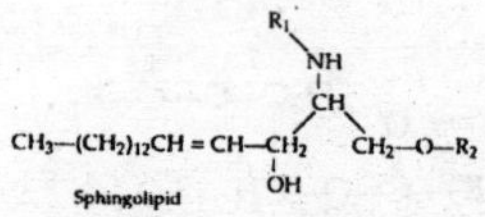

Sphingolipid

The myelin sheath surrounding many nerve cells is particularly rich in the sphingolipid *sphingomyelin.* In sphingomyelin, the amino group of the sphingosine skeleton is linked to a fatty acid and the hydroxyl group is esterified to phosphorylcholine.

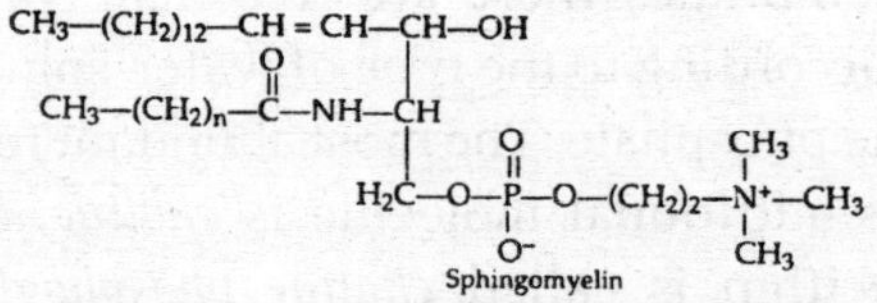

Sphingomyelin

Glycolipids : Glycolipids (and glycoproteins) occur in the plasma membrane. The former play vital roles in immunity, blood group specificity and cell-cell recognition. The lipid portion of a glycolipid is similar to sphingosine with the amino group of the sphingosine skeleton acylated by a fatty acid (as in sphingomyelin) and the hydroxyl group associated with the carbohydrate.

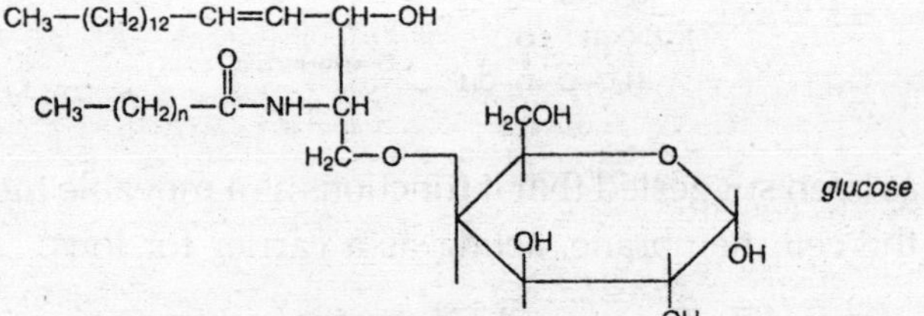

Cerebroside (a glycolipid)

The simplest glycolipids are the *cerebrosides,* which, as the name suggests, are abundant in the brain tissue where they occur in the myelin sheaths and may account for as much as 20% of the sheath's dry weight. The sugar of a cerebroside is either glucose or galactose. In the *gangliosides* of the nerve tissue, the carbohydrate portion of the molecule consists of a chain of sugar molecules usually including glucose, galactose, and neuraminic acid.

Steroids : These are fat-soluble dervatives of *cyclopentanoperhydrophenanthrene* and play important roles in metabolic activities of the organism. They include cholesterol, vitamin D, cholic acid and a number of sex hormones:

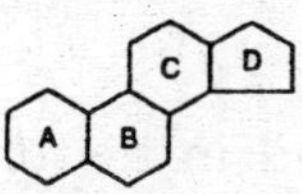

Basic unit of all steroids

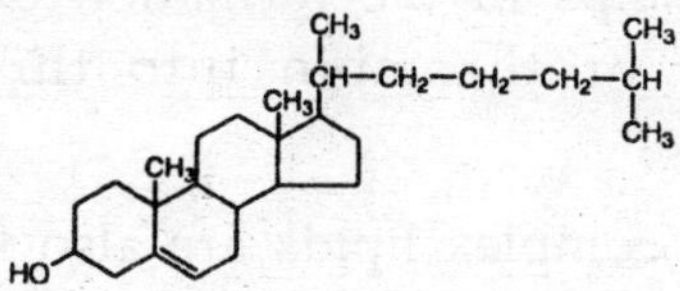

Cholesterol

The most common animal steroid is *cholesterol,* found in all cell membranes, in the myelin sheath of nerves and in many lipids. It is also the metabolic precursor of most steroid hormones. In its structure there is phenanthrene nucleus (rings A, B and C) and the attached cyclopentane ring (D) characteristic of all steroids. Cholesterol gives rise to three basic types of steroid hormones:

(a) Androgens (C_{19} compounds)

(b) Estrogens (C_{18} compounds)

(c) Progesterones and corticosterones (C_{21} compounds).

The blood contains cholesterol (215 mg/litre) as free cholesterol and cholesterides (esters of cholesterol with fatty acids). Cholesterol is also present in bile (0.6 to 1.7 g/litre). The concentration of the bile which occurs in the gall bladder may lead to a further rise in the cholesterol concentration and crystallization. Cholesterol crystals in turn form nuclei for the crystallization of bile pigments and calcium salts (bile stones).

Function and Uses of Lipids

Fats and fatty acids are important *food storage compounds* in most organisms. The fat is stored in the adipose tissue.

Like carbohydrates and proteins, fats also play significant roles as *structural components* of cells. For example, certain phospholipids form an important component of the cell membranes and of enzyme systems in the mitochondria. Phosphatides have also been considered essential for the formation of one of the blood clotting factors. Phospholipid *cephalin* helps in the formation of *prothrombinase* which converts prothrombin into thrombin during blood coagulation.

Some complex lipids are also found in the brain and nerve tissue and in the heart and skeletal muscles. Besides, the oxidation of fats provides a large amount of energy to body cells.

The animal fats are the most important sources of some of the vitamins, *i.e.,* A and D. The steroids are also of great physiological importance and the cholesterol is the main precursor of steroid hormones, *(e.g.,* estrogen progesterone, corticosterone) which affect cellular activities by influencing gene expression.

Some steroids are vitamins *(e.g.,* vitamin D_2) and influence the activities of certain cellular enzymes. Other steroids are regular constituents of membranes, where they influence membrane structure, permeability and transport.

Nucleotides

A nucleotide is a molecular complex of three units : a nitrogen base, a pentose sugar, and a phosphate group. The nitrogen base of a nucleotide may be a purine (adenine, guanine) or a pyrimidine (cytosine, thymine and uracil). The phosphate groups are derivatives of phosphoric acid (H_3PO_4), and the pentose sugar in a nucleotide is ribose or deoxyribose. The nucleotides are the building blocks of the following three types of compounds.

(i) *High Energy Compounds:* The most important high-energy molecule is the ATP, which is synthesized in the mitochondria of living cells.

(ii) *Co-enzymes*: Co-enzymes are also chemical derivatives of nucleotides. A coenzyme is a carrier molecule that functions in conjuction with a particular enzyme. Important hydrogen-carrying cellular co-enzymes are NAD, NADP, FMN and FAD.

(iii) *Nucleic Acids* : Nucleic acids (DNA and RNA) are polymers of nucleotides. They are an especially significant group of compounds found in the nucleus (DNA) and in the cytoplasm (RNA). In the nucleus, they are associated with the chromosomes and are important in transmitting information from the nucleus to the cytoplasm. In the cytoplasm, they are most closely concerned with the synthesis of proteins. The DNA is the material that forms the genes, the cellular components that have long been recognised as the carriers of heredity, but now they are known to be the ultimate controllers of all metabolism, or the 'secrets of life'.

Hormones and Vitamins

A hormone may be defined as a specific product (organic substance such as protein, steroid, etc.) secreted by the cells of an endocrine gland into the bloodstream which carries it to some part of the body where it produces a definite physiological effect. Hormones function in *growth, reproduction, metabolism* and *digestion.*

Vitamins are a specific group of organic compounds that are essential for the normal functioning of an organism. There are two main categories of vitamins : fat soluble and water soluble. Fat soluble vitamins are A, D, E and K. Another class of vitamins includes vitamin B complex and vitamin C, which are water soluble. The deficiency of vitamins causes a number of diseases. In digestion, the vitamins act as the enzymes. Some vitamins also function as the structural components of the co-enzymes (*e.g.*, NAD contains the vitamin nicotinamide).

Biological Characteristics of Protoplasm **:** Since protoplasm is the living matter, all attributes of life lie in the protoplasm. These are : irritability, metabolism, digestion and assimilation, respiration, growth, movement, adaptation, excretion, and reproduction. These characteristics distinguish the living from the non-living and are the criteria by which we measure the dynamic aspects of the protoplasmic system. They also represent the adaptive nature of the living systems, as revealed in their mechanisms of survival.

Superficially, the protoplasm appears the same in all the organisms, but its submicroscopic and physicochemical organization vary widely with different species and with different cells in the same organism. This organization accounts for the morphological and functional differences between the organisms.

Biological Significance of Protoplasm **:** All the organisms are composed of protoplasm, which is the *'physical basis of life'* (*Huxley*, 1868). Later biologists have given the special name *nucleoplasm* to the protoplasm in the nucleus and identified the protoplasm outside the nucleus as *cytoplasm*. In fact, all phenomena of life are performed by the protoplasm and the properties of the protoplasm are the basic expressions of life.

8

Endocrine Glands

Most glands of the body deliver their secretions by means of ducts. These are the exorcrine glands. Other glands manufacture chemical substances which they secrete into the blood stream for transmission to various "target" tissues. These are the endocrine (Gr., *endon* = within ; *krinein* = to secrete) or ductless glands. The secretions which they produce *do not leave* the gland through a duct or a canal but pass *directly* from the cells into the blood streams. Study of these glands and their hormones is called Endocrinology.

The secretions produced by the endocrine glands are called Hormones. The term *Hormone* was introduced by William Bayliss and Earnest Starling in 1904, which is derived from the Greek *hormon,* meaning to excite, to arouse or to set in motion. These secretions are prompted by a stimulus and have an activating influence on the neighbouring or distant tissues. They constitute a mode of communication between cells, a mode of regulation of cell machinery and a channel for integration and coordination in multicellular organisms.

Enzymes vs. Hormones — Hormones act as body catalyst resembling enzymes since they are required only in very small amounts and are not used during their catalytic action. However, they differ from enzymes in the following ways :

(1) They are produced in an organ other than that in which they ultimately perform their action.

(2) They are secreted into the blood in response to body requirements.

(3) Structurally, they are not always proteins. They include proteins with molecular weights of 30.000 or less, small polypeptides, single amino acids an steroids.

The principal role of hormones in the body is to maintain constant, within a narrow range, the characters of the environment of the cells in our body. The great French biologist Claude Bernard (1885) called it constancy of the *milieu interior* (internal environment): the English physiologist, Walter Cannon (1920) described it as *homoeostasis.* Our body cells can live and function normally only if the fluid that surrounds them and constitutes their environment' remains more or less constant both in physical character and chemical composition. This fluid is in a constant dynamic flux : substances are being added and removed, physical characteristic like alkalinity or temperature are being altered continually. These are the hormones which are responsible for homeostasis in the wake of such changing situations.

The internal environment of the body is controlled partly by the autonomic nervous system and partly by the endocrine glands. The hormones secreted by the ductless glands are mainly excitatory in nature stimulating activity in other organs.

The endocrine system consists of the following glands :

1 Pituitary gland

1 Thyroid gland

4 Parathyroid glands

2 Adrenals or Supra-renal glands

The islets of langerhans in the pancreas

2 Testes in the male

2 ovaries in the female

Placenta in the female

1 Thymus

1 Pineal gland or body.

Pituitary Gland (Hypophysis)

The human pituitary is a reddish-gray oval structure, about 10 mm in diameter, located in the brain just behind the optic chiasma as an extension from the floor or the hypothalamus. It is lodged in a depression, the *sella* turcica or hypophyseal fossa of the sphenoid bone. In man it weighs about 0.5 gm.

Structure : The pituitary gland has a dual origin, both components being derived from the *ectoderm.* These are adenohypophysis and neurohypophysis. The adenohypophysis arises from the roof of the buccal cavity in the form of an outgrowth called "Rathke's pouch". The neurohypophysis originates as a downgrowth from the floor of the diencephalon in the fore brain. Adenohypophysis soon loses its connection with the buccal cavity whereas neuronypophysis remains attached to the brain by an *infundibula or pituitary stalk.*

The Adenohypophysis : This is composed of three parts– the pars intermedia, the pars distalis and, the pars tuberailis (the two latter being the anterior lobe' of the older terminology). The adenohypophysis consists of cells (chromosomes and chromophils) which themselves secrete with the exception of the pars tuberalis which is simply a supporting structure for the gland and its blood vessels.

The Neurohypophysis : It comprises Infundlbulum or Neural Stalk and pars nervosa or neural lobe. The neural lobe (Infundibular process) consists of non-myelinated nerve fibres and non-secretary supporting cells. The hormones of neurohypophysis are now clearly established to be formed in the supra-optic and paraventricular nuclei of the hypothalamus of brain. Two tracts of non-medullated nerve fibres arise from these nuclei and supply the posterior lobe. These tracts contain motor fibres only. It is believed that these hormones are released from the nerve terminals which contain them when impulses pass down the axon, which means that the hypothalamic neurons concerned are both nervous and secretory in function.

The production of hormones by nerve cells is known as neuro-secretion. Thus, the neural lobe or infundibular process itself is simply a storage depot for hypothalamic secretions and does not contain its own secretory cells.

The blood supply to the pituitary is by the superior and inferior hypophyseal arteries which are branches from the internal carotid arteries.

Venous drainage is into one of the venous sinuses of the brain.

Functions : The pituitary gland is often described as the *master gland* of the endocrine system because it secretes a number of hormones which regulate and control the activity of other endocrine glands.

Anterior Pituitary or Adenohypophysis Hormones

(I) Growth Hormone (GH) or Somatotropic Hormone, (STH) : This hormone stimulates growth of bones, cartilage, muscles, viscera and the body as a whole. Growth hormone promotes protein anabolism, the absorption of calcium from the intestine and the conversion of glycogena to glucose. (Growth hormone has direct actions on cellular activity and metabolism).

Hyposecretion

(a) *During Childhood* : Due to lack of the growth hormone, there is stunted growth of the skeleton resulting in dwarfism. There are two types of dwarfs :

(i) *The Lorain Type* : A minute individual, fairly well proportional who is mentally and physically normal.

(ii) *Frohlich's Type* : In this type of dwarfism growth is stunted. There is considerable obesity, mental deficiency and retarted sexual development.

(b) *During Adult Life* : This condition is very rare and is probably due to atrophy of the anterior lobe of the gland. This condition is exhibited by early *senility*. The skin becomes dry and wrinkled, the hair grey and sparse, the body emaciated, and there is degeneration of the sexual organs with cessation of menstruation in the female. This condition is known as Simmond's disease.

Hypersecretion

(a) *During Childhood* : Due to increased secretion of the growth hormone gigantism occurs. There is excessive skeletal growth before the epiphyses have closed and the individual may becomes 8 or 9 feet tall.

(b) *During Adult Life* : The hyperactive gland in the adult (after union of epiphyses of bones) results in a condition known as acromegaly. There is excessive growth of the bones of the face, specially the frontal bone and the mandible. The hands and feet become large and spade-like. There is also thickening of the skin on the face and hands.

The Pituitary Tropins : Pituitary hormones which influence the activities of other endocrine glands (*e.g.* Adrenal cortex, thyroid and Gonads), principally those involving reproduction or stress; are called tropic hormones. They are carried by the blood to other target glands and aid in maintaining these glands and stimulating production of their respective hormones.

Adrenocorticotropic Hormone (ACTH) : This hormone stimulates the *cortex* of the *adrenal gland* to produce Glucocorticoids Production of CTH is particularly important during emotional and physical stress.

Thyroid Stimulating Hormone (TSH) or Thyrotropic Hormone (TTH) : This hormone controls the growth and activity of the *thyroid gland*. It influences the uptake of iodine, the synthesis of thyroid hormones (T_3 and T_4) by the thyroid gland and the release of stored hormones into the blood stream.

Gonadotropic Hormones : These regulate the development and functions of the gonads.

Follicle Stimulating Hormone (FSH)

(a) *In the Female* : The target organs are the ovaries where FSH stimulates the development and ripening of the ovarian follicle. During its development the ovarian follicle secretes its own hormone *estrogen*.

(b) *In the Male* : It stimulates the development of seminiferous tubules and spermatogenesis.

Due to it's action on both male and female gametes, FSH is also called Gametokinetic factor.

Luteinizing Hormone (LH)

(a) *In the Female* : This hormone promotes the final maturation of the ovarian follicle, ovulation and formation of the *corpus luteum* which secretes the second ovarian hormone known as *progesterone*.

(b) *In the Male* : It stimulates the interstitial cells of testes causing them to release male sex hormones (androgens)

into the blood stream. Hence, in the male, it's also known as *Interstitial Cell Stimulating Hormone (ICSH).*

Feedback Control of Tropic Hormones : The tropic hormone are usually subject to Feedback inhibition at the pituitary or hypothalamic level by the hormone product of the final target gland. Thus when the blood level of Adrenocorticosteroids, sex *steroids* and thyroid hormones becomes high they inhibit the release of their respective tropic hormones. If the concentration of these hormones in the blood falls, the hypothalamus and the anterior lobe of pituitary secrete more of the tropic hormones. Through such control systems the output of tropic pituitary hormones is automatically self-adjusting.

The blood level of the hormones is thus maintained and their production related to needs. This type of regulatory system of hormonal secretion is called "feedback", "plus-minus" or "pull and push" control mechanism that operates between the hormone and the target organ.

Lactogenic Hormone (Prolactin, PL, Mammotropin, Leuteotropic Hormone LH) : The lactogenic hormone has a direct effect upon the *breasts* or *mammary glands* immediately after the delivery of a baby and the expulsion of the placenta. In conjunction with other hormones like growth hormone, ACTH and TSH, prolactin stimulates the breasts to secrete milk.

During pregnancy, the ovarian hormones inhibit the secretion of prolactin. Besides, prolactin induces maternalism, strong emotional attachment to the young and drive to bring the offspring to a state of self-sufficiency. In males, prolactin promotes protective attitudes towards mates and offspring, intensive food gathering for the family and expressions of Paternalism in general.

Melanocyte Stimulating Hormone (MSH) or Intermedin : MSH is produced in the pars intermedia. Though found in all vertebrate groups, this hormone appears to be functional only in poikilothermic ones. Its secretion causes a dramatic darkening of the skin of many fishes, amphibians and reptiles. The darkening occurs because of the spread of granules of *melanin* through the branches of specialized *melanocyte cells* in the epidermis. When the melanin is concentrated in the centre of the melanocytes, the skin has a light appearance. This mechanism is of great value in enabling the animal to blend in with its surroundings.

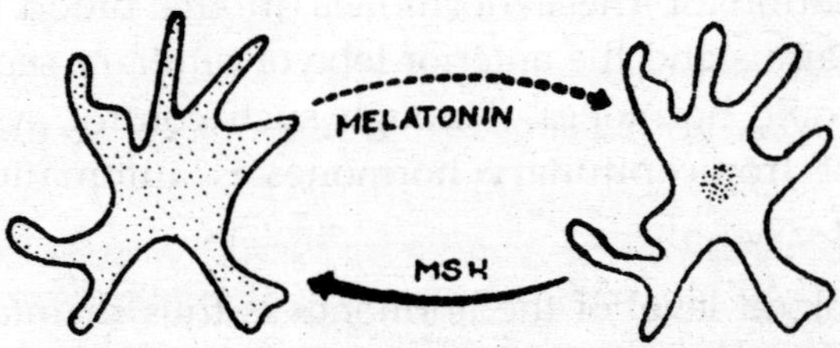

A generalized vertebrate melanocyte. Dispersed condition (Skin dark) Concentrated Condition (Skin Light).

Although MSH does not seem to play an important role in the normal behaviour of human melanocytes, but under certain conditions such as pregnancy, an increase in its secretion does cause some darkening of the skin. Normally, large number of melanocytes are present in the human skin, where they are responsible for moles and freckles.

Posterior Pituitary or Neurohypophysis Hormones : Whereas adenophypophysis secretes hormones on its own, neurohypophysis is not secretory in function. It consists of nonmyelinated nerve fibres and supporting cells and receives its quota of hormones from neurosecretory cells in the hypothalamus of brain.

(i) *Infundibulum* : This part sends a number of release factors or hormones that travel by portal circulation to the anterior pituitary and trigger the secretion of a series of tropic hormones :

Release Factors

No.	*Hypothalamic Hormone or Release Factor*	*Abbreviation*
1.	Corticotropin (ACTH)—release factor	CRF
2.	Thyrotropin (TSH)—release factor	TRF
3.	Luteinizing hormone (LH)—release factor	LH-RF
4.	Follicle-stimulating hormone (FSH)—release factor	FSH-RF
5.	Growth hormone (GH)—release factor	GH-RF
6.	Prolactin—release factor	PRF
7.	Prolactin release—inhibiting factor	PIF
8.	Melanocyte—stimulating hormone (MSH) release—inhibiting factor	MIF
9.	Somatostatin or GH Inhibiting factor	GHIF

These release factors (hormones) may be built up of 3 to 20 amino acids units. For example, TRF is a tripeptide, whereas GH-RF is a decapeptide. The study of these release factors was considered so important that the Noble prize in physiology or Medicine for 1977 has been awarded to Andrew V. Schally and Roger C.L. Cuillemin in U.S.A. for isolating, synthesizing and characterizing the polypeptides from the hypothalamus which control the anterior pituitary function.

(ii) *The Infundibular Process* : Extract of the posterior lobe is called Pituitrin which contains 2 hormones :

Oxytocin (Pitocin) ("Birth Hormone" or "Milk Ejection Hormones") : *(i)* Produces *contraction of the uterine muscles (a)* to facilitate the ascent of spermatozoa after coitus and (b) for delivery of the foetus. In late pregnancy, the uterus becomes very sensitive to oxytocin. The amount is increased during labour, *(ii)* Facilitates the *flow of milk* from mammary glands, by contraction of the myoepithelial cells of the lactating breast, squeezing milk into the large ducts behind the nipple. Sucking plays an essential part in reflex stimulation of the pituitary gland. When sucking stops location will continue only for a short time.

Antidiuretic Hormone (ADH), Vasopressin or Pitressin : This has two important functions :

(i) Influences water-balance of the animal by reducing its output of urine. It exerts a direct effect upon the facultative reabsorption of water by the kidney tubules. ADH may be increased by a variety of stimulators of neural activity. Emotional and physical stress, electrical stimulation, acetylcholine, nicotine, and morphine increase ADH secretion.

If the antidiuretic hormone is lacking, reabsorption of water is reduced and large amounts of dilute urine are excreted. This ailment is called *diabetes insipidus.* It differs from the more common *Diabetes mellitus* in that the urine is free from sugar (*mellitus* is 'sweet' diabetes, while *insipidus* is 'tasteless' diabetes, the common feature in both being the excessive flow or "syphoning off or urine).

(ii) Under the influence of ADH, the involuntary smooth muscle of the intestine, gall bladder, urinary bladder and blood vessel contracts. Contraction of the muscle layer in the blood vessel walls raise the blood pressure.

Thyroid

It is a bilobed gland situated in the lower part of the neck, ventral to the trachea and immediately behind the larynx. The two lobs of the thyroid are joined together by a narrow bridge of tissue called *isthumus* which gives the entire gland a more or less H-shaped appearance.

Histologically, it consists of *acini* or *vesicles* lined with cuboidal epithelium. The lumen of these acini is filled with a proteinaceous colloidal secretion, *thyroglobulin* (an iodized glycoprotein).

Synthesis of Thyroid Hormones : The thyroid gland produces hormones which are iodinated derivates of the amino acid throsine. The formation of these hormones takes place in several stages :

(1) The first stage in their production is the accumulation of *inorganic iodide* in the aceiner cells of the thyroid from the blood plasma. This accumulation is accelerated by the *thyrotropic hormone (TSH)* of the pituitary.

(2) As iodide is absorbed by the aciner cells, it is oxidised enzymatically to iodine $2I^- == I_2 + 2e$.

(3) The iodine is immediately taken up by the amino-acid tyrosine of the gland cells to form *Mono* - and *Di-iodotyrosine (I Tyr I_2 Tyr).*

(4) These then undergo *oxidative coupling (condensation)* to form *derivatives of Thyronine.* One Mono-and one Di-iodotyrosine units combine to form a *Tri-iodothyronine (I_2 Thn),* whereas two Di-iodotyrosine units form Tetra-iodothyronine (I_4 Thn) which is also referred to as *Thyroxine.* It was first crystallised by Kendall in 1919.

(5) The Thyroxine and *l_3 Thn* produced in the gland are still in peptide linkage within the thyroglobulin molecule and are stored as such in the colloid.

(6) Under simulation by TSH the thyroglobulin is broken down by a *hydrolytic protease enzyme* to yield *Thyroxine* (80%) and *I_3 Thn.* (20%) which pass into the blood. The enzyme is secreted by the aciner cells and derived from their lysosomes.

(7) In the blood, I_4 Thn and I_3 Thn are transported in association with two plasma proteins — *thyroxine-binding globulin* and *thyroxine binding prealbumin.* These act as specific carrier agents for the hormones and represent the Protein Bound iodine (PBI) of the plasma.

(8) Finally, they enter the cells of tissues where about 50% of the *Thyroxine* is dehalogenated (one I_2 removed) by a dehalogenating enzyme to I_3 *Thn* and *inorganic iodide* which is returned to the thyroid gland.

(9) Both I_4 *Thn* (65 - 90%) and I_3 *Thn* (10 - 35%) are active hormones but I_3 Thn is 3-5 times more active than I_4 *Thn* and has a more rapid onset of action.

Functions of Thyroid Hormones

(1) An adequate supply is essential for normal development and growth of body.

(2) They are responsible for the maintenance of healthy hair and skin.

(3) They accelerate O_2 consumption by the cells, energy production and BMR (Calorigenic action).

(4) They influence heat production during the ketabolism of food-stuff in the cells.

(5) They increase the intestinal absorption of glucose ; glycogenolysis, deamination and gluconeogenesis in liver cells.

(6) They enhance rate of heart-beat and blood pressure.

(7) They are associated with nerve stability controlling the excitability of nerve fibres.

(8) They are essential for metammorphosis in amphibia.

(9) They regulate reproduction and cellular physiology.

Abnormalities of Thyroid Function

Hyposecretion : A deficiency of thyroid hormones produces a number of clinical disorders depending upon the degree of the deficiency and the age at which it occurs :

(a) In Infancy : Cretinism

(1) Lack of skeletal development resulting in deformed bones and stunted body growth.

(2) Mental deficiency due to subnormal development of the nervous system.

(3) There is a large protruding tongue from a sagging mouth and pot-bellied abdomen.

(4) The BMR is lowered accompanied by slow pulse and respiration rate.

(5) Sex organs and secondary sexual characters becomes retarded.

(b) In Adult Life : Myxoedema (Gull's Disease)

(1) The BMR, body temperature and pulse rate are lowered, and there is undue sensitivity to cold.

(2) There is considerable general oedema particularly manifested by the puffiness of face and hands due to thickening of the skin by the deposit of an albuminium semifluid material underneath.

(3) The skin becomes coarse and dry and there is loss of hair.

(4) Anaemia and slowing of physical and mental reactions.

(5) Sexual development is retailed.

Simple (Endemic or Colloid) Goitre : It is a deficiency disease caused by an inadequate supply of iodine in the diet. Simple Goitre is fairly prevalent in certain countries such as Switzerland, the Himalayan regions and in the Pyrenees, where the iodine content of water and soil is low. In this disease there is a pathological enlargement of the gland itself, which protrudes, as a marked swelling in the neck. This enlargement is believed to be due to overestimation of the gland by the thyrotropic hormone from the anterior pituitary in the absence of "braking" or feedback inhibition of the thyroid hormone.

Since only small amounts of iodine are necessary for the formation of thyroid hormones, treatment with iodine or with iodized salt (NaI_2) is adequate for cases of simple goitre.

Hypersecretion : This condition generally occurs in adult life, and is also referred to as *thyrotoxicosis* or *hyperthyroidism*.

Exophthalmic Goitre or Graves' Disease or Basedow's Disease (Plummer's disease) : The chief symptoms are :

(1) The BMR is greatly increased and the individual appears to be literally "burning himself up".

(2) There is a raised nervous excitability, restlessness and anxiety, accompanied by fatigue, muscular weakness, tremors and loss of weight.

(3) There is an increase in the heart rate and pulse rate with a corresponding increase in respiration rate and copius sweating.

(4) Increased body temperature with excessive sweating.

(5) A characteristic protrusion of the eye-balls (exophthalmus) with a staring look, usually accompanies hyperthyroidism. This protrusion is thought to be due to increased post-orbital pressure and possible weakening of the extrinsic muscles of the eye, as a result of excessive TSH.

Hyperthyroidism can be treated by thyroid removed by *surgery* or by *radioactive iodine* (^{131}I). It is also treated by *antithyroid drugs (goitrogens)*.

Antithyroid Drugs : These are used experimentally to produce hypothyroidism (as in the amphibian metamorphosis), and clinically in the treatment of hyperthyroid conditions. These inhibit the production of thyroxine both at the organification and at the coupling steps.

Examples of these antithyroid drugs are goitrogens—thiouracil, propylthiouracil, methylthiouracil carbimazole, thiourea, and methimazole.

Thyrocalcitonin (TCT) is a non-iodinized calcium-lowering hormone originating from the thyroid gland. The release of TCT is stimulated by high levels of ionized calcium in the serum. The hormone lowers the level of calcium in blood and increases calcium excretion by the kidneys.

Parathyroids

The parathyroids are four small glands placed on the thyroid, two on each lobe. In man, the glands are reddish or yellowish-brown egg shaped bodies; the 4 glands together weigh about 0.1 to 0.2 g. The glands were first recognised by Raynard (1835) and their structure was described by Sandstrom (1880).

Hormone of the Glands : They secrete a hormone, called *Paratharmone* or *Parathyroid hormone (PTH)* OR *Collip's Hormone* after its discoverer Phillip Collip (1925). It is under the feed-back control of blood Ca^{++} level.

Functions

(1) The primary function of paratharmone is to maintain the concentration of ionized calcium (Ca^{2+}) in the plasma, within the narrow range characteristic of this electrolyte, despite wide variations in calcium intake, excretion, and deposition in bone.

(2) Lowers the serum phosphorus by increasing urinary excretion of phosphate but decreases excretion of calcium.

(3) Increases the concentration of Ca^{++} in the blood plasma by mobilising calcium from bones, particularly if the dietary intake of calcium is inadequate.

(4) Enhances the rate of absorption of calcium from the intestine.

The growth of bones, muscle-tone and normal-nervous activity is dependent on a fixed Ca-P balance. Normally in man there is 12 mg of Ca and 4.5 mg of inorganic phosphate per 100 ml of blood.

Secretion of PTH is subject to control by a negative feedback mechanism relating to the levels of ionized Ca^{++} in the plasma.

Abnormalities of Parathyroid Activity

Hyposecretion : Removal of the parathyroid glands in error during the operation of thyroidectomy, or due to some local condition of the glands themselves causes a decrease in the secretion of paratharmone. In this state the level of free calcium in the blood is reduced, because much of the calcium in the blood is bound to other substances. As a result the muscles go into a type of spasm known as tetany (sustained contractions of the muscles of larynx, face, hands and feet).

Characteristic of Tetany :

(1) Muscle twitching leading to muscle spasm of the hands and feet termed carpo-pedal spasm.

(2) Increased neuromuscular irritability and spasm of the eye muscles.

(3) In infants and children spasm of the muscles of the larynx occurs accompanied by cyanosis.

(4) Serum level of calcium falls by more than 50% and phosphorus shoots high, resulting in abnormally greater excitation of the nerve fibres and muscles.

Hypoparathyroidism occurs only very rarely.

Hypersecretion : This is rare and is probably due to a tumour of one mere of the parathyroid glands. In case of hypersecretion of Paratharmone osteoclasts (the cells which destroy bone) become overactive so that there is an excessive destruction of bone and non-utilization of calcium. This results in softening of the bones and a high blood calcium. Eventually calcium is deposited in the walls of arteries and lungs. In the areas of destruction in the bones fibrous cysts develop. This condition is called osteitis rosa cystica. The bones becomes painful and fractures occur frequently.

Adrenal or Suprarenal Glands

The adrenal glands are situated one on each side of the vertebral column, closely applied to the upper poles of the kidneys ; hence the name supra-renals. The adrenal glands are surrounded by a capsule of areolar tissue containing fat.

Structure : The glands are composed of two distinct parts which differ both anatomically and physiologically. The outer part (derived from mesoderm) is known as the cortex and the inner part (derived from neural crest) as the medulla.

Cortex : It is composed of three layers of cells :

(i) The outer layer zona glomerulosa, is made up of comparatively small thickly set ovoid groups of cells. These secrete mineralocorticoids which regulate salt and water balance.

(ii) The middle layer, zona fasciculata is the widest layer which is made up of polyhedral cells arranged in radiating columns. This layer secretes glucocorticoids affecting carbohydrate metabolism.

(iii) The inner layer, zona reticularis is made up of an irregular network of rows of cells. These secrete sex hormones.

Medulla : The medulla is completely surrounded by the cortex. In man this region is thin, comprising only 5-10 per cent of the gland. The cells of the medulla assemble in rounded groups or short anastomosing columns with many intermingling blood capillaries. The medulla is composed of sympathetic gang-lion cells which have no postganglionic fibres. Medullary cells are essentially neurosecretory cells because they are stimulated by preganglionic sympathetic fibres to produce the hormones *adrenatine* and *moradrenaline.*

Medullary cells have a special affinity for chromium compounds such as $K_2Cr_2O_7$ and Chromic acid, and they are, therefore, also referred to as *chromaffine cells.*

Functions

Cortex : The adrenal cortex produced a number of potent hormones all of which are steroid derivatives :

The Mineralocorticoids : These are associated with the maintenance of electrolyte balance in the body and distribution of water in tissues.

Specific examples are *aldosterone* and *deoxycorticosterone.* The most potent mineralocorticoid is aldosterone. Deoxycorticosterone is only 4% as potent as aldosterone.

Functions of Aldosterone

(1) It is a "salt retaining hormone" which promotes the reabsorption of sodium and chloride ions by renal tubules. Retention of these in the blood keeps its osmotic pressure high, as a result of which water is also reabsorbed. This in turn assures normal blood volume and pressure.

(2) Increases excretion of potassium ions.

The amount of aldosterone produced is influenced by the sodium level in blood. When the sodium blood level falls, more aldosterone is secreted and more sodium reabsorbed.

Aldosterone secretion is stimulating by the angiotensin produced by kidneys.

Hyposecretion of aldosterone causes Addison's disease. This disease was first described by Thomas Addison *(Father of Endocrinology)* in 1855. It is characterized by :

(1) Excessive loss of sodium chloride in the urine and elevated levels of potassium in the serum.

(2) Lowering of blood pressure, body temperature and BMR.

(3) Muscular weakness.

(4) Hypoglycemia.

(5) Loss of appetite, vomitting and gastro-intestinal disturbances.

(6) Brownish pigmentation or bronzing of the skin especially the exposed parts, for example, the face and hands due to deposition of melanin.

The Glucocorticoids : These primarily affect metabolism of carbohydrate, protein and lipids.

Specific examples are *corticosterone, hydrocortisone* or *cortisol* and *cortisone.* Their secretion is stimulated by ACTH from the anterior lobe of pituitary.

Functions

(1) They are metabolically antagonistic to insulin. Under their influence glycogen is changed to glucose in the liver, raising the blood sugar level (hyperglycemic action).

(2) During stress, they promote the conversion of fats and proteins to glycogen that is ultimately converted to glucose (gluconeogenesis).

(3) They have anti-inflammatory and anti-allergic properties. Cortisol is used in the treatment of rheumatoid arthritis when it is even injected inside of

the diseased joint. It is a life saving drug in anaphylactic or adverse reactions to certain drugs- like reaction to the injection of penicillin and in severe attacks of asthma. This property of cortiocosteroids is called immunosuppressive whereby they suppress the normal immune reaction of body against antigens and attack of micro-organisms. For this very reason, cortisol in often used for treatment of allergy. It is also useful in the form of ointments for skin and eye diseases.

(4) In recent years, this hormone, has been utilized to suppress the phenomenon of rejection in transplantation of organs.

(5) Together with ACTH, the glucocorticoids serve to counteract the harmful effects of long-lasting stress such as starvation, cold, heat, burns, toxins, infections, forced exercises etc; by helping the tissues to develop resistance and adapt them to meet new conditions. The endocrinilogist Hans Selye calls these responses "the general adaptation syndrome".

Hypersecretion of cortisol in adults results from one of the two causes :

(1) Adrenocortical hyperplasia initiated by increased production of ACTH due to hyperactivity of basophil cells in the pituitary gland.

(2) Benign or malignant tumours of the adrenal cortex. The term *Cushing's disease* has been restricted to those cases which are of pituitary origin whereas *Cushing's Syndrome* denotes adrenocortical hyperfuction directly involving the adrenal glands.

The signs of hyperadrenocorticism (Cushing's disease and Syndrome, 1932) include :

(1) Hyperglycemia and glycosuria (diabetogenic effect).

(2) Excessive deposition of fat over the face, back of the neck and abdomen.

(3) Muscular weakness due to potassium depletion through the kidneys.

(4) Wasting of the muscles due to protein anti-anabolic effect and glyconeogenesis.

(5) (a) *Virilism* in the female *i.e.,* a tendency to develop male characteristics viz., (i) there is excessive growth of hair on the chest and public region with an increase and darkening of the facial hair (hirsutism) (ii) cessation of the menstrual flow (amenorrhoea) and (iii) gradual atrophy of the breasts (iv) overgrowth of clitoris, (b) *Feminism* in the male *i.e.,* a tendency to develop female sex characteristics, *i.e.,* enlarged mammary glands (gynaecomastia) and retailed growth of bread.

The Anrogens or Estrogens : These primarily affect secondary sex characteristics in their specific target organs.

The primary adrenal androgens are *Dehydroepiandrosterone (DHEA) and Androstenedione.* These cause retention of nitrogen (a Protein-anabolic effect), phosphorus, potassium, sodium and cloride. If present in excessive amounts, they also lead to masculinization in the female. Adrenal estrogens influence the development and maintenance of the secondary sex characters in females.

Medulla

The adrenal medulla is a derivative of the symphathetic part of the autonomic nervous system. When stimulated by preganglionic symphathetic nerve fibres, it secretes two hormones : *adrenaline (epinephrine)* (80%) and *norarenaline (norepinephrine)* (20%). These hormones belong to a class of substances known as catecholamines that are derivatives of amino acid tyrosine.

Functions of Noradrenaline : It is a precursor of adrenaline which does not possess a methyl group (—CH_3), as is found in the adrenaline. Nor adrenaline produces a vasoconstriction of peripheral organs like skin ; digestive and reproductive organs. This effect raises blood pressure.

Functions of Adrenaline :

(1) Causes vasodilation of the arterioles of the skeletal muscles increasing the supply of oxygen and nutritional materials for a sustained muscle activity.

(2) Causes vasoconstriction of the arterioles of the skin, mucous membranes and splanchnic viscera, directing the blood away from them to the skeletal muscles, brain, coronary artery and liver.

(3) Dilates trachea and bronchi to permit a greater volume of air to enter the lungs at each inspiration.

(4) Increases the rate of heart beat, thus increasing blood pressure.

(5) Slows down peristalsis in the digestive tract and limits the flow of saliva.

(6) Increases the tone of the anal and urethral sphincter muscles, thus inhibiting micturition and defaecation.

(7) Stimulates the conversion of liver glycogen stored in liver and muscles to glucose that is ready for oxidation.

(8) Decreases *insulin*, thereby preventing the glucose from being taken up by peripheral tissues and preserving it for the CNS.

The net of such changes is that an increased quantity of blood is pumped faster to heart, muscles and brain ; it is well oxygenated and rich in nutrients (stress or alarm reactions). These preparations are valuable for the body in providing a rapid physiologic response to emergencies such as fatigue, shock, fear, excitement and danger. In this sense, it mobilises what has been termed the "fight or flight" mechanism *i.e.*, the emergency function, as suggested by Cannon.

Pancreas

The pancreas is a rather diffuse gland which is a mixture of two histologically and functionally separate tissues. The bulk of the gland is exocrine in function and forms clusters of glandular cells called acini or lobules lined with cuboidal epithelium. Scattered among the exocrine cells are patches of tissues named Islets of Langerhans after their discover and these consists of endocrine cells that discharge their secretion directly in the blood. These cells are small, prism-shaped, closely packed together and without a lumen ; they collectively form an endocrine gland.

In mammals, two distinct types of cells are recognizable in the islet tissue, the α-cells which contain alcohol-soluble granules, and the β-cells, whose granules are not soluble in alcohol. These two cell types each secrete a different hormone, *glucagon* being secreted by α-cells and *insulin* by the β-cells.

Insulin

Insulin is the hypoglycemic antidiabetic factor and the protein hormone which regulates blood glucose.

History : Banting and Best extracted insulin in the year 1921 for which they were awarded Noble Prize in 1923. Abert prepared pure crystalline insulin in 1926.

Chemical Structure : Insulin is a relatively small protein of 51 amino acids disposed in two parallel chains A and B, interconnected by two disulfide bonds. A chain has 21, and the B chain 30 amino acid units ; thus one molecule of Insulin has 51 amino acids in all. Its molecular weight is 6,000.

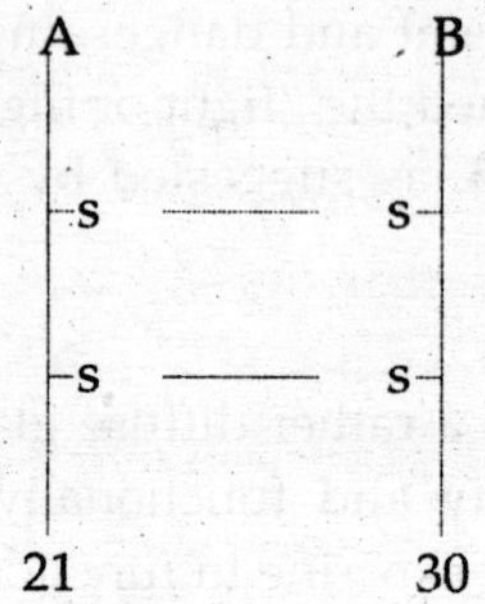

The Insulin Structure

Functions

1. *Increases Combustion of Sugar in tissues* : Insulin accelerates the phosphorylation of glucose by the enzyme Glucokinase or Hexokinase. Glucose-6-phosphate thus formed enters the respiratory cycle to liberated chemical energy.

$$\text{Glucose} \xrightarrow[\text{Glucokinase}]{\text{Insulin}} \text{Glucose-6-phosphate}$$

2. It helps in the *transport of glucose* into the cell.

3. *Stimulates Glycogenesis* : Insulin promotes the formation of glycogen from glucose in the liver and muscles.

4. *Prevents Gluconeogenesis* : It reduces the production of glucose from non-carbohydrate sources such as proteins and fats.

5. *Prevents formation of Ketone Bodies (Antiketogenic Effect):* When the supply of insulin is inadequate fat metabolism is also affected, because fat must be used instead of carbohydrate for the release of energy, through cellular oxidation. The large increase in the amount of $CH_3CO.CoA$ which accumulates from the breakdown of fat is too great to be dealt with by the cellular TCA Cycle. Instead of being combined with oxaloacetic acid in the first stage of that cycle, it passes to the liver where it is condensed to form acetoacetic acid (a 4C)

molecule). The liver is unable to make use of acetoacetic acid ; and, therefore, the aid is delivered to the body tissues by circulation. In the blood, the acetoacetic acid undergoes further dissimilation forming β-*Hydroxybutyric acid* and *acetone.* These degradation products are known as Ketone bodies.

Acetyl CoA
(2C)
↓

6. *Stimulates protein-synthesis and growth e.g.,* nitrogen-retension, bone formation etc. Increases the incorporation of amino acids into peptides.
7. *Decreases Pyaemia i.e.,* accumulation of excess fat in blood and liver.

Diabetes Mellitus

Insulin deficiency causes diabetes mellitus. This is a disorder of carbohydrate metabolism characterised by :

(1) *Hyperglycemia* : It is a condition in which blood sugar increases above the mammal level (80 to 120 mg/100 ml).

(2) *Glycosuria* : It is a condition in which glucose is present in the urine in such quantities as will reduce Benedict's or Fehling's reagent. It occurs when the blood sugar level exceeds the renal threshold which is 180 mg.

(3) *Polyuria* : The high glucose level in the nephric filtrate increased osmotic pressure, sharply reducing the reabsorption of water back into the blood, hence victims of the disease urinate large volumes of urine frequently. This is known as polyuria.

(4) *Dehydration* : The continued loss of water and electrolytes in the urine leads to increasing dehydration.

(5) *Polydipsia* : Because of polyuria, the patient is thirsty inspite of drinking large amounts of fluid.

(6) *Weight Loss* : Inspite of *polyphagia* (increased appetite) the patient loses weight because of the excessive breakdown of protein.

(7) *Abnormal Metabolism of Fat* : Muscles are unable to utilize the glucose in the absence of Insulin, therefore the diabetic feels weak and tried. Since carbohydrate cannot be used as fuel, fat is used instead to provide the energy.

(i) *Lipaemia* : Fat is mobilized from the body stores and transferred to the liver, hence fat content of the liver and blood rises.

(ii) *Ketosis* : The disproportionate metabolism of fat in diabetes results in the overproduction of ketone bodies.

(iii) *Acidaemia or Acidosis* : As acetoacetic acid and P-hydroxybutyric acids are produced faster than they can be metabolised, the patient develops acidosis.

(8) *Diabetic Coma* : The keto acidosis is associated with increasing drowsiness and if untreated, the patient may become unconscious (diabetic coma) and die from a combination of hyperglycemia, ketosis, acidaemia and dehydration.

(9) Reduced Resistance to infections due to altered immunoligical response and altered metabolism etc.

Stimulus to Insulin Secretion

It is a high blood sugar level. There are no secretory nerve fibres to the islets ; when the blood sugar rises, this rise, is all the stimulus the islets need to increase their output.

Therapy : Insulin is now available in large quantities from the glands of slaughtered livestock (cattle an pigs). The molecular structure of these animal insulins is so close to that of human insulin that they may be used to treat diabetes. Insulin is protein in nature therefore, it is inactive when given by mouth, since it is destroyed by the digestive enzymes.

In recent year some types of "modified", insulin have appeared. Some are combined with Protamine, which is another protein and some with zinc salts :

1. IZS (Insulin Zinc Suspension)

2. PZI (Protamine Zinc Insulin)

The purpose of such modification is to render the insulin less soluble so that a given dose will be more slowly absorbed.

Insulin Reaction or Shock : Insulin injected after a period of exercise or long after a meal may drive the blood sugar level down to abnormally low levels, 43 mg 100ml. The victim becomes irritable, fatigued, and may lose consciousness. This condition is of insulin shock.

Glucagon

It is a polypeptide consisting of 29 amino acids arranged in a single chain. The effect of glucagon is to cause a rise in the blood glucose level by mobilizing the glycogen in the liver, which is broken down to glucose.

The pancreatic hormones are secreted into the hepatic portal vein, so that the liver forms a barrier between the hormones and the body cells.

The liver destroys glucagon before it can get into the body circulation, and its metabolic action is exerted solely upon the liver cells. Therefore, the ability of the liver to destroy the hormone has resulted in failure of its effect on muscle glycogen.

Alimentary Canal

Historically, some of the hormones concerned with the digestion of food were among the first hormones to be discovered. They are 'local' hormones, being produced by a localized area of the gut and acting upon the gut itself or one of its accessory glands.

Stomach

Gastrin : Gastric distension and protein in the stomach stimulates the gastric mucosa to secrete a hormone gastrin, which has a strong stimulating effect on the acid-secreting parietal cells of the stomach.

Duodenum : The duodenal wall secretes several hormones when food is passed into it from the stomach.

(i) *Secretin :* It was the first hormone to be discovered (in 1902) which circulates in the blood, and when it reaches the pancreas, it causes the exocrine cells of the gland to secrete the fluid base of the pancreatic juice (water and electrolytes like Na^+ HCO_3 and Cl^- with smaller concentration of K^+ and Ca^+).

(ii) *Pancreozymin (PZ) :* A second duodenal hormone, pancreozymin, follows the same path to evoke the production of the pancreatic enzymes.

(iii) *Cholecystokinin (CCK) :* Stimulates the gall bladder to rhythmic contractions and these result in an increase in the rate of flow of bile into the duodenum.

Small Intestine

It secretes two hormones :

(i) *Enterocrinin :* The hormone stimulates the intestinal mucosa to secrete the intestinal juice.

(ii) *Enterogastrone :* The presence of fats in the small intestine stimulates the intestinal mucosa to produce another hormone *enterogastrone* which eventually stops the secretion of gastric juice in the stomach and cairns down peristalsis there.

Kidney : Erythropoietin is formed by kidney that acts on haemopoietic organs for the stimulation of the formation of blood cells, especially in response to anemia. Erythropoietin acts on bone marrow to increase its production of R.B.C's.

Gonads : The testes and ovaries, in addition to their cytogenic function of producing spermatozoa or ova, manufacture steroid hormones which control secondary sex characteristics, the reproductive cycle, and the growth and development of the accessory reproductive organs.

Male Hormones : In mammalian testes, the Interstitial (Leydig) cells lying scattered in between the seminiferous tubules synthesize male hormones, the androgens. The principal male hormone is *testosterone,* which is synthesized from cholesterol through pregnenolone, progesterone, and hydroxyprogestrone. The latter is then converted to the C-19 ketosteroid-androstenedione, the immediate precursor of testosterone.

Of the synthetic products, *methyl testosterone* and *testosterone prepionate* are important members.

The principal metabolites of testosterone are *androsterone* and *etiocholanolone,* the major 17-ketosteroids in the urine.

Functions of Testosterone

(1) Promotes the growth and functions of the epididymis, vas deferens, prostate, seminal vesicles and penis.

(2) Stimulates the development of male secondary sex characters — (i) the voice deepens (ii) hair begins to grow on the face and on the upper lip ; in the axillae and in the pubic region. (iii) there is a greater skeletal and muscular growth, (iv) aggressive nature and sex urge. In brief, it determines libido.

(3) Influences spermatogenesis and seminal plasma secretions.

(4) Exerts a feedback inhibitory effect on pituitary LH secretion.

(5) It has a general anabolic effect on the body by stimulating protein synthesis.

Female Hormones : Two main types of female hormones are secreted by the ovary: the *estrogens* or *follicular hormones* produced by the cells of developing graafian follicle and the *progesterone* derived from the corpus luteum that is formed in the ovary from the ruptured follicle ovulation.

Function of Estrogens : The principal estrogenic hormone in the blood circulation is *estradiol* which is secreted by theoa internal cells of the graafian follicle. The estradiol is excreted as *estrone* or *estriole*.

(1). Promotes the development of female secondary sex characteristics. At the start of the adolescence (puberty),

(i) the mammary glands (breasts) develop chiefly by proliferation of ducts.

(ii) the pelvis becomes broader,

(iii) voice stays high pitched,

(iv) the genitalia enlarge and there is growth of hair around them and in the axillae,

(v) development of fat deposits in subcutaneous tissues and other particular regions to make a typical feminine body (characterised by rounded body contours).

(2) Prepares the uterine mucosa for the later action of progesterone *i.e.*, proliferative growth of the lining of the endometrium, deepening of uterine glands, and increased vascularity.

(3) Influences growth of fallopian tubes and stratification of vaginal epithelium.

(4) Suppresses the production of pituitary FSH.

(5) Influences behaviour and psychic patterns in females.

Functions of Progesterone : After ovulation, the discharged follicle cells become transformed into a characteristic body known the corpus luteum, a temporary endocrine structure that secretes the hormone, *progesterone.* It is a steroid which is synthesised from its immediate precursor, pregnenolone.

(1) Causes extensive development of the endometrium, preparing the uterus for the reception of the embryo and for its nutrition.

(2) The hormone suppresses estrus, ovulation and the production of pituitary LH, which originally stimulated corpus luteum formation.

(3) Stimulates the mammary glands.

(4) Maintains the foetus by forming placenta and inhibits uterine muscles to contract during pregnancy.

The chief excretory product of progesterone is *pregnanediol* which is present in the urine of pregnant women. Progesterone is metabolized to pregnanediol in liver by a series of reductive steps.

Relaxin : In addition to progesterone, corpus luteum of some mammals like guinea pig, mouse, rabbit, dog and cow, reduce a second hormone termed *relaxin.* It relaxes pubic symphysis and thus facilitates birth.

The Placenta

It secretes a hormone *Chorionic Gonadotropin* (CG) which is a product of the very early placenta (foetal part). Scientists have known for some time that Human Chorionic Gonadotropin exerts a protective influence on the unborn child, because deficiency or absence of the hormone in the early stages of pregnancy has been associated with abortion. This hormone has a major promoting influence on the production of progesterone from the ovary in the first six weeks. It is well known that progesterone is the gestational hormone which is essential for the continuation of pregnancy.

The Thymus

Thymus is partly an endocrine gland and partly a lymphoid structure. Located in the upper part of the thorax, the thymus is quite large in the young animal, but as the animal grows, the gland atrophies. By the time adulthood is reached, it is no more than a pair of small lobes lying on either side of trachea. Each lobe has numerous lobules and every lobule consists of a peripheral cortex densely packed with lymphocytes and a central medulla in which lymphocytes are less numerous.

The medulla also contains reticular cells and the unique *"Hassall's Corpuscle"* which are spherical structures composed of concentric layers of spindle-shaped cells whose function is unknown.

Functions

(i) The thymus normally secretes a polypeptide hormone—thymosin or thymin which depresses neuro-muscular transmission (by regulating the amount of Arch at the neuro-muscular junction of skeletal muscles). An excess production of thymin by the abnormal thymus results in a disease myasthenia gravis in which the skeletal muscles are weak and there is a derangement of neuromuscular transmission.

(ii) The thymus in the infant mammal plays a major role in setting up the lymphocyte-producing machinery of the lymph nodes, thus providing the basis for the development of antibodies, in this way it keeps the owner immune from certain diseases.

The Pineal Body

The Pineal gland is a small, pea-sized structure lying deep in the groove between the cerebellum and the cerebral hemispheres in rabbit. Histologically, it consists of Parenchyme cells and Neuroglia cells.

In lower vertebrates, it produces a hormone called melatonin. When injected into frogs, it causes a marked lightening of the skin. Its action on the melanophores is thus just the reverse of that of MSH. In the Laboratory rat, melatonin inhibits the gonads both as gamete producers and as endocrine glands.

As for humans, it atrophies at the age of 7 years and its role is uncertain.

Endocrine Glands in Insects

It has been established that the moulting and metamorphosis in insects is controlled by hormones. The insects have a well-developed compliment of endocrine organs consisting primarily of (1) the brain (protocerebrum) (2) the prothoracic glands and (3) the corpora allata. The glandular cells of all three organs show secretory activity prior to each moult, and the three types of secretion are essential for the normal course of larval moults.

1. *The Brain (protocerebrum)* : There are four groups of *neurosecretory cells* in the brain. Of these, two groups lie near the midline and one group on each side. The neurosecretory cells secrete a hormone which serves to activate the prothoracic gland as a result of which the moulting is initiated. The secretion of the brain, therefore, cannot act directly but only through the activation of the prothoracic gland.

2. *The Prothoracic Gland* : It is an irregular branching mass of glandular cells located in the thorax. The prothoracic gland lies in close association with the tracheal tubes (Fig.). The hormone produced by the prothoracic gland is called the growth and moulting hormone or ecdysone. It has following functions :

   (i) It initiates the process of moulting in the epidermis as a result of which the epidermal cells undergo growth and proliferation.

   (ii) It is responsible for the shedding of the old cuticle and the production of the new one.

Butenandt and Karlson (1951) prepared ecdysone in a chemically pure form silkworm pupae. 25 mg of the pure substance was isolated from five hundred kilograms of silkworm pupae. It was found that 0.0057 mg. of ecdysone thus isolated was enough to cause the pupation of a fly larva. Karlson and hoffmeister (1963) carried out the chemical analysis of the ecdysone and gave the empirical formula as $C_{27}H_{41}O_6$. Chemically, it is a steroid. Its structural formula is as follows :

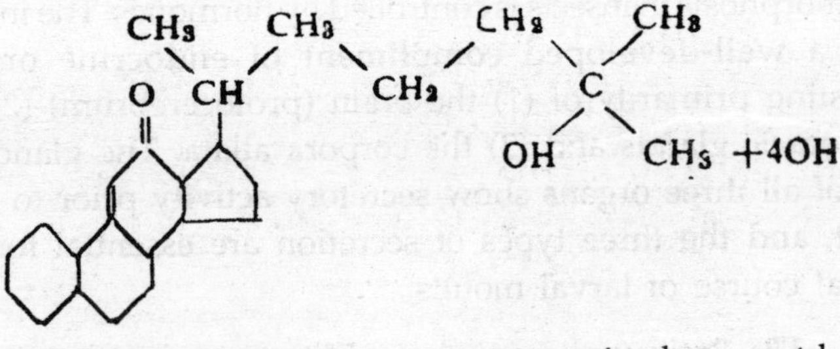

So far two ecdysones are now recognized : α -ectdysone and β -ecdysone.

3. *The Corpora Allata* : In most insects there are found two pairs of bodies connected by nerve strands to the protocerebrum and lying alongside to the dorsal aorta. First of these bodies on each side is the corpora cardiaca which is of the nature of nerve ganglia. Located more

posteriorly is the corpora allata which consists of secretory cells.

The secretion of the corpora allata controls the nature of the change that is going to take place at the time of moulting. The hormones secreted by the protocerebral neurosecretory cells and the thoracic gland induce an insect to moult but they do not determine whether it will be a larval moult, the pupal moult, or the imaginal moult. On the other hand, the secretion of the copora allata controls the qualitative changes in the body of the insect during metamorphosis.

The presence of the corpora allata is necessary to prevent metamorphosis and to keep the insect in the larval state. Therefore, the secretion of the corpora allata has been called the juvenile hormone. At every larval moult the cells of the corpora allata show signs of secretory activity but no such activity is observed during the pupal or imaginal moult. It is inferred, therefore, that at the time metamorphosis takes place, the corpora allata do not produce their hormone or at least are less active. It is actually the absence of the juvenile hormone that is the necessary condition for metamorphosis.

The juvenile hormone is chemically related to ecdysone and also a steroid or a substance of the terpene group.

The Physiology

The various physiological activities of the body are controlled by two systems : *nervous* and *endocrine.* In humans, as in all other mammals, the *endocrine system* consists of about ten different glands located in particular parts of the body. These glands have no ducts and liberate their secretions, known as the *hormones,* directly into the bloodstream. For this reason the endocrine glands are also referred to as the *ductless glands.*

Definition : A *hormone* may be defined as a *specific product (organic substance) of an endocrine gland secreted into the blood* *which carries it to some part of the body where it produces a definite physiological effect.* This effect may be either *excitatory* or *inhibitory* in its action.

Hormones are not only the *"chemical messengers"* that pass from the glands via the bloodstream to the target organ or some process ; but they are important regulating substances that control virtually every aspect of the metabolism in living cells. There are also chemical coordinators other than hormones, such as CO_2 and other metabolites. Such substances are often called *Parahormones.*

Characteristics of Hormones

Hormones possess the following characteristics, which distinguish them from other chemicals :

(a) Hormones are produced by a specialized gland and are secreted directly into the bloodstream.

(b) Hormones exert their effects not where they are locally produced, but on other tissues of the body.

(c) Hormones act upon their target cells by regulating the rates of specific metabolic reactions.

(d) Hormones are required in very small amounts and their effect may be excitatory or inhibitory depending upon their concentration and the physiological state of the responding tissue.

(e) As against enzymes which are proteins, hormones have a vaired chemical nature, including amino acid derivatives, polypeptides, steroids, etc.

Chemical Nature of Hormones

All the hormones are organic substances of varying structural complexity. Chemically, they may belong to any of the following categories:

(a) *Amino acid derivative: e.g.*, thyroxine.

(b) *Catecholamine: e.g.*, adrenaline, noradrenaline.

(c) *Polypeptide* or *protein : e.g.*, oxytocin, vasopressin, prolactin, adrenocorticotropic hormone, somatotropin (growth hormone), insulin, glucagon, secretin, relaxin.

(d) *Glycoprotein : e.g.*, thyrotropin, follicle-stimulating hormone, luteinizing hormone.

(e) *Steroid : e.g.*, cortisone, aldosterone, testosterone, estrogen, progesterone.

Kinds of Hormones

On the basis of their control and influence over physiological processes, the hormones may be divided into five categories:

(a) hormones concerned with the metabolism, e.g., insulin,

(b) hormones of digestion e.g., gastrin, secretin,

(c) hormones that regulate the growth and development, e.g., growth hormone or somatotropin ;

(d) hormones that control reproduction, *e.g.*, gonadotropic and sex hormones; and

(e) hormones that control other endocrine glands, *e.g.*, thyrotropin.

Some hormones are called *local hormones;* for example, acetylcholine, secretin and cholecystokinin. These hormones have specific local effects. Other hormones are the so-called *general hormones,* which are secreted by specific endocrine glands. For example, epinephrine and norepinephrine, etc. Hormone concentrations in the blood are measured by *radioimmunoassay.*

Mechanism of Hormone Action : The following mechanisms have been suggested :

The cAMP Mechanism

A number of peptide hormones perform their action through cyclic AMP (3'-5'-adenosine mono-phosphate) and *adenyl cyclase*. The Nobel Prize-winning work of E.W. Sutherland demonstrated that cyclic AMP, called as *second messenger*, mediates the action of many hormones and consequently regulates cellular activity in all forms of life. Cyclic AMP is produced by the catalytic action of *adenyl cyclase* on ATP.

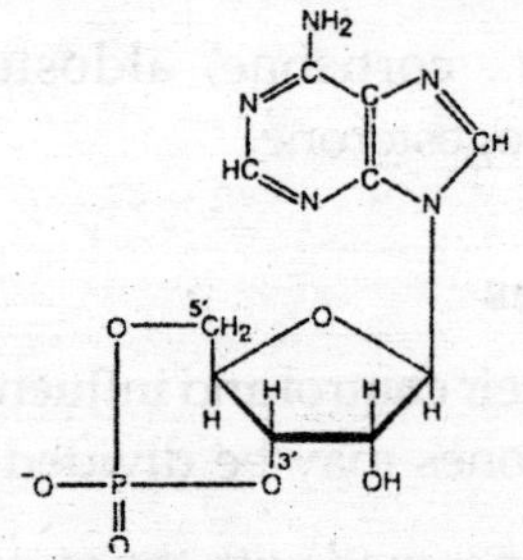

The structure of cyclic AMP.

$$\text{ATP} \xrightarrow{\text{adenyl cyclase}} \text{cAMP} + 2\ \text{Pi}$$

Once the hormone-activated reactions have been completed, cAMP is hydrolyzed in the presence of Mg^{2+} to 5' AMP by an enzyme *phosphodiesterase*. All cyclic AMP-dependent reactions are turned off, because 5 AMP is ineffective on these targets.

Five components are involved in the mechanism by which ligands bring about changes in the intracellular concentration of cyclic AMP. These are : a catalytic unit (adenyl cyclase), which catalyzes the conversion of ATP to cyclic AMP; stimulatory and inhibitory receptors; and stimulatory and inhibitory G proteins (G_s & Gi that link the receptor to the catalytic unit, like the receptors *(receptor proteins)*, adenyl cyclase is a transmembrane protein and eight isoforms of this enzyme have been described.

Hormones affect target cells by binding to specific receptor proteins on the outer surface of the cell membrane. After the receptor binds its hormone, there is stimulation of adenyl cyclase (AC) associated with the inner surface of the membrane. Coupling of the function of the receptor to the adenyl cyclase is accomplished by another protein that is also bound to the inner surface of the membrane. This protein, known as the *G-protein* strongly binds the nucleotide GTP when the receptor has bound hormone. With GTP bound, the a subunit of G-protein interacts with and stimulates or inhibits the activity of adenyl cyclase. When the appropriate ligand binds to a stimulatory receptor, a Gs α subunit activates the adenyl cyclase, and when it binds to the inhibitory receptor, a Giα subunit inhibits adenyl cyclase. The G-protein also has a GTPase activity. Since adenyl cyclase is stimulated when a G_s-protein binds GTP, but not GDP, the GTPase activity tends to stop the stimulation of adenyl cyclase.

It appears that one hormone-receptor complex can activate several G-proteins to bind GTP and thus stimulate or inhibit several adenyl cyclases.

Two bacterial toxins (cholera toxin and pertussis toxin) have important effects on adenyl cyclase that are mediated by G-proteins. The drug *forskolin* stimulates adenyl cyclase activity by a direct action on the enzyme.

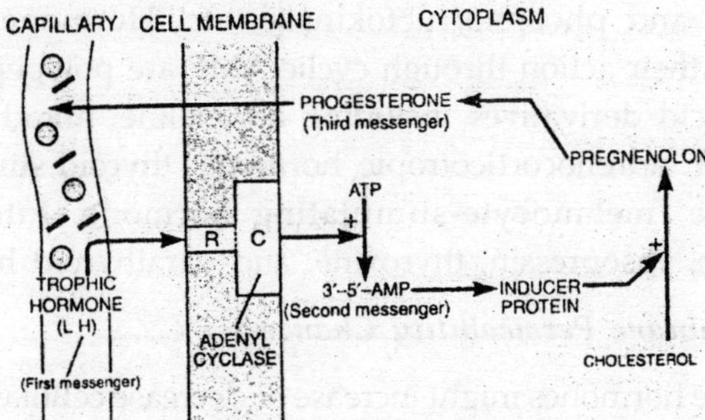

Mechanism of trophic hormone action on the cell to stimulate the production of cyclic-AMP, which then acts as a 'second messenger' to stimulate steroid secretion.

One interesting example of hormone-enzyme interaction is found in the liver cells where glucagon and epinephrine activate adenyl cyclase on the membrane of liver cell resulting in increased intracellular cAMP level.

The increased cAMP concentration activates protein kinase which then converts by phosphorylation nonactive enzyme phosphorylase *b* kinase into active phosphorylase *b* kinase. The phosphorylase *b* kinase is in turn phosphorylated to phosphorylase *a* which converts glycogen to glucose-1-phosphate (Fig.).

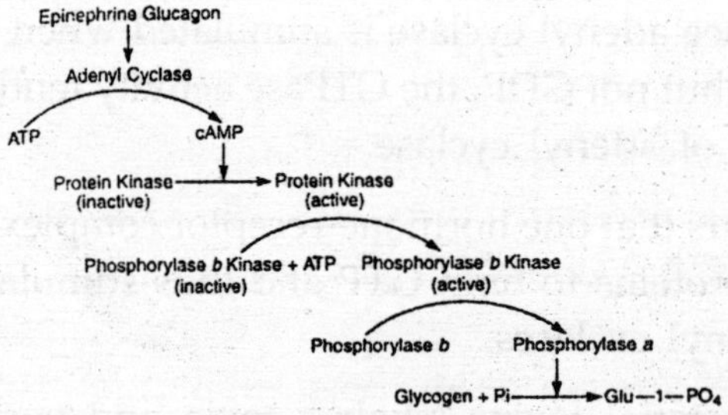

Activation of liver phosphorylase by hormones.

Other enzymatic processes that are known to be directly affected by the cyclic-AMP are glycogen synthetase, protein kinases, and phosphofructokinase, etc. Hormones which mediate their action through cyclic-AMP are polypeptides or amino acid derivatives including adrenaline, noradrenaline, glucagon, adrenocorticotropic hormone, thyroid-stimulating hormone, melanocyte-stimulating hormone, luteinizing hormone, vasopressin, thyroxine, and parathyroid hormone.

(ii) Membrane Permeability Change

Some hormones might increase or decrease cellular activity by affecting the permeability of the cell membranes. All the neurotransmitter substances (local hormones) combine with receptors in the postsynaptic membrane. This causes a conformational change in the protein structure of the receptor, usually opening or closing a channel for one or more ions like Na, K or Ca, etc. Then, it is the altered movement of these ions through the channels that causes the subsequent effects on the postsynaptic cells. Some general hormones like epinephrine and nonepinephrine have especially powerful effects in opening or closing membrane ion channels for Na, K, or both, thus changing the membrane potentials of specific smooth muscle cells and causing excitation or inhibition.

(iii) Hormone-Gene Interaction

It is considered that some hormones act by controlling the gene activity in the target cell so that the DNA, through the genetic code mechanism, induces protein or enzyme synthesis that may produce physiological responses. *A steroid hormone* molecule released in the bloodstream combines with a receptor molecule present in the cytoplasm of the target cell (intracellular receptor). The receptor molecule for the steroid hormone consists of two subunits and each subunit can bind to a single hormone molecule.

The receptor and the hormone molecule together form the hormone-receptor complex. This complex enters the nucleus and attaches itself to the target cell chromatin. The binding site occurs on a specific portion within the nonhistone protein known as the AP_3 fraction. Thus, on entering the nucleus the hormone-receptor complex binds to AP_3 fraction with the second subunit (designated as Y in Fig.) attached to it. This combination determines what genes in the DNA molecule are going to be activated. The subunits then dissociate and the first subunit (designated as X in Fig.) binds to the DNA enabling a molecule of RNA polymerase to occupy an initiation site at the DNA molecule. A segment of DNA is then transcribed producing a strand of messenger RNA that serves as a template for the synthesis of a protein. The latter in turn performs the necessary function of the hormone (Fig.).

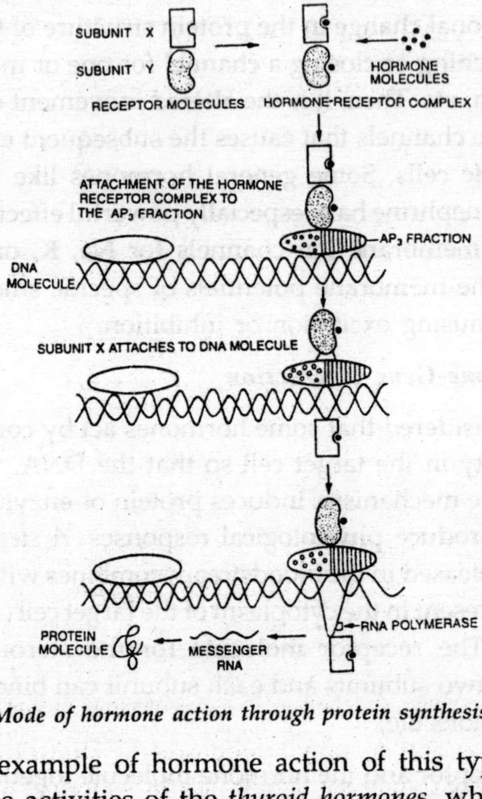

Mode of hormone action through protein synthesis.

As an example of hormone action of this type we may consider the activities of the *thyroid hormones,* which play an important role in temperature regulation in homeothermic vertebrates. Administration of small amounts of thyroid hormone to a mammal increases its metabolic rate. It is typical of this and other effects of thyroid hormones that the response to the treatment is delayed for several hours. This indicates that the calorigenic effect is secondary to some other effects of the hormones.

A study of the time course of events after the injection of a small, physiological dose of triiodothyronine into a thyroidectomized rat showed that increase in respiratory rate lagged behind an increase in protein synthesis in ribosomes and protein accumulation in mitochondria, which again lagged behind the synthesis of mRNA. It was, moreover, shown that treatment with actinomycin D, which is believed to suppress the formation of mRNA, completely prevented the calorigenic effect of the thyroid hormones. Treatments with other substances known to interfere with RNA or protein synthesis were also found to suppress or strongly reduce the calorigenic as well as the growth-stimulating effects of the thyroid hormones. The locus of action of these hormones thus appears to be at the level of the genes that produce the mRNA.

Although there are a number of instances wherein the ultimate hormonal effects turn out to be exerted on the level of synthesis of proteins, this mode of action does not appear to be universal.

Regulation of Hormone Secretion

The secretion of hormones by the endocrine glands is controlled by a complex of *chemical* and *nervous factors*. The secretory activity of most of the endocrine glands seems to be regulated by certain substances carried by the blood. In some cases these chemical factors are hormones themselves. In other cases they are relatively simple substances such as the sugars (*e.g.*, blood glucose stimulates the flow of *insulin* by the pancreas) and acids (*e.g.*, the secretion of *secretin* by the small intestine is stimulated by the acid chyme entering from the stomach). In a few glands (*e.g.*, the *adrenal medulla*) the secretory activities are regulated directly by the visceral nervous system.

Metabolic Clearance Rate

Two factors can increase or decrease the concentration of a hormone in the blood. One of these is the *rate of secretion* of the hormone . The other is the *rate of removal* of the hormone from the blood, which is called the *metabolic clearance rate.* It is calculated by the following formula :

$$\text{Metabolic Clearance Rate} = \frac{\text{Rate of disapperance of hormone from the plasma}}{\text{Concentration of hormone in each millilitre of plasma}}$$

Hormones are "cleared" from the plasma in several ways, including

(1) binding with the tissues,

(2) metabolic destruction by the tissues,

(3) excretion by the liver into the bile, and

(4) excretion by the kidneys into the urine.

Endocrine Glands in Invertebrates : Most of our knowledge of invertebrate endocrine glands is based on the investigations that have been recently made in crustaceans and insects. In general, it is difficult to homologize the invertebrate endocrine glands with those of the vertebrates. But many hormonal functions are analogous in the two groups and there is sufficient reason to believe that the hormones have an equally important role among the invertebrates.

The sows *gland* in the eye stalk was formerly regarded a gland of internal secretion among the decapod crustaceans. But it is now definitely known that a cluster of neurosecretory cells, called the *X-organ* located near the sinus gland, is actually the source of internal secretion. The axons of the neurosecretory cells of the X-organ convey the secretions to the sinus gland which functions as a storage and release centre for the hormones. The latter are released into the blood, which distributes them to the responsive tissues in the body. Several hormones are known to be definitely secreted by the X-organ. These are : (a) three *chromatotropic hormones* for the regulation of chromatophores or pigment cells, (b) *moult-inhibiting* hormone, and (c) the hormone for the control of eye pigment.

The process of metamorphosis in insects, like the process of moulting in crustaceans, is under the control of particular hormones produced by the neurosecretory cells and transported by the blood. Certain neurosecretory cells in the corpus brain of insects (Fig.) secrete the cardiacum brain hormone *(prothoracicotropic hormone)* which stimulates the so-called *prothoracic gland.* In *Periplaneta,* this endocrine gland is a X-shaped strand of endodermal cells in the ventral part of the prothorax. The prothoracic gland then secretes *ecdysone,* a growth and differentiation hormone identified as a steroid. By initiating the complex process of growth and differentiation leading to moulting and the transformation from the larval to the pupal stage, *ecdysone* thereby initiates the post-embryonic growth and development of insects. Another hormone termed the *juvenile hormone,* secreted by a third endocrine gland, called the *corpus allatum,* lying near the brain, stimulates larval development but inhibits metamorphosis.

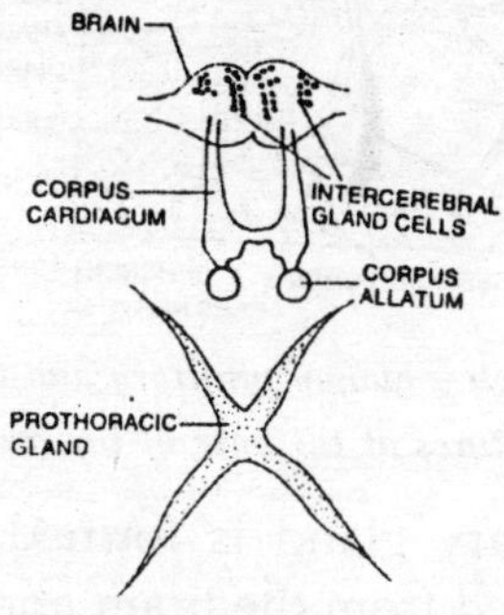

Endocrine glands in Periplaneta americana.

There is some evidence that the secondary sex characteristics among invertebrates are under the hormonal control. When parasitic castration by the parasite *sacculina* occurs in male crayfish, sexual changes toward femaleness are known to occur.

According to their location in the body, the principal endocrine glands of the vertebrates belong to the following four categories :

1. *Cranial*: *e.g.,* the Pituitary gland.
2. *Pharyngeal: e.g.,* the Thyroid and the Parathyroid glands.
3. *Abdominal : e.g.,* the Islets of Langerhans, Adrenal, and Gastrointestinal.
4. *Sexual: e.g.,* Testes, Ovaries and Placenta.

The Pituitary Gland or Hypophysis : *Location and Structure :* The pituitary gland is well protected, for it lies between the roof of the mouth and the floor of the brain in a depression of the sphenoid bone. It is a small rounded body about the size of a large pea weighing approximately 0.5 g in man but usually becomes much larger in a pregnant woman.

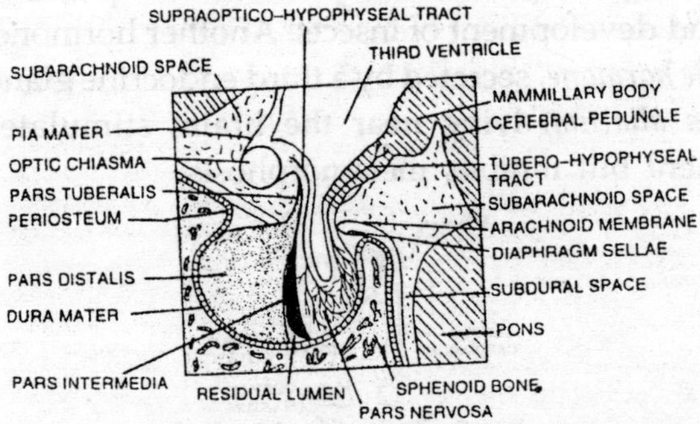

Section through a human pituitary and the adjacent structures at the base of the brain.

The adult pituitary gland is formed of two embryonic structures, one derived from the brain and the other from the epithelium of the mouth. Thus, it is a gland of double embryonic origin, consisting of two lobes *i.e.*, the anterior and posterior lobes (Fig.). The two lobes constituting the pituitary are essentially two different endocrine glands, as are the adrenal medulla and adrenal cortex, and the thyroid and parathyroid glands.

(a) *Anterior Lobe* : The upper or anterior lobe, now usually called the *adenohypophysis*, is derived from the epithelial tissue of the mouth. In the course of embryonic development the *neurohypophysis* moves downward, becoming posterior in position to the adenohypophysis which has developed upward to assume an anterior location relative to the neurohypophysis. The adenohypophysis in a mammal consists of the following parts :

(i) *Pars Tuberalis :* The pars tuberalis forms a collar of cells 25 to 60 μ thick around the neural stalk. It is thickest anterior to the stalk and frequently incomplete on the posterior aspect. The cells are arranged in short cords or globular clusters. They are polyhedral, with centrally placed nuclei and the cytoplasm contains fine granules, and is weakly basophilic. Nests of squamous cells are often found around the pars tuberalis. They are important because they may give rise to tumours (craniopharyngiomas).

(ii) *Pars Intermedia (middle or intermediate lobe) :* In most species this portion of the adenohypophysis is quite distinct. However, in others *(e.g.,* the whale) and in birds, it is lacking. In man, the pars intermedia is rudimentary and its cells often surround colloid-filled cysts, and merge imperceptibly with those of the pars distalis. They are either *chromophobes* or *basophils.* The cells lining the follicles may be ciliated. A unique feature in humans is the invasion of the parsnervosa by basophil cells of the pars intermedia.

(iii) *Pars Distalis* : It forms about 75% of the hypophysis. The cells in the pars distalis are arranged mainly in cords between which are large-bore capillaries. On the basis of the staining properties, three types of cells are distinguished in it, *acidophils, basophils,* and *chromophobes.* These are said to constitute about 40, 10, and 50 per cent of the anterior lobe (pars distalis) cells respectively.

(b) *Posterior Lobe* : The posterior lobe or the *neurohypophysis* has two subdivisions, *pars nervosa* or *infundibular process,* and *infundibulum* (median eminence and infundibular stem). While developing from the floor of the diencephalon behind the optic chiasma, the *infundibulum* and the *pars nervosa* at first contain the continuation of the cavity of the third ventricle.

This cavity is obliterated during the development except for cystic remnants lined by the ependymal cells. Following this obliteration, the hilar or central portion of the pars nervosa is formed by a densely packed bundle of nonmyelinated fibres known as the *hypothalamohypophysial tract*. In man, however, the *median eminence* is not prominently developed, and it does not have the characteristic neurohypophysial structure seen in the other vertebrates, including the lower primates.

Hormones of Adenohypophysis

1. *Thyrotropin or Thyroid-Stimulating Hormone (TSH)*: This hormone has been purified to a considerable extent and is shown to be a *glycoprotein* with a molecular weight of less than 10,000. It controls various aspects of the thyroid gland function including the development and maintenance. Thyrotropin also stimulates iodine accumulation, the conversion of iodine into the thyroid hormones, and their release from the gland into the circulation. It is thought that certain nerve secretions from specific areas of the hypothalamus are transmitted by way of the pituitary stalk to regulate the secretion of thyrotropin. If there is an excess of secretion, there is an increase in the size of the thyroid gland. But excess of secretion is prevented by *thyroxine*.
2. *Adrenocorticotropic Hormone (ACTH)* : The ACTH regulates the development and secretion of the adrenal cortex in the same way as TSH controls the thyroid gland. The ACTH molecule has been well characterised as a *polypeptide* of molecular weight 4500, containing 39 amino acids. The secretion of ACTH and adrenal

cortical hormones are mutually affected by a negative feedback mechanism. In addition, as in the case of TSH secretion, there is evidence of nerve secretions from certain hypothalamic areas of the brain which influence the secretion of ACTH. Such a mechanism suggests a basis for the release of ACTH by several different stimuli including the emotional stress, extreme temperatures, poisons and drugs, etc.

3. *Growth Hormone or Somatotropin (STH)*: This hormone is a protein. The growth hormone of humans has a molecular weight of 21500 and consists of 190 amino acids. The growth hormone stimulates growth of all the body cells including the skeletal and muscular tissues. Under-secretion of the growth hormone during the years of skeletal growth of an individual results in *dwarfism,* whereas an over-secretion of the growth hormone prior to adulthood result in *gigantism.* Over-secretion of the hormone after the attainment of full skeletal growth gives rise to the condition called *acromegaly* (*acros* = extremity, *megale* = large). It is manifested by a characteristic enlargement or overgrowth of the bones of the hands, feet, and in particular the jaws, cheeks and face as well as an increase in the overlying soft tissues.

4. *Gonadotropic Hormones (GTH)*: The three gonadotropic hormones secreted by the adenohypophysis are proteins. Of them, the FSH and LH are glycoproteins with a molecular weight of about 30,000 and with 15 to 18 per cent carbohydrate while the LTH is a polypeptide. Their secretion is apparently controlled by feedback mechanisms and by nerve secretions from the hypothalamus. These are :

   (a) *Follicle-Stimulating Hormone (FSH):* It is so-called because it stimulates the development and maturation of the ovarian follicles producing the egg cell and one of the female sex hormones, *estrogen*. In the male, the FSH induces spermatogenesis in the testes.

(b) *Luteinizing Hormone (LH) or Interstitial Cell-Stimulating Hormone (ICSH):* In the female, LH is involved in the further development of the egg cell and its release by rupturing of the ovarian follicle. It is also responsible for stimulating the development of the *corpus luteum* that in turn secretes *progesterone*. The progesterone and estrogen are the two principal female sex hormones responsible for sexual drive and the development of the secondary sexual characters (*e.g.*, mammary glands, etc). In the male, LH stimulates the secretion of the male sex hormone, *testosterone*, by specialized cells in the testes. Testosterone is primarily responsible for the sexual drive and development of the secondary sexual characteristics of the male (*e.g.*, hair on the body, deeper voice, etc).

(c) *Lactogenic* (Luteotropic) *Hormone LTH* or *Prolactin* : Prolactin (PRL) is a polypeptide. It has been termed by *Nicoll* and *Bern* (1971) as "jack-of-all-trades", with more than eighty different hormonal activities ascribed to it. According to them, the many functions of prolactin belong to the five major categories : reproduction, osmoregulation, growth, integument, and a group of actions in which prolactin has a synergistic effect with the steroid hormones.

In mammals, this hormone induceds the milk secretion (lactogenesis) in the mamary gland of the mother shortly after giving birth. It is now known that prolactin also functions together with estrogen in promoting the development of the mammary glands. It also serves together with LH to maintain the corpus *luteum* (and therefore *progesterone* secretion) which is so vital for the maintenance of pregnancy. In the rat prolactin has luteotropic functions, *i.e.*, it promotes and maintains the secretion of progesterone. In bony fish, prolactin enhances production of a dilute urine and helps in osmoregulation by both gills and kidney.

5. *Intermedin or Melanocyte-Stimulating Hormone (MSH)*: Two distinct hormones known as the melanocyte-stimulating hormones have been purified from several vertebrates. They stimulate the distribution and concentration of the *melanin* pigment granules in the epidermal cells of the skin. In the frogs and tadpoles, MSH is responsible for the generalized blackening of the skin. Removal of the pituitary in these animals results in a concentration of the pigment near the centre of the cells, thus lending a lighter or silvery colour to the skin. However, the role of MSH in pigmentation in humans is not known.
6. *Diabetogenic Hormone* : It is also produced in the anterior lobe and affects the metabolism of carbohydrates.
7. *Ketogenic Hormone* : It causes an increased breakdown of fats to ketone bodies which appear in the blood and urine.

***Hypothalamic Regulation of the Adenohypophysis* :** Many of the functions of the adenohypophysis depend on its connection with the hypothalamus. The hypothalamus produces three important substances, *gonadotropic release factor, corticotropic release factor* and *thyrotropic release factor* (or thyrotropin-releasing hormone, TRH), which govern the release of various tropic hormones on reaching the adenohypophysis by the hypophyseal portal system or by the general circulation. In return, these tropic hormones react back *(feedback)* and influence the functions of the nervous system. Thus, this plan involves a three stage hormonal system : releasing factors from the hypothalamus regulate tropic hormones from the adenohypophysis, which in turn regulate hormone secretions of the respective target glands, *e.g.*, gonads, thyroid, etc. (Fig.). This mechanism of feedback, whereby a target organ can influence its own production positively or negatively by sending back information to an early stage in its synthetic pathway, is not clear in all instances. But it is thought that whenever a feedback occurs it is mediated through the nervous system and reaches the hypophysis by way of the hypothalamus. Apparently some pituitary hormones are not subject to feedback control and are secreted independently of the central nervous system. This general scheme shows that parts of the nervous system function as the specialized endocrine organs.

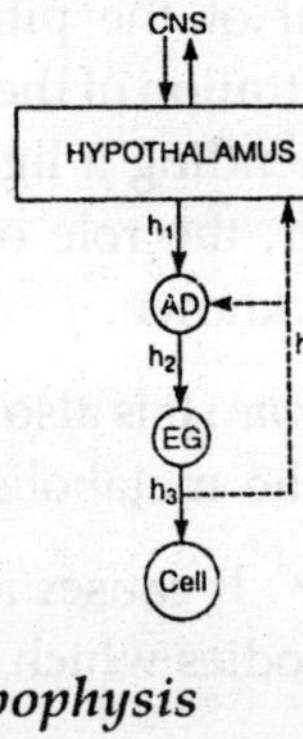

Hormones of Neurohypophysis

The neurohypophysis of most mammals, including man, releases *oxytocin* and *vasopressin*. Both vasopressin and oxytocin, are almost identical in structure with *isotocin, mesotocin* and *vasotocin*, the hormones released from this posterior lobe of the pituitary gland in the fish, frogs, and other nonmammalian vertebrates respectively.

However, the hormones of the neurohypophysis are secreted in the secretory nerve cells of the hypothalamus and migrate through the axons to the neurohypophysis where they are stored. It is generally believed that the nerve impulses control the synthesis of these hormones in the hypothalamus as well as their release from the neuro-hypophysis. Oxytocin and vasopressin are polypeptide hormones and they have seven amino acids in common. The similarity suggested that they arose through mutations in duplicate copies of a single gene. Vaso-pressin differs from oxytocin in having phenylalanine replacing isoleucine at position 3 in the ring, and arginine replacing leucine at position 8 in the side chain and it may be called 8-arginine-vasopressin. The structure is given in Fig.

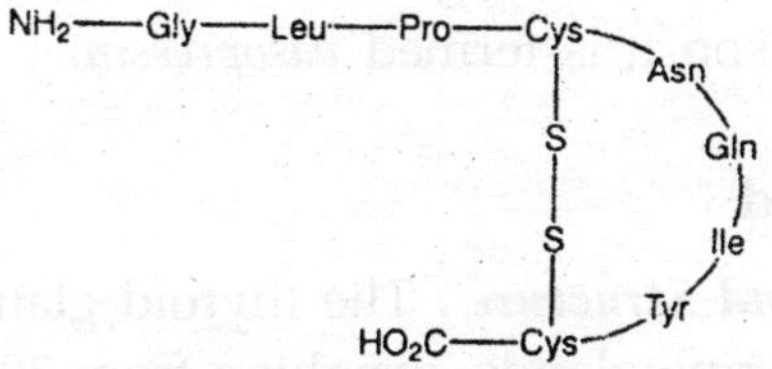

Oxytocin (Condensed Structure)

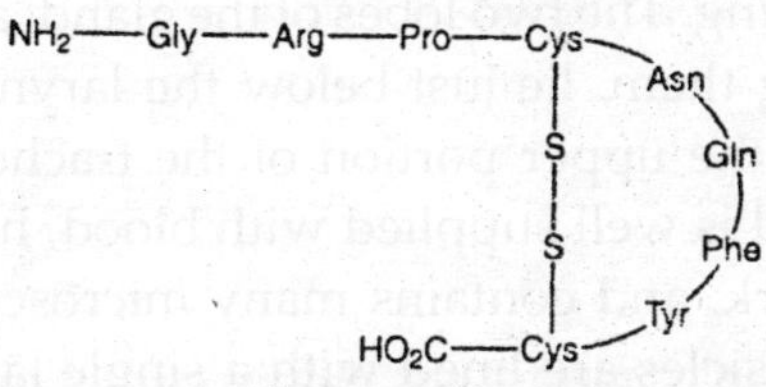

Vasopressin (Condensed Structure)

1 *Oxytocin* : It has a unique effect in producing the contraction of uterine muscles and uterus. It also promotes the ejection of milk from the ducts of the lactating breast. Several observations suggest that oxytocin may play a role in the reabsorption of sodium by the kidney, which may help to explain why the proportion of vasopressin and of oxytocin is the same in both sexes. Oxytocin is sometimes frequently injected during and after birth to stimulate contraction of the uterus and constriction of blood vessels, thus facilitating the birth process (parturition) and impeding the loss of blood.

2. *Vasopressin or Antidiuretic Hormone (ADH)* : It has an important role in regulating water reabsorption by the renal tubules. Apparently changes in the osmotic concentration of the blood and body fluids stimulate particularly sensitive neurons in the hypothalamus resulting in the release of the hormone from the neurohypophysis. An under secretion of vasopressin results in the disease *Diabetes insipidus,* in which a huge amount of urine is excreted. Vasopressin also causes the constriction of the musculature of the arterioles and capillaries, including the coronary and pulmonary vessles, thus causing some rise in blood pressure. For this reason it is termed *vasopressin.*

Thyroid Gland

Location and Structure : The thyroid gland is the largest of all the endocrine glands, weighing from 20 to 30 grams in adult human being. The two lobes of the gland, and the isthmus of tissue joining them, lie just below the larynx or voice box, just anterior to the upper portion of the trachea or windpipe (Fig.). The gland is well supplied with blood, has a connective tissue framework, and contains many microscopic cavities or vesicles. The vesicles are lined with a single layer of cuboidal epithelial cells which are the actual secretory cells responsible for the production of the thyroid hormones. The hollow centre of each vesicle is filled with a gelatinous colloidal material made up mostly of a large complex glycoprotein called *thyroglobulin.*

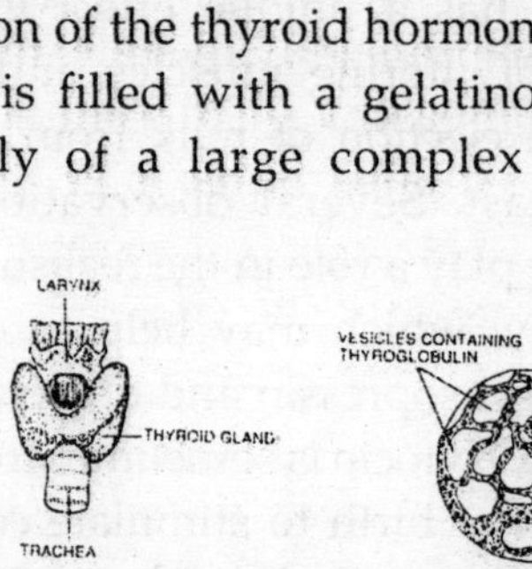

Gross and microscopic structure of the thyroid gland in man. Thyroid Hormones and their Physiological and Biochemical Effects

It is now known that the thyroid hormone consists of atleast four physiologically active iodinated organic compounds collectively called *iodothyronines* or thyroid hormones (Fig.). The most important and abundant (about 93%) of the four is called *thyroxine* or *tetraiodothyronine* (T_4) which until recently was believed to be the only physiologically active component or hormone secreted by the gland.

In human beings, thyroid glands also secrete *calcitonin,* which is a large polypeptide with a molecular weight of about 3400 and a chain of 32 amino acids. The so-called *parafollicular cells,* or C-cells, in the interstitial tissue between the follicles of the human thyroid gland which constitute only 0.1 percent of the thyroid gland, secrete *calcitonin.* These cells are remnants of the ultimobranchial glands of lower animals. Thyroid secretion is controlled primarily by thyroid-stimulating hormone (TSH) secreted by the adenohypophysis.

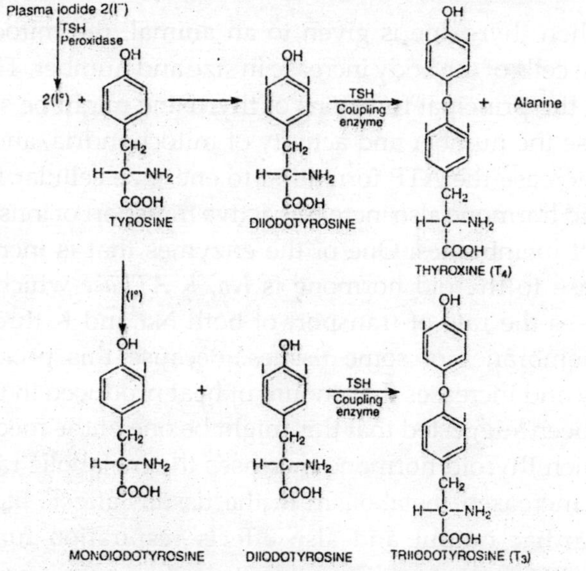

The synthesis of thyroid hormones (T_3 and T_4) from amino acid tyrosine. Before the iodide can be incorporated, it must be oxidised (I°) by a peroxidase system of enzymes, which is under the influence of TSH, thyroid stimulating hormone.

The structure of *thyroxine* ($C_{15}H_{11}O_4I_4N$) is the iodinated derivative of the essential amino acid *L-tyrosine*. Its potency as a hormone is in some way associated with its four iodine atoms. If two of them are removed or, substituted by other atoms, potency is sharply reduced. If all the four iodine atoms are removed, potency disappears altogether. Being an amino acid, thyroxine is not digested in the gut, and it can, therefore, be taken by mouth.

Functions of the Thyroid Hormones

1. *Increased Matabolic Activity* : The thyroid hormones stimulate the rate of oxygen consumption or *oxidative metabolism* and resulting heat production (calorigenesis) of the various cells and tissues of the body. Thyroid hormone, in the homeotherms, stimulates many phases of *carbohydrate metabolism*, from the absorption of sugars to their utilisation in the tissues.

When thyroxine is given to an animal, the mitochondria in most cells of the body increase in size and number. Therefore, one of the principal functions of thyroxine might be simply to increase the number and activity of mitochondria, and they in turn increase the ATP formation to energize cellular function. Thyroid hormone also increases active transport of ions through the cell membranes. One of the enzymes that is increased in response to thyroid hormone is *Na, K-ATPase* which in turn increases the rate of transport of both Na and K through the cell membranes of some tissues. Because this process uses energy and increases the amount of heat produced in the body, it has been suggested that this might be one of the mechanisms by which thyroid hormone increases the metabolic rate of the body. Increased metabolism in the tissues affects blood flow and cardiac output and also affects respiration function of muscles and sleep.

2. *Regulatory Role* : Thyroid gland also has a second important function, namely a *regulatory role* (which it

performs by its effects on the cellular metabolism) in general process of growth and development of all cells and tissues. These include the sexual development, maturation of bones and teeth, mental development, and energy metabolism, etc. The neo-natal requirement for thyroid hormones is, of course, well established, notably in human cretinism where thyroid deprivation results in reduced skeletal and somatic growth, retarded sexual development, mental deficiency, low basal metabolic rate, and increased risk of infection. However, the growth-promoting action of the thyroid hormones involves interaction with other hormones, especially somatotropin.

The effect of thyroid hormones on the nervous system is also quite clear. For example, if rats are made thyroid-deficient at birth, the cerebral cortex fails to develop normally. The behaviour of rats is also affected. Such rats are sluggish and perform poorly in learning experiments. However, the defects in the cerebral cortex can be corrected by thyroxine treatment, but only if the rats are treated within a critical period of about 15 days after birth. If the treatment is delayed, the defects are irrversible, as they are also in the humans.

3. *Effect on Other Endocrine Glands* : Increased thyroid secretion increases the rates of secrection of most other endocrine glands, but it also increases the needs of the tissues for hormones. For instance, increased thyroxine secretion increases the rate of glucose metabolism in the body and therefore a corresponding need for increased insulin secretion by the pancreas. Thyroid hormone also increases many metabolic activities related to bone formation and, as a consequence, increases the need for parathyroid hormone. Finally, thyroid hormone increases the rate at which adrenal glucocorticoides are inactivated by the liver. This leads

to feedback increase in adrenocorticotropic hormone production by the adenohypophysis, and therfore, increased rate of secretion of glucocorticoid by the adrenal gland.

4. *Effect of Calcitonin on Blood Calcium* : Calcitonin, secreted by thyroid gland in human beings, tends to reduce the blood calcium ion concentration (an effect which is opposite to that of parathyroid hormone). In fish, amphibians, reptiles and birds, calcitonin is secreted by the *ultimobranchial* glands and plays an important role in controlling the blood calcium ion concentration when these animals change their habitat from freshwater to seawater (which contains more calcium).

However, it should be noted that calcitonin mechanism acts only weakly and only as a short-term regulator of Ca ion concentration because it is rapidly overridden by the much more powerful parathyroid control mechanism.

5. *Effect on Development* : Thyroid hormone also controls some developmental processes, at least in amphibia and teleosts. In these vertebrates the hormone is specifically required in the *metamorphosis.* If thyroid function in an embryo or a young tadpole is in some way inhibited (*e.g.*, by excising the thyroid gland), then the animal remains a tadpole permanently. Conversely, if a young tadpole is given an excess of thyroxine, then the larva metamorphoses prematurely into a tiny froglet. Metamorphosis takes place normally when the thyroid gland has developed to maturity and produces appropriate amounts of the hormone. The developmental role of the thyroid hormone here resembles the role of the insect hormones.

Diseases of the Thyroid-Hypothyroidism : The phenomenon of under-secretion by the thyroid gland is called *hypothyroidism*. The occurrence of hypothyroidism (which may by due to a variety of factors including atrophy of the gland, or deficient dietary supply of iodine, etc.) in adults produces the condition called *myxedema*, a disease characterized by a low metabolic rate, increased plasma cholesterol, thickness and puffiness of the skin, coarseness and brittleness of the hair and finger nails, and a general physical and mental lethargy. There is usually a weight-gain, loss of hair, slower pulse rate, reduced blood pressure, and a decreased body temperature. The administration of adequate dose of the thyroid extracts quickly restores normal function to individuals with myxedema.

Cretinism : When *hypothyroidism* is present from the birth, the condition is called *cretinism*. It results from congenital lack of a thyroid gland *(congenital cretinism)*, from failure of the thyroid gland to produce thyroid hormone, or from iodine deficiency in the diet *(endemic cretinism)*. Cretinism is typified by retarded physical, sexual and mental development and low metabolic rate. Such children are stunted and of low intelligence. An early administration of the thyroid hormones by mouth or injection is necessary to prevent permanent effects of cretinism.

Goitre : Thyroxine deficiency may be accompained by thyroid enlargement or *goitre*. Here the number of thyroid cells increases and more hormone may then be secreted by the enlarged gland. This may or may not suffice to compenstate for an original hormone deficiency. If it does, the body functions will be normal but a *simple goitre* will be in evidence. If the deficiency is not compensated for, then cretinism in the young and myxedema in the adult may develop in spite of and in addition to the goitre. One of the causes of *simple goitre* is dietary deficiency of iodine and so the addition of iodine to certain foods (*e.g.*, table salt) and drinking water, or the administration of adequate amounts of thyroxine fronvan external source reduces its incidence.

Hyperthyroidism: The phenomenon of oversecretion by the thyroid gland is called hyperthyroidism or *thyrotoxicosis*.

It is accompained by physiological effects and symptoms which are the opposite of those observed in hypothyroidism. One of the most severe conditions of hyperthyroidism is called *Graves' disease* or *Exophthalmic goitre.*

It is characterised by an enlargement of the entire gland, an elevated basal metabolic rate, weight-loss, profuse perspiration, higher pulse rate, and a typical protrusion of the eyeballs called *exophthalmos.* The condition may be treated by surgical removal of a portiori of the thyroid gland, administration of radioactive iodine which accumulates in the thyroid destroying some of its tissues, or the use of certain thyroid-inhibiting drugs which interfere with the synthesis of thyroid hormones. Hyperthyroidism may also be caused by a tumor or growth of the thyroid gland.

Control of Thyroid Secretion

The various aspects of the thyroid gland function are controlled by *thyrotropin.* There are two main factors which influence the secretion of thyrotropin. They are as follows :

(a) The thyrotropin-releasing factor (TRF) of the hypothalamus, and

(b) The blood level of the thyroid hormones.

It is believed that the hypothalamus acts by secreting the TRF which is transmitted via the hypophysial portal system to stimulate the anterior pituitary cells. Thus, the secretion of thyroid hormones is apparently controlled by the thyroid-stimulating Hormone (also known as the thyrotropin or *thyrotropic hormone TTH)* secreted by the anterior lobe of the hypophysis. The thyrotropic hormone is produced in larger quantities as a result of low levels of thyroxine in the blood. The rise of thyroxine in the bloodstream in turn inhibits, by a negative feedback mechanism, the production of the thyrotropic hormone, thus decreasing stimulation of the thyroid.

The regulation of the level of thyroid hormones once secreted into the bloodstream is affected in part by the liver through enzymatic mechanisms involving the removal of iodine and the amino group from the hormones.

The Parathyroid Glands Location and Structure : In human, the parathyroid glands consist of two pairs of small oval glands attached to the thyroid gland. These glands are absent in fishes, but are present in amphibians, reptiles, birds and mammals in either one or two pairs, depending on the species.

In adult human beings, the parathyroid gland consists of mainly *chief cells* and a small to moderate number of *oxyphil cells,* but oxyphil cells are absent in many animals and in young human beings. The chief cells are believed to secrete most of the parathyroid hormone. However, the function of oxyphil cells is not certain.

Parathyroid Secretion and its Function : The parathyroid hormone is a polypeptide with 84 amino acids and has a molecular weight of about 9500. Smaller ompounds with as few as 34 amino acids adjacent to the N terminus of the molecule have also been isolated from the parathyroid glands that exhibit full parathyroid hormone activity.

The parathyroid hormone tends to elevate the blood calcium level and decreases the phosphate level. The rise in Ca concentration is caused by two effects:

(1) an effect of parathyroid hormone in causing calcium and phosphate absorption from the bone *(osteolysis)* and

(2) a rapid effect of parathyroid hormone in decreasing the excretion of Ca by the kidneys. The decline in phosphate concentration is caused by a strong effect of parathyroid hormone on the kidneys in causing excessive renal phosphate excretion.

Parathyroid hormone also increases tubular reabsorption of Ca at the same time that it diminishes phosphate reabsorption. Parathyroid hormone greatly enhances both calcium and phosphate absorption from the intestine by increasing the formation in the kidneys of 1,25 dihydroxycholecalciferol from vitamin D. It should be mentioned that in the absence of vitamin D, the effect of parathyroid hormone in causing bone absorption is greatly reduced.

Hyperparathyroidism: It is produced by tumors on the gland and causes the calcium to be withdrawn from the bones, thus producing a high blood level of calcium. Under such conditions the bones may become soft and porous, or have only localized areas of calcification.

It should be noted that vitamin D is related in some unknown fashion to parathyroid function and the calcium and phosphate metabolism. For example, it can be shown that high doses of vitamin D can maintain normal blood calcium concentration in *Hypoparathyroidism,* because vitamin D promotes the rate of calcium absorption from the intestine. Thus, it can be noted that the bone cells withdraw calcium and phosphate from the blood for production of the bone, whereas parathormone withdraws these minerals from the bones. These two actions are antagonistic and tend to balance each other.

The Pancreas (The Islets of Langerhans)

Location and Structure : In human adults, the pancreas is an elongated gland weighing about three ounces. It lies behind the stomach close to the posterior abdominal wall, and functions both as a digestive gland and an endocrine gland. Besides the bulk of the glandular cells, which secrete a mixture of several digestive enzymes, the pancreas also contains scattered, rounded microscopic clusters or islets of tissue approximately 0.3 mm in diameter and about a million or so in number. These islets, discovered by *Langerhans* in 1869, are not connected with any ducts and perform an important endocrine function.

The *islets of Langerhans* in mammals consist of four known types of cells. The so-called *beta-cells* constitute majority (about 70-80%) in the islets and secrete the hormone *insulin.* About 20% of the islet cells are the *alpha-cells,* which are generally believed to secrete a second hormone called *glucagon.* However, the functions of the small percentage of remaining cells, called the C-cells, and D-cells are not known. D-cells may secrete the hormone *somatostatin.* In addition, at least one other type of cell, the PP cell, is present in small numbers in the islets and secretes a hormone called *pancreatic polypeptide.* Its function is uncertain.

Insulin and its Functions

The term insulin was given in 1916 and it was first extracted by Nobel Laureate *Banting* in 1922. Insulin is a proteinaceous hormone, consisting of 51 amino acid residues in the form of two linked peptide chains of 21 and 30 amino acids respectively with a moleuclar weight of 6000. The two chains are connected by two disulphide (-S-S-) linkages, while a third such linkage forms an intrachain bridge in the smaller chain. These disulphide bridges, which contribute to the tertiary structure of the molecule, are essential for its biological activity.

Insulin is rapidly digested by protein-digesting enzymes and must, therefore, be administered by injection and not by mouth to avoid its destruction in the digestive tract. The demonstration that insulin is actually present in the circulating blood (which is an essential condition for normal activity) depends in part upon the fact that plasma from normal animals can lower the blood-sugar levels of normal or diabetic ones. The most convincing evidence, however, rests upon radio-immunoassay which makes possible the measurement of the circulating levels.

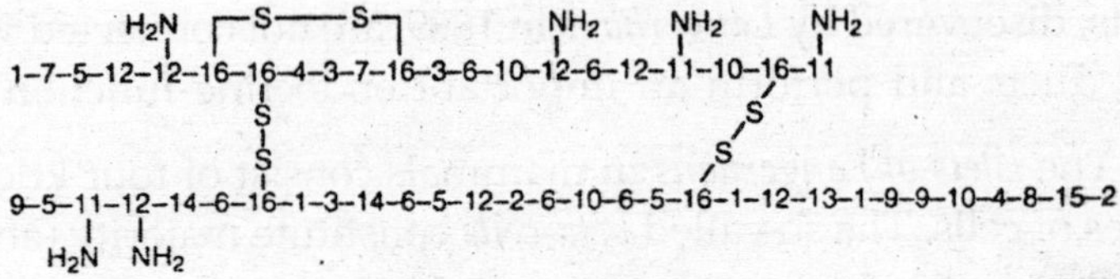

1. glycine 5. valine 9. phenylalanine 13. arginine
2. alanine 6. leucine 10. tyrosine 14. histidine
3. serine 7. isoleucine 11. aspartic acid 15. lysine
4. threonine 8. proline 12. glutamic acid 16. cysteine

Structure of the human insulin molecule.

The insulin is of great importance in the metabolism of fats and carbohydrates. Deficiency of insulin leads to excess sugar in the blood (300—1200 mg/dl) which is excreted in the urine. In this disease, known as *diabetes mellitus* meaning "a running through of sweet urine", much urine is excreted and the individual loses weight. When the blood glucose level rises to 300—500 mg/dl, which is common in people with untreated diabetes, 100 or more grams of glucose is lost into the urine each day.

As the diabetic can no more utilize the rich energy source of his high blood sugar, the person shifts from carbohydrate metabolism to fat metabolism for energy. As a result, there is an accumulation of incompletely oxidized ketone bodies from fat oxidation. These substances are toxic producing metabolic *acidosis*, and eventually lead to death of the patient.

Proper treatment of *diabetes mellitus* requires the administration of the right amount of insulin so that carbohydrate metabolism is balanced. Because insulin is a polypeptide and is not absorbed through the intestinal mucosa, it must be administered by intramuscular injection. Thus, insulin appears to be necessary for the storage of glucose or glycogen in the liver and for the oxidation of glucose.

But how does insulin stimulate the conversion of glucose to glycogen? One theory proposes that insulin stimulates the movement of glucose across the cell membrane into the cell by somehow increasing permeability to various sugars. Although there is appreciable evidence in favour of this view, there is also experimental support for another important theory. It proposes that insulin influences the early stage of glucose metabolism by possibly affecting hexokinase, which converts glucose to glucoses-phosphate (G-6-P) by phosphorylation utilizing phosphate group from ATP.

Insulin has three sites of action. The first important field of action of the insulin is the *muscular tissue* where it acts to promote carbohydrate metabolism and the storage of glycogen. The glycogen can later be used for energy by the muscle.

The second important field of action of insulin is the *liver*. Here, in contrast to its action upon muscle, it does not affect the cell membrane, for the hepatic cells are freely permeable to glucose. In liver cells it favours glycogenesis and glycolysis, decreases hepatic gluconeogenesis, favours the uptake of fatty acids, and also increases lipogenesis. When the quantity of glucose entering the liver cells is more than can be stored as glycogen or be used for local liver cell matabolism, insulin promotes the conversion of all this excess glucose into fatty acids. These fatty acids are subsequently packaged as triglycerides in very low density lipoproteins and transported in this form via blood to the adipose tissue and deposited as fat.

The third site of action of insulin is the *adipose tissue,* where it enhances the membrane transfer of glucose, which may then be stored as glycogen, or used immediately as the source of energy. It also enhances the uptake of fatty acids by adipose tissue, promotes lipogenesis, and decreases lipolysis. In all these three tissues insulin stimulates the uptake of amino acids and their incorporation into the tissue protein. It also facilitates the oxidation of glucose in most peripheral tissues.

Insulin also affects protein metabolism and growth. It promotes, protein formation and preventes degradation of proteins.

Elevated glucose levels in the blood apparently serve as the stimulus for insulin secretion by the beta-cells of the islets. A lowering in blood glucose concentration is accompanied by a decrease or cessation in insulin secretion. The available quantity of insulin is the resultant of the rate of secretion and the rate of inactivation. A proteolytic enzyme, termed *insulinase*, is present in the liver, which appears to be the major site of insulin inactivation.

Hyperinsulinism and Insulin Shock : Increased insulin production resulting from adenoma of an islet of langerhans is called *hyperinsulinism*. When insulin causes the level of blood glucose to fall to low values (hypoglycemia caused by excess insulin), the metabolism of the central nervous system (which normally derives all its energy from glucose metabolism, and insulin is not necessary for this use of glucose) is depressed. Consequently, in patients with hyperinsulinism or in patients with diabetes who administer too much insulin to themselves, the syndrome called *insulin shock* may occur.

Glucagon and its Functions

Glucagon or the *hyperglycemic factor* is another important hormone that is secreted by the *alpha-cells* of the islets of Langerhans. Glucagon (from the beef pancreas) is a relatively small protein molecule with a molecular weight of 3450, and consists of a single straight chain of 29 amino acid residues. A remarkable feature of the structure is that amino acid sequence of glucagon closely resembles that of secretin.

Whereas insulin in normally lacking in diabetes, glucagon is present in excessive amounts, and this excess of glucagon is probably as important a factor in diabetes as is insulin's lack. This hormone has the opposite effect of insulin, tending to increase the level of blood glucose by promoting the breakdown of liver glycogen (but not of muscle glycogen) to glucose *(glycogenolysis)*. The liver, therefore, appears to be the primary target organ for glucagon, which is believed to act by somehow enhancing the activity of the enzyme *(phosphorylase)* responsible for the breakdown of glycogen to glucose-1-phosphate. It possibly prevents enzyme inactivation or stimulates its activity. It also affects glucose metabolism by increased *gluconeogenesis* in the liver. Both these effects greatly enhance the availability of glucose to the other organs of the body. Thus glucagon provides an immediate defence against hypoglycemia and this defensive action seems to be one of its major functions.

Glucagon is also an important regulator of amino acid metabolism. The importance of amino acid stimulation of glucagon secretion is that glucagon then promotes rapid conversion of amino acids to glucose, thus making even more glucose available to the tissues.

The secretion of glucagon is regulated by a feedback mechanism similar to that which controls insulin secretion, but it works in the opposite direction. Low blood sugar stimulates the alpha-cells of the islets of Langerhans to secrete glucagon, whereas high blood sugar inhibits them.

Role of Somatostatin

Somatostatin or the *somatotropin-release inhibiting factor* has been isolated from sheep brain in the 1973 by *Brazeau* and his associates. It has recently been reported that somatostatin is not an exclusively hypothalamic peptide but may be synthesized in the pancreas by the *delta cells* of islets of Langerhans. *Dubois* (1975) has shown immunoreactive somatostatin in the endocrine cells of human foetal pancreas.

Besides inhibiting the release of somatotropin (growth hormone) somatostatin also acts on endocrine pancreas to inhibit the secretion or glucagon and insulin. It has been shown by several investigators that somatostatin reduces the concentration of insulin, glucagon and glucose in the blood of healthy individuals. Recent experiments with somatostatin strongly indicate that it is possible to alleviate the symptoms of diabetes by reducing the concentration of glucagon by somatostatin.

The Adrenal Glands

Location and Structure : Each adrenal or suprarenal gland in man adrenal medulla weighs about 6 grams, measures 25 to 50 mm in length, and lies at the top of each kidney (Fig.). The adrenal glands are somewhat pyramidal in shape and have a rich blood supply. They are innervated by the sympathetic nervous system, and consist of two parts that are embryologically, anatomically and functionally different. In the typical mammal, each gland consists of an external reddish brown *cortex* which surrounds an internal grayish *medulla.* The adrenal cortex has arisen from the same origin *(mesodermal)* as that of the sex glands, whereas the medulla has an ectodermal origin similar to that of the postganghonic fibre cells of the sympathetic nervous system.

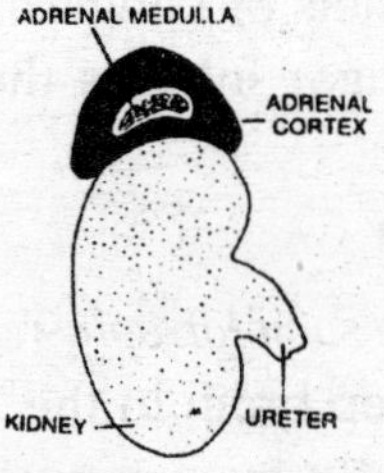

Showing the position and internal structure of the adrenal gland in man.

The cortex is made up of three zones, the cells of which differ in form and function (Fig.). The outermost *zona glomerulosa* is formed by closely packed, low columnar cells and it secretes mineralocorticoids, hormones concerned with water and electrolyte mechanism and the most important of which are aldosterone and deoxycorticosterone. The secretion of these hormones is mainly regulated by negative feedback through plasma Na^+ and K^+ levels, but there is some evidence that ACTH may also be involved in their regulation. The thick, middle *zona fasciculata* is formed by cords of cells filled with lipid, the amount of which varies with the secretory activity of these cells. The third or the innermost zone is the *zona reticularis*. The middle and inner zones secrete glucocorticoids, hormones that influence carbohydrate and protein metabolism. Cortisol is the most important glucocorticoid, and its secretion is regulated by ACTH.

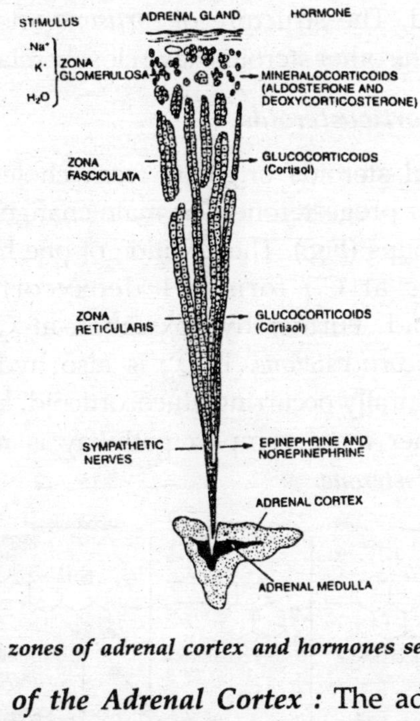

Showing three zones of adrenal cortex and hormones secreted by them.

Hormones of the Adrenal Cortex : The adrenal cortex is absolutely essential to life. When it is completely removed most animals will not live more than a week or so. The cortex has several hormones and it ranks next to the adenohypophysis in endocrine importance. Secretion of the cortical hormones is under the control of ACTH from the pituitary. All the *adrenocortical hormones* are steroid compounds and can be divided into three groups.

(i) *Glucocorticoids* such as *cortisone, hydrocortisone* (cortisol), and *corticosterone.*

(ii) *Mineralocorticoids,* such as *aldosterone* and *deoxycorticosterone.*

(iii) *Androgens,* sex hormones which have chiefly *androgenic* and to a minor extent *estrogenic* effects.

The adrenal cortex is rich in *ascorbic acid* (vitamin C) and *cholesterol,* the chief compound from which the steroid hormones are synthesized. The structure of *cortisone* and *aldosterone* is shown in Fig. The other steroids have closely related structures.

Synthesis of Corticosteroids

The adrenal steroids originate from cholesteral through pregnenolone or progesterone. The main changes are additions of hydroxyl groups (Fig.). The addition of one hydroxyl group to *progesterone* at C_{21} forms 11 deoxycorticosterone, a mineralocorticoid. Further hydroxylation at C_{11} produces a glucocorticoid, *corticosterone.* If Q_{17} is also hydroxylated, the most potent naturally occurring glucocorticoid, *hydrocortisone* is formed. A rather more complex pathway is needed for the synthesis of *aldosterone.*

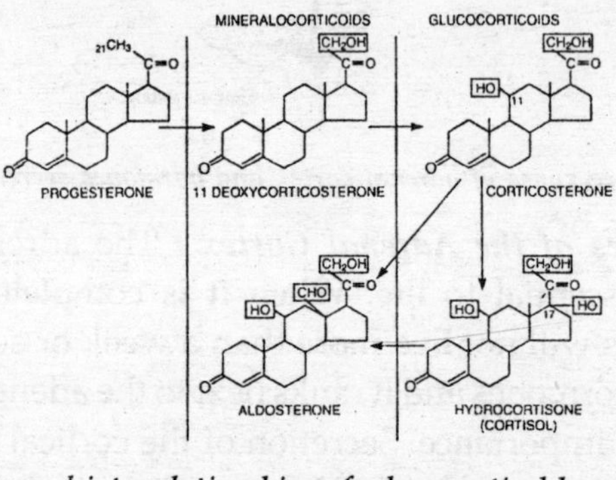

Origin and interrelationships of adrenocortical hormones.

Functions of the Adrenocortical Hormones

1. *Effect on Carbohydrate and Protein Metabolism* : The glucocorticoid hormones exert their influence primarily on carbohydrate metabolism. They increase the blood glucose level at the expense of liver glycogen and enhance the conversion of protein (and fat) to carbohydrates. Cortisol also affects fat metabolism.
2. *Effect on Mineral, Salt, and Water Metabolism*: The mineral corticoids have a regulatory effect on the relative concentrations of mineral ions (especially sodium and potassium) in the body fluids, and therefore on the water contents of the tissues. The insufficiency of the mineral corticoids results in an increased excretion into the urine of sodium ions, chloride ions, and water, accompained by a fall in blood sodium chloride, and bicarbonate. The administration of the mineral corticoid hormones has the opposite effects, resulting in an increase in blood sodium and chloride concentrations, more retention of water by the body, a decrease in potassium concentration, and a restoration of pH of the blood to its more normally alkaline level.

Aldosterone also promotes sodium retention in the salivary glands, sweat glands, and colon, and the excretion of *magnesium* in the urine. Thus *aldosterone* has almost the same effects on sweat glands and salivary glands that it has on the renal tubules. The effect on the sweat glands is important to conserve body salt in hot environments, and the effect on the salivary glands is necessary to conserve salt when excessive quantities of saliva are lost.

3. *Effect on Sexual Development* : Under certain circumstances the adrenal cortex appears to secrete steroids which are similar in their effects to male hormones. For example, adrenal cortical tumours in females may give rise to masculine characters, such as beard and deep

voice, etc. Although the female sex hormones have also been detected in cortical extracts, it is believed that they are simply intermediates in the formation of other adrenocorticosteroids.

4. *Anti-allergy, Anti-inflammatory and other Effects* : Several of the cortical hormones *(e.g., cortisol* and *cortisone)* have a marked influence in preventing the appearance of an inflammation. Some cortical hormones *(e.g., cortisol)* also counteract the symptoms of allergic or hypersensitive reactions. Accordingly, they have been used in the treatment of several kinds of acute inflammatory and allergic diseases especially of the eye. The two basic anti-inflammatory effects of *cortisol* are :
   (1) it can block the early stages of the inflammation process before inflammation even begins, and
   (2) if inflammation has already begun, it causes rapid resolution of the inflammation and increased ripidity of healing. They are also used to control the symptoms of particular arthritic conditions and rheumatic fever.

The steroid secretions of the adrenal cortex also exercise several other influences on the body functions including a role in decreasing the level of red and white blood cells and stimulating the secretions of the digestive tract.

Hypoadrenalism

Insufficiency of the cortical hormones results in *Addison's disease.* In this disease the adrenal cortex degenerates, resulting in a number of symptoms, including a characteristic bronzing of the skin, muscular weakness, low blood pressure, and digestive disturbance. The sufferers of this disease are cured by a proper administration of cortical hormones from some external source.

Hyperadrenalism

Hyperfunction or oversecretion by the adrenal cortex as a result of excessive secretion of ACTH, gives rise to *Cushing's disease.* The symptoms include high blood pressure, salt retention, swelling of the tissues with water, demineralization of the bones, and loss of sexual function. Oversecretion may also arise from a tumor in the cortical cells and the resulting symptoms will depend on the kinds of hormones secreted by the tumor. If glucosteroids are predominantly produced, metabolism is affected. Treatment in Cushing's disease consists of removing an adrenal tumour if this is the cause, or decreasing the secretion of ACTH, if possible. If more concentrations of male-like hormones are secreted, the females begin to attain masculine characteristics and males show a greater accentuation of the male sexual characteristics.

Hormones of the Adrenal Medulla : The adrenal medulla in man consists of two distinct types of cells. These are *adrenaline-secreting* cells and the *noradrenaline-secreting* cells which are responsible for the production of the two hormones, *adrenaline* or *epinephrine* and *norepinephrine* or *noradrenaline* respectively (Fig.). In human beings, these catecholamines of the medulla comprise about 80% adrenaline and 20% noradrenaline.

They are secreted in response to sympathetic stimulation. In turn, these hormones cause almost the same effects as direct stimulation of the sympathetic nerves in all parts of the body.

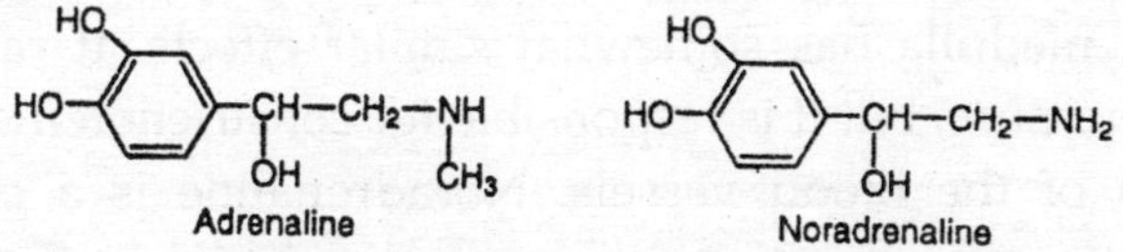

The structure of adrenal medulla hormones.

Adrenaline (Epinephrine) : It is one of the simplest hormones known and can be produced synthetically. It was the first hormone to be isolated in crystalline form in 1901 and was first synthesized in 1904.

It functions as the so-called "emergency" hormone for the body because the general secretion of this hormone centres around emergency functions of the body, such as those of fear, fight, and rage, etc. During emotional states epinephrine is secreted in great amounts. It has a decided effect in elevating the glucose level in the bloodstream by somehow enhancing enzymatic conversion of liver glycogen to glucose.

The rise in blood glucose is accompained by an increase in oxygen consumption, body temperature and heat production. Adrenaline also tends to increase the flow of blood in skeletal muscles, heart and viscera by dilating the blood vessels. In this way it serves to direct the circulation of blood where necessary during exertion or increased activity.

It also increases blood pressure, heart beat, inhibits peristaltic movements of the gut, stimulates the tiny muscles which attach to hair and feathers (appendages which are raised as a consequence producing "goose flesh" in man), dilates the pupils of the eye, increases muscular power and resistance to fatigue, promotes faster coagulation of blood, and generally promotes faster and sharper responses to external stimuli.

Together, most or all these effects are sometimes called the *alarm reaction.* If the action of adrenaline is somehow inhibited, a vertebrate cannot adjust rapidly to emergencies.

Noradrenaline (Norepinephrine) : This hormone of the adrenal medulla has somewhat similar effects. It raises the blood pressure, but it is responsible for constriction instead .of dilation of the blood vessels. Noradrenaline is a chemical transmitter at sympathetic nerve endings and in the central nervous system, but is also present in the adrenal medulla, both as a hormone and also as a precursor of adrenaline.

The following differences between the actions of epinephrine and norepinephrine must be noted.

1. Epinephrine, because of its greater effect in stimulating the beta receptors, has a greater effect on cardiac stimulation than does norepinephrine.

2. Epinephrine causes only weak constriction of the blood vessels in the muscles, in comparison with much stronger constriction caused by norepinephrine. This difference is of special importance because norepinephrine greatly increases the total peripheral resistance and elevates arterial pressure, whereas epinephrine raises the arterial pressure to a lesser extent but increases the cardiac output considerably more because of its excitatory effect on the heart.

3. Their effects on tissue metabolism are also different. Epinephrine has 5 to 10 times as great a metabolic effect as norepinephrine. Epinephrine can increase the metabolic rate of the whole body often to as much as 100 per cent above normal, thus increasing the activity of the body. It also increases the rate of other metabolic activities such as glycogenolysis in the liver and muscle, and glucose release into the blood.

Gastrointestinal Hormones

Several hormones are associated with the activities of the digestive tract. All of these hormones are produced in the mucosa lining and help regulate the volume and character of the secretions. Chemically, these hormones are polypeptides or polypeptide derivatives.

Gastrin : When food enters the stomach, *gastrin* (a hormone produced in the pyloric wall) stimulates the gastric glands to secrete gastric juice. Two closely related peptides (gastrin I and II) are present in the stomach of various mammalian species, including man, sheep, pig, and dog, etc. These are heptadecapeptide amides, gastrin II having a sulphate group attached by ester linkage to the single tyrosyl.

Secretin : The secretion of pancreatic juice, like that of the gastric juice, is regulated by both neural and hormonal factors. The entrance of acid chyme from the stomach into the duodenum causes the release of *secretin* from the cells of the duodenal wall into the blood circulation.

Secretin is a single-chain polypeptide of low molecular weight with 27 amino acid residues and bearing a strongly basic charge due to the arginine and histidine residues, and to the amidation of two of the glutamic acid ones. The arrival of *secretin* at the pancreas by way of the bloodstream stimulates the secretion of copious volumes of pancreatic juice low in digestive enzymes. *Secretin* also stimulates the flow of bile from the gall bladder and intestinal juice from the glands in the wall of the small intestine.

Pancreozymin : A second intestinal hormone called *pancreozymin,* stimulates the secretion of pancreatic juice rich in digestive enzyme activity.

Cholecystokinin : The entry from the stomach of fatty foods into the upper part of the duodenum promotes the release into the bloodstream of an intestinal hormone called *cholecystokinin.* It causes the gall bladder to contract discharging its contents into the cystic and common bile ducts.

Cholecystokinin/pancreozymin is a single-chain polypeptide with 33 amino acids and with a sulphated tyrosyl. As with gastrin and secretin, it has a number of actions additional to its primary ones upon the gall bladder and the secretion of pancreatic enzymes. It stimulates secretion of gastric acid and of the duodenal succus entericus, it inhibits gastric motility, and it increases duodenal motility.

Enterocrinin and Duocrinin : There is some evidence that the secretion of intestinal juice is under the control of two specific intestinal hormones called *enterocrinin* and *duocrinin,* which are secreted by the intestinal lining into the bloodstream.

Enterogastrone : The entry of lipids into the duodenum apparently causes the liberation by the intestinal wall of a hormone called enterogastrone which inhibits gastric juice secretion.

Gastric Inhibitory Polypeptide : GIP is a well characterized product of the small intestine, inhibiting both gastric mobility and gastric secretion. In has 43 amino acids, the sequence resembling with secretin and glucagon. Fifteen of the first 26 amino acids at the NH_2-terminal end are in the same position as in porcine glucagon, while 9 of the first 26 amino acids are in the same position as in secretin. It has been suggested that it is identical with enterogastrone.

However, no deficiency diseases have been associated with the under-secretion of the gastrointestinal hormones and some endocrinologists consider them of minor importance.

The Gonads

Location and structure : The specialized organs which eventually produce the reproductive cells (germ cells) are termed the *gonads.* In higher animals, the male gonad produces the sperm cells and is called the testes. The female gonad which produces the egg cells or ova is called the *ovary.* The mature gonads of both sexes are not only the centres for production of the germ cells but they also secrete the *sex hormones* which control the maturation and function of the reproductive system and certain other tissues of the body.

The Testes

Each of the two testes in the adult male is an ovoid body about the size of a walnut. Both are located outside the body cavity within a skin covered pouch called the *scrotum.* The wall of each testis consists largely of connective tissue which is continuous with wall-like partitions dividing the interior of the testis into approximately 250 small compartments called *lobules.*

Each lobule contains from one to three tiny convoluted seminiferous tubules within which the sperms are formed. The special cells located among the *seminiferous tubules* are collectively called the *interstitial tissue,* which performs the important function of producing and secreting the *male sex hormone.* The interstitial cells are not present during childhood but are found after puberty.

The Ovaries

Each of the two ovaries in the adult female is about the size and shape of large almond and is separately located in the lower or pelvic region of the abdominal cavity. Each ovary consists of a large number of *ovarian follicles.* The ovarian fallicles produce *the female sex hormones* while undergoing maturation.

Male Sex Hormones and Their Physiological Effects : The male sex hormones are usually referred to as androgens, including *testosterone, dihydrotestosterone* and *androstenedione.* The term "androgen" means any steroid hormone that has masculinizing effects. The male sex hormones have not only been obtained from the testes but from the adrenal cortex, placenta, and the ovaries as well.

The principal male sex hormone is *testosterone* (Fig.). It is a steroid, and is secreted by the interstitial cells of the testes of most mammals, including man. Much of the testosterone that becomes fixed to the tissues is converted within the cells to *dihydrotestosterone,* especially in certain target organs such as the prostate gland in the adult. Some important target tissues do not have the reductase enzyme in their cells to convert testosterone into dihydrotestosterone. In these tissues, testosterone functions directly. For instance, this direct effect of testosterone is essential in the male fetus for the development of the epididymis, vas deferens, and seminal vesicles. The direct effect is also responsible for much of the effects of testosterone to promote spermatogenesis.

The production of testosterone is stimulated by such gonadotropins as the luteinizing (or the interstitial cell-stimulating) hormone. The basic intracellular mechanism of action of testostrone is that it greatly stimulates production of proteins virtually everywhere in the body, although increasing in particular those proteins in target organs responsible for the development of secondary sexual characteristics.

Testosterone is responsible for the secondary male sex characteristics, such as the beard of the face, the distribution of hair on the body, and the

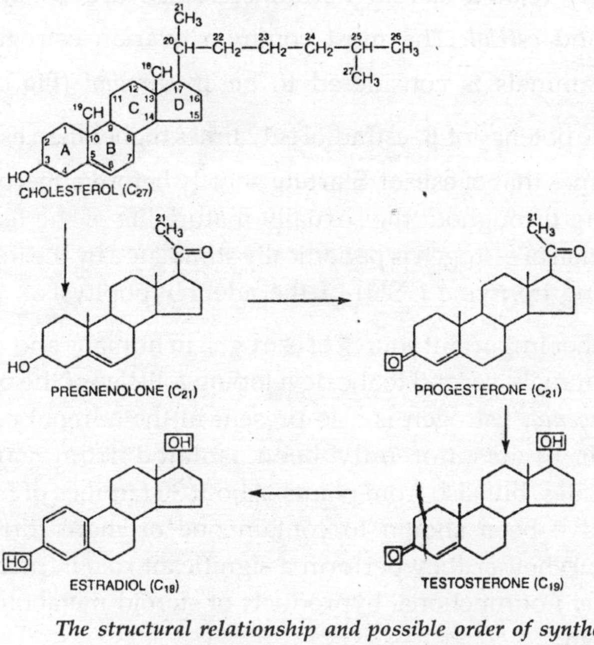

The structural relationship and possible order of synthesis of steroid hormones from cholesterol.

character of the voice, etc. It is also concerned with the growth and development of the accessory sex characters, penis, seminal vesicle and prostate gland. When injected into a female, secondary male sex characteristics are often developed. Testosterone is also responsible for the sex urge and for sexual behaviour. Testes-removal or *castration* in the human male prior to the onset of puberty results in the failure of the secondary sexual characteristics to develop. After puberty, castration causes an attenuation of the sexual characteristics.

Female Sex Hormones and Their Physiological Effects : The ovary secretes two types of female sex hormones called *estrogens* and *progestins.* By far the most important of the estrogens is the hormone *estradiol,* and by far the most important progestin is *progesterone.*

Estrogen : Estrogen is a collective name for three different but closely related steroid hormones. These are β-*estradiol, estrone,* and *estriol.* The most common ovarian estrogen of many mammals is considered to be β-*estradiol* (Fig.). The estrogenic potency of β-estradiol is 12 times higher than estrone and 80 times that of estriol. Starting shortly before puberty and continuing throughout the sexually mature life of the female, the secretion of estrogen is periodically stimulated by the follicle-stimulating hormone (FSH) of the adenohypophysis.

Another important source of estrogen in humans and many other mammals, second to the developing follicles of the ovary, is the *placenta.* Estrogen is also present in the adrenal cortex. However, it has not only been isolated from various invertebrates, but also from plants. About 30 families of higher plants have been shown to contain one or more forms of estrogen. Whether they perform a significant role in plants or are simply nonfunctional byproducts of steroid metabolism is not known.

Estrogen is responsible for the mature growth, development and maintenance of the female reproductive tract, the secondary sex organs (breasts, vagina, uterus) and other secondary sexual characteristics. It also contributes to the sexual desire of the female. Estrogen promotes repair of the uterine lining following menstruation in preparation for possible implantation of a fertilized egg. Although the hormone helps to control the development and functioning of the mammary gland, it appears to inhibit lactation by inhibiting the secretion of the lactogenic hormone by the pituitary gland. In general, estrogen tends to decrease the secretion of the gonadotropic hormones of the pituitary.

After the monopaure almost no estrogens are secreted by the ovaires. This estrogen deficiency leads to diminished osteoblastic activity in the bones, decreased bone matrix, and decreased deposition of bone calcium and phosphate. In some old women, this effect is extremely severe, and the resulting condition is called *osteoporosis* of bones.

Estrogens also affect protein and fat deposition, hair distribution, skin, and electrolyte balance.

Progesterone : Progesterone is the second principal female sex hormone. It is also a steroid somewhat related in chemical structure to estrogen. It is secreted during the later half of the menstrual cycle in human females chiefly by the temporary endocrine tissue, the *corpus luteum,* which rapidly develops in the ruptured follicle shortly after the release of the egg cell.

Progesterone further promotes the secretory changes in the uterine endometrium already initiated by the estrogen during the first half of the menstrual cycle, in preparation for implantation of the fertilized ovum. At this particular phase of the female menstrual cycle, it takes over the general functions of estrogen (when the latter is being secreted at a minimum level) in maintaining the mature female reproductive tract and secondary sexual characteristics. Thus, together with estrogen, progesterone is responsible for the sequence of changes of the repeating menstrual cycles that occur in the female reproductive tract starting at puberty and extending through the reproductive life of the female.

Progesterone also affects fallopian tubes and breasts. It promotes development of the lobules and alveoli of the breasts, causing the alveolar cells to proliferate, enlarge and become secretory in nature. However, it does not cause the alveoli actually to secrete the milk which is secreted under the stimulus of prolactin.

Like some of the adrenal cortical hormones, progesterone also seems to promote retention of sodium salts and water (electrolyte balance), presumably by affecting kidney function. Progesterone is also found in the mammalian placenta, the adrenal cortex, and the testes of some mammals (*e.g.*, the pig). The occurrence of the hormone in the adrenal cortex and testes is probably due to its role as an intermediate in the synthesis of other steroid hormones.

Relaxin **:** Another female sex hormone called *relaxin* is produced by the corpus luteum of the ovary and by the placenta. It is a polypeptide, having a molecular weight of about 9000. It specifically causes the relaxation of pelvic ligaments in child birth. It has also been claimed that relaxin softens the cervix of the pregnant woman at the time of delivery.

The Placenta

The human placenta, which has been extensively studied, has been described as an 'incomplete endocrine gland' because it is deficient in many of the essential enzymes necessary for complete synthesis of steroid hormones from acetate. The importance of the human foetus in the production of placental hormones has been recognised by the use of the term 'foeto-placental unit' or the 'foeto-placento-maternal unit'. The latter term appears more appropriate as both the maternal and foetal liver and adrenal glands supply preformed precursors to the placenta, which converts them into active hormones. The placenta of some mammals secretes the following protein and steroid hormones :

Progesterone : Human placenta probably converts cholesterol or pregnenolone from the maternal blood to progesterone which is secreted by the placenta in large amounts and is important in maintaining pregnancy. Besides inducing widespread adaptations in maternal physiology, progesterone reduces the contractility and excitability of uterine muscle and thereby helps to prevent premature explusion of the foetus from the uterus.

Androgens and Estrogens : Progesterone is converted by the foetal adrenal to adrenocortical steroids, as well as a variety of Q_{19} androgens such as dehydroepiandrosterone. An additional hydroxyl group is added to this weak androgen by the foetal liver, and the 16-hydroxylated compound is then passed back to the placenta where it is converted to *oestriol*, the main estrogen secreted by the human placenta. As the human placenta lacks the enzyme 16-hydroxylase, it is incapable of converting dehydroepiandrosterone directly to oestriol. The human placenta also synthesizes and secretes smaller quantities of estradiol-17 b and oestrone from other C_{19} precursors such as androstenedione.

Like ovarian estrogens, placental estrogens promote the growth and development of the breast and the uterine muscle in pregnancy, in addition to inducing widespread changes in maternal metabolism which are reflected by alterations in proteins synthesis, carbohydrate metabolism, and adrenal and thyroid function.

Human Chorionic Gonadotrophin (HCG) : The human placenta also secretes large quantities of HCG, a protein hormone. It is a glycoprotein of molecular weight around 30,000. The chemical structure and properties of HCG are very similar to those of the pituitary LH, although it has a higher carbohydrate content (8.5% sialic acid). It is composed of at least two subunits one of which is common to FSH, LH, and TSH, accounting for the immunological similarity of these protein hormones. It was first demonstrated by Aschheim and

Zondek in 1927 that pregnancy in the human female is characterised by the appearance in the urine of HCG which reaches its maximum concentration between the 60th and 80th days of pregnancy.

This hormone has almost the same effects on the sexual organs as LH. Its most important function is to prevent the normal involution of the corpus luteum at the end of the female sexual cycle. Instead, it causes the corpus luteum to secrete even larger quantities of estrogens and progesterone for the next few months. These sex hormones prevent menstruation and cause the endometrium to continue growing and to store large amounts of nutrients rather than being shed in the menstruum.

Human chorionic gonadotropin also exerts an *interstitial cell-stimulating effect* on the testes, thus resulting in the production of testosterone by fetal tests. This small secretion of testosterone causes the fetus to grow male sex organs. At the end of pregnancy, the testosterone secreted by the fetal testes also causes the testes to descend into the serotum.

Human Chorionic Somatomammotrophin (HCS) : Another protein hormone of the human placenta is HCS. It was formerly known as human placental lactogen (HPL). It is synthesized by the syncytiotrophoblast cells of the placental villi, and is secreted almost exclusively into the maternal circulation at a rate of about 1 g per day. It is similar immunologically and chemically to human growth hormone having a molecular weight of around 38,000.

It promotes development of the breast, and together with HCG is probably responsible for the maintenance of the corpus luteum of pregnancy. It also produces changes in carbohydrate and protein metabolism similar to those of the growth hormone. It causes decreased insulin sensitivity and decreased utilization of glucose in the mother, thereby making more quantities of glucose available to the fetus for its growth. Furthermore, the hormone promotes release of free fatty acids from the fat stores of the mother, thus providing an alternative source of energy for her metabolism.

Pregnant Mare's Serum Gonadotropin (PMSG) : This protein hormone is secreted by the endometrial cups of the pregnant uterus in horse and donkey. The molecular weight of PMSG is larger than that of any other gonadotrophin (about 70,000). The high molecular weight together with high carbohydrate content (49% with 10.4% sialic acid) impairs its clearance from the circulation and accounts for its long half-life and high concentration. It has properties of both the FSH and LH, but the former predominates. However, its physiological function is not fully known and its secretion is regulated by the genotype of the foetus.

Other Endocrine Organs

Pineal Gland : The pineal is situated on a stalk between the two halves of the brain. In some vertebrates, for example, lampreys and tuataras, the pineal is a quite large structure. Here it is located directly under the skin of the head, and the skin at that point is transparent. In these animals the pineal is actually the third functional eye situated on the top of the head.

The pineal gland plays an important regulatory role in sexual and reproductive function at least in some lower animals. The pineal gland is photosensitive and is controlled by the amount of light or "time pattern" of light seen by the eyes each day. The *melatonin,* secreted by the pineal gland, is believed to pass either by way of the blood or through the fluid of the third ventricle to the adenohypophysis to suppress gonadotropic hormone secretion. This occurs during the early winter months. Thereafter, the gonadotropic hormone secretion breaks through the inhibitory effect of the pineal gland and the gonads become functional.

In humans, tumors often occur in the region of the pineal gland. Such tumors are often associated with serious hypogonadal or hypergonadal function. Therefore, it appears that pineal gland plays atleast some role in controlling sexual drive and reproduction in humans.

Thymus Gland

It lies in the neck and the upper chest regions. In many mammals it persists throughout the life, but in man and some other mammals the organ gradually decreases in size during youth and disappears altogether at the time of puberty.

The most important function of the thymus gland is that it preprocesses the *T-lymphocytes.* The T-lymphocytes, after their origin in the bone marrow, first migrate to the thymus gland. Here they divide rapidly and develop extreme diversity for reacting against different specific antigens. The different types of processed T-lymphocytes now leave the thymus and spread throughout the body in the lymphoid tissue. The thymus also makes certain that any T-lymphocytes leaving the thymus will not react against proteins or other antigens that are present in the body's own tissues. Most of the processing of the T-lymphocytes in the thymus occurs shortly before birth of a baby and for a few months after birth. However, removal of the thymus several months before birth can prevent development of all cell-mediated immunity.

9

The DNA

The nucleic acids were discovered over 100 years ago, but their role in heredity and in the control of cellular metabolism has been described only during the past 50 years. There are two nucleic acids, deoxyribonucleic acid (DNA) and ribonucleic acid (RNA). The DNA is the predominant nucleic acid in the nucleus, while the RNA is the predominant nucleic acid in the cytoplasm, although all cytoplasmic RNA is of nuclear origin.

Deoxyribonucleic Acid

In 1968, *F. Miescher* extracted from the nuclei of pus cells (white blood cells) a substance which he called *nuclein.* However, Miescher was not aware that the substance he isolated was the hereditary material. It was later on renamed *deoxyribonucleic acid* (DNA) and is now recognised as being the chemical structure that stores the cell's hereditary information.

In eucaryotic cells, two distinct types of DNA are present. The nuclear DNA of eucaryotes differs fundamentally from other DNA in that it is complexed with certain proteins called *histones,* with *non-histone* proteins, and with *small amounts* of *RNA* to form the structures called *chromosomes.*

The DNA of self-replicating organelles such as mitochondria and chloroplasts resembles closely that of procaryotes (a) in its relatively smaller size, *(b)* in not being found in combination with histone proteins, and (c) in its circular form.

Molecular Structure of DNA

DNA is composed of:

1. *A Pentose Sugar, called deoxyribose ("de" = lacking, "deoxy"* = lacking in an oxygen atom found in the ribose at no. 2 position), Fig.

2. *Nitrogen Bases* : There are four heterocyclic bases in DNA. Two of the bases are *purines* : Adenine (A) and Guanine (G), and two are *pyrimidines* : Cytosine (C) and Thymine (T). As separate chemicals these bases exist largely in their *enol* forms ; but when these purines and pyrimidines are bound into nucleic acids, they are in their *keto* forms (as shown in Fig.). In the DNA molecule, two hydrogen bonds exist between adenine and thymine and three hydrogen bonds join guanine and cytosine.

3. *Phosphoric Acid* : or H_3PO_4 (Fig.) links the nucleotides by joining the pentoses of two consecutive nucleotides with an ester-phosphate bond. This bond links carbon 3' in one nucleoside with carbon 5' in the next. In this way phosphoric acid uses two of its three groups. The remaining acid group enables the molecule to form ionic bonds with basic proteins (histones and protamines). This group makes nucleotides highly basophilic *i.e.,* they stain readily with basic dyes.

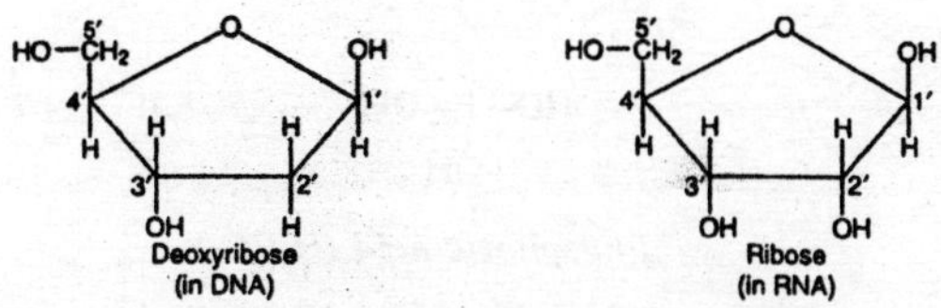

The pentose sugars of nucleic acids.

In DNA, each of the bases is bound to a deoxyribose sugar. Linkage between a purine base and the sugar occurs between group 9 (base) and group 1' (sugar), while pyrimidines link at group 3. The resulting combinations of base and sugar are known as *nucleosides* and are designated as deoxyadenosine, deoxyguanosine, deoxycytidine and thymidine respectively.

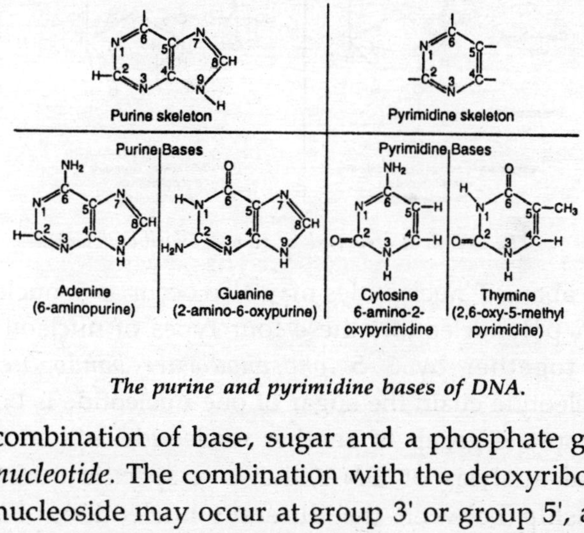

The purine and pyrimidine bases of DNA.

A combination of base, sugar and a phosphate group is called *nucleotide.* The combination with the deoxyribose part of the nucleoside may occur at group 3' or group 5', and the resulting nucleotides are called 3' nucleotides and 5' nucleotides respectively. The four nucleotides of DNA, (Fig.) called *deoxyribonucleotides,* are as follows:

(i) Deoxyadenylic acid or Deoxyadenosine-3'-monophosphate.

(ii) Deoxyguanylic acid or Deoxyguanosine-3'-monophosphate.

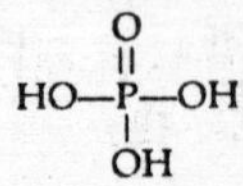

Phosphoric acid (H_3PO_4)

(iii) Deoxycytidylic acid or Deoxycytidine-3'-monophosphate.

(iv) Deoxythymidylic acid or Deoxythymidine-3'-monophosphate.

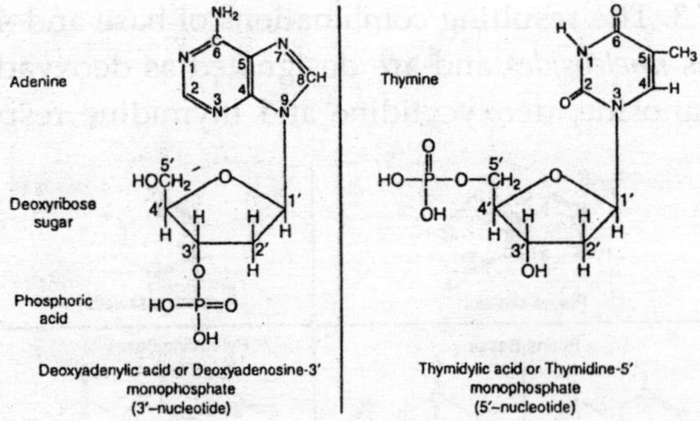

The chemical structure of two deoxyribonucleotides.

The above 3' nucleotides may also occur as 5' nucleotides. In DNA polynucleotide, these four types of nucleotides are linked together by 3'-5' *phosphodiester bonds, i.e.,* in a polynucleotide chain the sugar of one nucleotide is bound to the phosphate group of another nucleotide and so on. The chemical analysis of DNA shows a specific quantitative relationship between the different nucleotides. For each nucleotide of adenine there is one of thymine (A = T), and for each nucleotide of guanine there is one of cytosine (G = C).

Watson and Crick Model of DNA

In 1950, *E. Chargaff* pointed out that in DNA, the molar amount of Adenine equals the molar amount of thymine and the molar amount of guanine equals the molar amount of cytosine. This observation has become a universal biological law which holds true for every form of DNA found in nature. While the work on the base composition of DNA was in progress in Chargaff's laboratory, R. *Franklin* and *M. Wilkins* used X-ray diffraction to obtain precise measurements of DNA. In 1951, *J.D.Watson* and *F.H. Crick* met in Cambridge (England) and decided to collaborate on the study of DNA.

In 1953, *J.D. Watson* and *F.H. Crick* proposed a model of the structure of the DNA molecule that has been widely accepted. For this work these two men shared with *Wilkins,* the 1962 Nobel Prize in medicine and physiology. According to this model, DNA usually consists of two very long, thin, polymeric, antiparallel chains twisted about each other in the form of a regular *double helix* (Fig.). Each chain is a polynucleotide or a collection of many nucleotides in which the sugar of one nucleotide is linked by a phosphate group to the sugar of the adjacent nucleotide.

The diameter of the helix is about 20 Å and each chain makes a complete turn every 34 Å. As there are 10 nucleotides on each chain for every turn of the helix, the distance per nucleotide base is thus 3-4 Å.

There are four nucleotides, each of them containing a deoxyribose sugar (D), a phosphate group (P) and a purine or pyrimidine base. There are two *purines,* adenine (A) and guanine (G), and two *pyrimidines,* cytosine (C) and thymine (T). In the polynucleotide chain the joining together of the sugar and phosphate groups always involves the same chemical groups. Hence, this part of the molecule, called the backbone, is very regular.

In contrast, the order of the purine and pyrimidine bases along the chain is highly irregular and varies from one molecule to another. Both the purine and pyrimidine bases are flat, relatively water-insoluble molecules which tend to stack above each other perpendicular to the direction of the helical axis.

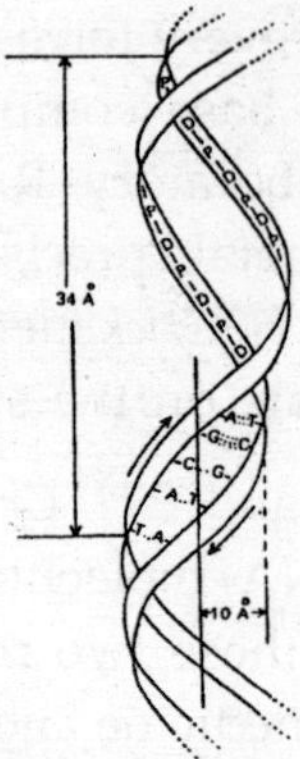

The DNA double helix, Watson and Crick model.

The two polynucleotide chains are joined together by *hydrogen bonds* between the pairs of bases (Fig.). According to the base-pairing rules *adenine* (purine) is always joined to *thymine* (pyrimidine) by two hydrogen bonds, and *guanine* (purine) is always bounded to *cytosine* (pyrimidine) by three hydrogen bonds. A hydrogen bond takes place from the sharing of a single hydrogen atom by two atoms, one of which acts as a hydrogen donor and the other as a hydrogen acceptor. These donor and acceptor atoms are found in each of the bases of DNA. Although the hydrogen bonds are weaker than the ordinary chemical bonds, the great number of these bonds contributes to the stability of the double helical structure.

Thus, in the DNA double helix hydrogen bonds exist only between adenine and thymine (A and T), and between guanine and cytosine (G and C) because other arrangements distort the helical structure of DNA. Hence, only these arrangements are possible for two purines (A and G) would occupy too much space to allow a regular helix and, correspondingly, two pyrimidines (C and T) would occupy too little. The existence of the base pairs means that the sequence of bases and, hence of nucleotides along the two chains are not identical but *complementary*. For example, if we have a sequence ATGCT on one chain, the opposite chain must have the sequence TACGA. According to *Watson*, a stereochemical consequence of the formation of the A...T and G...C base pairs is that the two polynucleotide chains run in opposite directions. Thus, if the helix is inverted by 180°, it superficially looks the same.

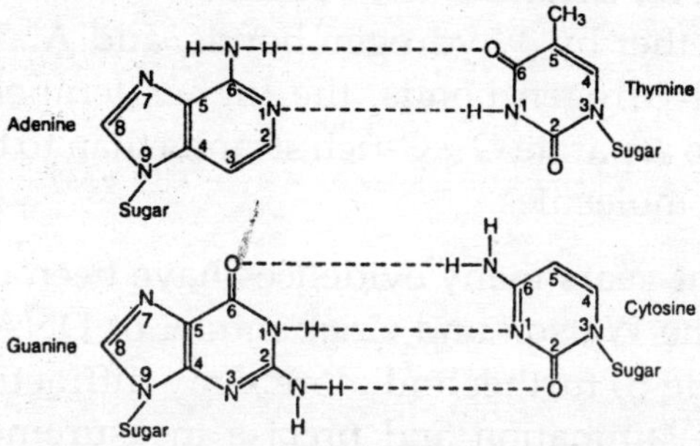

The pairing of purine and pyrimidine bases by means of hydrogen bonds.

An important chemical feature of the DNA structure is the position of the hydrogen atoms (Fig.) in the purine and pyrimidine bases, for the H atoms are found at precise or fixed locations. The N atoms attached to the purine and pyrimidine rings are usually in the *amino* (NH_2) form and only very rarely assume the imino (NH) configuration. Similarly, the *oxygen atoms* attached to the C-6 atoms of guanine and thymine normally have the *keto* (C = O) form, and only rarely take up the *enol* (COH) configuration. These relatively stable locations are quite essential to the biological function of DNA.

In contrast to the ratios of A/T and G/C, which must always be 1-1 to satisfy the base-pairing rules, there are wide variations in the A + T/ G + C contents of different DNA molecules. Higher plants and animals all have an excess of A + T over G + C in their DNA, whereas, among the viruses, bacteria and lower plants, there is much more variation ; and both A + T-rich and G + C-rich species occur. However, these variations are not purely random, and the base ratios of taxonomically related organisms are quite similar.

DNA molecules with a high G + C content are more resistant to thermal collapse than A + T- rich molecules. When DNA molecules are heated above physiological temperatures (to near 100°C), their hydrogen bonds break and the complementary strands often separate from one another (DNA denaturation). Because the G...C base pair is held together by 3 hydrogen bonds, and A...T base pair by only 2 hydrogen bonds, the higher temperatures are necessary to separate G + C-rich strands than to break apart A + T-rich molecules.

In recent years many evidences have been collected in favour of the Watson and Crick model of DNA structure. They include (i) further and more x-ray diffraction studies, (ii) direct examination and precise measurements of the DNA molecule under the electron microscope, (iii) physico-chemical data on the so-called "melting point" of DNA, which is the temperature at which the hydrogen bonds break causing the stiff DNA molecules to collapse, (iv) enzymatic digestion studies in which the kinetics observed follow a pattern predicted from a double-stranded structure, and (v) enzymatic synthesis of DNA in the test tube by *Kornberg* (1962).

Replication of DNA

According to *Watson* (1970), DNA carries all the specificity involved in its *self-replication* and it is important for the linear DNA molecules to be able to circularize while replicating. Initiation of replication begins at a fixed point on the circular chromosome, with growth of many chains starting by elongation of pre-existing strands. However, several enzymes are involved in replication. These are endonuclease, exonuclease, DNA polymerase and polynucleotide-ligase. Enzymes which break the internal phosphodiester bonds are called *endonucleases*. Thus, according to *Watson* (1970), replication must start with an endonuclease cut. *Exonuclease* is the enzyme that digests DNA for 3' or 5' ends of strands. *DNA polymerase* catalyzes the formation of DNA from deoxyribonucleoside triphosphates, using DNA as a template. The *polynucleotide-ligase* covalently joins the backbone chains of DNA. Some of these enzymes probably also function in the repair of damaged DNA molecules. Thus, replication is an enzymatic process, and several enzymes participate in it.

The replication of DNA has been explained by the following hypotheses :

(a) *Template Hypothesis* : This is the hypothesis of *Watson* and *Crick* and it can be explained in their own words " Now, our model for DNA is, in effect, *a pair of templates*, each of which is complementary to the other. We imagine that prior to duplication the hydrogen bonds are broken, and the two chains unwind or separate (Fig.). Each chain then acts as a template for the formation on to itself of a new companion chain, so that eventually we have *two pairs* of chains, whereas we only had one before. Moreover the sequence of the pairs of bases will have been duplicated exactly. This is the so called *template hypothesis* for DNA replication, and it could account for accurate and specific gene duplication as well as occasional modification of the DNA. It can be imagined that on rare occasions a nucleotide pair containing the wrong base is somehow incorporated into the newly forming complementary polynucleotide strand, thus, resulting in a mutation."

(b) *Hypothesis of Levinthal and Crane: A* precise geometrical model (Fig.) whereby the duplication of DNA molecule could occur in a continuous manner without requiring the prior separation of strands was suggested by C. *Levinthal* and *H.R.Crane* (1956). According to this hypothesis, the unwinding and rewinding of the DNA can occur simultaneously. They suggested that duplication could begin at one end, separating the strands, and the free energy released in forming the polymer would provide the energy for the rotation of the two lengthening daughter strands as well as the shortening parent strand.

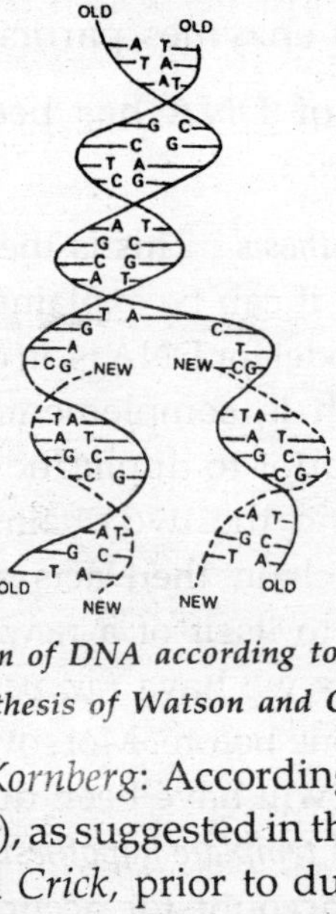

The replication of DNA according to the template hypothesis of Watson and Crick.

(c) *Hypothesis of Kornberg*: According to the hypothesis of *Kornberg* (1962), as suggested in the template hypothesis of *Watson* and *Crick*, prior to duplication two strands of DNA separate by opening of hydrogen bonds. Then two new strands, having a complementary base composition with respect to the parent DNA strands are synthesized.

However, in this mechanism of the cellular duplication of DNA, which takes place in a medium of nucleotide pool, the two strands do not completely separate before the synthesis of the new strands. Instead, duplication goes hand in hand with strand separation. The monomeric precursors for DNA synthesis are the *deoxyribonucleoside triphosphates*, which are enzymatically joined together in the presence of the DNA template.

$$\begin{matrix} \text{dATP} \\ + \\ \text{dTTP} \\ + \\ \text{dGTP} \\ + \\ \text{dCTP} \\ \text{the four deoxyribo-} \\ \text{nucleoside triphosphates} \end{matrix} \xrightarrow[\substack{\textit{DNA polymerase} \\ \text{(Kornberg)}}]{\text{DNA template}} \text{DNA + diphosphate (2Pi)}$$

Selection of the correct nucleotides by hydrogen bonding to the correct template bases is a very accurate process. *Kornberg* (1969) has recently proposed what he calls a *"speculative model for helix replication, in vivo"* and he suggests that the enzyme DNA *polymerase* participates in achieving a sequential and practically simultaneous replication of both helical strands of DNA.

Mechanisms for DNA Replication

There are three possible mechanisms for the replication of DNA. These are conservative, semiconservative, and dispersive. In the *conservative* mechanism, the strands of the parent double helix come back together again after serving as templates for the two new strands. The *dispersive* mechanism implies that the original strands break up and then somehow recombine with the new DNA strands.

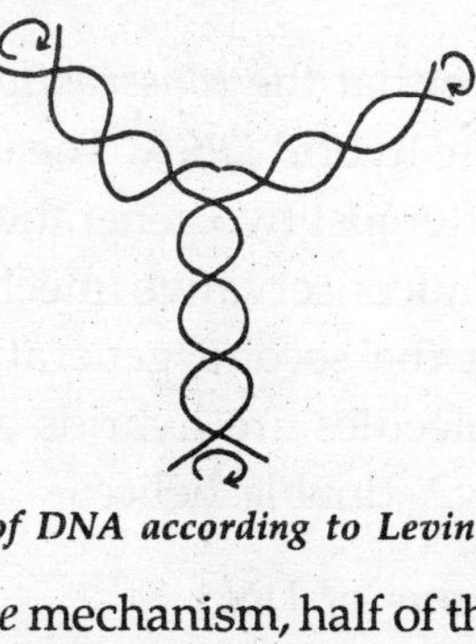

The replication of DNA according to Levinthal and Crane.

In *semiconservative* mechanism, half of the DNA is conserved at the molecular level. Each of the two strands of the parent molecule serves as a template for the synthesis of a new complementary strand. Thus, in the newly formed DNA molecules one strand is always new and one is always old. This mechanism has been demonstrated both at the molecular and cellular levels.

The most convincing evidence for the *semiconservative* mechanism has been given by *M. Meselson* and *F.W. Stahl* in 1958. Working with the common colon bacillus, *Escherichia coli*, they grew some of this bacteria in a medium containing N^{15} (a heavy isotope of nitrogen) as the only source of nitrogen. As a result, all the nitrogen of the bases in the DNA of freshly grown bacteria were of the heavier N^{15}. The density of this DNA was greater than that of the ordinary DNA of this species made with N^{14}.

The analytical data were obtained in this experiment by measuring densities. Some of these bacteria grown in the N^{15} medium were then transferred to a new medium where N^{14} was the only source of nitrogen. By the time a new generation had developed, the density of the new DNA indicated that each DNA molecule must consist of one strand of "heavy" (N^{15}) DNA, and one strand of "light" (N^{14}) DNA. We may call this *"isotopic hybrid DNA"*. In the second generation, half the DNA molecules were of the isotopic hybrid type, and half were light (Fig.).

Thus, we can see that the *conservative* mechanism would not yield any isotopic hybrid DNA. The *dispersive* mechanism would not, atleast after just two generations, yield any wholly light DNA. The semiconservative mechanism is consistent with the results. At the second generation (grand daughter molecules), two molecules are hybrids and two are not, *i.e.*, they are "light" DNA double helices.

Three Structural Forms of DNA

There are three structural forms of DNA. These are : *B-form*, *A-form* and *Z-form*, which are described as follows :

(a) *B-form* : B-form or B-DNA exists in solutions of purified DNA. It is the naturally occurring double helical DNA existing as right-handed form.

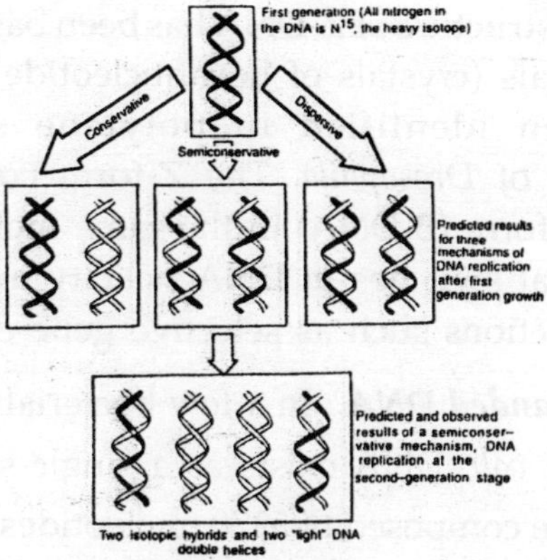

The Meselson-Stahl experiment. Dark strands contain N^{15}. Light strands contain N^{14}.

(b) *A-form* : A-form or A-DNA is also double stranded molecule in the form of a right-handed helix. It occurs when DNA fibres are dried to about 75% relative humidity. The A-form has also been detected in double-stranded RNA molecules, and in single crystals of double-stranded oligonucleotides, and the mixed DNA/RNA oligonucleotides. In A-form, the two strands are antiparallel in their 5'-3' directions.

(c) *Z-form* : The Z-Form or Z-DNA is double-stranded in the form of a left-handed helix (*Rich*, 1979). The Z-form appears to be favoured by repeating dC - dG runs as shown below :

dC dG dC dG dC dG
dG dC dG dC dG dC

In the left-handed form, the phosphate-sugar backbone of each nucleotide follows a zig-zag pattern around the axis of the helices, hence the molecule is called Z-DNA. In the Z-helix, the conformations of the guanine and cytosine deoxyribonucleotides differ considerably from one another, so that the repeating unit becomes a dinucleotide, unlike the B-helix in which repeating unit is a single nucleotide.

While the structure of Z-DNA has been based upon studies of DNA crystals (crystals of hexanucleotide d), Z-DNA has recently been identified in polytene salivary gland chromosomes of *Drosophila*. The Z-form co-exists with the right handed form (B-DNA) in the same molecule. Now it is considered that switches in DNA helicity may be related to biological functions such as selective gene expression.

Single-stranded DNA : In a few bacterial viruses, *e.g.*, the $\phi \times 174$ of E. *coli*, DNA exists as a single-stranded circular DNA molecule composed of 5370 nucleotides. It contains nine genes most of which code for the structural proteins of the virus. In single-stranded DNA, the amounts of adenine and thymine, as well as guanine and cytosine, are not equal. During reproduction of these viruses, the single-stranded DNA (referred to as the "+" strand) is injected into the host bacterial cell, where it acts as a template for the reproduction of a complementary polynucleotide chain (called the "-" strand). These two polynucleotides combine to form a conventional double helix, which then serves as the template for the production of additional + strands. The newly formed + strands are then encapsulated in the viral protein coats to form new virus particles (*Kornberg*, 1979).

Satellite DNA : In human DNA, four satellite DNAs have been observed that constitute 6% of the chromosomal DNA, or a total of 174,000 kb per genome. In *Drosophila melanogaster*, satellite DNAs have been separated from main-band DNA (buoyant density of 1.701g/cm^3.) by CsCl density-gradient centrifugation (*Peacock et al*, 1973). This is most readily accomplished by shearing the DNA into double-stranded segments of sizes from 20 to 60 kb before centrifugation. Under these conditions, the DNA distributes into a main band and a set of smaller bands called *satellite bands* having densities of 1.672 and 1.688 g /cm^3.

The predominant satellite DNAs generally contain very

simple sequences that are repeated many times. For example, in the satellite DNA with a buoyant density of 1.672 g / cm^3 the sequence

—A C A A A C T—
—T G T T T G A—

is repeated 12 × 10^6 times. The satellite DNAs are generally found in the centromeric region of chromosomes and may act in pairing and segregation.

Biological Function of DNA

DNA is the primary genetic material of the cell. The 20 × 10^{-12} gram of DNA present in the human fertilized egg has the ability to determine all the hereditary characteristics of the mature human individual. The main functions of DNA are as follows :

1. *Replication* : DNA, as the primary genetic material, has the ability to replicate. Formation of new molecules of daughter DNA is called the replication of parent DNA. The replication of DNA must occur at every cell division to ensure the equal distribution of DNA in daughter cells. Thus, the most important function of DNA (gene) is the control of its own duplication.
2. *Transcription or Synthesis of RNA* : All the three types of RNA (messenger RNA, ribosomal RNA, and transfer RNA), that in turn take part in protein synthesis, are synthesized under the control of DNA by the process known as transcription.
3. *Translation or Synthesis of Proteins and Enzymes* : Translation is the guiding of amino acid sequence in proteins. Thus, DNA controls protein synthesis via RNA. Due to this ability of DNA, specific proteins and enzymes are synthesized by the cell. The relation of DNA to RNA and protein is usually summarized by

the following formula, which is called the "central dogma" of molecular biology.

$$DNA \xrightarrow{transcription} RNA \xrightarrow{translation} Protein$$

4. *DNA and Mutation* : The genetic material or DNA possesses properties, which determine the development of many different phenotypes. All types of mutations are almost certainly related to changes in the structure of the genetic material. From what we have learnt of the Watson-Crick theory, a change in the nucleotide sequence of a DNA molecule must necessarily affect the amino acid sequence of an enzyme. Of the important physical agencies for altering DNA, atomic radiations and X-ray produce the most serious changes.

5. *DNA and Viruses* : When a *virus* (a chemical aggregate made of protein and nucleic acid) encounters its host cell, *its nucleic acid core penetrates the cell.* Once inside the living cell, the viral nucleic acid directs the making of more of itself and of protein overcoats for the new cores. When enough new particles are made, the host cell bursts, and neighbouring host cells are invaded by the cores of the newly made virus particles. Thus, the viral infection (*e.g.*, the tobacco mosaic virus infecting the leaves of the tobacco plant) spreads. One virus, that has been extensively studied, is labelled $f \times 174$, and has *E. coli* as its host cell. The nucleic acid in $f \times 174$, which is DNA, has 5500 nucleotide units and exists as a single- stranded, closed ring. This DNA core material can be isolated from the virus and yet retain the natural infectivity of the virus.

6. *DNA as the Thread of Life* : However, the biological significance of the DNA can be judged from the statement: "DNA is the secret of life", or "*Life is the secret of DNA.*" It has been called "the thread of life" by *Kendrew* (1967).

10

The RNA

Ribonucleic acid or RNA is the predominant nucleic acid in the cytoplasm, although all cytoplasmic RNA is of nuclear origin. In the nucleus it is present in the nucleolus, nucleoplasm and chromatin, and in the cytoplasm RNA is the main component of the ribosomes. In some viruses, *e.g.*, the tobacco mosaic virus (TMV), the RNA is present in the centre, surrounded by a protective coat of protein and serving as the genetic component of the virus. Almost all RNA molecules have a linear shape. However, recently only one circular replicating form of RNA of the mouse encephalomyocarditis (EMC) virus has been reported by *Agol et al.* (1970).

There is a great variation in the size of the linear RNA molecules. Some of them contain only a small number of nucleotides (about 75), while others have over 10,000. Independent of their length, they generally do not have a completely random shape. Some parts of the RNA molecules are double helical, while others are non-helical. In the double helical regions, the bases are hydrogen bonded to each other using DNA-type base pairing.

Chemical Composition of RNA

RNA is composed of:

(i) *A Pentose Sugar* : called ribose (Fig.). It has a pentagonal ring with 5 carbons, two of which (3' and 5') are linked to phosphoric acid and a third one (carbon 1') to the base.

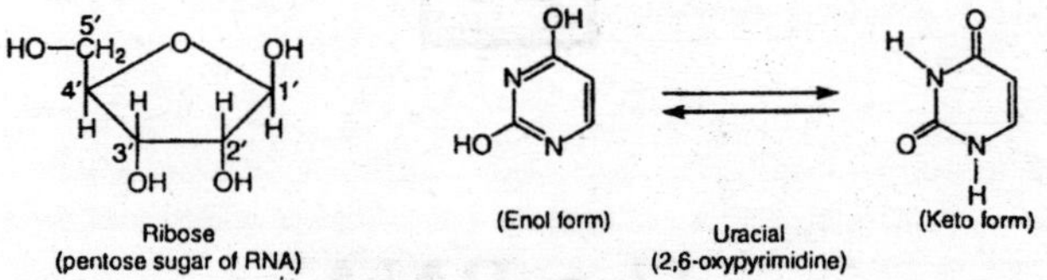

The chemical structure of Ribose and Uracil.

(ii) *Nitrogen Bases* : Usually there are four heterocyclic bases in RNA. Two of the bases are *purines.* These are adenine (A) and guanine (G). *Pyrimidine bases* of the RNA are cytosine (C) and uracil (U). There is no thymine (T) in RNA. The uracil of RNA (Fig.) can pair with adenine. In addition to these four usual bases, some RNA molecules (*e.g.*, tRNA) may have some " unusual " bases.

(iii) *Phosphoric Acid* : or H_3PO_4 links the nucleotides by joining the riboses of two consecutive nucleotides with an ester-phosphate bond. This bond links carbon 3' in one nucleotide with carbon 5' in the next. Linkage between a purine base and the sugar (ribose) occurs between group 9 (base) and group 1' (sugar), while pyrimidines link at group 3. The resulting combinations of the base and sugar are called *nucleosides* and are designated as : adenosine, guanosine, cytidine and uridine.

When these ribosides combine with H_3PO_4, the nucleotides of the RNA are formed. Thus, like the DNA, the RNA is also made up of nucleotides which are called *ribonucleotides* (Fig.). These are as follows:

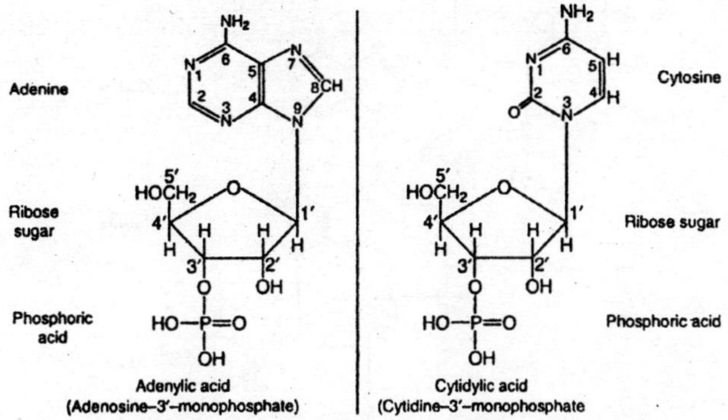

The chemical structure of two ribonucleotides.

(a) Adenosine-3'-monophosphate(Adenylic acid)

(b) Guanosine-3'-monophosphate(Guanylic acid)

(c) Cytosine-3-monophosphate(Cytidylic acid)

(d) Uridine-3'-monophosphate (Uridylic acid)

The above 3' nucleotides also occur as 5' nucleotides. In RNA polynucleotide, these four types of nucleotides are linked together by 3'-5' *phosphodiester bonds.*

Structure of RNA

Like the DNA, RNA is also a long unbranched molecule consisting of four types of nucleotides linked together by 3'-5' phosphodiester bonds (Fig.) But there are several important differences between DNA and RNA (Table). The first is that the sugar of DNA is deoxyribose, whereas RNA contains *ribose.* The latter is identical to deoxyribose except for the presence of an additional OH (hydroxyl group). The second difference is that RNA contains no thymine but instead contains *uracil,* and the third difference is that RNA is not a regular double helix.

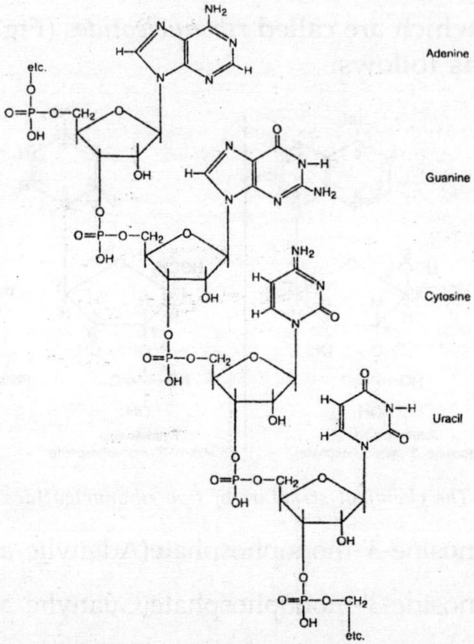

A small segment of RNA chain formed by 3'-5' linkage of ribonucleotides.

In RNA, the amount of adenine is not equal to the amount of uracil. Similarly, the amount of guanine and cytosine also usually differ from each other This suggests that RNA does not possess a regular hydrogen-bonded structure. But unlike double-helical DNA, RNA is usually single stranded (Fig.). However, the RNA strand may be twisted around itself to produce DNA-like helices at several places. In the regions of helix formation, the bases are hydrogen-bonded, *i.e.*, adenine pairing with uracil, and guanine with cytosine. Non-helical regions of RNA contain many non-bonded bases. However, the significance of non-bonded bases is not known.

The RNA-Type and their Functions

On the basis of functions and physical properties there are several kinds or classes of RNA. The principal types of RNA are :

(1) Viral RNA (normally wrapped in a protein coat), (2) ribosomal RNA, (3) messenger RNA, (4) transfer RNA, and (5) small nuclear RNA.

Viral RNA

In most plant viruses and a number of bacterial and animal viruses the hereditary material is single-strand RNA. But two groups of animal viruses called *revoviruses* contain double-strand RNA. The RNA of tobacco mosaic virus (TMV) is one of the most extensively studied and best-known molecules of RNA, which has been used for the study of virus assembly, protein synthesis, the mechanism of mutagenesis, and infectivity. Each TMV virus particle of molecular weight 4×10^7 contains a single strand of RNA of molecular weight 2.2×10^6. There is less stacking of bases in TMV RNA than in DNA. The presence of *intramolecular RNA loops* is an important part of the assembly of TMV from the RNA and protein sub-units.

Function: It was first thought that viral RNA was structurally incapable of forming a double helix. But now it is well known that in different RNA viruses a variety of mechanisms are used for viral RNA replication. They include various ways of using the RNA genome directly, and the unusual mechanism of transcribing RNA into DNA with the help of an enzyme called *reverse transcriptase* and forming double-strand DNA. This observation created a sensation when it was first discovered as it modified the "central dogma" of molecular biology from

$$\text{DNA} \longrightarrow \text{RNA} \longrightarrow \text{protein, to}$$

$$\text{DNA} \rightleftarrows \text{RNA} \longrightarrow \text{protein}$$

The RNA to DNA arrow is shown dashed because reverse information flow from RNA to DNA does not occur very frequently.

The use of RNA, in RNA viruses, for the storage of genetic information may represent an earlier evolutionary stage of information transfer.

The self-replicating RNA viruses lack protein coats. These *viroids* are surprisingly small; the one that causes spindle tuber disease of potatoes is among the largest and has only 359 nucleotides. They are capable of coding only for approximately 100 amino acids, far too few for the minimum number and weight of enzymes necessary for RNA replication, but not too few for a short peptide that could alter the activity of host cell enzymes. Alternatively, the RNA itself may affect the synthesis or processing of normal cellular RNAs.

Similarities and differences between DNA and RNA

DNA	*RNA*
1. DNA occurs mainly in the nucleus as the chromosomal DNA, but also in mitochondria and other self-reproducing organelles.	1. RNA occurs in the cytoplasm (as the main component of ribosomes), nucleolus, and chromosomes.
2. The acid component of DNA is the phosphoric acid.	2. Same
3. The pentose sugar of DNA is *deoxy-ribose*, hence the name DNA.	3. The pentose sugar of RNA is *ribose*, hence the name RNA
4. The purine bases of DNA are *adenine* and *guanine*.	4. Same
5. The pyrimidine bases of DNA are *cytosine* and *thymine*.	5. The pyrimidine bases of RNA are *cytosine* and *uracil*.
6. DNA contains only the usual bases (A, G, C and T).	6. In addition to the four usual bases (A, G, C and U), some RNA molecules may have some 'unusual' bases.
7. In DNA, adenine pairs with thymine by hydrogen bonds, and cytosine with guanine.	7. In RNA, adenine pairs with uracil by hydrogen bonds, and cytosine with guanine.
8. DNA is usually a *double helix* so that the ratios of A-T and G-C are always 1-1 to satisfy the base-pairing rules.	8. RNA is usually single-stranded having both helical and non-helical regions. Hence amount of adenine is not equal to the amount of uracil, and the amount of guanine and cytosine also differ from each other.
9. The hydrolyzing enzyme of DNA is *deoxyribonuclease*.	9. The hydrolyzing enzyme of RNA is *ribonuclease*.

Contd...

DNA	*RNA*
10. Cellular DNA regulates its own synthesis , *i.e.*, DNA is a self-replicating molecule.	10.Cellular RNA is not a self-replicating molecule,
11. DNA is synthesized in the nucleus in the presence of DNA template, four deoxyribonucleoside triphosphates and the enzyme DNA *polymerase*.	11. RNA is synthesized in the nucleus upon a DNA template in the presence of the four ribonucleoside triphosphates and the enzyme *RNA polymerase*.
12. DNA is Feulgen-positive, the deoxyribose is responsible for the reaction.	12.RNA is Feulgen-negative. It is detected by basophilic dyes with ribonuclease treatment.
13. DNA is the primary genetic material of the cell.	13. RNA is not the primary genetic material of the cell. It is the genetic component of some viruses (e.g., TMV) only.
14. DNA regulates the synthesis of protein via RNA because all the three types of RNA are coded on the DNA template.	14. All the three types of RNA play their role in protein synthesis,

Ribosomal RNA (rRNA) : About 75 to 80% of all the RNA of a cell is found in the ribosomal particles. Each sub-unit of the ribosome contains single-strand RNA (fig.). The large subunit (50 S) of the *E. coli* ribosome (70 S ribosome) has 23 S ribosomal RNA and 5 S ribosomal RNA, which have a molecular weight of about 1.2×10^6 and 4×10^4 respectively. The small subunit (30 S) of the E. *coli.* ribosome has 16 S ribosomal RNA whose molecular weight is about 0.6×10^6.

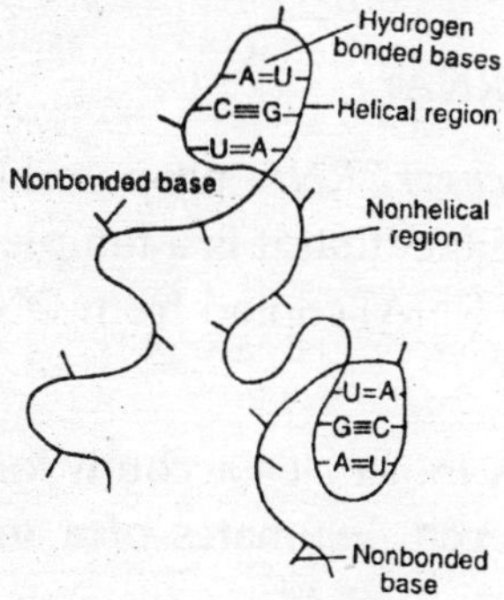

In 80 S ribosomes (with 60 S and 40 S sub-units), there are 28 S and 18 S ribosomal RNAs, with a molecular weight of 1.3×10^6 and 6×10^5 respectively. The length of a large unfolded RNA molecule has been estimated to be about 7000 Å by *Hart* (1965).

In both sub-units, the RNA molecules are associated with protein whose chief function is to help bring the template RNA and the amino acid-transfer RNA precursors together correctly. Because ribosomes easily attach basic dyes, it is thought that RNA is exposed at the surface of the subunit, and the protein may be in the interior in relation to non-helical parts of the RNA. About 60% of the ribosomal RNA is helical or double stranded and contains paired bases. The helical part of the ribosomal RNA is supposed to be on the surface of the sub-units.

The base composition of the ribosomal RNAs of different sources varies considerably and it does not follow a rule of A = U and G = C. The most abundant bases are adenine and guanine. However, ribosomal RNA also contains some methyl groups.

Function : The function of ribosomal RNA is not known and at present it is difficult to say why ribosomes contain rRNA as well as protein. It is believed that unpaired bases in ribosornal RNA are in some way involved in the binding of transfer RNA and messenger RNA to ribosomes.

Messenger RNA (mRNA)

The name *"messenger" RNA*, proposed by *Jacob* and *Monod* in 1961, refers to the fact that it is a template molecule (hence also called template RNA) copied from DNA and has a rapid turnover.

Messenger RNA molecules account for about 5 to 10% of the total RNA in a cell. Estimates of a molecular weight of mRNA so far indicate that it is a polydisperse material ranging in molecular weight from 5×10^4 to 5×10^5 or more. Another characteristic of mRNA is that it is *heterogeneous.* The size of the molecule varies considerably since it is adapted to the size of the polypeptide chain for which it will code. In E. *coli,* the average size of mRNA cistron is 900 to 1500 nucleotides, thus corresponding to peptide chains consisting of 300 to 500 amino acids. Sometimes messenger RNAs are much longer if several adjacent cistrons are copied at the same time.

These are the so-called *polygenic* or *polycistronic* messengers. For example, the 10 specific enzymes involved in the metabolism of histidine may be coded in the same mRNA molecule. In higher cells, although messenger RNAs of very high molecular weight are found, some mRNAs may be *monocistronic.* For example, haemoglobin is coded by two (α and β) structural genes.

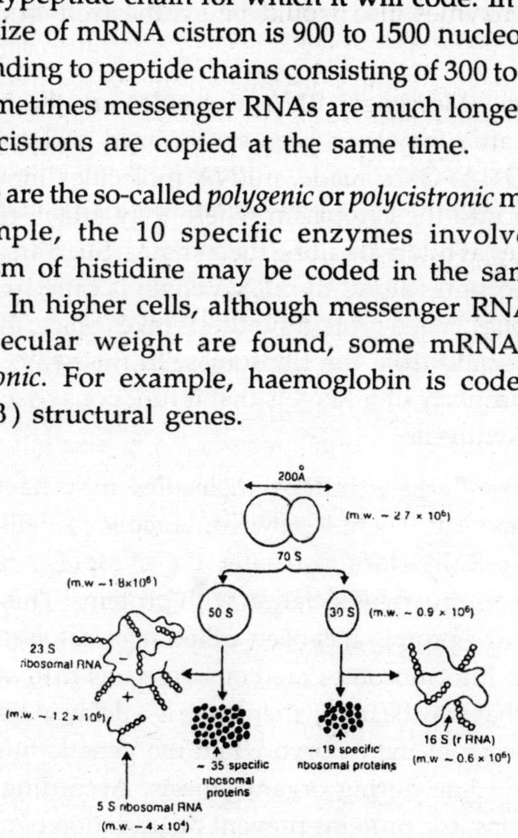

The structure of E. coli ribosome and their RNA.

Messenger RNA molecules are synthesized in the nucleus by DNA in the presence of the enzyme *RNA polymerase* and the triphosphate forms of the nucleotides (ATP, UTP, CTP and GTP) by a process that is known as *transcription.* The sequence of bases in mRNA is determined by complementary sequence on DNA. Thus, A in DNA is replaced by U in messenger RNA,

C is replaced by G; G is replaced by C, and T is replaced by A. However, it appears that only either of the two strands of the DNA is used in RNA synthesis. The reason why only one strand of the double-stranded DNA is a *"coding strand"* is that mRNA otherwise would always be formed in complementary pairs and enzymes also would be synthesized in matching pairs.

Function : Messenger RNA molecules are the bearers *of genetic code* after they have been synthesized under the direct control of DNA. Once made, mRNA molecules move out of the nucleus into the cytoplasm where they attach ribosomes to themselves at intervals along their chains. Such an assembly of many ribosomes along an mRNA chain is called a *polysome* or *polyribosome.* When protein synthesis takes place, the mRNA cannot dissociate from the ribosomes. In this way, the most important property of mRNA is that it functions as a template in protein synthesis.

In eukaryotic cells mRNA molecules may have a high degree of stability. There is now convincing evidence that in these cells mRNA does not enter the cytoplasm as naked strands of RNA but in association with proteins. This complex of mRNA with protein has been called *informosome* by *Spirin et al.* (1965). Informosomes are considered as mRNA-protein complexes that may be used when there is a delay in translation. This may occur in the embryo when the genetic information is manifested late during organogenesis. According to other interpretations, the proteins prevent degradation of mRNA by intracellular ribonucleases, thus playing a protective role. It has been suggested that the protein may also be required, for recognition of the ribosomes, because when mRNA is purified, it tends to lose its template properties. The protein may also reversibly mask the mRNA and in this way control the synthesis at the level of translation. This may result in a regulation or modulation of protein synthesis.

Transfer RNA (tRNA) or Soluble RNA (sRNA) : Because this RNA has a much lower molecular weight than rRNA or mRNA, it is more soluble in the media used to isolate the higher-formula-weight form. Thus, the designation "soluble" refers to a physical property and says nothing about the function. When its function is to be described, various authors refer to sRNA as *transfer RNA* (tRNA) or as *amino acid adapter RNA* (aaRNA).

In *E. coli,* tRNA accounts for 10 to 15% of the total RNA of the organism. Each molecule of tRNA consists of a single chain of about 70 to 80 nucleotides that is folded back on itself.

The first nucleotide sequence of a specific tRNA molecule (from yeast) was worked out by *Robert Holley* in 1964-65, who suggested a cloverleaf diagram for the molecule. For this he earned the Nobel Prize in 1968. This tRNA specifically attaches to *alanine* and therefore is called *alanine-tRNA.* It contains 77 nucleotides arranged in a particular manner (Fig.), and has a molecular weight of about 25,000.

The most striking feature of the alanine -tRNA is the presence of several "unusual" bases i.e., the bases other than adenine, guanine, cytosine, or uracil. These unusual bases differ from the normal bases by the presence of one or more methyl (CH_3) groups. Most of the methyl groups are enzymatically added after the nucleotides are linked together by 3'-5' phosphodiester bonds. However, the function of the unusual bases is not clear. They cannot form the usual base pairs and may thus have the function of disrupting the double -helical regions, thereby exposing the free keto and amino gr[illegible] which can then form secondary bonds to mRNA, ribosome, or to the enzyme required to attach a specific amino acid to its specific tRNA molecule.

The important aspects of the suggested *cloverleaf model* (Fig.), which shows the two-dimensional folding pattern, for tRNA molecules are as follows.

The 3' end of the molecule consists of ACC (adenylic acid, cytidylic acid, cytidylic acid) sequence and a variable fourth nucleotide. The amino acid always attaches to the 3' terminal A. The terminal nucleotide of the 5' end is usually guanylic acid. On moving along the backbone away from the 3' end we find the acceptor stem, T stem, and the T loop or the first loop of the cloverleaf. The T loop is so named because it includes thymine (T). It is made up of 7 unpaired bases and always contains the sequence 5'-T pseudo UCG-3', *i.e*, towards the 5' end the first base is *ribothymidine* and it is followed towards the 3' end by pseudouridine, cytosine, and guanine, etc. The first loop binds to the ribosomal surface during protein synthesis.

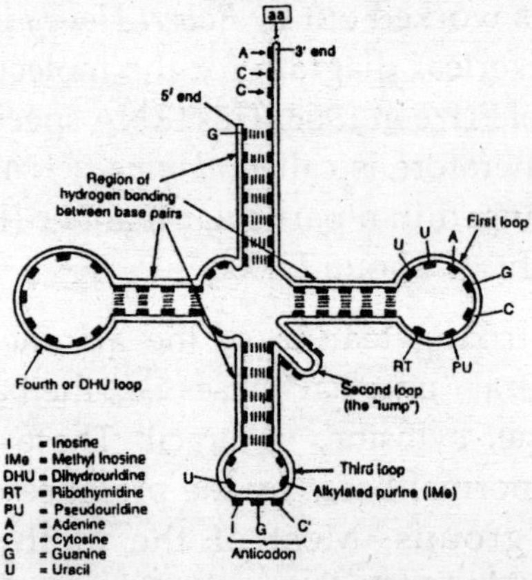

The cloverleaf model of tRNA molecule.

The first loop is followed by a loop of highly variable size often called "the lump" or the extra arm or the variable loop, which in different tRNAs has different numbers of nucleotides ranging from 4 to 21. The third loop or the anticodon loop consists of 7 unpaired bases, of which three form the *anticodon* and bind to the mRNA codon during protein synthesis. The anticodon is bracketed on its 3' end by an alkylated purine base (methyl inosine) and on its 5' end by uracil. The fourth non-hydrogen-bonded section of the molecule or the fourth loop consists of 8 to 12 unpaired bases and is relatively rich in *dihydrouridine*. Hence, it is called DHU loop or D loop, and may be involved in binding to the specific activating enzyme (amino-acyl synthetase).

Besides the above non-hydrogen-bonded sections the alanine-tRNA molecule consists of four regions of hydrogen bonding between the base pairs. Thus, a majority of the bases are hydrogen-bonded to each other because the number of adenine is equal to the number of uracil, and similarly, the number of guanine is equal to the number of cytosine.

Thus, in a given tRNA molecule we may recognise four specific sites. These are : *(a)* the amino acid attachment site, *(b)* the codon recognition site (anticodon), *(c)* the ribosome recognition site, and *(d)* the specific site for recognizing the amino acid-activating enzyme.

An interesting feature of the cloverleaf diagram is the presence of nucleotide sequences that are constant in all the 100 of the tRNA sequences that have been determined so far. The number of base pairs on the stem regions is also constant; 7 in the acceptor stem, 5 in the T stem, 5 in the anticodon stem, and 3 or 4 in the D stem. These features are maintained in tRNA molecules from bacteria, plants and animals. Indeed, the pattern of loops, stems and constant nucleotides found in the tRNA cloverleaf appears to have the same universality as the genetic code.

In 1974, the three-dimensional structure of the yeast phenylalanine-tRNA has been described by *Alexander Rich* and *Sung Hou Kim* from an X-ray crystallographic study at a resolution of three angstroms. This L-shaped map confirmed the earlier finding that the molecule is organized into two columns of nucleotide bases stacked at right angles to each other. These columns have both helical and non-helical regions corresponding to the stems and loops of the cloverleaf model. The map further revealed the two helical regions, each consisting of about 10 base pairs corresponding to one turn of the double helix, and possessing the same type of hydrogen bonding between complementary nucleotide bases as found in the double helix of DNA.

Function : The function of tRNA is to attach to itself a particular amino acid for which it is coded and to carry it to the site of protein synthesis (specially an mRNA site in contact with a ribosome), and there give it up to the growing end of a polypeptide chain at the particular moment called for by the genetic code. For each amino acid, there is a specific activating enzyme which activates the carboxyl group of that amino acid for covalent bonding with the 3' terminal adenosine of the tRNA. The first step of the reaction uses the energy of ATP.

$$\underset{\text{(amino acid)}}{AA + ATP} \xrightarrow{\textit{amino acyl synthetase}} \underset{\text{(amino acid adenylate)}}{AA\text{-}AMP + P\text{-}P}$$

The AA-AMP so formed remains bound to the enzyme until the proper tRNA arrives to form the AA-tRNA complex.

$$AA - AMP + tRNA \xrightarrow{\textit{amino acyl synthetase}} AA\text{-}tRNA + AMP$$

For this purpose the enzyme must have two different sites : one to recognise the amino acid and another to recognize the specific tRNA. Every cell needs at least 20 different kinds of activating enzymes and at least 20 kinds of tRNA molecules. There must be at least one of each for every amino acid. It was formerly known that only 20 different tRNA molecules were used, but now it is clear that there may be more than one tRNA molecule for the same amino acid. The amino acid which is used first of all in protein synthesis is formyl methionine and the tRNA which attaches to it is called F-met-tRNA.

Thus, tRNA acts as a vehicle for transferring amino acids into the assembled chain of the protein. In other words, it plays a key role in translating the genetic code into the sequence of amino acids in a protein. Without the tRNA molecule genetic information could not be expressed in the synthesis of protein. *Alexander Rich* and *Sung Hou Kim* (1978) have stated that tRNA participates in a variety of other processes also, some of which are of great importance. For example, tRNA molecules can donate amino acids to preformed protein molecules or to the molecular structure of the cell wall in bacteria independently of the ribosome.

Another process in which tRNA participates is the control of gene expression. Certain tRNAs with an amino acid attached to them are known to determine whether genes (segments of DNA) will be expressed by regulating their transcription into messenger RNA or not. However, the details of the mechanism are not known, but in some systems the control function is associated with a particular modified nucleotide in the tRNA molecule, say, a uracil that has been converted into a pseudouracil. It is thought that tRNA helps to control the expression of many different genes, although the exact number is not known. Some other observations suggest that tRNA may be involved in still more types of biochemical regulation. For example, in embryonic development one kind of modification of certain nucleotides in a tRNA gives way to another kind. Similarly, when a normal cell becomes cancerous, the kinds of modifications of nucleotides in its tRNA molecules change substantially. However, it is not yet known whether these transformations are associated with the regulatory functions of tRNA.

Now it is known that tRNA is associated not only with the synthesis of polypeptide chains but also with that of polynucleotide chains. This synthesis is carried out by special enzymes such as *reverse transcriptase*. The latter, discovered a few years ago as a constituent of several rumor viruses, synthesizes a strand of DNA from a template of single-strand RNA, a direction of information flow that is the reverse of the normal one. For this a specific type of tRNA first binds to the enzyme and to the viral RNA and signals the synthesis of the

DNA copy to begin. But why a tRNA serves this purpose is not known at present.

Small, Nuclear RNA

The small, nuclear RNA (snRNA) is a metabolically stable RNA molecule found in the nucleus of the cell. It occurs predominantly in ribonucleoprotein particles. It is a small molecule consisting of 100 to 300 bases. The RNA found in the cytoplasm of eucaryotic cells is synthesized in the nucleus. However, much of this RNA never leaves the nucleus and thus never participates directly in protein synthesis.

RNA Polymerases

In *E. Coli,* the protein that catalyzes the synthesis of RNA using DNA as a template, the *RNA polymerase* is structurally well known. The holoenzyme is composed of five non-covalently bonded polypeptides two α sub-units with MW 40;000; P with MW 150,000; β ′ with MW 160,000-and α (sigma) subunit with MW 90,000.

In eucaryotic cells, three RNA polymerases are found, each having great subunit complexity. They can be distinguished from one another by their sensitivity to the fungal toxin α -amanitin and by their cellular location. RNA polymerase 1 (also known as A) occurs in the nucleolus, is resistant to α -amanitin, and transcribes the large ribosomal RNAs. RNA polymerase 2 (also known as B) occurs in the nucleoplasm, is sensitive to low concentrations of α -amanitin, and is the main enzyme responsible for synthesizing the heterogeneous nuclear RNA that later becomes processed into mRNA. RNA polymerase 3 (also known as C) is also in the nucleoplasm, is sensitive only to high concentrations of α -amanitin, and synthesizes tRNAs and 5S rRNA. There are also RNA polymerases found in mitochondria and chloroplasts.

11

Homeostasis

The term 'homeostasis' consists of two parts : *homeo* which means "like" or "similar", and stasis meaning "a standing" or "a state of equilibrium." Thus, *homeostasis* is defined as the tendency to maintain stability or equilibrium within an organism or a social group. The term homeostasis is used by physiologists to mean maintenance of constant or static conditions in the internal environment.

However, we should differentiate between homeostasis and *cybernetics.* The term 'cybernetics' was coined by *Norbert Wiener* (1948) and is defined as "the science of controls." Homeostasis includes the mechanisms responsible for the maintenance of the internal environment in the living organisms and also the tendency to maintain equilibrium between the internal and the external environment of the organism or social groups.

It is also necessary to differentiate between homeostasis and haemostasis. The term *haemostasis* means the "control of bleeding" and depends in part on appropriate vascular responses at the site of injury.

Brief History : The French Physiologist *Claude Bernard* (1813-1878) was the first to observe the importance of homeostasis in animal function. As a physiologist he was concerned with the *internal environment* ("milieu interieur") of the living organism and described that internal environment resisted change and seemed to remain fixed ("La fixite").

According to him, "all the vital mechanisms, however varied they may be, have only one object, that of preserving constant conditions of the life in the inner environment". In other words, *"constancy of the internal environment"*, as *Bernard* (1859) phrased it, has been found to be a universal phenomenon in living animals and plants, allowing them to survive in different habitats including stressful and varying external environments.

In 1929, the American physiologist *Walter B. Cannon* published an article entitled "Organization for Physiological Homeostasis". He suddenly was struck by the remarkable constancy of the internal environment. But in the French edition of his book, *The Wisdom of the Body, Cannon* made it clear that *Bernard* deserves priority for emphasizing the role of the inner environment in the establishment and maintenance of steady states in the body. In other words, it was the Bernard's original idea of the constancy of the internal environment that *Cannon* adopted, developed, and then neatly packaged under the label of *homeostasis*.

The following lines of *Cannon* (1932) may be quoted. "The coordinated physiological processes which maintain most of the steady states in the organism are so complex and so peculiar to living beings— involving, as they may, the brain and nerves, the heart, lungs, kidneys and spleen, all working cooperatively—that I have suggested a special designation for these states, *homeostasis*. The word does not imply something set and immobile, a stagnation. It means a condition—a condition which may vary, but which is relatively constant."

Langley (1965) in his brilliant monograph "Homeostasis" writes : "*Bernard* and *Cannon* and many other physiologists were impressed by the self-regulating negative feedback systems which serve to maintain the constancy of the internal environment of living organisms. This is termed homeostasis. But homeostasis is not limited to animals, it is applicable to all living things. Engineers have long used such systems. The principle may even be applicable in sociology, economics, and ecology".

Homeostatic Mechanism

The regulatory processes are based primarily on the principle of feedback or *feedback mechanism* or *servomechanism* (as the engineers call it). The essential components of such a mechanism include : (1) a sensor, (2) a circuit, (3) an effector, and (4) a shut off device (Fig.). In servomechanism, the terms *input* and *output* are used. If we take an example of thermostatic regulation of room temperature, the working of the basic elements of homeostasis may be explained as follows :

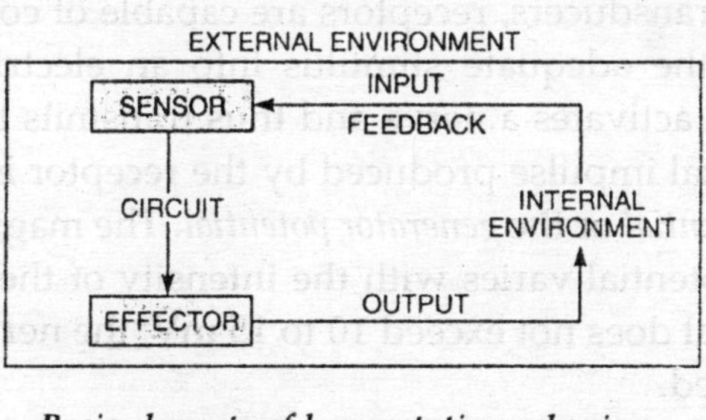

Basic elements of homeostatic mechanism.

If we want to keep the house temperature between 22°C and 25°C, the *sensor* (thermostat) must respond when the temperature of the house rises above 25°C, or falls below 22°C. In the first case when it rises above 25°C the sensor responds by closing a circuit which activates the air conditioner. As a result, air at a temperature considerably below 22°C is blown into the house. The internal temperature of the house will, accordingly, fall. Similarly, if the temperature of the house falls below 22°C, the thermostat responds by closing a circuit which activates the furnace to produce hot air (Fig.). In this example, the temperature of the room is the *input,* the heat produced by the furnace (or the cold air generated by the air conditioner) is the *output.*

Each living organism maintains its internal environment in a similar manner. There are *sensors* called *receptors* ; there are *circuits* which in some instances are composed of *nerves,* but in other cases utilize different mechanisms, for example, the circulation of hormones in the blood, and there are *effectors* which can alter the internal environment and return it to the normal narrow range despite wide variations in the external environment. In the living organism receptors are *biological transducers (Langley,* 1965). A receptor is specialized nervous tissue sensitive to a specific change in its environment.

Receptors generally respond only to a specific type of stimulus. For example, a receptor may respond to changes in temperature, but not to changes in light or pressure. As biological transducers, receptors are capable of converting the energy of the adequate stimulus into an electrical impulse which then activates a nerve and thus transmits the message. The electrical impulse produced by the receptor is termed the *receptor potential* or the *generator potential.* The magnitude of the receptor potential varies with the intensity of the stimulus. If the potential does not exceed 10 to 15 mV, the nerve is usually not activated.

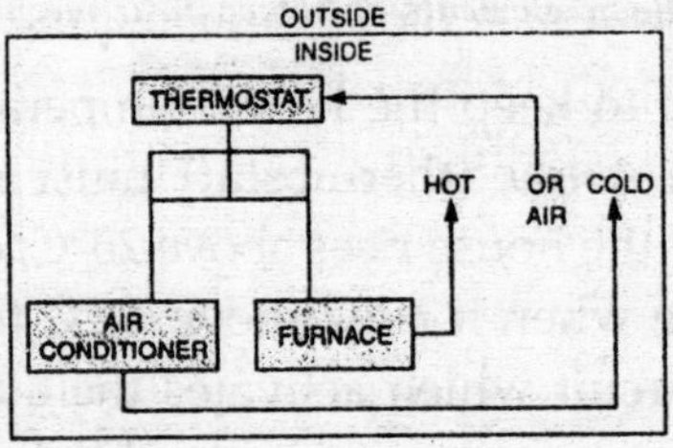

Control of house air temperature by simple feedback mechanism.

In homeostatic mechanisms we deal with messages. The first message which indicates a difference between the external and internal environment influences a transducer. The latter converts the energy of that message to a convenient form. In that form a second message is transmitted to a device which can do something to oppose the influence of the external environment *(Langley,* 1965).

Characteristics of Homeostatic Mechanisms : All the homeostatic control mechanisms have certain characteristics in common, which are as follows :

1. *Negative Feedback :* Most control systems of the body act by negative feedback. For example, in the regulation of CO_2 concentration, a high concentration of CO_2 in the extra-cellular fluid increases pulmonary ventilation. This in turn decreases CO_2 concentration because the lungs then excrete greater amounts of CO_2 out of the body. In other words, the high concentration causes a decreased concentration, which is negative to the initiating stimulus.

Therefore, in general, when some factor becomes too high or too low, a control system initiates *negative feedback,* which consists of a series of changes that return the factor towards a certain mean value, thus maintaining homeostasis.

2. *Gain of a Control System :* The degree of effectiveness with which a control system maintains homeostasis is determined by the *gain* of the negative feedback. The gain of the control system is calculated by the following formula :

Gain = Correction divided by error.

For example, a large volume of blood is transfused into a person whose barorecepter pressure control system is not functioning, and the arterial pressure rises from the normal level of 100 mm Hg to 175 mm Hg. Then, if the same volume of blood is injected into the same person when the beroreceptor system is functioning and this time the pressure rises only 25 mm Hg. Thus, the feedback control system has caused a "correction" of -50 Hg, that is from 175 mm Hg to 125 mm Hg. There remains an increase in pressure of +25 mm Hg, called the "error", which means that the control system is not 100 per cent effective in preventing change.

The gain of the baroreceptor system for control of arterial pressure is -50 divided by +25, or -2. That is, an extraneous factor that decreases or increases the arterial pressure does so only one third as much as would occur if this control system were not present. But the gains of some other control systems are much higher than that of the barorecepter system. For example, the gain of the system controlling body temperature is about -33, which indicates that the temperature control system is much more effective than the barorecepter system.

3. Adaptive control system : In some cases more complex types of control systems are used. For instance, some movements of the body occur so rapidly that there is not enough time for nerve signals to travel from the peripheral parts of the body to the brain and then back to the periphary again in time to control the movements. Therefore, the brain uses a principle called *feed-forward control* to cause the required muscle contractions. Then, sensory nerve signals from the moving parts apprise the brain in retrospect whether or not the appropriate movement as envisaged by the brain actually has been performed correctly. If not, the brain corrects the feed-forward signals that it sends to the muscles the next time the movement is required. Then, once again, if still further correction needs to be made, this, too, will be done for subsequent movements. This is called *adaptive control*, which, in a sense, is delayed negative feedback.

Examples of Homeostatic Mechanisms

Numerous examples of homeostatic mechanisms may be given right from subcellular and cellular level to organism, population and ecosystem level. The following examples from human physiology are important.

1. *Regulation of Body Fluids* : The volume and composition of body fluids are homeostatically maintained with almost complete precision. The body fluids consist of *intracellular fluid* (fluid within the cells) and *extracellular fluid* (blood and interstitial fluid). As blood passes through the capillaries, continuous exchange of extracellular fluid also occurs between the plasma portion of the blood and the interstitial fluid that fills the intercellular spaces, the spaces between the cells.

Blood contains a wide variety of substances—glucose, potassium, sodium, calcium *etc.*, each of which is regulated by a feedback mechanism. Although a number of factors work for the regulation of the volume of the body fluids, two major factors are : *(a)* the control of urine formation, and *(b)* the control of fluid intake..

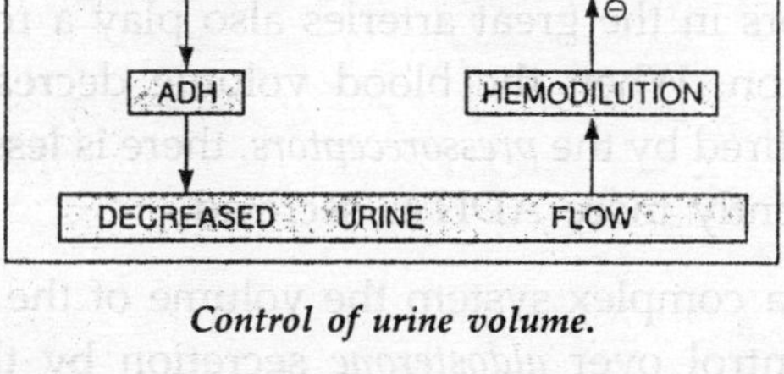

Control of urine volume.

(a) *Control of Urine Output* : There are two types of sensors that are necessary to fluid volume control—*osmoreceptors* (that evaluate the concentration) and *volume receptors* (that evaluate the change in volume). The former are located in the hypothalamus. When the organism is dehydrated, the osmotic concentration of the blood increases and therefore hypothalamic osmoreceptors respond. As a result, messages are sent to the posterior lobe of the hypophysis and a hormone called the *antidiuretic hormone* (ADH) is secreted (Fig.). It is carried by the blood to the kidneys where it acts to diminish the formation of urine. Thus urine volume is reduced, and if fluid intake remains normal, fluid will accumulate in the body thereby returning the osmotic state of the blood to normal. Thus, there is a simple but quite effective negative feedback mechanism, which regulates the volume of urine vis-a-vis osmotic state of the blood.

It should be noted that in this feedback mechanism the sensor is sensitive only to the concentration (osmotic pressure) of the blood. But there is also a feedback mechanism that responds to the volume of the blood. For example, following haemorrhage there is a sharp decrease in urine formation. Such volume receptors have been located in the walls of the atria and great veins. As they are stretch receptors, they respond when the volume of blood increases. These messages are inhibitory and cause a reduction in the release of ADH. Pressoreceptors in the great arteries also play a role in body fluid regulation. When the blood volume decreases, fewer impulses are fired by the *pressoreceptors,* there is less inhibition and consequently more ADH is secreted.

Through a complex system the volume of the blood also excercises control over *aldosterone* secretion by the adrenal cortex. The aldosterone influences the excretion of Na^+ and water by the kidneys. The sensors are thought to be the specific cells in the blood vessels of the kidneys. These cells themselves secrete a substance called *renin.* In this case, then, the receptors are also effectors. That is, instead of sending message to other cells that respond, they respond themselves. The renin secreted into the blood reacts with a constituent of the blood called *angiotensinogen* to form *angiotensin.* Angiotensin then is carried by the blood to the adrenal cortex where it stimulates the secretion of *aldosterone.*

The aldosterone is carried by the blood back to the kidney where it decreases the excretion of Na^+ and water. The more the kidney cells are stretched, the less renin they secrete. This negative feedback mechanism works as shown in Fig. As the blood volume decreases, the volume receptors secrete more renin, more angiotensin is formed, the adrenal cortex secretes more aldosterone, the kidneys retain more Na^+ and water, and therefore the blood volume increases, thus completing the cycle.

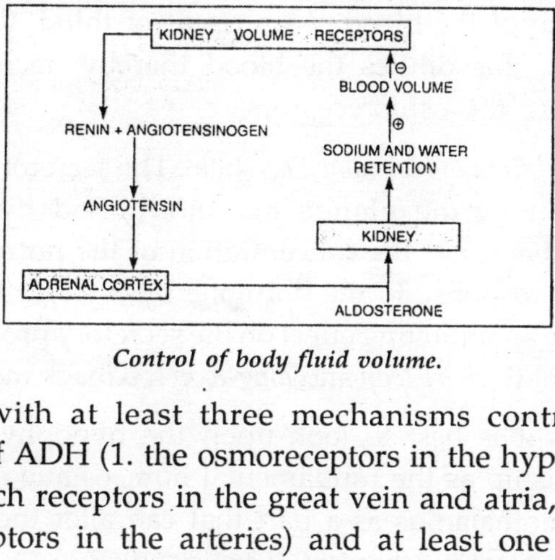

Control of body fluid volume.

Thus with at least three mechanisms controlling the secretion of ADH (1. the osmoreceptors in the hypothalamus, 2. the stretch receptors in the great vein and atria, and 3. the pressoreceptors in the arteries) and at least one regulating aldosterone secretion, it is no wonder that the urine output is so precisely attuned to body fluid needs.

(b) *Control of Fluid Intake :* We are familiar with the fact that in hot weather, due to the loss of water through sweating, we feel thirsty and accordingly our fluid intake is increased. The sensation of *thirst* has at least two components: *(i)* a local response represented by drying of the mucous membrane of the mouth and throat, and *(ii)* the hypothalamic drinking centre. The cells that make up this part of the hypothalamus an the sensors of the controlling feedback mechanism of thirst (Fig.)

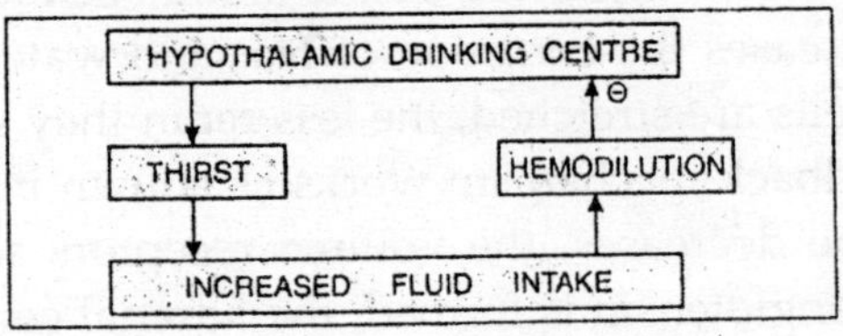

Control of fluid intake.

They are osmoreceptors and monitor the osmotic concentration of the blood. As a result of thirst, fluid intake is increased; this dilutes the blood thereby decreasing the stimulation to the osmoreceptors.

2. *Regulation of Hormone Secretion* : The secretory activities of endocrine glands are modulated by negative feedback, *i.e.,* the concentration of the hormone itself, or a response to the hormone by a target tissue, will have an inhibiting effect on the secretory process. There are both *short-loop* and *long-loop* feedback mechanisms.

Perhaps it is best to look upon the hypophysis-thyroid interrelationship as the fundamental homeostatic mechanism and the hypothalamus as a part that can alter the setting of that fundamental mechanism *(Langley,* 1965).

The anterior lobe of the hypophysis secretes the *thyroid stimulating hormone* (TSH), which stimulates the thyroid gland to secrete *thyroxin.* But thyroxine is inhibitory to the anterior lobe of hypophysis. As a result, less TSH is secreted. Through this short loop negative feedback mechanism, the thyroid secretion is controlled. Similarly, the level of blood calcium regulates the activity of the parathyroid gland.

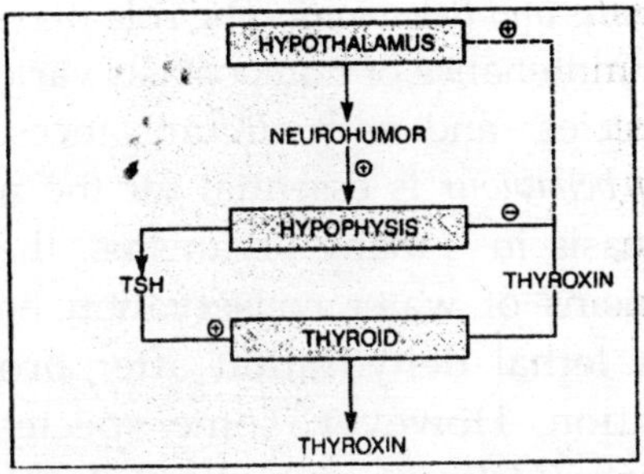

Control of thyroid secretion by a simple feedback mechanism.

Negative feedback control is also illustrated by the secretion of ADH in response to increased osmotic concentration of the blood. The ADH upon reaching the kidney increases water reabsorption thus counteracting the loss of fluid. As a result, plasma osmolarity is stabilized, and, in turn, so is the secretory activity of the ADH-producing tissue of the posterior lobe of the hypophysis.

The negative feedback signal acting to slow the output of an endocrine tissue may be a physiological response to the hormone secreted by that tissue (*e.g.*, reduced osmolarity in the ADH loop ; reduced blood glucose levels in the insulin loop) or it may be a rise in concentration of a second hormone, the secretion of which is stimulated by the first hormone (*e.g.*, high *estrogen* levels inhibit production of the *gonadotropic hormone* that stimulates estrogen production). Thus the negative feedback signal may be simply the reduction of a stimulus to the endocrine tissue, or it may be an active inhibition of secretion of the hormone.

3. Other Examples : Many other examples of homeostatic mechanisms may be found in physiological processes such as (i) control of blood pressure, (ii) control of respiration, (iii) control of blood glucose level, and (iv) control of body temperature, etc. Details of these regulatory mechanisms may be found in different chapters of this book.

4. *Homeostasis and Behaviour :* The role played by *behaviour* in the maintenance of homeostasis varies considerably with species and with circumstances. For example, *drinking behaviour* is essential for the maintenance of homeostasis in many organisms, the physiological mechanisms of water conservation being unable to prevent lethal dehydration after prolonged water deprivation. However, some species such as the budgerigar *(Melopsittacus undulatus)* and the mongolian gerbil *(Meriones unguiculatus)* are able to survive indefinitely without water. Because of the great efficiency of their water conservation mechanisms they can live on the water contained in the seeds they eat *(McFarland,* 1981). Others, such as aquatic species need no special behaviour to obtain the water necessary for the maintenance of homeostasis.

In some aspects of homeostasis the role of behaviour is negligible. Experiments involving surgical manipulation with the physiological mechanisms responsible for homeostasis show that animals often possess the capacity for appropriate behaviour, even though it is not manifested in their normal lives. For example, when the thermal homeostasis of rats is interfered with by removal of the thyroid gland, they respond if given the opportunity by the experimenter, by building warmer nests and by showing other forms of behavioural thermoregulation *(Curt Richter,* 1943).

In a similar way, removal of the adrenal gland, which is involved in salt balance, causes rats to shift their food preferences towards more salty food and water. In other cases there may be no special organs (such as thyroid and adrenal glands mentioned above) involved. Rats fed on a vitamin-deficient diet are able to select food containing the required vitamins, even though they are unable to taste the presence of vitamins in the food. It seems that they are able to learn which food made them feel better *(Rozin* and *Kalat,* 1971). But how does the rat know which food to select at a particular time ? The answer is that for some constituents of diet such as water, sodium, and possibly glucose, there is different monitoring by taste and by sensory processes in the brain that detect the presence of each of these substances in the blood. For other dietary elements such as the vitamins and minerals, there is no direct monitoring and the animal learns what to eat in order to avoid sickness.

It may be mentioned that *homeostatic requirements* place behavioural demands upon animals. The traditional view of *motivation* is built upon a simple feedback principle. A change in the animal's internal state is sensed by the brain and leads to a drive to perform the appropriate behaviour. The classical view of hunger and thirst as homeostatic drives implies that feeding and drinking are initiated as a result of monitored changes in the animal's physiological state. Feeding and drinking are said to be under negative feedback control because food and water intake serve to diminish the deviations in physiological state that initiated the behaviour.

However, a distinction has been made between homeostatic aspects of behaviour, such as feeding, drinking, and temperature regulation, and non-homeostatic aspects of behaviour, such as aggression and sexual behaviour. According to *McFarland* and *Nunez* (1978) and *Toates* (1980), the control of sexual behaviour is not fundamentally different than that of feeding or drinking. The interactions among the physiological processes that give rise to behaviour are complex.

Physiological processes may become attuned to the normal routine. In humans, for example, the liver may cease to mobilize glycogen just before a meal is due. This leads to a fall in blood sugar level in anticipation of the increase that will occur when the meal is digested. Experiments have shown that such physiological adjustments can undergo conditioning in relation to the time of day.

It may be concluded that the concept of homeostasis has had great influence upon the development of thought in animal behaviour research. This is partly because an animal's physiological requirements and interactions among physiological processes are bound to have a great effect upon its *motivation* and *motivational state,* and partly because homeostasis is an excellent example of the *feedback principle,* by which the behaviour is partly controlled by its own consequences.

12

The Osmoregulation

The maintenance of an optimal concentration of water and salts in the tissues and body fluids of animals is called *osmoregulation.* The term was first introduced by Hober (1902).

Isotonic, Hypotonic and Hypertonic Animals : Animals whose body fluids have the same concentration as that of the surrounding medium are called *Isotonic Animals.* Such animals never face the problem of osmoregulation. For example, most marine animals have their body fluids isotonic with sea water.

Animals which live in a medium of lower salt concentration are called *hypotonic animals.* In these animals, the water continuously enters into their bodies and dilutes their body fluids. Hence, they have to develop special mechanisms to get rid of the excess water that enters into their bodies. Examples of such animals are fresh water protozoans and crustaceaus that expel their excess water through contractile vacuoles and excretory organs respectively.

Animals which live in a medium of higher salt concentration, *e.g.*, marine bony fishes, are called *hypertonic animals*. These animals face a danger of dehydration as the water is continually lost from their bodies. As such, they have to drink sea water and to evolve regulatory mechanisms by which excessive loss of water from their body is prevented.

Poikilosmotic and Homoiosmotic Animals : Aquatic animals which have entered media of different salinities may be either *poikilosmotic* or *homoiosmotic*.

Poikilosmotic Animals : Most of the aquatic invertebrate animals are evolved with tissues which tolerate dilutions of their body fluids and can adjust their internal medium to that of the external medium. Such osmotetbile or osmotically variable animals are called poikilosmotic.

Stenohaline : Poikilosmotic animals that can tolerate only slight changes in salinity are known as *stenohaline*. For example, the marine spider crab, *Maia*, when transferred to 80% sea water (80% sea water mixed with 20% distilled water), puts up weight up to an certain extent due to quick entry of water into its body. Although the excretory organs of the crab work actively to eliminate the excess water in the form of a dilute urine, the crab loses large quantities of salts also along with the urine making the body fluid more dilute. Since the crab is not able to replenish the salts, it dies within 18 hours. *Arenicola* is another example of stenohaline condition.

Euryhaline : Poikilosmotic animals which can tolerate and withstand wide ranges in salinity are known as *euryhaline*. They are found in estuaries or river mouths or near sea shores where the salinity may fluctuate to a great extent. For example, the sea mussel, *Mytilus* can tolerate the dilution of sea water upto 4% and lives unaffected. Sipunculid worms such as *Phascolosoma* ; *Aurelia* and *Aplysia* are also *euryhaline*.

Homoiosmotic Animals : These are *osmostable i.e.*, they can regulate and maintain the osmotic concentration of their body fluid irrespective of the outside concentration.

Homoiosmotic animals living in fresh water adopt different methods for osmoregulation which are as follows :

(1) Presence of an impermeable cuticle.

(2) Restriction of semi-permeable surfaces to relatively small areas of respiratory and digestive systems.

(3) Development of excretory organs (nephridia or kidney) to produce hypo tonic urine.

(4) Possession of special cells (on the gills, in the gut or in the kidney) to absorb salts from the surrounding fluid to make good the salt loss through urine.

(5) Tolerance of dilute body fluids to minimise the expenditure of metabolic energy.

Mechanism of Osmoregulation

I. Osmoregulation in Fresh Water Animals : Osmoregulation in fresh water animals is effected by pumping out of excess water from their bodies. The salt loss through the excretion of water is made good by salt absorbing gills, skin, and various parts of the alimentary canal.

1. *Protozoa :* In fresh water protozoans such as *Amoeba, Euglena* and *Paramecium,* which live in a hypotonic medium, outside water enters into the body and dilutes the body fluid. This excess water is eliminated by the rhythmic pulsation of the contractile vacuole.

   Parasitic and urine protozoa do not posses a contractile vacuole.

2. *Crustacea :* The blood of fresh water crustaceans like *Palaemon* is hypertonic to the surrounding water. The latter, therefore, continuously diffuses into the blood through highly permeable gills. The excess water is excreted in urine by the antennary or green glands. Salt loss through urine is made good by the active uptake

of chloride, sodium and potassium from the surrounding medium by the chloride cells of gills even if the concentration of salts in the medium is extremely low.

3. *Mollusca* : Fresh water mussels like *Unio* and *Anodonta* carry on osmoregulation through kidneys. Osmotic pressure of these animals is higher than the surrounding aqueous medium and as such water diffuses into the tissues. This water is eliminated by the kidneys during excretion.

4. *Fresh Water Teleosts (Bony Fishes)* : These are *hyperosmotic* to the surrounding medium *i.e.,* their blood is more concentrated than the external medium. As a result, outside water enters through the thin permeable skin, the smooth surface of the gills and lining of oral and pharyngeal cavities. This water is quickly transported to the well-developed glomerular kidneys which reabsorb desirable solutes and secrete a dilute urine. Thus the blood concentration is restored. However, alongwith the urine, some salts also pass out. The salt loss is made good by chloride cells present on the gill epithelium. These cells are capeble of absorbing the small quantities of salts present in the fresh water. In this way, kidney of fresh water fishes serves primarily as filter and reabsorber of solutes.

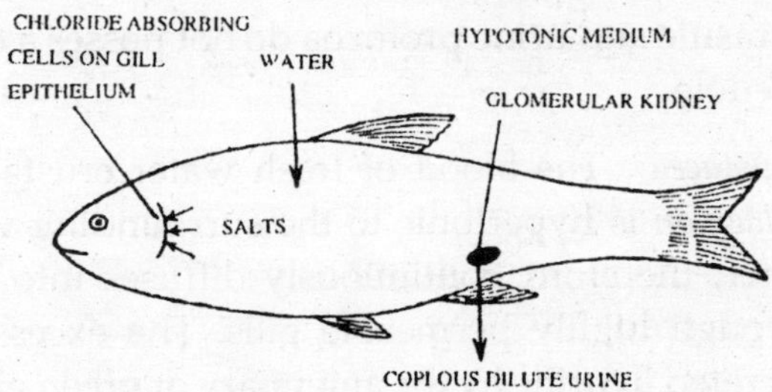

Osmoregulatin in a fresh water teleostean fish.

5. *Fresh Water Elasmobranchs (Cartilaginous Fishes) : Pristis* (sawfish) and *Characharíasgiganticus* (Shark), which often enter the large rivers and other fresh water habitats, have a tendency to retain some urea in their blood and tissues. As a result, large quantity of water enters the body by osmosis. This water is eliminated by the large glomerular kidneys.

6. *Amphibia* : Fresh water which contains a small percentage of salts diffuses into the body through smooth skin of amphibians. The well developed glomerular kidneys serve to remove the excess water which is excreted as dilute urine. The salts lost through the urine are replenished by reabsorption of ions from the urine.

Osmoregulation in Marine Animals

1. *Hagfish* : A few marine animals such as the hagfish, *Myxine,* maintain body fluids about as salty as the sea waters of the surrounding ocean (isosmotic) and so do not tend to lose water by osmosis. However, the glomerular kidney of the hagfish is an ion-regulating structure; it removes Ca^{++}, Mg^{++} and SO_4^- and reabsorbs K^+ and Cl^-.

2. *Marine Teleosts* : In the marine bony fishes e.g., *Opsanus* and *Lophius,* the body fluids are hypotonic to the marine environment, having a solute concentration only about 1/3 that of sea water. Consequently, they lose water through the smooth surfaces of the body like gills and oral epithelia. Thus they are constantly in danger of losing so much water to their environment that the solutes in their body fluids may become extremely concentrated and the fish may die of osmotic dessication. Therefore, to compensate for their osmotic water loss, they *drink* sea water. This restores their

water content but leads to a new problem—how to eliminate the excess salt ingested. This problem has been solved by the evolution of special "chloride secreting cells" in the gills that excrete excess salt. Hence these fish can take in salt water and still remain hypotonic. Water loss through urine is minimised by the development of *aglomerular* kidneys. Further, the distal convoluted tobule of the nephron is also absent. Consequently, urine flow in *Opsanus* is only 2.5 ml/kg per day as compared to 300 ml/kg per day in a fresh-water teleost.

3. *Marine Elasmobranchs :* Marine cartilaginous fish such as *Scoliodon* and *Stegostoma* have body fluids isotonic with sea water. They achieve their isotoniciry in a different way. Because they have an unusual tolerance for urea, instead of constantly, pumping this waste out, as do all other fish, sharks retain a higher concentration of it in their blood equal to that of sea-water. Though the osmotic imbalance is removed in this way, the ionic concentration of the body fluids and the surrounding sea-water remains different. The marine elasmobranchs achieve their ionic regulations by actively removing the excess salts with the help of a *rectal gland.* They do not have chloride secreting cells in their gills.

4. *Migratory Fishes :* Catadromous fishes like the eel, *Anguilla bengalensis* migrate from fresh-water to sea, while anadramous fishes like Salmon, *Salmofario,* migrate from the sea to fresh-water for spawning. Such fishes have to regulate their blood tonicity in the two contrasting media. It has been observed that the blood concentration of eel is hypertonic when it lives in the fresh-water and it shows the same pattern of osmoregulation as fresh-water teleosts. On the other

hand, blood concentration of the spawning eel in the marine medium is hypotonic to the sea-water. Therefore, it maintains itself against an osmotic dessication (a) by drinking sea-water and *(b)* eliminating the excess salts by the chloride secreting cells of the gills.

5. *Marine Reptiles* : Some reptiles like turtles, *sea-snakes* and the *iguana* are able to maintain the osmotic and ionic concentration of their body fluids by the same mechanisms as met within marine teleosts. They drink sea-water and consume salty prey. Thus they are able to absorb the water but eliminate the salt through special "salt glands".

   Turtles have special glands in their heads (lachrymal glands and premaxillary glands) that can excrete sodium chloride at a concentration about twice that of sea water. Since ancient times, turtle watchers have reported that these great armored reptiles come ashore, with tears in their eyes, to lay their eggs, but it is only recently that biologists have learned that this is not caused by an excess of sentiment — as is the case with Lewis Carroll's mock turtle—but is, rather, a useful solution to the problem of excess salt from ingestion of sea water.

6. *Marine Birds* : Sea birds such as *petrels, cormorants, penguins* and *herring gulls* eliminate their excess salts, chiefly sodium chloride, in the form of a salty fluid through their nostrils. The salt glands are located in their nasal chambers. In the herring gull, the concentration of sodium chloride thus got rid off is about five times its concentration in the blood.

7. *Marine Mammals* : Whales, dolphins, porpoises and seals do not drink water, hence they depend upon food

for their water requirements. These mammals never sweat and in this way avoid loss of water by perspiration from their skin.

Osmoregulation in Terrestrial Animals : Since terrestrial animals do not always have automatic access to either fresh or salt water, they must regulate water content in other ways, balancing off gains and losses. Terrestrial animals gain water by drinking fluids, by eating water-containing foods, and as an end product of the oxidative processes that take place in the mitochondria. When 1 gm of glucose is oxidized, 0.6 gm of water is formed. When 1 gm of protein is oxidized, only about 0.3 gm of water is produced.

Oxidation of 1 gm of fat, however, produces 1.1 gm of water because of the high hydrogen content of fat (the extra oxygen comes from the air).

The main problem of terrestrial animals is the loss of body water content. It is solved by a combination of some of the following adaptations:

1. Presence of a highly impermeable cuticle (arthropods) or skin covered with scales (reptiles), feathers (birds) or fur (mammals) and deposited with wax or keratin. Such a water proof cover prevents loss of water by evaporation.

2. Elimination of nitrogenous wastes, as urea or uric acid with little or no loss of water. In reptiles and birds, loss of water is restricted due to uricotelism. They excrete a semisolid urine in the form of uric acid crystals. Reabsorption of water from faeces by the rectal glands and cloacal wall further minimises water loss. In mammals and in some birds there is a water absorbing column called Henle's loop in the uriniferous tobule of the kidney which minimises the water loss through urine.

3. Ability of water absorption through the body surface. The skin of the desert lizard, *Moloch horridus* absorbs water like a blotting paper when the atmosphere is humid or the animal is drenched in rain.

4. Production of metabolic water by the oxidation of food stuffs, particularly the fat. Some animals living in extremely dry habitats depend upon metabolic water for their vital activities. The Kangaroo rat, Dipodomys, of the American desert for example, may spend its entire life without drinking water. It is not surprising that it prefers a diet of fatty seeds, which yield a large amount of water on oxidation. Analysis shows that the Kangaroo rat is highly conservative in its water expenditures. It has no sweat glands, and being nocturnal, it searches for food only when the external temperature is relatively cool. Its faeces have a very low water content, and its urine is highly concentrated. Its major water loss is through respiration, and even this loss is reduced by the animal's long nose in which some cooling of the expired air takes place, with condensation of water from it.

If the Kangaroo rat is fed high-protein seeds, such as soyabeans—the oxidation of which produces a large amount of nitrogenous waste and a relatively small amount of water—it will die of dehydration unless some other source of water available.

On the average, a human takes in about 2,300 milliliters of water (about 2.5 quarts) a day in food and drink, and gains an additional 200 milliliters a day by oxidation of nutrient molecules.

Water is lost from the lungs in the form of moist exhaled air, is eliminated in the faeces, is lost by evaporation from the skin, and is removed from the blood and excreted as urine. The latter is usually the major route of water loss. In a normal human adult, the rate of water excretion in urine averages 1,500 milliliters a day, but the actual amount produced may vary from 500 to 2,300 milliliters.

The *cleidoic eggs* of reptiles and birds are provided with a large amount of yolk and enclosed by a hard shell. The eggs are laid on land. The developing embryo has limited supply of water, chiefly *metabolic* in nature. The nitrogenous wastes are converted into non-toxic uric acid crystals which are stored in an *allantoic sac* to be excreted at a later stage.

5. *Tolerance of Dehydration and Fluctuation in Internal Temperature* : The *camel* "ship of the desert", can tolerate the loss of more than 25 per cent of their body weight in water, going without drinking for as long as an entire week in the summer months, three weeks in the winter. Besides, the camel can tolerate a fluctuation in internal temperature of 5° to 6°C. It is estimated that the camel saves as much as 5 litres of water a day as a result of these internal temperature fluctuations. Its body temperature can vary from 34° to 40.5° C. Without taxing its regulatory machinery like sweat glands (Man cannot tolerate a temperature fluctuation of 1°C). The camel never sweats till the body temperature rises to 41° C. Therefore, much water loss is saved by this method. Moreover, the camel excretes a much more concentrated urine, in other words, it does not need to use so much water to dissolve its waste products.
6. Behavioral adaptations tending to confine the animals to regions of high humidity and to burrow into the soil.
7. Restriction of activity to those periods when the temperature is not severe.
8. Regulatory mechanisms to check water loss (diuresis) through nervous and chemical coordination (secretion of antidiuretic hormone, ADH, from the posterior lobe

of the pituitary gland and Aldosterone from the adrenal cortex). ADH increases the reabsorption of water from kidney tubules whereas *aldosterone* is concerned with ionic balance. In mammals, aldosterone increases the reabsorption of sodium and promotes the renal excretion of potassium.

Osmoregulation

With the help of their several structures all the living organisms maintain a proper amount of water and salts in their bodies. This regulation of the water and salt contents of the body is termed as *Osmoregulation*. However, the *osmotic regulation* and *ionic regulation* are sometimes treated as the different mechanisms. But in most animals the osmoregulatory mechanism includes both of them because the regulation of one (water) always affects the other (salts).

Marine, Brackishwater, and Freshwater Fauna : It is generally considered that life originated in the sea. Part of the evidence for this is the extreme variety and richness of the present-day marine fauna and there is marked impoverishment as one proceeds into brackishwater and freshwater.

Many marine organisms are stenoheline, highly intolerant to dilution of the external medium, and some groups (Cephalopoda, Echinodermata) are virtually absent from brackishwater and freshwater. But from stenohaline marine fauna many forms have managed to adapt to and colonize brackish and ultimately fresh water. The abundance of freshwater insects and branchiopod Crustacea and their rarity in the sea is noteworthy.

There is also evidence that some groups have secondarily returned from freshwater to the sea, particularly fish groups—elasmobranchs, teleosts, and coelacanths. Thus, movement from sea to freshwater and *vice versa* has been possible through osmotic and ionic regulation. The successful penetration of marine animals into fresh water required the conquest of an osmotic problem which was the opposite of any which might exist in the sea. Thus, osmoregulatory devices have been developed in accordance with the evolutionary trends of animals.

The concentration of sea water and brackish water can be defined by the salinity (g solute per kg solution, written S%o) or be given directly in terms of the concentration of Na^+ and Cl— ions. The concentration of open sea water varies from 30 to 40%o. Water more dilute than this would be regarded as brackish water. Usually a salinity < 0.5%o (7.3 mM Cl^-) is regarded as freshwater.

Isotonic, Hypotonic and Hypertonic Animals : Animals whose body fluids have the same concentration as that of their surrounding medium *(isotonic* or *isosmotic animals)* never face the problem of osmoregulation as long as they live in such a medium.. For example, most marine animals are isotonic, as they have their body fluids isotonic with sea water. Animals which live in a medium of low-er salt concentration *(hypotonic* or *hyposmtoic animals)* have to face the difficulty of hydration and hence they have evolved special mechanisms to get rid of the excess water that usually enters their body and dilutes their body fluids.

For example, fresh water protozoans and crustaceans are hypotonic and eliminate the excess water by their contractile vacuoles and excretory organs respectively. In a similar way animals which live in a medium of higher salt concentration *(hypertonic* or *hyperosmotic animals)* have to evolve regulatory mechanisms by which excessive loss of water from the body is prevented. Since these animals *(e.g.,* bony fishes of the sea) are in constant danger of dehydration, they compensate for the water loss by drinking the surrounding water or by other means.

Osmosis and Osmotic Pressure

In order to understand the physiological reactions taking place in the mechanism of osmoregulation, we must have some knowledge of the behaviour of substances in solution. When a weak and a strong solution are separated by a semipermeable membrane, there is always a transfer of the solvent from the weaker solution to the stronger one. This process of selective transmission of a liquid in preference to another or a solvent in preference to the solute through a semipermeable membrane is termed *osmosis.* This process continues until the hydrostatic pressure due to the flow of the liquid or solvent has attained a value sufficient to stop further flow.

The excess pressure which is just sufficient to stop the flow through the membrane is called the *osmotic pressure* of the stronger solution. The osmotic pressure is proportional to the concentration of the solution, or in other words, the greater the concentration of a solution the greater would be the osmotic pressure of it. Thus, the magnitude of the osmotic pressure or the *osmotic concentration* depends on the number of particles dissolved per volume of the solvent and not on their size or nature. If several different substances are dissolved, the osmotic concentration depends on the sum of all the dissolved particles.

One solution is said to be *isosmotic* with another when they have equal osmotic concentration. Similarly a *hyposmotic* solution is more dilute and a *hyperosmotic* one more concentrated than a comparison solution. The osmotic pressure of a dilute solution is given by the Van't Hoff Equation :

$$\pi = RTC$$

where π = the osmotic pressure

C = the total concentration of nonelectrolytes and ions present

R = gas constant

T = temperature (in Kelvin)

Dimensionally, it is correct to express C as moles and g-ions per litre of solvent. However, it is more usual to express concentration as moles and g-ions per kg of solvent *(molal scale)*. This is numerically correct in above equation for an *aqueous* solution. This total concentration is sometimes, incorrectly, termed *osmolality*.

It should be noted that above equation is only approximate in the upper range of concentration found in many biological situations. However, if the osmotic pressure of a solution is known, the ratio π /RT defines a value C that is *osmolality* . Osmotic pressure is frequently expressed using this *osmolal scale*. It is difficult to measure the osmotic pressure of a solution directly, but other properties releated to osmotic pressure can be more easily determined. These properties, called the *colligative properties*, are all determined by *water activity* in the system. Van't Hoff equation shows that osmotic pressure of a solution at any given temperature T depends only upon the molar concentration C, *i.e.*, on the number of moles (and hence on the number of molecules) present per unit volume, quite independent of the nature of the solute. Thus, osmotic pressure of a solution is a colligative property.

Animal Response to Osmotic Conditions of the Medium: There are two extreme patterns of response to osmotic conditions of the environment. Animals may be osmotically labile (dependent) and their body fluid concentration may change with the medium. These are *osmoconformers*. Other animals are osmotically stable (independent) and when the medium changes, the internal concentration of the body fluid remains constant or unchanged. These are *osmoregulators*. The terms *poikilosmotic* and *homoiosmotic* are often applied respectively to osmoconforming and osmoregulating animals. Osmoconformers can tolerate wider variation in their internal osmotic concentration, while osmoregulators can withstand a wider environment range. However, there are all gradation between the extremes of lability and constancy. An animal may conform osmotically in one concentration range and regulate in other.

Euryhaline and Stenohaline Animals : Poikilosmotic animals which can tolerate wide ranges in salinity are known as *euryhaline.* For example, *Mytilus* when left in sea water diluted to a great extent, can adjust the salinity of its haemolymph to exactly the same concentration and lives unaffected. Similarly *Aplysia,* left in 75% sea water increases in weight by 20% in 2 hours, but loses salts, and water in the next 4 hours. In this way its original volume is regained and it is able to live with this diluted blood in the changed water content.

On the other hand poikilosmotic animals which cannot tolerate wide ranges in salinity are known as *stenohaline.* For example, a marine crustacean, spider-crab *(Maia)*, if left in 80% sea water increases in its weight up to a certain extent due to the quick entry of water into its body. The excretory organs of the animal activate to get rid of the excess water in the form of dilute urine. However, along with the excreted water some salts also escape out, making the body fluid more dilute, and as a consequence the animal dies in 2 to 4 hours.

Osmoregulation in Freshwater Animals : All freshwater invertebrates, as well as fishes and amphibians, are hypertonic (hyperosmotic) to the surrounding water. Since they are never completely impermeable, they have steady inflow of water and some loss of salt. They eliminate the excess amount of water as urine, and by elaborating a urine of very low concentration, they keep salt loss in the kidney to a minimum. However, some salt is lost and must be made up by food or by active absorption from the very dilute water. Numerous freshwater invertebrates can absorb salt actively and probably all of them have some salt absorbing mechainsm.

Freshwater protozoa living in *hypotonic* medium, eliminate excess water by their contractile vacuoles. The pulsation rate of the vacuole increases when more water penetrates into the body. In freshwater crustaceans the excess water is eliminated by their excretory organs, the antennary or green glands.

The small amount of salt lost with the dilute urine is restored by active absorption of salts by the gills from the freshwater. Freshwater molluscs, *e.g., Unio* and *Anodonta* also excrete a large amount of dilute urine and absorb salts to make up the loss. Similarly, in freshwater fishes, the water that enters the body is eliminated by the kidney. The urine is quite low in salt and is highly hypotonic, as it is in most freshwater animals. The salts loss is balanced by active absorption of salt in the gills.

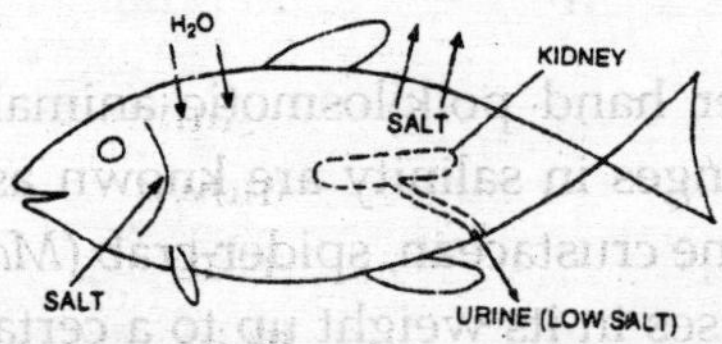

Osmoregulation in a freshwater bony fish.

Aquatic insects vary considerably in the permeability of the cuticle to water. Gills are more permeable than body surfaces, and larvae and nymphs tend to be more permeable than adults. Aquatic insects in air tend to lose more water by evaporation than do terrestrial insects. Larvae of aquatic insects conserve or extract salts not with their gills but with their guts. Either the rectal cuticle or special anal papillae developed by some larvae, such as *Aedes aegypti,* can absorb salts and the gut wall is histologically differentiated in these areas. In *A. aegypti,* the anal papillae are responsible for about 90% of salt conservation.

The fresh water crayfish, *Astacus,* secretes uirne equal to about 4% of its body weight in 24hr, the osmotic concentration of which is about one-fifth that of its blood. The antennary organ of this animal has a long canal between labyrinth and bladder. The canal functions as a water-secreting segment. Salt loss is compensated for by the active uptake of salts from the medium, even when the external salt concentration is very low, Cl^-, Na^+ and K^+ all being absorbed by the gills. *Astacus* can extract Na^+ from water containing as little as 0.02 to 0.09 mM/litre, and Cl^- from a medium hardly less dilute.

Amphibians are very similar to freshwater fish, although the adult animals have no gills and the absorption of salts takes place from the whole skin surface. The common frog can exctract Cl^- from a medium containing only 0.01 mM/litre. As they have a low salt concentration, they have not been able to adapt to a marine life, for in salt water they would become dehydrated and salt would enter the body.

Osmoregulation in Brackish Water Animals : Brackishwater is generally defined as dilute sea water that ranges in concentration from 1.5% to 90% of that of pure sea water. Most brackish waters contain both marine animals which can tolerate low salinity, and true brackishwater animals which are not found in the sea. Typical marine animals in brackish water include the shore crab *Carcinus maenas* and the mussel *Mytilus edulis*.

These marine forms tend to disappear from the brackish water fauna when the water becomes equivalent to about 15% sea water concentration. An example of a true brackish water animal is the prawn *Palaeomonetes varians*, which can live in water rangning from 0.5% to 100% sea water, but does not in fact live in genuinely marine habitats.

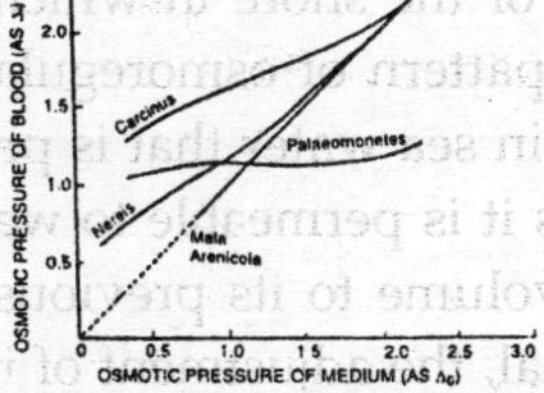

Osmotic pressures of the body fluids of some marine or brackish water animals plotted against the osmotic pressure of external medium.

When we plot the osmotic pressures of the body fluids of a number of brackish water or marine animals against the osmotic pressure of the external medium, different types of curves are obtained (Fig.). The greater the deviation of the curve from a linear curve that passes back through zero if it is extrapolated back, the greater is the ability of the animal concerned to regulate its blood concentration despite the osmotic effects of the diluted medium. Linear curves are characteristic of the spider crab *Maia* and the lung-worm *Arenicola marina*. In *Nereis* and *Carcinus*, the osmotic pressure of the blood is at first linear, but as the dilution of the medium is progressively increased the internal osmotic pressure follows the external osmotic pressure to a decreasing extent, the blood becoming increasingly hyperosmotic to the medium.

The ability to regulate is greater in *Carcinus* than *Nereis diversicolor* and therefore it is no wonder that *Carcinus* can penetrate further into brackish water than *Nereis*. The curve given by *Palaeomonetes* is quite different from the others, reflecting the true brackish water nature of this prawn. Even in 100% sea water its blood concentration is equivalent to only 70% sea water. But it can maintain a blood concentration equal to 50% sea water when the external medium is diluted to 5% sea water. Below this level some further dilution of the blood occurs, and the animal dies in media more dilute than 0.5% sea water.

In the planarian *Procerodes ulvae*, which lives in tidal estuaries and areas of the shore in which fresh water enters the sea, a different pattern of osmoregulation is exhibited. If *Procerodes* is placed in sea water that is progressively diluted, the animal swells as it is permeable to water. But eventually, it adjusts its body volume to its previous level. By chemical analysis of the animal, the adjustment of its body volume can be shown to be achieved by the simple expedient of actively secreting ions outwards, which is done to a striking extent. In fresh water, *Procerodes* possesses an internal salt concentration that is only one-tenth of that in sea water. The limitations of such a method are obvious, and the fact that the blood concentration reached in fresh water is dangerosusly low is illustrated by the observation that *Procerodes* cannot live indefinitely in fresh water. Such a method of adjustment is only suitable for an animal living in tidal conditions.

Osmoregulation in Marine Animals

Most invertebrates living in the sea have an osmotic concentration in their body fluids which equals that of the surrounding seawater but they can regulate their ionic composition. When a marine invertebrate is placed in dilute sea water, the water enters the animal due to the higher osmotic concentration in the animal. But after sometime a state of osmotic equilibrium is attained with the new environment. Most animals that are exclusively marine can tolerate only a limited change in the concentration of their medium. If the medium becomes too dilute, they would die.

The shore carb, *Carcinus,* can actively resist changes in its osmotic concentration because it has the power of osmoregulation. *Carcinus* can live in brackish water. In normal sea water it is in osmotic equilibrium, but in dilute sea water it maintains a higher osmotic concentration in the blood than that in the surrounding water. In this situation, the internal osmotic concentration in the crab causes a steady inflow of water, which must be eliminated again.

Since the crab has a kidney, the so-called green gland or antennary gland, the water can be excreted. But curious fact is that the urine formed is not dilute and has the same osmotic concentration as the blood. A high salt concentration in the crab is, therefore, maintained by the gill's ability to secrete salt into the blood. The salt naturally tends to diffuse from the concentrated blood to the more dilute water. For example, in

40% sea water *Carcinus* loses 4.4 mM/Kg body weight/hr of sodium ions through diffusion from its permeable sufraces, and 3.8 mM/ kg/hr in its urine. This total net loss of 8.2 mM/ kg/hr is recouped entirely through the active uptake of Na^+ ions from the medium against the concentration gradient.

The ability of a marine animals to penetrate into a brackish or esturaine environment depends on three different abilities, or a combination of them. These are:

(i) the animal may temporarily exclude dilute sea water, for example, a clam or oyster may close its shells when the changing tides bring highly diluted sea water ;

(ii) the animal may tolerate the decrease occuring in the concentration of its body fluids as the salinity of the water decreases temporarily;

(iii) the animal may actively take up salt from the water and maintain a higher osmotic concentration than the surrounding water. This form of osmotic regulation will enable the efficient animal to live in dilute water, and may even permit him to penetrate into freshwater. The Chinese mitten crab, *Eriocheir,* can live in the rivers of Central Europe, but it must return to the sea to breed, for the eggs do not tolerate fresh water. This crab, then shows the transition in osmoregulation that is necessary for life in freshwater.

Marine animals placed in more concentrated sea water will usually show increased blood concentrations equal to that of the water. But some including the mitten crab, can maintain concentrations lower than the surrounding water by actively transporting salt from the blood to the water. Such regulation becomes fairly important in the animals living in salt lakes, such as the fairy shrimp, *Artemia.* It can exist in salt solutions containing as little as 0.26 per cent NaCl and in concentrated brine containing as much as 30 per cent salt. The osmotic concentration of the blood is relatively independent of the medium and increases only slightly as the salt content of the water increases.

The myxinoids (hagfishes) are exclusively marine. They are osmotic with the medium and show considerable ionic regulation by varying the total osmoconcentration largely by charges in amino acids. Lampreys show both hyposmotic and hyperosmotic regulation. When marine lampreys *(Petrotmyzon marinus)* enter freshwater they stop drinking, but if returned to brackishwater they drink and extrarenal extraction of Na^+ and Cl^- is activated. In freshwater, they can actively absorb chloride.

The *bony fishes* in the sea are not in osmotic equilibrium with their *hyperosmotic* medium, for their osmotic concentration is about one-third that in sea water. Since they are not impermeable, they lose water to the more concentrated sea water and are actually in constant danger of dehydration. They compensate for the water loss by drinking the very sea water and by absorbing water and salt (Na^+ and Cl^-) from the intestine. The fish kidney is unable to produce a concentrated urine to help eliminate the salt. Instead, the gills secrete the excess salt ions (Na^+, K^+ and Cl^-) into the sea water. Since this secretion is from a lower concentration in the blood to a higher one outside, it is an active transport and needs energy. Some ions are lost in the faeces and urine (Mg, SO_4, Ca, K and PO_4).

The *elasmobranchs* (cartilaginous) fishes have solved their osmotic problem with the help of their kidney. Their blood contains sufficient amounts of urea and trimethylamine oxide, thereby maintaining an osmotic concentration equal or more to that of the sea water. Thus, for osmoregulation, their kidney possesses a special mechanism for pumping urea back into the blood almost as fast as it passes through the glomerular filter. Even the young elasmobranch is provided with its store of urea, either directly from the parent as in the viviparous dogfish

Squalus, or by its being placed in an egg case impermeable to urea. By the time they hatch or are born, the young animals have developed their own urea conserving mechanism.

As the osmotic concentration of the blood is often 10 to 50 mOsm/kg more than that of the sea water, marine elasmobranchs can, therefore, use mechanisms similar to those of freshwater fish for the excretion of water. There is a tendency for osmotic inflow and excess water is removed by the kidneys. However, there is also an inward diffusion of salt which must be eliminated. Excess salt is excreted not by the gills but by specialized organs, the *rectal glands*. For example, the dogfish *Squalus acanthias* secretes about 2 ml/kg/hour of solution through its rectal glands; this is a volume about equal to the urine output of the animal. The rectal glands are responsible for eliminating excess monovalent ions, whereas the kidneys eliminate divalent ions. The gut plays little or no role in water or solute balance in elasmorbranchs. Some urea which diffuses through the gut wall is eliminated in the faeces.

The maintenance of a high level of urea is clearly an adaptation to marine conditions, and is presumed by most authorities to indicate that elasmobranchs too were originally a fresh water group, but that all returned to the sea at some stage and were enabled to live there more easily through the development of urea conservation. On this view, fresh water elasmobranchs have returned once more to their habitat of origin. Since they possess a blood urea concentration only little below that of their marine relatives, it seems that they have been able to reverse this evolutionary change to only a limited extent. The presence of this amount of urea in the blood is additional evidence for a return to fresh water from the sea, since it would not have arisen in fresh water.

Marine turtles, marine iguanas, and sea snakes all have plasma osmotic and ionic concentration similar to those of bony fishes. Their interguments are highly impermeable, and they also lack the exposed permeable surfaces of gills because they are air breathers. Thier major problem is a continuing water loss because they must produce sufficient urine to eliminate metabolic wastes and they must lose some water in respiration. However, they are not known to drink water and the physiological mechanisms used to control water are not known. The urine is produced at a low rate, and the kidneys of these animals have a low glomerular filtration rate. The salt intake occurs, in part at least through the food (*Schmidt-Nielsen* and *Davis*, 1968).

In marine birds such as sea gulls, cormorants, and pelicans that drink sea water, the excess salt is eliminated as NaCl by nasal glands of their heads. Marine mammals do not drink water to any great extent and rely on their food as the sole source of water. Whales are a good example. They have a slightly higher blood osmotic concentration than terrestrial mammals and this also aids in osmoregulatory problems.

Most marine mammals feed on teleosts rather than marine invertebrates. This is an advantage because teleosts have already performed the osmotic work of eliminating excess salts from their tissues. Whether marine mammals possess mechanisms for extrarenal salt loss is not known, but they can produce a hyperosmotic urine when stressed.

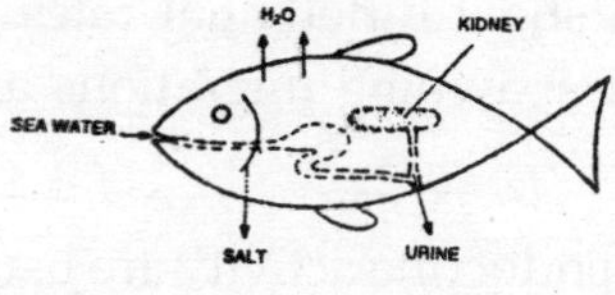

Osmoregulation in a marine bony fish.

Osmoregulation in Migratory Fishes

When an eel (*Anguilla*) reaches the freshwater habitat on its upstream migration, it faces problem of salt depletion and over-hydration as compared with the tendency towards dehydration and salt excess in the ocean. A young salmon

(Oncorhynchus) migrating downstream to its ocean habitat comes up against these conditions in reverse. It follows then, that both *anadromous* (marine fishes that migrate from sea to rivers for spawning, *e.g.*, Salmon) and *catadromous* (freshwater fishes that migrate to sea for spawning, *e.g., Anguilla)* fishes must be versatile in their osmotic adjustment, have glomerular kidneys which can adjust to differences in urine volumes due to different salinities and possess gills and oral membranes capable to cope both with the uptake and the secretion of certain ions against existing diffusion gradients.

Experiments with eel have suggested that the chloride cells can function both as secreting and absorbing units. However, it should be noted that *diadromons fishes* are adjusted either to the freshwater or to the saltwater phase of their life cycles and that such adjustments are primarily dependent on genetically determined phsiological changes.

The sea lamprey *(Petromyzon)*, for example, after having entered freshwater in order to spawn cannot adjust to salinities higher than 17 parts per thousand, and young salmons *(Oncorhynchns, Salmo salar)* cannot successfully enter the sea before their chloride cells are well developed. The case of transition depends on anatomical pecularities such as the gill to body surface ratio, well illustrated in different races of *Gasterosteus*, which show differential rates of adjustments to freshwater on their spawning migrations along the coasts of western Europe.

Changes in the endocrine activity are usually simultaneous with, or precede, changes in salts and water balance mechanism.The pituitary, the thyroid, and the gonads are primarily concerned with changes in physiological adjustment prior to and during migration. An increase in thyroid activity has been reported for salmon on downstream migration, possibly to facilitate the energy requiring process of salt excretion in sea water.

Osmoregulation in Terrestrial Animals : All terrestrial animals lose water to their environment and therefore preserve water to get rid of the danger of dehydration. The only truly successful land animals are higher arthropods and higher vertebrates. All other animals depend more or less on a moist environment. If an animal has to maintain its normal water content, all the water it loses must be exactly made up by the water it gains. If the total gain in water is equal to the loss, then the animal is said to be in *water balance;* and if the water loss is not covered by an equal gain, the animal is said to be in *negative water balance.* This situation is dangerous and is avoided by the animals.

There are several ways by which water is gained or lost in land animals. The water gain includes drinking water, water in food, and the oxidation water formed inside the body in the combustion of food materials, while water loss occurs through evaporation from the lungs and skin, urine formation, and faeces.

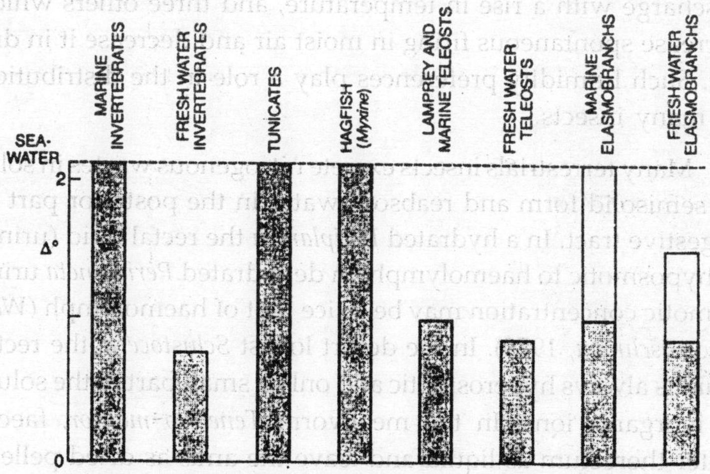

Comparative histogram of the osmotic pressures of the body fluids of invertebrates and fishes. The proportion due to urea is represents by clear areas.

To prevent the loss of water from the body, the land arthropods (arcahnids, myriapods and insects) have a chitinous coat. The insects have been the most successful in minimizing surface evaporation. They have a thin waxy cuticle of such unique properties that it not only prevents evaporation but can also take up water from a moist environment.

A unique property of the cuticle of insects and ticks is the capacity to absorb water from an atmosphere which is humid but less than saturated. Water absorption from the atomsphere has been demonstrated in ticks, mites, fleas, *Tenebrio* larvae, and lepidoptera. Another adaptive mechanism is behavioural selection of "preferred" humidity. Slightly dessicated animals tend to go to higher humidities and hydrated ones to drier air. For example, adult *Drosophila* from a normal culture prefer drier air, but they reverse this preference if their water content is reduced by 12 to 16 per cent.

In some lepidopteran caterpillars, the third antennal segment has three receptor neurons which increase their discharge with a rise in temperature, and three others which increase spontaneous firing in moist air and decrease it in dry air. Such humidity preferences play a role in the distribution of many insects.

Many terrestrials insects excrete nitrogenous wastes in solid or semisolid form and reabsorb water in the posterior part of digestive tract. In a hydrated *Periplaneta* the rectal fluid (urine) is hyposmotic to haemolymph; in dehydrated *Periplaneta* urine osmotic concentration may be twice that of haemolymph *(Wall* and *Oschman*, 1970). In the desert locust *Schistocera*, the rectal fluid is always hyperosmotic and only a small part of the solute is inorganic ions. In the mealworm *Tenebrio molitor*, faeces enter therectum as liquid and leave the anus as dried pellets. The animal has perirectal malpighian tubules, which can take K^+ from the haemolymph/ and transport out of rectum through an osmotic gradient.

Of all the invertebrates, only insects can produce markedly hyperosmotic urine and are able to conserve water. However, the mechanisms by which they accomplish this are quite different from those in mammals. In terrestrial insects, K^+ is cycled and water can be moved via osmotic gradients in restricted intercelluar channels back to the haemolymph, to leave hyperosmotic excreta.

In other higher vertebrates, protection against surface drying is afforded by horny scales (reptiles), feathers (birds), and hair (mammals) of the skin. Another means of water conservation in birds and mammals is that their kidney excretes a concentrated urine. This ability of their kidney depends upon the *loop of Henle,* the function of which is the resorption of further water. Those animals *(e.g.,* desert rat) which can produce the most concentrated urine have the longest loops of Henle.

In mammals, the volume of urine is controlled by the antidiuretic hormone, (ADH), that is secreted by the posterior lobe of the pituitary gland. The hormone plays an important role in regulating water reabsorption by the renal tubules. An under-secretion of ADH causes a disease in which a huge amount of urine is excreted (diabetes insipidus). The osmoregulatory mechanisms of terrestrial animals, thus, operate by reducing the loss their water content to minimum and involve several organ systems, including the endocrine system. However, the lack of such protection keeps some groups confined to a moist environment, *e.g.,* Onychophora and Amphibia.

Hormonal Control of Osmoregulation : Endocrine secretions control osmotic and ionic balance in both invertebrates and vertebrates chiefly by action on gills, skin and excretory organs, and on tissues in relation to extracellular ions. The role of following hormones is considered important.

Neurohypophyseal Hormones : The neurophypophyseal polypeptides are best known for their function in water and salt balance of vertebrates. These are produced by the supraoptic nucleus of the hypothalamus, pass along the infundibular stalk to the neurohypophysis, where they are stored and released on demand. Those with a basic amino acid such as arginine or lysine in the 8-position are called *vasopressins* and have marked antidiuretic (ADH) or water-conserving activity. Those with neutral 8-amino acids are called *oxytocins* and have weak ADH activity.

Most mammals have *arginine vasopressin* (AVP), while others such as wild pig and hippopotamus have only *lysine vasopressin* (LVP), and some Suiformes have both. In birds, amphibia, reptiles, and most fishes the active polypeptides are oxytocin, and arginine vasopressin. Other polypeptides are *arginine vasotocin,* AVT *(e.g.,* in lampreys, amphibians, teleosts, reptiles, and birds), *mesotocin* or 8-isoleucine oxytocin (in *Protopterus,* amphibians and birds), *isotocin* or 4-serine, 8-isoleucine oxytocin (in *Polypterus,* holosteans and teleosts), and *glumitocin* or 4-serine, 8-glutamine oxytocin (in elasmobranch, *Raja).*

However, the mode of action of vasopressins is different in the various vertebrate classes. In birds and mammals, the hormone increases the permeability of the renal tubules to water and thus permits more water reabsorption and reduced urine production. In amphibians and reptiles, AVT is also antidiuretic, but in addition to a small tubular effect the hormone reduces the glomerular filtration rate, probably by constriction of the glomerular arterioles. In anurans, an additional action is to increase permeability of the skin to Na^+ and water ; this, in combination with a sodium pump, favours salt uptake. Another function of AVT is to increase water reabsorption from the bladder. In freshwater fishes (goldfish, eel), AVT increases urine flow and glomerular filtration rate, (Jones *et al,* 1968). Thus, the action on kidneys of freshwater fish is opposite to that in tetrapods, namely to elicit diuresis. Sodium uptake by the gills is increased. In the lungfish *Protopterus,* AVT increases glomerular filtration rate as in freshwater fish, but in addition it enhances Na^+ excretion, *i.e.*, it is natriuretic (*Sawyer*, 1970). Possibly the hormone is important as a diuretic and natriuretic when the lungfish emerges from aestivation. In lamprey, AVT also promotes Na^+ excretion.

Prolactin: In bonyfish, prolactin from the anterior pituitary enhances production of a dilute urine. This hormone is critical for euryhaline fish in migration between fresh and salt water. A hypophysectomized euryhaline fish *Fundulus heteroclitus*, fails to survive well when transferred from sea-to freshwater; but survival is furthered by injection of prolactin. Hypophysectomized freshwater *Fundulus, Gasterosteus* and *Xiphophorus* lose Na^+ rapidly mostly by the gills, but injection of prolactin reduces the rate of Na^+ loss to one-fourth and the fish survive. According to *Lam* and *Leatherland* (1970), prolactin enhances freshwater osmorgulation by both gills and kidneys.

Adrenal Cortical Hormones : Several hormones of the adrenal cortex also function in n intaining salt balance. Aldosterone increases Na^+ reabsorption by the bladder of a toad *Bufo marinus* (*Sharp* and *Leaf*, 1966). In mammals, aldosterone favours Na^+ retention and K^+ excretion ; Na^+ reabsorption in the distal tubules is enhanced. In a Na^+-depleted dog, low plasma Na^+ increases secretion of aldosterone and corticosterone. Corticosterone enhances K^+ excretion and decreases Na^+ excretion, in salivary glands as well as in kidney. Aldosterone appears to act on toad bladder by inducing the synthesis of some mediator or transport protein, since its effect is blocked by inhibitors of protein synthesis. The action of vasopressin on Na^+ transport by bladder is not so blocked. In birds, glucocorticoids from the adrenal cortex stimulate salt output by the salt glands.

Renin : In mammals, elevated blood Na^+ causes the release from juxtaglomerular cells of the kidney a hormone *renin*, which enzymatically converts angiotensin precursor (angiotensinogen) in the blood to angiotensin I; this is converted to angiotensin II, which then acts on the adrenal cortex to increase the secretion of *aldosterone*. The latter stimulates reabsorption of Na^+ in the kidney, which in turn favours water retention. Renin seems to be absent from lampreys and elasmobranchs ; it is found more in freshwater fishes than in marine species. When an eel is transferred from fresh to sea water, the renin level in the kidney decreases over a three-week period.

Thyrocalcitonin : Another hormone active in ion regulation is thyrocalcitonin, which is released from the thyroid in the presence of high serum Ca^{2+}. It lowers plasma Ca^{2+} by inhibiting bone resorption. In fish, cells of the ultimobranchial bodies release a polypeptide which lowers blood-Ca^{2+} level of eel and catfish, but not of *Fundulus*.

Neuroendocrine Regulation : Neuroendocrine regulation of salt and water balance has also been observed in crustaceans and insects. In crayfish, ligation of the eyestalks increases water influx and urine production when in fresh water, and increases blood osmoconcentration when in 50% sea water; a brain extract counteracts these effects (*Kamemoto* and *Ono*, 1969). A brief review of what has been discussed above is given again :

(a) Marine invertebrates are in osmotic equilibrium with sea water, but can regulate their ionic composition.

(b) Marine elasmobranches have a low salt content, but are in osmotic equilibrium with sea water because they are capable of retaining urea in their blood.

(c) Marine teleosts are hypotonic and make up for the water loss by drinking sea water and excreting the excess salts by the gills.

(d) Freshwater invertebrates are hypertonic to the water

and have mechanisms for actively absorbing salt from the very low concentration in fresh water.

(e) Freshwater fishes and amphibians are in a similar situation and also absorb salt from the water.

(f) Reptiles, birds and mammals are all terrestrial and have impermeable integuments and some other devices for the conservation of water.

(g) Some hormones control osmotic and ionic balance in both inverterbates arid vertebrates.

Bibliography

Alcock, John : *Animal Behavior : An Evolutionary Approach*, Sinauer, 1998.

Altmann, S. A. and Altmann, J. : *Baboon Ecology : African Field Research*, Oxford, New York, 2000.

Asdell, S. A. : *Patterns of Mammalian Reproduction*, Comstock, New York, 1946.

Aslam, M. S. : *Genetics-Behavioural and Molecular*, Campus Books, New Delhi, 2003.

Aubrey, Manning : *An Introduction to Animal Behaviour*, ELBS Publication, New York, 2001.

Bahr, G. F. : *Chromosomes and Chromatin Structure*, Academic Press, New York, 2001.

Barnett, S. A. : *Study in Behaviour*, Science Paper Backs, London, 2001.

Bendall, J. R. : *Muscles, Molecules and Movement*, Heinemann Educational Books, London 1969.

Benjamin, Gutfreud : *Enzymes : Physical Principals*, John Wiley, London, 1972.

Bernhard, : *The Structure and Function of Enzymes*, Routledge, New York, 1968.

Bhaskar, Harsh Vardhan : *Animal Ecology*, Campus Books, New Delhi, 2004.

Bogert, L. J. and Calloway D.H. : *Nutrition and Physical Fitness*, Saunders, London, 1973.

Bolhuis, J. J. and Hogan, J.A. : *The Development of Animal Behaviour : A Reader*, Blackwell, 1999.

Boulenger, E. G. : *An Introduction to Animal Behaviour*, Discovery Publishing, New Delhi, 2003.

———— : *Animal Mysteries*, Discovery Publishing, New Delhi, 1993.

Boyd, J. D. and Hamilton, W.J. : *The Human Placenta*, Cambridge, London, 1978.

Buffon's, B. : *Natural History of Birds, Fishes, Insects and Reptiles*, Discovery Publishing, New Delhi, 2003.

Carpenter, K. : *Dinosaur Eggs and Babies*, Cambridge, New York, 1996.

Castillo, J. : *Progress in Biophysics and Biophysical Chemistry*, Prentice Hall, London, 1956.

Chauvin, Remy : *Ethology : The Biological Study of Animal Behaviour*, International Universities Press, New York, 1999.

Damron W. Stephen : *Introduction to Animal Science : Global, Biological, Social and Industry Perspectives*, Prentice Hall, London, 2000.

David, M. : *The Marvels of Animal Behaviour*, National Geographic Society, Washington, 2000.

David, McFarland : *Animal Behaviour : Psychobiology, Ethology and Evolution*, ELBS Publication, New York, 2000.

Dickerson and Geis : *The Structure and Action of Proteins*, Academic Press, New York, 1973.

Ernster and Drahota : *Mitochondrial Structure and Function*, Academic Press, New York, 1969.

Filley, G. F. : *Acid-Base and Gas Regulation*, Lea & Febiger, New York, 1971.

Finn, Frank : *Classification of Birds*, Discovery Publishing, New Delhi, 1994.

Florey, E. : *An Introduction to General and Comparative Animal,* Oxford University Press, New York, 1989.

Freeman, W. H. : *Animal Societies and Evolution,* A Scientific American Publication, New York, 2001.

Gene and Reynolds : *Water-soluble Vitamins,* John Wiley, London, 1963.

Gorbman, A. and Bern, H.A. : *A Text Book of Comparative Endocrinology,* John Wiley and Sons, New York, 1962.

Greenberg, Rodwell : *Amino Acid Metabolism,* Cambridge University Press, Cambridge, 1964.

Gyorgy, Pearson : *The Vitamins,* Academic Press, New York, 1967.

Houssay, Frederic : *Industries of Animals,* Discovery Publishing, New Delhi, 1999.

Huxley, H. E. : *The Mechanism of Muscular Conduction,* Spring Harbor, Harbor, 1972.

James, F. : *Animal Social Behaviour,* Wittenberger Duxbury Press, Boston, 1999.

Jencks, : *Catalysis in Chemistry and Enzymology,* McGraw Hill, New York, 1961.

Jennifer, Owen : *Feeding Strategy,* Oxford, New York, 2002.

John, Alcock : *Animal Behaviour: An Evolutionary Approach,* Sinaver Associates, New York, 2002.

Johnson, Christopher B., Franz, Chlodwig : *Breeding Research on Aromatic and Medical Plants,* The Howorth Herbal, London, 2002.

Johnston, R. G. : *Introduction to Sheep Farming,* Granada, 1983.

Jorge, Peter and Ewert Verlag : *Neuroethology : An Introduction to the Neuropathy Sociological and Fundamentals of Behaviour,* Oxford, New York, 2001.

Kamra, D. N. and Pathak, N. N. : *Nutritional Microbiology of Farm Animals,* Vikash Publishing House, New Delhi, 1995.

Kay, Ian : *Introduction to Animal Physiology*, BIOS, London, 1998.

Khanna, D. R. : *Animal Distribution*, Discovery Publishing, New Delhi, 2004.

Konrad Lorenz : *King Solomon's Ring*, MacMillan, London, 2001.

Konrad, Z. and Lorenz, Verlag : *Messengers to the Brain : Our Fantastic, Five Senses*, National Geographic Society Publication, Washington, 2001.

Konrad, Z. and Lorenz, Verlag : *The Foundations of Ethology*, New York, 1999.

Krishnan, M. : *The Hand Book of India's Wildlife*, Oxford, New York, 2002.

Kumar, Ashok : *Animal Physiology*, Discovery Publishing, New Delhi, 2003.

———— : *Aviculture*, Discovery Publishing, New Delhi, 2003.

———— : *Movement and Locomotion in Animals*, Discovery Publishing, New Delhi, 2003.

Kumar, Bipin : *Histology*, Campus Books, New Delhi, 2001.

Kung, S. D. and Yang, S.F. : *Discoveries in Plant Biology*, MacMillan, London, 2000.

Linville and Kelly : *A Text Book in General Zoology*, Discovery Publishing, New Delhi, 1993.

Lister, Harry : *An Introduction to Zoology*, Discovery Publishing, New Delhi, 1999.

Lubibiecki, A. S. : *Large-Scale Mammalian Cell Culture Technology*, Dekker, 1990.

Maheswari, S. K. : *A Textbook of Life Science*, Dominant Publishers, New Delhi, 2004.

Mary, E. Torrence and Issacson : *Microbial Food Safety in Animal Agriculture*, Blackwell, London, 2003.

McCance and Widdowson : *The Composition of Food*, Her Majesty's Office, London, 1960.

Merler, E. : *Immunoglobulins : Biological Aspects and Clinical Uses*, Natural Academy of Sciences, Washington, 1970.

Michael, A. M. : *Irrigation : Theory and Practice*, Vikash Publishing House, New Delhi, 1997.

Morgan, S. J. : *Animal Cell Culture : Introduction to Biotechniques*, BIOS, London, 1993.

Nama, Shinde and Jadhav : *The Nematodes of Indian Mammals*, Dominant Publishers, New Delhi, 2004.

Openheimer, J. H. and Others : *Nuclear Receptors and the Initiation of Thyroid Jormone Action*, Little Brown and Company, Boston, 1976.

Pathak, Nityanand : *Textbook of Feed Processing Technology*, Vikash Publishing House, New Delhi, 1997.

Price, M. R. S. : *Animal Re-Introduction: The Arabian Oryx in Oman*, Cambridge, 1989.

Prosser, Brown, : *Comparative Animal Physiology*, Saunder, Philadelphia, 1961.

Quirk, R. P. : *Chemistry of Muscular Contraction*, Academic Press, New York, 1970.

Ranjhan, S. K. : *Animal Nutrition and Feeding Practices*, Vikash Publishing House, New Delhi, 1995.

——— : *Animal Nutrition in the Tropics*, Vikash Publishing House, New Delhi, 1998.

——— : *Textbook on Buffalo Production*, Vikash Publishing House, New Delhi, 1996.

——— : *Textbook on Quality Meat Production*, Vikash Publishing House, New Delhi, 1999.

Rao, D. Bhaskara : *Animal Kingdom*, Discovery Publishing, New Delhi, 1995.

Robert, Burton : *Animal Behaviour*, Time Life International, Nederland, 1999.

Roger, Tory : *The Birds : Life Nature Library*, Peterson, London, 1999.

Ronald, H. Bailey : *The Role of the Brain*, Time-Life Books, New York, 2001.

———— : *The Secret World of Animals*, A National Geographic Society Publication, Washington, 2002.

Saharia, V. B. : *Wildlife in India*, Natraj Publishers, Dehradun, 1999.

Samuel, K. Wasser : *Social Behaviour of Female Vertebrates*, Academic Press, New York, 2002.

Samuel, Israel and Toby, Sinclair : *Indian Wildlife*, Hans Johannes Hoefer, Singapore, 2001.

Sandhu, G. S. : *A Textbook of Fish and Fisheries*, Dominant Publishers, New Delhi, 2005.

———— : *Analytic Genetics of Animal Development*, Campus Books, New Delhi, 2004.

———— : *An Introduction to Birds*, Campus Books, New Delhi, 2003.

Sarkar, Amira : *Development of Animal Behaviour*, Discovery Publishing, New Delhi, 2003.

———— : *Fundamentals of Animals Behaviours*, Discovery Publishing, New Delhi, 2003.

———— : *Social Behaviour in Animals*, Discovery Publishing, New Delhi, 2003.

Schmidt, Nelson K. : *Animal Behaviour*, Prentice Hall, New Jersey, 1970.

Sela, M. : *New Perspectives in Biology*, American Elsevier, New York, 1963.

Shreeve, W. W. : *Physiological Chemistry of Carbohydrates in Mammals*, Saunders, New York, 1974.

Singh, Gurdarshan, Bhaskar, H. : *An Introduction to Fishes*, Campus Books, New Delhi, 2003.

Spater, P. J. B. : *An Introduction to Ethology*, Cambridge University Press, New York, 1999.

Tikader, B. K. : *Threatened Animals of India*, Zoological Survey of India Publication, Calcutta, 1999.

Tortora and Anagnostrakes : *Principles of Anatony and Physiology*, Harper and Row Publishers, New York, 2002.

Tullts, L. P. : *Blood Cells and Plasma Proteins : their State in Nature*, Academic Press, New York, 1953.

Underwood, E. J. : *Trace Elements in Human and Animal Nutrition*, Academic Press, New York, 1971.

Van, Grvisen : *Wildlife of India*, Guide Book Company Ltd., New York, 2002.

Wadsworth, Cecie Starr : *Basic Concepts in Biology*, Wadsworth, New York, 1997.

Whelan : *Control of Glycogen Metabolism*, Academic Press, New York, 1968.

Whitfield, Dr. P. : *Longman World Guide to Mammals : A Visual Who's Who of the Mammal Kingdom*, Longman, London, 1985.

Williams, C. A. and Chase, M. W. : *Methods of Animal Behaviour*, Academic Press, New York, 1968.

Williams, S. R. : *Animal Behaviour*, MacMillan, London, 1973.

Yadav, Manju : *Applied Entomology*, Discovery Publishing, New Delhi, 2003.

Yadav, P. R. : *Environmental Biodiversity*, Discovery Publishing, New Delhi, 2003.

Yapp, W. B. : *An Introduction to Animal Behaviour*, Clarendon Press, New York, 1960.

Zhukov, M. F. : *Organic Chemistry of Biological Compounds*, Englewood Cliffs, Prentice Hall, 1971.

Zumdahl, S. S. : *An Introduction to General and Comparative Animal Behaviour*, Oxford University Press, New York, 1981.